ATOMIC DIFFUSION IN SEMICONDUCTORS

ATOMIC DIFFUSION IN SEMICONDUCTORS

Edited by

D. Shaw

Department of Physics
The University
Hull
England

PLENUM PRESS • LONDON AND NEW YORK • 1973

Plenum Publishing Company Ltd
Davis House
8 Scrubs Lane
London NW10 6SE
Telephone 01-969 4727

U.S. Edition published by
Plenum Publishing Corporation
227 West 17th Street
New York, New York 10011

ISBN: 0-306-30455-4

Library of Congress Catalog Number: 74-178779

Printed in Great Britain by
The Whitefriars Press Ltd., London and Tonbridge

PREFACE

The diffusion or migration of atoms in matter, of whatever form, is a basic consequence of the existence of atoms. In metals, atomic diffusion has a well established position of importance as it is recognized that there are few metallurgical processes which do not embody the diffusion of one or more of the constituents. As regards semiconductors any thermal annealing treatment involves atomic diffusion. In semiconductor technology diffusion processes provide a vital and basic means of fabricating doped structures. Notwithstanding the importance of diffusion in the preparative processes of semiconductor structures and samples, the diffusion based aspects have acquired an empirical outlook verging almost on alchemy. The first attempt to present a systematic account of semiconductor diffusion processes was made by Boltaks [1] in 1961. During the decade since Boltaks' book appeared much work germane to understanding the atomic mechanisms responsible for diffusion in semiconductors has been published. The object of the present book is to give an account of, and to consolidate, present knowledge of semiconductor diffusion in terms of basic concepts of atomic migration in crystalline lattices. To this end, exhaustive compilations of empirical data have been avoided as these are available elsewhere [2, 3] : attention has been limited to considering evidence capable of yielding insight into the physical processes concerned in atomic diffusion. As a consequence consideration is restricted to diffusion in dilute solutions (typically less than 1% atomic) and, where diffusion occurs in non-equilibrium (chemical) conditions, to situations where departure from equilibrium is caused by thermal and/or external phase changes. Thus diffusion processes in ion implanted layers or in the annealing of radiation damage is not discussed. A further area, which is not specifically treated is that of diffusion controlled reactions (e.g. precipitation phenomena). The inclusion of any discussion of diffusion technology for device fabrication was also felt to be inappropriate especially as adequate accounts are available elsewhere.

The presentation assumes some familiarity with the elementary ideas of atomic diffusion in crystalline solids together with a basic knowledge of semiconductor theory, including defect chemistry. Chapter 1 presents an introduction to the general concepts essential to interpreting atomic diffusion in semiconductors in which the Fermi level has a prominent role. Ideally one should be able to calculate the diffusion coefficient of a particular atomic species from first principles and Chapter 2 discusses what has been achieved in this direction. During the past decade it has become increasingly clear that the diffusion processes in a semiconductor specimen are crucially affected by conditions in the phase external and adjacent to the specimen. The role of these conditions on point defect concentrations within the semiconductor is the substance of Chapter 3. It is very obvious that an interpretation of a diffusion situation is only as good as the available experimental data and in Chapter 4 an account and assessment is given of the various techniques currently used for determining a diffusion coefficient. The remaining four chapters are concerned with presenting and interpreting the available evidence in the elemental semiconductors germanium and silicon (Chapter 5), in the III-V compounds (Chapter 6), in the II-VI and IV-VI compounds (Chapter 7) and finally in oxide semiconductors (Chapter 8).

1. B. I. Boltaks, Diffusion in Semiconductors, Infosearch Ltd., London (1963).
2. B. L. Sharma, Diffusion in Semiconductors (Trans. Tech. Publications, Clausthal, Germany, 1970).
3. Diffusion Data (F. H. Wöhlbier, ed.) Trans. Tech. Publications, Clausthal, Germany. A continuous compilation of reference data on diffusion phenomena (3 issues per year).

D.S.

PRINCIPAL SYMBOLS

a_o — lattice constant
A — solvent or solute species, acceptor impurity
$\widetilde{A}$ — radioisotope of A
B — solvent or solute species
$\widetilde{B}$ — radioisotope of B
c — concentration, number/cm^3
$\tilde{c}$ — radioisotope concentration, number/cm^3
$\mathbf{c}$ — concentration, mole fraction
$\tilde{\mathbf{c}}$ — radioisotope concentration, mole fraction
C — solvent or solute species
$\widetilde{C}$ — radioisotope of C
D — donor impurity
$\widetilde{D}$ — radioisotope of D
$D(B)$ — diffusion coefficient for isothermal diffusion of B species when no driving forces (section 1.3.2) or couplings with other component fluxes are present
$\mathbf{D}(B)$ — the chemical or interdiffusion coefficient for the isothermal diffusion of B
$D_e(\widetilde{B})$ — the self or isoconcentration tracer diffusion coefficient for the isothermal diffusion of component B
$D^i(B)$ — the value of $D(B)$ under intrinsic electrical conditions. Similarly for $\mathbf{D}(B)$ and $D_e(B)$
D_o — the pre-exponential factor in the Arrhenius expression for a diffusion coefficient
e — electronic charge
e' — free electron (conduction band)
eV — electron volt

E_c	energy of the conduction band edge	referred to a zero below the valence band edge.
E_F	the Fermi level	
E_v	energy of the valence band edge	

f — correlation factor
$F_{½}(y)$ — Fermi-Dirac integral, $= (2/\sqrt{\pi}) \int_0^\infty x^{½}(1 + e^{x-y})^{-1}\,dx$

G	Gibbs free energy, gaseous state
h	free hole (valence band)
H	enthalpy
J	flux of a species
k	Boltzmann's constant
K	equilibrium constant
L	liquid state
m_e	free electron mass
m_n	electron effective mass
m_p	hole effective mass
M	electropositive solvent species
$\tilde{M}$	radioisotope of M
n	free electron concentration (conduction band)
n_i	intrinsic free electron concentration
p	free hole concentration (valence band)
p_A	partial vapour pressure of species A
P	total vapour or gas pressure
Q	activation energy for diffusion occurring in the Arrhenius expression for a diffusion coefficent. Where numerical values for a diffusion coefficient are given in the form of an Arrhenius expression the activation energies are in electron volts
S	entropy, solid phase
t	time
T	absolute temperature
V	lattice vacancy
x	distance
X	electronegative solvent species
$\tilde{X}$	radioisotope of X
X_A, $X(A)$	mole fraction of species A
z	absolute electric charge of an ion in units of the magnitude electron charge
Z	coordination number of a lattice species
ϵ	dielectric constant
ϵ_g	energy band gap of a semiconductor
γ	activity coefficient

NOTATION FOR SOLUTE AND SOLVENT SPECIES

Charge Notation for Defects

Kröger's [1] system is followed in which the charge relative to the normal solvent lattice site is indicated rather than the absolute charge of the defect. Thus the superscripts $\times$, $\cdot$, $'$ to a defect species symbol denote respectively the neutral (un-ionized) state, the loss of one electron (singly ionized donor), the gain of one electron (singly ionized acceptor). A doubly ionized donor or acceptor would be designated as $D^{\cdot\cdot}$ or A'' respectively.

If a solvent lattice species has an absolute charge $\pm ze$ (z is not necessarily integral) the charge states of a substitutional impurity can be compared as follows:

Defect	Absolute Charge	Kröger Notation
Neutral donor	$D^{z\pm}$	$D^{\times}$
Singly ionized donor	$D^{(z\pm 1)\pm}$	$D^{\cdot}$
Neutral acceptor	$A^{z\pm}$	$A^{\times}$
Singly ionized acceptor	$A^{(z\mp 1)\pm}$	A'

Concentrations of Lattice Species

Square brackets [ ] around the species symbol or c denote the concentration in number/cm^3. The concentration in mole fraction for a species A is expressed as $X(A)$, X_A or [A] ; where there is no need to specify the species c can be used. The activity of a species is represented by $\{\,\}$ brackets enclosing the species symbol.

Suffixes

Subscripts are used with the species symbol to denote the lattice site occupied by the species e.g. a solute species A substituting for a solvent lattice species M is represented by A_M, or again, V_X, denotes a lattice vacancy in the X sub-lattice of the solvent. The subscript i denotes the occupancy of an interstitial site e.g. M_i for a self-interstitial or D_i for an interstitial donor impurity. The i subscript is also used outside the concentration bracket to show that the value of the concentration corresponds to intrinsic electrical conditions, e.g. $[A_M]_i$. The subscript e is used to denote values appropriate to conditions of chemical equilibrium e.g. $[A_M]_e$. Thus $[A'_M]_{e,i}$ denotes the equilibrium concentration of a singly ionized acceptor impurity substituting at an M lattice site under intrinsic conditions.

F. A. Kröger, The Chemistry of Imperfect Crystals, p. 198 (North-Holland Publishing Co., Amsterdam, 1964).

CONTRIBUTORS

J. C. Brice	Solid State Physics Division, Mullard Research Laboratories, Redhill, Surrey, England.
H. C. Casey, Jr.	Bell Telephone Laboratories, Mountain Avenue, Murray Hill, N.J. 07974, USA.
S. M. Hu	IBM, Thomas J. Watson Research Center, P.O. Box 218, Yorktown Heights, N.Y. 10598, USA.
D. Shaw	Department of Physics, The University, Hull, Yorkshire, England.
D. A. Stevenson	Department of Materials Science, Stanford University, Stanford, California 94305, USA.
R. A. Swalin	University of Minnesota, Institute of Technology, Minneapolis, Minnesota 55455, USA.
J. Bruce Wagner, Jr.	Department of Materials, Science and Materials Research Center, Technological Institute, Northwestern University, Evanston, Illinois, USA.
T. H. Yeh	IBM, East Fishkill Facility, Route 52, Hopewell Junction, New York 12533, USA.

CONTENTS

GENERAL FEATURES OF DIFFUSION IN SEMICONDUCTORS 1

D. SHAW

CONTENTS

1.1 INTRODUCTION

Atomic diffusion in a crystalline lattice is the phenomenon by which atoms wander through the lattice, occupying, as a consequence of their thermal energy, successively different lattice sites. Most lattice sites are usually occupied so this poses the basic problem: how can an atom actually leave one site and occupy another? Diffusion studies in metals and ionic crystals have led to the general recognition and acceptance of several basic mechanisms for atomic diffusion [1]. In semiconductor crystals very few definitive results have so far been achieved and in consequence the ideas that have evolved from metals and ionic crystals dominate the interpretation of semiconductor diffusion experiments. The essential feature distinguishing a semiconductor from a metal or ionic crystal is the wide energy range available to the Fermi level (E_F) in a given semiconductor which leads to a given lattice defect appearing in a variety of ionized states. Although this complicates the situation it also provides an important means for investigating semiconductor diffusion processes.

The diffusion of solute and solvent atoms in a semiconductor can occur under one or other of two basic experimental regimes. These are: diffusion in a crystal at chemical equilibrium, i.e. uniform chemical and native defect composition; and diffusion entailing a net chemical flux, i.e. the system is not in chemical equilibrium and the chemical potential gradients set up chemical fluxes. The former regime is invariably studied using radiotracers, therefore involving a net isotopic flux, and gives rise to what is known as self-diffusion for solvent species and to isoconcentration diffusion for solute species. Diffusion in a gradient of chemical potential is known as chemical diffusion or interdiffusion. At high solute concentrations the distinction between solute and solvent becomes blurred as we now have an alloy or mixed crystal system. Solute diffusion work in semiconductors has been predominantly concerned with solutes having a limited solubility in the solvent, typically $\lesssim 1\%$ mole fraction, because such solutes confer important electrical properties on the semiconductor. The interdiffusion of one impurity into a semiconductor already doped by diffusion, but not necessarily uniformly, with a second impurity has been described as double or sequential diffusion: double diffusion has also been used to describe the simultaneous interdiffusion of two impurities. It is also important to know whether intrinsic or extrinsic conditions obtain in the semiconductor at the diffusion temperature. Extrinsic behaviour can of course arise from native defects, especially from departures from stoichiometry, as well as from the presence of foreign impurities. If interdiffusion results in the formation of a second crystalline phase this is referred to as reactive diffusion and is beyond the scope of this chapter [1, 2].

For one-dimensional diffusion the flux, J, of a species is related to the concentration gradient of the species by Fick's first law

$$J = -D\frac{\partial c}{\partial x} \tag{1.1}$$

which defines the diffusion coefficient D of the species. Equation 1.1 is purely phenomenological: it enables D to be obtained from experimental measurements but reveals nothing about the diffusion process at the atomic level. Use of random walk analysis and an atomic jump model permits D from equation 1.1 to be related to the parameters of the model [1]. Chemical diffusion results in a net flow of matter which generally leads to an expansion or contraction of the diffusion region. In this case it becomes important to specify the reference plane to which the flux of equation 1.1 is referred, as the diffusion coefficient will depend on the choice of reference plane. In metals D is known as the intrinsic diffusion coefficient when J is referred to local lattice planes and as the chemical diffusion or interdiffusion coefficient when J is referred to an interface, or to a plane beyond, the diffusion region. Obviously the choice of reference plane is immaterial in self and isoconcentration diffusion. A fuller discussion of these coefficients in relation to semiconductor diffusion is given in section 1.3.6.

The first task in semiconductor diffusion is to identify the diffusing species i.e. to determine its position in the lattice and its charge state(s). This information can in principle be obtained by observing how D varies (i) with the known chemical activity (e.g. partial vapour pressure) of the species in an adjacent phase (i.e. the adjacent phase is a reservoir for the diffusion of the species into the semiconductor) and (ii) with the position of the Fermi level in the semiconductor. Identifying the diffusing species may also identify the jump mechanism (e.g. an interstitial). Otherwise the properties of D must be compared with the predictions of an appropriate atomic model. The temperature dependence of D is generally described by [1]

$$D = D_0 \exp(-Q/kT) \tag{1.2}$$

where D_0 is a temperature independent factor and Q is the activation energy for the atomic jump mechanism. The magnitude of the experimental Q can also assist in identifying the diffusing species (see sections 2.3 and 2.4).

The presence of concentration gradients in chemical diffusion introduces many additional and complicating features which are absent in self and isoconcentration diffusion. The simpler situation of isothermal diffusion under conditions of chemical equilibrium will therefore be used in section 1.2 as a background for introducing the basic ideas germane to semiconductor diffusion. The effects of gradients are then considered in sections 1.3 and 1.4. Section 1.5 discusses the role and importance of dislocations in semiconductor diffusion. Surface diffusion is a relatively unexplored field and is not considered further: an account of this area for solids has been given by Adda and Philibert [3].

1.2 ISOTHERMAL DIFFUSION IN CONDITIONS OF CHEMICAL EQUILIBRIUM

1.2.1 Some Properties of the Atomic Diffusion Coefficients D_e

D_e, the self-diffusion or isoconcentration diffusion coefficient for a diffusant in a crystal lattice, is a symmetric second order tensor [4]. The off-diagonal elements however become zero for symmetries other than tri- and mono-clinic if, as is customary, D_e is referred to the principal crystallographic axes. For cubic symmetry (e.g. the diamond and zinc blende lattices) the diagonal terms are all equal so that diffusion in a cubic lattice is isotropic and can be described by a scalar diffusion coefficient. In a hexagonal lattice (e.g. wurtzite) two independent diagonal elements are required: one for diffusion perpendicular to the basal plane and one for diffusion (isotropic) in the basal plane. Tetragonal and trigonal symmetries are similar to the hexagonal case whereas D_e has three independent elements for orthorhombic symmetry. D_e for mono- and tri-clinic crystals has four and six independent elements respectively. It is simpler however with these two symmetries to work with the three principal diffusion coefficients and the appropriate direction cosines [4]. Experimental results for diffusion in Sb [5], Se [6], Te [7] and anisotropic metals [8] indicate that lattice anisotropy causes comparatively small differences in the diagonal elements of D_e.

Random walk theory can be applied to atomic diffusion [9, 10] in a crystal lattice and yields equations analogous to Fick's first and second laws. The latter is expressed for one-dimensional diffusion by

$$\frac{\partial c}{\partial t} = D_e \frac{\partial^2 c}{\partial x^2} \tag{1.3}$$

and identity with the random walk analysis of the same situation with only nearest neighbour jumps allowed requires

$$D_e = \Gamma_x d^2/2 \tag{1.4}$$

where Γ_x is the average jump rate of an atom with jump distance component d along the x direction.

In a cubic lattice $\Gamma_x = \Gamma_y = \Gamma_z = s\Gamma$ where Γ is the average jump rate of an atom, Γ_y and Γ_z are the average jump rates parallel to the y and z axes and $s = 1/3, 2/3, 1$ for the simple cubic, *fcc* and *bcc* structures respectively. Equation 1.4 then becomes

$$D_e = \Gamma a^2/6 \tag{1.5}$$

for a cubic lattice if a is the nearest neighbour separation. The diffusion coefficients perpendicular and parallel to the basal plane in an *hcp* lattice are respectively [10]

$$D_{e\perp} = \Gamma_{\perp} c^2/8 \tag{1.6}$$

$$D_{e/\!/} = (3\Gamma_{/\!/} + \Gamma_{\perp})a^2/12 \tag{1.7}$$

where $\Gamma_{\perp}$ and $\Gamma_{/\!/}$ are the average jump rates out of and in the basal plane respectively: c is the height of the unit cell and a the nearest neighbour separation in the basal plane.

The temperature dependence of D_e is represented by equation 1.2. D_e is also pressure dependent but no experimental data appear to be available for semiconductors: metals have received most attention experimentally [11].

The simplest experimental situation for studying diffusion is planar diffusion into an effectively semi-infinite lattice ($x \geqslant 0$) for which equation 1.3 is directly applicable. The two most common boundary conditions are (i) the infinite or replenished source at $x = 0$ which gives for $t \geqslant 0$

$$c = c_o \operatorname{erfc}(x/2\sqrt{D_e t}) \tag{1.8}$$

where c_o is the value of c at $x = 0$ and (ii) the thin film or non-replenished source at $x = 0$ which for $t \geqslant 0$ yields

$$c = W \exp(-x^2/2\sqrt{D_e t})/\sqrt{\pi D_e t} \tag{1.9}$$

where W is the initial concentration per unit area at $x = 0$ (see the Appendix for other solutions to equation 1.3).

1.2.2 The Fermi Level and Its Evaluation

Main Features

Although a space charge layer exists at the surface of a semiconductor [12] (which is normally neglected in diffusion studies) there will be zero space charge within the volume of a homogeneous semiconductor. The electro-neutrality condition is given by [13]

$$p + \begin{pmatrix}\text{Total relative charge}\\ \text{per unit volume of}\\ \text{ionized donor levels}\end{pmatrix} = n + \begin{pmatrix}\text{Total relative charge}\\ \text{per unit volume of}\\ \text{ionized acceptor levels}\end{pmatrix} \tag{1.10}$$

Electronic or ionization equilibrium is defined in terms of the Fermi level and the distribution of electrons at defects in the band gap.† The Fermi

† We neglect levels due to dislocations.

level E_F is given by

$$p = N_v F_{1/2}(\eta_v/kT) \tag{1.11}$$

$$n = N_c F_{1/2}(\eta_c/kT) \tag{1.12}$$

where

$$\eta_v = E_v - E_F, \quad \eta_c = E_F - E_c$$

and

$$F_{1/2}(y) = (2/\sqrt{\pi}) \int_0^\infty x^{1/2} \{1 + \exp(x-y)\}^{-1} dx$$

is the Fermi-Dirac integral. The distribution of electrons over the multivalent states of a donor D is described by [14]

$$\begin{aligned} g_{D^\cdot}[D^\times] &= g_{D^\times}[D^\cdot] \exp\{(E_F - E_{D^\times})/kT\} \\ g_{D^{\cdot\cdot}}[D^\cdot] &= g_{D^\cdot}[D^{\cdot\cdot}] \exp\{(E_F - E_{D^\cdot})/kT\} \end{aligned} \tag{1.13}$$

etc.

where the g factors denote the spin degeneracy of the indicated donor state. Similar equations apply to a multivalent acceptor A,

$$\begin{aligned} g_{A^\times}[A'] &= g_{A'}[A^\times] \exp\{(E_F - E_{A^\times})/kT\} \\ g_{A'}[A''] &= g_{A''}[A'] \exp\{(E_F - E_{A'})/kT\} \end{aligned} \tag{1.14}$$

etc.

In the case of native defects $[D^\times]$ and $[A^\times]$ will (in equilibrium) be solely functions of temperature in an elemental semiconductor. In a compound semiconductor the neutral native defect concentrations will be determined both by the temperature and by the activity of one or more of the components in an adjacent phase (section 3.2.1). The concentration of a solute in an elemental semiconductor will be controlled by temperature and by the activity of the solute in an adjacent phase. In a compound semiconductor the activity of one or more of the host components in an external phase is also needed to determine the solute concentration (section 3.2.2). It is clearly essential in experimental diffusion work that the composition of the phase adjacent to the semiconductor be controlled and specified.

Equations 1.10 to 1.14 define electronic equilibrium in a semiconductor and E_F, in principle, can be obtained from these in terms of the neutral defect concentrations, the various energy parameters and temperature. In general an analytical solution is not possible and alternative methods must be used. Two methods are available, one of which is a graphical approach, due to Shockley and Moll [15], whose application to binary compound semiconductors has been described by Reynolds [16]. The alternative but approximate method of solution originated with Brouwer [17] who showed that equation 1.10 will, for a given composition of the phase adjacent to the semiconductor, usually be dominated by only two or three terms (this is also evident from the Shockley-Moll method). If the dominant terms can be identified E_F can be easily obtained, sometimes in analytical form. Kroger and associates [18] have shown that Brouwer's [17] approximation can be used with great effectiveness to portray in graphical form (Brouwer diagrams) the point defect structure of a semiconductor.

Whenever η_c or $\eta_v \leqslant -2kT$ equations 1.11 and 1.12 simplify to $p = N_v \exp(\eta_v/kT)$ and $n = N_c \exp(\eta_c/kT)$ respectively. The error never exceeds 5% and decreases rapidly with increasing negative argument. In the non-degenerate range these simplifications greatly facilitate the determination of E_F.

The zero space charge condition expressed by equation 1.10 has been shown to be valid in an n-type non-degenerate semi-conductor provided that relative changes in n over the screening length $r_s = (\epsilon kT/4\pi e^2 n)^{1/2}$ are small [19]. A corresponding condition exists for a p-type crystal. Typically $r_s < 10^3$ Å for n, $p > 10^{16}$ cm^{-3} and usual diffusion temperatures [19]. We therefore expect equation 1.10 to be valid within the bulk of a (nominally) homogeneous semiconductor. The characteristic depth of the space charge layer at a semiconductor surface is also $\sim r_s$ [12] so that violation of equation 1.10 will only be important in experiments where diffusion is restricted to a surface region of depth $\sim r_s$. Diffusion depths generally exceed r_s by many orders of magnitude.

Heavily Doped Semiconductors

It is implicit in equations 1.13 and 1.14 that the defects provide spatially discrete, or isolated, electron levels within the lattice. This assumption ceases to be valid for hydrogenic centres at concentrations much in excess of 10^{16} cm^{-3}, due to overlap of the defect wave functions. Overlap means that the electrons occupying a given set of defect levels will no longer be localized at the parent centres. A narrow band of energy levels results, formed by the overlap of the donor or acceptor defects, whose width increases with defect concentration. The condition for the formation of an impurity band from hydrogenic centres at a concentration c is $c^{1/3} r_B \simeq 0.2$ where r_B is the Bohr radius of the ground state of the defect [20]. As c increases, one edge of the impurity band will merge with either the conduction (donors) or valence (acceptors) band. In GaAs the donor impurity band forms and merges at concentrations of $\sim 2 \times 10^{16}$ cm^{-3} and $\sim 3 \times 10^{17}$ cm^{-3} respectively [21], whereas the acceptor impurity band only merges with the valence band at a concentration of $\sim 2 \times 10^{19}\,cm^{-3}$ [22]. Donor bands merge with the conduction bands of Si and Ge at concentrations of $\sim 6 \times 10^{18}\,cm^{-3}$ and $\sim 2 \times 10^{17}\,cm^{-3}$ respectively, while in Si an acceptor band merges at an acceptor concentration of $\sim 1.5 \times 10^{18}\,cm^{-3}$ [23].

The presence of impurity banding will affect equations 1.10 to 1.14 through changes in the energy level spectrum for free carriers and also through the disappearance of neutral or unionized states of the defect constituting the band: these defects will be totally ionized. Impurity bands can form in compensated and uncompensated crystals. The former case still resists a theoretical description whereas the latter has been extensively studied theoretically, primarily for homopolar semiconductors [24, 25]. When a donor impurity forms a merged impurity band in an uncompensated crystal the donors are fully ionized so that the electroneutrality condition is expressed by $n = [D^{\cdot}]$. In this situation the change in the distribution of electron states causes the conduction band edge to disappear as the electron states now penetrate, with a density which decreases rapidly with penetration, into what was the band gap of the pure crystal. Additional to the modified density of electron states is a displacement of the electron energy levels, ΔE, because of the Coulomb interactions of the free electrons with the ionized donors and with each other. If E_1 and E_2 respectively denote these interaction energies then the total shift is [26]

$$\Delta E = E_1 + E_2 \tag{1.15}$$

Although it might appear that in these circumstances the evaluation of E_F would be difficult it turns out, rather surprisingly, that, if a conduction band edge defined by

$$E_c = E_c^0 + \Delta E \tag{1.16}$$

is used, then E_F relative to E_c can be found using equation 1.12 with an error of $<15\%$: the error decreases with increasing n [24, 25]. E_c^0 is the band edge in the pure crystal. Thus equation 1.12, despite the modified density of states due to impurity banding, can still be used to obtain E_F with not too great an error. Similarly equation 1.11 gives E_F for the case of heavy acceptor doping.

The results discussed above apply specifically to a homopolar crystal. Keldysh and Kopaev [27] have studied the effect of impurity band formation on the density of states in a heteropolar semiconductor. Their results can be used to again show that E_F values obtained from equations 1.11 and 1.12 differ by only a few per cent from values based on the modified density of states.

Zero space charge requires any variations in n or p to be small over a distance equal to the screening length r_s [19]. In a fully degenerate semiconductor with n or $p = c$, the impurity concentration, then [25] $r_s = (\pi r_B^3/3c)^{1/6}/2$ which for typical values gives $r_s < 10^2$ Å so that provided $d(\ln c)/dx < \sim 10^6$ cm^{-1} electroneutrality will obtain.

The Electron and Hole Activity Coefficients

Defect equilibria in semiconductors are described by means of the law of mass action (section 3.2.2) in which e' and $h^{\cdot}$ appear as chemical species. To retain the validity of this description under degenerate conditions requires the introduction of activity coefficients for e' and $h^{\cdot}$. Departure from ideality of the free carriers arises because firstly, the carriers obey Fermi-Dirac rather than Maxwell-Boltzmann statistics and secondly, carrier-impurity and carrier-carrier interactions occur. The two effects can be represented as separate factors in the carrier activity coefficient. Thus the activity coefficient for both electrons and holes can be written as $\gamma = \gamma(FD)\gamma(\text{int})$ where $\gamma(FD)$ and $\gamma(\text{int})$ are respectively the factors representing the contributions to non-ideality due to Fermi-Dirac statistics and carrier interactions [28]. Rosenberg [29] and Hwang and Brews [26] have obtained the following expressions for $\gamma(FD)$ and $\gamma(\text{int})$ respectively:

$$\left.\begin{aligned} \gamma_n(\text{FD}) &= \exp(\eta_c/kT)/F_{1/2}(\eta_c/kT) \\ \gamma_p(\text{FD}) &= \exp(\eta_v/kT)/F_{1/2}(\eta_v/kT) \end{aligned}\right\} \tag{1.17a}$$

$$\left.\begin{aligned} \gamma_n(\text{int}) &= \exp\{(E_{1n} + E_{2n})/kT\} \\ \gamma_p(\text{int}) &= \exp\{(E_{1p} + E_{2p})/kT\} \end{aligned}\right\} \tag{1.17b}$$

E_1 and E_2 are defined in equation 1.15. The n or p subscript refers to the conduction and valence bands respectively. Explicit expressions are

$$\begin{aligned} E_{1n} &= -4\pi e^2 n r_s^2/\epsilon \\ E_{2n} &= 2e^2(3n/\pi)^{1/3}/\epsilon \end{aligned} \tag{1.17c}$$

with parallel expressions for E_{1p} and E_{2p}. Equation 1.17a assumes the validity of equations 1.11 and 1.12.

In formulating the equilibrium conditions it is important to note that the energy shift given by equation 1.15 will be different for the conduction and valence bands in a given doping situation.† The band associated with the minority carrier will have $E_2 \sim 0$ because of the low carrier concentration. E_1 will only be the same for the two bands if

† C. J. Hwang (private communication). I am grateful to Dr. Hwang for the following comments.

$m_n = m_p$, or when $m_n \neq m_p$, if r_s is small relative to the Bohr radii of electrons and holes. E_1 has not as yet been evaluated for the minority carrier band when neither of these conditions is fulfilled.

If, for a donor doped crystal, equation 1.12 is valid then from equations 1.11, 1.12, 1.17a and 1.17b we obtain

$$\gamma_n n = N_c \exp \left\{(\eta_c + \Delta E_n)/kT\right\} = N_c \exp \left\{(E_F - E_c^0)/kT\right\} \quad (1.18a)$$

$$\gamma_p p = N_v \exp \left\{(\eta_v + \Delta E_p)/kT\right\} = N_v \exp \left\{(E_v^0 - E_F)/kT\right\} \quad (1.18b)$$

where $\Delta E_n (= E_{1n} + E_{2n})$ is the shift in the conduction band edge relative to E_c^0 (equation 1.16) and is calculable from equation 1.17c, as given equation 1.12, then [26]

$$r_s^{-2} = (4\pi e^2/\epsilon kT) N_c F_{-\frac{1}{2}}(\eta_c/kT) \quad (1.19)$$

where the term $F_{-\frac{1}{2}}(\eta_c/kT)$ is the differential of $F_{\frac{1}{2}}(\eta_c/kT)$.

ΔE_p is the associated shift in the valence band edge ($E_v = E_v^0 - \Delta E_p$) and is not generally calculable. Equations 1.18a and 1.18b yield

$$\gamma_n n \gamma_p p = N_c N_v \exp \left\{(E_v^0 - E_c^0)/kT\right\} = n_i^2 \quad (1.20)$$

Once the values of γ_n and n are known then the value of $\gamma_p p$ is determined. The determination of the minority carrier concentration p calls however for a knowledge of ΔE_p.

1.2.3 Diffusion Mechanisms

Basic Mechanisms

Adequate accounts of the essential concepts involved in the basic mechanisms already exist [1, 30]. These mechanisms are listed in Table 1.1 and apply equally to the diffusion of solute and solvent atoms. Several if not all the mechanisms will be simultaneously operative for a given atomic species but, under given experimental conditions, the isotopic diffusion flux will usually be dominated by one mechanism. Thus the exchange and ring mechanisms are held to be ineffective competitors against the defect dependent processes because of the significantly lower Q values for the latter [1, 31]. A reduced Q for an exchange mechanism may however be possible in a compound semiconductor, say MX, for M atoms surrounding a V_X and vice versa.† Diffusion by the direct interstitial mechanism is favoured by atoms, generally of a solute, which are substantially smaller than the solvent atoms and occupy interstitial sites. According to the lattice type there may be more than one type of stable

† Such a process has been invoked in the self-diffusion of Cd in CdS (see section 7.4.2) and of O in Cu_2O (see section 8.5.2).

TABLE 1.1

Basic diffusion mechanisms in crystals

Defect independent processes	1. Direct exchange mechanism 2. Ring mechanism	
	Defect involved	Mechanism
Defect dependent processes	Interstitial	1. Direct interstitial 2. Interstitialcy (indirect interstitial) 3. Crowdion 4. Dissociative
	Vacancy	1. Vacancy 2. Relaxion
	Dislocation	1. Pipe diffusion 2. Grain boundary diffusion

interstitial site [32, 33].† With increasing size of the interstitial species the indirect interstitial, or interstitialcy mechanism, provides an easier mode for diffusion of both solute and solvent species. A variant of this for the latter species is the dumb-bell interstitialcy or split interstitial [34, 35]; a further variation for solute species has also been proposed [36] which involves an interstitital solute atom displacing a solvent atom at a lattice site which in turn displaces a solute atom at a nearest neighbour lattice site into an interstitial site. The interstitialcy mechanism has been established as the mode of Ag diffusion in the silver halides [37], in which a striking feature is the very small migration enthalpy, a value of 0.008 eV having been found for collinear jumps of Ag in AgCl. The crowdion is another solvent interstitial configuration by which it is possible for atoms to migrate through the lattice [34]. The dissociative mechanism was specifically introduced by Frank and Turnbull [38] to account for the diffusion of Cu in Ge. This is a solute diffusion mechanism in which the solute occupies both substitutional and interstitial sites but is only mobile, by the direct interstitial mechanism, when in the latter sites. The diffusion of Cu and Ni in Ge and Si, of Au in Si (section 5.4.2) and of Cu, Zn and Cd in GaAs (section 6.3) has been interpreted in terms of this mechanism.

In terms of Q the vacancy or relaxion mechanisms are the most probable diffusion modes after interstitial ones at least as far as volume diffusion is concerned. Variants of the relaxion model have been described [39, 40, 41]. The concepts of extended vacancies and interstitials have

† Bourgoin and Corbett [33a] have proposed that if the equilibrium site of an interstitial depends upon its charge state then diffusion of the interstitial can occur due to successive changes in the charge state.

been introduced as a basis for explaining the large values of D_o (in equation 1.2) observed in self-diffusion in Ge and Si [39, 40].

Stable associations between point defects also occur which are capable of diffusing as a single entity, such as the di-interstitial [34] and di-vacancy [34, 42]. The di-vacancy has also been considered as the mechanism for self-diffusion in Si [42] and a full discussion of this aspect is given in section 5.3. Diffusion of an interstitial solute-vacancy pair has also been described [43]. No new concepts are needed however to describe the movements involved.

Diffusion in dislocations is discussed separately in section 1.5.

A prominent aspect of defects in semiconductors is the ability of a particular defect to be present in various states of ionization (equations 1.13 and 1.14). It is reasonable to expect that the jump rate of a defect will depend on its charge state. This is an important relationship but little information, either experimental or theoretical, is available. In Si there is evidence that a V'' is considerably more mobile than a $V^{\times}$ [44] and in the II-VI compounds evidence is emerging that the native defects which are dominant electrically are not necessarily dominant in self-diffusion (section 7.4.4).

A solute atom can also occupy both substitutional and interstitial sites with each site providing multivalent states. In a compound semiconductor there will be more than one sub-lattice providing substitutional and interstitial sites for a solute. Although the solubility on one site may be orders of magnitude greater than on others, low solubility is coupled to a high mobility [45], so that solute diffusion could be determined by the occupancy of the low solubility sites.

Correlation Effects

The jump kinetics on which equations 1.4 to 1.7 are based assume that successive atom jumps are wholly random i.e. the directions of consecutive jumps are quite independent of each other. Whether this assumption is valid depends upon the diffusion mechanism. Thus the jumps of a tracer atom are related to each other in a vacancy mechanism [1]. This dependence is known as correlation and can be simply allowed for by introducing a correlation factor f into the right-hand side of equations 1.4 and 1.5. LeClaire [46] has given a recent and comprehensive review of correlation effects and what follows is drawn largely from this review and the account by Manning [47].

The value of f depends upon the diffusion mechanism, the lattice type and whether solute or solvent diffusion is being considered. In non-dilute alloys f may also be a function of composition. When $f = 1$ the jumps are uncorrelated, which is the situation for the direct interstitial and dissociative mechanisms, provided the interstitial concentration is small enough to avoid interference effects between neighbouring interstitials. In the non-defect mechanisms correlation effects may or may not occur depending on the number of atomic species participating in the mechanism. Correlation effects are present in the vacancy and interstitialcy mechanisms. They may arise with a relaxion mechanism depending on the structure of the relaxion. A crowdion mechanism is expected to show strong correlation effects, i.e. $f \ll 1$. It is worth noting that although for vacancy diffusion in a pure solvent the tracer jumps are correlated the vacancy jumps are not. When f becomes dependent on composition it is necessary to introduce a vacancy correlation factor which will also be a function of composition.

Values of f have been calculated for vacancy and interstitialcy tracer self-diffusion in a variety of lattices [46]. In an anisotropic lattice the jumps will generally be different along different crystal axes so that it is necessary to introduce directional correlation factors e.g. $f_\perp$ and $f_\parallel$ into equations 1.6 and 1.7. f values for self-diffusion by divacancies and bound vacancy pairs, (V_M, V_X), have also been calculated. Miller [43] has obtained an expression for the solute correlation factor in diffusion by an interstitial solute-vacancy pair.

The most important aspect of correlated jumps appears in isoconcentration diffusion where correlation effects can lead to $f \ll 1$, and hence small diffusion coefficient, for a solute atom. Correlation factors in isoconcentration diffusion by both vacancy and interstitialcy mechanisms have been calculated. In the former case

$$f = H/(2w_2 + H) \tag{1.21}$$

where H depends upon the lattice type and the various possible vacancy jump rates w_0, w_1, w_2, w_3, w_4, w_5 which are defined in figure 1.1. Table 1.2 lists expressions for H obtained by Manning [47]. The F factors are functions of w_0, w_4 and w_5 and figure 1.2 shows the values of the coefficient of w_3, in the expressions for H of Table 1.2, as a function of the appropriate jump rate ratio. The various vacancy jump rates arise of

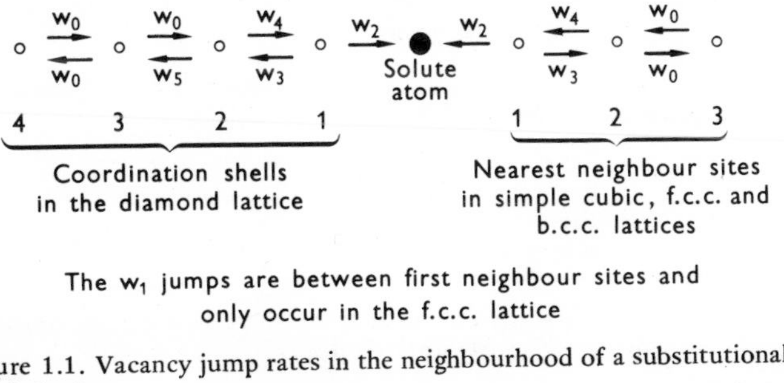

w_0 w_0 w_4 w_2 w_2 w_4 w_0

w_0 w_5 w_3 Solute atom w_3 w_0

4 3 2 1 1 2 3

Coordination shells in the diamond lattice

Nearest neighbour sites in simple cubic, f.c.c. and b.c.c. lattices

The w_1 jumps are between first neighbour sites and only occur in the f.c.c. lattice

Figure 1.1. Vacancy jump rates in the neighbourhood of a substitutional impurity in a cubic lattice.

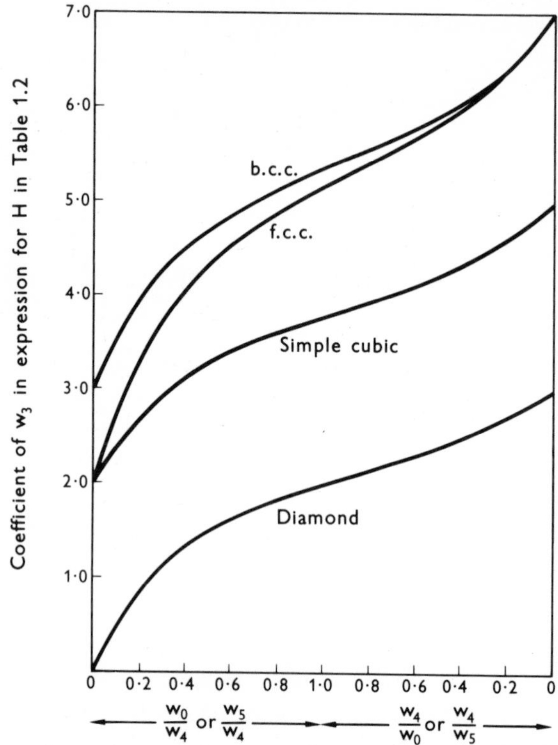

Figure 1.2. The variation of the coefficient of w_3 in the expressions for H (Table 1.2) with the ratios of w_0 and w_5 to w_4. The data for the simple cubic lattice were kindly made available by Dr. J. R. Manning. The remaining data are from ref. 47.

course from solute-vacancy interactions (section 1.2.4). When all the jump rates are equal f corresponds to self-diffusion. f will vary with temperature for isoconcentration diffusion, and also for self-diffusion, if there is more than one type of jump contributing to diffusion in a given direction (e.g. collinear and non-collinear in the interstitialcy mechanism). Thus the observed Q can include a contribution from f.

TABLE 1.2.
Expressions for H in $f = H/(2w_2 + H)$

Crystal lattice	H
Simple cubic	$5Fw_3$
Body centred cubic	$7Fw_3$
Face centred cubic	$2w_1 + 7Fw_3$
Diamond†	$3Fw_3$

† More detailed calculations of f for solute diffusion by a vacancy mechanism in a diamond lattice, with specific applications to Si, have recently been made [47a, 47b].

Tracer Diffusion Coefficients

The tracer diffusion coefficients for vacancy and interstitial mechanisms involve the concentrations of either type of defect. This dependence is examined below for diffusion in the cubic M sub-lattice of a compound semiconductor in which only nearest neighbour jumps are considered.

Vacancy Mechanism. In order for a tracer atom $\tilde{M}_M$ to jump to an adjacent site that site must be vacant. The probability that such a site will be vacant is given by $[V_M]/[M_M]$ provided $[V_M] \ll [M_M]$. If we consider first self-diffusion involving only $V_M^{\times}$ then the jump rate of $\tilde{M}_M$ is given by

$$\Gamma(\tilde{M}_M) = Z w_0(V_M^{\times})[V_M^{\times}]/[M_M] \tag{1.22}$$

where Z is the number of nearest neighbour sites, and $w_0(V_M^{\times})$ is defined in figure 1.1. Equation 1.22 is easily extended to include ionized states of V_M. Thus for acceptor states

$$\Gamma(\tilde{M}_M) = (Z/[M_M])\left\{w_0(V_M^{\times})[V_M^{\times}] + w_0(V_M')[V_M'] + \cdots\right\} \tag{1.23}$$

Using equation 1.5 and taking correlation effects into account we then have for the tracer diffusion coefficient

$$D_e(\tilde{M}) \equiv D_e(\tilde{M}_M) = \Gamma(\tilde{M}_M) f_V a^2/6 \tag{1.24}$$

where f_V is the vacancy correlation factor and $\Gamma(\tilde{M}_M)$ is given by equation 1.23. It also follows that the average jump rate of a $V_M^{\times}$ is $\Gamma(V_M^{\times}) = Z w_0(V_M^{\times})$ and similarly for other charge states so that equation 1.24 can be rewritten as

$$D_e(\tilde{M}) = (f_V/[M_M])\left\{D_e(V_M^{\times})[V_M^{\times}] + D_e(V_M')[V_M'] + \cdots\right\} \tag{1.25}$$

Turning now to the diffusion of a substitutional solute B_M a similar analysis gives

$$D_e(\tilde{B}) = (Za^2/6[M_M]\left\{f_B(V_M^{\times}) w_2(V_M^{\times})[V_M^{\times}] + f_B(V_M') w_2(V_M')[V_M'] + \cdots\right\} \tag{1.26a}$$

where the f factors are given by equation 1.21 and the w_2 jump rates by figure 1.1. The vacancy concentrations refer to nearest neighbour sites of B_M. If B_M is present in more than one charge state then each charge state will have a diffusion coefficient given by an expression like equation 1.26a. The tracer coefficient will then be given by

$$D_e(\tilde{B}) = D_e(\tilde{B}_M^{\times}) + \sum_{r,s}\left\{D_e(\tilde{B}_M^{r\prime}) + D_e(\tilde{B}_M^{s\cdot})\right\} \tag{1.26b}$$

Although from section 1.2.2 we might expect only one V_M charge state to be dominant in a given instance, it does not follow that this charge state will be dominant in either equation 1.25 or 1.26 as the various vacancy jump rates are also involved.

Suppose that only V_M' is dominant in diffusion so that from equations 1.25 and 1.26a for diffusion in the same crystal at the same temperature

$$\frac{D_e(\widetilde{B})}{D_e(\widetilde{M})} \equiv \frac{D_e(\widetilde{B}_M)}{D_e(\widetilde{M}_M)} = \frac{f_B(V_M')w_2(V_M')[V_M']_1}{f_V w_0(V_M')[V_M']_0} \tag{1.27}$$

where the subscripts 1 and 0 on $[V_M']$ denote concentrations at first neighbour sites to, and beyond the influence of B_M respectively. That $[V_M']_1 \neq [V_M']_0$ is a consequence of the differing jump rates because, in equilibrium, relations like $[V_M']_1/[V_M']_2 = w_4(V_M')/w_3(V_M')$ must exist (the subscript 2 denoting a second nearest neighbour site to B_M). With reference fo figure 1.1 we then have†

$$\frac{[V_M']_1}{[V_M']_0} = \frac{w_4(V_M')w_0(V_M')}{w_3(V_M')w_5(V_M')} \quad \text{or} \quad \frac{w_4(V_M')}{w_3(V_M')}$$

for the diamond and other cubic structures respectively. For the latter, equation 1.27 then becomes

$$\frac{D_e(\widetilde{B}_M)}{D_e(\widetilde{M}_M)} = \frac{w_2(V_M')w_4(V_M')f_B(V_M')}{w_3(V_M')w_0(V_M')f_V} \tag{1.28}$$

Comparison of $D_e(\widetilde{B}_M)$ and $D_e^i(\widetilde{M}_M)$ at the same temperature gives

$$\frac{D_e(\widetilde{B}_M)}{D_e^i(\widetilde{M}_M)} = \frac{f_B(V_M')w_2(V_M')[V_M']_1}{f_V w_0(V_M')[V_M']_{i,0}} \tag{1.29}$$

with the subscript 'i, 0, denoting $[V_M']_0$ under intrinsic conditions i.e. $[B_M] = 0$. We also have that

$$\frac{[V_M']_1}{[V_M']_{i,0}} = \frac{[V_M']_1}{[V_M']_0}\frac{[V_M']_0}{[V_M']_{i,0}}$$

The ratio $[V_M']_0/[V_M']_{i,0}$ is governed by the electroneutrality condition in the extrinsic crystal: it follows from equation 1.14 that this ratio will be $\gg 1$ in a strongly n-type crystal and $\ll 1$ in a strongly p-type crystal. Little is known about the relative values of the jump rates contained in equations 1.27, 1.28 and 1.29 other than if B_M and V_M attract or repel each other then $w_4(V_M') \gtrless w_3(V_M')$ respectively. It is also evident that $D_e^i(\widetilde{B}_M)/D_e^i(\widetilde{M}_M)$ follows from equation 1.27. Recent experiments in Si show that [48] $\mathbf{D}^i(\widetilde{\text{As}})/D_e^i(\widetilde{\text{Si}}) = 6.7 \times 10^{-3} \exp(0.93/kT)$. Although $\mathbf{D}^i(\widetilde{\text{As}})$ was actually measured it is not unreasonable, because of intrinsic conditions, to equate $\mathbf{D}^i(\widetilde{\text{As}})$ with $D_e^i(\widetilde{\text{As}})$. Hu [49] has given an analysis of vacancy jump kinetics

† Neglecting variations in the effective coordination numbers at the different sites.

for substitutional solute diffusion in a diamond lattice with special reference to Si.

LeClaire [46] has pointed out that when

$$w_2(V'_M) \gg w_1(V'_M) \sim w_3(V'_M) \sim w_4(V'_M) \sim w_0(V'_M)$$

then

$$D_e(\tilde{B}_M) \sim a^2 w_0(V'_M)[V'_M]_0/[M_M]$$

and the experimentally measured Q will not contain the activation enthalpy for B_M jumps.

Interstitial Mechanisms. Consider the diffusion of a native interstitial tracer $\tilde{M}_i^{\times}$, then by Fick's first law

$$J(\tilde{M}_i^{\times}) = -D_e(\tilde{M}_i^{\times})\partial[\tilde{M}_i^{\times}]/\partial x \tag{1.30}$$

Local isotopic equilibrium will however ensure that, provided isotopic mass effects can be neglected, $[\tilde{M}_i^{\times}]/[\tilde{M}_M] = [M_i^{\times}]/[M_M]$ and usually this ratio will be $\ll 1$. Experimentally only the total tracer concentration $[\tilde{M}] = [\tilde{M}_M] + [\tilde{M}_i^{\times}]$ can be measured so that equation 1.30 can be written as

$$J(\tilde{M}_i^{\times}) = -\mathbf{D}_e(\tilde{M}_i^{\times})\frac{\partial}{\partial x}([\tilde{M}_i^{\times}] + [\tilde{M}_M]) \tag{1.31a}$$

which defines an effective diffusion coefficient $\mathbf{D}_e(\tilde{M}_i^{\times})$ for $\tilde{M}_i^{\times}$, or as

$$J(\tilde{M}_i^{\times}) \equiv J(\tilde{M}) = -D_e(\tilde{M})\frac{\partial}{\partial x}[\tilde{M}] \tag{1.31b}$$

which defines the tracer coefficient $D_e(\tilde{M})$ for self-diffusion of $\tilde{M}$ in MX. Clearly $D_e(\tilde{M}) \equiv \mathbf{D}_e(\tilde{M}_i^{\times})$. If however there is more than one interstitial species contributing to the tracer flux then $D_e(\tilde{M})$ will now be equal to the sum of the effective diffusion coefficients of the participating species. Now $[\tilde{M}] = [M]\,[\tilde{M}_i^{\times}]/[M_i^{\times}]$, where $[M] = [M_M] + [M_i^{\times}]$ and we also expect $[M]/[M_i^{\times}]$ to be constant, so substitution for $[\tilde{M}]$ in equations 1.31a and 1.31b yields after comparison with equation 1.30

$$\mathbf{D}_e(\tilde{M}_i^{\times}) \equiv D_e(\tilde{M}) = ([M_i^{\times}]/[M])D_e(\tilde{M}_i^{\times}) \tag{1.32}$$

In a dissociative mechanism involving a solute B, which is present as two species B_M and B_i each of which is in a single charge state, a similar approach leads to

$$D_e(\tilde{B}) \equiv \mathbf{D}_e(\tilde{B}_i) = D_e(\tilde{B}_i)\left\{1 + \frac{[B_M]}{[B_i]}\right\}^{-1} \tag{1.33}$$

where $D_e(\widetilde{B})$ is the isoconcentration diffusion coefficient. $\mathbf{D}_e(\widetilde{B}_i)$ is the effective diffusion coefficient for the mobile tracer species $\widetilde{B}_i$. The case of the direct interstitial follows if $[B_M] = 0$. For neutral centres the ratio $[B_M^\times]/[B_i^\times]$ will depend on $[V_M^\times]$ through the mass action relation $[B_i^\times]\,[V_M^\times] = K_B\,[B_M^\times]$. $[B_i^\times]$ will also be directly proportional to the chemical activity of $B^\times$ in an adjacent phase (e.g. vapour). In the event of ionized states of B_M and B_i being present the electroneutrality condition must be known in order to determine the concentrations of the various charge states. If B_i is present in more than one charge state then $D_e(\widetilde{B})$ will be the sum of the effective diffusion coefficients for each charge state (see for example equation 1.40).

If differences due to isotopic mass are neglected then with reference to equations 1.32 and 1.33 we can expect $D_e(\widetilde{M}_i^\times) = D_e(M_i^\times)$, $D_e(\widetilde{B}_i) = D_e(B_i)$. Generally however the mass of a tracer isotope differs from that of the non-tracer isotope and the jump rate of a diffusing atom does depend on its mass, so that it is not obvious that isotopic mass effects can be neglected. LeClaire [46] has estimated that isotopic mass differences will lead to a difference in jump rates for tracer and non-tracer isotopes of the same element of ~2% and so for most purposes the effect of isotopic mass variations on a diffusion coefficient can be neglected.

With self-diffusion by an interstitialcy mechanism it can be shown that the tracer diffusion coefficient for collinear jumps is [50]

$$D_e(\widetilde{M}) = f_M\,[M_i^\times]\,D_e(M_i^\times : M_M)/2\,[M_M] \tag{1.34}$$

assuming a single interstitial species $M_i^\times$. f_M is the interstitialcy correlation factor for collinear jumps.

The Isotope Effect

The atomic masses of tracer and non-tracer atoms of the same element are generally unequal. As a consequence the jump rates of different isotopes of the same element will differ. This difference is the source of the isotope effect which has been reviewed by LeClaire [46]. The effect is manifest in both solute and solvent diffusion and when $f = u/(w + u)$, where w is the jump rate of the diffusant and u contains only the jump rates of the host atoms, its 'strength' E_β^α is given by

$$E_\beta^\alpha = \frac{(D_e^\alpha - D_e^\beta)/D_e^\beta}{\left\{\dfrac{m_\beta + (s-1)m_\gamma}{m_\alpha + (s-1)m_\gamma}\right\}^{1/2} - 1} = f^\alpha \Delta K \tag{1.35}$$

where m_α and m_β are the isotopic masses of two radio-isotopes α and β of the same element; s is the number of atoms involved in an elementary jump ($s = 1$ for vacancy and direct interstitial jumps, $s = 2$ for interstitialcy jumps); m_γ is the average mass of the non-tracer atoms. f^α is the correlation factor and ΔK is the fraction of the total translational kinetic energy possessed by the tracer atom in the diffusion direction at the saddle point, so that $\Delta K \leqslant 1$. The quantities in the middle of equation 1.35 are all measurable so that E^α_β can be obtained.

The isotope effect is generally small and, other than for light elements, $(D^\alpha_e - D^\beta_e)/D^\beta_e < 10\%$. Equation 1.35 will also apply to chemical diffusion provided the diffusion coefficient is independent of concentration. Very few measurements of the isotope effect in semiconductors have been reported. In Si, Pell [51] found $E^\alpha_\beta = 0.94 \pm 0.25$ for Li6 and Li7. The isotope effect has also been observed for Co and Fe diffusion in cobaltous and ferrous oxides respectively: these two cases are discussed more fully in section 8.5.2. The experimental technique commonly used is to diffuse the radioisotopes α and β simultaneously; the effect can be enhanced if an electric field is also applied [52].

It should also be noted that equation 1.35 is based on transition rate theory in which irreversibility of an atomic jump is necessarily assumed. An alternative theory by Prigogine and Bak, based on a jump model with 'built in' irreversibility, has been used to show that when m_α, $m_\beta \ll m_\gamma$ a very much stronger isotope effect than predicted by rate theory should be found [53]. For m_α, $m_\beta \gg m_\gamma$ the effect given by rate theory is greater. The application of the theory to experiment is at present only at a qualitative stage.

1.2.4 Defect Interactions

This section deals with the ways in which interactions of defects with the solvent lattice and with other defects can affect both solute and solvent diffusion. Magnetic interactions are excluded. Interactions between defects and dislocations are discussed in section 1.5.1.

Interactions with the Solvent Lattice

The primary interaction of a point defect with an otherwise pure solvent lattice is a thermodynamic one which determines the solubility of the defect. This forms the substance of Chapter 3. Here we shall be concerned with changes in the parameters of the solvent lattice due to the presence of point defects.

The incorporation of point defects, native or foreign, changes the lattice constant of the solvent lattice [54, 55, 56, 57]. This is due to the size misfit between a defect and solvent atom which results in a local contraction or dilatation of the lattice surrounding the defect. The resultant average lattice strain is given by βc where β is the lattice contraction or dilatation coefficient and c is the defect concetration. The fractional change in the lattice constant of a cubic lattice is equal to βc [56]. A change Δa_o in the lattice constant a_o can change (i) the band gap, (ii) the concentrations of native defects through changes in formation enthalpies and (iii) the atomic jump rates because of changes in the activation enthalpies.

A change of $\Delta\epsilon_g$ in ϵ_g can be important in that n_i will be affected and hence the concentrations of ionized defects. In order to estimate $\Delta\epsilon_g$ we assume that the average internal stress is equivalent to an externally applied hydrostatic stress. This then gives [58]

$$\Delta\epsilon_g = -\frac{3}{\chi}\frac{\Delta a_o}{a_o}(\partial\epsilon_g/\partial P)_T = -\frac{3\beta c}{\chi}(\partial\epsilon_g/\partial P)_T \tag{1.36}$$

where χ and $(\partial\epsilon_g/\partial P)_T$ are the lattice compressibility and band gap variation with applied pressure, P, respectively. Typical values are $\beta = \pm 2 \times 10^{-24}\,\text{cm}^3/\text{atom}$, $\chi = 10^{-12}\,\text{cm}^2\,\text{dyne}^{-1}$ and $(\partial\epsilon_g/\partial P)_T = 10^{-11}\text{eV cm}^2\,\text{dyne}^{-1}$, and substitution in equation 1.36 yields $\Delta\epsilon_g = \pm 6 \times 10^{-23}\,c$ eV. Even with $c = 10^{21}\,\text{cm}^{-3}$, $\Delta\epsilon_g$ is only ± 0.06 eV so that in dilute solutions $\Delta\epsilon_g$ is unlikely to be large enough to have a significant effect on ionization equilibrium except perhaps for small band gap materials. In concentrated solutions (i.e. alloys or mixed crystals) a_o and ϵ_g both vary continuously with composition and the magnitudes of the observed variations are comparable to those obtained by extrapolation of the above estimates [59, 60]. Account must now of course be taken of the compositional dependence of all aspects of a diffusion mechanism.

Girifalco and Welch [61] have considered theoretically the effect of lattice strain on diffusion and the concentration of native defects in Cu. The situation is a complex one and in semiconductors, because of a lack of knowledge of the relevant parameters, only an order of magnitude estimate can be attempted. Adopting the same premise on which equation 1.36 is based gives

$$\Delta(\ln D_e) = -\frac{3\beta c}{\chi}(\partial \ln D_e/\partial P)_T \tag{1.37}$$

for the fractional change in the diffusion coefficient. Taking $\beta = \pm 2 \times 10^{-24}\,\text{cm}^3/\text{atom}$, $\chi = 10^{-12}\,\text{cm}^2\,\text{dyne}^{-1}$ and a value, based on self-diffusion under pressure in metals [11], of $10^{-10}\,\text{cm}^2\,\text{dyne}^{-1}$ for $(\partial \ln D_e/\partial P)_T$ yields $\Delta(\ln D_e) = \pm 6 \times 10^{-22}\,c$. For $c > 10^{20}\,\text{cm}^{-3}$ significant changes, say $\gtrsim 10\%$, can clearly occur in D_e. The reliability of this estimate is obviously uncertain but it does suggest that any dependence of D_e on c, at least in dilute solution is more likely to be due to changes in defect formation and jump activation enthalpies rather than to effects consequent on a change in ϵ_g. Changes in jump activation enthalpies will also cause a change in the correlation factor for mechanisms which involve more than one type of jump.

Interactions between Defects

Two broad categories can be recognized under this heading: (i) the electronic interaction which determines E_F and the solubility of ionized defects [62], and (ii) those interactions which act directly between defects with a strength related to the separation of the defects [62, 63].

Longini and Greene [64] first pointed out that with a vacancy mechanism the diffusion coefficient may vary with E_F due to the varying concentrations of ionized vacancies, as shown by equations 1.25 and 1.26. The varying solubility of V_M' was also utilized in the discussion of equation 1.29. Such effects can be quite large: a shift in E_F of 0.5 eV with $kT \sim 0.1$ eV will change concentrations by a factor $\sim 10^2$. Consider now diffusion, by the dissociative mechanism, of a fully ionized monovalent acceptor B_M' with singly and doubly ionized interstitials $B_i^{\cdot}$ and $B_i^{\cdot\cdot}$. The total tracer flux is then given by

$$J(\widetilde{B}) = -D_e(\widetilde{B}_i^{\cdot})\frac{\partial[\widetilde{B}_i^{\cdot}]}{\partial x} - D_e(\widetilde{B}_i^{\cdot\cdot})\frac{\partial[\widetilde{B}_i^{\cdot\cdot}]}{\partial x}$$

which leads to the tracer diffusion coefficient (isoconcentration)

$$D_e(\widetilde{B}) = \frac{D_e(\widetilde{B}_i^{\cdot})[B_i^{\cdot}] + D_e(\widetilde{B}_i^{\cdot\cdot})[B_i^{\cdot\cdot}]}{[B_M'] + [B_i^{\cdot}] + [B_i^{\cdot\cdot}]} \tag{1.38}$$

Ionization equilibrium is described by

$$B_i^{\cdot} + V_M^{\times} \rightleftharpoons B_M' + 2b^{\cdot}; \quad B_i^{\cdot\cdot} \rightleftharpoons B_i^{\cdot} + b^{\cdot}$$

which yield by the law of mass action

$$[B_M'](\gamma_p p)^2 = K_{\text{Diss}}[V_M^{\times}][B_i^{\cdot}]; \quad [B_i^{\cdot\cdot}] = K_B[B_i^{\cdot}]\gamma_p p \tag{1.39}$$

where K_{Diss} and K_B are the appropriate mass action constants. Eliminating the defect concentrations between equations 1.38 and 1.39 gives

$$D_e(\widetilde{B}) = D_e(\widetilde{B}_i^{\cdot})\left\{1 + K_B\gamma_p p + \frac{K_{\text{Diss}}[V_M^{\times}]}{(\gamma_p p)^2}\right\}^{-1}$$

$$+ D_e(\widetilde{B}_i^{\cdot\cdot})\left\{1 + \frac{1}{K_B\gamma_p p} + \frac{K_{\text{Diss}}[V_M^{\times}]}{K_B(\gamma_p p)^3}\right\}^{-1} \tag{1.40}$$

The electroneutrality condition in this case is

$$n + [B_M'] = p + [B_i^{\cdot}] + 2[B_i^{\cdot\cdot}] \tag{1.41}$$

The situation summarized by equations 1.40 and 1.41 is obviously complex but not physically unrealistic. Simplification can be achieved only if one of the ionized components dominates equation 1.41 and if one of the interstitial species is effectively immobile. It is interesting to note however that even though ionized states of V_M may be present only the neutral state is involved in equation 1.40.

Kroger [63] has discussed interactions in the second category which can be attractive or repulsive due to either Coulomb, strain, electron orbital or localized lattice vibrational mode (around a defect) interactions. These various interactions are the physical bases for the different jump frequencies shown in figure 1.1. Such interactions will cause a given defect to attract or repel defects of other species so that the distributions of the latter cease to be random in the vicinity of the given defect. Thus the local vacancy and/or native interstitial concentrations surrounding a solute atom can be enhanced or depressed relative to the concentrations at large distances from the solute atom. Although Kroger [63] considered Coulomb and covalent effects to be paramount in any interaction between defects it seems probable that the strain interaction can be of comparable magnitude [65, 66]. Considering only the Coulomb and strain interactions, then, if ΔH_c and ΔH_s denote the respective interaction energies between a substitutional solute atom and, say, a V_M' at a nearest neighbour site then with reference to equation 1.27 [65]

$$[V_M']_1 = [V_M']_0 \exp\left\{-(\Delta H_c + \Delta H_s)/kT\right\} \tag{1.42}$$

The specific charge state of V_M is clearly not important in determining the form of equation 1.42.†

Coulomb interactions beyond the second nearest neighbour separation can be neglected [63] and this would appear to be valid also for the strain interaction [66]. Extending the condition of Reiss *et al.* [67] on ΔH_c we can say that if the total interaction energy between two defects is $\leqslant -2kT$ then the two defects can be regarded as a bound pair, or associate. Associates can form between defects in the same or different sub-lattice and can be treated as a separate defect species [62]. Triplet and higher multiple associates are also feasible [63, 68].

The existence of defect associates in semiconductors has been well established through electrical measurements [69], luminescence [70], infra-red absorption [71] and ESR [33]. Their role however in diffusion in semiconductors appears to have been ignored perhaps because they are essentially a low temperature phenomenon and so can be neglected at the

† Generally in this chapter, even though $[V_M']_1 \neq [V_M']_0$, the vacancy concentrations which are used do not differentiate explicitly between sites. It should be clear from the context which particular concentration is involved.

higher temperatures where tracer diffusion experiments are usually made. An important and special case however is the pairing of a substitutional solute atom, B_M, with a vacancy in its own sub-lattice. If B_M only diffuses by a vacancy mechanism then clearly it must catch a vacancy and form a pair, however transient, before it can move. Such pairing is important in the alkali halides [72]. Associations between different solute species however are likely to be immobile relative to the isolated solute atoms so leading to an effective diffusion coefficient for a species which is smaller than that for the isolated species.

The influence, through correlation effects, of a solute-vacancy interaction on the diffusion coefficient of a solute diffusing by a vacancy mechanism has already been noted in section 1.2.3. If the solvent atoms also diffuse by a vacancy mechanism then it is possible that the solute-vacancy interaction will modify the solvent diffusion coefficient. Howard and Manning [73] investigated this situation for dilute Ag alloys (*fcc*). Their results however can be applied directly to the intrinsic semiconductor case to yield

$$D_e^i(\tilde{M}_M) = D_e^i(\tilde{M}_M, 0)\,(1 + \alpha\,[\mathbf{B}_M]) \tag{1.43}$$

which is valid for $[\mathbf{B}_M] \lesssim 2\%$ mole fraction. The first term on the right-hand side of equation 1.43 is the solvent diffusion coefficient when $[\mathbf{B}_M] = 0$; α is a function of the vacancy jump frequencies defined in figure 1.1 and can take positive or negative values. Fitting to experimental data in Ag shows α varies with the solute and that $-8 < \alpha < 90$ [73] Equation 1.43 should also apply to the extrinsic range for $V_M^\times$. For ionized vacancies, say V_M', equation 1.43 is still applicable if a factor $[V_M']/[V_M']_i$ is introduced on the right-hand side. Generally the jumps of a solvent atom are likely to be altered in the vicinity of a solute atom whatever lattice site is occupied by the solute and whatever the diffusion mechanism. Thus $D_e(\tilde{M}_M)$ should vary with $[B]$. The magnitude of the variation is rather uncertain but the results of Howard and Manning [73] suggest significant effects for $[B] \gtrsim 10^{20}$ cm^{-3}. LeClaire [74] has discussed the application of equation 1.43 to dilute solutions in *bcc* metals.

Briefly summarizing this section we see that defect interactions with both the lattice and other defects give rise to a variety of effects which according to circumstances may oppose or reinforce each other in the overall influence on solvent and solute diffusion. The strongest effect is probably the electronic interaction which governs the mutual solubility of ionized defects, but which will only be operative under extrinsic conditions. The other effects will still be present in intrinsic crystals but it is then probable that $[B]$ will be too low for them to be of significance.

The experiments that have been done in semiconductors which permit interpretation in terms of the effects described in this section have all relied for interpretation upon the electronic interaction. Self-diffusion in Ge and Si under intrinsic and extrinsic conditions (section 5.3), the isoconcentration diffusion of Zn and Cd in the III-V compounds (section 6.3.7, tables 6.4, 6.5, 6.6), self-diffusion under extrinsic conditions in the chalcogenides (Chapter 7) and oxides (Chapter 8) are the principal areas where experimental work has been done.

1.2.5 Identifying the Diffusion Mechanism

In any real situation a variety of mechanisms can be expected to be simultaneously active. Unfortunately there is no one experimental technique which enables these mechanisms to be scanned: the available methods are all in some degree selective (Chapter 4). Although the tracer method observes the total tracer flux, and therefore the combined mechanisms, only those mechanisms giving the highest flux gradients will be 'seen'. Attention in this section will be limited to the tracer method because of its wide applicability and use in self and isoconcentration diffusion experiments. The necessary use of an infinite source boundary condition ensures that in the experimental profiles, volume diffusion, given by $c = c_0 \operatorname{erfc}(x/2\sqrt{D_e t})$, can be distinguished from type B diffusion involving dislocations (section 1.5.3). It will be assumed that only one mechanism is operative.

Self-Diffusion

Two general approaches are possible. The first is based on the isotope effect and a determination of its strength. From equation 1.35 $E^{\alpha}_{\beta} = f^{\alpha} \Delta K$ and as $\Delta K \leqslant 1$ then $E^{\alpha}_{\beta} \leqslant f^{\alpha}$ so that mechanisms with $f^{\alpha} < E^{\alpha}_{\beta}$ can be immediately excluded. Whether or not a firm identification can then be made depends upon the particular circumstances [46]. The method calls for high experimental precision and the availability of suitable radioisotopes.

The second approach studies the dependence of self-diffusion, at a given temperature, on E_F and also, for compounds, on the degree of non-stoichiometry. E_F can be varied by impurity doping and by control of non-stoichiometry. In short, solvent diffusion is studied as a function of the concentrations of ionized and neutral defects in the semiconductor. If direct proportionality of the solvent diffusion coefficient to a particular defect concentration is found (e.g. equations 1.22 and 1.31) this identifies the mechanism and the defect. An invariance to E_F means that if a point defect is involved it must either be neutral or, if ionized, require very specific electroneutrality condition(s): the existence of the latter can be confirmed (or otherwise) by measurements of electrical conductivity versus doping level or non-stoichiometry. Such electrical measurements are preferably performed at the same temperature as the diffusion experiments rather than at room temperature following a high temperature anneal. Self-diffusion in Ge and Si has been studied in terms of background impurity doping levels (section 5.3) whereas in the oxide and chalcogenide compound semiconductors component self-diffusion has been investigated mainly as a function of non-stoichiometry.

The more complicated case of non-stoichiometric compounds is best illustrated by an example. Consider the M sub-lattice which is in equilibrium with an ambient monatomic vapour of partial pressure p_M. Let $M_i^\times, M_i^\cdot, M_i^{\cdot\cdot}, V_M^\times, V_M', V_M''$ be present together with a totally ionized donor $D^\cdot$ whose presence is only important in the electroneutrality condition. Any other ionized defects are neglected and non-degeneracy is assumed. The electroneutrality condition is then expressed by

$$p + [M_i^\cdot] + 2[M_i^{\cdot\cdot}] + [D^\cdot] = n + [V_M'] + 2[V_M''] \tag{1.44}$$

Defect and ionization equilibria are described by

$$\begin{aligned} p_M[V_M^\times] &= K_R(M); \quad [M_i^\times][V_M^\times] = K_F(M) \\ n[M_i^\cdot] &= K_1[M_i^\times]; \quad p[V_M'] = K_3[V_M^\times] \\ n[M_i^{\cdot\cdot}] &= K_2[M_i^\cdot]; \quad p[V_M''] = K_4[V_M'] \\ np &= n_i^2 \end{aligned} \tag{1.45}$$

From equations 1.25 and 1.31 we then have by substitution from the set of equations 1.45 that

$$\begin{aligned} [M_M]D_e(\tilde{M}) = (K_R(M)f_V/p_M)\left\{D_e(V_M^\times) + \frac{K_3}{p}D_e(V_M') + \frac{K_3K_4}{p^2}D_e(V_M'')\right\} \\ + (K_F(M)p_M/K_R(M))\left\{D_e(\tilde{M}_i^\times) + \frac{K_1}{n}D_e(\tilde{M}_i^\cdot) + \frac{K_1K_2}{n^2}D_e(\tilde{M}_i^{\cdot\cdot})\right\} \end{aligned} \tag{1.46}$$

Use of the set 1.45 in equation 1.44 leads to an equation containing only the variables n (or p) and p_M. The variation of each term with p_M in equation 1.46 is then defined. Brouwer's [17] approximation applied to equation 1.44 allows a considerable simplification and in Table 1.3 the dependence on p_M for each term in equation 1.46 is shown for different electroneutrality conditions. As an illustration suppose it was found that $D_e(\tilde{M}) \propto p_M^{-1/3}$. Table 1.3 shows that the diffusing defect could be a V_M', V_M'' or $M_i^{\cdot\cdot}$ depending on the electroneutrality condition. If at the diffusion temperature the electrical conductivity varies as $p_M^{1/3}$ (n-type) or $p_M^{-1/3}$ (p-type) this identifies the diffusing defect as a V_M''. On the other hand

TABLE 1.3

Dependence of free carriers and native defect concentrations on p_M and $[D^\cdot]$ for various electroneutrality conditions.

Electroneutrality condition \ Species	$V_M^\times$	V_M'	V_M''	$M_i^\times$	$M_i^\cdot$	$M_i^{\cdot\cdot}$	n	p
$n = [M_i^\cdot]$ $p = [V_M']$ $[M_i^{\cdot\cdot}] = [V_M'']$	p_M^{-1}	$p_M^{-1/2}$	const.	p_M	$p_M^{1/2}$	const.	$p_M^{1/2}$	$p_M^{-1/2}$
$n = 2[M_i^{\cdot\cdot}]$ $p = 2[V_M'']$	p_M^{-1}	$p_M^{-2/3}$	$p_M^{-1/3}$	p_M	$p_M^{2/3}$	$p_M^{1/3}$	$p_M^{1/3}$	$p_M^{-1/3}$
$[V_M'] = [M_i^\cdot]$	p_M^{-1}	const.	p_M	p_M	const.	p_M^{-1}	p_M	p_M^{-1}
$[V_M'] = 2[M_i^{\cdot\cdot}]$ $[M_i^\cdot] = 2[V_M'']$	p_M^{-1}	$p_M^{-1/3}$	$p_M^{1/3}$	p_M	$p_M^{1/3}$	$p_M^{-1/3}$	$p_M^{2/3}$	$p_M^{-2/3}$
$n = [D^\cdot]$	p_M^{-1}	$[D^\cdot]p_M^{-1}$	$[D^\cdot]^2 p_M^{-1}$	p_M	$p_M[D^\cdot]^{-1}$	$p_M[D^\cdot]^{-2}$	$[D^\cdot]$	$[D^\cdot]^{-1}$
$[V_M'] = [D^\cdot]$	p_M^{-1}	$[D^\cdot]$	$p_M[D^\cdot]^2$	p_M	$[D^\cdot]^{-1}$	$p_M^{-1}[D^\cdot]^{-2}$	$p_M[D^\cdot]$	$p_M^{-1}[D^\cdot]^{-1}$
$2[V_M''] = [D^\cdot]$	p_M^{-1}	$[D^\cdot]^{1/2} p_M^{-1/2}$	$[D^\cdot]$	p_M	$p_M^{1/2}[D^\cdot]^{-1/2}$	$[D^\cdot]^{-1}$	$p_M^{1/2}[D^\cdot]^{1/2}$	$p_M^{-1/2}[D^\cdot]^{-1/2}$

if the electrical conductivity varies as $p_M^{2/3}$ (n-type) or $p_M^{-2/3}$ (p-type) the defect is either a V_M' or $M_i^{\cdot\cdot}$. In this case by donor doping and observing the dependence of $D_e(\tilde{M})$ on $[D^{\cdot}]$ and p_M, for any of the three electroneutrality conditions involving $[D^{\cdot}]$ in Table 1.3, permits the V_M' or $M_i^{\cdot\cdot}$ to be identified. This type of approach is basic to self-diffusion studies in the chalcogenides and oxides described in Chapters 7 and 8.

Isoconcentration Diffusion

The same methods can be applied here as for self-diffusion. Use of the isotope effect is only viable if $f_B = 1$ (i.e. direct interstitial or dissociative mechanisms) as with other defect mechanisms $0 < f_B < 1$ so precluding any basis for identification. The technique has been used for Li in Si [51].† The most profitable strategy is to vary the solute concenttration and, where appropriate, p_M or p_X and observe the dependence of D_e. If the electroneutrality condition is known then an interpretation in terms of equations such as 1.26 and 1.40 can be made. The diffusion of As in Si (section 5.4.3) and of Zn and Cd in the III-V compounds (section 6.3.7) has been studied in this way.

1.3 DIFFUSION INVOLVING A NET CHEMICAL FLUX

1.3.1 The General Situation

The existence of a net flow of a chemical species generally involves, but not always, a chemical concentration gradient of the species and arises in a system trying to attain equilibrium. It is this aspect of atomic diffusion which makes it of such practical importance in all forms of matter and not least in semiconductors. Manning [35] has pointed out that the so-called driving forces for diffusion can be split into real driving forces and non-driving forces on the basis of a simple but general definition of a driving force. Figures 1.3(a), (b) and (c) show the three basic diffusion situations considered by Manning in terms of the energy barriers a diffusing atom must surmount. A net flow can arise in figure 1.3(a) because in a concentration gradient, although the jump rates are symmetrical and independent of position, there are more atoms jumping from the left than from the right. In figure 1.3(b) the jump rates are still symmetrical for a given barrier but depend now on position. This situation represents a diffusion coefficient gradient and a net flux can again occur when there are more atoms jumping over a given barrier from the left than from the right i.e. a concentration gradient exists. The asymmetric jump

† The actual experiments were based on the chemical diffusion of Li but as intrinsic conditions obtained throughout the diffusion region it is plausible to equate such conditions to those of isoconcentration diffusion.

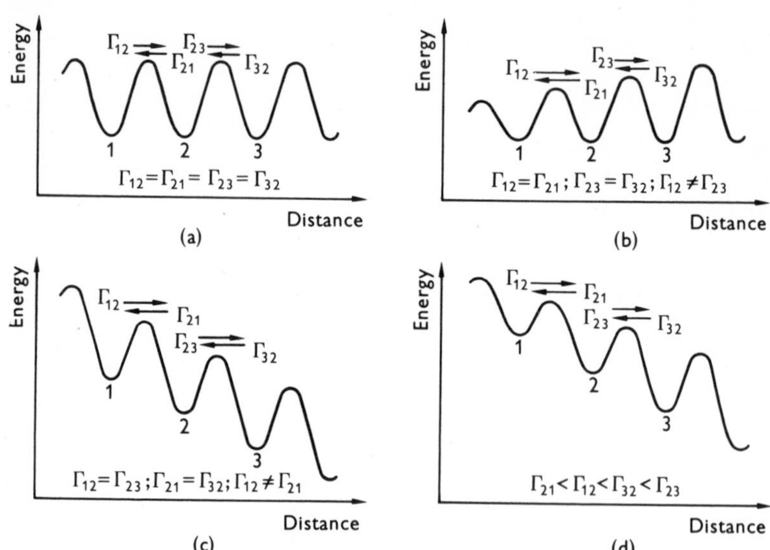

Figure 1.3. Representative energy barrier diagrams for diffusion. The Γ factors denote the jump rates over the indicated barrier.

rates over a given barrier in figure 1.3(c) mean however that a net flux will occur even in the absence of a concentration gradient. The asymmetry in jump rates is due to the variation with position of the energy minima. Such a situation can arise in a variety of ways. The most obvious one is that of an ion in an electric field. A stress field can also create asymmetry in the jump rates of a diffusing particle. Chemical potential gradients are also sources of asymmetry if the solution is non-ideal and/or if temperature gradients are present. The most general situation is depicted in figure 1.3(d). Manning [35] defines a driving force as any effect causing the energy minima to vary with position. A driving force can therefore produce a net chemical flux without a chemical concentration gradient whilst a net chemical flux can arise because of a concentration gradient even though there is no driving force. This chapter will follow Manning's classification.

It should perhaps be noted that the alternative approach is in terms of thermodynamic driving forces. A thermodynamic driving force is any factor giving rise to a net flux. Such a definition does not distinguish between diffusion in a situation represented by figure 1.3(a) from a situation represented by figure 1.3(c), i.e. the difference in jump kinetics is not recognized. In the thermodynamic description of diffusion the ordinate scales in figure 1.3 would be the free energy of the diffusant whereas in Manning's representation the ordinate scales used in figure 1.3 are the free energy minus the entropy of mixing. If the diffusant forms an ideal solution and the situation depicted in figure 1.3(a) applies, there will be a free energy gradient due to the gradient in the entropy of mixing term: this gradient constitutes a thermodynamic driving force of magnitude $(kT/c)\ \partial c/\partial x$. More generally a thermodynamic driving force is any effect which causes the free energy minima (of the diffusant) to vary with position.

1.3.2 Driving Forces and Diffusion Fluxes

To provide a basis for describing all circumstances in which a diffusion flux can occur, Fick's first law has been generalized to [72]

$$\mathbf{J}_r = \sum_{j=1}^{s} L_{rj}\mathbf{X}_j + L_{rq}\mathbf{X}_q \tag{1.47}$$

where

$$\mathbf{X}_j = \mathbf{F}_j - T\,\nabla\,(\mu_j/T), \quad \mathbf{X}_q = -\,\nabla \ln T$$

$\mathbf{J}_r$ is the flux of the rth species in a system containing s species; $\mathbf{F}_j$ is the external driving force acting on particles of the jth species whose chemical potential is μ_j. The term $L_{rq}\mathbf{X}_q$ describes the flux of the rth species due to a temperature gradient ∇T: the flux created by ∇T is known as thermal diffusion or, alternatively, as the Soret effect. A set of s equations, each one similar to equation 1.47, describes the fluxes of all the species present in the system. These equations are purely phenomenological and simply state that the flux of any species is linearly coupled by the L coefficients to all the driving forces and chemical potential gradients present in the system. The L coefficients are undetermined in this representation. Irreversible thermodynamics shows that generally, in the absence of a magnetic field. $L_{rj} = L_{jr}$. Beyond this, however, the L coefficients are quite empirical. The nature of the couplings they represent only emerges when specific systems and interaction models are considered. In this way it can be shown that equation 1.47 is fully consistent with kinetic descriptions of specific diffusion systems [72]. Manning [75] has identified the L coefficients, for diffusion in multicomponent alloys, by considering the kinetic diffusion equations. Hu [49] has discussed solute diffusion in a semiconductor by a vacancy mechanism in terms of equation 1.47.

Driving forces will be present when there are gradients in electric potential (for charged species), temperature, mechanical stress and chemical composition (if the species forms a non-ideal solution) [35]. A centrifugal acceleration field also constitutes a driving force and has been used to observe the movement of Au atoms in potassium [76]. Non-ideality of a defect, native or foreign, in a semiconductor has been exclusively identified with the ionization of the defect [67]. While justified at dilute concentrations the situation in concentrated solutions will be more complex. The consequence of non-ideality of a species, in a dilute concentration gradient, is the appearance of an internal electric field provided the particular species makes the semiconductor extrinsic (section 1.4.2). Mention must also be made of the enhancement of a chemical flux during irradiation of the semiconductor crystal with photons (optical, X- and γ-rays) and nucleons. Kaneev [77] found that γ radiation (Co60) would transport Cd through CdS at room temperature. In Si Niyazova and associates have observed a substantial penetration of Au under a neutron flux [78a], under X-ray irradiation (50 kV and 100 kV energies at room temperature) [78b] and also under illumination from an incandescent lamp (with the sample maintained at the temperature of liquid nitrogen) [78c].† Boron has been transported in Si by a proton flux with an incident energy of 10 or 50 keV [79]. In this case the Si crystal was held at a temperature between 500 and 700°C during irradiation. The movement of the boron appeared to be independent of temperature over this interval. The enhancement mechanism is unclear in all of these cases but a correlation with dislocation content has been noted [77].‡ It is also important to note that enhancement can occur when the irradiating particle energy is less than the atomic displacement threshold in the solvent lattice [78b]. Even when displacements do occur the magnitude of the observed enhancement would seem to be far too large to be attributed simply to a vacancy supersaturation. Such enchancement is an intriguing situation which obviously demands further investigation.

The diffusion of inclusions of a second phase (gas, liquid, or solid) can also occur in a crystal lattice [80, 81]. Theoretical treatments of their migration in metals under various driving forces have been given [80, 81]. Nichols [82] has given a unified treatment of the diffusion of inclusions of arbitrary shape for any driving force and transport mechanism.

The setting up of stress gradients, through size misfit, and of an internal electrical field, provides two examples of how the flux of one species can interact with other species. Manning [83] has described a more subtle coupling which appears in the chemical diffusion of a species by a vacancy (or interstitialcy) mechanism in a driving force. The flow of the species is enhanced through the creation of a counter flow of vacancies (or interstitialcies). These separate effects illustrate some of the origins of the cross coefficients in equation 1.47.

Despite its obvious importance in crystal growth and thermoelectric phenomena, and the potential area of solid state zone refining [84],

† Enhanced Au mobility in Si has also been found during irradiation with 100 keV electrons [78d].

‡ It has been suggested that electronic excitation is the origin of the enhancement [79a, 79b].

thermal diffusion in semiconductors has attracted conspicuously little attention. Apart from oxides (section 8.5.8) only one investigation, in $Bi_2 Te_3$, has been reported [85]. Oriani [86] has given a recent account of thermal diffusion in metals.

Consider now the planar, isothermal chemical diffusion in a cubic lattice of a neutral dilute solute, $B^{\times}$, which, due to size misfit, introduces an elastic strain energy density G. The associated stress is $\beta[B^{\times}]\Upsilon/(1-\nu)$ provided $[B^{\times}]$ averaged throughout the crystal is negligible [87]; β, Υ and ν are the solute-lattice dilatation/contraction coefficient, Young's modulus and Poisson's ratio respectively. Thus $G = \beta^2[B^{\times}]^2\Upsilon/2(1-\nu)$. Assuming that the flux of $B^{\times}$ has only two components, one due to diffusion down the concentration gradient and the other due to the driving force of the strain energy gradient, then relative to the local lattice planes [88]

$$J(B^{\times}) = -D(B^{\times})\frac{\partial}{\partial x}[B^{\times}] - D(B^{\times})[B^{\times}]\frac{\partial G}{\partial x}/kTS \qquad (1.48)$$

where S is the total concentration of lattice sites and

$$\mathbf{D}(B^{\times})\frac{\partial}{\partial x}[B^{\times}]$$

is the flux due to the concentration gradient (corresponding to figure 1.3(a) or (b)). The second term on the right-hand side of equation 1.48 is the flux arising from the strain energy gradient. The overall situation is then represented by figure 1.3(c) or (d). Substitution for G in equation 1.48 yields an effective diffusion coefficient

$$\mathbf{D}(B^{\times}) = \{1 + (2G/kTS)\}D(B^{\times}) \qquad (1.49)$$

Taking

$$\beta = 10^{-24}\ \text{cm}^3/\text{atom},$$
$$\Upsilon/(1-\nu) = 10^{12}\ \text{dyne cm}^{-2},$$
$$[B^{\times}] = 10^{21}\ \text{cm}^{-3},$$
$$S = 10^{22}\ \text{cm}^{-3}$$

and

$$kT = 0.1\ \text{eV}$$

gives a value for $2G/kTS$ of 0.06. This is an upper value for a dilute solution and it is therefore clear that the strain energy gradient makes a negligible contribution to the solute flux, provided the solution is dilute. This situation can be contrasted with the flow of solute to a dislocation where the stress field of the dislocation provides a significant contribution to the solute flux [89].

Suppose $B^{\times}$ is a substitutional solute, diffusing by a vacancy mechanism in a cubic lattice, then from equation 1.26 we have

$$D(B_M^{\times}) = (Za^2/6[M_M])\{f_B(V_M^{\times})w_2(V_M^{\times})[V_M^{\times}] + f_B(V_M')w_2(V_M')[V_M'] + \cdots\}$$

The various vacancy jump rates (w_0, w_1, etc.) and hence the correlation factors are unaffected by driving forces but may vary with composition [83], but such a dependence should be negligible in dilute solutions. If the solute ionizes then in the extrinsic range the ionized vacancy concentrations, by virtue of equations 1.10 to 1.14, will vary with position so causing $D(B_M)$ to be position dependent. Strain effects may also cause variations in the local concentrations of native defects.

1.3.3 Isothermal Diffusion in an Electric Field

An external electric field, $\mathbf{E}_{ex}$, will impart a net drift velocity to all electrically charged species in a semiconductor due to:

(a) the direct interaction of $\mathbf{E}_{ex}$ with a charged particle;
(b) the drag of atomic particles by free electrons and holes—the 'electron wind';
(c) a counter flow of vacancies (interstitialcies) in the case of a vacancy (interstitialcy) diffusion mechanism.

If υ is the drift mobility of an ion then (a) induces an average drift velocity of $\upsilon\mathbf{E}_{ex}$ per ion. The electron wind modifies the ion mobility to an effective value

$$\upsilon_w = \upsilon(1 \pm nl\sigma) \tag{1.50}$$

in n-type material [90]. A similar equation will apply to a p-type crystal. The + or − refers to negatively or positively charged ions respectively; l and σ are respectively the mean free path of an electron and the scattering cross section for electrons by the particular ion species. It is feasible for $nl\sigma$ to be >1 so that for positive ions υ_w can change sign due to the electron wind. For ions of the same polarity as the free carrier clearly $\upsilon_w \geqslant \upsilon$. Neutral centres are also susceptible to the electron wind. υ_w should start to differ significantly from υ when $n, p \gtrsim 10^{17}$ to 10^{19} cm^{-3} and Zn has been transported in GaAs by this means [91]. The effect however has been more intensively studied in metals [92].

Manning [83] has given a detailed treatment of (c). The analysis is complicated but essentially he shows that for an ion B with absolute charge $z_B e$

$$\frac{\upsilon(B)}{D(B)} = \frac{z_B e}{kT}\lambda \tag{1.51}$$

where λ is a parameter which includes the magnitudes of the charges of all the species involved in the diffusion mechanism and the various vacancy (interstitialcy) jump rates. The important conclusion is drawn that λ in general will not be equal to unity and hence the Nernst-Einstein relation given, by equation 1.51 for $\lambda = 1$, is not generally valid for ions in a crystal lattice. Temkin [92a] has also recently investigated this point. Only in the case of a direct interstitial mechanism will λ always be equal to unity. An expression equivalent to the Nernst-Einstein equation, for the drift mobility of a species in a stress field, is the basis of equation 1.49. It therefore seems likely that this also is not generally valid.

The chemical flux of a dilute solute ion B in the presence of both internal, $\mathbf{E}$, and external, $\mathbf{E}_{ex}$, electric fields is given by

$$J(B) = -D(B)\frac{\partial}{\partial x}[B] + (\mathbf{E}_{ex}v_w(B) + \mathbf{E}v(B))[B] \qquad (1.52)$$

It should be noted that $\mathbf{E}$ does not lead to any flow of free carriers (section 1.4.2).

1.3.4 Summary of Isothermal Interdiffusion

We can conclude that for solute diffusion of a dilute species only two factors will contribute to a net chemical flux: the concentration gradient and the drift in an electric field as described by equation 1.52. It is possible that $D(B)$ may vary with position due to varying concentrations of native defects arising from mutual solubility effects and other interactions (section 1.2.4).

With non-dilute solutions the non-ideality of the solid solution will give rise to a driving force, additional to $\mathbf{E}$, which can be treated in an analogous manner to interdiffusion in metal alloys [93] and alkali halides [94]. Referring to equation 1.49 we see that for $[B] \sim S/2$ then $(2G/kTS) \sim 1.5$ so that strain energy gradients could also be significant provided that plastic flow does not occur first. An additional contribution to $\mathbf{E}$ will occur if there is a band gap gradient [95]. A species diffusion coefficient will also be a function of composition and therefore of position. The case also arises of solute diffusion in a solvent (mixed crystal) lattice which has a composition gradient, so that the solute diffusion coefficient will be a function of composition of the solvent lattice [95]. Strain effects could also be important here.

1.3.5 Some Consequences of Interdiffusion

Consider the diffusion of a substitutional solute, B_M, by any mechanism, in the M sub-lattice of a binary compound semiconductor MX. We assume the presence of Schottky disorder and that there is no change in the lattice constant, that intrinsic conditions obtain and that B_M and V_M occur in only one charge state each. If S is the concentration of lattice sites on the M sub-lattice then [1]

$$\begin{aligned}
&[B_M] + [V_M] + [M_M] = S \\
&\frac{\partial}{\partial t}([B_M] + [V_M] + [M_M]) = 0 \\
&J(B_M) + J(V_M) + J(M_M) = 0 \\
&\frac{\partial}{\partial x}([B_M] + [V_M] + [M_M]) = 0
\end{aligned} \tag{1.53}$$

where the origin of the x axis is now a point beyond and outside the crystal surface and fixed relative to the position of the surface at $t = 0$. The set of equations 1.53 can only be generally consistent, without demanding unique relationships between $D(B_M)$, $D(V_M)$ and $D(M_M)$ if

$$\begin{aligned}
J(B_M) &= -D(B_M)\frac{\partial}{\partial x}[B_M] + u[B_M] \\
J(V_M) &= -D(V_M)\frac{\partial}{\partial x}[V_M] + u[V_M] \\
J(M_M) &= -D(M_M)\frac{\partial}{\partial x}[M_M] + u[M_M]
\end{aligned} \tag{1.54}$$

where u is the velocity of lattice planes parallel to the x axis at position x. The fluxes of the set of equations 1.54 are now relative to the x axis rather than a lattice plane. From the sets of equations 1.53 and 1.54 we obtain

$$uS = \{D(B_M) - D(M_M)\}\frac{\partial}{\partial x}[B_M] + \{D(V_M) - D(M_M)\}\frac{\partial}{\partial x}[V_M] \tag{1.55}$$

so that for dilute concentrations of B_M the contribution $u[B_M]$ to $J(B_M)$ can be neglected (assuming $(\partial/\partial x)[V_M]$ is either zero or small). $J(M_M)$ however is given exactly by

$$J(M_M) = D(B_M)\frac{\partial}{\partial x}[B_M] - D(M_M)\frac{\partial}{\partial x}[V_M] \tag{1.56}$$

A simple interpretation of the above analysis is possible when $[V_M]$ is constant as then an increase in $[B_M]$ in a given volume element of the lattice must be balanced by an equal decrease in $[M_M]$. If there is no loss of M from the crystal to the ambient phase then the total number of sub-lattice sites must increase at a rate ($u_{\text{surface}}S$) per unit area of surface, whatever the value of $[V_M]$; u_{surface} will be given by equation 1.55 evaluated at the crystal surface. The diffusion of B_M is therefore strictly a moving boundary problem but for dilute concentrations a fixed boundary is a satisfactory approximation [96]. We can therefore equally well place the origin of the x axis at the crystal surface for chemical diffusion in dilute solution.

The maintenance of a single crystalline phase also requires any increase in the total number of M lattice sites to be matched by an equal increase of X lattice sites. Depending on the ambient phase this can be achieved by incorporating X atoms and/or V_X. In the case of an elemental semiconductor the lattice simply grows as the solute is incorporated.

1.3.6 The Various Types of Lattice Diffusion Coefficient in Isothermal Chemical Diffusion

The literature on diffusion in metals distinguishes between an intrinsic diffusion coefficient and a chemical or interdiffusion coefficient [97]. The former is obtained by referring the diffusant flux to the local lattice planes whereas the latter is operationally defined by reference to some interface or fixed set of axes in the diffusion system and is the coefficient obtained by tracer measurements. Thus in the situation considered in section 1.3.5 the chemical diffusion coefficient of B_M is equal to

$$-J(B_M)/\frac{\partial}{\partial x}[B_M];$$

in the situation described by equation 1.48 the intrinsic diffusion coefficient of B is

$$-J(B^\times)/\frac{\partial}{\partial x}[B^\times].$$

Because in dilute solutions u is very small and can be neglected it follows that the intrinsic and chemical diffusion coefficients become identical in such solutions. In non-dilute solutions however the analysis of experimental data must naturally be in terms of a set of equations similar to 1.54.

In this chapter we have chosen to use an effective chemical diffusion coefficient $\mathbf{D}$ for a species obtained from the application of equation 1.1. As long as dilute solutions only are considered it is immaterial which choice of reference plane is made and $\mathbf{D}$ is then equal to the intrinsic and chemical diffusion coefficients defined above. The other lattice chemical diffusion coefficient which has been introduced (e.g. $D(B^\times)$ in equation 1.48, $D(B)$ in equation 1.52) describes the flux component, relative to local lattice planes, due solely to diffusion down the concentration gradient i.e. flux components due to driving forces and couplings with other fluxes that may be present are excluded. As driving forces and couplings between fluxes are absent in self and isoconcentration diffusion it might be expected that $D_e = D$ at the same species concentration. This equality is generally unlikely to hold in isoconcentration experiments as the native defects will not necessarily be in local chemical equilibrium with the solute in the corresponding interdiffusion experiment. In the case of solvent diffusion by native defects the equality will hold only if the defects are again in local chemical equilibrium during interdiffusion.

1.4 ISOTHERMAL CHEMICAL DIFFUSION OF POINT DEFECTS IN DILUTE SOLUTION

1.4.1 Introduction

Experimental chemical diffusion in semiconductors has concentrated on the isothermal diffusion of electrically important defects which have limited, dilute, solubilities in the solvent lattice. The latter is initially homogeneous and the diffusant is contained in an ambient phase (vapour, liquid or solid). In these conditions equation 1.52 is applicable, with $D(B)$ possibly varying with position due to varying concentrations of native defects. The latter may originate in mutual solubility effects via the electronic interaction of ionized defects and/or strain effects (section 1.2.4). We shall here be concerned with the diffusion situation in which

$$J(B) = -D(B)\frac{\partial}{\partial x}[B] + v(B)[B]\,\mathbf{E} \tag{1.57}$$

describes the flux of any point defect B, native or foreign.

For completeness mention needs to be made of diffusion of defects to or from dislocations when the driving force, due to the dislocation stress field, contributes an additional flux component to the right-hand side of equation 1.57. Bullough and Newman [89] have discussed this situation and it will not be considered further here. It may be noted however that they neglected any effects due to **E** and assumed an expression, equivalent to the Nernst-Einstein relation, for the defect stress field mobility.

1.4.2 Electroneutrality and the Internal Electric Field

Although chemical equilibrium does not obtain in chemical diffusion it is assumed that electronic equilibrium is achieved in all parts of the diffusion region [18]. Equations 1.10 to 1.14 describe the local fulfilment of electronic equilibrium, in which the concentrations are now position dependent. We shall also assume that $(r_s/n)\nabla n \ll 1$ so as to ensure the validity of the electroneutrality condition (section 1.2.2). The existence of concentration gradients leads to the appearance of an internal electric field, **E**. Zaromb [98] was the first to consider the role of **E** in semiconductor diffusion but limited his analysis to near intrinsic conditions. If the electroneutrality condition is

$$p + \sum_{s,r} s[D_r^{s\cdot}] = n + \sum_{s,q} s[A_q^{s\prime}] \tag{1.58}$$

then by differentiation

$$\nabla p + \sum_{s,r} s\nabla[D_r^{s\cdot}] = \nabla n + \sum_{s,q} s\nabla[A_q^{s'}] \tag{1.59}$$

As E_F is the electrochemical potential of the free electrons [13] we can write quite generally

$$E_F = \text{constant} + kT \ln(\gamma_n n) - eV_s \tag{1.60}$$

where V_s is the macroscopic electrostatic potential due to local space charge and/or an external source. Consider the situation where there is no external potential source so that the condition for electronic equilibrium is expressed by $\nabla E_F = 0$. Equation 1.60 then yields for this situation

$$e\nabla V_s = kT\nabla \ln(\gamma_n n)$$

where ∇V_s is the gradient in the local space charge potential. Clearly this gradient is manifest as an internal electrostatic field **E** given by

$$\mathbf{E} = -\frac{kT}{e}\nabla \ln(\gamma_n n) \tag{1.61}$$

E originates because equation 1.58 is not strictly exact: the very much more mobile free carriers tend to diffuse ahead of the ionized parent defects. This segregation establishes a small space charge and hence **E**, such that **E** opposes any further increase in the space charge density. The space charge density however required to sustain **E** is quite negligible relative to the concentrations in equation 1.58 which for diffusion purposes can be treated as exact.

If γ_n and γ_p are known functions of the free carrier and defect concentrations then by means of equations 1.20, 1.59 and 1.61 **E** can be evaluated. At present **E** can only be obtained in this way for an uncompensated heavily doped semiconductor and for a lightly doped semiconductor (i.e. non-degenerate, so that classical statistics are applicable) with a variable degree of compensation. Both cases are discussed in the next section.

1.4.3 Solute Diffusion

Consider the planar diffusion of a fully ionized monovalent donor $D^\cdot$ in an otherwise intrinsic semiconductor. The electroneutrality condition is $n = [D^\cdot]$ and we confine attention to the degenerate region $\eta_c \geqslant 0$. Thus equation 1.59 simplifies to

$$\partial[D^\cdot]/\partial x = \partial n/\partial x$$

and substitution, in conjunction with equations 1.12, 1.18a, 1.19, into equation 1.61 yields

$$e\mathbf{E} = -\frac{kT}{N_c}\left[\frac{F_{1/2}(\eta_c/kT)F_{-3/2}(\eta_c/kT)}{F^3_{-1/2}(\eta_c/kT)} + \frac{2e^2}{3\epsilon kT}\left(\frac{3N_c}{\pi}\right)^{1/3} F_{1/2}^{-2/3}(\eta_c/kT)\right]\partial[D^{\cdot}]/\partial x \tag{1.62}$$

The result [99] $dF_j(y)/dy = F_{j-1}(y)$ has been utilized in deriving equation 1.62. If the absolute charge of the ionized donor is $+z_D \cdot e$ then by equation 1.51 $v(D^{\cdot}) = z_D \cdot e\lambda D(D^{\cdot})/kT$. Substituting this expression together with equation 1.62 into equation 1.57 gives

$$\mathbf{D}(D^{\cdot}) = D(D^{\cdot})(1 + z_D \cdot \lambda\vartheta) \tag{1.63}$$

for the effective diffusion coefficient of $D^{\cdot}$ where

$$\vartheta = 1 + \frac{F^2_{1/2}(\eta_c/kT)F_{-3/2}(\eta_c/kT)}{F^3_{-1/2}(\eta_c/kT)} + \frac{2e^2}{3\epsilon kT}\left(\frac{3N_c}{\pi}\right)^{1/3} F^{1/3}_{1/2}(\eta_c/kT)$$

It is evident that $\mathbf{E}$ enhances the flow of $D^{\cdot}$ so that $\mathbf{D}(D^{\cdot}) > D(D^{\cdot})$ as shown by equation 1.63. The value of λ is obviously important and experimental confirmation that $\lambda = 1$ for the direct interstitial mechanism has been obtained for Li in Ge and Si [100] and for Li and Cu in GaAs [101, 102]. We therefore expect $\lambda = 1$ for the interstitial in the dissociative mechanism. The variation of $\mathbf{D}(D^{\cdot})/D(D^{\cdot})$ with $[D^{\cdot}]$ is shown in figure 1.4, according to equation 1.63, for representative values of the various parameters. Values of $F_j(\eta_c/kT)$ are tabulated in reference [99]. In the remainder of this chapter it will be assumed that $\lambda = 1$ and that the absolute charge of a defect is identical to its relative charge (i.e. relative to the host atoms (ions) which are substituted for by the particular species).

Equation 1.63 simplifies appreciably under fully degenerate conditions as then $F_{1/2}(\eta_c/kT)$ can be replaced by [99] $(4/3\sqrt{\pi})(\eta_c/kT)^{3/2}$. Equation 1.63 becomes, with $z_{D^{\cdot}} = \lambda = 1$,

$$\mathbf{D}(D^{\cdot}) = D(D^{\cdot})\left[1 + \frac{2}{9}\frac{\eta_c}{kT} + \frac{2e^2}{3\epsilon kT}\left(\frac{4N_c}{\pi^{3/2}}\right)^{1/3}\left(\frac{\eta_c}{kT}\right)^{1/2}\right]$$

This differs from an earlier expression due to Shockley [103] because non-ideality of the conduction electrons in that case was attributed solely to the reign of Fermi-Dirac statistics i.e. $\gamma_n = \gamma_n(FD)$. The statement is sometimes met in the literature that $\mathbf{D}(D^{\cdot})/D(D^{\cdot})$ has a maximum value of 2; this is erroneous and is a consequence of using classical statistics.

Exactly parallel equations and remarks apply to the diffusion of a monovalent ionized acceptor in the degenerate range.

Turning now to planar diffusion in the lightly doped non-degenerate situation represented by equation 1.58, equations 1.11 and 1.12 simplify

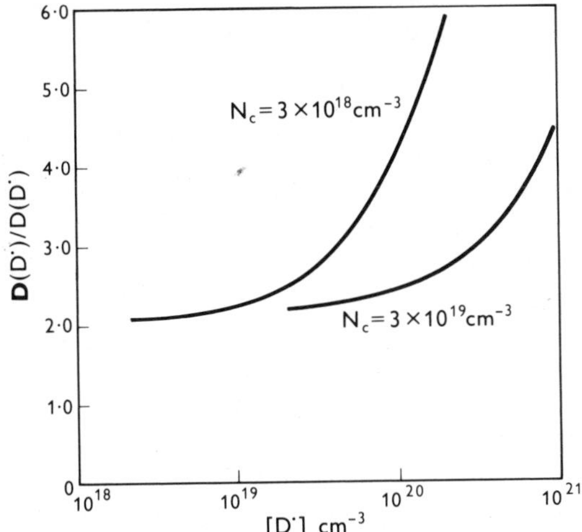

Figure 1.4. The enhanced diffusion of an ionized donor according to equation 1.63. It was assumed that $z_{D^{\cdot}} = \lambda = 1$, $\epsilon = 12$ and $kT = 0.10$ eV. Two values of N_c equal to 3×10^{19} and 3×10^{13} cm^{-3} were used corresponding to electron effective mass ratios of 0.29 and 0.063 respectively.

to $p = N_v \exp(\eta_v/kT)$ and $n = N_c \exp(\eta_c/kT)$ respectively. As $\gamma_n = \gamma_p = 1$ equation 1.61 becomes

$$\mathbf{E} = -\frac{1}{e}\,\partial\eta_c/\partial x.$$

Substituting these simplifications into equation 1.59 yields

$$e\mathbf{E} = -kT\left[\frac{\sum_{q,r,s} s\partial([D_r^{s\cdot}] - [A_q^{s'}])/\partial x}{(n+p)}\right] \tag{1.64}$$

The effective diffusion coefficient of any of the defect species can then be derived by substitution of equation 1.64 into equation 1.57. In this way the effective diffusion coefficient of an ionized monovalent donor $D^{\cdot}$ diffusing into an otherwise intrinsic semiconductor is readily shown to be

$$\mathbf{D}(D^{\cdot}) = D(D^{\cdot})\left(1 + \frac{[D^{\cdot}]}{([D^{\cdot}]^2 + 4n_i^2)^{1/2}}\right) \tag{1.65}$$

The diffusion of a completely ionized monovalent donor into a uniformly doped p-type crystal has also been investigated with the assumptions that the acceptors are fully ionized and immobile [104, 105]. It is found that again $\mathbf{D}(D^{\cdot}) \geqslant D(D^{\cdot})$, similar to equation 1.65, but now $\mathbf{D}(D^{\cdot})$ also has a maximum value of

$$D(D^{\cdot})[1 + (1 + [A']^2/4n_i^2)]^{1/2}$$

when

$$[D^{\cdot}] = [A'] + 4n_i^2/[A']$$

where for simplicity compensation is provided by a monovalent acceptor. At the p-n junction

$$\mathbf{D}(D^{\cdot}) = D(D^{\cdot})(1 + [A']/2n_i)$$

The effect of the internal electric field has also been studied in the simultaneous and sequential diffusion of substitutional donors and acceptors [106, 107, 108, 109]. Enhancement or retardation can now occur depending on the relative magnitudes of the donor and acceptor concentration gradients. In the dissociative mechanism, if the interstitial and substitutional species have opposite polarity, then the direction of the internal field will be determined by the larger gradient. If this is the substitutional species then the internal field will retard the interstitial flow [110].

Kendall [111] has described in terms of solubility effects the redistribution of a dopant, initially of uniform concentration, during the in-diffusion of an impurity of opposite polarity.

Solute chemical diffusion can be followed by various methods (chapter 4). In the tracer method the ratio $[\widetilde{B}]/[B]$ ($\ll 1$) is constant, if the isotope effect is neglected, although both $[\widetilde{B}]$ and $[B]$ vary with position.† This contrasts with the isoconcentration situation where $[B]$ is constant and $[\widetilde{B}]/[B]$ varies with position.

1.4.4 The Diffusion of Native Defects

The chemical diffusion of native defects is important in establishing changes in the degree of non-stoichiometry of compound semiconductors. Consider planar diffusion of fully ionized monovalent vacancies $V_X^{\cdot}$ and V_M' in a binary compound MX. The departure from stoichiometry, Δ, is defined by

$$\Delta = [M_M] - [X_X] = [V_X^{\cdot}] - [V_M'] \tag{1.66}$$

which utilizes $S = [M_M] + [V_M'] = [X_X] + [V_X^{\cdot}]$ where S is now the concentration of lattice sites on each sub-lattice. Electroneutrality requires

$$[V_X^{\cdot}] + p = [V_M'] + n \tag{1.67}$$

Ionization equilibrium is fully described by equation 1.20 and

$$[V_X^{\cdot}]\,[V_M'] = K_s \tag{1.68}$$

Note that equation 1.68 also implies local equilibrium between $V_X^{\cdot}$ and V_M'. A change in non-stoichiometry can then be represented as a flow in Δ given by

$$J(\Delta) = -\mathbf{D}(\Delta)\,\frac{\partial \Delta}{\partial x} = J(M_M) - J(X_X) \tag{1.69}$$

† Carrier free radioisotopes are rarely used in semiconductor experiments.

which defines the effective diffusion coefficient, $\mathbf{D}(\Delta)$, for the change. Continuity imposes

$$J(M_M) + J(V_M') = J(X_X) + J(V_X^{\cdot}) = 0$$

so that

$$J(\Delta) = J(V_X^{\cdot}) - J(V_M') \tag{1.70}$$

We also have

$$\begin{aligned} J(V_X^{\cdot}) &= -D(V_X^{\cdot})\left(\frac{\partial [V_X^{\cdot}]}{\partial x} - \frac{e\mathbf{E}}{kT}[V_X^{\cdot}]\right) \\ J(V_M') &= -D(V_M')\left(\frac{\partial [V_M']}{\partial x} + \frac{e\mathbf{E}}{kT}[V_M']\right) \end{aligned} \tag{1.71}$$

The vacancy concentrations are invariably low enough to ensure that $D(V_X^{\cdot})$ and $D(V_M')$ do not vary with composition, and also the validity of Maxwell-Boltzmann statistics, so that through equations 1.20, 1.64 and 1.67 we obtain

$$\mathbf{E} = -(\Delta^2 + 4n_i^2)^{-1/2}\frac{\partial \Delta}{\partial x}$$

Substituting for $\mathbf{E}$ in equations 1.71 followed by substitution of equations 1.71 into equation 1.70 and using equations 1.20 and 1.68 finally yields

$$2\mathbf{D}(\Delta) = \left\{D(V_M') + D(V_X^{\cdot}) + \frac{\Delta(D(V_X^{\cdot}) - D(V_M'))}{(\Delta^2 + 4K_s)^{1/2}}\right\}\left\{1 + \left[\frac{\Delta^2 + 4K_s}{\Delta^2 + 4n_i^2}\right]^{1/2}\right\} \tag{1.72}$$

This result was first obtained by a thermodynamic analysis and shows that $\mathbf{D}(\Delta)$ varies with composition [112]. On the M rich side of the MX existence region $\Delta \approx [V_X^{\cdot}]$ so that from equations 1.25 and 1.72 we see that

$$\mathbf{D}(\Delta)/D_e(\widetilde{M}_M) = 2D(V_X^{\cdot})S/f_V D(V_M')[V_M']$$

and

$$\mathbf{D}(\Delta)/D_e(\widetilde{X}_X) = 2S/f_V[V_X^{\cdot}].$$

Similarly for the X rich side

$$\mathbf{D}(\Delta)/D_e(\widetilde{M}_M) = 2S/f_V[V_M']$$

and

$$\mathbf{D}(\Delta)/D_e(\widetilde{X}_X) = 2D(V_M')S/f_V D(V_X^{\cdot})[V_X^{\cdot}].$$

Normally $[V_X^{\cdot}], [V_M'] \lesssim 10^{18}\ \text{cm}^{-3}$ and $S \sim 10^{22}\ \text{cm}^{-3}$ so that $\mathbf{D}(\Delta)$ will generally be many orders of magnitude greater than $D_e(\widetilde{M}_M)$ and $D_e(\widetilde{X}_X)$ at the same vacancy concentrations. Thus $D_e(\widetilde{M}_M)$ and $D_e(\widetilde{X}_X)$ provide little information by themselves about the progress of a change in Δ: they can only provide lower limits on $\mathbf{D}(\Delta)$.

A special case arises under extrinsic conditions when $\Delta = 0$. This defines the position of the *p-n* junction whose diffusion coefficient $\mathbf{D}(0)$ will satisfy

$$\frac{\mathbf{D}(0)}{D_e^i(\widetilde{M}_M)} = \frac{S}{2f_V}\left(1 + \frac{D(V_X^{\cdot})}{D(V_M')}\right)\left(\frac{1}{n_i} + \frac{1}{K_s^{1/2}}\right)$$
$$\frac{\mathbf{D}(0)}{D_e^i(\widetilde{X}_X)} = \frac{S}{2f_V}\left(1 + \frac{D(V_M')}{D(V_X^{\cdot})}\right)\left(\frac{1}{n_i} + \frac{1}{K_s^{1/2}}\right) \tag{1.73}$$

and clearly $\mathbf{D}(0) \gg D_e^i(\widetilde{M}_M), D_e^i(\widetilde{X}_X)$. At the junction the crystal will of course be intrinsic.

Similar results would have been obtained with native interstitials rather than vacancies so that the preceding conclusions are generally valid for the diffusion of native defects [112]. An important corollary is that, in the volume diffusion of solvent tracer atoms by a defect mechanism, the native defects will have equilibrated in the tracer diffusion region long before the arrival of the tracer atoms. This therefore precludes the use of tracers to follow the chemical diffusion of native defects. The foregoing analysis can be equally well applied to an elemental semiconductor and similar conclusions obtained.

Information on $\mathbf{D}(\Delta)$ can be obtained from the movement of *p-n* junctions [113] and colour boundaries [114]. Such measurements however are made at room temperature, following a high temperature diffusion anneal, and rely for interpretation on two crucial assumptions: (i) the high temperature defect structure is 'frozen in', i.e. there is no precipitation or association during quenching of the defects being monitored; (ii) no defect species are added to, or removed from, the electroneutrality condition due to the shift in E_F arising from the quench. The justification of either assumption at present calls for more knowledge of a system than is usually available. $\mathbf{D}(\Delta)$ for CdTe has been obtained by Zanio [115] who avoided quenching problems by measuring the high temperature relaxation time of the electrical conductivity as a function of non-stoichiometry.

1.4.5 The Introduction of Defects by Solute Diffusion

Point Defects

The various interactions discussed in section 1.2.4 show that a solute gradient will generally have associated concentration gradients of native defects, neutral and ionized. The supply (or removal) of such defects must accompany the assimilation of the solute in the solvent crystal. In a compound semiconductor a solute gradient in one sub-lattice can in this way set up coupled concentrations of defects in all the sub-lattices.

Furthermore a solute, diffusing by the dissociative mechanism, consumes vacancies in order to build up the local substitutional solute concentration. Whether or not the native defects will be in local thermodynamic equilibrium, in addition to ionization equilibrium, with the solute therefore depends on the operating efficiency of sources and sinks for, and the diffusion of, native defects. Sources and sinks for native defects are dislocations and the crystal surface [116]. Evidently if the vacancy supply or removal rate is restricted then the diffusion of solute will also be restricted. This is of particular importance in the dissociative mechanism and was first stressed by Frank and Turnbull [38]: a vacancy sub-saturation can generally be expected. Various analyses have been made of chemical dissociative diffusion in which explicit consideration is given to the vacancy supply [117, 118, 119]. A basic difficulty to a quantitative approach however is knowing the relative importance of surface and dislocation vacancy sources. The widely held view that a deviation of 1% from the thermal equilibrium vacancy concentration would be adequate to activate a dislocation source (or sink) has recently been questioned by experiments in metals which showed that a 15% vacancy supersaturation was insufficient to cause dislocation climb (the surface acting as the sink) [120].

Hu and Mock [96] have investigated the diffusion of a substitutional solute by a vacancy mechanism in which there is a strong solute-vacancy association. They find that vacancy diffusion should be fast enough from the surface to maintain thermal equilibrium vacancy concentrations. Local defect equilibrium should therefore obtain. Hanneman and Anthony [121] have shown however that, in quenching from the diffusion temperature, very substantial redistributions of solute can occur near the crystal surface due to the vacancy flux to the surface. The redistribution of a solute during quenching has obvious and important implications in experimental diffusion work. It is an aspect however that so far has received little attention in semiconductors. Williams [122] has interpreted the observed distribution of Cu in Ga diffused Ge as arising from a redistribution of Cu, due to solubility effects in the diffused layer, during cooling.

Dislocations

Prussin [87] showed theoretically that the chemical diffusion of boron and phosphorus in Si could set up stresses sufficient to create dislocations (edge type). The resulting dislocation density ρ_d created by plastic flow was shown to be

$$\rho_d = \frac{\beta \partial [B]}{b \partial x} \tag{1.74}$$

where b is the component of the Burgers vector parallel to the surface and

β is as defined in section 1.2.4. A separate approach by Shockley [123], based on energy considerations, expresses the condition for the occurrence of plastic flow in terms of the total number of solute atoms per unit surface area, $W = \int [B]\, dx$. If W exceeds a critical value, which can be calculated, dislocations will be created. The two approaches can be regarded as complementary: the W criterion reveals whether plastic flow occurs, if it does then equation 1.74 gives the dislocation density. Although both theories necessarily oversimplify the problem (e.g. the Prussin model is a static one whereas the situation is a dynamic one) the broad features of dislocation generation appear to be well established with order of magnitude agreement between experiment and equation 1.74 being found [124, 125] and substantially better quantitative agreement with the W criterion [126]. Both theories apply equally well to over- and under-sized solute atoms so that by the simultaneous diffusion of over- and under-sized solutes it has proved possible to keep, by strain compensation, the ensuing stresses below the yield stress and so avoid plastic flow [57]. The bulk of the experimental work on diffusion induced dislocations has been confined to Si but the same phenomenon should be found in any lattice when the W criterion is met. Thus high values of ρ_d have been reported in Zn diffused GaAs [54, 127] and InSb [128].

Parker [129] suggested that the diffusion by a vacancy mechanism of a solute, which had caused plastic flow, would be enhanced by the excess vacancies created by the glide and climb, under the residual stress, of the diffusion induced dislocations. This is the situation for an under-sized solute such as boron and phosphorus in Si. The edge dislocations created by an over-sized solute (e.g. Sn in Si [57]) will however be of opposite sign so that climb in this case, by the same argument, should absorb vacancies and hence retard diffusion. Parker [129], Hu and Yeh [130] and Thai [131] have given different analyses of solute enhanced diffusion in Si all based on a vacancy supersaturation set up by dislocation motion.† Thai also included the effect of the internal electric field. A further origin of enhancement, which does not appear so far to have been considered, could arise if, due to the large ρ_d, the diffusion region corresponded to Harrison's type A regime (section 1.5.3). Vacancy generation by the motion of diffusion induced dislocations has been used to explain the chemical diffusion of Zn in the III-V compounds [131c].

It has also been found in Si that a solute concentration gradient which is insufficient to generate misfit dislocations can create dislocation loops and stacking faults [132].

† A more recent treatment of this problem [131a] in terms of a moving dislocation network with vacancy equilibrium has been shown to be inconsistent with experimental data [131b].

1.4.6 Precipitation

Diffusion has an important role in precipitation phenomena and valuable information can be obtained from precipitation kinetics about the diffusion coefficients involved [39, 133]. Bullough and Newman [89] have given a recent review of the migration kinetics of point defects to dislocations. These authors also point out the possibility of simultaneous diffusion and precipitation of a solute under isothermal conditions. The precipitation is due to the formation of a second phase inclusion by a chemical interaction between the diffusant and an impurity already present. The growth of such inclusions is a problem in reactive diffusion. Recent treatments of the growth kinetics of spherical [134] and plate-like [135] precipitates have been given.

1.4.7 Determination of the Diffusion Mechanism

The foregoing sections have shown that in the chemical diffusion of a solute both $D(B)$ and $\mathbf{D}(B)$ will in general vary with position. Before any elucidation of the diffusion mechanism can be made these variations need to be known. Given an experimental diffusion profile it is first necessary to separate the volume, or type A, diffusion regime from the types B and C regimes (section 1.5.3), all of which may simultaneously contribute to the solute profile [136]. If $\mathbf{D}(B)$ is solely a function of $[B]$ and if

$$\frac{\partial[B]}{\partial t} = \frac{\partial}{\partial x}\left(\mathbf{D}(B)\,\frac{\partial[B]}{\partial x}\right) \tag{1.75}$$

then $[B]$ must be a function of $x/\sqrt{t}$. When the infinite source boundary condition is used $\mathbf{D}(B)$ can be obtained from the type A diffusion profile by Matano's method [137] to yield

$$\mathbf{D}(B) = -\frac{1}{2t}\left(\frac{\partial x}{\partial[B]}\right)_{[B]} \int_0^{[B]} x\,d[B] \tag{1.76}$$

The values of $\mathbf{D}(B)$ can then be compared to those of an appropriate model but it must be noted that the Matano method is only valid when equation 1.75 is applicable and, experimentally, the infinite source boundary condition is used. If either of these conditions is not met recourse must be made to integration of the continuity equations of an appropriate model with matching of the computed and experimental profiles.

Kucher [138] has given a theoretical analysis of volume diffusion by a solute occupying two types of site and diffusing simultaneously by two different mechanisms. Account was taken of an interchange (non-steady state) of solute atoms between the two types of site. The total solute concentration profile was obtained by integration of the continuity equations and involved four parameters; it would seem that profile matching would not necessarily yield a unique set of values for these four characteristic parameters. The same problem has been considered by Laskar [139] who obtained simple expressions for the solute diffusion profile under various conditions by making suitable approximations in the flow equations. Boktor and Watt [140] have dealt with the simpler situation which arises when the impurity is immobile in one of the two sites. Numerical solutions of the continuity equations describing dissociative diffusion under intrinsic conditions have been obtained by Yoshida and Kimura [140a]. Their results have specific application to the diffusion of Au in Si. Vaskin and Uskov [141] have considered the problem of the chemical diffusion of one impurity into a crystal uniformly doped with another impurity with which pairing takes place.

The room temperature analysis of high temperature diffusion experiments generally assumes that the parameter being used to follow the diffusion is unchanged by the quench to room temperature. This is an assumption that needs careful consideration in any experimental investigation.

1.5 DIFFUSION BY SHORT CIRCUIT PATHS

1.5.1 The Role of Short Circuit Paths in Atomic Diffusion

There is ample evidence that grain boundaries, sub-grain boundaries and isolated dislocations all provide high mobility routes for atom migration relative to lattice diffusion [142, 143]. The generic name for such high mobility routes is short circuit paths: pipe diffusion is the generic name for diffusion along a dislocation.

Bullough and Newman [89, 144] have described the numerous ways in which a point defect can interact with a dislocation. The situation in semiconductors is further complicated by the electrical interaction between ionized centres and charged dislocations for which a complete theory is lacking [89]. Dislocations can also exhibit both donor- and acceptor-like properties in a semiconductor [145, 146, 147, 148]. The primary result of a defect-dislocation interaction is an enhanced defect concentration surrounding the dislocation core due to a binding energy, estimated at a few tenths of an eV, between the defect and dislocation [144]. Studies in heavily doped Si have shown that n_i is increased by a factor $\sim 10^2$ in the vicinity of a dislocation relative to the undisturbed lattice [149]. A further long range effect of the dislocation strain field is a change in atomic jump frequencies through lattice distortion. A detailed analysis, in Cu, of edge and screw dislocations concluded that in the region where linear elasticity theory holds in the dislocation strain field (i.e. outside the dislocation core) any change in the jump frequencies for self-diffusion by point defects is quite negligible [61]. Separate calculations by Koehler [150], also for Cu, show that the vacancy migration energy outside the core can change by ~10%. These results suggest that any change in atomic mobility, directly due to lattice strain, outside the core are likely to be small and commensurate with the experimental error. A grain boundary however does not have a long range strain field [142], so that the only long range interaction possible with centres in the adjacent lattice will be electrical. Little information appears to be available on the electrical properties of grain boundaries, as distinct from dislocations, in semiconductors [151].

Atoms diffusing in short circuit paths will not be 'trapped' or 'locked' in these paths but can jump out into the surrounding lattice. In this way short circuit paths enhance the overall penetration of a diffusant. Such enhancement occurs not only with a static network of short circuit paths but also with a moving network such as in plastic deformation [152], polygonization and grain boundary migration [153], and cyclic straining [154]. Generally the enhancement due to movement will be extra to that of a stationary network, at least in the case of dislocations, because of (a) an increased lattice vacancy concentration (and hence lattice diffusion coefficient) due to climb components in the dislocation motion, and (b) direct transport of the diffusant while it is in the moving pipe. An extensive analysis of these processes has been made [155, 156].

Short circuit paths are also important as sources and sinks for point defects (including solute) and can play an important part in the equilibration of a lattice by dislocation climb [116], by acting as static funnels to a free surface [157, 158] and by serving as sites for precipitation [89]. The efficacy of quenching in retaining a vacancy population in Ge and Si has been shown to depend on the dislocation density [159]. Gutnikova and Lyubov [160] have also found that if there is a strong vacancy-dislocation interaction then the dislocation strain field will be sufficient to maintain a vacancy flux to a dislocation at temperatures where the flux due to a concentration gradient alone is negligible. These authors have also investigated the relaxation time of an impurity atmosphere surrounding a dislocation [161].

1.5.2 Characteristics of Short Circuit Paths

Girifalco and Welch [61] concluded that a diffusion pipe could be identified with the core of a dislocation. They further found that a vacancy diffusion mechanism in the core, similar to that in the bulk lattice, was incompatible with experimental data. Lothe [162] had shown earlier that a vacancy mechanism operating along a linear atomic chain, which was equated to a dislocation core, led to unrealistic vacancy parameters. Luther [163] sought to relate the high mobility of a pipe to the vacancy atmosphere surrounding a dislocation. An unrealistically high binding energy of the vacancy to a dislocation was however then required for consistency with experimental data. Despite these various conclusions against a vacancy mechanism being operative in the dislocation core Balluffi [164], in a recent review of experimental self-diffusion data along dislocations in *fcc* metals, has concluded that most of these data are not inconsistent with a vacancy mechanism. A satisfactory atomic model of pipe diffusion is still lacking.

Experimentally, however, it is known that the activation energy for diffusion along edge dislocations is less than that for screw dislocations [61, 165, 166]. Diffusion in grain boundaries is also anisotropic, being much greater in directions parallel to the dislocation cores than perpendicular to them [143, 167]. There is evidence from experimental work in Ni that the activation energy for self-diffusion along isolated edge dislocations is the same as that in a tilt boundary [168]. Previous reports that the activation energy for diffusion varied with misorientation in a simple tilt boundary are believed to be in error due to the use of an oversimplified method of analysis of the experimental data [168]. The lack of a suitable atomistic model for pipe diffusion has forced the use of a phenomenological model in which the pipe is taken to be a tube, of radius r_0, having a homogeneous structure and uniform diffusion coefficient. A similar situation applies to grain boundaries which are taken to be parallel sided layers of uniform thickness δ. D_s will denote the diffusion coefficient of an atomic species in a pipe or boundary; in the latter it will refer to the direction of easy flow. In general terms it can be expected that r_0 will depend on the Burgers vector of the dislocation as well as on its edge or screw character. Because of the existence of impurity atmospheres around dislocations a dependence of r_0 upon doping level is quite feasible. These same factors, furthermore, may also exert a significant influence on D_s. Very few experimentally based values of r_0 and δ are available. Table 1.4 summarizes experimental values for the phenomenological parameters of short circuit paths in semiconductors and alkali halides. The most striking aspect of the data in Table 1.4 are the large values of r_0 and δ which are substantially greater than a dislocation core radius, which typically has a value of $\sim 10^{-7}$ cm. The wide range in the r_0 values is also notable. There is little doubt that our present knowledge of short circuit paths is still very rudimentary and a great deal more work is desirable both to confirm the existing data and to explore other semiconductors.

1.5.3 The Relative Contributions of Lattice and Short Circuit Path Diffusion

Short circuit paths provide a ubiquitous element in all diffusion studies and it is essential to be able to recognize the relative effects of such

TABLE 1.4

Experimental values of short circuit path diffusion parameters D_{so} and Q_s where $D_s = D_{so} \exp(-Q_s/kT)$. The volume diffusion parameters D_o and Q are given in brackets.

Crystal	Diffusant	Ref.	$D_{os}(D_o)\text{cm}^2\text{s}^{-1}$	$Q_s(Q)eV$	$r_o$cm
			(a) Pipe Diffusion		
Te	Te	166	Edge: 9.7×10^{-6}; Screw: 7.1×10^{-3} (//c: 1.3×10^2; ⊥c: 3.9×10^4)	Edge[a]: 0.65; Screw: 0.98 (//c: 1.75; ⊥c: 2.03)	Edge[a]: 4.5×10^{-5}; Screw: 1.3×10^{-5}
Se	Se	6	(//c: 0.2; ⊥c: 10^2)	//c: ≈ 0.5(1.2); ⊥c: ≈ 0.7(1.4)	—
Ge	In	169	6.8×10^6 (10)	2.76(2.75)	3×10^{-7} (assumed)
Si	In	170	10^4 (16.5)	3.31(3.87)	10^{-6}
	Sb	171	450(5.6)	3.0(3.91)	3×10^{-7} (assumed)
	P	172	1.1×10^4 (10.5[b])	3.38(3.65[b])	—
	Sb	172	7.0×10^4 (5.6[b])	3.61(3.91[b])	—
	B	172	190(10.5[b])	3.24(3.65[b])	—
	A1	172	140(8.0[b])	3.01(3.44[b])	—
NaCl	Na	165	—	Edge: 0.61; Screw: 1.1; (1.8)	—
		173	—	—	2×10^{-6}
KI	T1	165	—	Edge: 0.51; Screw: 0.63; (1.17)	2×10^{-6}
LiF	Na	173a	Edge and Screw: 0.01	Edge and Screw: 0.3	—
			(b) Grain Boundary Diffusion		
Te	Te	166	7.5×10^{-4} (//c: 1.3×10^2; ⊥c: 3.9×10^4	0.87 (//c: 1.75; ⊥c: 2.03)	1.49×10^{-5}

[a] Mimkes and Wuttig [190] have re-analyzed Ghoshtagore's [166] results using a more accurate theory and find for edge dislocations: $Q_s = 0.80$ eV, $r_o = 1.5$ μm.

[b] Data from ref. 100.

diffusion paths, in combination with lattice diffusion, in a given situation. The relevant parameters are D_s, the effective lattice diffusion coefficient $\mathbf{D}_l$, r_0, δ, the dislocation density, ρ_d, and the mean linear dimension of a grain, L. Harrison [174] has shown that, when there is simultaneous diffusion through the lattice and over short circuit paths, three regimes can be distinguished which he designated as types A, B and C. He also pointed out that different experimental methods (e.g. sectioning, isotopic exchange, residual activity or radiographic methods) generally observe quite different aspects of a given regime. These regimes are applicable to interdiffusion and to diffusion in an isotopic gradient. We now consider diffusion into a semi-infinite crystal located at $x \geqslant 0$.

Type A

This regime arises when in any plane normal to the x axis, the diffusant concentration in the lattice is uniform between pipes and grain boundaries. In effect a local dynamic equilibrium exists between atoms of the diffusant in short circuit paths and in the lattice i.e. there is no net flow out of, or into, such paths. If t_s and t_l are the average times a diffusant atom spends in a short circuit path and in the lattice respectively then the effective diffusion coefficient is [175]

$$\mathbf{D} = (1 - t_s/t_l)\mathbf{D}_l + t_s D_s/t_l \tag{1.77}$$

For solvent diffusion t_s/t_l is equal to the ratio of the total volume of short circuit paths to lattice volume. For pipe diffusion

$$t_s/t_l = \pi r_0^2 \rho_d/(1 - \pi r_0^2 \rho_d) \approx \pi r_0^2 \rho_d$$

so that

$$\mathbf{D} = \mathbf{D}_l + \pi r_0^2 \rho_d D_s \tag{1.78}$$

whereas for grain boundaries $t_s/t_l \approx 3\delta/L$. In the case of solute diffusion it has been proposed that the above expressions for t_s/t_l should be multiplied by a factor c_d/c_l where c_d and c_l are the solute concentrations in the short circuit paths and lattice respectively [175]. We would expect $c_d > c_l$. The data of Table 1.4 suggest that $\mathbf{D}$ can be significantly greater than $\mathbf{D}_l$ for $\rho_d \gtrsim 10^6$ cm^{-2}, $L \lesssim 1$ cm.

Harrison [174] has summarized the various criteria which have been proposed to define the conditions for type A diffusion. The least restrictive condition† is $\rho_d^{-1/2}$, $L \ll (\mathbf{D}_l t)^{1/2}$ where t is the diffusion time. The

† $\rho_d^{-1/2}$ and L give respectively the scale of the pipe and grain boundary networks. Even though a criterion may be explicitly obtained in terms of, say, L it is customary to assume that it is equally valid for a pipe network and vice versa.

most restrictive one, due to Harrison, is ρ_d^{-1}, $L^2 \leqslant 10^{-5} (\mathbf{D}_l t)$. Both of these criteria however assume a stationary pipe or grain boundary network. If the network is moving then type *A* diffusion can arise even if neither criterion is satisfied [155]. This could account for the observation of type *A* diffusion in Cu despite $\rho_d^{-1/2} \ll (\mathbf{D}_l t)^{1/2}$ not being met [176].

Planar type *A* diffusion will be described by

$$\frac{\partial c}{\partial t} = \frac{\partial}{\partial x}\left(\mathbf{D}\,\frac{\partial c}{\partial x}\right) \tag{1.79}$$

where c is the diffusant concentration averaged over the lattice and short circuit paths, and $\mathbf{D}$ is given by equation 1.78. The *A* regime can then be recognized if $c(x, t)$ is a solution of equation 1.79. The magnitude of $\mathbf{D}/\mathbf{D}_l$ will obviously depend on ρ_d or L for a given system; the dependence on ρ_d has been used to obtain $r_0^2 D_s$ in NaCl [165]. Widmer [177] found a slight enhancement of self-diffusion in Ge with $\rho_d = 2 \times 10^6$ cm^{-2} and $(\mathbf{D}t)^{1/2} \sim 1.5\,\mu$m, whereas Heldt and Hobstetter [178], with $\rho_d = 2 \times 10^6$ cm^{-2} but unspecified $(\mathbf{D}t)^{1/2}$ reported no enhancement of In or Sb diffusion in Ge. Although Widmer's results violate the type *A* criterion, the *A* regime could still be operative if the pipe or boundary network was moving. No firm conclusion can be drawn about the null result of ref. 178 as $(\mathbf{D}t)^{1/2}$ is unknown and it is also possible that the solute concentration will induce a $\rho_d > 2 \times 10^6$ cm^{-2} which will override the initial ρ_d i.e. no effect would be expected.

Type B

Type *B* diffusion occurs when, as in the *A* regime, both lattice and short circuit path diffusion contribute to the flow of the diffusant, but now there is a net flow of diffusant out of grain boundaries and pipes into the lattice (for an in-diffusing diffusant). The concentration of diffusant between short circuit paths, in a plane normal to the x axis, is no longer constant as in the *A* regime. The diffusion problem presented by this situation has attracted the attention of many workers dealing either with grain boundaries [179, 180, 181, 182, 183, 184] or pipes [163, 171, 180, 185]. Only the treatments of Whipple [181] (infinite source) and Suzuoka [184] (thin film) yield exact general solutions which embrace the *A*, *B* and *C* regimes. The other treatments are approximate and deal only with the *B* regime: the essential approximation which is made is that lattice diffusion is normal to either a grain boundary or pipe. The exact solutions of Whipple [181] and Suzuoka [184] have been computed and compared with the approximate solutions for the different experimental methods [168, 186, 187, 188, 189]. Confining attention to the widely used sectioning method the approximate solutions for an infinite source

[171, 179, 180, 183, 187] give $\ln c \propto x^r$ where c is as defined in equation 1.79 and $1 \leqslant r \leqslant 4/3$. The value of r is set by the particular approximation. Whipple's [181] solution can be used to show that

$$r = 4/3 \quad \text{if} \quad x(2\mathbf{D}_l/\delta D_s\sqrt{\mathbf{D}_l t})^{1/2} \gtrsim 10$$

or

$$r = 6/5 \quad \text{if} \quad 2 < x(2\mathbf{D}_l/\delta D_s\sqrt{\mathbf{D}_l t})^{1/2} < 10$$

and in both cases $(\delta D_s/2\mathbf{D}_l\sqrt{\mathbf{D}_l t})^{1/2} \geqslant 1$ [187]. Significant errors can arise in the activation energy for D_s if Fisher's analysis ($r = 1$) is used [168, 187, 188]. The solutions for pipe diffusion [163, 171, 180] yield $r = 1$ and are basically Fisher type solutions. As the Ge and Si data of Table 1.4 are based on $r = 1$ caution is required in their use. All the approximate theories assume a negligible overlap of lattice diffusion from adjacent short circuit paths. This condition requires $L/(D_s t)^{1/2} \geqslant 7$ or 10 for $(\delta D_s/\mathbf{D}_l\sqrt{\mathbf{D}_l t})^{1/2} = 1$ or 10 respectively [184]. It also turns out that the above remarks for $r = 6/5$ are equally applicable to diffusion with the thin film boundary condition [188]. It has been pointed out that all the analyses described above apply when r_0 and δ are small and of the order of a few Burgers vectors: for larger values of r_0 and δ account must be taken of the diffusant distribution within the pipe or grain boundary [190]. This becomes particularly important in sectioning experiments when, with large r_0 or δ, a significant proportion of the diffusant can be in the short circuit paths. Solutions of the general diffusion equations appropriate to this situation have been obtained [190]. Stark [191] has considered a variant of the more usual grain boundary problem by finding the diffusant flux which passes through a crystal of finite thickness containing grain boundaries.

Harrison [174], using the Whipple [181] solution, has derived the criterion $\mathbf{D}_l t \ll \delta^2, r_0^2$ for which, in the B regime, a particularly simple time dependence of the mass flow occurs in an isotopic exchange experiment. A more general criterion for the B regime does not appear to be available, other than if the criteria for types A and C are violated then presumably type B conditions prevail.

Type B diffusion involving pipe diffusion of impurities has been reported for Zn in GaAs [192] and also for impurities from groups III and V in Si [170, 171, 172] and Ge [169]. Ghoshtagore [166] has given details of self-diffusion in Te along both grain boundaries and pipes under type B conditions but appears to have used L rather than δ in Harrison's various criteria.

Type C

The C regime appears when the diffusant is only mobile in the short circuit paths i.e. lattice diffusion can be neglected. Harrison [174] has used

Whipple's [181] asymptotic solution to show that this situation requires $\delta^2, r_0^2 \gg \mathbf{D}_l t$. An alternative, less restrictive condition is

$$(\delta D_s/\mathbf{D}_l)^2 > 4 \times 10^3 (\mathbf{D}_l t) [183].$$

The sectioning method gives D_s directly under type C conditions for infinite source or thin film boundary conditions [174]. Type C self-diffusion, for both pipes and grain boundaries, in Te has been reported in which neither of the above two criteria is satisfied although the least restrictive condition is nearly met [166].

Harrison [174] has noted that the A, B and C regimes will follow in sequence in any diffusion experiment having appropriate boundary conditions (e.g. infinite source) and extending for a long enough time. Initially, near the surface the penetration will be type C, developing into type B and then maturing into type A. A mature penetration profile should then correspond to type A conditions near the surface, merging into type B with increasing depth and finally becoming type C at the deepest penetrations. Examples of profiles showing the A and B regimes have been obtained from self-diffusion experiments in Te mono-crystals [166] in which the tracer concentration in the B regime exhibits a $\ln c \propto x$ dependence. A further example of this situation, obtained by the self-diffusion of Cd in CdTe [193a], is given in figure 1.5 (profile 1). The discussion of Harrison's categories has implicitly assumed that δ, r_0, L and ρ_d are independent of position. If these parameters do vary with position then $\mathbf{D}$ (equation 1.79) will vary with position in the A regime, assuming no position dependence of $\mathbf{D}_l$, which could erroneously be interpreted as a concentration dependent lattice diffusion coefficient $\mathbf{D}_l$ (it is also conceivable that D_s may depend on δ or r_0). Circumstances giving rise to a continuous variation of ρ_d with penetration have been described in section 1.4.5. Irregular variations in ρ_d will give correspondingly irregular profiles, as seen by sectioning experiments, in the B and C regimes. An example of a profile showing such characteristics is shown in figure 1.5 (profile 2) for the chemical diffusion of Zn in AlSb [193b].

Type C diffusion can, in principle, be distinguished from type A through a dependence of the diffusion coefficient on ρ_d. A type A profile will also not change its x dependence with time whereas, at a given depth, a C regime will give way to a B regime. The B regime can be identified if $\ln c \propto x^r$ with $r \sim 1$ but the absence of such a dependence does not necessarily rule out type B diffusion. Separating lattice from short circuit path diffusion can be a difficult problem when $\ln c$ is not $\propto x^r$ and has been discussed in connection with diffusion in tungsten [194]. Unusual type B diffusion has been found in Bi self-diffusion [195]. The effect of

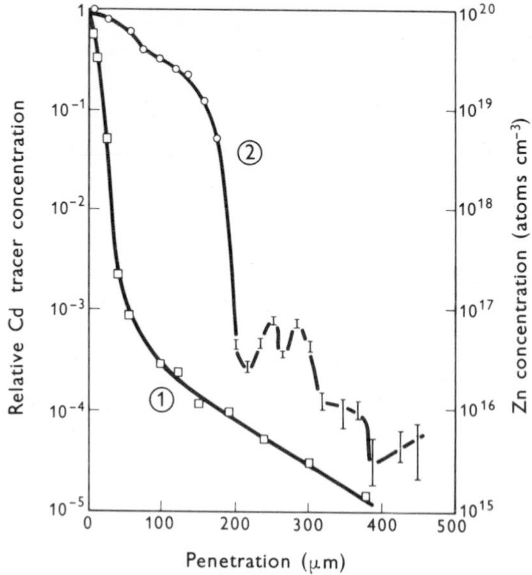

Figure 1.5. Curve 1 shows the Cd tracer distribution in CdTe following self-diffusion at 796° C (after Whelan and Shaw [193a]). Beyond 100 μm $\ln c \propto x$ which is a feature of the type *B* regime. Curve 2 is the Zn distribution observed after chemical diffusion in AlSb at 775° C (after Showan and Shaw [193b]). The region between the surface and 200 μm is identified as the type *A* regime; beyond 200 μm is attributed to the type *B* regime with variable ρ_d. Note that in both diffusions infinite source boundary conditions were used.

an external hydrostatic pressure on short circuit path diffusion relative to lattice diffusion has also been considered [196].

1.6 CONCLUSIONS

This chapter has been concerned with diffusion in dilute solutions and it has been shown that generally in the extrinsic range (n or $p > n_i$) the various diffusion coefficients of a species will vary with the species concentration. The variation arises because the common diffusion processes are sensitive to the point defect structure of the semiconductor. It is possible to claim that a satisfactory theoretical framework exists for the interpretation of bulk self and isoconcentration diffusion experiments in terms of atomic jump kinetics. Quantitative limits on interpretation are however imposed by the lack of the necessary precision in the measured diffusion coefficients and also by uncertainty in determining E_F for degenerate materials. The discussion of section 1.2.2 dealt in terms of a single parabolic conduction or valence band together with total ionization of a randomly distributed impurity. In reality as degeneracy increases it is to be expected that more and more sub-bands will contribute to the density of states. With increasing impurity concentration there is also the increasing possibility of local clustering of the impurity centres. Such clusters are ordered structures and have been observed at solute concentrations as low as 1% mole fraction [197] (see also section 5.5.2). The electronic properties of such clusters are unknown. Any calculation of E_F in the degenerate case must have regard to the detailed band structure, and the associated density of states, of the particular semiconductor and to the occurrence of clusters. It may be that the onset of significant clustering will provide an effective demarcation between dilute and non-dilute solutions.

With chemical diffusion further problems are introduced due to the concentration gradient. At all diffusant concentrations the important question arises as to whether local chemical equilibrium exists along the diffusant profile. If equilibrium cannot be assumed to exist the situation becomes considerably more difficult to interpret on a quantitative basis. At high impurity concentrations there is the further complication arising from the generation of misfit dislocations. The feedback effect of such dislocations on the diffusion of the impurity is a complex dynamical problem. Several attacks have been made on this problem but a satisfactory quantitative model is still lacking.

The sensitivity of diffusion to the defect structure in the semiconductor can be exploited in the control of a diffusant flux and also in the study of defects in a semiconductor. The latter, particularly when combined with electrical measurements, is a rich area but one in which relatively little prospecting has so far been done. It must however be emphasized that in all diffusion experiments meaningful conclusions can only be drawn if the composition and defect structure of the semiconductor are under experimental control.

Looking to the future it seems likely that diffusion in non-dilute semiconductor systems will become of increasing interest. A further area which promises to have considerable basic and technological interest is the enhancement of diffusion processes during irradiation with photon and other particle fluxes (section 1.3.2). Although few results are available they are sufficient to show many exciting possibilities.

ACKNOWLEDGMENTS

The author wishes to express his appreciation to Dr. A. D. LeClaire, Dr. R. F. Peart and Professor L. M. Slifkin for reading and commenting on an earlier draft of the chapter. Special thanks are due to Drs. A. D.

LeClaire, J. R. Brews and C. J. Hwang for providing copies of their papers prior to publication and also for clarifying correspondence.

APPENDIX

Solutions of $\dfrac{\partial c}{\partial t} = D\dfrac{\partial^2 c}{\partial x^2}$

Solutions of equation 1.3 for a wide range of different boundary conditions are already available [198, 199, 200]. The following solutions have particular relevance to diffusion in semiconductors.

A.1 Planar Diffusion in a Semi-Infinite Crystal ($x \geqslant 0$) with Depletion of the Reservoir

In most situations the diffusant is present in an external phase at a uniform concentration $c_R(t)$ and occupies a volume V_R : this constitutes the source or reservoir. $c_R(t)$ decreases with time as a consequence of diffusion into the crystal. Assuming that diffusion within the reservoir is fast enough to ensure no gradients in $c_R(t)$ and that the surface area of the crystal is A, then [201]

$$c(x, t) = c(0, 0) \exp\left(\frac{Ax}{\vartheta V_R} + \frac{A^2 Dt}{\vartheta^2 V_R^2}\right) \operatorname{erfc}\left(\xi + \frac{A\sqrt{Dt}}{\vartheta V_R}\right) \tag{1.80}$$

where

$$\xi = x/2\sqrt{Dt}, \quad \vartheta = c_R(t)/c(0, t)$$

and

$$\operatorname{erfc} y = 1 - \frac{2}{\sqrt{\pi}}\int_0^y e^{-s^2}\, ds$$

Useful properties of $\operatorname{erfc} y$ to note are

$$d(\operatorname{erfc} y)/dy = -\frac{2}{\sqrt{\pi}} \exp(-y^2)$$

and for large y, $\operatorname{erfc} y \to \exp(-y^2)/\sqrt{\pi}y$. Equation 1.80 shows that the reservoir content varies with time according to

$$c_R(t) = \vartheta c(0, t) = c_R(0) \exp\left(\frac{A^2 Dt}{\vartheta^2 V_R^2}\right) \operatorname{erfc}\left(\frac{A\sqrt{Dt}}{\vartheta V_R}\right) \tag{1.81}$$

Two practically important limiting cases for equation 1.80 arise, corresponding to negligible and extreme depletion of the reservoir. For the former this means a very large reservoir, i.e. $\vartheta V_R \gg A\sqrt{Dt}$, so yielding

TABLE 1.5[a]

Values[b] of erfc ξ and $\frac{2}{\sqrt{\pi}} \exp(-\xi^2)$

(a) $\operatorname{erfc} \xi = 1 - \frac{2}{\sqrt{\pi}} \int_0^{\xi} \exp(-s^2\, ds)$

ξ	0	0.1	0.2	0.3	0.4	0.5	0.6	0.7	0.8	0.9
0	1.000	8.875(−1)	7.773(−1)	6.714(−1)	5.716(−1)	4.795(−1)	3.961(−1)	3.222(−1)	2.579(−1)	2.031(−1)
1.0	1.573(−1)	1.198(−1)	8.969(−2)	6.599(−2)	4.772(−2)	3.390(−2)	2.365(−2)	1.621(−2)	1.091(−2)	7.210(−3)
2.0	4.678(−3)	2.980(−3)	1.863(−3)	1.143(−3)	6.885(−4)	4.070(−4)	2.360(−4)	1.343(−4)	7.501(−5)	4.110(−5)

(b) $\frac{2}{\sqrt{\pi}} \exp(-\xi^2)$

ξ	0	0.1	0.2	0.3	0.4	0.5	0.6	0.7	0.8	0.9
0	1.128	1.117	1.084	1.031	9.615(−1)	8.788(−1)	7.872(−1)	6.913(−1)	5.950(−1)	5.020(−1)
1.0	4.151(−1)	3.365(−1)	2.673(−1)	2.082(−1)	1.589(−1)	1.189(−1)	8.722(−2)	6.271(−2)	4.419(−2)	3.053(−2)
2.0	2.067(−2)	1.371(−2)	8.922(−3)	5.689(−3)	3.556(−3)	2.178(−3)	1.308(−3)	7.699(−4)	4.442(−4)	2.512(−4)

[a] Compiled from 'Tables of the Error Function and Its Derivative', Applied Mathematics Series 41, National Bureau of Standards (U.S. Dept. of Commerce, 1954).

[b] The bracketed digits denote the power of ten to which an entry must be multiplied.

$c_R(t) = c_R(0) = \vartheta c(0, 0)$ and $c(x, t) = c(0, 0)$ erfcξ which is equation 1.8. Extreme depletion occurs when the reservoir is very small, i.e.

$$\vartheta V_R \ll A\sqrt{Dt},$$

hence

$$c(x, t) = W \exp(-\xi^2)/\sqrt{\pi Dt}$$

where $W = \vartheta V_R\, c(0, 0)/A$. This is equation 1.9. Values of $(2/\sqrt{\pi})\exp(-\xi^2)$ and erfcξ are given in Table 1.5. The difficulty with equation 1.9 when $t \gtrsim 0$ has been discussed by Weymann [202]. Malkovich [203] has given a useful discussion of the experimental application of equations 1.8 and 1.9.

A.2 Planar Diffusion into a Semi-Infinite Evaporating Crystal

(a) Infinite Source. If the crystal surface is receding at a rate u by evaporation then [204]

$$c(x, t) = (1/2)c(0, 0)\left\{\operatorname{erfc}\left(\xi + \frac{ut}{2\sqrt{Dt}}\right) + \exp\left(-\frac{ux}{D}\right)\operatorname{erfc}\left(\xi - \frac{ut}{2\sqrt{Dt}}\right)\right\} \tag{1.82}$$

where $c(0, 0) = c(0, t)$ and the crystal surface at time t defines the x origin.

(b) Thin Film Source. In this case [205]

$$c(x, t) = W\left\{\frac{\exp\left[-\left(\xi + \frac{ut}{2\sqrt{Dt}}\right)^2\right]}{\sqrt{\pi Dt}} - \frac{u}{2D}\operatorname{erfc}\left(\xi + \frac{ut}{2\sqrt{Dt}}\right)\right\} \tag{1.83}$$

where x, u and W are defined above.

A.3 Planar Diffusion into a Finite Crystal (Infinite Source)

When $\sqrt{Dt} > 0.2l$ where l is the total thickness of the crystal, and the diffusant penetrates from both faces, then [199]

$$c(x, t) = c(0, 0)\left\{1 - \frac{4}{\pi}\sin\frac{\pi x}{l}\exp\left(-\frac{\pi^2 Dt}{l^2}\right)\right\} \tag{1.84}$$

REFERENCES

1. P. G. Shewmon, Diffusion in Solids, Chapters 2, 4, 5. (McGraw-Hill, New York, 1963)
2. B. I. Boltaks, Diffusion in Semiconductors, Chapter 2. (Infosearch, London, 1963)
3. Y. Adda, J. Phillibert, La Diffusion dans les Solides, Vol. 2, Chapter 13. (Presses Universitaires de France, Paris, 1966)
4. L. A. Girifalco, Atomic Migration in Crystals, Appendix D. (Blaisdell Publishing Company, New York, 1964)

5. H. B. Huntington, P. B. Ghate and J. H. Rosolowski, *J. appl. Phys.*, **35**, 3027 (1964)
6. P. Brätter and H. Gobrecht, *phys. stat. sol.*, **37**, 869 (1970); **41**, 631 (1970)
7. R. N. Ghoshtagore, *Phys. Rev.*, **155**, 598 (1967)
8. Ref. 3, Vol. 1, Chapter 9
9. F. H. Ree, T. S. Ree and H. Eyring, *Adv. chem. Phys.*, **4**, 1 (1962)
10. J. R. Manning, Diffusion Kinetics for Atoms in Crystals, Chapter 2. (D. van Nostrand Co. Inc., Princeton, 1968)
11. Ref. 3, Vol. 2, Chapter 14
12. F. A. Kroger, The Chemistry of Imperfect Crystals, Chapter 19. (North Holland Publishing Co., Amsterdam, 1964)
13. V. I. Fistul, Heavily Doped Semiconductors, Chapter 2. (Plenum Press, New York, 1969)
14. J. S. Blakemore, Semiconductor Statistics, Chapter 3. (Pergamon Press, Oxford, 1962)
15. W. Shockley and J. L. Moll, *Phys. Rev.*, **119**, 1480 (1960)
16. R. A. Reynolds, Thesis, Stanford University (1965)
17. G. Brouwer, *Philips Res. Rep.*, **9**, 366 (1954)
18. Ref. 12, Chapter 10
19. A. K. Jonscher, Principles of Semiconductor Device Operation, p. 154. (Bell, London, 1960)
20. N. F. Mott and W. D. Twose, *Adv. Phys.*, **10**, 107 (1961)
21. C. Hilsum, *Prog. Semicond.*, **9**, 153 (1963)
22. H. C. Casey, F. Ermanis and K. B. Wolfstirn, *J. appl. Phys.*, **40**, 2945 (1969)
23. Ref. 13, Chapter 1
24. V. L. Bonch-Bruevich, The Electronic Theory of Heavily Doped Semiconductors. (Elsevier Publ. Co., New York, 1966)
25. C. J. Hwang, *J. appl. Phys.*, **41**, 2668 (1970)
26. C. J. Hwang and J. R. Brews, *J. Chem. Phys.*, **54**, 3263 (1971); *Physics Chem. Solids*, **32**, 837 (1971)
27. L. V. Keldysh and Y. V. Kopaev, *Soviet Phys. Solid St.*, **5**, 1141 (1963)
28. W. W. Harvey, *Physics Chem. Solids*, **23**, 1545 (1962)
29. A. J. Rosenberg, *J. Chem. Phys.*, **33**, 665 (1960)
30. Ref. 3, Vol. 1, Chapter 1
31. A. D. LeClaire, *Prog. Metal. Phys.*, **4**, 300 (1953)
32. K. Weiser, *Phys. Rev.*, **126**, 1427 (1962)
33. J. W. Corbett, Electron Radiation Damage in Semiconductors and Metals, p. 178, 200. (Academic Press, New York, 1966); Radiation Damage in Semiconductors, p. 8. (Ed. F. L. Vook, Plenum Press, New York, 1968)
33a. J. C. Bourgoin and J. W. Corbett, *Phys. Lett.*, **38A**, 135 (1972)
34. A. C. Damask and G . J. Dienes, Point Defects in Metals, p. 14. (Gordon and Breach, New York, 1963)
35. Ref. 10, Chapter 1
36. V. V. Voronkov, G. I. Voronkova and M. I. Iglitsyn, *Soviet Phys. Solid St.*, **11**, 1344 (1969)
37. M. D. Weber and R. J. Friauf, *Physics Chem. Solids*, **30**, 407 (1969)
38. F. C. Frank and D. Turnbull, *Phys. Rev.*, **104**, 617 (1956)
39. A. Seeger and K. P. Chik, *phys. stat. sol.*, **29**, 455 (1968)
40. A. Seeger, *Comments Solid State Physics*, **2**, 55 (1969)
41. G. Borelius, *Phys. Lett.*, **30A**, 267 (1969); *Physica Scripta*, **1**, 141 (1970)
42. D. L. Kendall and D. B. de Vries, Semiconductor Silicon, p. 358. (The Electrochemical Society, New York, 1969)
43. J. W. Miller, *Phys. Rev.*, **188**, 1074 (1969)
44. G. D. Watkins, Radiation Effects in Semiconductors, p. 67. (Ed. F. L. Vook, Plenum Press, New York, 1968)
45. J. C. Brice, *Solid-St. Electron.*, **6**, 673 (1963)

46. A. D. LeClaire, 'Physical Chemistry–An Advanced Treatise', Solid State, Vol. 10, Chapter 5, to be published. (Academic Press, New York)
47. Ref. 10, Chapter 3
47a. H. Mehrer, *Zeit. Naturforsch*, **26a**, 308 (1971)
47b. M. Yoshida, *J. appl. Phys. Japan*, **10**, 702 (1971)
48. J. M. Fairfield and B. J. Masters, *J. appl. Phys.*, **38**, 3148 (1967); **40**, 2390 (1969)
49. S. M. Hu, *Phys. Rev.*, **180**, 773 (1969)
50. Ref. 1, Chapter 5
51. E. M. Pell, *Phys. Rev.*, **119**, 1014 (1960)
52. Ref. 10, p. 130
53. B. Pegel, *phys. stat. sol.*, **22**, 223, K45 (1967)
54. J. Black and P. Lublin, *J. appl. Phys.*, **35**, 2462 (1964)
55. H. R. Potts and G. L. Pearson, *J. appl. Phys.*, **37**, 2098 (1966)
56. B. G. Cohen, *Solid-St. Electron.*, **10**, 33 (1967)
57. T. H. Yeh and M. L. Joshi, *J. electrochem. Soc.*, **116**, 73 (1969)
58. Ref. 13, p. 34
59. O. Madelung, Physics of III-V Compounds, Chapter 6. (John Wiley and Sons, New York, 1964)
60. R. K. Willardson and A. C. Beer (Editors), Semiconductors and Semimetals, Vol. 4, Chapter 7.(Academic Press, New York, 1968)
61. L. A. Girifalco and D. O. Welch, Point Defects and Diffusion in Strained Metals, Chapter 3. (Gordon and Breach, New York, 1967)
62. N. B. Hannay (Editor), Semiconductors, Chapter 5. (Reinhold, New York, 1959)
63. Ref. 12, Chapter 9
64. R. L. Longini and R. F. Green, *Phys. Rev.*, **102**, 992 (1956)
65. R. A. Swalin, *J. appl. Phys.*, **29**, 670 (1958)
66. J. R. Hardy and R. Bullough, *Phil Mag.*, **15**, 237 (1967)
67. H. Reiss, C. S. Fuller and F. J. Morin, *Bell Syst. tech. J.*, **35**, 535 (1956)
68. M. Martens, L. Mehrkam and F. Williams, *Phys. Rev.*, **186**, 757 (1969)
69. C. S. Fuller and K. B. Wolfstirn, *J. appl. Phys.*, **38**, 4339 (1967)
70. F. Williams, *phys. stat. sol.*, **25**, 493 (1968)
71. R. C. Newman, *Adv. Phys.*, **18**, 545 (1969)
72. R. E. Howard and A. B. Lidiard, *Rep. Prog. Phys.*, **27**, 161 (1964)
73. R. E. Howard and J. R. Manning, *Phys. Rev.*, **154**, 561 (1967)
74. A. D. LeClaire, *Phil Mag.*, **21**, 819 (1970)
75. J. R. Manning, *Metall. Trans.*, **1**, 499 (1970)
76. L. W. Barr and F. A. Smith, *Phil Mag.*, **20**, 1293 (1969)
77. M. A. Kaneev, *Soviet Phys. Solid St.*, **10**, 726 (1968)
78a. A. I. Koifman and O. R. Niyazova, *phys. stat. sol. (a)*, **10**, 59 (1972)
78b. O. A. Klimkova and O. R. Niyazova, *Soviet Phys. Solid St.*, **12**, 1760 (1971)
78c. L. N. Zyuz, A. E. Kiv, O. R. Niyazova and F. T. Umarova, *J.E.T.P. Lett.*, **12**, 147 (1970)
78d. M. A. Zaikovskaya, O. A. Klimkova and O. R. Niyazova, *Soviet Phys. semicond.*, **5**, 802 (1971)
79. D. G. Nelson, J. F. Gibbons and W. S. Johnson, *Appl. Phys. Lett.*, **15**, 246 (1969)
79a. A. E. Kiv and F. T. Umarova, *Soviet Phys. semicond.*, **4**, 474 (1970); (with Z. A. Iskanderova) **4**, 1543 (1971)
79b. V. S. Yakhot, *phys. stat. sol. (b)*, **48**, 141 (1971)
80. E. F. Koch and K. T. Aust, *Acta metall.*, **15**, 405 (1967)
81. A. B. Lidiard and R. S. Nelson, *Phil Mag.*, **17**, 425 (1968)
82. F. A. Nicholls, *Acta. metall.*, **20**, 207 (1972)
83. Ref. 10, Chapter 4
84. D. S. Kamenetskaya, I. B. Piletskaya and V. I. Shiryaev, *Soviet Phys. Dokl.*, **13**, 245 (1968)

85. H. P. Dibbs and J. R. Tremblay, *J. appl. Phys.*, **39**, 2976 (1968)
86. R. A. Oriani, *Physics Chem. Solids*, **30**, 339 (1969)
87. S. Prussin, *J. appl. Phys.*, **32**, 1876 (1961)
88. Ref. 1, p. 24
89. R. Bullough and R. C. Newman, *Rep. Prog. Phys.*, **33**, 101 (1970)
90. V. B. Fiks, *Soviet Phys. Solid St.*, **1**, 1212 (1960); **5**, 2549 (1964)
91. B. I. Boltaks, T. D. Dzhafarov, V. I. Sokolov and F. S. Shishiyanu, *Soviet Phys. Solid St.*, **6**, 1181 (1964)
92. Ref. 3, Vol. 2, Chapter 16
92a. D. E. Temkin, *Soviet Phys. Solid St.*, **13**, 2840, 2864 (1972)
93. Ref. 3, Vol. 2, Chapter 10
94. R. Lindstrom, *Physics Chem. Solids*, **30**, 401 (1969)
95. K. A. Arseni, B. I. Boltaks and T. D. Dzhafarov, *phys. stat. sol.*, **35**, 1053 (1969)
96. S. M. Hu and M. S. Mock, *Phys. Rev.*, **IB**, 2582 (1970)
97. Ref. 10, Chapter 5
98. S. Zaromb, *IBM Jl Res. Dev.*, **1**, 57 (1957)
99. Ref. 14, Appendix C
100. Ref. 62, Chapter 6
101. R. N. Hall and J. H. Racette, *J. appl. Phys.*, **35**, 379 (1964)
102. B. I. Boltaks and T. D. Dzhafarov, *phys. stat. sol.*, **19**, 705 (1967)
103. W. Shockley, *J. appl. Phys.*, **32**, 1402 (1961)
104. D. Shaw and A. L. J. Wells, *Br. J. appl. Phys.*, **17**, 999 (1966)
105. V. V. Vas'kin, V. A. Uskov and M. Y. Shirabokov, *Soviet Phys. Solid St.*, **7**, 2703 (1966)
106. T. Klein and J. R. A. Beale, *Solid-St. Electron.*, **9**, 59 (1966)
107. V. V. Vas'kin, V. S. Metrikin, V. A. Uskov and M. Y. Shirabokov, *Soviet Phys. Solid St.*, **8**, 2779 (1967)
108. V. V. Vas'kin, *Soviet Phys. semicond.*, **2**, 84 (1968)
109. S. M. Hu and S. Schmidt, *J. appl. Phys.*, **39**, 4272 (1968)
110. H. C. Casey, Jr., M. B. Panish and L. L. Chang, *Phys. Rev.*, **162**, 660 (1967)
111. Ref. 60, Chapter 3
112. R. F. Brebrick, *J. appl. Phys.*, **30**, 811 (1959)
113. R. W. Broderson, J. N. Walpole and A. R. Calawa, *J. appl. Phys.*, **41**, 1484 (1970)
114. H. H. Woodbury, *Phys. Rev.*, **134**, A492 (1964)
115. K. Zanio, *J. appl. Phys.*, **41**, 1935 (1970)
116. J. W. Mitchell, *J. appl. Phys.*, **33**, 406 (1962)
117. F. A. Huntley and A. F. W. Willoughby, *Solid-St. Electron.*, **13**, 1231 (1970)
118. R. S. Malkovich, *Soviet Phys. Solid St.*, **9**, 1676 (1968)
119. M. Yoshida, *Jap. J. Phys.*, **8**, 1211 (1969)
120. B. S. Bokshtien, S. Z. Bokshtein, A. A. Zhukhovitskii, S. T. Kishkin, L. G. Kornelyuk and Y. S. Nechaev, *Soviet Phys. Solid St.*, **11**, 194 (1969)
121. R. E. Hanneman and T. R. Anthony, *Acta metall.*, **17**, 1133 (1969)
122. R. L. Williams, *J. appl. Phys.*, **40**, 2932 (1969)
123. Quoted by H. J. Queisser, *J. appl. Phys.*, **32**, 1776 (1961)
124. M. L. Joshi and S. Dash, *IBM Jl Res. Dev.*, **10**, 446 (1966)
125. M. L. Joshi, C. H. Ma and J. Makris, *J. appl. Phys.*, **38**, 7255 (1967)
126. W. Czaja, *J. appl. Phys.*, **37**, 3441 (1966)
127. E. D. Jungbluth, *Metall. Trans.*, **1**, 575 (1970)
128. R. L. Mozzi and J. M. Lavine, *J. appl. Phys.*, **41**, 280 (1970)
129. T. J. Parker, *J. appl. Phys.*, **38**, 3471, 3475 (1967); **39**, 2043 (1968); **41**, 424 (1970); *J. electrochem. Soc.*, **115**, 1290 (1968): Proceedings of the Thomas Graham Memorial symposium, Strathclyde, 1969, p. 503. (Gordon and Breach. London, 1971)
130. S. M. Hu and T. H. Yeh, *J. appl. Phys.*, **40**, 4615 (1969), **42**, 2153 (1971); S. M. Hu, *J. appl. Phys.*, **40**, 4684 (1969)

131. N. D. Thai, *Solid-St. Electron.*, **13**, 165 (1970); *J. appl. Phys.*, **41**, 2859 (1970)
131a. V. I. Pashkov and P. V. Pavlov, *Soviet Phys. Solid St.*, **13**, 867 (1971)
131b. D. Shaw, *phys. stat. sol. (a)*, **11**, No. 1 (1972)
131c. D. Shaw and S. R. Showan, *phys. stat. sol.*, **32**, 109 (1969)
132. S. Dash and M. L. Joshi, *IBM Jl Res. Dev.*, **14**, 453 (1970)
133. N. S. Rytova, V. I. Fistul and P. M. Grinshtein, *Soviet Phys. semicond.*, **5**, 1700, 1762 (1972)
134. F. V. Nolfi, P. G. Shewmon and J. S. Foster, *Trans Metall. Soc. A.I.M.E.*, **245**, 1427 (1969)
135. R. Trivedi and G. M. Pound, *J. appl. Phys.*, **40**, 4293 (1969)
136. E. Nebauer, *phys. stat. sol.*, **36**, K63 (1969)
137. Ref. 1, p. 29
138. T. I. Kucher, *Soviet Phys. Solid St.*, **6**, 623 (1964)
139. A. L. Laskar, *Proc. Nat. Inst. Sci. India,* **A30**, 4800 (1964)
140. S. A. Boktor and L. A. K. Watt, *J. appl. Phys.*, **41**, 2844 (1970)
140a. M. Yoshida and K. Kimura, *J. appl. Phys. Japan,* **10**, 566 (1971)
141. V. V. Vas'kin and V. A. Uskov, *Soviet Phys. Solid St.*, **11**, 1429 (1970)
142. J. Friedel, Dislocations, Chapter 10. (Pergamon Press, Oxford, 1964)
143. Ref. 3., Vol. 2, Chapter 12
144. R. Bullough and R. C. Newman, *Prog. Semicond.*, **7**, 99 (1963)
145. R. H. Glaenzer and A. G. Jordan, *Solid-St. Electron.*, **12**, 247 (1969)
146. W. Schroter and R. Labusch, *phys. stat. sol.*, **36**, 539 (1969)
147. U. Baitinger, J. Arndt and D. Schnepf, *J. Mater. Sci.*, **4**, 396 (1969)
148. Y. A. Osip'yan and Y. I. Fedaev, *JETP Lett.*, **9**, 21 (1969)
149. A. D. Paddock and S. H. Carpenter, *Metall. Trans.* **1**, 651 (1970)
150. J. S. Koehler, *Phys. Rev.*, **181**, 1015 (1969)
151. H. F. Mataré, Defect Electronics in Semiconductors (Wiley–Interscience, New York, 1971), section 12.4
152. J. E. Lawrence, *Br. J. appl. Phys.*, **18**, 405 (1967)
153. C. A. Pampillo and N. W. De Reca, *J. Mater. Sci.*, **4**, 985 (1969)
154. A. C. Damask, G. J. Dienes, H. Herman and L. E. Katz, *Phil Mag.*, **20**, 67 (1969)
155. R. W. Balluffi and A. L. Ruoff, *J. appl. Phys.*, **34**, 1634, 1848, 2862 (1963)
156. A. L. Ruoff, *J. appl. Phys.*, **38**, 3999 (1967)
157. T. E. Volin, K. H. Lie and R. W. Balluffi, *Acta metall.*, **19**, 263 (1971)
158. I. R. Sanders and P. S. Dobson, *Phil Mag.*, **20**, 881 (1969)
159. J. Melngailis and S. O'Hara, *J. appl. Phys.*, **33**, 2596 (1962)
160. G. M. Gutnikova and B. Y. Lyubov, *Soviet Phys. Solid St.*, **11**, 1011 (1969)
161. G. M. Gutnikova and B. Y. Lyubov, *Soviet Phys. Dokl.*, **14**, 165 (1969)
162. J. Lothe, *J. appl. Phys.*, **31**, 1077 (1960)
163. L. C. Luther, *J. chem. Phys.*, **43**, 2213 (1965); **45**, 1080 (1966)
164. R. W. Balluffi, *phys. stat. sol.*, **42**, 11 (1970)
165. Y. E. Geguzin and E. R. Dobrovinskaya, *Soviet Phys. Solid St.*, **7**, 1660, 2866 (1965)
166. R. N. Ghoshtagore, *Phys. Rev.*, **155**, 603 (1967)
167. K. T. Aust and B. Chalmers, *Metall. Trans.*, **1**, 1095 (1970)
168. R. F. Canon and J. P. Stark, *J. appl. Phys.*, **40**, 4361, 4366 (1969)
169. V. A. Panteleev, *Soviet Phys. Solid St.*, **7**, 734 (1965)
170. P. V. Pavlov, L. V. Lainer, V. A. Sterkhov and V. A. Panteleev, *Soviet Phys. Solid St.*, **8**, 580 (1966)
171. P. V. Pavlov, V. A. Panteleev and A. Maiorov, *Soviet Phys. Solid St.*, **6**, 305 (1965)
172. G. V. Dudko, M. A. Kolegaev and V. A. Panteleev, *Soviet Phys. Solid St.*, **11**, 1097 (1969)
173. L. W. Barr, I. M. Hoodless, J. A. Morrison and R. Rudham, *Trans. Faraday Soc.*, **54**, 697 (1960)

173a. R. Tucker, A. Lasker and R. Thomson, *J. appl. Phys.*, **34**, 445 (1963)
174. L. G. Harrison, *Trans. Faraday Soc.*, **57**, 1191 (1961)
175. Ref. 1, Chapter 6
176. H. M. Morrison, *Phil Mag.*, **12**, 985 (1965)
177. H. Widmer, *Phys. Rev.*, **125**, 30 (1962)
178. L. A. Heldt and J. N. Hobstetter, *Acta metall.*, **11**, 1165 (1963)
179. J. C. Fisher, *J. appl. Phys.*, **22**, 74 (1951)
180. R. Smoluchowski, *Phys. Rev.*, **87**, 482 (1952)
181. R. T. P. Whipple, *Phil Mag.*, **45**, 1225 (1954)
182. R. C. Exterman (Editor), Radioisotopes in Scientific Research, Vol. 1, p. 212. (Pergamon Press, London, 1958)
183. H. S. Levine and C. J. MacCullum, *J. appl. Phys.*, **31**, 595 (1960)
184. T. Suzuoka, *Trans. Japan Inst. Metals*, **2**, 25 (1961)
185. J. P. Stark, *J. appl. Phys.*, **36**, 3938 (1965)
186. V. E. Wood, A. E. Austin and F. J. Milford, *J. appl. Phys.*, **33**, 3574 (1962)
187. A. D. LeClaire, *Br. J. appl. Phys.*, **14**, 351 (1963)
188. T. Suzuoka, *Jap. J. Phys.*, **19**, 839 (1964)
189. J. Kučera, *Can. J. Phys.*, **46**, 1511 (1968)
190. J. Mimkes and M. Wuttig, *J. appl. Phys.*, **41**, 3205 (1970) (grain boundary); *Phys. Rev. B*, **2**, 1619 (1970) (pipe)
191. J. P. Stark, *J. appl. Phys.*, **40**, 3101 (1969)
192. D. A. Granning and A. H. Herzog, *J. appl. Phys.*, **39**, 2783 (1968)
193a. R. C. Whelan and D. Shaw, II-VI Semiconducting Compounds, p. 451. (Editor D. G. Thomas). (W. A. Benjamin, Inc., New York, 1967)
193b. S. R. Showan and D. Shaw, *phys. stat. sol.*, **32**, 97 (1969)
194. R. E. Pawel and T. S. Lundy, *Acta metall.*, **17**, 979 (1969)
195. W. P. Ellis and N. H. Nachtrieb, *J. appl. Phys.*, **40**, 472 (1969)
196. R. F. Peart, *phys. stat. sol.*, **20**, 545 (1967)
197. M. E. Fine, The Chemistry of Extended Defects in Non-Metallic Solids, p. 575. (Editors L. Eyring and M. O'Keeffe). (North Holland Publishing Co., Amsterdam, 1970)
198. J. Crank, The Mathematics of Diffusion (Oxford University Press, 1956)
199. Ref. 2., Chapter 4
200. Ref. 3, Vol. 1, Chapter 3
201. P. M. Borsenberger, Thesis, Stanford University, 1967
202. H. D. Weyman, *Am. J. Phys.*, **35**, 488 (1967)
203. R. S. Malkovich, *Soviet Phys. Solid St.*, **1**, 548 (1959)
204. T. L. Kucher, *Soviet Phys. Solid St.*, **3**, 401 (1961)
205. R. N. Ghoshtagore, *phys. stat. sol.*, **19**, 123 (1967)

THE CALCULATION OF DIFFUSION COEFFICIENTS IN SEMICONDUCTORS 2

R. A. SWALIN

CONTENTS

2.1 INTRODUCTION

The intent of the present chapter is to discuss the current status of theoretical work which has been performed in connection with predicting rates of atomic motion in semiconducting materials. This theoretical work, aside from being of intrinsic interest, has some utility for individuals interested in experimental phenomena. Firstly, the calculations provide a means for estimating diffusion properties for systems which have not been studied or for systems for which it is difficult to perform measurements. Secondly, some calculations can often be used as a test of the accuracy of experimental data inasmuch as data for systems which do not conform to calculations based on reasonable models are suspect, and those systems become targets for future work in order to verify or reject the early measurements. As an example, in metal systems theoretical calculations have played a major role confirming the accuracy or inaccuracy of a given set of experimental data. Thirdly, inasmuch as the process of atomic diffusion involves ionic interactions which deviate substantially from equilibrium conditions, experimental data interpreted through various models, in principle, can provide information about the nature of very short-range as well as very long-range forces in crystals.

In order to provide a background, a general review of the nature of activated process will be first presented. An attempt will be made to show some of the underlying assumptions of the various theoretical approaches which have been employed in calculating D_o and Q in the diffusion equation. Subsequently, the approaches employed in calculating D_o and Q for self and solute diffusion for various types of crystals will be discussed. Since detailed discussion of data and implied diffusion mechanisms will be given in later chapters in this monograph, the principal emphasis here will be on the methods of calculation of various diffusion parameters under various assumed mechanisms. Also, since phenomenological equations and random walk considerations are presented in Chapter 1, the emphasis in this chapter will be on calculation of terms associated with the elemental jumping process itself.

2.2 GENERAL THEORETICAL CONSIDERATIONS OF THE ACTIVATED PROCESS

2.2.1 Application of Activated State Theory and the Arrhenius Equation

It is observed empirically that most diffusion processes in solids exhibit a temperature dependence described by the Arrhenius equation

$$D = D_o \exp(-Q/kT) \tag{2.1}$$

where D_o and Q are constants for a particular system. This equation is in fact so generally applicable that when experimental data are found which deviate from it, considerable effort is generally devoted to rationalize the discrepancy. Many factors can be involved in causing deviations from the Arrhenius equation. Among these are the existence of short circuiting diffusion paths at low temperature, multiple diffusion mechanisms, and impurity effects. The burden is always on the investigator in cases where deviation exists to show that the data are correct and that the effect is not due to improper experimental technique. There are several instances for semiconductors where deviations do exist, however, and these cases provide interesting situations for interpretation.

The approach used in chemical rate theory was first employed to explain the applicability of equation 2.1. This approach is based on absolute rate theory [1]. Expositions of the treatment have been given by many authors [2, 3, 4, 5, 6, 7, 8] and, thus, only a brief summary will be presented here.

In the application of absolute rate theory, a principal assumption is that in isothermally moving from one equilibrium site to another a particle moves *reversibly* from its initial position over a potential energy maximum along the reaction coordinate between the initial and final site. The potential energy barrier is schematically illustrated in Figure 2.1. The

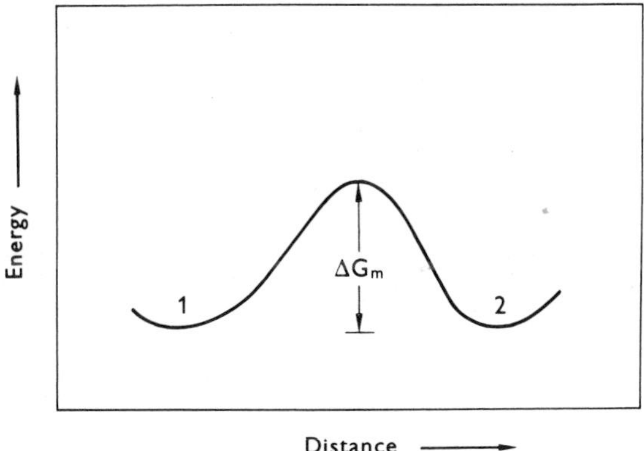

Figure 2.1. Schematic illustration of diffusion potential energy barrier for solids.

moving particle is considered to be in thermal equilibrium with the lattice at all times and, thus, in a thermodynamic sense the motion must occur at an infinitesimal velocity. If these assumptions are employed, standard methods of statistical mechanics may be employed to calculate the probability of motion of particles over the potential energy maximum. The rate of 'reaction' is obtained by a calculation of the number of particles in the ensemble at the potential energy maximum and then calculating the velocity at which particles move over the barrier into the other equilibrium site. By employment of these procedures Zener [6] obtained for the jump frequency, Γ

$$\Gamma = n\bar{\nu} \exp(-\Delta G^*/kT) \tag{2.2}$$

or

$$\Gamma = n\bar{\nu} \exp\left(\frac{\Delta S^*}{k}\right) \exp\left(-\frac{\Delta H^*}{kT}\right) \tag{2.3}$$

In equation 2.2 n is the number of paths by which a particle can move from an equilibrium site to another, $\bar{\nu}$ is the vibrational frequency of the particle in the initial state, and ΔG^* represents the isothermal work associated with the motion of the particle from state A to state B. In obtaining equation 2.2 the vibrational degree of freedom along the diffusion direction has been removed and, thus, ΔG^* is associated with the two vibrational degrees of freedom normal to the diffusing direction.

Equation 2.3, when substituted into the basic expression for the diffusion coefficient D, as obtained by random walk consideration, is consistent with the empirically observed equation 2.1 if ΔH^*, ΔS^*, $\bar{\nu}$ and interatomic distance are relatively temperature insensitive. As a consequence, most theoretical calculations of Q and D_0 employ tacitly the concepts employed in the absolute reaction rate theory. Nonetheless, the basic assumptions are open to question since it is difficult to believe that the motion of the diffusing particle from a potential well to the maximum can occur in a reversible fashion. In addition, there is some serious concern that the lifetime of the activated state exists for a duration of time which is too short to allow equilibrium statistical mechanical concepts to be employed. Further, the correct value of the frequency $\bar{\nu}$ which is employed is undefined. For want of a suitable value, either the Debye frequency or Einstein frequency is generally employed although there is no fundamental reason why either of these is applicable. The absolute rate theory, in addition, focuses on the diffusing particle and ignores the rest of the lattice. Thus, it tends to give an incorrect picture of the solid state diffusion process.

2.2.2 Many Body Approach

Vineyard's Approach

In an attempt to circumvent some of the problems mentioned above and in particular the use of the thermodynamics approach with the implied condition that the diffusing particle moves infinitely slowly, Vineyard [9] has used a more generalized approach to calculate Γ. To illustrate this approach, consider a defect lattice with N atoms and $3N$ degrees of freedom. In figure 2.2 is shown schematically a configuration space for the system. A and B denote the initial and final positions of a jumping atom, respectively. Two-dimensional contours of constant potential energy are shown. At S is shown the saddle surface and illustrates the possible configurations associated with a diffusing atom at the maximum in potential energy. The jump rate Γ is the ratio of probabilities of all points in the saddle surface S which have positive velocities in the diffusion direction to the sum of all points in configuration space around A. Thus,

Vineyard obtains

$$\Gamma = n\sqrt{\left(\frac{kT}{2\pi}\right)} \frac{\int_S \exp\left[-G/kT\right] dS}{\int_A \exp\left[-G/kT\right] dV} \tag{2.4}$$

where G represents potential energy. The symbols S and V after the integral signs indicate that integrations are to be carried out over the hyper-surface S and regions to the left of S respectively.

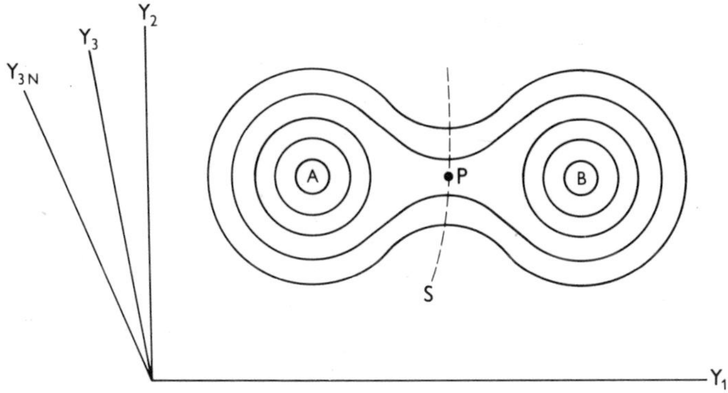

Figure 2.2. Configuration space of 3N dimensions. Solid lines represent constant potential energy surfaces. Dotted line labeled S represents constraining hypersurface. P represents the minimum along S. After Vineyard [9].

Expanding the potentials in a Taylor series expansion about A and the saddle point P and neglecting terms of higher order than the harmonic term, the following is obtained

$$\Gamma = \frac{n \prod_{i=1}^{3N} \nu_i \exp\left(-\Delta G'/kT\right)}{\prod_{i=1}^{3N-1} \nu_i'} \tag{2.5}$$

where ν_i are the 3N normal frequencies of a system of atoms in their equilibrium states and ν_i' are the $3N-1$ normal frequencies associated within the activated complex. $\Delta G'$ is the free energy difference between an atom at the saddle point and in an equilibrium position with the exclusion of contributions associated with vibrational degrees of freedom. Equation 2.5 rests on the assumption that an activated state exists and that when an atom moves over the potential energy barrier to state B, the probability of return is very small. Further, it is assumed that the diffusing atom behaves as an harmonic oscillator. The principal conceptual advantages of Vineyard's approach is that the frequency term is more appropriately defined in terms of the frequencies of the host atoms in the lattice (although it is not readily calculated) and that the need for the assumption of thermodynamic reversibility is obviated.

As a consequence, equation 2.5 has the advantage over equation 2.2 in that some of the simplifying and possibly incorrect assumptions inherent in the classical theory of absolute rates have been avoided. On the other hand, the problem of defining $\bar{\nu}$ has not been eliminated since it is required that detailed knowledge be available concerning the frequency terms at the saddle point and equilibrium positions. It might be mentioned at this point that all treatments discussed here neglect quantum effects. These effects are probably very small except for very light atoms diffusing at very low temperatures.

Dynamical Theory of Diffusion

Rice [10] has avoided the use of equilibrium statistical mechanics and has considered the derivation of Γ in terms of a cooperative model by employing normal mode analysis. If one considers, for example, an atom next to a vacancy, diffusion can occur if (1) the atom has a sufficiently large amplitude of motion in the direction of the vacancy and (2) the atoms at the saddle point move sufficiently to reduce closed-shell repulsion between the diffusing atom and atoms forming the 'barrier.' (In the diamond-cubic lattice the second term is relatively unimportant for self-diffusion as shall be seen later.) The jump frequency is then

$$\Gamma = n \sum \frac{[V]}{S} P(\{\delta\}) \tag{2.6}$$

where $P(\{\delta\})$ is the frequency of occurrence of the appropriate configuration in which the central atom has a sufficiently large amplitude of vibration in the right direction for diffusion to occur and there is appropriate out of phase motion of surrounding atoms. $[V]$ represents the concentration of vacancies, S is the concentration of lattice atoms and the symbol $\{\delta\}$ refers to the critical configuration. The summation is taken over all nearest neighbors in the crystal. If the fluctuations in a subvolume containing the diffusing atom are random

$$P(\{\delta\}) = \bar{\nu} \exp(-U_o/kT) \prod_j \exp(-U_j/kT) \prod_{k>l} g_{kl}^{(2)} \tag{2.7}$$

where $\bar{\nu}$ is some weighted mean frequency, U_o is the energy required for the diffusing atom to reach the critical vibrational amplitude, U_j is the energy required to shift the barrier atoms from their equilibrium positions, and $g_{kl}^{(2)}$ are pair correlation functions for atoms k and l [8].

Equation 2.7 is very general, but in order to obtain a useful expression some simplifying assumptions must be made. The following assumptions were made by Rice:

(a) The required critical amplitude of vibrational motion for diffusion is one-half the interatomic distance. (This assumes that the basic philosophy underlying the activated state concept is applicable.)

(b) This large amplitude is achieved with much less frequency than smaller motions required by barrier atoms.

(c) Oscillators including the diffusing atom are harmonic.

(d) The frequency of thermal fluctuations in the region surrounding a diffusing atom is much greater than that of a diffusion jump.

(e) The number of normal modes involved in the process whereby the diffusing atom reaches the critical amplitude is much smaller than the total number of the crystal.

The following equation for Γ is obtained by employing these conditions

$$\Gamma = n\bar{\nu} \exp\left[-\left(U_o + \sum_j U_j + \sum_{k>l} W_{kl}\right)/kT\right] \tag{2.8}$$

where W_{kl}, the potential of the mean force between atoms k and l, is defined by

$$g_{kl}^{(2)} = \exp(-W_{kl}/kT)$$

Since for a harmonic crystal U_o and U_j are temperature independent and since

$$W_{kl} = \Delta H_{kl} - T\Delta S_{kl}$$

equation 2.8 becomes

$$\Gamma = n\bar{\nu} \exp\left(\sum_{k>l} \Delta S_{kl}/k\right) \exp\left[-\left(U_o + \sum_j U_j + \sum \Delta H_{kl}\right)/kT\right] \tag{2.9}$$

By comparison with equation 2.3 we see that the activation enthalpy associated with a diffusive jump is given by

$$\Delta H^* = U_o + \sum_j U_j + \sum_{k>l} \Delta H_{kl} \tag{2.10}$$

Thus, the treatment given by Rice is consistent with the general form of absolute rate theory but provides greater physical insight into the actual mechanics of the diffusion process. Like the Vineyard treatment, the condition of thermodynamic reversibility is eliminated. A principal weakness, of course, is the assumption of harmonicity. The error introduced might be relatively small in connection with the host atoms surrounding the saddle point, but certainly the assumption of harmonic behavior for a diffusing atom displaced a distance equal to one-half the interatomic distance introduces considerable uncertainty into the final result. The principal utility of the approach is the focus which is provided on the cooperative many-body aspect of the diffusion process.

Synthesis of Vineyard's and Rice's Approaches

By employing normal coordinates in conjunction with random variable theorems, Rice assumed that he had thereby avoided the use of equilibrium statistical mechanics with the necessity of assuming that the lifetime of the activated state was sufficiently long.

The developments of Vineyard and Rice, however, are basically equivalent. Glyde [11] and Franklin [12] recently published papers which show this equivalence for an harmonic lattice. Both Vineyard's and Rice's treatments rest on two assumptions; namely, that the lattice may be considered to be at equilibrium and that the independence of the chosen units is maintained up to the point of critical fluctuation during the jump. Franklin considers in some detail the use of the harmonic approximation as employed in earlier treatments and points out the obvious problem associated with using this approximation for a diffusing atom which becomes displaced a distance of the order of magnitude of the interatomic distance. Thus, in principle, one needs to consider anharmonic terms; and the importance of these is indicated by the fact that anharmonic terms amount to about 75% of the harmonic term. Franklin adds anharmonic terms to the potential employed in the Vineyard treatments and interestingly enough finds that the jump frequency Γ varies as $m^{-1/2}$ where m is the atomic mass. This is the same mass dependence as one obtains by use of the theory of absolute rates and indicates that a study of the mass dependence of diffusion provides information concerning the validity of employment of equilibrium statistics. A deviation from the predicted mass dependence indicates a failure of a principal underlying assumption. Franklin employs, in calculating the jump frequency, the harmonic, cubic, and quartic force constants. As Franklin points out, one interesting use of his equation would be to calculate the fourth-order energy term in the activation energy for crystals where the experimental activation energy is available and where second and third order elastic constants are available.

2.2.3 Implication of Theoretical Development for Diffusion Calculations

In the brief survey presented above, some of the basic premises associated with calculation of the temperature dependence of the jump frequency Γ have been examined. We have started with the semi-phenomenological theory of absolute reaction rates which has been used as a successful model for chemical reactions. Various investigations since have attempted to improve upon this approach as employed for solid state diffusion and in doing so have clearly illustrated the nature of the assumptions which must be made in order to make the problem tractable. All treatments make the assumption of a critical jump amplitude for a diffusing atom. Furthermore, all treatments assume the existence of a single low potential energy diffusion path and, of importance, all assume that the lattice is in equilibrium during the diffusion process. The more recent developments are useful in that they point out the cooperative nature of atomic motions that must occur among the diffusing atoms and atoms at the saddle point for diffusion to occur. Vineyard's approach to calculating Γ is simpler and more general than that used by Rice and co-workers and, thus, should be a suitable framework for future calculations. Franklin illustrates clearly the dangers in using the harmonic approximation and two-body potentials in calculations of the jump frequency.

These recent developments create a dilemma for investigators interested in making calculations of diffusion constants since we do not have suitable information to employ in a straightforward fashion in the various theoretical equations. For example, there is no clear-cut way of obtaining the various frequencies employed in equation 2.5 nor is there any accurate method of obtaining exactly the atomic displacements which occur on the saddle point configuration since exact interatomic potentials are not available for atomic displacements of the magnitude which exist. In general, we obtain energy relationships for very small displacements, from, say, elastic constant data and then extrapolate to larger separations by use of whatever form captures the individual's fancy. As a consequence of these problems, all numerical calculations of diffusion parameters no matter how seemingly exact must be considered as approximate. For elemental semiconductors such as germanium and silicon, the situation is somewhat better than it is for close-packed metals. The open nature of the diamond-cubic lattice means that distortions of atoms about the saddle point are probably much smaller than is the case for close-packed crystals. Thus, the barrier is somewhat more easily handled. Furthermore, for valence crystals the calculation of effective pair potentials is easier than is the case for metals since electron-phonon interactions may be considered small for non-metals.

In the next section calculations of the diffusion coefficients for various non-metals will be presented which employ tacitly the assumption underlying equations 2.2 and 2.5 with the understanding that the important frequency term $\bar{\nu}$ in equation 2.2 and

$$\prod_{i=l}^{3N} \nu_i / \prod_{i=l}^{3N-l} \nu_i'$$

must be crudely approximated.

It seems important to present some of the important assumptions underlying rate calculations since too often absolute rate theory is employed without a clear attempt to present the assumptions and approximations which are inherent in its use. Occasionally, because of this, too much has been expected from these calculations in terms of expecting exact agreement with experimental data. Much more information is needed concerning the lattice dynamics of a given material before anything approaching an exact treatment may be made.

2.3 SELF-DIFFUSION IN ELEMENTAL COVALENT CRYSTALS

2.3.1 Brief Summary of Available Data

A detailed discussion of diffusion in germanium and silicon is presented in Chapter 5 of this volume. In addition a detailed discussion of data obtained in semiconductors prior to 1960 has been given by Boltaks [13]. Most of the self-diffusion studies of elemental semiconductors have been performed on germanium, so for this material an excellent opportunity exists for comparison of calculations with experiment. The data from three separate investigations [14, 15, 16] agree well enough to define a single straight line on a plot of log D versus T^{-1} over a span of temperature of 300°C. These compiled data result in the following equation for the tracer self-diffusion coefficient [17] in pure germanium:

$$D_e(\tilde{Ge}) = 10.8 \exp(-2.99/kT) \text{cm}^2 \text{ s}^{-1} \tag{2.11}$$

For silicon, results are not as satisfactory because of experimental difficulties in studying the system. The principal difficulty is possibly due to the fact that radioactive isotopes of silicon are very short-lived. The investigations which have been performed [18, 19, 20] yield data which are reasonably close to one another when plotted as log D versus T^{-1} but the data appear to define two straight lines represented by the following equations:

$$D_e(\widetilde{\text{Si}}) = 1800 \exp(-4.86/kT)\,\text{cm}^2\,\text{s}^{-1} \tag{2.12}$$

$$D_e(\widetilde{\text{Si}}) = 9000 \exp(-5.14/kT)\,\text{cm}^2\,\text{s}^{-1} \tag{2.13}$$

Equation 2.12 is fit closely by data from two separate investigators [18, 21] whereas equation 2.13 is fit by only one group of investigators [19, 20]. Also, Peart's data [21] were the only data obtained by the use of a regular sectioning technique so preference is shown here toward equation 2.12 at this time.

Valenta and Ramasastry [22] investigated the effect of *n*- and *p*-type doping on self-diffusion in germanium. These data indicate a substantial effect of doping particularly at lower temperatures. It was found that an *n*-type impurity (As) resulted in an enhancement of the self-diffusivity, whereas a *p*-type impurity (Ga) had a retarding effect. A similar study, although more limited in extent, has been made for silicon [18, 20] and the results are somewhat conflicting. At 1178°C, for example, Ghostagore [18] found that a concentration of $8.6 \times 10^{19}\,\text{cm}^{-3}$ of *n*-type impurity (P) resulted in approximately a doubling of the self-diffusion coefficient of Si, whereas a similar concentration of a *p*-type impurity (B) resulted in a decrease of the self-diffusion coefficient by about 25%. Fairfield and Masters [20], on the other hand, found that both *n*- and *p*-type impurities increase the self-diffusion coefficient.

To date no data are available for diamond; thus, only the results of theoretical calculations can be used to provide information about possible diffusion characteristics of this substance.

2.3.2 Effect of Fermi Level Position on Self-Diffusion

It has been demonstrated both theoretically and experimentally that defects in germanium and silicon can exist in the charged state. If this is true, then the concentration of the charged species can be strongly influenced by the position of the Fermi level which in turn is dependent on the relative concentration of donors and acceptors as well as temperature. If the concentration of defects responsible for the self-diffusion process can be changed, of course, the self-diffusion constant will be also altered if it is assumed that self-diffusion occurs by a vacancy mechanism. It has generally been assumed that self-diffusion in germanium and silicon occurs by a vacancy mechanism (although this premise has been challenged by Seeger and Chik [17]). There is extensive evidence that vacancies exist principally as acceptors in lightly doped silicon and germanium [23, 24]. Thus, if we assume that a vacancy mechanism applies, we may use the following equations for D_e:†

$$D_e = (1/6)fd^2\Gamma \tag{2.14}$$

where f is the correlation coefficient and d is the interatomic distance. For the diamond lattice $f = 1/2$ and n in equation 2.5 is given by

$$n = Z[V]/S \tag{2.15}$$

where Z is the coordination number and S is the number of atomic sites cm^{-3}. Equation 2.5 may be written as:

$$\Gamma = n\bar{\nu} \exp(\Delta S'/k) \exp(-\Delta H'/kT) \tag{2.16}$$

and upon substitution:

$$D_e = \frac{a_o^2 \bar{\nu} [V]}{16S} \exp(\Delta S'/k) \exp(-\Delta H/kT) \tag{2.17}$$

where a_o is the lattice parameter. Upon doping the semiconductor with relatively small quantities of n- or p-type impurities, we might reasonably assume that all quantities in equation 2.17 are unaffected except for $[V]$. Under isothermal conditions, therefore,

$$D_e \propto [V] \tag{2.18}$$

For illustrative purposes let us assume that the vacancies may exist as singly charged acceptors. (This restriction is not necessary in the calculations, and the treatment may be readily extended to include other charge states.) As a consequence:

$$[V] = [V^\times] + [V'] \tag{2.19}$$

The term $[V^\times]$ is independent of the position of the Fermi level, and the temperature dependence of the concentration is given by:

$$[V^\times] = S \exp(\Delta S_V/k) \exp(-\Delta H_V/kT) \tag{2.20}$$

where ΔS_V and ΔH_V are the entropy and enthalpy of vacancy formation, respectively. The calculation of the variation of $[V']$ with position of the Fermi level was first given by Longini and Greene [25] although a simpler treatment will be employed here to show the effect. Let us assume that each vacancy introduces an acceptor level E_V with a spin degeneracy $g = 1/2$. According to Fermi-Dirac statistics, the probability $f(E_V)$ that state E_V is occupied is:

$$f(E_V) = \frac{1}{1 + g^{-1} \exp(E_V - E_F)/kT} \tag{2.21}$$

† In this chapter D_e will denote a self-diffusion coefficient and D_e(imp) will denote the isoconcentration diffusion coefficient of an impurity.

Thus, the concentration of charged vacancies will be given by:

$$[V'] = [V]f(E_V) = \frac{[V]}{1 + 2\exp(E_V - E_F)/kT} \tag{2.22}$$

There is much confusion in the literature about the spin degeneracy term g. Many treatments of semiconductor problems neglect this term entirely, but if it is assumed that each state may be occupied by one electron of either spin but once occupied cannot be occupied further, the term must be specifically included. Now since $[V^\times]$ is not influenced by the position of the Fermi level, upon employing equation 2.19 we find:

$$[V] = [V^\times]\{1 + (1/2)\exp(E_F - E_V)/kT\} \tag{2.23}$$

As a consequence, the ratio of the vacancy concentration in a crystal with Fermi level E_F to that of the intrinsic crystal with Fermi level E_F^i is:

$$\frac{[V]}{[V]_i} = \frac{1 + (1/2)\exp(E_F - E_V)/kT}{1 + (1/2)\exp(E_F^i - E_V)/kT} \tag{2.24}$$

If one makes the assumption that neutral and charged vacancies are equivalent in the diffusion process, then:

$$\frac{D_e}{D_e^i} = \frac{[V]}{[V]_i} = \frac{1 + (1/2)\exp(E_F - E_V)/kT}{1 + (1/2)\exp(E_F^i - E_V)/kT} \tag{2.25}$$

This assumption could well be invalid and is probably certainly invalid for the case of charged substitutional impurities which diffuse by a vacancy mechanism. Seeger and Chik [17] have shown that in the case where differences in diffusivity of charged and uncharged vacancies must be considered, the following expression can be used for the effective vacancy diffusivity:

$$\mathbf{D}_e(V) = (1/2)D_e(V^\times)\,\frac{[V^\times]f}{S}\left\{1 + \frac{D_e(V')}{2D_e(V^\times)}\exp(E_F - E_V)/kT\right\} \tag{2.26}$$

The $D_e(V)$ terms are related to the self-diffusivity D_e by the following:

$$D_e = f\{D_e(V^\times)[V^\times] + D_e(V')[V']\}/S \tag{2.27}$$

Let us, however, ignore this complication for the present and utilize equation 2.25.

Since for n-type material where non-degenerate statistics are applicable:

$$\frac{n}{n_i} = \exp(E_F - E_F^i)/kT \tag{2.28}$$

We find upon substitution that

$$\frac{D_e}{D_e^i} = \frac{[V]}{[V]_i} = \frac{1 + \frac{n}{2n_i} \exp (E_F^i - E_V)/kT}{1 + (1/2) \exp (E_F^i - E_V)/kT} \tag{2.29}$$

If $(E_F - E_V) \gg kT$ then

$$\frac{D_e}{D_e^i} \cong \frac{n}{n_i} \tag{2.30}$$

For p-type material under the same condition

$$\frac{D_e}{D_e^i} \cong \frac{p}{p_i} \tag{2.31}$$

If $|E_F - E_V| < kT$

$$\frac{D_e}{D_e^i} \cong \frac{2}{3}\left(1 + \frac{n}{2n_i}\right) \text{for } n\text{-type material} \tag{2.32}$$

and for p-type material

$$\frac{D_e}{D_e^i} \cong \frac{2}{3}\left(1 + \frac{p}{2p_i}\right) \tag{2.33}$$

For the case where $(E_F - E_V) \ll 0$, $D_e/D_e^i = 1$ for both n- and p-type material.

As has been pointed out by other writers, equation 2.29 can in principle be used to determine the position of the vacancy acceptor level in a given material.

As mentioned in an earlier section, self-diffusion measurements have been made for both silicon and germanium at various doping levels. For silicon, Ghoshtagore [18] found at 1178°C, for example, that the self-diffusion coefficient of silicon increased from 4.6×10^{-14} cm^2 s^{-1} for nearly intrinsic material to 7.3×10^{-14} cm^2 s^{-1} for n-type silicon which contained 8.6×10^{19} atoms cm^{-3} of phosphorus, thus resulting in a value for $D_e/D_e^i = 1.6$. Upon the introduction of 10^{20} atoms cm^{-3} of boron in the pure crystal, the diffusivity dropped to 3.3×10^{-14} cm^2 s^{-1} and, thus, $D_e/D_e^i = 0.72$. The position of the vacancy acceptor level for silicon is not known with precision but is thought to lie above the Fermi level for intrinsic silicon. If it is assumed that $(E_F - E_V) \sim 0$, equations 2.32 and 2.33 can be utilized for n- and p-type material, respectively. It is found that the predicted ratios of D_e/D_e^i and 1.8 and 0.75 for n- and p-type doping, respectively. These are very close to the measured ratios and offer some support for the vacancy mechanism of diffusion.

In Table 1 of reference 26, theoretical ratios of D_e/D_e^i are calculated which utilize forms of equations 2.32 and 2.33 which are based on a g

value of unity and, thus, deviate considerably from the values of D_e/D_e^i calculated above.

For germanium, the evidence is that the vacancy acceptor level lies close to the valence band and, therefore, $(E_F - E_V) \gg kT$. Under this condition equations 2.30 and 2.31 are applicable. Valenta and Ramasastry [22], for example, find for germanium that the presence of 6×10^{18} As atoms cm^{-3} resulted in a ratio of D_e/D_e^i of 1.40 at 800°C whereas 5×10^{19} Ga atoms cm^{-3} resulted in a ratio of D_e/D_e^i of 0.17. By use of the data of Morin and Maita [27] concerning values of n_i, these investigators find reasonable agreement between calculation and experiment.

The data for both silicon and germanium appear to be basically consistent with the vacancy mechanism of self-diffusion although the evidence is not as clear for silicon as for germanium.

2.3.3 Theoretical Calculations of Vacancy Formation and Migration Energies

Application of Pair Potential Concepts

Examination of the diamond cubic lattice shows that, because of the openness of the lattice, no close-packed configuration of saddle atoms surrounding the saddle point exists. As a consequence, the closed shell repulsion term is not expected to be particularly important in determining the activation energy. This is rather unlike the situation for metals and simplifies the procedure somewhat. In view of this lack of overlap, it is tempting to attempt to calculate the energy of formation and migration of a vacancy by use of a valence bond approach. This was initially done by Swalin [28] in 1961. In this calculation, a Morse potential was employed as a representation of the energy of elongation of nearest neighbor valence bonds.† This potential has the form:

$$\Delta U = S_D\{1 + \exp[-2\alpha(d - d_o)] - 2\exp[-\alpha(d - d_o)]\} \qquad (2.34)$$

where S_D represents the enthalpy of dissociation of the chemical bond and α is obtained from elastic constant data [29, 30]. The energy of formation of a vacancy is considered to consist of three parts: (1) energy needed for rupture of the four bonds surrounding an atom, (2) energy gained upon deposition of the atom on the surface, thereby reforming two bonds, and (3) energy gained upon reforming dangling bonds between next nearest

† Since the nature of the interatomic potential is unknown except in the immediate vicinity of the potential well, the appropriate form at large strains can only be surmised; and, consequently, the form actually employed is largely a matter of personal choice. No *a priori* method is available for determining the correct form.

neighbors around the vacancy and allowing relaxation to occur. This process is illustrated schematically in figure 2.3. The activation energy associated with the motion of the vacancy is then obtained by calculating the difference in bond enthalpies between an atom in the saddle point position and normal position in the relaxed state. These respective states are shown schematically in figure 2.4.

In calculating the energies of vacancy formation and motion, equation 2.34 was used to calculate the changes in bond distances. Several implicit assumptions are made in these calculations: (a) Angular contributions to the appropriate energy calculations were considered unimportant. Thus, it was assumed that the bond energies were only a function of the bond stretching term represented by equation 2.34. (b) In calculating relaxations of atoms in a crystal upon vacancy formation, it was arbitrarily assumed that strains in the second neighbor shell were one-half of those in the first neighbor shell and that strains in the third

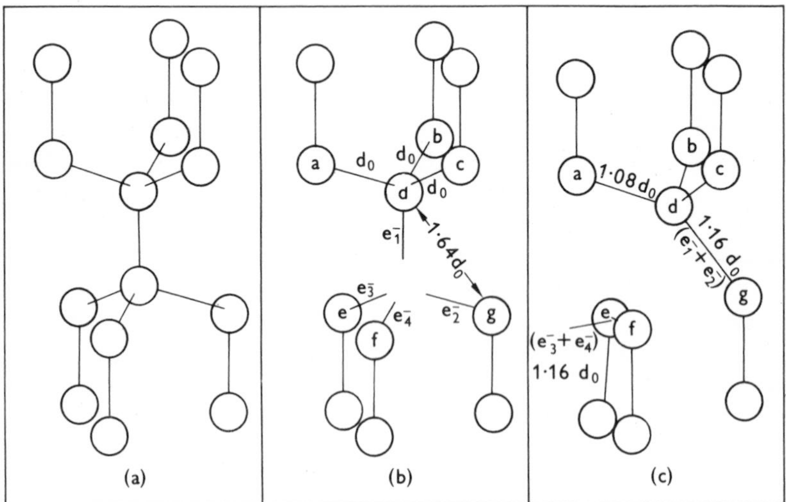

Figure 2.3. (a) Arrangement of atoms in the diamond lattice. (b) Unrelaxed vacancy. (c) Vacancy after relaxation has occurred. From reference 28.

neighbor shell were one-half those in the second shell, etc. (c) In the use of equation 2.34 the value for the bond energy S_D was obtained from sublimation data as mentioned. This experimentally obtained value does not represent solely the dissociation energy associated with the sp^3 electronic configuration but rather represents a conversion to the s^2p^2

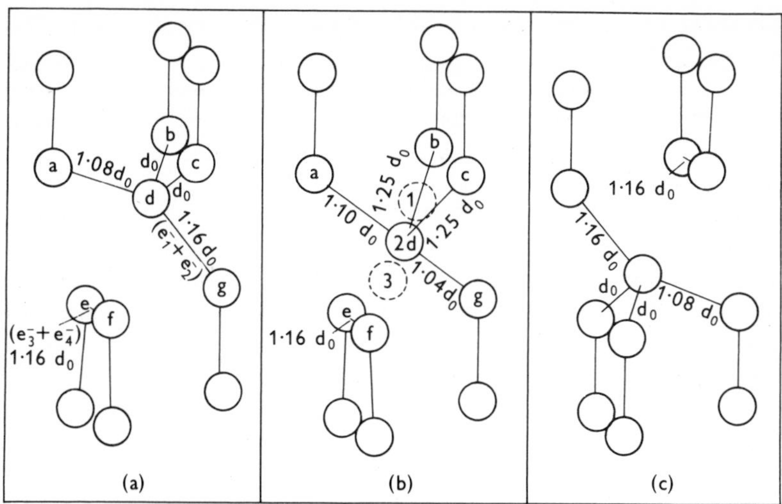

Figure 2.4. (a) Arrangement of vacancy and nearest neighbors in the diamond lattice. (b) Diffusing atom in saddlepoint position. (c) Crystal after exchange of atom with a vacancy. From reference 28.

configuration characteristic of the atoms in the gaseous state. The results obtained from the calculations are nonetheless interesting and are shown in Table 2.1. For germanium the calculated value for ΔU_V of 2.02 eV is very close to experimental values of about 2.0 eV obtained for the enthalpy from quenching experiments [24, 31, 32, 33, 34]. Further, the calculated energy of migration of 0.95 eV for germanium compared with values of 1.3 eV deduced from electron bombardment experiments [35] and 1.2 eV obtained from a study of the annealing kinetics of quenched-in acceptors [34]. The sum $(\Delta U_m + \Delta U_V)$, of course, represents the calculated activation energy of self-diffusion and compares very well with the experimental value for germanium. The calculated values are in only fair agreement with the data for silicon, however. The data obtained from quenching experiments refers to vacancies in the charged state whereas calculations of ΔU_V refer to neutral vacancies. It is, therefore, of interest to examine the difference between $\Delta U_V(V')$ and $\Delta U_V(V^{\times})$.

In quenching experiments, a sample is held at an elevated temperature for a time sufficiently long to achieve equilibrium; the sample is then rapidly quenched to a lower temperature where the concentration of defects is measured electrically. At the elevated equilibration temperature, T_Q, both charged and uncharged vacancies will exist according to equations 2.20 and 2.23. Upon quenching to the measurement temperature T_M, virtually all vacancies will be charged if, as expected for

TABLE 2.1

Calculated values of vacancy formation and migration energies, diffusion frequency factors, and activation energies for elemental semiconductors

Quantity	Ge (eV)	Si (eV)	Diamond (eV)	Reference
ΔU_V	2.07	2.32	4.16	28
ΔU_V	1.97	2.35	–	37
ΔU_V	1.91	2.13	3.68	40
ΔU_m	0.95	1.06	2.02	28
ΔU_m	0.98	1.09	1.85	40
$Q = (\Delta U_V + \Delta U_m)$	3.02	3.38	6.18	28
$Q = (\Delta U_V + \Delta U_m)$	2.89	3.22	5.53	40
Q (experiment)	2.99	4.86	–	
D_0 (theor)	5.4 cm² s⁻¹	8.9 cm² s⁻¹	11.6 cm² s⁻¹	28
D_0 (experiment)	10.8 cm² s⁻¹	1800 cm² s⁻¹	–	

germanium, E_V is very close to the valence band edge and if E_F is close to the center of the gap. Thus,

$$[V]_{T_Q} = [V^{\times}]_{T_Q} + [V']_{T_Q} \cong [V']_{T_M} \tag{2.35}$$

The quantity $\Delta U_V\,(\exp)$ is obtained from the slope of the plot of $\ln\,[V']_{T_M}$ versus $1/T_Q$ which is equivalent to the slope of $\ln\,[V]_{T_Q}$ versus $1/T_Q$. The value for $\Delta U_V\,(\exp)$ will, therefore, be equal to the following from equation 2.23:

$$\Delta U_V\,(\exp) = -k\frac{\partial \ln\,[V]}{\partial(1/T)} = \Delta U_V(V^{\times}) - \frac{(E_F - E_V)\exp\,(E_F - E_V)/kT}{2 + \exp\,(E_F - E_V)/kT} \tag{2.36}$$

Since the calculated value actually refers to the neutral vacancy, the following relation exists between $\Delta U_V\,(\exp)$ and $\Delta U_V\,(V^{\times})$ if $(E_F - E_V)_{T_Q} \gg kT$

$$\Delta U_V(V^{\times}) \cong \Delta U_V\,(\exp) + (E_F - E_V)_{T_Q} \tag{2.37}$$

If $(E_F - E_V) \ll kT$

$$\Delta U_V(V^{\times}) = \Delta U_V\,(\exp) + 1/3(E_F - E_V)_{T_Q}$$

For germanium, at the annealing temperature, if E_V is close to the valence band edge, the correction term $(E_F - E_V)$ could be equal to about 0.1– 0.2 eV.

In reference 28, values for D_o were calculated by utilization of the Delye frequency for ν. Also, the entropy term in D_o associated with the possible orientation of reformed bonds around a vacancy was calculated. No attempt was made to calculate a vibrational entropy term. Calculated and experimental D_o values are listed in Table 2.1. For germanium, the agreement between theory and experiment is quite close but is inadequate for silicon.

Seeger and co-workers have attempted to improve the calculations of Swalin by improving on some of the assumptions made in the calculations of ΔU_V. Scholz [36] and Seeger and Scholz [37] considered the relaxation contribution to ΔU_V in much more detail. Rather than assuming a geometric decrease in strain as one goes from the first neighbor shell to the second, etc., atomic displacements for 14 neighbor shells were considered in a more rigourous fashion. Somewhat surprising, the resulting values for ΔU_V differ only slightly from those calculated in reference 28. These values are shown in Table 2.1. Seeger and Swanson [38] also have generalized the Morse potential by substitution of two values for α and by correcting S_D in order to take into account the promotion energy in converting the sp^3 configuration to s^2p^2. Thus, in reference 38 the enthalpy associated with bond dissociation is represented by

$$\Delta H = S'_D \left\{ \frac{\alpha_2}{\alpha_1 - \alpha_2} \exp\left[-\alpha_1 (d - d_o)\right] - \frac{\alpha_1}{\alpha_1 - \alpha_2} \exp\left[-\alpha_2 (d - d_o)\right] \right\} \tag{2.38}$$

where S'_D is given by

$$S'_D = (\Delta H_p + \Delta H_s)/2 \tag{2.39}$$

In equation 2.39, ΔH_s is the experimental sublimation enthalpy and ΔH_p is the enthalpy associated with the promotion from the s^2p^2 to sp^3 electronic configuration. It is interesting to note that these corrections appear to result in a small net correction presumably because of compensation of one correcting term by another.

At the present time no rigorous attempt has been made to take into account the energy associated with angular distortion of the covalent bonds. Schmid *et al.* [39] suggest that this energy E_b, can be represented by the 'ansatz'.

$$E_b = \frac{C_0}{1 + 4\beta} \left[(1 + \beta) - \cos\phi - \beta \cos 2\phi \right] \tag{2.40}$$

where ϕ is the angle change, C_0 is equal to $\Omega_0(C_{11} - C_{12})/4$ where Ω_0 is the atomic volume and C_{11} and C_{12} are the usual elastic constants. β is an adjustable parameter. No suitable test of this 'ansatz' can be made since β cannot be obtained directly from experimental data.

Application of Pseudopotential Concepts

It appears to this writer that the approximate approach utilized in bond energy calculations has been pushed about as far as desirable. In order to improve on first order calculations, the introduction of more parameters is necessary. These corrections, although potentially capable of yielding better results, result in a loss of generality of the approach. Bennemann [40] points out further that one is on dangerous ground by assuming that the parameters used in the pair potential as calculated for atoms in an undisturbed state are the same for atoms in the immediate vicinity of the lattice defect. If the redistribution of valence electrons due to the presence of a point defect is large, serious errors could result. Bennemann points out that calculations of the energies associated with self-interstitial formation and migration are particularly affected. Bennemann, therefore, attempted to provide a more fundamental base for calculations of defect formation and migration energies. In this approach, the ions (nuclei plus tightly bound core electrons) are considered to be situated in a static lattice. Each valence electron then moves in the potential field produced by the other valence electrons plus the ions. Because of the open nature of the diamond cubic lattice, overlapping of the tightly bound core electrons is negligible. The distribution of valence electrons in the lattice which includes a defect is determined by use of the t-matrix approximation [41]. Following the approach used in pseudopotential theory, the crystal potential is considered to be strong only within the core region of the ions. The valence electron wave functions are orthogonalized to the wave functions of the closed shell electrons. An effective Hamiltonian is calculated for the effective wave function (original function minus orthogonalization terms) which now represents the crystal potential as a pseudopotential [42] which is generally smooth and quite weak. A self-consistent crystal potential is calculated by considering the interaction of valence electrons with the pseudopotential. Since the scattering of valence electrons depends upon lattice displacement, energy terms due to these displacements may be calculated. The formation energy of a defect is then determined by subtracting the ground state energy of a perfect crystal from the ground state energy of a crystal with the defect present. In making these calculations, Bennemann neglects completely the contribution to the formation energy from displacement of the lattice ions so a possible error of some significance is introduced. In similar fashion, the migration energy is calculated by subtracting the ground state energy of the crystal with defect from the energy of a crystal with the defect in the saddlepoint position.

The energy of formation of a vacancy is in principle given by the sum of five terms

$$\Delta U_V = \Delta U_1 + \Delta U_2 + \Delta U_3 + \Delta U_4 + \Delta U_5 \tag{2.41}$$

where

ΔU_1 = Coulomb interaction of an ion with the uniform gas of valence electrons,

ΔU_2 = change in electrostatic energy of ions due to lattice distortions associated with the point defect,

ΔU_3 = the energy associated with a change in volume of the crystal $\Delta\Omega$ resulting from formation of the lattice vacancy,

$\Delta U_4, \Delta U_5$ = two types of contributions from that part of the energy of the valence electrons which is sensitive to the lattice configuration.

As mentioned above, Bennemann neglects terms due to lattice distortions and, therefore, it is assumed that

$$\Delta U_V \cong \Delta U_1 + \Delta U_3 \tag{2.42}$$

In the term ΔU_3, it is assumed that the volume change of the crystal upon formation of a vacancy is Ω_0. By a similar approach ΔU_m was calculated by Bennemann and the results of calculation are listed in Table 2.1. It is interesting to note in spite of the approximations how well the calculated quantities obtained by use of this approach agree with those obtained by use of the pair potential approach. It should be pointed out, however, that in addition to the neglect of terms dependent upon lattice distortions several approximations were made in the development of expressions for ΔU_1 and ΔU_3 in order to obtain numerical quantities for ΔU_V and ΔU_m. The pseudopotential approach is basically more fundamental than the use of empirical pair potential concepts and provides a better overall framework for the development of future calculations.

2.3.4 Theoretical Calculations of Divacancy Formation and Migration

For a crystal in thermal equilibrium, the concentration of divacancies, V_2, is given by the following expression:

$$[V_2] = \frac{Z}{2S} \exp(-U^B_{V_2}/kT) \exp(S^B_{V_2}/k)[V]^2 \tag{2.43}$$

where Z is the coordination number and $U^B_{V_2}$ and $S^B_{V_2}$ are the binding energy and entropy associated with pairing of vacancies to form a divacancy, respectively. In equation 2.43 a negative value for $U^B_{V_2}$ indicates an attractive interaction and, thus, a reduction in energy for the system. For self-diffusion by a divacancy diffusion mechanism

$$D_e = D_e(V_2) f_{V_2} g^{SD}_{V_2} [V_2]/S \tag{2.44}$$

where f_{V_2} and $g^{SD}_{V_2}$ are the correlation coefficient and a geometric factor, respectively. The divacancy diffusion coefficient $D_e(V_2)$ is given by [17]

$$D_e(V_2) = g^M_{V_2} a_o^2 \nu(V_2) \exp \frac{\Delta S_m(V_2)}{k} \times \exp\left(-\frac{\Delta U_m(V_2)}{kT}\right) \tag{2.45}$$

where $g^M_{V_2}$ is the appropriate geometric factor. Since in the diamond cubic lattice no common nearest neighbors exist to a divacancy, dissociation of the divacancy would probably occur as part of the process of diffusion (in this case $g^{SD}_{V_2} = 8$ and $g^M_{V_2} = 1/32$). Another possibility, although it does not appear as likely, is a jump of a divacancy through a distance equal to that between next nearest neighbors (in this case $g^{SD}_{V_2} = 8$ and $g^M_{V_2} = 1/8$). In equations 2.44 and 2.45 $\nu(V_2)$ is the appropriate frequency term (undoubtedly different from the ν term used in the single vacancy case), f is the correlation coefficient (equal to 0.78) [17], and $\Delta S_m(V_2)$ and $\Delta U_m(V_2)$ are the entropy and energy, respectively, associated with migration of divacancies.

The approaches discussed in section 2.3.3 to calculate formation and migration energies for single vacancies in principle can be used to calculate these properties for divacancies. No quantitative calculations, however, have been performed as yet. Scholz and Seeger [43], however, have used the Morse potential approach for calculation of energy of formation of divacancies ΔU_{V_2} and the binding energy $U^B_{V_2}$ [37]. The calculated formation energies are in the vicinity of 4-4.7 eV for germanium and silicon and 9.2 eV for diamond. The quantity $U^B_{V_2}$ is estimated to be about −1.2 eV for germanium and silicon and −2.4 eV for diamond.

The energy of migration has not been calculated directly, but one would expect $\Delta U_m(V_2) > \Delta U_m(V)$ since the divacancy must either dissociate first in order to diffuse or else undergo an unusually wide angular reorientation in order to facilitate a jump equal in magnitude to the separation between next nearest neighbors.

In view of the fact that detailed calculations have not been made, we can only make an estimate of the diffusion coefficient for divacancies as compared to that for vacancies.

If we consider, for example, that diffusion of divacancies occurs through dissociation and if differences in the frequency and migration entropy terms for the divacancy and single vacancy are negligible, we find from equation 2.45 and the discussion in section 2.3.3 that

$$\frac{D_e(V_2)}{D_e(V)} \cong (1/4)\exp\,[-(\Delta U_m(V_2) - \Delta U_m(V))/kT] \tag{2.46}$$

Since dissociation of the divacancy must occur with an energy increase of $-U^B_{V_2}$, we estimate that the energy of divacancy diffusion is given by

$$\Delta U_m(V_2) = \Delta U_m(V) - U^B_{V_2} \tag{2.47}$$

and therefore

$$\frac{D_e(V_2)}{D_e(V)} \cong \tfrac{1}{4}\exp\,(U^B_{V_2}/kT) \tag{2.48}$$

If the binding energy of vacancies in germanium is in the vicinity of -1 eV, we see that at 1100° K, for example,

$$\frac{D_e(V_2)}{D_e(V)} \cong 10^{-5}$$

As a consequence, it appears that divacancies diffuse much slower in the diamond cubic lattice than single vacancies. This is due, of course, to the fact that the divacancy must dissociate before motion can occur. For more close-packed structures where this dissociation need not occur, divacancies can move considerably faster than single vacancies.

2.3.5 Theoretical Calculations of Self-Interstitial Formation and Migration Energies

By use of the *t*-matrix approach discussed in 2.3.3, Bennemann [40] has calculated the energy associated with formation and migration of self-interstitials. In the diamond cubic lattice, two types of interstitial positions exist. In what is called the tetrahedral position, an atom at the site center is surrounded by four nearest neighbors at distance 0.433 a_0 and six next nearest neighbors at distance 0.500 a_0. In the so-called hexagonal position, an interstitial atom is surrounded by six nearest neighbors at 0.415 a_0 and eight next nearest neighbors at distance 0.648 a_0. The hexagonal position is usually considered to represent the saddle point position and bisects a line from the center of two neighboring equilibrium tetrahedral interstitial positions which lie along ⟨100⟩ [44, 45]. This diffusion path is shown schematically in figures 2.5 and 2.6.

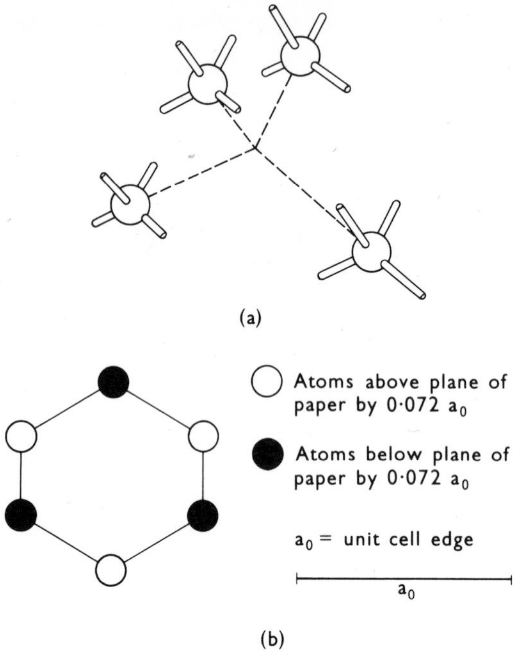

Figure 2.5. Interstitial sites in the diamond lattice: (a) Tetrahedral site, (b) Hexagonal site; plane of paper normal to ⟨111⟩. From reference 45.

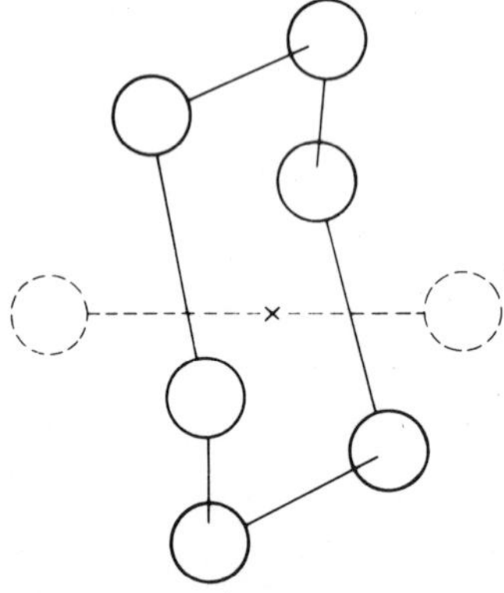

Figure 2.6. Hexagonal saddlepoint position which connects two tetrahedral sites.

In making the numerical calculations for interstitials, Bennemann made the same basic assumptions as in the case of the vacancy calculations. The results of these calculations are shown in Table 2.2 for the energy of

TABLE 2.2
Theoretical calculations of the energy of self-interstitial formation and migration and activation energy of self-diffusion by a self-interstitial mechanism

Energy (eV)	Ge	Si	Diamond	Reference
ΔU_i	0.93	1.09	1.76	40
$\Delta U_m(I)$	0.44	0.51	0.85	40
Q (calc)	1.37	1.60	2.61	40
$\Delta U_m(I)$	0-0.25	0-0.22	–	46

interstitial formation ΔU_i, the energy of interstitial migration $\Delta U_m(I)$ and the self-diffusion activation energy

$$Q\text{ (calc)} = \Delta U_i + \Delta U_m(I)$$

The values calculated by Bennemann for the self-diffusion activation energy on the basis of an interstitial mechanism are surprising in the sense that they are much lower than values calculated for a vacancy mechanism. Further, in the case of germanium and silicon at least, the calculated values are less than one-half of the values determined experimentally. This indicates that a very serious error exists in the method of calculation and that a need exists for a closer examination of some of the simplifying assumptions used in employment of the *t*-matrix approach.

Hasiguti has made approximate calculations of the migration energies associated with singly charged self-interstitials in germanium and silicon [46]. The approach used was similar to that developed by Weiser [45] for diffusion of interstitial impurities. In calculating the migration energy, two potential terms are considered: namely, an attractive term due to the polarization of the host atoms and a repulsive term due to the overlapping of nonbonding electrons. This approach will be discussed in more detail in the next section and will be only outlined here. The polarization contribution to the migration activation energy is the difference between the terms for hexagonal and tetrahedral sites. Thus

$$U_{\text{pol}} = U_{\text{pol}}^H - U_{\text{pol}}^T \tag{2.49}$$

where the superscripts H and T refer to the hexagonal and tetrahedral sites, respectively. The value of U_{pol} is assumed to be 0.75 eV for both germanium and silicon. The repulsive term was determined by use of a Born-Mayer potential.

$$U_{\text{rep}} = A \exp\left[(d_o - d)/\rho\right] \tag{2.50}$$

where A and ρ are appropriate constants. Thus, in similar fashion the repulsive contribution to the activation energy is

$$U_{\text{rep}} = U_{\text{rep}}^H - U_{\text{rep}}^T \tag{2.51}$$

Consequently, the migration energy $\Delta U_m(I)$

$$\Delta U_m(I) = U_{\text{pol}} - U_{\text{rep}} \tag{2.52}$$

Upon calculating various values for A and ρ from compressibility data and considering nearest and next nearest neighbors in computing U_{rep}^H and U_{rep}^T, values for $\Delta U_m(I)$ from 0 to -0.25 eV and 0 to -0.22 eV were obtained for germanium and silicon, respectively. The minus sign implies that the saddlepoint is represented by the tetrahedral position rather than the hexagonal position as assumed and, therefore, the equilibrium position for a self-interstitial is calculated to be the hexagonal site. This is just the opposite of Bennemann's results. There is a serious question, however, in Hasiguti's calculation concerning the use of the Born-Mayer potential for covalent crystals such as germanium and silicon since the overlap of wave functions of closed shell core electrons is assumed to be small. As Hasiguti showed in his paper, the value obtained for $\Delta U_m(I)$ is very sensitive to the choice of values for A and ρ.

2.3.6 Relation Between Calculations and Experimental Data

In some respects, the results of theoretical calculations are gratifying. For example, the calculated influence of Fermi level position on self-diffusion for both germanium and silicon based on a vacancy mechanism of diffusion yields results which are close to the observed effect. This tends to lend credence to the assumption that self-diffusion occurs by a vacancy mechanism although it is possible that other mechanisms could be conceived which would yield the same results.

Further, there is a surprising agreement, with regard to calculations of ΔU_V, ΔU_m, and Q for elemental semiconductors, between the approach based on pair potential concepts as applied by several investigators and the approach based on pseudopotential theory. The calculations are not only consistent among themselves but for germanium, at least, show excellent agreement with measurements of ΔU_V, ΔU_m, and Q as determined from quenching experiments and self-diffusion measurements. In addition, the value of D_o measured for germanium of about 10 cm^2 s^{-1} is consistent with the calculated value of about 5 cm^2 s^{-1} based on entropy contributions from bond rearrangements under the assumption that a vacancy mechanism is operative [28]. The vibrational contributions would be expected to yield a positive contribution to the entropy terms ΔS_V and ΔS_m and, thus, increase the calculated value of D_o. In order to obtain a value for D_o of about 10 cm^2 s^{-1}, a vibrational entropy contribution of only 1.4 entropy units is needed which is well within reason. This conclusion is different from that of Seeger and Chik [17] who believe that a D_o value of 10 cm^2 s^{-1} is not consistent with a vacancy mechanism.

Their conclusion is partly based on a comparison with metal systems in which D_o normally is between 0.1 $cm^2\ s^{-1}$ and 1.0 $cm^2\ s^{-1}$. In metals, the entropy term is due principally to vibrational contributions and would be expected to be smaller than the total entropy term for directionally bonded semiconductors.

In spite of the optimistic picture just presented, there are several unresolved problems. First, the calculations of D_o and Q for silicon are not in agreement with experiment. Since experiments with silicon are much more difficult to perform than with germanium, it is possible that although the self-diffusion coefficients are probably reasonably accurate, serious errors exist in the calculation of the slope of the line (which gives Q) and, hence, the intercept of the line extrapolated to $(1/T) = 0$ (which gives D_o). As it now stands and as clearly pointed out [17], the very large experimental D_o value for silicon cannot be explained in terms of a simple vacancy mechanism. In addition the calculated value for Q is more than one electron volt lower than that obtained experimentally from the slope of the log D versus $1/T$ plot. The only way to explain this problem away and still maintain the vacancy model is to prove experimentally that the measured slope is too high. To illustrate how sensitive the calculated value of D_o is to change in slope, a line whose slope is equivalent to an activation energy of 4.1 eV (compared to the experimental value of 4.8 eV) drawn through the high temperature experimental points yields a value of D_o equal to 10 $cm^2\ s^{-1}$ instead of the experimentally deduced value of $1.8 \times 10^3\ cm^2\ s^{-1}$. In view of this high sensitivity of D_o to the deduced slope, it is preferable that the question of mechanism be considered as open until one is absolutely certain as to the validity of the experimental data. It perhaps is relevant that in the case of diffusion of Group III and Group V impurities in silicon, the experimentally obtained values for D_o and Q are typically in the region of 10-20 $cm^2\ s^{-1}$ and 3.5-4.0 eV, respectively.

In spite of the rationalization presented above, some serious questions remain. The calculated values of Q based on a self-interstitial mechanism are very much lower than that for a vacancy mechanism as illustrated by the summaries in Tables 2.1 and 2.2. If the methods employed in these calculations have validity, one would be forced to draw the conclusion that the predominant self-diffusion mechanism is one of creation and migration of self-interstitials and not vacancies. It would appear, however, that there must be a serious deficiency in the theoretical procedures employed since the calculated Q values are also much lower than the experimental values measured.

The most serious challenge to the results of calculations based on a vacancy model came from some experimental data based on electron spin resonance experiments (ESR). By employing ESR, centers have been found in silicon which are identified as vacancies. These centers are mobile at very low temperatures. One center identified as a neutral vacancy has a migration energy of 0.33 eV [23] in *p*-type silicon and another type of center which is identified as a doubly charged vacancy, V'', has a migration energy of 0.18 eV in heavily doped *n*-type silicon [47]. If, indeed, these centers are the ones which participate in self-diffusion, great difficulties of interpretation are created. If, for example, it is assumed that in silicon ΔU_m is 0.33 eV for a vacancy, then a value for ΔU_V of about 4.5 eV is obtained since Q has a value of bout 4.8 eV. To date, this problem has not been resolved although Seeger and Chik [17] put forth a model of self-diffusion for silicon based on the concept of an extended interstitial in order to rationalize the various discrepancies.

It is interesting to note that neutral vacancies in silicon were found by ESR techniques to have much higher energies of mobility than doubly charged negative vacancies. The relative mobility of defects with different charge states represents a problem that has received little theoretical attention and should be a fruitful area for future work.

2.4 CALCULATION OF SOLUTE DIFFUSION COEFFICIENTS IN ELEMENTAL COVALENT CRYSTALS

Since the presentation of detailed data and a discussion of mechanisms of impurity diffusion will be presented in Chapter 5, this section will consist of the presentation of various theoretical approaches which have been employed within the context of various assumed mechanisms of diffusion. As Chapter 5 will show, there is a rich variety of diffusion mechanisms which exist for various impurities. These include the common mechanisms of the simple vacancy mechanism, the simple interstitial mechanism, but also more complex mechanisms such as dissociative diffusion and for the case of diffusion of oxygen, an interstitial mechanism in which the oxygen appears to be tightly bound to two nearest neighbor host atoms.

Theoretical work has only been performed on the first two mechanisms named above and, thus, these will of necessity be the focal points of this section.

2.4.1 Diffusion of Group III and Group V Impurities

Under this topic we shall be concerned with two aspects of the diffusion of group III and group V impurities. Firstly, we shall discuss the factors which govern the magnitude of the impurity diffusion coefficient in intrinsic material and secondly, the influence of position of the Fermi level will be discussed. The discussion of complicating structural features such as dislocations and concentration dependent terms such as field enhanced diffusion will be treated in Chapter 5.

Impurity Diffusion in Intrinsic Germanium and Silicon

Examination of experimental data indicates that an interesting trend exists with regard to the diffusivities of *n*-type and *p*-type impurities. In germanium, for example, *n*-type impurities diffuse about 100 times faster than *p*-type impurities. In silicon, on the other hand, *n*-type impurities diffuse about one-fifth as fast as *p*-type impurities. By use of a vacancy mechanism approach, we attempted to understand these generalizations by consideration of three basic terms [48]. One term pertains to the relative position of the vacancy E_V acceptor level with regard to the position of the Fermi level E_F. Thus, the total vacancy concentration which is the sum of neutral vacancies and singly charged vacancies is influenced markedly by the relative position of E_F^i and E_V as shown in equation 2.23. Since evidence at the time indicated that $(E_F^i - E_V) > kT$ for germanium and $(E_F^i - E_V) \ll 0$ for silicon, it would appear that in germanium most of the vacancies are charged whereas in silicon most of the vacancies are neutral. The second term considered was the Coulombic interaction, ΔG_c, between charged impurities and charged vacancies in these materials. One would conclude from this consideration that *n*-type impurities would be attracted to charged vacancies whereas *p*-type impurities would be repulsed. The third term and the most difficult to quantitatively assess is a strain term ΔG_s due to overlap between the closed shells of impurities and host atoms. According to equation 2.14 the diffusion coefficient is given by

$$D_e^i = (1/6)d^2 f\Gamma \tag{2.14}$$

where Γ is given by equation 2.16
The number of paths n in equation 2.16 is given by

$$n = p_V \tag{2.53}$$

where p_V is the probability that a vacancy is a nearest neighbor to a given impurity atom. For self-diffusion p_V is simply equal to $Z[V]/S$ if neutral and charged vacancies are treated equally. (As suggested in section 2.3, this assumption might be in error.) For impurity diffusion, because of Coulombic interaction one must consider neutral and charged vacancies separately. Thus

$$p_V = p_V^\times + p_V' \tag{2.54}$$

where $p_V^\times$ and p_V' are the respective probabilities that an impurity ion is next to a neutral and a charged vacancy. The term $p_V^{\times}$ is given by

$$p_V^{\times} = \frac{Z[V^{\times}]}{S} \exp(-\Delta G_s/kT) \tag{2.55}$$

and p_V' is given by

$$p_V' = \frac{Z[V']}{S} \exp[-(\Delta G_s + \Delta G_c)/kT] \tag{2.56}$$

From equations 2.22 and 2.23 we obtain for $[V']$ the following:

$$[V'] = (1/2)\,[V^{\times}] \exp[(E_F - E_V)/kT] \tag{2.57}$$

and upon appropriate substitution we find, if ΔS_c and ΔS_s are considered small,

$$D_e^i(\text{imp}) = (1/8)a_0^2\tilde{\nu} f \exp[(\Delta S_V + \Delta S_m)/k] \exp[-(\Delta H_V + \Delta H_m + \Delta H_s)/kT] \times \{1 + (1/2)\exp[(E_F^i - E_V - \Delta H_c)/kT\} \tag{2.58}$$

If equation 2.58 is compared with the expression for self-diffusion, equation 2.17, we obtain for the ratio of the impurity diffusion coefficient to the self-diffusion coefficient if f and $\tilde{\nu}$ are assumed to be the same for the two processes

$$\frac{D_e^i(\text{imp})}{D_e^i} = \exp(-\Delta H_s/kT)\frac{\{1 + (1/2)\exp(E_F^i - E_V - \Delta H_c)/kT\}}{\{1 + (1/2)\exp(E_F^i - E_V)/kT\}} \tag{2.59}$$

Equation 2.58 differs from equation 16 in reference 48 since the spin degeneracy term was not included in reference 48. The Coulombic term is given by

$$\Delta H_c \cong \Delta U_c = z_i z_V e^2/\epsilon d \tag{2.60}$$

where $z_i e$ and $z_V e$ are the charges on impurity and vacancy, respectively, and ϵ is the dielectric constant.

The strain term is difficult to estimate with any accuracy, but following an approach employed by Friedel [49] the following term was calculated in reference 48:

$$\Delta H_s = \frac{6\pi(r_s - r_s')^2 r_s}{Z(1+\alpha)\chi} \tag{2.61}$$

where

$$\alpha = \frac{(1+p)\chi r_s}{2(1-2p)\chi' r_s}$$

In the above equations, r_s and r_s' are appropriate solvent and solute ion radii respectively, χ and χ' are appropriate solvent and solute compressibilities and p is Poissan's ratio for the host lattice.

It is interesting to compare the predicted values of $D_e^i(\text{imp})/D_e^i$ with experimental data for n- and p-type impurities in germanium and silicon. The term ΔH_c is calculated to be ±0.36 eV and ±0.46 eV for singly charged impurities in germanium and silicon respectively. Values of ΔH_s as calculated by equation 2.61 are presented in reference 48. In order to estimate the quantity $(E_F^i - E_V)$, values measured at room temperature for quenched centers were employed, and it was assumed that at elevated temperatures the change in $(E_F^i - E_V)$ is proportional to the change in band gap. Thus, $(E_F^i - E_V)$ was calculated to be equal to 0.05 eV for germanium at 800°C and −0.29 eV for silicon at 1200°C.

A comparison between calculated values of $D_e^i(\text{imp})/D_e^i$ and experiment is shown in Table 2.3. The values used for experimental ratios of $D_e^i(\text{imp})/D_e^i$ are the same as those given in reference 17 as representative values. It should be pointed out that there is some uncertainty in several of these values.

TABLE 2.3

Comparison between calculated and experimental[a] values of the ratio of impurity diffusivity to self-diffusivity, $D_e^i(\text{imp})/D_e^i$

		Ge(T = 800°C)		Si(T = 1200°C)	
	Solute	Theory	Experiment	Theory	Experiment
	$Sb^{\cdot}$	23	180	2.9	9.4
n-type	$Bi^{\cdot}$	24	61	3.1	4.5
	$As^{\cdot}$	46	150	4.2	7.8
	$P^{\cdot}$	150	63	14	5.0
	Ga'	2.2	1.2	4.2	56
p-type	In'	3.1	1.5	6.1	27
	B'	8.7	5.5	7.0	41
	Al'	18	—	34	310

[a] Only experimental data for $D^i(\text{imp})$ are available (i.e. chemical diffusion) but as the solutions are dilute and the crystals are intrinsic it is plausible to expect $D^i(\text{imp}) = D_e^i(\text{imp})$.

Seeger and Chik [17] calculated values of $D_e^i(\text{imp})/D_e^i$ by the same approach as discussed above although they included a term that is based on the assumption that the energy of migration of a singly charged vacancy is 0.1 eV less than the migration energy of a neutral vacancy. This refinement would result in some changes in the calculated values of $D_e^i(\text{imp})/D_e^i$ as presented in Table 2.3.

Examination of Table 2.3 shows that the general trends exhibited by experimental data are predicted by equation 2.59. It is predicted for example that n-type impurities in germanium diffuse 5 to 10 times faster than p-type impurities. Experimental results indicate that n-type impurities diffuse 10 to 150 times faster. For silicon, it is predicted that n-type impurities diffuse generally about one-half as fast as p-type impurities whereas experimentally the ratio is about one-fifth. Given the uncertainties in calculating ΔH_s and $(E_F^i - E_V)$, the results are in reasonable accord with the general vacancy mechanism of diffusion upon which equation 2.59 is based. From a physical point of view, the difference between n- and p-type impurities in germanium and silicon is due to the supposed difference in the quantity $(E_F^i - E_V)$. As pointed out, most of the vacancies will be charged in germanium. Thus, Coulombic interaction between vacancies and impurities will favor rapid diffusion of n-type impurities as compared with p-type impurities. In silicon, on the other hand, because of the negative value of $(E_F^i - E_V)$, most vacancies will be neutral and, thus, Coulombic interaction is not as important as in germanium. Consequently, the dominating term in equation 2.61 is the strain energy term.

Effect of Fermi Level Position on Impurity Diffusion

The effect of Fermi level position on the diffusivity of impurities can be readily seen by examining equation 2.58. If one assumes that $\tilde{\nu}, f,$ the ΔS terms, ΔH_V, ΔH_m, and ΔH_s are independent of doping level then

$$\frac{D_e(\text{imp})}{D_e^i(\text{imp})} = \frac{1 + (1/2)\exp[(E_F - E_V - \Delta H_c)/kT]}{1 + (1/2)\exp[(E_F^i - E_V - \Delta H_c)/kT]} \tag{2.62}$$

For n-type impurities in germanium and silicon, equation 2.62 becomes

$$\frac{D_e(\text{imp})}{D_e^i(\text{imp})} \cong \exp[(E_F - E_F^i)/kT] \tag{2.63}$$

For a non-degenerate semiconductor

$$\frac{n(\text{doped})}{n_i} = \exp[(E_F - E_F^i)/kT] \tag{2.64}$$

and thus

$$\frac{D_e(\text{imp})}{D_e^i(\text{imp})} = \frac{n(\text{doped})}{n_i} \tag{2.65}$$

For p-type material, equation 2.62 must be employed directly.

The influence of solute doping level on solute diffusion in germanium has been studied by Valenta (data reported in reference 17) and the agreement between experiment and theory is quite good for both n- and p-type solutes.

Recently, a study has been made of the diffusivity of As in silicon as a function of As doping level [50]. A radiotracer technique was employed and homogeneously doped samples were used and, thus, any effect due to a concentration gradient was eliminated. The comparison between experimental and calculated values is shown in Table 2.4. The agreement is quite good particularly if it is borne in mind that the experiment error associated with D/n is estimated to be ±50% [50]. Millea has studied the influence of doping level on the diffusivity of antimony, phosphorus, and indium [51]. Seeger and Chik [17] employ these data and show reasonable agreement between theory and experiment for diffusion of n-type solutes but poor agreement for p-type indium. Inasmuch as these were not isoconcentration experiments, the results could be influenced by effects due to the concentration gradient.

TABLE 2.4

Influence of doping level on diffusion of arsenic in silicon (data from reference 50)

T(°C)	[As] in x10^{20} atoms cm^{-3}	$D_e(\text{As})/D_e^i(\text{As})$ (calc)	$D_e(\text{As})/D_e^i(\text{As})$ (expt)
950	0.71	11	8.9
950	0.97	15	30
950	1.68	26	34
950	2.70	41	78
1150	1.68	7.8	4.1
1150	2.40	11	16

As a footnote, the strong effect which doping level has on impurity diffusion rates in semiconductors might be pointed out. At 950°C, for example, the presence of 0.1% As increases the diffusion coefficient of As by an order of magnitude when compared to a crystal with 0.01% As.

2.4.2 Diffusion of Interstitial Impurities

As illustrated in section 2.3, the structure of silicon and germanium is sufficiently open so that when an interstitial moves, for example, from a tetrahedral interstitial site to the hexagonal saddle point, virtually no geometric barrier exists. As a consequence, it would appear that calculations of the activation energy Q for interstitial impurities is considerably simpler in Ge and Si than for the case of more close-packed crystals. In calculating the activation energy for interstitial diffusion, Swalin [44] assumed that the principal contribution was from Coulombic interaction. Thus, the interaction energy of an interstitial impurity i with valence z_i whose center is situated at distance d_{ij} from the center of

TABLE 2.5

Comparison between calculated and experimental values of diffusion parameters for the interstitial mechanism. ( ) denotes reference number in this chapter

Solute		Germanium				Silicon			
		D_o(calc) $cm^2\ s^{-1}$	D_o $cm\ s^{-1}$ (exp)	Q(calc) eV	Q(exp) eV	D_o $cm^2\ s^{-1}$ (calc)	D_o $cm^2\ s^{-1}$ (exp)	Q(calc) eV	Q(exp) eV
Li	(44)			0.55	0.51			0.76	0.66
Li^H	(45)	1.7×10^{-3}	2.5×10^{-3}	0.57	0.51	1.6×10^{-3}	2.3×10^{-3}	0.53	0.66
H	(44)			0.33	0.38			0.46	0.48
He	(44)			0.66	0.70			0.92	1.26
Na^H	(45)	4×10^{-4}		0.22		2×10^{-4}		0.07	
Cu^H	(45)	3×10^{-4}		0.21		2×10^{-4}		0.06	
Ag^T	(45)	9×10^{-4}		0.60		1.1×10^{-3}		1.05	
Au^T	(45)	9×10^{-4}		1.08		1×10^{-3}	2.44×10^{-3}	1.55	0.38

NOTE: Superscripts H and T denote the calculated equilibrium position to be H-site and T-site, respectively.

solvent ion j with effective valence z_j is represented as

$$\Delta U_c = \frac{z_i z_j e^2}{\epsilon d_{ij}} \tag{2.66}$$

The activation energy is calculated by moving the interstitial from the center of a tetrahedral site to the center of the hexagonal site. Upon consideration of the interaction of the solute with 55 neighboring solvent ions, it is found that

$$Q = \Delta U_c^H - \Delta U_c^T = \frac{8.90 z_i z_j}{\epsilon d_o} \text{ eV} \tag{2.67}$$

where d_o is the nearest neighbor distance expressed in Angstrom units. The effective valences z_i and z_j were determined from some rather empirical screening arguments as put forth by Pauling [52]. The calculated effective valences for germanium, silicon, hydrogen, helium, and lithium were calculated on this basis to be 2.4, 2.4, 0.6, 1.2, and 1.0, respectively. As shown in Table 2.5, the agreement between calculation and experiment is rather remarkable given the empirical nature of the valence calculations. Such agreement is particularly interesting in view of the fact that no arbitrary proportionality constants are introduced.

A more elaborate theory has been developed by Weiser [45] for calculating D_o and Q for interstitial diffusion. As indicated in figures 2.5 and 2.6, two interstitial positions exist. Weiser points out that it cannot be assumed *a priori* that the saddlepoint position is the configuration shown in figure 2.6. The argument proceeds in the following way: The interaction energy of an ionized interstitial is the sum of two principal contributions. One contribution is attractive and is due to the interaction between the impurity ion and induced dipoles in neighboring solvent ions. (There is an additional van der Waals term which is calculated to be much smaller and, therefore, is ignored.) The second contribution is repulsive and is due to the closed-shell interaction between solvent and impurity. In calculating the attractive polarization term U_{pol}, Weiser basically used the approach employed by Mott and Littleton [53] as modified by Rittner *et al.* [54]. Numerical calculations by Weiser indicate for germanium and silicon that the polarization energy term itself is approximately 5 eV in magnitude. Further, the polarization energy associated with the hexagonal site (H-site) is about 0.75 eV more negative than the energy associated with the tetrahedral site (T-site).

For calculating the repulsive energy U_{rep}, the standard Born-Mayer approach was employed. In considering the interaction, nearest neighbor and next nearest neighbors to an impurity atom were considered. In the Born-Mayer expression

$$U_{\text{rep}} = A \exp\left[(r_L + r_i - d)/\rho\right] \tag{2.68}$$

r_L and r_i represent effective radii of solvent and impurity ions respectively, d is the distance between impurity and solvent ion, and ρ is a measure of how rapidly the repulsion falls off with distance. For r_L the covalent radii were used by Weiser and for ρ a value of 0.33×10^{-8} cm was chosen. The difference in repulsive energy for an interstitial ion in H-site and T-site configuration is calculated to be

$$\Delta U_{\text{rep}} = U_{\text{rep}}^H - U_{\text{rep}}^T = \Delta U_{\text{rep}} = A \exp (r_i/\rho) \qquad (2.69)$$

where A is calculated to be 0.030 eV and 0.038 eV for germanium and silicon respectively. Equation 2.69 shows that when an impurity ion has a large radius, the energy associated with an impurity ion in an H-site increases and hence the impurity ion tends to prefer a T-site. As indicated above

$$\Delta U_{\text{pol}} = U_{\text{pol}}^H - U_{\text{pol}}^T \cong -0.75 \text{ eV} \qquad (2.70)$$

Thus, we see that the polarization energy term favors an H-site over the T-site as the equilibrium site for an interstitial. For interstitials of large radius, however, the term $(U_{\text{rep}}^H - U_{\text{rep}}^T)$ is large and thus the sum of equations 2.69 and 2.70 will be positive in sign. As a consequence, according to the theory, the equilibrium site for large interstititals will be the T-site and, consequently, the saddlepoint for diffusion will be represented by an H-site. On the other hand, for small impurities $(U_{\text{rep}}^H - U_{\text{rep}}^T)$ will be small and thus the sum of equations 2.69 and 2.70 will be negative in sign. Consequently, the equilibrium site is the H-site and the saddlepoint is the T-site. The activation energy for diffusion in either case will be given by

$$Q_{\text{calc}} = |\Delta U_{\text{rep}} + \Delta U_{\text{pol}}| \qquad (2.71)$$

and will be large for both very small impurities and very large impurities. At an intermediate size, the value of Q will be close to zero and thus, the activation energy will be low and consequently diffusion will be very rapid. Calculated values of Q based on a simple interstitial mechanism are shown in Table 2.5. Copper, silver, and gold are believed to diffuse by a dissociative mechanism and thus values of Q for the interstitial part must be deduced indirectly from experiment. It is interesting to note that for the diffusion of copper in silicon a very low activation energy is calculated which is in general agreement with experimental results.

By use of the Wert-Zener approach, values for D_o were calculated and the following equation was derived

$$D_o = B a_o \left(\frac{Q_{\text{calc}}}{m} \right)^{1/2} \qquad (2.72)$$

where B is equal to 0.22 and 0.11 for the tetrahedral equilibrium site and hexagonal equilibrium site respectively and m is the solute atomic mass. The calculated values of D_o are tabulated in Table 2.5. Weiser's treatment is important since it provides a mechanism for explaining the rapid diffusion of copper in silicon which is observed and also because it indicates that one cannot *a priori* specify that the tetrahedral site, for example, is the normal equilibrium site for all interstitial impurities. The theory is approximate in the sense that the Mott-Littleton and Born-Mayer approaches are used to calculate the polarization energy and repulsive energy respectively. It is also assumed in the approach that the interstitial impurity resides in the center of its cage and that no distortion of the lattice occurs in the region of the solute. The agreement between theory and experiment is on the whole quite reasonable although significantly it is predicted that the activation energy for diffusion of lithium in silicon and germanium is basically the same. Experimental results, however, show substantial difference. This problem for lithium is similar to that for calculations of the activation energy for self-diffusion in germanium and silicon as discussed earlier. Most theoretical calculations indicate that germanium and silicon have very similar properties with regard to the calculation of diffusion parameters.

2.5 DIFFUSION IN IONIC CRYSTALS

From both experimental and theoretical viewpoints, diffusion investigations in ionic crystals are in a primitive state compared with the situation in elemental semiconductors. At the present time, no completely accepted theoretical techniques are available for calculating energies of defect formation and migration although some progress is being made for alkali halides. In no case have suitable calculations of D_o been made. In fact, experimentally one observes a very wide range of D_o values in various ionic systems. It is not clear at this point how much of this range is due to experimental errors. For an investigator interested in estimating diffusion activation energies and energies of defect formation, two theoretical approaches can be employed. One is phenomenological in nature and the other rests on the application of the Born model. These will be discussed in turn.

2.5.1 Phenomenological Approach

Self-diffusion of both cation and anion has been extensively studied in the alkali halides. Laurent and Benard [55, 56] examined the relation between cohesive energy and activation energies for cation and anion diffusion in alkali halide systems with potassium as a common cation. Since the cohesive energy can be expressed in the form

$$U_{coh} = \frac{\alpha e^2}{d}\left(1 - \frac{1}{\beta}\right) \tag{2.73}$$

where α, d, and β are the Madelung constant, interionic distance, and a characteristic constant, respectively, a correlation was sought between diffusion activation energies and $1/d$. A plot of this type for some alkali halides is shown in figure 2.7. Inasmuch as polarization effects make an

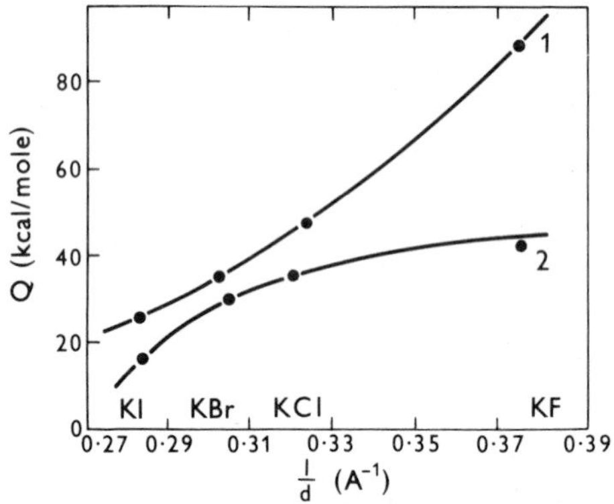

Figure 2.7. Relation between activation energy for self-diffusion and interionic spacing. Curve 1 is for anion diffusion and Curve 2 is for cation diffusion. From reference 56.

important contribution to the vacancy formation energy, one would expect that a plot of Q versus U_{coh} or $1/d$ would show only a rough trend rather than a simple straight-line relation since the polarization contribution will be different from one material to another in the series. Laurent and Benard deduced that this effect would cancel if the sum of activation energy for cation and anion are considered. A plot of $(Q^+ + Q^-)$ versus $1/d$ is shown in figure 2.8, and a straight-line relationship is obtained. It would appear that this type of relation might be employed in some other systems. Laurent and Benard also observed a linear relation between $(Q^+ + Q^-)$ and the melting point of salts with a common cation.

In similar fashion, Swalin [57] suggested that there might be a relation between a modified heat of formation of the salt $\Delta H_f'$ and the activation energy for diffusion. The experimental heat of formation available in the literature is modified so that each pure element is in the gaseous form in a monatomic state and the standard enthalpy is obtained for the following reaction.

$$M(G, 1\text{ atm}) + X(G, 1\text{ atm}) = MX(S, 1\text{ atm});\ \Delta H_f'$$

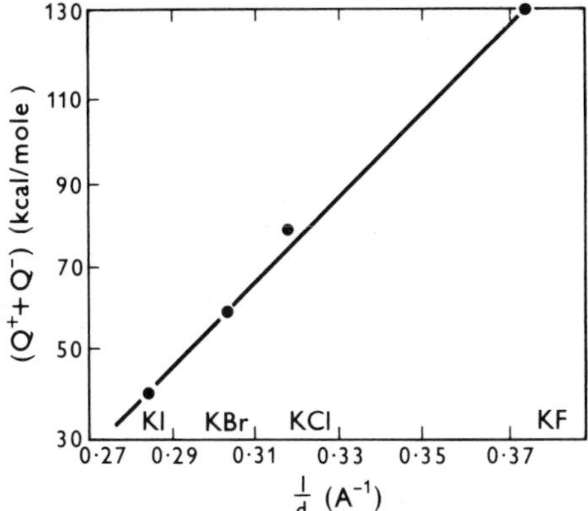

Figure 2.8. Relation between the sum of activation energies for cation and anion self-diffusion and interionic distance. From reference 56.

Since for most halides and oxides the metallic constituent is in the solid state in its pure form and the non-metallic constituent is in the diatomic state, $\Delta H_f'$ is related to the standard heat of formation of the compound by the following relation

$$\Delta H_f' = \Delta H_f^0 - \Delta H_S - \Delta H_D \tag{2.74}$$

where ΔH_S and ΔH_D refer to the heat of sublimation of the metallic substance and dissociation enthalpy of the diatomic species, respectively. A definite trend exists between the activation energy for cation self-diffusion and $\Delta H_f'$ as shown in figure 2.9. The line through points is arbitrarily placed through zero since Q would be expected to be close to zero when the heat of formation of the compound is very small. A similar attempt was made in reference 57 to correlate the activation energy of anion diffusion with $\Delta H_f'$, but no clear-cut trend was observed in this case.

Insofar as the writer knows, virtually no other correlations have been examined for diffusion in salts although one would expect to find some other possible empirical correlations. Such correlations would be useful, not for ascertaining diffusion mechanisms but for rationalizing experimental data and enabling useful estimates to be made for systems which are of practical interest but which are difficult to study experimentally.

2.5.2 Application of the Born Model

Discussion of the Born model and its applicability in calculations of the energy of formation and migration of defects has been given in considerable detail for alkali halides [58, 59, 60] and for oxides [61]. As a consequence, the substance of these reviews will not be presented here in

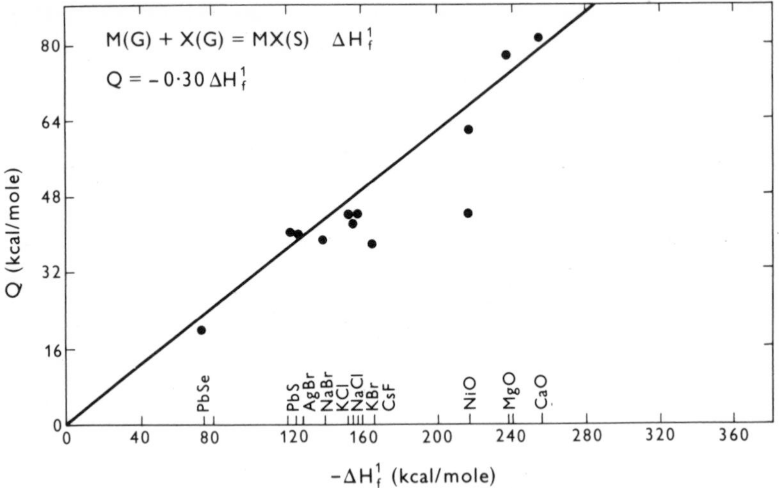

Figure 2.9. Relation between activation energy for cation diffusion and modified heat of compound formation for some materials with the NaCl structure. From reference 57.

detail. Rather the reader is referred to the references given above. In brief, most of the calculations are based on the concepts developed by Mott and Littleton [53]. In the use of the Born model, of course, one assumes that a static crystal exists which contains separate charged ions. The energy of creating a vacancy, for example, is calculated by removing the ion and in calculating three contributions to the energy. One term U_c is due to the change in Coulombic energy of crystal as caused by the removal of a charged ion. The second term U_p is due to the polarization of ions which surround the vacancy and the third U_{rep} is due to short-range closed shell interactions. Thus

$$\Delta U_V = U_c + U_p + U_{\text{rep}}$$

Since all three terms are functions of the displacement of ions surrounding the vacancy, all one needs to do in principle is to remove the ions from the interior of the crystal to the surface and minimize ΔU_V with regard to displacement of ions which surround the vacancy. Most papers since Mott and Littleton have been concerned with methods of calculating appropriate ionic radii, appropriate displacement fields, and appropriate repulsive potentials. Generally a Born-Mayer type potential has been used in calculating U_{rep}. For alkali halides, good agreement is obtained between calculated and experimental energies of Schottky defect formation ΔU_S although most calculated values generally tend to be on the low side. For NaCl, for example, calculated values of ΔU_S range from 1.3 to 2.17 eV as compared to an experimental value of about 2.12 eV. For KCl, calculated values range from 1.81 to 2.28 eV as compared with an experimental value of about 2.25 eV. By using the same general approach as used to calculate defect formation energies, activation energies of defect migration were calculated [62]. In the process of diffusion, for this calculation, the ion passes through a rather severely constricted saddlepoint which results in significant dilation. From a fundamental point of view, it is doubtful that the repulsive energy can be adequately calculated by use of a Born-Mayer potential since the parameters are obtained from macroscopic compressibility data. Compressibility data refer to small strains whereas we need data for very extensive core overlap. To compensate for this, a modified potential was employed in reference 62 which in effect stiffens the potential at very short interionic distances. With this stiffening included, good values were obtained for ΔU_m^+ and ΔU_m^-. For example, experimental values of ΔU_m^+ and ΔU_m^- for NaCl are about 0.80 and 1.06 eV respectively, whereas calculated values of about 0.85 and 0.90 eV were obtained. For KCl the experimental values of ΔU_m^+ are in the range 0.6-0.85 eV and the value for ΔU_m^- is about 0.95 eV. Calculated values are 0.85 and 0.82 eV respectively.

Oxide materials are of more interest than alkali halides to individuals interested in semiconducting materials. Since calculations for the alkali halides have been made by employing the Born model, it is tempting to use the same approach for calculating defect properties for oxides. As pointed out by Boswarva and Franklin [61], however, serious fundamental problems arise in the case of oxides. In order to apply an ionic model such as the Born model, one makes several assumptions. It is assumed that ionic properties such as short-range interaction parameters can be assigned to individual ions and further it is assumed that these parameters are independent of displacement. Also, it is assumed that the additivity rule holds. This rule states that properties such as ionic radii can be determined and that these properties remain constant for a given ion in different compounds within the same family. As pointed out by several investigators, calculated parameters for oxygen ions appeared to vary unsystematically from one compound to another, and this lack of consistency was generally attributed to failure of the additivity rule. Boswarva and Franklin, however, made a systematic study of alkaline earth chalcogenerates and found that the additivity rule appeared to work. Inasmuch as past calculations of vacancy formation energies in oxides have been very unsatisfactory, a fundamental problem must exist. There is evidence that the problem is associated with assumptions concerning the nature of the oxygen-oxygen potential in oxide crystals. One normally assumes that the potential between nearest neighbor oxide ions is repulsive. There is substantial evidence, however, that the short-range interactions between neighboring oxygen ions are attractive rather than repulsive [61]. If this is correct, the use of the Born-Mayer approach is inappropriate and, therefore, the general techniques employed successfully for alkali halides cannot be employed for oxides. As a consequence, all calculated values of ΔU_S which have been made in the past are suspect and little reliance should be placed on them until the problem concerning the corrected nature of the ionic interactions is resolved.

2.6 DIFFUSION IN COVALENT COMPOUND SEMICONDUCTORS

The status of experimental work and proposed mechanisms of diffusion will be discussed in some detail in Chapters 6 and 7 of this monograph. For our purposes here, it will suffice to point out that the diffusion mechanisms appear to be complex and that most of the past theoretical work has been devoted toward deciphering the experimental diffusion results in terms of general atomistic mechanisms of atomic transport. These efforts have been concerned with determining the path or paths through which diffusion occurs as well as clarifying the state of charge of the diffusing species. As a consequence, very little opportunity has existed for detailed quantitative calculations of diffusion parameters. Furthermore, such calculations will prove complicated inasmuch as the compounds have both covalent and ionic character in varying degrees. Thus, neither the Born approach nor the use of pair potential concepts will form by themselves a suitable framework for calculations. It appears that a synthesis of these concepts is needed as a basis for adequate calculations. One expects that this will tend to make calculations more approximate and, hence, probably less useful than is presently the case for elemental semiconductors on one hand and more nearly ionic solids such as the alkali halides on the other.

An illustration of the problem is provided by some calculations which were made by Bailly [63]. Bailly attempted to calculate energies of metal vacancy formation in some II-VI telluride compounds, and the general approach used was similar in concept to that employed by Swalin [44] for elemental semiconductors. In this calculation a metal atom M is removed from the interior of the crystal (thus breaking four bonds with tellurium nearest neighbors) and deposited on the surface (reforming two bonds). Prior to rupture the M–Te bond is characterized by a certain ionicity λ and each M ion, therefore, has an effective charge $+z_1 e$ and each Te ion has an effective charge $-z_1 e$. The electrons remaining after removal of the M atom rearrange themselves among the four telluride ions which are nearest neighbors of the vacancy; subsequently relaxation of ions surrounding the vacancy occurs in order to minimize the energy of vacancy formation. Thus, in principle, the set of operations is identical to those discussed in reference 44. In order to make calculations, Bailly assumed that the metal-telluride bond energy, ΔU_{el} could be represented by a Morse type potential

$$\Delta U_{el} = \Delta' \{1 - \exp[-\alpha(d - d_o)]\}^2 \tag{2.75}$$

where Δ' is the bond energy of the M–Te bond in its normal state. For the Te–Te bond after removal of a metal ion it is postulated that

$$\Delta U'_{el} = W\{1 - \exp[-\alpha'(d' - d'_o)]\}^2$$

where W is an appropriate bond energy, α' is a parameter characteristic of the Te–Te bond and d'_o and d' represent the separations between Te ions before and after relaxation, respectively. In addition to covalent bond changes a Coulombic energy term must be included inasmuch as the M and Te ions are assumed to have effective charges.

Thus we see that several parameters must be obtained for calculations involving compound semiconductors which were not needed for elemental semiconductors. For example, the degree of ionicity must be calculated in order to obtain $+z_1e$ and $-z_1e$. The bond strength Δ' must be calculated for the M–Te bond. The terms W and d'_o cannot be determined from experimental data alone since in the perfect II-VI crystal Te–Te bonds are assumed not to exist. Thus, in addition to the many approximations inherent in the application of pair potential concepts to defect energy calculations, several additional parameters which are difficult to determine must be utilized.

2.7 CONCLUDING REMARKS

The present status of diffusion calculations for semiconducting materials is quite mixed. With the exception of simple interstitial impurities in germanium and silicon, attempts to calculate D_o are primitive. This is due in part to uncertainty in knowing what effective vibrational frequency to use in the calculation and in part to inability in making accurate calculations of the entropy term associated with defect formation and the change in entropy associated with motion of an atom from its normal position to the saddlepoint. Since the entropy terms enter as an exponential term in D_o, a major influence on calculated values of D_o results and, hence, relatively small errors in calculation will result in large errors in D_o. Coupled with this is the problem of knowing accurately the experimental value of D_o to compare with calculation. When experimental values of log D are plotted versus the reciprocal of absolute temperature, a relatively small error in slope will magnify greatly the errors in D_o.

In the case of activation energy calculations, the situation is considerably better. For germanium and silicon, in fact, quite good agreement between various types of calculation and experimental results for solute diffusion is obtained. Similarly good agreement with regard to self-diffusion in germanium is obtained. Experimental self-diffusion parameters for silicon, on the other hand, are found to deviate substantially from calculated values. At the present time the reason for this is unclear.

In the case of alkali halides, use of the Born approach appears to be fruitful in calculations of energies of defect formation and migration. Unfortunately, the same approach is not proving fruitful for the more interesting case of semiconducting oxide materials. To date, no satisfactory approach has been developed which yields insight into the problem although there is considerable evidence that short-range interactions between oxide ions in crystals are attractive rather than repulsive. If this proves to be the case, the applicability of models employed successfully with alkali halide crystals will be fruitless. Although highly non-stoichiometric oxides were not discussed in this chapter, it might be mentioned that further complications exist with these materials since there is considerable evidence that defects in these materials have some form of order. If this is the case, this order would have substantial influence on calculated diffusion properties.

A class of materials for which little theoretical work has been performed are the semiconductor covalent compounds such as III-V and II-VI materials. For these open-structured materials there is increasing evidence that diffusion occurs mostly by some form of interstitial or interstitialcy mechanism. Theoretical work on these materials is sparse, and in many respects these materials are more complicated to deal with since neither the covalent bond assumptions utilized in treating elemental covalent materials nor the Born model used for materials which are mostly ionic may be employed directly. Some form of synthesis of these approaches perhaps must be used in future calculations.

REFERENCES

1. S. Glasstone, K. J. Laidler and H. Eyring, The Theory of Rate Processes. (McGraw-Hill, New York, 1941)
2. R. M. Barrer, Diffusion In and Through Solids, p. 291. (Cambridge University Press, London, 1941)
3. W. Jost, Diffusion in Solids, Liquids, Gases (Revised), p. 172. (Academic Press, New York, 1960)
4. C. A. Wert and C. Zener, *Phys. Rev.*, **76**, 1169 (1949)
5. C. A. Wert, *Phys. Rev.*, **79**, 601 (1950)
6. C. Zener; W. Shockley (Editor), Imperfections in Nearly Perfect Crystals, p. 289. (Wiley, New York, 1952)

7. A. D. LeClaire; B. Chalmers (Editor), Progess in Metal Physics, Vol. 4, p. 265. (Pergamon, London, 1953)
8. D. Lazarus; F. Seitz and D. Turnbull (Editors), Solid State Physics, Vol. 10, p. 81. (Academic Press, New York and London, 1960)
9. G. Vineyard, *J. Phys. Chem. Solids,* **3,** 121 (1957)
10. S. A. Rice, *Phys. Rev.,* **112,** 804 (1958)
11. H. R. Glyde, *Rev. Mod. Phys.,* **39,** 373 (1967)
12. W. Franklin, *J. Phys. Chem. Solids,* **28,** 829 (1967)
13. B. I. Boltaks, Diffusion in Semiconductors. (Academic Press, New York, 1963)
14. H. Letaw, L. Slifkin and W. M. Portnoy, *Phys. Rev.,* **102,** 636 (1956)
15. H. Widmer and G. R. Gunther-Mohr, *Helv. Phys. Acta,* **34,** 635 (1961)
16. M. M. Valenta and C. Ramasastry, *Phys. Rev.,* **106,** 731 (1957)
17. A. Seeger and K. P. Chik, *Phys. Stat. Sol.,* **29,** 455 (1968)
18. R. N. Ghoshtagore, *Phys. Rev. Letters,* **16,** 890 (1966)
19. B. J. Masters and J. M. Fairfield, *Appl. Phys. Letters,* **8,** 280 (1966)
20. J. M. Fairfield and B. J. Masters, *J. Appl. Phys.,* **38,** 3148 (1967)
21. R. F. Peart, *Phys. Stat. Sol.,* **15,** K119 (1966)
22. M. W. Valenta and C. Ramasastry, *Phys. Rev.,* **106,** 73 (1957)
23. G. D. Watkins, *J. Phys. Soc. Japan,* **18,** Supp. II, 22 (1963)
24. R. A. Logan, *Phys. Rev.,* **101,** 1455 (1956)
25. R. L. Longini and R. F. Greene, *Phys. Rev.,* **102,** 992 (1956)
26. A. F. W. Willoughby, *J. Mat. Science,* **3,** 89 (1968)
27. F. J. Morin and J. P. Maita, *Phys. Rev.,* **94,** 1525 (1954)
28. R. A. Swalin, *J. Phys. Chem. Solids,* **18,** 290 (1961)
29. J. Waser and L. Pauling, *J. Chem. Phys.,* **18,** 747 (1950)
30. R. A. Swalin, *Acta metall.,* **7,** 736 (1959)
31. S. Mayburg, *Phys. Rev.,* **95,** 38 (1954)
32. S. Ishino, F. Nakazawa and R. R. Hasiguti, *J. Phys. Soc. Japan,* **20,** 817 (1965)
33. A. D. Belyaev, L. I. Datsenko and S. S. Malogolevets, *Ukr. Fiz. Zh.,* **13,** 654 (1967)
34. A. Hiraki and T. Suita, *J. Phys. Soc. Japan,* **18,** Supp. III, 254 (1962)
35. G. Bemski and W. M. Augustyniak, *Phys. Rev.,* **108,** 645 (1957)
36. A. Scholz, *Phys. Stat. Sol.,* **3,** 42 (1963)
37. A. Seeger and A. Scholz, *Phys. Stat. Sol.,* **3,** 1480 (1963)
38. Reported in reference 17
39. G. Schmid, K. P. Chik and A. Seeger; L. Vook (Editor), Radiation Effects in Semiconductors, p. 60. (Plenum Press, New York, 1968)
40. K. H. Bennemann, *Phys. Rev.,* **137,** A1497 (1965)
41. K. H. Bennemann, *Phys. Rev.,* **133,** A1045 (1964)
42. W. A. Harrison, Pseudopotentials in the Theory of Metals. (Benjamin, New York and Amsterdam, 1966)
43. A. Scholz and A. Seeger; P. Baruch (Editor), Radiation Damage in Semiconductors, p. 315. (Dunod, Paris, 1965)
44. R. A. Swalin, *J. Phys. Chem. Solids,* **23,** 153 (1962)
45. K. Weiser, *Phys. Rev.,* **126,** 1427 (1962)
46. R. R. Hasiguti, Calculation of the Properties of Vacancies and Interstitials, p. 27. (U.S. Government Printing Office, Washington, D.C., 1966)
47. G. D. Watkins, Symposium in Radiation Damage in Semiconductor Components, as reported in reference 17. (Conference in Toulouse, 1967)
48. R. A. Swalin, *J. Appl. Phys.,* **29,** 670 (1958)
49. J. Friedel, *Adv. in Phys.,* **3,** 446 (1954)
50. B. J. Masters and J. M. Fairfield, *J. Appl. Phys.,* **40,** 2390 (1969)
51. M. F. Millea, *J. Phys. Chem. Solids,* **27,** 315 (1966)
52. L. Pauling, The Nature of the Chemical Bond (2nd ed.). (Cornell Press, Ithaca, 1940)
53. N. F. Mott and M. J. Littleton, *Trans. Faraday Soc.,* **34,** 485 (1938)

54. E. S. Rittner, R. A. Hutner and F. K. Dupre, *J. Chem. Phys.*, **17**, 1981 (1949)
55. J. F. Laurent and J. Benard, *J. Phys. Chem. Solids,* **3**, 7 (1957)
56. J. F. Laurent and J. Benard, *J. Phys. Chem. Solids,* **7**, 218 (1958)
57. R. A. Swalin, Thermodynamics, Vol. II, p. 197. (International Atomic Energy Agency, Vienna, 1966)
58. M. P. Tosi, Calculations of the Properties of Vacancies and Interstitials, p. 1. (Nat'l. Bureau of Standards, Misc. Pub. 287, 1966)
59. I. M. Boswarva and A. B. Lidiard, Calculations of the Properties of Vacancies and Interstitials, p. 7. (Nat'l. Bureau of Standards, Misc. Pub. 287, 1966)
60. A. B. Lidiard, Calculations of the Properties of Vacancies and Interstitials, p. 61. (Nat'l. Bureau of Standards, Misc. Pub. 287, 1966)
61. I. M. Boswarva and A. D. Franklin, Mass Transport in Oxides, p. 15. (Nat'l. Bureau of Standards Special Pub. 296, 1968)
62. R. Guccione, M. P. Tosi and M. Asdente, *J. Phys. Chem. Solids,* **10**, 162 (1959)
63. F. Bailly; R. Hasiguti (Editor), Lattice Defects in Semiconductors, p. 231. (University of Tokyo and Pennsylvania State University, Tokyo University, College Park and London, 1968)

PHASE RELATIONSHIPS IN SEMICONDUCTORS 3

J. C. BRICE

CONTENTS

3.1 INTRODUCTION

The distribution of solutes in semiconductors has a significant effect on device performance and has therefore been extensively studied. A considerable part of this work has been concerned with the distribution of solutes between phases both during crystal growth and in the course of subsequent processing. As a result of this, the theory of equilibrium processes is now relatively well understood but because of difficulties with the analysis of the very dilute solutions involved, the amount of reliable experimental data is still rather limited.† Studies of non-equilibrium processes have been hindered by the problems of measuring all the parameters involved but a considerable body of semiempirical data has been amassed.

This chapter considers the factors which control the composition of the surface layers of a semiconductor in contact with another phase. The

† The increasing use of the solid source mass spectrometer (see for example reference 116) has shown that semiconductor materials are in general much dirtier than was previously believed. Typical total impurity levels are several parts per million (i.e. $\sim 10^{17}$ atoms cm^{-3}). Although most of these impurities are electrically inactive or only partially active, there are usually sufficient different solutes present at levels which make the determination of purity by the use of electrical measurements somewhat dubious.

treatment is in three main parts. The first of these (section 3.2) considers static equilibria and deals first with the composition with respect to major components and then with the problems arising when small quantities of additional solutes are present. The second part (section 3.3) extends the treatment to dynamic equilibria in which the interface between the phases moves with respect to a fixed point in the crystal. The final part (section 3.4) considers non-equilibrium cases about which less is known but which are more important practically.

In sections 3.2 and 3.3 the treatment will assume that the solid is uniform so that transport processes in the solid may be neglected. In these sections it will also be implicitly assumed that the interfaces between phases are sharp (i.e. the transition from one phase to another takes place over a single atomic layer). This rather naive assumption is useful and it is shown in section 3.2.3 that when comparing phases in equilibrium, the nature of any material separating them is completely irrelevant. The exact nature of the interface only becomes important when the kinetics of the various processes are considered or if the phases are not in equilibrium. Such cases form the subject of section 3.4.

For simplicity the interfaces between phases are taken to be flat and to lie usually in the $y - z$ plane at $x = 0$: a one-dimensional treatment of the processes at this interface is therefore usually sufficient. Table 3.1 gives the principal symbols used in this chapter. Note in particular the use of K_{ij} for mass action constants and $\mathbf{K}_{kl}$ for segregation coefficients.

Inevitably a relatively short account of a complex subject must omit some detail and assume a fair amount of background knowledge. Table 3.2 lists some references for detailed discussion of peripheral topics and general background information and some sources of numerical data. This list is not comprehensive but represents a reasonable selection of the available works.

3.2 STATIC EQUILIBRIA

3.2.1 Major Components

The equilibrium relations between vapour, liquid and solid phases can be represented by a diagram. If the effects of applied hydrostatic pressure can be ignored, in the case of an element or other material of invariant composition this diagram can be plotted in two dimensions and is simply the plot of vapour pressure as a function of temperature. Figure 3.1 gives some examples. To a first approximation the curves can be described in the form

$$\ln P = A + \Delta H_v / RT \tag{3.1}$$

TABLE 3.1

Notation

Symbol	Meaning
A, B, C	used as general atomic symbols and constants
C_p	specific heat
D	diffusion coefficient
G	free energy
H	enthalpy
K	mass action constant
K	segregation coefficient
k	distribution coefficient, Boltzman's constant
N	number of atoms per unit volume
P	gas pressure
R	gas constant
S	entropy
T	absolute temperature
t	time
V	volume, reaction velocity
W	energy
w	interaction energy
X	atom fraction
x	distance normal to an interface
Z	number of nearest neighbours
γ	activity coefficient
δ	diffusion layer thickness
Δ	used to denote a change
κ	thermal conductivity
Subscripts	
B	bulk
f	fluid
L	liquid
m	melting
o	equilibrium or starting value
s	surface
t	transition
Superscripts	
L	refers to the second phase in equilibrium with the crystal
O	refers to a reference state

where *P* is the vapour pressure, *A* is a constant, ΔH_v is the enthalpy of evaporation, *R* is the gas constant and *T* the absolute temperature. ΔH_v is a slowly varying function of temperature.

A binary system composed of non-volatile constituents can also be described in two dimensions, but in the case of a binary system with volatile components, three dimensions are required to represent composition (*X*), equilibrium partial pressures (*P*) and temperature (*T*). Two-dimensional representation is normally achieved by using projections on to the $T-X$, $P-T$ and $P-X$ planes. Figure 3.2 shows these

TABLE 3.2

Some general references

Phase theory:	A simple account of the principles is given by Gordon [1] and a more detailed account by Kroger [2]. Zernike [45] gives a comprehensive account of phase diagrams. Alper [122] discusses their application to electronic systems.
Crystal growth:	General discussions have been given by Buckley [3] and Gilman [4]. Semiconductor growth is described by Tanenbaum [5] and Lawson and Nielsen [6]. For particular methods see Brice [7] and Zief and Wilcox [8] (melt growth), Schaefer [9], Francombe and Sato [10] and Hirth and Pound [11] (vapour growth).[a]
Transport in fluids:	For general description see Goldstein [12] and Knudson and Katz [13]. Stability conditions are given by Chandrasekhar [14].
Data:	For phase diagrams see Hansen [15], Elliott [16a] and Shunk [16b]. Vapour pressures of the elements are given by Honig and general thermodynamic data by Stull and Sinke [17] and N.B.S. [18]. Data for particular semiconductors is given by Goryunova [19] and Runyan [20].[b]

[a] Other recent books on crystal growth include: R. F. Strickland-Constable, Kinetics and Mechanisms of Crystallization (Academic Press, New York, 1968); R. A. Laudise, The Growth of Single Crystals (Prentice Hall, Englewood Cliffs, 1970) and J. C. Brice, The Growth of Crystals from Liquids (North Holland, Amsterdam, 1972).

[b] Data on the properties of materials are also given by EPIC, Handbook of Electronic Materials (Plenum, New York); and Y. S. Touloukian, Thermophysical Properties of High Temperature Solid Materials (Macmillan, New York, 1967).

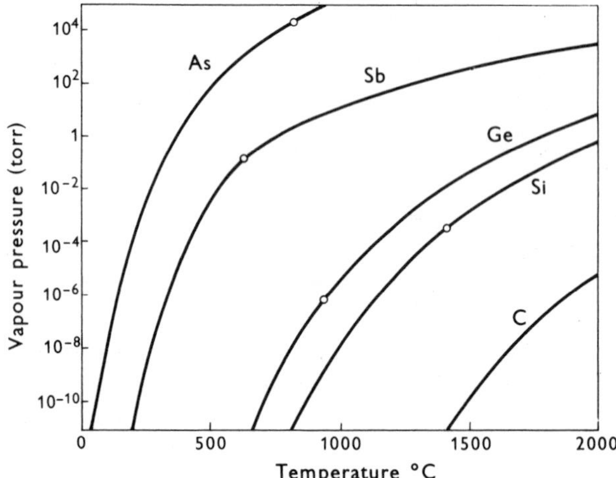

Figure 3.1. The vapour pressure of some elements as a function of temperature. ⊙ denotes the melting point.

projections for the gallium-arsenic system. This system is probably the most thoroughly investigated semiconductor system. Data exists to plot

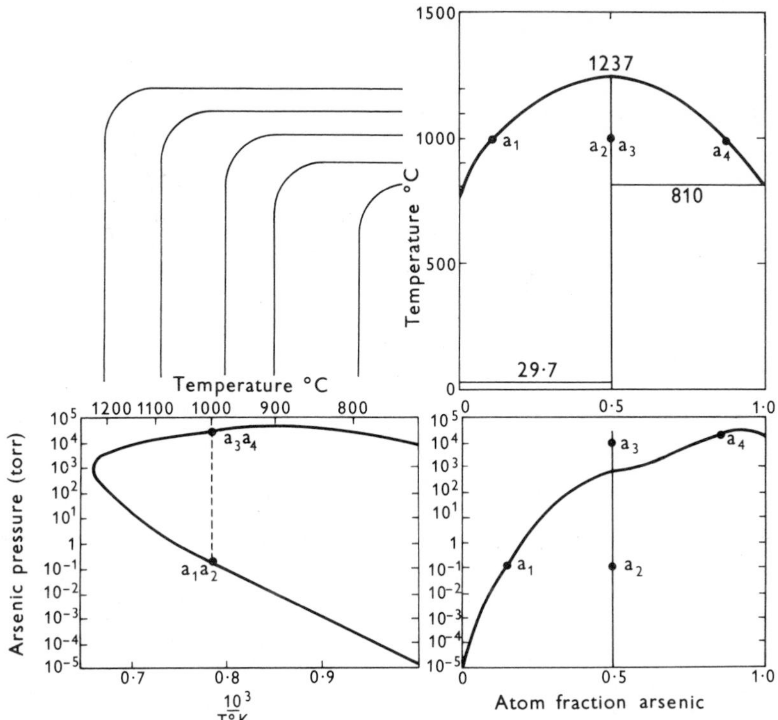

Figure 3.2. The phase diagram of the Ga-As system [21-25]. The thick lines on these diagrams give the parameters of solid, liquid and vapour in equilibrium. For example, the point a_1 represents a liquid at 1000° C which is in equilibrium with the vapour and the solid a_2. Similarly the liquid a_4 is in equilibrium with the vapour and the solid a_3. At 1000° C, liquids containing less arsenic than a_1 or more arsenic than a_4 cannot be in equilibrium with solid gallium arsenide, and a crystal with a composition between a_2 and a_3 can exist in equilibrium only with a vapour phase.

similar diagrams for the systems Ga – P [25], Ga – Sb [26, 27], and Ga – Te [28, 29]. The most important information on these diagrams is the position of the three-phase lines on which solid, liquid and vapour are in equilibrium. On the $P - T$ projection a single continuous line suffices to give all the relevant information. On the other two projections two lines are required to give the compositions of the solid and liquid. Note that although a single line is used to denote the composition of the solid, a really accurate representation would require a double line at all positions except the maximum melting point. Consider, for example the set of points marked a_1, a_2, a_3, and a_4 which represent the two sets of equilibria which can occur at 1000° C. The points a_1 and a_4 describe the liquids which are in equilibrium respectively with the solids a_2 and a_3. It is known from the work of Straumanis and Kim [30] that a_2 and a_3 correspond to compositions of 49.998 and 50.009 atom % arsenic respectively and that the deviations from the ideal formula GaAs are achieved by having vacant lattice sites.† Logan and Hurle [126] have calculated that at higher temperatures the arsenic content can fall to 49.990%.

The presence of lattice vacancies has a pronounced effect on the properties of crystals. In the case of GaAs, it is known that vacancies affect the electrical properties [31], the formation of dislocations [32, 33] and the incorporation of solutes (a subject discussed in section 3.2.2). The occurrence of vacancies is not limited to crystals with more than one type of atom. For example vacancies have been shown to be associated with the shallow etch pits found in germanium [35] which have a pronounced correlation with the hardness of the crystals [36].

In a semiconductor near its melting point, one site in about 10^5 might be vacant.‡ Under equilibrium conditions (of temperature and pressure) there is a fixed concentration of vacancies. In the case of an element the fraction of vacant sites ϕ in a lattice of A atoms is given by

$$\phi = \frac{[V]}{[A]} = \alpha \exp(-W/kT) \tag{3.2}$$

where $[V]$ is the concentration of vacancies, W is the energy to form a vacancy and α is a constant which allows for effects due to the distortion of the lattice. For germanium $\alpha = 13.5$ [35, 41] and for silicon $\alpha = 1$ [43, 44]. Kroger [2] suggests that $W \sim 0.5\ \Delta H_v$. For germanium $W = 2.18$ eV atom^{-1} $= 0.56\ \Delta H_v$ and for silicon $W = 2.32$ eV atom^{-1} $= 0.49\ \Delta H_v$ [35, 41] or $W = 2.53$ eV atom^{-1} [43]. Swalin [44] has discussed the derivation of W from material properties.

In a binary material, the situation is more complex. In a material AB, both A and B vacancies (V_A, V_B) are possible. Using symbols with the superscript L to represent atoms in the phase in equilibrium with the crystal, there are the following relations and reactions to be considered:

(a) Conservation of lattice sites

$$[A] + [V_A] = [B] + [V_B] = N_0 \tag{3.3}$$

where N_0 is half the total number of lattice sites per unit volume.

(b) Reactions involving vacancies

$$nA \rightleftharpoons A_n^L + n V_A \tag{3.4}$$

$$mB \rightleftharpoons B_m^L + m V_B \tag{3.5}$$

† Possible alternative explanations of deviation from stoichiometry are that interstitital atoms are present or that atoms can sit on the wrong sites (e.g. As in Ga positions). However in this case these explanations are not consistent with the X-ray and density data.

‡ For comparison in a metal as many as one in 10^3 sites might be vacant.

where n and m are the states of association of A and B in the second phase. (For example if B = As, in the vapour arsenic exists mostly as As_4 giving $m = 4$). When this second phase is a gas, it is usual to replace A_n^L by the pressure P_{A_n} to which it is proportional.

(c) The formation of AB

$$mA_n^L + n\,B_m^L \rightleftharpoons (m+n)AB \tag{3.6}$$

Applying the law of mass action (with the assumption that the activity coefficients are constant and so may be incorporated in the reaction constants†) gives

$$\frac{[V_A]^n\,[A_n^L]}{[A]^n} = K_1 \tag{3.7}$$

$$\frac{[V_B]^m\,[B_m^L]}{[B]^m} = K_2 \tag{3.8}$$

$$\frac{[A]^{m+n}\,[B]^{m+n}}{[A_n^L]^m\,[B_m^L]^n} = K_3 \tag{3.9}$$

If $[V_A] \ll [A]$ and $[V_B] \ll [B]$ then from (3.3)

$$[A] = [B] = N_0$$

and 3.7 and 3.8 become

$$[V_A] = N_0 K_1^{1/n}\,[A_n^L]^{-1/n} \tag{3.10}$$

$$[V_B] = N_0 K_2^{1/m}\,[B_m^L]^{-1/m} \tag{3.11}$$

Using 3.10 and 3.11 with 3.9 gives

$$[V_A]\,[V_B] = N_0^2 (K_1^m K_2^n K_3 / N_0^{2(m+n)})^{1/mn} \tag{3.12}$$

The relations allow the calculation of $[V_A]$ and $[V_B]$ if the condition of the second phase is known. Thurmond [25] has used these relations to deduce the ratio of maximum to minimum vacancy concentrations in GaAs and shows that at 1000°C the ratio is about 20. Using this figure with the data of Straumanis and Kim [30] gives maximum vacancy concentration of $1.3 \times 10^{18}\ \mathrm{cm^{-3}}$ on Ga sites and $3.7 \times 10^{17}\ \mathrm{cm^{-3}}$ on As sites. In the case of II-VI compounds, in which vacancies are electrically active, this type of relation is very important, see for example references 2 and 48a, b.

† This assumption will usually only be valid for extremely small changes in composition but it produces a useful first approximation. Some more elaborate arguments are given in the discussion of equations 3.18 to 3.22.

A similar sequence of arguments may be used to derive the concentrations, $[A_i]$, $[B_i]$, of interstitial atoms. If, for example, there are N_i interstitial sites available for A atoms and $[V_i]$ of these are unoccupied, then

$$nA_i \rightleftharpoons A_n^L + nV_i \qquad (3.13)$$

but, in general, $[A_i] \ll N_i$ so that $[V_i] \sim N_i$

and we can write

$$[A_i] = [A_n^L]^{1/n} K_i \qquad (3.14)$$

where K_i is the appropriate mass action constant.

The calculation of the equilibrium state of a phase in equilibrium with a crystal has been widely attempted. In particular there have been many attempts to calculate the composition of a liquid containing only the major components in equilibrium with a solid as a function of temperature. The starting point of such calculations is to consider a random mixture of atoms. If the atoms do not interact then a random mixture of A and B atoms has a free energy

$$G = X_A G_A + X_B G_B + RT(X_A \ln X_A + X_B \ln X_B)$$

where X_A and X_B are the atom fractions of A and B which on their own would have free energies G_A and G_B. The term $R(X_A \ln X_A + X_B \ln X_B)$ is the entropy of mixing. The free energies can be written

$$G = H - TS$$

where H is the enthalpy and S the entropy and

$$H = H^0 + \int_{T_0}^{T} C_p \, dT + \Delta H_t \qquad (3.15)$$

$$S = S^0 + \int_{T_0}^{T} \frac{C_p}{T} \, dT + \frac{\Delta H_t}{T_t} \qquad (3.16)$$

where C_p is the specific heat, ΔH_t is the enthalpy of a transition at T_t which lies between the actual temperature T and the reference temperature T_0 at which $H = H_0$ and $S = S_0$.

A similar calculation can be carried out for the solid (omitting, of course, the entropy of mixing, since the solid is not a random mixture). In equilibrium the solid and the liquid must have the same (minimum) free energy. At any temperature, this minimum value can be found by differentiation with respect to X_A or X_B (note $X_A + X_B = 1$). Equating the free energies enables us to express the enthalpies and entropies in terms of changes in these parameters (ΔH_m, ΔS_m) on melting and hence [38] obtain

$$-\ln\,(4X_A^L X_B^L) = \frac{2\Delta S_m}{R}\left(\frac{T_m}{T} - 1\right) - \frac{2\Delta C_p}{R}\left\{\frac{T_m}{T} - 1 - \ln\left(\frac{T_m}{T}\right)\right\} \quad (3.17)$$

where ΔC_p is the difference between the specific heats of the solid and the liquid and T_m is the melting point of the compound. Table 3.3 gives values of the thermodynamic quantities involved for some binary semiconductors and for comparison gives the same parameters for germanium and silicon. In general the second term on the right-hand side of equation 3.17 can be neglected because:

(a) ΔC_p is small compared with ΔS, and

(b) if $T_m = T + \Delta T$, expanding the logarithm shows that the term in curly brackets becomes

$$\frac{\Delta T^2}{2T^2} - \frac{\Delta T^3}{3T^3} + \frac{\Delta T^4}{4T^4} \text{ etc.}$$

which will also usually be small.

TABLE 3.3

Some thermodynamic quantities [25, 26, 42, 47]

	T_m °K	ΔH_m k cal (g atom)$^{-1}$	ΔS_m cal (g atom deg.C)$^{-1}$	C_p cal (g atom deg.C)$^{-1}$ solid	liquid
InSb	797	5.707	7.16	6.70	8.0
GaSb	985	7.78	7.90	6.79	7.50
AlSb	1330	9.796	7.37	6.79	8.90
GaAs	1510	12.59	8.32	7.35	~9.2
Ge	1210	7.6	6.28	6.86	7.0
Si	1685	12.1	7.18	6.77	7.4
InAs	1210	8.79	7.26	7.47	8.0

For semiconductors with $T_m > 700$°K, the following empirical relations can be used:

(1) $\Delta S_m = \Delta S_{mix} + 5.1 + 9 \times 10^{-3}\ T_m \pm 0.6$ cal (g atom deg C)$^{-1}$ where ΔS_{mix} is the entropy of mixing. For a binary material $A_r B_s$ where $r + s = 1$,

$$\Delta S_{mix} = R(r \ln r + s \ln s)$$

(2) For the liquids

$$C_p = 6.0 + 1.7 \times 10^{-3}\ T_m \pm 0.5 \text{ cal (g atom deg C)}^{-1}$$

(3) For the solids at the melting point

$$C_p = 7.0 \pm 0.4 \text{ cal (g atom deg C)}^{-1}$$

(4) For the solids at $T > 700$°K

$$\frac{dC_p}{dT} = C_p^2 \frac{T}{T_m}\ (2.7 \pm 0.4) \times 10^{-2} \text{ cal/g atom/(deg C)}^2$$

In each of these relations, the limits given include at least 90% of the available data.

Equation 3.17 gives a reasonable first approximation to the shape of the $T - X$ phase diagram. A better picture can be obtained by assuming that the liquid solution is not ideal. To do this we introduce activity coefficients γ_A and γ_B which are defined so that for example γ_A^L is the ratio of the vapour pressure of A over the solution to the product of X_A and the vapour pressure of pure A. If the solution is regular [37] (i.e. there is an enthalpy of mixing but no excess entropy of mixing) then

$$RT \ln \gamma_A^L = w(X_B^L)^2 \quad (3.18)$$

and

$$RT \ln \gamma_B^L = w(X_A^L)^2 \quad (3.19)$$

where w is an interchange energy. Carrying out the same analysis [38] gives

$$-R/2) \ln (4X_A^L X_B^L) = (\Delta H_m - T_m \Delta C_p) \left(\frac{1}{T} - \frac{1}{T_m}\right) + \Delta C_p \ln \left(\frac{T_m}{T}\right) + \frac{w}{T} (X_B^L - 0.5)^2 \quad (3.20)$$

Neglecting the terms in ΔC_p this equation has been used by Vieland [38] to describe the Ga-As system ($w = 5.0 \pm 0.8$ kcal g atom^{-1}) and by Schottky and Bever [26] to describe the systems In-Sb ($w = 3.96 \pm 0.80$ kcal g atom^{-1}) and Ga-Sb ($w = 0.67 \pm 0.55$ kcal g atom^{-1}). Hall [46] has examined these systems (and Ga-P, In-As and In-P) and gives very different values of w and ΔS_m. A detailed analysis of the Ga-As and Ga-P systems by Thurmond [25] shows that w varies with temperature: for Ga-As

$$w = 9.960 - 0.0115\,T \text{ kcal mole}^{-1} \quad (3.21)$$

and for Ga-P

$$w = 7.900 - 0.00700\,T \text{ kcal mole}^{-1} \quad (3.22)$$

The form of these equations suggests that the solutions are not regular and that there is an excess entropy of mixing.

The use of equation 3.20 can give a good description of the $T - X$ plot and data about vapour pressures can be derived from equations 3.18 and 3.19 if the vapour pressures over pure A and B are known. These three equations therefore enable us to give a complete description of a binary phase diagram composed of two volatile components. A complete graphical description of such a system would require the addition of extra lines (representing the second vapour species on the $P - T$ and $P - X$ plots of the type shown on figure 3.2). This approach is widely used to describe II-VI compounds with two volatile components.

The description of systems involving three major components is in general more difficult. Some systems can be described as pseudo-binary e.g. equilibria between solids and liquids of $Hg_xCd_{1-x}Te$ can be described as mixtures of HgTe and CdTe. However there is now considerable interest in the growth of semiconductor crystals from relatively dilute solution in metals. Such systems cannot be described in pseudo-binary terms and a graphical description is usually given in terms of a triangular diagram which uses the relation that for a system composed of *A*, *B* and *C*,

$$X_A + X_B + X_C = 1 \tag{3.23}$$

Figure 3.3 gives the liquidus diagram of the Ga-GaAs-GaP diagram. The

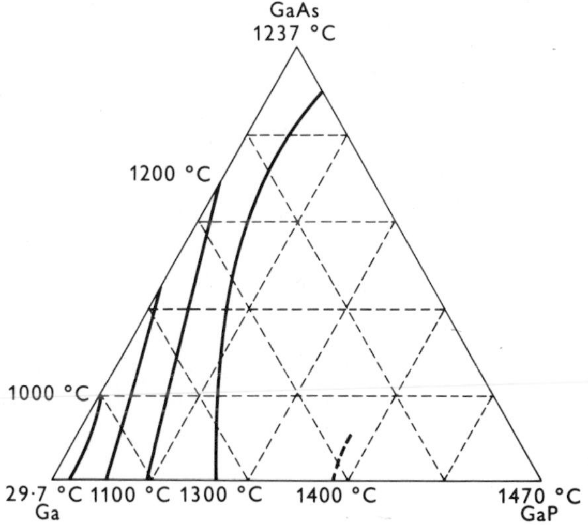

Figure 3.3. Liquidus curves for the system Ga-GaAs-GaP. (After Panish [39].)

representation of the results in the solid can be in terms of the molar fraction X_{GaAs}. Figure 3.4 gives this data.

Results obtained experimentally on this type of system can be represented analytically by considering the various possible reactions. Considering for example the system Ga-Al-P and using unmarked symbols for the solid and symbols with the superscript *L* for the liquid, the relevant reactions are

$$GaP \rightleftharpoons Ga^L + P^L \tag{3.24}$$

$$AlP \rightleftharpoons Al^L + P^L \tag{3.25}$$

Applying the law of mass action (assuming that γ_{Ga} and γ_{Al} may be

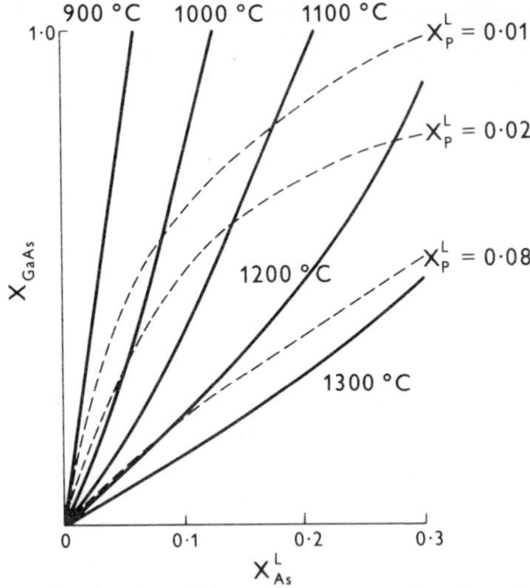

Figure 3.4. The molar fraction of GaAs(X_{GaAs}) as a function of the molar fraction of arsenic in the liquid (X_{As}^L) (full lines). Broken lines are lines of equal concentration of phosphorus in the liquid. The concentrations are given in terms of the molar fraction of phosphorus X_P^L.

incorporated in the mass action constants) gives

$$K_{GaP} = \frac{X_{Ga}^L X_P^L \gamma_P^L}{X_{GaP}} \tag{3.26}$$

and

$$K_{AlP} = \frac{X_{Al}^L X_P^L \gamma_P^L}{X_{AlP}} \tag{3.27}$$

Applying the restrictions

$$X_{Ga}^L + X_{Al}^L + X_P^L = 1 \tag{3.28}$$

and

$$X_{GaP} + X_{AlP} = 1 \tag{3.29}$$

gives

$$K_{AlP}(X_{Ga}^L - (X_{Ga}^L)^2 - X_{Ga}^L X_{Al}^L) + K_{GaP}(X_{Al}^L - (X_{Al}^L)^2 - X_{Ga}^L X_{Al}^L) = \frac{K_{AlP} K_{GaP}}{\gamma_P^L} \tag{3.30}$$

If it is assumed that γ_P^L behaves in the same way as in the GaP system, then

$$\ln \gamma_P^L = \frac{7900 - 7.00T}{RT} (1 - X_P^L)^2 \tag{3.31}$$

A complete description of the system can thus be given in terms of K_{AlP}, and K_{GaP}. The experimental results of Panish, Lynch and Sumski [40] can be approximately represented by

$$\log K_{GaP} = 11.05 - 1.56 \times 10^4/T \tag{3.32}$$

and

$$\log K_{AlP} = 4.55 - 1.21 \times 10^4/T \tag{3.33}$$

in the range $1500°K > T > 1300°K$.

Such arguments can be extended to more complex systems if necessary and account may be taken of more interactions.† For a discussion of the means of representing more complex systems see Zernike [45].

3.2.2 Minor Components

Segregation Coefficients

When two phases are in equilibrium, it is frequently observed that the ratio of the concentrations of a solute in the phases is roughly constant. This ratio is usually called the segregation coefficient of the solute and is an extremely useful quantity in the prediction of the distribution of the solute. For a solute C which substitutes for A atoms in a crystal, the segregation coefficient is defined as

$$\mathrm{K}_C = \frac{X_C}{X_C^L} = \frac{[C]([C^L] + [A^L])}{([C] + [A])[C^L]} \tag{3.34}$$

For semiconductors $[C] \ll [A]$ so that

$$\mathrm{K}_C = \frac{[C]\,[A^L]}{[C^L]\,[A]} \tag{3.35}$$

If the second phase is a stoichiometric melt so that $[A^L] \simeq [A]$, the segregation coefficient becomes

$$k_C = \frac{[C]}{[C^L]} \tag{3.36}$$

and is known as the distribution coefficient. Table 3.4 gives some examples of values of distribution coefficients.

Values of distribution coefficients have been correlated with tetrahedral radii [49] heats of sublimation [49] and diffusion coefficients [50] but perhaps the most useful correlation is that due to Millett [51] who has shown that distribution coefficients (and in some cases the more general segregation coefficients) correlate with the fractional misfit volume of the

† Thus G. M. Blom and T. S. Plaskett (*J. Electrochem. Soc.* **118**, 1831 (1971)) and G. M. Blom (Ibid., page 1834) discuss the systems In-Ga-Sb and In-Ga-P in terms of all the possible interactions.

TABLE 3.4

Some values of distribution coefficients[a]

Solute	Germanium [49]	Silicon [49]	Gallium Arsenide [7, 34]	Indium Antimonide [7]
K	–	–	1	–
Ca	–	–	0.8	4.16
Zn	4×10^{-4}	1×10^{-5}	1.9	2.3
Cu	1.5×10^{-5}	4×10^{-4}	$<2 \times 10^{-3}$	5×10^{-4}
Cd	1×10^{-5}	1×10^{-8}	$<2 \times 10^{-3}$	0.73
Al	7×10^{-2}	2×10^{-3}	1	–
Ga	9×10^{-1}	8×10^{-3}	–	3
In	1×10^{-3}	4×10^{-4}	<0.1	–
C	–	–	0.2	–
Si	5.5	–	2	–
Ge	–	0.33	1×10^{-2}	–
O	–	–	0.3	–
S	1.1×10^{-6}	–	0.5	0.1
Se	1×10^{-6}	–	0.1	0.3
Te	1×10^{-6}	4×10^{-6}	5×10^{-2}	0.5
P	8×10^{-2}	3.5×10^{-1}	2	0.16
As	2×10^{-2}	0.3	–	5.4
Sb	3×10^{-3}	2.3×10^{-2}	–	–

[a] Further data is given by J. C. Brice, The Growth of Crystals from Liquids (North Holland, Amsterdam, 1972) Chapter 2.

atom in the lattice. This misfit volume is calculated for semiconductors using tetrahedral covalent radii. The relation takes the form

$$\ln \mathbf{K}_C = \beta + \phi \Delta V/V_A \qquad (3.37)$$

where $\Delta V = V_A - V_C$ and V_A and V_C are the volumes of the solvent and solute atoms. Some values of β and ϕ are given in Table 3.5.

TABLE 3.5

Typical values of constants in equation 3.37

Solvents	Solutes	β	ϕ
Si and Ge	Group IV	0	1.43
Si and Ge	Groups III and V	−0.59	1.43
III-V Compounds	Groups II and III	0	1.43
III-V Compounds	Groups IV, V and VI	−0.59	1.43

Although there is much data about melt grown crystals, there is considerably less about solution and vapour grown semiconductors. Such information as there is however suggests that we can define a general segregation coefficient. Table 3.6 illustrates the constancy of $\mathbf{K}_{Sn}$ for the vapour growth of GaAs. Similarly, Silvestri [59] has shown that for

TABLE 3.6

Segregation coefficients of tin in GaAs grown from the vapour on to (100) faces at rates of $0.5 \pm 0.1\ \mu m\ min^{-1}$

Atom fraction of tin in source	$\mathbf{K}_{Sn}$
0.05	0.023
0.15	0.025
0.17	0.021
0.20	0.024
0.24	0.026

From data supplied by M. J. King.

germanium grown by the hydrogen reduction of $GeCl_4$, the segregation coefficient for boron is approximately 1×10^{-2} for $10^{-5} < X_B^L < 10^{-2}$ and $973°K < T < 1100°K$ provided that the flow rate is constant. In the case of growth of GaAs from solution in excess gallium, Harris and Snyder [60] report that $\mathbf{K}_{Sn}$ is constant at 1×10^{-4} for $3 \times 10^{-5} < X_{Sn}^L < 10^{-2}$. This value is again relatively independent of temperature (in the range 800-860° C) and can be compared with the value of $\mathbf{K}_{Sn}$ for growth from a melt which is $\sim 3 \times 10^{-3}$.

Phase Theory

The thermodynamic approach to the analysis of segregation coefficients is very similar to that used to calculate the binary phase diagrams in section 3.2.1. In both phases we assume random replacement of A atoms by the solute C and write the free energy in the form

$$G = X_C H_C + X_A H_A - T(X_C S_C + X_A S_A) + RT(X_C \ln X_C + X_A \ln X_A) \tag{3.38}$$

For an elemental solvent

$$X_A + X_C = 1 \tag{3.39}$$

The minimum free energy will occur when

$$\frac{dG}{dX} = 0 \tag{3.40}$$

i.e. when

$$H_C - H_A - T(S_C - S_A) + RT(\ln X_C - \ln X_A) = 0 \tag{3.41}$$

For a dilute solution $X_A \to 1$ and $\ln X_A \to 0$ so that

$$\ln \mathbf{K}_C = \frac{1}{RT}\left\{H_A - H_A^L - H_C + H_C^L\right\} - \frac{1}{R}\left\{S_A - S_A^L - S_C + S_C^L\right\} \tag{3.42}$$

The differences of enthalpy and entropy can obviously be expressed in terms of the thermodynamic parameters of the system. For example in the case of growth from a melt if pure A melts at T_A and pure C at T_C we have

$$\ln \mathbf{K}_C = \frac{\Delta H_A}{R}\left[\frac{1}{T} - \frac{1}{T_A}\right] - \frac{\Delta H_C}{R}\left[\frac{1}{T} - \frac{1}{T_C}\right] \tag{3.43}$$

If the solution is very dilute, by applying equation 3.38 to both the crystal and the melt, it is easily shown that

$$\begin{aligned}\frac{\Delta H_A}{R}\left(\frac{1}{T} - \frac{1}{T_A}\right) &= \ln X_A - \ln X_A^L \\ &= \ln(1 - X_C) - \ln(1 - X_C^L) \\ &\sim X_C^L(1 - \mathbf{K}_C)\end{aligned} \tag{3.44}$$

The published values of $\mathbf{K}_C$ for semiconductor systems (see for example Table 3.4) show that $\mathbf{K}_C < 10^{-1}$ in 80% of cases and $\mathbf{K}_C < 10^{-2}$ in more than 50% of cases so that in general

$$X_C^L \sim \frac{\Delta H_A}{R}\left(\frac{1}{T} - \frac{1}{T_A}\right) \tag{3.45}$$

If we put $T = T_A$ in equation 3.43 we get

$$\ln \mathbf{K}_{CO} = -\frac{\Delta H_C}{R}\left[\frac{1}{T_A} - \frac{1}{T_C}\right] \tag{3.46}$$

substituting this value into 3.43 and using 3.45 gives

$$\ln \mathbf{K}_C = \ln \mathbf{K}_{CO} - X_C^L\left[1 + \frac{\Delta H_C}{\Delta H_A}\right] \tag{3.47}$$

This relation is illustrated by figure 3.5. Thurmond and Struthers [55] give other examples.

Empirically [53] it is found that k_{CO} is proportional to the maximum solid solubility for germanium and silicon. Following Statz [54] and Kroger [2], the maximum solid solubility can be found by finding the maximum value of $k_C X_C^L$. From equation 3.47

$$\ln X_C = \ln(k_C X_C^L) = \ln k_{CO} - X_C^L\left[1 + \frac{\Delta H_C}{\Delta H_A}\right] + \ln X_C^L \tag{3.48}$$

and

$$\frac{d \ln X_C}{d X_C^L} = 0$$

when

$$X_C^L = \left(1 + \frac{\Delta H_C}{\Delta H_A}\right)^{-1} \tag{3.49}$$

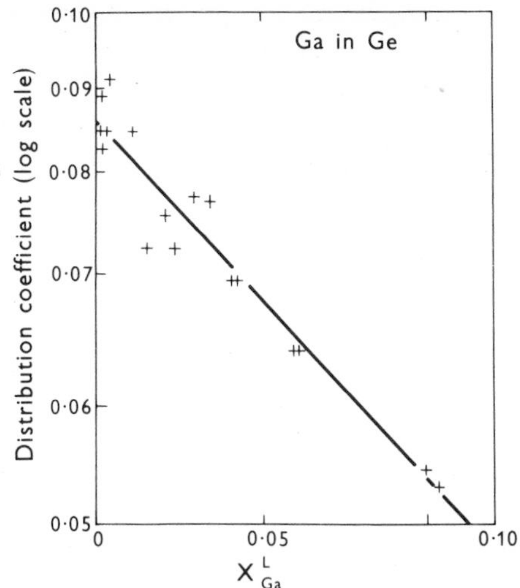

Figure 3.5. The distribution coefficient of gallium in germanium as a function of concentration in the liquid [52].

so that

$$\ln X_{C\,\max} = \ln k_{CO} - 1 - \ln\left(1 + \frac{\Delta H_C}{\Delta H_A}\right) \tag{3.50}$$

Or

$$\begin{aligned} X_{C\,\max} &= \frac{k_{CO}}{e}\left(1 + \frac{\Delta H_C}{\Delta H_A}\right)^{-1} \\ &= (0.28 \pm 0.09)k_{CO} \qquad \text{for germanium} \\ &= (0.18 \pm 0.06)k_{CO} \qquad \text{for silicon} \end{aligned} \tag{3.51}$$

A more rigorous analysis [55] would add an enthalpy of mixing to equation 3.42 in the same way as was done to give equation 3.20 and an even better description of the results can be obtained by using an enthalpy of mixing w in the form

$$w = A + BT$$

Thurmond and Kowalchik [56] have examined the data for solutes in germanium and silicon and deduced the values of A and B.

This type of analysis can readily be extended to more complex systems (see for example the work of Berkes and White [56b]) and by analogy with 3.42, we expect to be able to write a segregation coefficient of a solute C in the form

$$\mathbf{K}_C = \exp\{-\Delta G_C/RT\} \tag{3.52}$$

where ΔG_C is the difference in free energy of the solute in the two phases.

For the case of growth from a melt

$$\Delta G_C = \Delta H_A - \Delta H_C - T(\Delta S_A - \Delta S_C) - w$$

where the ΔH's and ΔS's are enthalpies and entropies of melting.

If other solutes are added to the system and these do not interact, then Saidov [57] has shown that there will be competition for incorporation in the crystal and the segregation coefficient becomes

$$\mathbf{K}_C = \frac{\mathbf{K}_{CO}}{1 + \Sigma(\mathbf{K}_q - 1)X_q^L} \tag{3.53}$$

where K_{CO} is the segregation coefficient for $X_q = 0$ and the $\mathbf{K}_q$ and X_q refer to the qth impurity.

For most semiconductor systems $|(\mathbf{K}_q - 1)X_q^L| \ll 1$ so that this effect is negligible. However the addition of another solute will in general affect the equilibrium temperature and hence the temperature in equation 3.52. In the case of growth from a melt it is readily shown that

$$\ln k_C = \ln k_{CO} + \frac{\Delta G_C + T(\Delta S_A - \Delta S_C)}{\Delta H_A} \Sigma(k_q - 1)X_q^L \tag{3.54}$$

Once again, in general, $(k_q - 1)X_q^L$ will be very small but in the case of a crystal grown from a non-stoichiometric melt, the excess of one of the major constituents can be considered to be a solute and this can produce an appreciable effect. Figures 3.6 and 3.7 illustrate the effect for some

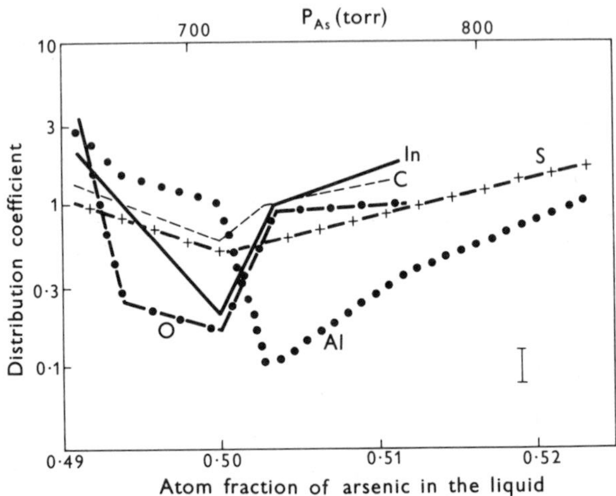

Figure 3.6. The distribution coefficients of In, C, O, S and Al in GaAs as a function of the atom fraction of arsenic in the liquid.

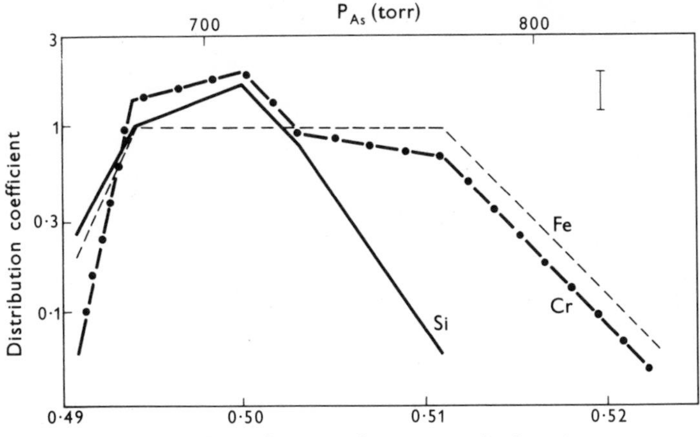

Figure 3.7. The distribution coefficient of Fe, Cr and Si in GaAs as a function of the atom fraction of arsenic in the liquid.

solutes in GaAs. Note that on these figures, the solutes giving maxima are those with the greatest values of ΔH_C. This would be expected from equation 3.54. This type of effect accounts for some of the variability of distribution coefficients in compound materials.†

The Use of the Law of Mass Action

The use of the law of mass action often provides a convenient method for predicting the variation of segregation coefficients.

The simplest case involves reactions of the type

$$A + C^L \rightleftharpoons C + A^L \tag{3.55}$$

where the symbols with superscripts refer to the second phase and symbols without superscripts refer to the crystal. From equation 3.55 applying the law of mass action gives

$$\frac{[C]\,[A^L]}{[A]\,[C^L]} = K \tag{3.56}$$

where

$$K = \frac{\gamma_{C^L}\gamma_A}{\gamma_{A^L}\gamma_C}\exp\left(-\frac{\Delta G}{RT}\right) \tag{3.57}$$

but

$$\mathbf{K}_C = \frac{[C]}{[C^L]} \tag{3.58}$$

† J. V. DiLorenzo and G. E. Moore (*J. Electrochem. Soc.* **118**, 1823 (1971)) show that the ratio of gallium to arsenic in the gas phase also affects the incorporation of impurities during vapour growth.

hence

$$\mathbf{K}_C = \frac{K[A]}{[A^L]} \tag{3.59}$$

This type of analysis can for example readily explain the variation of $\mathbf{K}_{\mathrm{Ga}}$ in InSb grown from melts with varying indium contents. In this case Merten, Vos and Hatcher [58] have shown that

$$\mathbf{K}_{\mathrm{Ga}} = \frac{1.5}{[X_{\mathrm{In}}^L]} \tag{3.60}$$

for $0.7 > [X_{\mathrm{In}}^L] > 0.3$.

The law of mass action becomes particularly useful in the discussion of the incorporation of charged atoms [61]: at high temperatures most shallow donors and acceptors will be completely ionized. Consider for example a solute C substituting on A sites to give a donor centre:

$$C^L + V_A^{\times} \rightleftharpoons C^{\cdot} + e' \tag{3.61}$$

hence

$$\frac{[C^{\cdot}]\,n}{[V_A^{\times}]\,[C^L]} = K_1 \tag{3.62}$$

where n is the electron concentration, and

$$\mathbf{K}_C = \frac{K_1 [V_A^{\times}]}{n} \tag{3.63}$$

$[V_A^{\times}]$ can then be expressed in terms of $[A^L]$ or the equivalent partial pressure of a volatile major component in the manner described in section 3.2.1. Figure 3.8 illustrates the effect on k at low doping levels. It frequently happens that at relatively high doping levels, $[C^{\cdot}] = n$ and Lorenz and Blum [62] have used this approach to account for their results on CdTe and InAs. Rearranging equation 3.63 on this basis gives

$$\mathbf{K}_C [C^{\cdot}] = K_1 [V_A^{\times}] \tag{3.64}$$

or

$$\mathbf{K}_C^2 = K_1 [V_A^{\times}]\,[C^L] \tag{3.65}$$

so that $\mathbf{K}_C \propto [C^L]^{1/2}$.

A more general approach (see section 1.2.2) is to remember that

$$\gamma_p \gamma_n p n = n_i^2 \tag{3.66}$$

and

$$[C^{\cdot}] + p = n \tag{3.67}$$

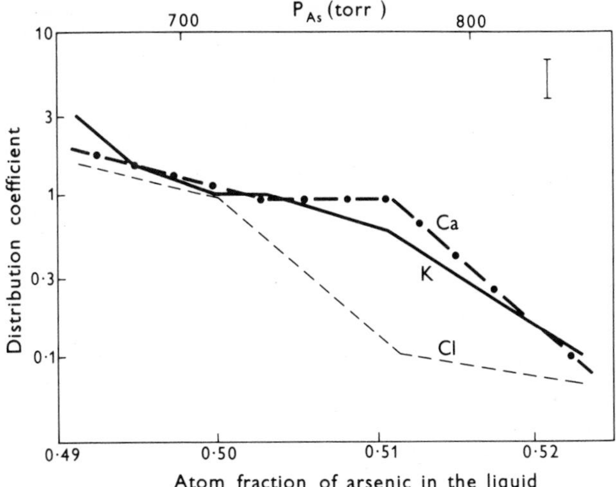

Figure 3.8. The variation of the distribution coefficients of Ca, K and Cl in GaAs with (X_{As}^L) from data at very low solute concentrations.

so that in equation 3.63 we should put with $\gamma_p = \gamma_n = 1$

$$n = (1/2)\{[C^{\cdot}] + \sqrt{[C^{\cdot}]^2 + 4n_i^2}\} \tag{3.68}$$

Similar arguments can be applied to acceptors or even amphoteric impurities, and this result has been used by Foster and Scardefield [63] to account for their results on Zn substituting in GaP grown from solution at 1144°C: up to $[\text{Zn}] \sim 2 \times 10^{17}\ \text{cm}^{-3}$ $K_{\text{Zn}} \sim 0.5$ but for $[\text{Zn}] > 10^{18}\ \text{cm}^{-3}$, $K_{\text{Zn}} \alpha [\text{Zn}^L]^{1/2}$. Note that these arguments assume that the bulk of the crystal is in equilibrium with the fluid. This will be true for solutes which diffuse rapidly in the solid. If, however, diffusion is slow, then the exchange processes will be governed by the density of surface states so that K_C may remain constant for much greater values of $[C^L]$. This has been demonstrated for Te in GaAs grown from solution in gallium [34]

The mass action constants K are given by

$$K = \exp - (\Delta G^{\circ}/RT) \tag{3.69}$$

Knowing ΔG°, it is possible to apply the various relations at different temperatures and so to allow for the change of equilibrium constant with temperature. This is particularly useful when treating the result of interactions between various solutes. This subject has been treated exhaustively by Kroger [2, Chapter 9] and Fuller [64]. If interaction takes place between imperfections A and B to form an associate AB we can consider the reaction

$$A + B \rightleftharpoons AB \tag{3.70}$$

and write

$$\frac{[AB]}{[A]\,[B]} = K_{AB} = \exp\left\{\frac{-\Delta G^\circ}{RT}\right\} \tag{3.71}$$

where

$$\Delta G^\circ = \Delta H - T\Delta S \tag{3.72}$$

In this relation ΔH is the interaction energy (for example $-e^2/\epsilon r$ for particles of charge e separated by a distance r in a medium of dielectric constant ϵ) and ΔS is formed of two parts one due to vibration and the other due to the configuration. The first of these entropies is given by

$$k \ln \left\{\prod_q \frac{\nu_q}{\nu_q^*}\right\}$$

where the ν_q are the vibrational frequencies of the free imperfections and the ν_q^* of the associate. The second entropy term is $k \ln (Z/\sigma)$ where Z is the number of possible equivalent arrangements and σ is a geometrical factor: $\sigma = 1$ for a binary AB associate, $\sigma = 2$ for an AA associate, $\sigma = 3$ for a triangular AAA associate etc. These interactions are of interest to us because they affect the apparent segregation coefficient. Normally we measure the segregation coefficient of a solute A as the ratio of total concentration in the solid to the concentration in the liquid so that

$$\mathbf{K}_{\text{measured}} = \frac{[A] + [AB]}{[A^L]} = \frac{[A](1 + K_{AB}[B])}{[A^L]} \tag{3.73}$$

hence

$$\mathbf{K}_{\text{measured}} = \mathbf{K}_A(1 + K_{AB}[B]) \tag{3.74}$$

For example Merten, Vos and Hatcher [58] have shown that k_{Cu} in InSb is affected by the presence of Zn in the crystal. Their results are shown on figure 3.9 where it can be seen that for small X_{Zn}, k_{Zn} follows the sort of relation expected from equation 3.47, but at larger X_{Zn}, the effect predicted by equation 3.74 predominates and $k_{Cu} \propto X_{Zn}$. A relation similar to equation 3.74 will, of course, also apply to B so that we can expect coupled substitution of interacting species to increase the total solute concentrations. The growth of pure crystals is, therefore, best carried out by low temperature processes: at low temperatures the solutes will not be completely ionized and the degree of interaction will therefore be less.

3.2.3 Some Consequences of Equilibrium

The most important, but frequently neglected, consequence of equilibrium involves equilibria between several systems. Consider systems 1, 2 and 3; if

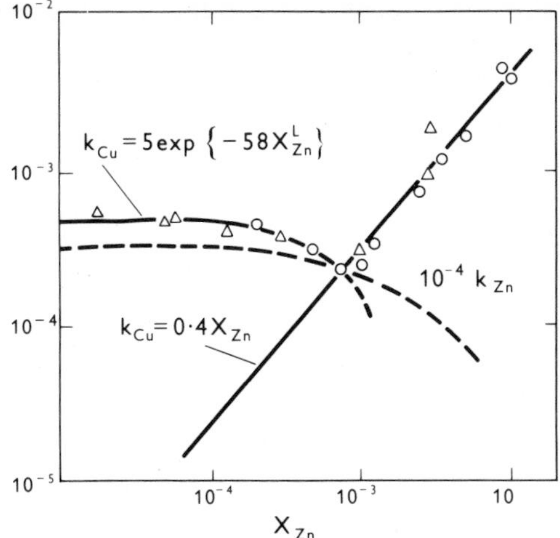

Figure 3.9. The variation of the distribution coefficient of copper in InSb as a function of the concentration of zinc in the solid. (After Merten, Vos and Hatcher [58].)

1 is in equilibrium with 2 and 2 is in equilibrium with 3, then 1 and 3 are also in equilibrium. (Any alternative allows perpetual motion machines to be made.) If these systems are a solid (1), a liquid (2) and a vapour (3), then it is clear that they must all be at the same temperature and that the partial pressures of all the components must be the same i.e.

$$\gamma_{C1}[C_1] = \gamma_{C2}[C_2] = \gamma_{C3}[C_3] \tag{3.75}$$

where C is any component, whose activity coefficient in the ith phase is γ_{Ci}.

Hence

$$\mathbf{K}_{12} = \frac{[C_1]}{[C_2]} = \frac{\gamma_{C2}}{\gamma_{C1}} \tag{3.76}$$

$$\mathbf{K}_{23} = \frac{[C_2]}{[C_3]} = \frac{\gamma_{C3}}{\gamma_{C2}} \tag{3.77}$$

and

$$\mathbf{K}_{13} = \frac{[C_1]}{[C_3]} = \frac{\gamma_{C3}}{\gamma_{C1}} \tag{3.78}$$

hence

$$\mathbf{K}_{12}\mathbf{K}_{23} = \mathbf{K}_{13} \tag{3.79}$$

If we consider two solids 1 and 2 in equilibrium with a gas phase 3, equations 3.75 to 3.79 still apply and may be used to evaluate the various segregation coefficients.

Use has already been made of these ideas to interpret the phase diagrams of multicomponent systems in section 3.2.1.

We can also consider phase 2 to lie between phases 1 and 3. This makes no difference to the relations between phases 1 and 3, so that in equilibrium we can ignore the nature of the material separating two phases. Consider the case of a crystal grown from a melt which is exposed to an atmosphere containing some solute. If it is convenient we can relate the amount of solute appearing in the crystal to the partial pressure in the vapour phase. For example in the case of germanium about 2×10^{16} cm^{-3} oxygen atoms are incorporated in a crystal for each torr of oxygen in the atmosphere to which the melt is exposed: actual mean values are 1.9 ± 0.3, 2.0 ± 0.2, 2.2 ± 0.1 and $2.3 \pm 0.5 \times 10^{16}$ cm^{-3} torr^{-1} at 2, 5, 10 and 28 torr respectively.

3.3 DYNAMIC EQUILIBRIA

The equilibria considered in section 3.2 were completely independent of any transport processes. When the interface between two phases† moves, concentrations at a point fixed relative to the interface may also be constant. However these concentrations are dependent on transport processes. The evaluation of these concentrations forms the subject of this section. The analysis will involve the concepts of interface and effective segregation coefficients, so that we can separate effects due to atomic kinetics and transport. These coefficients are most easily defined by reference to figure 3.10. The interface segregation coefficient $\mathbf{K}_I$ is given by

$$\mathbf{K}_I = \frac{[C]}{[C_I^L]} \tag{3.80}$$

and the effective segregation coefficient $\mathbf{K}_{\text{eff}}$ is given by

$$\mathbf{K}_{\text{eff}} = \frac{[C]}{[C_\infty^L]} \tag{3.81}$$

For the remainder of this section it will be assumed that exchanges of atoms between the solid and the fluid take place randomly. (Cases in which this assumption is not justified are considered in section 3.4.1.) Even with this assumption it is evident that processes at the interface will depend on the nature of the interface. The simplest cases are those in which the interface is sharp, i.e. the transition from one phase to another takes place over a single atomic spacing as shown on figure 3.11.

† In this section it is assumed that the second phase is a fluid.

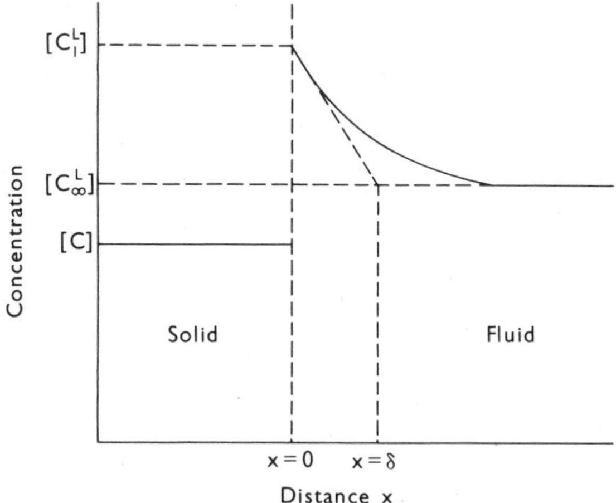

Figure 3.10. Concentrations at a solid-fluid boundary. Boundary layers of this type have been clearly revealed in metal alloy systems [125].

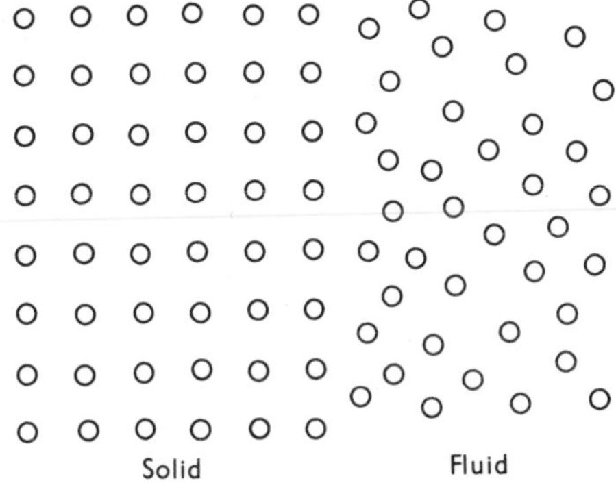

Figure 3.11. A sharp solid-fluid boundary.

3.3.1 Perfectly Stirred Fluids

We shall consider first cases in which we can expect perfect mixing in both phases. Consider a stationary interface. If the reaction rates from phase 1 to phase 2 and from phase 2 to phase 1 are V_{12} and V_{21} then clearly in equilibrium

$$[C_1]\,V_{12} = [C_2]\,V_{21} \tag{3.82}$$

and

$$\mathbf{K}_{12} = \frac{[C_1]}{[C_2]} = \frac{V_{21}}{V_{12}} \tag{3.83}$$

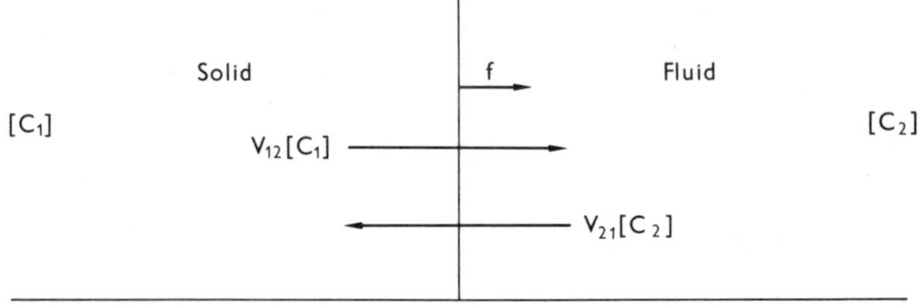

Figure 3.12. Reactions at a sharp solid-fluid boundary.

If the interface is advancing from phase 1 into phase 2 at a rate f then, in the absence of the reactions, solute C would be incorporated at a rate $[C_2]f$, but it is in fact only incorporated at a rate $[C_1]f$ (see figure 3.12). The reactions at the interface must therefore account for the difference between these rates, so that

$$[C_2]f - [C_1]f = [C_1]V_{12} - [C_2]V_{21} \tag{3.84}$$

Hence

$$\mathbf{K}_I = \frac{f + V_{21}}{f + V_{12}} \tag{3.85}$$

which reduces to $\mathbf{K}_{12}$ for $f = 0$.
Using equation 3.83

$$\mathbf{K}_I = \frac{f + V_{21}}{f + V_{21}/\mathbf{K}_{12}} \tag{3.86}$$

An exact evaluation of the reaction rates is difficult but to a first approximation we might expect

$$V_{12} \sim D_1/d_1$$

and

$$V_{21} \sim D_2/d_2$$

where the D's are diffusion coefficients and the d's are interatomic spacings. Since $d \sim 10^{-8}$ cm, if phase 2 is a fluid ($D_2 > 10^{-5}$ cm^2 sec^{-1}) then V_{12} would be expected to be $>10^3$ cm sec^{-1} and $\mathbf{K}_I$ would usually be $\sim\mathbf{K}_{12}$. However examination [124] of the available data [125] suggests that V_{12} can be much smaller.

This analysis assumes that equilibrium reaction rates may be applied to a non-equilibrium case. This will not in general be allowed. For example a finite rate of growth requires that the interface temperature is below the equilibrium temperature (see section 3.4.1). A rigorous analysis therefore requires the use of non-equilibrium thermodynamics. Baralis [65] has performed the necessary calculations in the case of melt grown materials and shown that the fractional change in distribution coefficient ($\Delta k_C/k_C$) associated with an interface undercooling of ΔT is

$$\frac{\Delta k_C}{k_C} \sim (\Delta H_A - \Delta H_C)\frac{\Delta T}{RT^2}. \tag{3.87}$$

where ΔH_A and ΔH_C are recspectively the enthalpies of fusion of A and C. Since it is known [66] that the undercoolings ΔT are usually very small this effect will normally be negligible.

The same problem has also been treated by Jindal and Tiller [67a] on the basis of non-equilibrium thermodynamics. They show that k_C is given by

$$\frac{1+k_C}{1-k_C}\ \ln\left[\frac{k_C+1}{2k_{CO}^{1/2}}\right] = \frac{2f}{M_C RT} \tag{3.88}$$

where k_{CO} is the equilibrium value of k_C, f is the growth rate and M_C is the mobility of solute atoms given by

$$\frac{M_C}{d} \sim (1/2)(D_{SO} + D_{LO}) \exp -\left\{\frac{Q_S + Q_L}{2kT}\right\} \tag{3.89}$$

where

$$D_j = D_{jO} \exp -(Q_j/kT) \tag{3.90}$$

Expected values of M_C are 10-100 cm sec^{-1}. Note that for $f \to \infty$, $k_C \to 1$ but $f \to 0$ does not result in $k_C \to k_{CO}$.

The treatment of a diffuse interface, at which the transition from solid to fluid takes place over several atomic layers, is most easily achieved by generalizing the argument used to deduce equation 3.86 to involve many intermediate layers (reference 7, page 66). This gives

$$\mathbf{K}_C = \prod_{i=o}^{n} \frac{f + V_{i,\,i+1}}{f + V_{i+1,\,i}} \tag{3.91}$$

where the layers are numbered from $i = 0$ (definitely crystal) to $i = n$ (definitely fluid). There is evidence [67b] that in rapidly quenched alloys, $\mathbf{K}_C$ can be much larger than the equilibrium value.

3.3.2 Cases of Imperfect Stirring

All the analysis so far has assumed that the concentrations were constant within each phase. We now consider cases in which this assumption cannot reasonably be made. The most obvious of these cases is a crystal growing from a fluid containing a solute with a segregation coefficient not equal to unity. It is clear, in this case, that in the fluid at the growth interface, there will be a concentration gradient unless there is infinitely good stirring. For simplicity the treatment will be in terms of a boundary layer of width δ (see figure 3.10). In a rigorous treatment we should consider the width of this boundary layer to be a function of the rate of movement of the interface. However the work of Wilcox [68] shows that if D_L is the diffusion coefficient in the fluid then for $f\delta/D_L \lesssim 0.5$ the change in thickness is less than 10% and in most practical cases the change will only be a fraction of 1%, so that δ is effectively a constant.

In general the thickness of the boundary layer [12] is given by

$$\delta = aD_L^{1/3}\nu^{1/6}\left(\frac{V}{L}\right)^{-1/2} \tag{3.92}$$

where a is a constant in the range 1-10, ν is the ratio of viscosity to density. L is a typical dimension (e.g. the radius of a disc or the length from the edge of a plate) and V is the relative velocity of the fluid. In the case of a disc rotating at a rate ω in a fluid

$$\delta = 1.6D_L^{1/3}\nu^{1/6}\omega^{-1/2} \tag{3.93}$$

In the case of a large planar boundary a one-dimensional treatment is usually sufficient and the flow is governed by

$$D_L\frac{d^2[C^L]}{dx^2} + f\frac{d[C^L]}{dx} = 0 \tag{3.94}$$

which can be solved to give

$$\frac{[C^L]-[C_0^L]}{[C_\infty^L]-[C_0^L]} = \frac{1-\exp(-xf/D_L)}{1-\exp(-\delta f/D_L)} \tag{3.95}$$

Using the boundary condition

$$[C]f = [C^L]f + D_L\frac{d[C^L]}{dx} \quad \text{at } x = 0 \tag{3.96}$$

with

$$[C] = \mathbf{K}[C^L] \tag{3.97}$$

gives

$$\frac{[C]}{[C_\infty^L]} = \mathbf{K}_{\text{eff}} = \frac{\mathbf{K}}{\mathbf{K}+(1-\mathbf{K})\exp\left(\frac{-f\delta}{D_L}\right)} \tag{3.98}$$

This result was first derived by Burton, Prim and Slichter [69] and has subsequently been used by many authors with great success to account for results obtained in melt growth. Figure 3.13 gives some examples for melt

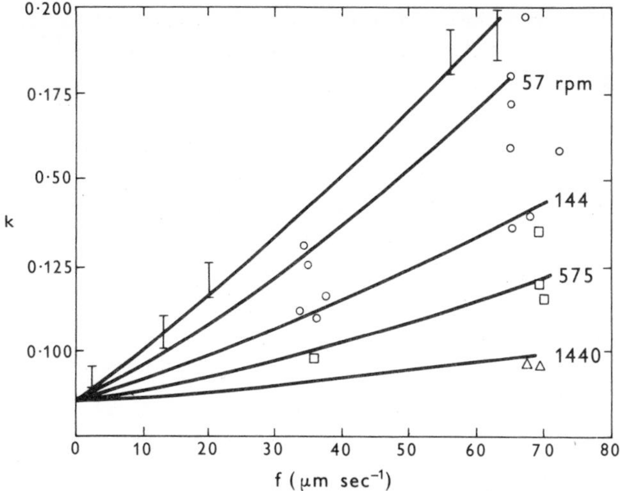

Figure 3.13. The effective distribution coefficient of gallium in germanium. The barred symbols refer to zone refining in a temperature gradient of about 30° C cm^{-1} (δ/D = 150 sec cm^{-1}). The other results are taken from Burton *et al.* [70].

Δ,	ω = 1440 r.p.m.,	δ/D = 27 sec cm^{-1}
□,	ω = 575 r.p.m.,	δ/D = 43 sec cm^{-1}
θ,	ω = 144 r.p.m.,	δ/D = 85 sec cm^{-1}
○,	ω = 57 r.p.m.,	δ/D = 134 sec cm^{-1}

growth. Other cases have been given by Merton *et al.* [58]. Figure 3.14 shows that similar analysis may be used in the case of vapour growth.

The principal difficulty with this type of analysis is finding values for the various parameters in equations 3.92 and 3.98. For most semiconductor melts $D_L \sim 10^{-4}$ cm^2 sec^{-1} and viscosities are about 10^{-2} poise. These figures should also apply to solutions in metals. For gases, diffusion coefficients at 300° K fall in the range 0.3 to 0.7 cm^2 sec^{-1} for mixtures containing hydrogen and 0.05 to 0.25 for other mixtures. If the diffusion coefficient at temperature T_1 and total pressure P_1 is D_1 then at a temperature T_2 and a pressure P_2 the diffusion coefficient is given approximately by

$$D_2 = \frac{D_1 P_1}{P_2}\left(\frac{T_2}{T_1}\right)^{1.8} \qquad (3.99)$$

Viscosities of gases are typically in the range 8×10^{-5} to 2.5×10^{-4} poise at 300° K and one atmosphere and increase linearly with pressure and the square root of temperature. Methods for estimating the properties of fluids are discussed by Reid and Sherwood [71].

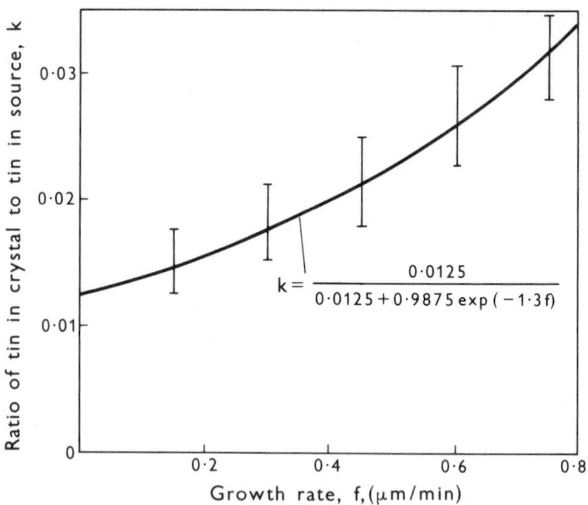

Figure 3.14. The segregation coefficient of tin in GaAs grown from the vapour. Plotted from results provided by M. J. King.

Equation 3.98 shows that the effective segregation coefficient can be controlled by the amount of stirring either deliberately applied or due to convection. This is true up to a point. However too much stirring may lead to instabilities [72] and in the case of growth from a solution (either in the liquid or gas phase), the stirring is frequently a controlling factor in the growth rate [73]. In general, if control over the segregation coefficient is necessary (for example to grow clean crystals from dirty materials), it is better to use the effect caused by changing the growth rate.

3.4 NON-EQUILIBRIUM CASES

3.4.1 The Facet Effect

There is a free energy per unit area associated with a phase boundary [74-77]. In the case of a solid this energy is anisotropic. A qualitative idea of the ratios of the surface free energies of various faces can be obtained by comparing the number of broken bonds per unit area of face. A more quantitative relation for the free energy per unit area σ is

$$\sigma = \frac{Z_B - Z_S}{Z_B} \Delta H_{vo} N_o^{2/3} \tag{3.100}$$

Where Z_B is the number of nearest neighbours to one atom in the bulk, Z_S is the number of neighbours in the solid of an atom on the surface, ΔH_{vo} is the enthalpy of evaporation per molecule at 0° K and N_o is the number of molecules per unit volume. In materials with the diamond and zinc blende structures, $\{111\}$ faces have the minimum free energy. A crystal of one of these materials in complete equilibrium would be bounded by $\{111\}$ faces. For germanium σ on $\{111\}$ faces from equation 3.100 is about 250 erg cm^{-2} and the experimental value is about 180 erg cm^{-2}. The addition of an extra atom or group of atoms to one of these low free energy planes makes a large increase in the total free energy and it can be shown that, in general, a large supercooling is required and that a cluster of atoms tends to be deposited at one time. Once the cluster has formed, it can spread laterally fairly rapidly (see figure 3.15). In germanium [78, 79] the growth rate on $\{111\}$ faces is given by

$$f = 2 \exp\left(\frac{-0.4}{\Delta T}\right) \text{mm min}^{-1} \qquad (3.101)\dagger$$

where ΔT is the supercooling (about 1.7°C for a growth rate of 1.6 mm min^{-1}). On other faces which have much higher free energies, the

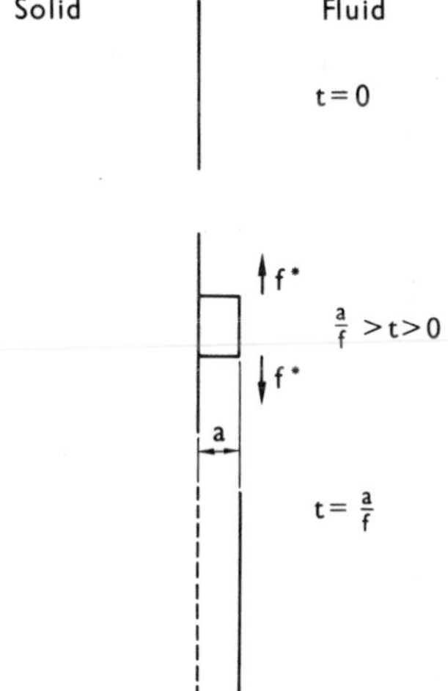

Figure 3.15. Growth on a singular face. At a time $t > 0$, a nucleus of height a forms on the face and expands laterally at a rate f^*. A second nucleus forms on the layer grown at some time $t > a/f$.

addition of an atom does not make a significant change in the free energy and atoms can be added randomly. For example in germanium such faces grow from the melt at rates given approximately by

$$f = 16\Delta T \text{ mm min}^{-1} \qquad (3.102)$$

† This expression represents experimental results over a small range of ΔT. The pre-exponential term may be a function of ΔT.

so that ΔT at 1.6 mm min^{-1} is only about 0.1°C. It is also possible for crystals to grow by the addition of atoms at defects (screw dislocations and twin planes) but such defects are not usually present in semiconductor crystals,† and we need only consider growth by the random addition of atoms (growth on a rough interface) and growth by a nucleation and lateral expansion process (growth on a singular interface). The processes occurring on a rough interface are exactly of the type that was discussed in section 3.3 in which we could take the interface segregation coefficient as being equal to the equilibrium one and then calculate the effective value on the basis of transport effects. These processes will not be considered further.

Jackson [81] has shown that the transition from a rough to a singular interface occurs when a parameter α exceeds 2. This parameter is given by

$$\alpha = \frac{Z_B \Delta H_f}{Z_S k T}. \tag{3.103}$$

where ΔH_f is the latent heat of fusion per atom, on the surface and k is Boltzmann's constant. (For germanium this factor exceeds 2 on $\{111\}$ but not on other faces, although there is evidence that $\{110\}$ faces are singular [66, 82].) The factors governing the size of the singular face in melt growth have been discussed by Brice [80].

Trainor and Bartlett [83] have examined the consequences of segregation on growth on a singular face for the case of melt growth including effects due to adsorption. With high lateral growth velocities, an absorbed layer will be incorporated in the growing crystal: at lower velocities, exchange between the layer and the fluid can occur. The model suggests that the ratio of interface to equilibrium distribution coefficients is

$$\frac{k_C}{k_o} = [1 - 2f^*/\{f^* + (f^{*2} + 4AD_C)^{1/2}\}(1 - k_o \tau_S/\tau_C)]^{-1} \tag{3.104}$$

where τ_S and τ_C are the sticking times for atoms of the solvent and solute on the surface‡; f^* is the transverse velocity and

$$A = \tau_C^{-1} + \Phi u_C n_C \sigma_o^{-1}$$

where Φ is a constant, u_C is the velocity of impurity atoms (10^3–10^4 cm sec^{-1}); n_C is the concentration of solute atoms, σ_o is the number of sites per unit area and D_C is the diffusion constant in the adsorbed layer ($D_C \sim 10^{-4}$ to 10^{-7} cm^2 sec^{-1}). Values of τ_C and τ_S

† For a discussion of the various mechanisms see references 78, 79, 80 and 7 (Chapter 2).

‡ The sticking time is the average time that an atom remains on the interface.

should be in the range 10^{-7} to 10^{-2} sec. This model has a number of unknown parameters but the values needed to fit the known data do not appear unreasonable and it is possible to show for example that the concentration dependence of the results is that expected on the basis of the model (reference 7, page 69). Hall [84] has proposed a somewhat similar relaxation model.

Table 3.7 gives some ratios of k_C/k_o found in melt growth. The

TABLE 3.7

Some values of $k_{(111)}/k_0$ [a]

Germanium [88]		Silicon [113]		Indium Antimonide [114]	
P	2.5	P	1.07-1.12	Zn	1.3
As	1.8	As	1.13-1.35	Cd	2.2
Sb	1.45	Sb	1.3 -1.45	Ge	1.6
Bi	1.65			Si	1.7
Ga	0.85	Ga	1.26	S	3
In	1.4	In	1.44	Se	3.6
Tl	1.2			Te	6

[a] The values should be treated with caution: the ratio tends to decrease with increasing concentration and decreasing growth rate.

effects found, however, are not confined to growth from the melt. Table 3.8 for example gives ratios of impurity contents found on various faces in vapour growth. In this case some allowance should be made for the variation of growth rate with orientation (see figure 3.16)† but the

TABLE 3.8

Ratios of segregation coefficients on various faces of GaAs grown from the vapour [115]

Solute	(111)*B* : (100)	(111)*B* : (111)*A*	(111)*B* : (110)
Zn	0.43	0.2	0.49
Te	7.4	20	15
Se	6.3		4.4
Sn	2.3		
S	1.4		

† For a discussion of factors concerned in growth from the vapour see Joyce [86].

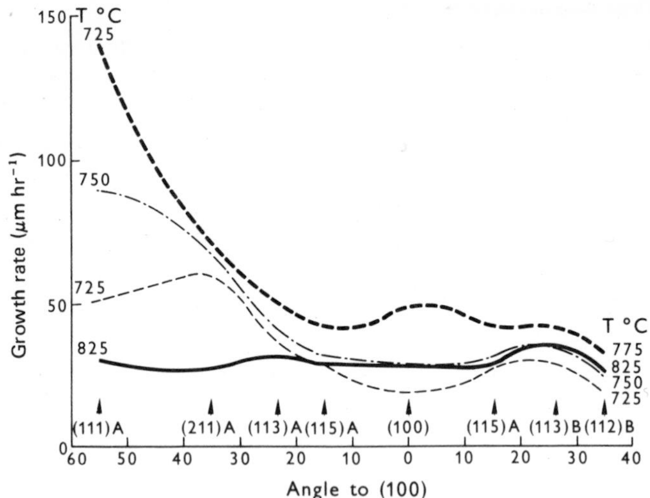

Figure 3.16. Growth rates on various faces of GaAs grown from the vapour. (After Shaw [85].)

discussion in the last section suggests that transport effects are not sufficient to account for the magnitudes found.

Taylor [87] has studied the effects of concentration on segregation of Se and Zn in vapour grown GaP. His results (see Table 3.9) show that the segregation coefficients can vary with concentration.

Equation 3.104 suggests that if $f^* \gg 4AD_C$ the ratio k_C/k_o will be independent of f^*. If f^* is given by a relation of the form of equation 3.102 and f is given by an equation of the form of equation 3.101 then, because the interface supercoolings in both equations are the same, at low growth rates f^* will decrease and the ratio of k_C/k_o will depend on f. This effect has been noted in InSb [58].

TABLE 3.9

Segregation in vapour grown GaP adapted from Taylor [87]

Solute	Face	Concentration (atom cm^{-3}) for an atom fraction X_C^L of solute in the vapour
Zn	(111)B	$1 \times 10^{20} X_C^L$
Zn	(111)A	$7 \times 10^{20} (X_C^L)^{0.84}$
Se	(111)B	$3 \times 10^{23} (X_C^L)^{1.12}$
Se	(111)A	$8 \times 10^{20} (X_C^L)^{0.85}$

3.4.2 Time Dependent Effects

Two classes of time dependent phenomena are of particular interest:

(a) the effects of small (frequently cyclic) changes of conditions on two phases which are very nearly in equilibrium, and

(b) the result of bringing two phases which are not in equilibrium into contact with one another.

In the first of these two classes, the subject which has received the most attention is the effect of cyclic changes of growth rate on the solute concentration in a growing crystal. The variations in growth rate may be deliberately imposed or arise from the experimental conditions (see references 88-96). The most pronounced effects are a result of temperature fluctuations due either to exceeding the critical Reynolds number in forced stirring [72] or to coupling between the conductive and convective modes of heat transport in fluids heated from one side or from below [97-100]. The effects have mostly been studied in melt grown crystals.† (Rates of growth from the vapour and from solution are usually so low that unless the fluctuations are very slow, the spacing between the parts of the crystal grown under different conditions are too small to allow examination.) The easiest method of analysis [96] is to differentiate equation 3.98 to give

$$\frac{d\mathrm{K_{eff}}}{df} = \frac{\delta \mathrm{K}_{\mathrm{eff}}^2(1-\mathrm{K})\exp(-f\delta/D_L)}{\mathrm{K}D_L} \tag{3.105}$$

This can then be multiplied by the change in growth rate Δf to give the change in segregation coefficient. For melt growth a change ΔT in the temperature of the bulk melt gives a change in growth rate of

$$\Delta f = \frac{\kappa_L \Delta T}{\delta_T \rho L} \tag{3.106}$$

where κ_L is the thermal conductivity of the melt which has a density ρ and a latent heat of fusion of L (calories per gram). This analysis has been tested for non-semiconductor systems with great success [96, 98]. For semiconductors the experimental difficulties of measuring ΔT have prevented a complete test. However it is clear from, for example, the results of Ueda [92] and Muller and Wilhelm [93] that a similar analysis will also account for results in semiconductors.

In the case of binary semiconductors, the growth rate can also change as a result of fluctuations in partial pressure over the melt (see for example

† In this case, it has been shown that the temperature oscillations may be suppressed by: (1) applying a magnetic field to the melt, (2) rotating the melt, or (3) putting baffles in the melt.

reference 117, which also discusses some other causes of inhomogeneity in GaAs).

The second class of phenomena–bringing two phases which are not in equilibrium into contact–reflect the conditions usually used for device diffusions [101-107]. Figure 3.17 shows the model considered here. This is the model considered by Barry and Olofsen [108] for the diffusion of boron from a doped oxide,† through an initially undoped oxide into an

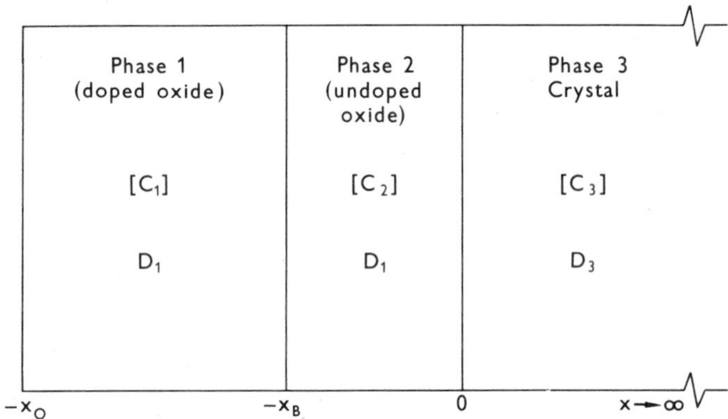

Figure 3.17. The model used by Barry and Olofsen [108].

initially undoped silicon crystal. The mathematics of this very general model can be readily made to yield answers for simpler systems. The one-dimensional equations required to describe the systems are:

$$\frac{\partial[C_1]}{\partial t} = D_1 \frac{\partial^2 [C_1]}{\partial x^2}, \quad -x_o < x < -x_B, \quad t > 0 \tag{3.107}$$

$$\frac{\partial[C_3]}{\partial t} = D_3 \frac{\partial^2 [C_3]}{\partial x^2}, \quad x > 0, \quad t > 0 \tag{3.108}$$

$$\frac{\partial[C_2]}{\partial t} = D_1 \frac{\partial^2 [C_2]}{\partial x^2}, \quad -x_B < x < 0, \quad t > 0 \tag{3.109}$$

With the boundary conditions:

$$[C_1] = [C_0], \quad [C_2] = [C_3] = 0 \text{ at } t = 0 \tag{3.110}$$

$$[C_1] = 0 \text{ at } x = -x_o \text{ for } t > 0 \tag{3.111}$$

(This is an assumption, but since in practice $-x_o \to -\infty$ it is not an important one.)

† H. J. Schnabel and F. Fleischer (*Phys. Stat. Sol.*, **8**, 71 (1971)) have recently shown that solid films of boron nitride can be used in the same way.

$$[C_3] \to 0 \text{ as } x \to \infty \tag{3.112}$$

$$[C_2] = [C_1] \text{ at } x = -x_B \text{ for } t > 0 \tag{3.113}$$

$$\frac{\partial [C_1]}{\partial x} = \frac{\partial [C_2]}{\partial x} \text{ at } x = -x_B \text{ for } t > 0 \tag{3.114}$$

These last two conditions imply equilibrium at the boundary between phases 1 and 2.

$$[C_3] = \mathbf{K}_{23} [C_2] \text{ at } x = 0 \text{ for } t > 0 \tag{3.115}$$

$$D_1 \frac{\partial [C_2]}{\partial x} = D_3 \frac{\partial [C_3]}{\partial x} \text{ at } x = 0 \text{ for } t > 0 \tag{3.116}$$

The last two conditions imply equilibrium at the boundary between phases 2 and 3. With these conditions, equations 3.107 to 3.109 can be solved to give

$$[C_1] = [C_0] \left[1 - \sum_0^\infty \frac{\alpha^n}{2} \left(\text{erfc} \left[\frac{2nx_o - x - x_B}{2\sqrt{D_1 t}} \right] \right.\right.$$
$$+ \alpha\, \text{erfc} \left[\frac{2nx_o - x + x_B}{2\sqrt{D_1 t}} \right] + 2\, \text{erfc} \left[\frac{(2n+1)x_o + x}{2\sqrt{D_1 t}} \right] - 2\alpha\, \text{erfc} \left[\frac{(2n+1)x_o - x}{2\sqrt{D_1 t}} \right]$$
$$\left.\left. - \text{erfc} \left[\frac{(2n+2)x_o + x - x_B}{2\sqrt{D_1 t}} \right] - \alpha\, \text{erfc} \left[\frac{(2n+2)x_o + x + x_B}{2\sqrt{D_1 t}} \right] \right) \right] \tag{3.117}$$

$$[C_2] = \frac{[C_0]}{2} \left[\sum_0^\infty \alpha^n \left(\text{erfc} \left[\frac{2nx_o + x_B + x}{2\sqrt{D_1 t}} \right] \right.\right.$$
$$- \alpha\, \text{erfc} \left[\frac{2nx_o + x_B - x}{2\sqrt{D_1 t}} \right] - 2\, \text{erfc} \left[\frac{(2n+1)x_o + x}{2\sqrt{D_1 t}} \right] + 2\alpha\, \text{erfc} \left[\frac{(2n+1)x_o - x}{2\sqrt{D_1 t}} \right]$$
$$\left.\left. + \text{erfc} \left[\frac{(2n+2)x_o - x_B + x}{2\sqrt{D_1 t}} \right] - \alpha\, \text{erfc} \left[\frac{(2n+2)x_o - x_B - x}{2\sqrt{D_1 t}} \right] \right) \right] \tag{3.118}$$

$$[C_3] = \frac{[C_0]\sqrt{D_1/D_3}}{(1+m)} \sum_0^\infty \alpha^n \left(\text{erfc} \left[\frac{2nx_o + x_B + m\mathbf{K}_{23}x}{2\sqrt{D_3 t}} \right] \right.$$
$$\left. - 2\, \text{erfc} \left[\frac{(2n+1)x_o + m\mathbf{K}_{23}x}{2\sqrt{D_3 t}} \right] + \text{erfc} \left[\frac{(2n+2)x_o - x_B + m\mathbf{K}_{23}x}{2\sqrt{D_3 t}} \right] \right) \tag{3.119}$$

where

$$m = \frac{1}{\mathbf{K}_{23}} \sqrt{D_1/D_3} \tag{3.120}$$

and

$$\alpha = \frac{1-m}{1+m} \tag{3.121}$$

If phase 1 has a large volume, i.e.

$$x_o - x_B > 4\sqrt{D_1 t} \tag{3.122}$$

then we need only consider terms in $n = 0$ and

$$[C_3] = \mathbf{K}_{23} \frac{[C_0]\sqrt{D_1/D_3}}{\mathbf{K}_{23} + \sqrt{D_1/D_3}} \operatorname{erfc}\left\{\frac{x_B + m\mathbf{K}_{23}x}{2\sqrt{D_3 t}}\right\} \tag{3.123}$$

For a case involving only two phases, we can put $x_B = 0$ and obtain a distribution with a surface concentration $[C_s]$ at $x = 0$ in the crystal given by

$$[C_s] = \frac{\mathbf{K}_{23}\sqrt{D_1/D_2}}{\mathbf{K}_{23} + \sqrt{D_1/D_2}} [C_o] \tag{3.124}$$

Figure 3.18 demonstrates the validity of this relation (see also reference 101). Note that if $D_1 \gg D_2$

$$[C_s] = \mathbf{K}_{23} [C_o] \tag{3.125}$$

This will in general be the situation if phase 1 is a gas.

A similar analysis [109, 110] can be used to consider a semiconductor with an initial concentration $[C_0]$ heated in a gas ambient with $[C_1] = 0$. The amount of solute lost per unit surface area in a time t is then

$$Q = 2[C_0]\left(\frac{D_3 t}{\pi}\right)^{1/2} \tag{3.126}$$

and the distribution in the solid is given by

$$[C_3] = [C_0] \operatorname{erf}\left\{\frac{x}{2\sqrt{D_3 t}}\right\}$$

If the crystal also evaporates at a linear rate v, the distribution in the solid becomes

$$[C_3] = C_0 - \frac{[C_0]}{2}\left[\operatorname{erfc}\left\{\left(\frac{(x+vt)^2}{4Dt}\right)^{1/2}\right\} + \exp\left(\frac{-vx}{D}\right)\operatorname{erfc}\left\{\left(\frac{(x-vt)^2}{4Dt}\right)^{1/2}\right\}\right] \tag{3.127}$$

These relations have been exhaustively tested by Dawson *et al.* [109, 110]. Note, however, that the derivation assumes that $[C_3] = 0$ at $x = 0$.

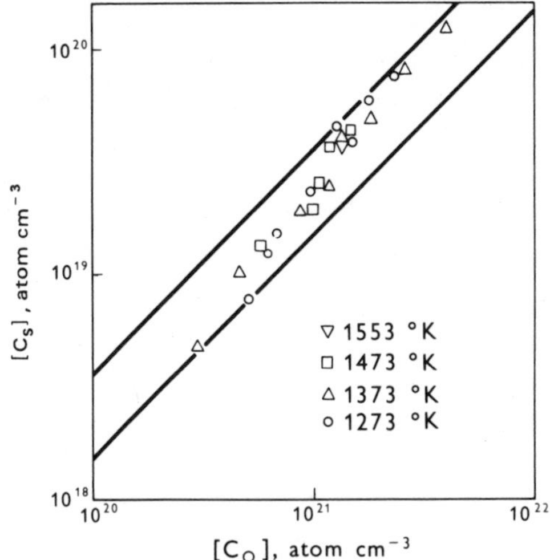

Figure 3.18. Some results obtained by Barry and Olofsen [108]. If equation 3.124 gave a completely accurate description of the processes, the results for each temperature should lie on a line parallel to the pair of lines drawn on the figure. These lines are separated by a distance equivalent to a factor of 2.4. The systematic divergence of the results suggests some concentration dependence –possibly because the $[C_S]$ values are near to the values of n_i (c.f. section 3.2.2–The Use of the Law of Mass Action and see references 111 and 112 for a discussion of the effects on n_i on diffusion coefficients).

The work reviewed by Fuller [64] suggests that in some systems, surface reactions occur, in which case $[C_3]$ will not be zero. Some cases involving surface reactions have been discussed by Jepsen and Somerjai [118].

In general, the rate of evaporation of the crystal can be calculated from its equilibrium vapour pressure. However, for some compound materials (e.g. CdS), it has been shown that the stoichiometry of the crystals is important and that the initial rate of evaporation can be much lower than the rate expected [119, 120]. In time, however, the excess constituent diffuses out and normal evaporation occurs. The effect may be modified by charge transfer, so that illumination of the sample may cause the rate of evaporation to change [122, 123].

In cases where the concentration of vacancies in the solid is not appropriate to the existing conditions, there will be a flux of vacancies to or from the surface. This flux can have a pronounced effect on the apparent diffusion coefficient. Hanneman and Anthony [121] give an analysis of the various cases which may occur.

3.5 CONCLUSIONS

This chapter has given a brief account of a very complex subject. Attention has been concentrated on the mathematical ways of describing the processes which occur, since these allow us to extrapolate available data to predict whether changes in conditions will produce useful results and to check whether available results are self-consistent. Attempts to do this are hindered by the lack of independent data to put in the various equations.

In the case of equilibria between major components it has been shown that even relatively complex phase diagrams can be described in terms of only a very few constants (melting points, entropies of fusion and heats of mixing). These constants are not available for many systems at present and attempts to fix the values from actual phase diagrams do not invariably get the same answer as a direct determination. The prediction of vacancy concentrations is more difficult: the effect of changing a partial pressure at a fixed temperature is readily predicted but the activation energies can really only be determined experimentally, which is not necessarily simple.

The analysis involved in predicting solute concentrations is in general simple and can readily be extended to cover the effects of several solutes being present, even if there are interactions. In particular the idea of a nearly constant segregation coefficient appears to be widely applicable.

In the case of a moving interface either in growth or evaporation, fairly simple transport theory with a velocity independent segregation coefficient is sufficient to describe the available data quite accurately. The orientation dependence of segregation coefficients can be understood qualitatively but at the moment quantitative predictions of effects are not possible.

Various time dependent effects can be described in terms of fairly simple mathematics and some of the effects giving rise to inhomogeneous solute distributions are reasonably well understood.

It seems therefore, that while further theoretical work might result in a quantitative understanding where at present we can only make qualitative predictions, the main need at present is for numbers to put into equations rather than even greater numbers of equations.

ACKNOWLEDGEMENTS

The Author is grateful for much data and pertinent comment provided by Dr. J. Burmeister, Mr. G. Frank, Mr. M. J. King, Mr. E. J. Millett and Mr. P. W. Whipps.

REFERENCES

1. P. Gordon, Principles of Phase Diagrams in Materials Systems. (McGraw-Hill, New York, 1968)
2. F. A. Kroger, Chemistry of Imperfect Crystals. (North Holland, Amsterdam, 1964)

3. H. E. Buckley, Crystal Growth. (Chapman and Hall, London, 1951)
4. J. J. Gilman, The Art and Science of Growing Crystals. (Wiley New York, 1963)
5. M. Tanenbaum, Semiconductors (Edited by N. B. Hannay). (Reinhold, New York, 1959)
6. W. D. Lawson and S. Nielsen, The Preparation and Purification of Semiconductor Crystals. (Butterworth, London, 1959)
7. J. C. Brice, The Growth of Crystals from the Melt. (North Holland, Amsterdam, 1965)
8. M. Zief and W. R. Wilcox, Fractional Solidification. (Arnold, London, 1967)
9. H. Schäfer, Chemical Transport Reactions. (Academic Press, New York, 1964)
10. M. H. Francombe and H. Sato, Single Crystal Films. (Pergamon, Oxford, 1964)
11. J. P. Hirth and G. M. Pound, Condensation and Evaporation. (Pergamon, Oxford, 1963)
12. S. Goldstein, Modern Development in Fluid Dynamics, Vols. 1 and 2. (Clarendon, Oxford, 1938)
13. J. G. Knudson and D. L. Katz, Fluid Dynamics and Heat Transfer. (McGraw-Hill, New York, 1958)
14. S. Chandrasekhar, Hydrodynamic and Hydromagnetic Stability. (Clarendon, Oxford, 1961)
15. M. Hansen, Constitution of Binary Alloys. (McGraw-Hill, New York, 1958)
16a. R. P. Elliott, Constitution of Binary Alloys, First Supplement. (McGraw-Hill, New York, 1965)
16b. F. A. Shunk, Constitution of Binary Alloys, Second Supplement. (McGraw-Hill, New York, 1969)
17. D. R. Stull and G. C. Sinke, Thermodynamic Properties of the Elements. (American Chemical Society, Washington, 1965)
18. National Bureau of Standard Circular 500, Selected Values of Chemical Thermodynamic Properties. (Washington, 1952)
19. N. A. Goryunova, The Chemistry of Diamond-like Semiconductors. (Chapman and Hall, London, 1965)
20. W. R. Runyan, Silicon Semiconductor Technology. (McGraw-Hill, New York, 1965)
21. J. Van Den Boomgard and K. Schol, *Philips Res. Repts.*, **12**, 127 (1957)
22. R. Goldfinger and J. Von Drowart, *J. Phys. Chem.*, **55**, 721 (1958)
23. V. J. Lyons and V. J. Sylvestri, *J. Phys. Chem.*, **65**, 1275 (1961)
24. D. Richman, R.C.A. Report No. 8 Oct.-Dec. 1961
25. C. D. Thurmond, *J. Phys. Chem. Solids,* **26**, 785 (1965)
26. W. F. Schottky and M. B. Bever, *Acta metall.*, **6**, 320 (1958)
27. J. C. Brice and M. J. King, *Nature,* **199**, 897 (1963)
28. P. C. Newman, J. C. Brice and H. C. Wright, *Philips Res. Repts.*, **16**, 41 (1961)
29. J. R. Dale, *Nature,* **197**, 242 (1963)
30. M. E. Straumanis and C. D. Kim, *Acta Cryst.*, **19**, 256 (1965)
31. J. C. Brice, *Solid-St. Electron.*, **10**, 335 (1967)
32. J. C. Brice and G. D. King, *Nature,* **209**, 1346 (1966)
33. J. C. Brice, *J. Cryst. Growth,* **7**, 9 (1970)
34. H. C. Casey, M. B. Panish and K. B. Wofstirn, *J. Phys. Chem. Solids,* **32**, 571 (1971)
35. A. G. Tweet, *J. Appl. Phys.*, **30**, 2002 (1959)
36. J. C. Brice and J. R. Dale, *Solid-St. Electron.*, **3**, 105 (1961)
37. C. Wagner, *Acta metall.*, **6**, 309 (1958)
38. L. J. Vieland, *Acta metall.*, **11**, 137 (1963)
39. M. B. Panish, *J. Phys. Chem. Solids,* **30**, 1083 (1969)
40. M. B. Panish, R. T. Lynch and S. Sumski, *Trans. Metall. Soc. A.I.M.E.*, **245**, 559 (1969)
41. A. G. Tweet, *Phys. Rev.*, **106**, 221 (1957)
42. B. D. Lichter and P. Sommelet, *Trans Metall. Soc. A.I.M.E.*, **245**, 1021 (1969)

43. D. L. K. Kendall and D. B. DeVries, Semiconductor Silicon. (Edited by R. R. Haberecht and E. L. Kern, Electrochemical Society, New York, 1969)
44. R. A. Swalin, *J. Phys. Chem. Solids,* **18**, 290 (1961)
45. J. Zernike, Chemical Phase Theory. (Kluwer, Antwerp, 1955)
46. R. N. Hall, *J. Electrochem. Soc.,* **110**, 385 (1963)
47. B. D. Lichter and P. Sommelet, *Trans. Metall. Soc. A.I.M.E.,* **245**, 99 (1969)
48a. M. Aven and J. S. Prener, Physics and Chemistry of II-VI Compounds. (North Holland, Amsterdam, 1967)
48b. W. Albers, and C. Haas, *Philips Tech. Rev.,* **30**, 82, 107 and 142 (1969)
49. F. A. Trumbore, *Bell System Tech. J.,* **39**, 205 (1960)
50. J. C. Brice, *Solid-St. Electron.,* **6**, 673 (1963)
51. E. J. Millett, unpublished (1964)
52. F. A. Trumbore, E. M. Porbansky and A. A. Tartaglia, *J. Phys. Chem. Solids,* **11**, 239 (1959)
53. S. Fischler, *J. Appl. Phys.,* **33**, 1615 (1962)
54. H. Statz, *J. Phys. Chem. Solids,* **24**, 699 (1963)
55. C. D. Thurmond and J. D. Struthers, *J. Phys. Chem.,* **57**, 831 (1953)
56a. C. D. Thurmond and M. Kowalchik, *Bell System Tech. J.,* **39**, 169 (1960)
56b. J. S. Berkes and W. B. White, *J. Cryst. Growth,* **6**, 29 (1969)
57. M. S. Saidov, *Fiz. Met. Metalloved,* **17**, 795 (1964)
58. U. Merten, K. D. Vos and A. P. Hatcher, *J. Phys. Chem. Solids,* **30**, 627 (1969)
59. V. J. Silvestri, *J. Electrochem. Soc.,* **116**, 81 (1969)
60. J. S. Harris and W. L. Snyder, *Solid-St. Electron.,* **12**, 337 (1969)
61. K. Lehovec, *J. Phys. Chem. Solids,* **23**, 695 (1962)
62. M. R. Lorenz and S. E. Blum, *J. Electrochem. Soc.,* **113**, 559 (1966)
63. L. M. Foster and J. Scardefield, *J. Eletrochem. Soc.,* **116**, 494 (1969)
64. C. S. Fuller, Semiconductors, Chapter 5 (Edited by N. B. Hannay). (Reinhold, New York, 1959)
65. G. Baralis, *J. Cryst. Growth,* **3, 4**, 627 (1968)
66. J. C. Brice and P. A. C. Whiffin, *Solid-St. Electron.,* **7**, 183 (1964)
67a. B. K. Jindal and W. A. Tiller, *J. Chem. Phys.,* **49**, 4632 (1968)
67b. J. C. Baker and J. W. Cahn, *Acta metall.,* **17**, 575 (1969)
68. W. R. Wilcox, *Mat. Res. Bull.,* **4**, 265 (1969)
69. J. A. Burton, R. C. Prim and W. P. Slichter, *J. Chem. Phys.,* **21**, 1987 (1953)
70. J. A. Burton, E. D. Kolb, W. P. Slichter and J. D. Struthers, *J. Chem. Phys.,* **21**, 1991 (1953)
71. R. C. Reid and T. K. Sherwood, The Properties of Gases and Liquids. (McGraw-Hill, New York, 1966)
72. J. C. Brice, *J. Cryst. Growth,* **2**, 395 (1968)
73. J. C. Brice, *J. Cryst. Growth,* **1**, 161 (1967)
74. J. W. Dunning, Chemistry of the Solid State, Chapter 6 (Edited by W. Garner). (Butterworth, London, 1955)
75. D. Turnbull, *Solid State Physics,* **3**, 226 (1956)
76. C. Herring, Metal Interfaces, Chapter 1. (American Society for Metals, 1952)
77. A. S. Shapski, *J. Chem. Phys.,* **16**, 389 (1948)
78. J. C. Brice, *J. Cryst. Growth,* **6**, 9 (1969)
79. J. C. Brice, *J. Cryst. Growth,* to be published
80. J. C. Brice, *J. Cryst. Growth,* **1**, 218 (1967)
81. K. A. Jackson, Liquid Metals and Solidification, p. 174. (American Society for Metals, 1958)
82. J. W. Faust and H. F. John, *J. Phys. Chem. Solids,* **23**, 1119 (1962)
83. A. Trainor and B. E. Bartlett, *Solid-St. Electron.,* **2**, 106 (1961)
84. R. N. Hall, *J. Phys. Chem.,* **57**, 836 (1953)
85. D. W. Shaw, *J. Electrochem. Soc.,* **115**, 405 (1968)
86. B. A. Joyce, *J. Cryst. Growth,* **3, 4**, 43 (1968)
87. R. C. Taylor, *J. Electrochem. Soc.,* **116**, 383 (1969)

88. J. A. M. Dikhoff, *Philips Tech. Rev.*, **25**, 195 (1964)
89. H. E. Bridgers, *J. Appl. Phys.*, **27**, 746 (1956)
90. H. E. Bridgers and E. D. Kolb, *J. Appl. Phys.*, **26**, 1188 (1955)
91. M. Tanenbaum, L. B. Valdes, E. Buehler and B. B. Hannay, *J. Appl. Phys.*, **26**, 686 (1955)
92. H. Ueda, *J. Phys. Soc. Japan*, **16**, 61 (1961)
93. A. Muller and M. Wilhelm, *Z. Naturf.*, **19a**, 257 (1964)
94. K. Morizane, A. F. Witt and H. C. Gatos, *J. Electrochem. Soc.*, **113**, 51 (1966)
95. A. F. Witt and H. C. Gatos, *J. Electrochem. Soc.*, **113**, 808 (1966)
96. J. C. Brice and P. A. C. Whiffin, *Brit. J. Appl. Phys.*, **18**, 581 (1967)
97. D. T. J. Hurle, *Phil. Mag.*, **13**, 305 (1966)
98. W. R. Wilcox and L. D. Fullmer, *J. Appl. Phys.*, **36**, 2201 (1965)
99. E. J. Harp and D. T. J. Hurle, *Phil. Mag.*, **17**, 1033 (1968)
100. J. R. Carruthers and J. Pavilouis, *J. Appl. Phys.*, **39**, 5814 (1968)
101. M. L. Barry and P. Olofsen, Solid State Technology, p. 39, October 1968
102. J. Scott and J. Olmstead, *R.C.A. Review*, **26**, 357 (1965)
103. C. S. Fuller and C. J. Frosch, Transistor Technology, Volume 3. (Van Nostrand, New York, 1958)
104. F. M. Smits, *Proc. I.R.E.*, **46**, 1049 (1958)
105. C. J. Frosch and L. Derick, *J. Electrochem. Soc.*, **104**, 547 (1957)
106. L. A. D'Asaro, *Solid-St. Electron.*, **1**, 3 (1960)
107. E. L. Jordan, *J. Electrochem. Soc.*, **108**, 478 (1961)
108. M. L. Barry and P. Olofsen, *J. Electrochem. Soc.*, **116**, 854 (1969)
109. D. K. Dawson and L. W. Barr, Proceedings of the British Ceramic Society No. 9, p. 171. July 1967
110. D. K. Dawson, L. W. Barr and R. A. Pitt-Pladdy, *Brit. J. Appl. Phys.*, **17**, 657 (1966)
111. F. M. Smits, *Proc. I.R.E.*, **46**, 1049 (1958)
112. A. D. Kurtz and R. Yee, *J. Appl. Phys.*, **31**, 303 (1960)
113. M. G. Mil'vidski and A. V. Berkova, *Soviet Phys. solid St.*, **5**, 517 (1963)
114. J. B. Mullin, Compound Semiconductors, Chapter 41 Vol. 1 (Edited by R. K. Willardson and H. L. Goering). (Reinhold, New York 1962)
115. F. V. Williams, *J. Electrochem. Soc.*, **111**, 886 (1964)
116. J. C. Brice, J. A. Roberts and G. Smith, *J. Mat. Sci.*, **2**, 131 (1967)
117. J. C. Brice, R. E. Hunt, G. D. King and H. C. Wright, *Solid-St. Electron.*, **9**, 853 (1966)
118. D. W. Jepsen and G. A. Somorjai, *J. Chem. Phys.*, **39**, 1665 (1963)
119. G. A. Somorjai and D. W. Jepsen, *J. Chem. Phys.*, **41**, 1389 (1964)
120. G. A. Somorjai and D. W. Jepsen, *J. Chem. Phys.*, **41**, 1394 (1964)
121. R. E. Hanneman and T. R. Anthony, *Acta metall.*, **17**, 1133 (1969)
122. G. A. Somorjai and J. E Lester, *J. Chem. Phys.*, **43**, 1450 (1965)
123. G. A. Somorjai and H. B. Lyon, *J. Chem. Phys.*, **43**, 1456 (1965)
124. J. C. Brice, *J. Crystal Growth*, **10**, 205 (1971)
125. M. Krumnacker and W. Lange, *Krist. Tech.*, **4**, 207 (1969)
126. R. M. Logan and D. T. J. Hurle, private communication

EXPERIMENTAL METHODS FOR DETERMINING DIFFUSION COEFFICIENTS IN SEMICONDUCTORS 4

T. H. YEH

CONTENTS

4.1 INTRODUCTION

Many different experimental methods have been developed in recent years to study atomic diffusion phenomena in semiconductors. All were developed for the purpose of either directly or indirectly establishing the distribution of the diffusing species in the semiconductors so that certain aspects of the diffusion phenomena could be properly understood and some theoretical predictions verified. Both direct and indirect methods of establishing the distribution of the diffusing species are numerous. Direct methods include radiochemical, chemical, ion mass spectroscopy, atomic absorption, X-ray, etc. The indirect methods involve the measurement of certain physical properties of the semiconductors that result from the diffusions, such as their electrical conductivity, resistivity, or their optical reflectivity in the infrared range, etc.

In this chapter, a comprehensive assessment will be made of the various methods of obtaining the diffusion coefficients in germanium and silicon, in particular, and also in compound semiconductors. Special attention will be paid to precisely what each method is capable of measuring and to comparing its advantages and limitations.

Three types of diffusion can be recognized experimentally: self, isoconcentration and chemical. The first two refer to diffusion of host and impurity atoms respectively in chemically homogeneous crystals. Experimentally such diffusions can only be followed using an isotope, stable or radioactive, differing from that in the crystal matrix. In isoconcentration diffusion the crystal is uniformly doped with the impurity under study and the diffusion of a different isotope of the impurity is then followed. There is no gradient of chemical potential in self and isoconcentration diffusion, only an isotopic gradient. Chemical diffusion can refer to both host and impurity atom diffusions. For the purpose of this chapter there is no need to distinguish further between self, isoconcentration and chemical diffusion beyond noting that only isotopic methods are available for the determination of self and isoconcentration diffusion coefficients. The symbol **D** will be used to denote the volume diffusion coefficient irrespective of species and of whether an isotopic or chemical gradient is present. Diffusion geometries will be limited to planar volume diffusion into a semi-infinite crystal.

Many of the experimentally determined impurity diffusion coefficients have been obtained by the *p-n* junction technique and surface concentration measurements. The former measures the depth of penetration of the diffused impurity, and the latter determines how much impurity is at the surface of the semiconductor after diffusion for a given time and temperature. The distribution of the diffusing impurity in the semiconductor is usually established after the junction depth and surface concentration are measured, by assuming that it obeys the complementary error function (erfc) or Gaussian distribution, depending upon which boundary conditions are ascribed to the partial differential equation

$$\frac{\partial c}{\partial t} = \mathbf{D}\frac{\partial^2 c}{\partial x^2}.$$

In this chapter two sections are devoted to discussing these two experimentally determined quantities.

On the other hand, several techniques have also been developed to establish a detailed impurity concentration profile in the semiconductor, such as differential resistivity, spreading resistance, and radiochemical. In many cases, these techniques found that the impurity profile is not necessarily an error-function complement or Gaussian distribution. In such situations the diffusion coefficient may then depend on the depth of diffusion. One section is devoted entirely to these techniques and the diffusion coefficients obtained from them. A discussion of the diode capacitance technique is also included. Separate sections are devoted to methods used for compound semiconductors, to the evaluation of experimental diffusion data, and to the importance of surface preparation of diffusion samples.

4.2 METHODS USED FOR GERMANIUM AND SILICON

4.2.1 *p-n* Junction Techniques

The *p-n* junction method is perhaps the most popular among all the experimental methods used to determine the diffusion coefficients of various impurities in semiconductors. It simply requires the diffusion of an impurity of the opposite type into the semiconductor in question to a depth where the semiconductor retains its original conductivity. The depth can be measured by various techniques described in detail below. Once the junction depth is determined, the diffusion coefficient of a particular impurity at a given temperature and time can be calculated from equation 4.1 for the experimental condition of constant supply of diffusion impurity source at the surface of the diffused semiconductor:

$$c(x_j) = c_o \operatorname{erfc} \frac{x_j}{2\sqrt{\mathrm{D}t}}, \tag{4.1}$$

where

$c(x_j)$ is the number of atoms of diffusant per cm^3 at the *p-n* junction located x_j cm from the surface,

c_o is the number of atoms of diffusant per cm^3 at the surface of the semiconductor,

$$\operatorname{erfc} \xi = 1 - \operatorname{erf} \xi = 1 - \frac{2}{\sqrt{\pi}} \int_0^{\xi} \exp(-y^2)dy$$

D is the diffusion coefficient in $\mathrm{cm}^2\ \mathrm{sec}^{-1}$, and

t is the time in seconds.

The value of $c(x_j)$ is known and should correspond to the impurity concentration of the original material. c_o usually is an experimentally determined value and will be discussed in detail in the following section.

When the supply of diffusing impurity at the surface of the diffused semiconductor is limited (for example, the diffusing impurity source is a thin layer of substance obtained by chemical deposition, evaporation, or other methods), the diffusion coefficient is calculated from:

$$c(x) = \frac{W}{\sqrt{\pi \mathbf{D}t}} \exp\left(-\frac{x}{2\sqrt{\mathbf{D}t}}\right)^2 \tag{4.2}$$

where $c(x)$ is the concentration of the diffusing impurity and is related to the total amount of impurity atoms in the sample, W, by the relation:

$$W = \int_0^\infty c(x)dx. \tag{4.3}$$

If the p-n junction depth x_j is accurately measured after the diffusion and the original impurity concentration is $c(x_j)$, then the diffusion coefficient can be calculated simply by the equation:

$$c(x_j) = \frac{W}{\sqrt{\pi \mathbf{D}t}} \exp\left(-\frac{x_j}{2\sqrt{\mathbf{D}t}}\right)^2. \tag{4.4}$$

Fuller's two-specimen method [1] consists of determining the p-n junction depths x_1 and x_2 in the specimens having resistivities of ρ_1 and ρ_2 respectively, after they are diffused together under identical conditions. Since the surface concentrations for the two specimens are the same, the following equation is true:

$$\frac{\rho_1\mu_1}{\rho_2\mu_2} = \frac{1-\operatorname{erf}\left(\frac{x_1}{2\sqrt{\mathbf{D}t}}\right)}{1-\operatorname{erf}\left(\frac{x_2}{2\sqrt{\mathbf{D}t}}\right)}, \tag{4.5}$$

where μ_1 and μ_2 are the drift mobilities [2]. Equation 4.5 is solved for $\mathbf{D}$ by the method of approximation, i.e., by

$$\mathbf{D} = \frac{1}{4t} \frac{(x_2^2 - x_2^2)}{\ln \frac{\rho_1\mu_1 x_1}{\rho_2\mu_2 x_2}}. \tag{4.6}$$

Several methods are available for measuring the p-n junction depth in a diffused sample. Two established ones are chemical staining and electrochemical displacement plating. A third and relatively new method, with definite advantages over staining or plating, is that of the scanning light spot. Attention will be limited to these three procedures.

Chemical Staining

Ever since Fuller and Ditzenberger [3] developed a staining procedure for measuring the diffused junction in silicon, it has been widely accepted by research workers for determining the depths of diffusion in semiconductors. The original staining procedure consisted of using a chemical solution of 0.1 to 0.5% by volume of concentrated (70%) HNO_3 in concentrated (48%) HF on an angle-lapped surface of the sample. If conditions are right, the *p*-region of the lapped surface turns dark, sharply defining the *p-n* junctions. By means of a traveling microscope fitted with a vertical illuminator, the width of the *p* or *n* diffusion layer can be accurately measured. The other important parameter to be measured is the exact angle of the lapped sample through the diffused layer.

The mechanism of the staining reaction, however, was not fully understood by Fuller and Ditzenburger [3]. They only stated that the formation of the dark stain does not involve appreciable etching and appears to be the result of the formation of SiO on the *p*-type regions by preferential oxidation. Subsequent studies indicated that the stained film might be silicon hydride [4] or H_2SiF_6 [5, 6]. Because of the uncertainty of the staining mechanism and the difficulties of delineating certain *p-n* junctions, many different etchants other than HF-HNO_3 combinations were also developed, such as replacing HNO_3 by sodium nitrite, hydrogen peroxide, or phosphoric acid in HF solution. All these new staining solutions were tried in an attempt to improve the contrast at the *p-n* junctions.

Electrochemical Displacement Plating

Another frequently used method for delineating the *p-n* junction is electrochemical displacement plating [7-9]. It involves immersing the sample in a metal salt solution such as $CuSO_4$, $AgNO_3$, or $AuCl_3$, with HF or hot alkalis, and observing the plating until the *p-n* junction is well delineated. The differences in plating presumably arise from the differences in electrode potential of the semiconductor regions at either side of the junction. Usually the plating occurs first on the *n*-side of the junction. The displacement of those metallic ions such as copper, silver, and gold by silicon occurs simply because the electromotive force of silicon is higher than that of metallic ions. Because the plating is literally by displacement, for each milligram-equivalent of metal depositing on silicon at cathode sites, 1 mg-equivalent of silicon dissolves at anodic sites. The total anodic current, therefore, equals the total cathode current. The rate of metal deposition by displacement can be controlled by regulating the rate of the anodic reaction.

Silicon dissolves anodically by first forming an oxide. Unless the electrolyte is capable of dissolving the oxide, the anode reaction will stop after the oxide becomes a few atomic layers thick. Displacement plating also will stop under these conditions. The hydrofluoric acid or hot alkalis in the metal salt solutions mentioned above are used to dissolve the oxide. Turner [7] reported a method of obtaining very sharp *p-n* junction delineation on silicon in each of the five different aqueous solutions containing simple metal salts such as Sb, Cu, Ag, Pt, and Au, plus different amounts of HF, as shown in Table 4.1. He was able to deposit the metals

TABLE 4.1

Solutions for junction delineation of Si (Turner [7])
Reprinted by permission of the Electrochemical Society

Metal deposited	Metal salt	g/l	cc (48%) HF/l
Sb	SbF_3	400	500
Cu	$CuSO_4 \cdot 5H_2O$	200	10
Ag	$AgNO_3$	40	4
Pt	$H_2PtCl_3 \cdot 6H_2O$	3	2
Au	$AuCl_3$	10	0.3

on silicon by displacement from the plating solutions because hydrofluoric acid dissolves SiO_2 by forming a soluble silicofluoride complex:

$$SiO_2 + 6HF \rightarrow H_2SiF_6 + 2H_2O. \tag{4.7}$$

The rate of silicon dissolution at anodic sites, and therefore the metal deposition rate at cathode sites, in acid fluoride displacement plating solutions can be regulated by controlling the rate of mass trasfer of HF from the solution to the anode sites. This is done easily by adjusting the HF concentration in the plating solution. The amount of HF used in each solution (Table 4.1) is believed to be optimum for the sharpest junction delineation. The best resolution of *p-n* junctions was obtained with Cu and Ag. Similarly, Silverman and Benn [8] were able to deposit Au on silicon by displacement from a gold cyanide plating solution, 10 g/litre $KAu(CN)_2$, because a hot, strong alkaline solution, 200 g/litre KOH, dissolves silicon dioxide and forms a soluble silicate complex:

$$2KAu(CN)_2 + 4KOH + Si \rightarrow 2Au + K_2SiO_3 + 4KCN + H_2 + H_2O. \tag{4.8}$$

Although the electrochemical action causes Au to plate on a silicon surface according to equation 4.8, it is only in the presence of light that one can obtain consistently predictable results in which the Au plates first on the more negative side of a junction. A brief explanation of the possible mechanism responsible for this phenomenon is discussed below.

If one considers the simplified energy diagram of a typical *p-n* junction, one can note the potential difference ΔV associated with the difference in carrier concentration on either side of the junction as indicated in figure 4.1. Neglecting possible complications due to surface states, one can, by the following equation, represent ΔV as a function of p and n; where p is the density of excess holes in the *p*-region, n is the density of

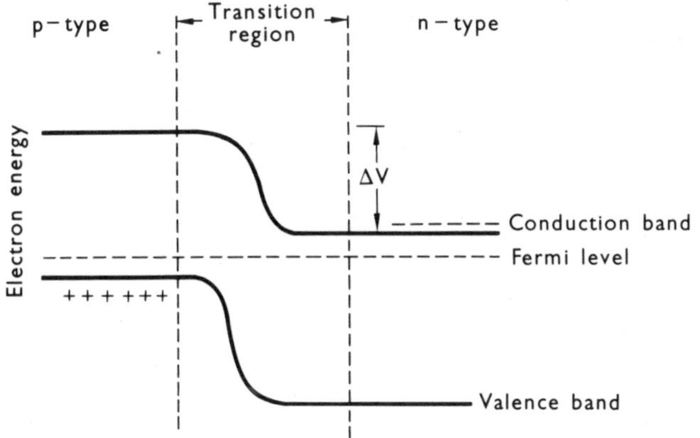

Figure 4.1. Potential distribution in a typical *p/n* junction.

excess electrons in the *n*-region, and n_i represents the density of intrinsic carriers:

$$\Delta V = V_n - V_p = \frac{kT}{e} \ln \frac{pn}{n_i^2}. \tag{4.9}$$

This equation then represents conditions in the vicinity of the junction which satisfy requirements of thermal equilibrium and charge neutrality throughout the specimen.

If one shines light on the junction, hole-electron pairs are created. The minority carriers in either region diffuse across the transition region under the influence of the built-in field created by ΔV. A photovoltage is then created on the surface near the junction. Positive Au ions resulting from dissociation of the $Au(CN)_2^-$ complex in the solution above the surface are attracted to the more negative region and plate on the *n*-side of the junction.

Whoriskey [5] illuminated a *p-n* junction in a dilute copper salt solution containing hydrofluoric acid and also plated copper on *n*-type silicon. Iles and Coppen [9] used silver fluoride for both diluted and stronger solutions. For diluted solutions, illumination was necessary and plating occurred again on *n*-type silicon. However, for stronger solutions, but far from saturation, the plating took place in the dark on the lower conductivity silicon regardless of conductivity type.

Although both chemical staining and electrochemical displacement plating are used extensively by research workers for locating the depths of diffused impurities in semiconductors, their reproducibility and accuracy depend on many considerations including

- The surface finish of the beveled area.
- The magnitude of the conductivity differences between the two layers.
- The optimum time for leaving the staining solution on the beveled surface or for leaving specimen in the plating solution to obtain the sharpest delineation of the junction.

Because of these difficulties related to the staining and electrochemical plating methods, a more recently developed scanning light spot method [10], which uses the photo-induced voltage at a junction for measuring the junction depth, deserves serious attention and is discussed below.

Scanning Light Spot

The photo-induced voltage at a junction as a means for measuring the junction depth [11-14] had not been developed sufficiently to use with confidence until recently. Tong, Schumann, and Dupnock [10], have perfected this technique for measuring *p-n* junction depths ranging from 0.5 to 10 μm in silicon. The method can locate the junction depth with a reproducibility to within ±2% of the mean depth. For this measurement, samples are beveled at an angle θ, usually one degree for a junction depth greater than 1 μm, and 30 min for junction depths less than 1 μm. Contacts are made either by thermal bonding or pressure contacts. Figure 4.2 shows a schematic setup. The open circuit voltage (or short circuit current) is fed first into a wide-band pre-amplifier with an amplification factor of 100. The output of the pre-amplifier is fed into a lock-in amplifier locked at the frequency of the light chopper. The light spot size can be adjusted to 10, 5, or 2.5 μm in diameter and is chopped at

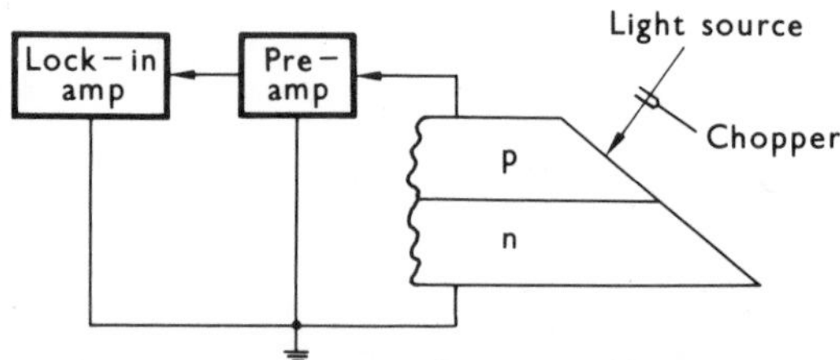

Figure 4.2. Experimental setup for the scanning light spot technique. The light spot is before the junction. (After Tong *et al.* [10].)

1000 Hz with a tuning fork. The sample is moved by a micrometer stage in 2.54 μm steps.

As shown by Tong *et al.* [10], when the light spot is above the junction the photo-induced short circuit current is given by the following equation:

$$I_n = \frac{2eG}{\left[\exp\left(\frac{l}{L_n}\right) + \exp\left(-\frac{l}{L_n}\right)\right] + \frac{L_n}{D_n}S_n\left[\exp\left(\frac{l}{L_n}\right) - \exp\left(-\frac{l}{L_n}\right)\right]} \tag{4.10}$$

where e = electron charge,

G = optical generation rate of free carriers,

L_n = minority carrier diffusion length (electrons),

D_n = diffusion constant of minority carrier (electrons),

S_n = surface recombination velocity (electrons), and

l = distance of the light spot from the junction.

When

$$\frac{l}{L_n} > 3/2, \exp\left(\frac{l}{L_n}\right) \gg \exp\left(-\frac{l}{L_n}\right),$$

equation 4.10 can be simplified to

$$I_n = \frac{2eG}{1 + \frac{L_n}{D_n}S_n} \exp\left(-\frac{l}{L_n}\right). \tag{4.11}$$

When $\ln I_n$ is plotted against l for equation 4.11, the slope gives L_n as shown in region 'a' of figure 4.3.

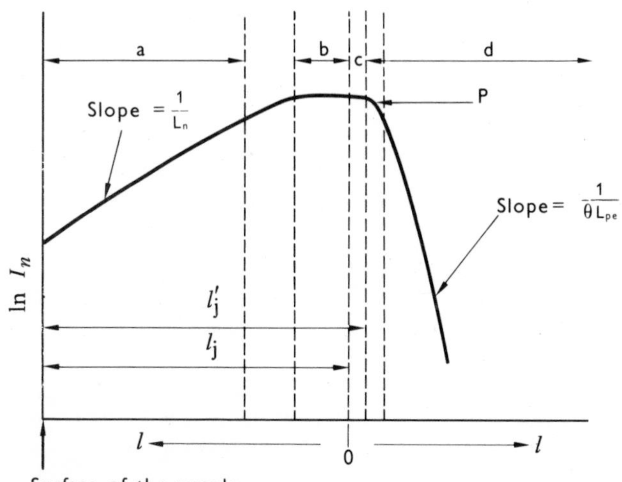

Figure 4.3. ln I_n of photo-response versus distance. (After Tong *et al.* [10].)

When

$$\frac{l}{L_n} \ll 1, \exp\left(\frac{l}{L_n}\right) \simeq 1 + \frac{l}{L_n}$$

and

$$\exp\left(-\frac{l}{L_n}\right) \simeq 1 - \frac{l}{L_n},$$

then equation 4.11 becomes

$$I_n = \frac{eG}{1 + \dfrac{l}{D_n} S_n}. \tag{4.12}$$

For $S_n \simeq 5 \times 10^3$ cm-sec^{-1} [15], $D_n \simeq 25$ cm^2 sec^{-1},†

and

$$l \simeq 1 \mu m = 1 \times 10^{-4} \text{ cm},$$

then

$$\frac{l}{D_n} S_n = 0.02 \ll 1.$$

Equation 4.12 then becomes $I_n = eG =$ constant, and is represented as region 'b' in figure 4.3.

When $l = 0$ (precisely at the junction), equation 4.12 also becomes $I_n = eG =$ constant. Therefore one knows that the true junction depth should be at the horizontal straight line portion of the curve when ln I_n is plotted against l.

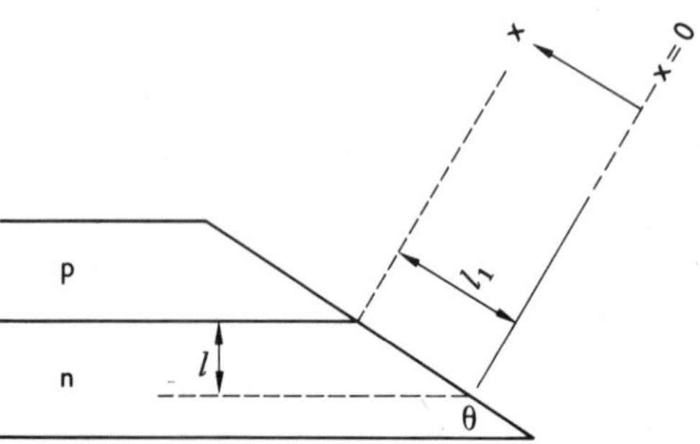

Figure 4.4. Schematic diagram of the sample when the light spot is below the junction. (After Tong *et al.* [10].)

When the light spot is below the junction, as shown in figure 4.4, the minority carriers generated do not diffuse vertically toward the junction, since it is no longer the shortest path. Instead, the minority carriers would

† This is calculated from Einstein's relationship $D_n = (kT/e)\mu_n$ for $\mu_n = 1000$ cm^2 V^{-1} sec^{-1}.

have an 'effective diffusion length' L_{pe} and would diffuse along the beveled surface toward the junction.

Following a similar mathematical analysis, the photo-induced current is

$$I_p = \frac{2eG}{\left[\exp\left(\frac{l_1}{L_{pe}}\right) + \exp\left(-\frac{l_1}{L_{pe}}\right)\right] + \frac{L_{pe}}{D_p} S_p \left[\exp\left(\frac{l_1}{L_{pe}}\right) - \exp\left(-\frac{l_1}{L_{pe}}\right)\right]} \tag{4.13}$$

where D_p = diffusion constant of minority carrier (holes) and
S_p = surface recombination velocity of holes.

When the bevel angle θ is small, $\sin\theta \simeq \theta$, and $l_1 = l/\theta$, equation 4.13 can be written as

$$I_p = \frac{2eG}{\left[\exp\left(\frac{l}{\theta L_{pe}}\right) + \exp\left(-\frac{l}{\theta L_{pe}}\right)\right] + \frac{L_{pe}}{D_p} S_p \left[\exp\left(\frac{l}{\theta L_{pe}}\right) - \exp\left(-\frac{l}{\theta L_{pe}}\right)\right]} \tag{4.14}$$

For

$$\frac{l}{\theta L_{pe}} > \frac{3}{2},\ \exp\left(\frac{l}{\theta L_{pe}}\right) \gg \exp\left(-\frac{l}{\theta L_{pe}}\right),$$

and equation 4.14 becomes

$$I_p = \frac{2eG}{1 + \frac{L_{pe}}{D_p} S_p} \exp\left(-\frac{l}{\theta L_{pe}}\right). \tag{4.15}$$

This is represented as region 'd' in figure 4.3.

For

$$\frac{l}{\theta L_{pe}} \ll 1$$

$$\exp\left(\frac{l}{\theta L_{pe}}\right) \simeq 1 + \frac{l}{\theta L_{pe}}$$

$$\exp\left(-\frac{l}{\theta L_{pe}}\right) \simeq 1 - \frac{l}{\theta L_{pe}}$$

equation 4.15 becomes

$$I_p = \frac{eG}{1 + \frac{l}{\theta D_p} S_p}. \tag{4.16}$$

Since $l/\theta \ll L_{pe}$, and assuming that $l/\theta \simeq 1$ μm and D_p and S_p have the same order of magnitude as D_n and S_n, then

$$\frac{l}{\theta D_p} S_p \ll 1$$

and equation 4.16 can be expressed as

$$I_p = eG = \text{constant} \tag{4.17}$$

and is represented as region 'c' in figure 4.3.

One can now estimate how big a region 'c' is in figure 4.3, since $l/\theta \sim 1$ μm, or $l \sim \theta$ μm, and if the sample is beveled at an angle of 30 min or $\theta = 0.0098$ radian, then region 'c' is approximately 0.0098 μm, which is negligible compared with the junction depths one intends to measure. This in turn makes it possible to assert that l_j' in figure 4.3 is not much different from l_j. Usually l_j' can be determined easily from figure 4.3; it corresponds to the point 'P' where the photo-response curve starts to fall off rapidly.

The minority carriers generated above the junction differ from those generated below the junction and cause the asymmetry of the photo-response curve. The minority carriers generated above the junction are diffused through the bulk of the material toward the junction, and those carriers below the junction are diffused through the surface. The asymmetry is characterized by the two different slopes of the photo-response curve as shown in figure 4.3: $-(1/L_n)$ in region 'a' above the junction, and $-(1/\theta L_{pe})$ in region 'd' below the junction. If one assumes a relatively constant L_{pe}, the slope in region 'd' is then inversely proportional to the bevel angle θ. Therefore, a small θ gives a steep fall-off and a more precise location of the junction depth. To substantiate the assumption that the fall-off of the photo-response curve is inversely proportional to the bevel angle θ, one sample was beveled at three different angles. The photo-responses are shown in figure 4.5. The values (slope x θ) listed in Table 4.2 show that the slope is a constant, indicating the validity of the relationship: the slope is proportional to $1/\theta$.

The mathematical analysis for this scanning light spot technique assumes that light is absorbed at the surface of the silicon. This is true when $e^{-\alpha x} \ll 1$, or that $\alpha x > 3$, where α is the absorption coefficient. For a light spot with wavelength of 0.4 μm, α is approximately 2×10^4 cm^{-1}. Therefore, $\alpha x > 3$ is true for $x > 1.5 \times 10^{-4}$ cm. In other words, the mathematical treatment holds true only for distances greater than 1.5 μm from the junction. But this is the case only when the light spot is descending from the surface toward the junction and is in the same direction as the minority carrier diffusion process. The situation is different, however, when the light spot is below the junction. The

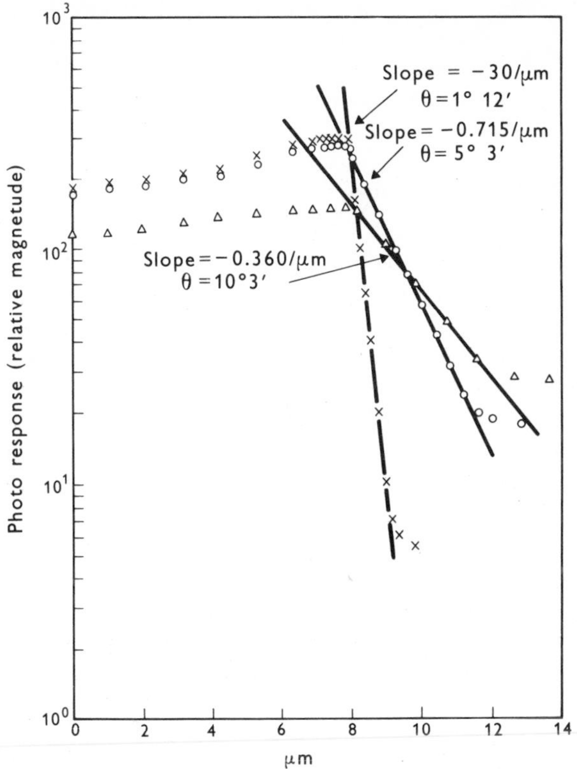

Figure 4.5. Photo-response with different bevel angles. (After Tong *et al.* [10].)

minority carrier diffusion process is on the surface toward the junction while any penetration of the light is into the material away from the junction. The fact that the junction depth is determined by the fall-off of the photo-response curve which is below the junction implies that the condition $e^{-\alpha x} \ll 1$ does not exist. Therefore, a junction depth thus determined is independent of α or the wavelength of the light used. This

TABLE 4.2

Slope x θ values for different θ (Tong *et al.* [10])
Reprinted by permission of the Electrochemical Society

θ	$\sin\theta \simeq \theta$ (in radians)	Slope	Slope x θ
1° 12′	0.0209	−3.000	−0.0627
5° 3′	0.0880	−0.715	−0.0630
10° 3′	0.1745	−0.360	−0.0628

technique is applicable to any junction depth, limited only by the uncertainty of the finite size of the light spot. For a 2.5 μm light spot with θ of 30 min, the uncertainty is approximately 0.0245 μm. This uncertainty can be reduced even further when the light spot is made smaller.

Problems in Beveling Techniques

In order to increase the depth of diffusion for accurate measurement, the diffused sample is generally beveled at a small angle θ, for example one to five degrees, in preparation for chemical staining or electrochemical plating. A stainless steel block, which has the desired angle θ machined onto one surface, is usually used. The diffused sample is then placed slightly above the beveled surface, as shown in figure 4.6, with the aid of some wax, while the steel block is heated. The block is then immediately cooled in cold water or other media to solidify the wax and anchor the sample in position. The block is turned upside down with the sample facing a glass plate in a mixture of fine alumina powder (0.05 to 0.1 μm particle size) suspended in water. The beveling action should be gradual so that the beveled surface is smooth, without deep grooves and scratch lines caused by the removed sample left on the glass plate.

Perhaps because of the extreme difficulty in defining the exact nature of the smoothness of the beveled surface, none of the staining or

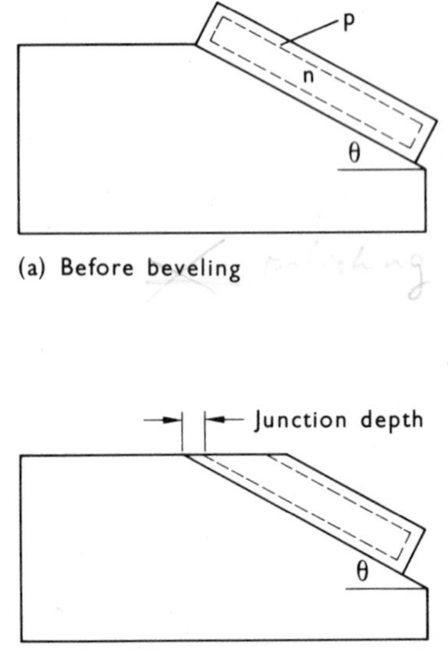

Figure 4.6. Bevel fixture for junction depth measurement.

electrochemical plating solutions can delineate a junction for a given sample with good reproducibility of data between research workers. This is evidenced by the poor 'round robin' test data gathered by the American Society for Testing and Materials (Document F110-69T) for the bevel and stain technique from different laboratories. For junction depths between 1 and 25 μm, the precision of the measurement is $\pm(0.15\ t + 0.5)$ μm (three sigma), where t is thickness in μm. Although part of the difference is due to the actual measurement, such as the beveled angle θ determination and the means of measuring the distance on a bevel from the surface to the junction, most of the discrepancy comes from the staining or electrochemical plating. It is the surface finish on the bevel that is considered to be the really critical parameter that has to be controlled.

This is best illustrated by the study conducted by Turner [7]. It consisted of measuring the potentials of individual electrodes of n- and p-type silicon with abraded and polished surfaces against a saturated KCl calomel reference electrode in an unstirred Cu solution. The electrodes were single-crystal 0.7 ohm-cm n-type and 1.5 ohm-cm p-type silicon. The results, as shown in figure 4.7, show considerable potential differences between abraded and polished silicon and also between polished n- and

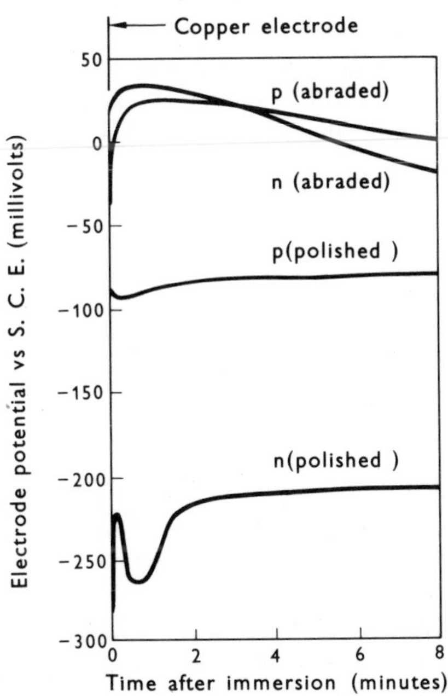

Figure 4.7. Electrode potential-time curves of abraded and polished n- and p-type Si in the $CuSO_4$ HF solution. (After Turner [7].)

p-type silicon in the Cu solution. The reversal in potential between *n*- and *p*-type silicon after 3 min is reproducible. On abraded silicon, metal begins to deposit more rapidly on the *n*-side of a *p-n* junction, but later the rate of metal deposition becomes greater on *p*-type than on *n*-type silicon.

Although the scanning light spot technique does not have the critical problem of surface finish on the bevel which affects the staining or electrochemical plating techniques, the accuracy of the measurement of the beveled angle θ and the exact position where the beveled surface really begins, do create problems. However, Tong *et al.* [10] have indicated that the reproducibility of the scanning light spot technique is within ±2% of the mean junction depth on the same sample and is far better than the bevel and stain or the electrochemical plating technique.

To avoid the controversial beveled surface discussed so far for the preparation of the junction depth measurement, it is possible to utilize one particular property of the semiconductor—brittleness—to obtain a cleavage surface through the *p-n* junction. Because the edges of the cleavage surface running through the *p-n* junction are very sharp, they can always be used as the reference lines for a direct measurement of the junction depth under a high-power metallographic microscope. The position of the diffused junction in the cleavage surface can usually be delineated without difficulty by chemical staining or electrochemical plating, as demonstrated by Jansen [16]. The only precaution for this method is that fracture lines should not parallel the diffused junction; otherwise, they may interfere with clear identification of the *p-n* junction, particularly if it is a very thin junction.

4.2.2 Radiochemical Techniques

Aside from the *p-n* junction method, perhaps radiochemical analysis, i.e., either radiotracer or neutron-activation techniques, is the next most popular experimental method used to determine the diffusion coefficients of various impurities in semiconductors. With the improved means of removing diffused layers in silicon [17], the entire profile of the diffused impurity can be established readily. The radiochemical method is now regarded by many workers as the most accurate method of obtaining the diffusion coefficients of various impurities in semiconductors.

However, radiochemical analysis requires careful calibration of the activities being measured and is very time consuming. Furthermore, it is only good for those elements for which a suitable radioactive tracer isotope is available, or which, upon activation, form a radioactive product of suitable half-life.

Neutron Activation

In the neutron activation technique, impurities are diffused into the semiconductor by normal processes with non-radioactive materials, and subsequently exposed to nuclear irradiation to convert the stable impurity nuclei into detectable radio-isotope species. This activation process is usually carried out with thermal neutrons. The (n, γ) nuclear reaction usually predominates for the thermal neutron activation. Generally, the specimen is annealed at a temperature of about 600° C to remove radiation-induced lattice damage. Successive layers of the structure are removed by anodization and etching for silicon [17] or by other chemical or mechanical processes. Anodic stripping is not feasible for germanium so that chemical or mechanical layer removal must be used. Non-radioactive isotopic carriers are used to measure the activation product; all solutions or suspensions involved in the sectioning process are retained for radioassay. However, it is necessary to separate the various radioactivities induced in the sample by means of additional instrumental or radiochemical procedures, such as decay-curve analysis and gamma-ray spectroscopy. A known amount of the impurity element of interest usually is carried through the same activation and counting procedure, to provide the necessary standardization.

A typical example is given below for the case of ^{32}P in silicon, which may get some interference from the activation of ^{31}P produced by ^{31}Si. In addition, a detailed procedure is described below for obtaining the radioactivities of both ^{32}P and ^{117m}Sn in the same silicon sample [18]. The tin or tin-and-phosphorus diffused wafers were irradiated in a nuclear reactor for 100 to 400 h. The neutron flux was 1 to 4×10^{13} neutron-cm^{-2}-sec^{-1} depending upon the tin's diffusion time and temperature. All irradiations were done in an aluminum capsule ¾ in. dia. and about 4 in. long. The standards used for the irradiation were 5-10 mg sample of Sn metal (99.9999%) and a single-crystal silicon wafer uniformly doped with P to about the 2×10^{20} atoms-cm^{-3} level. The isotopes used during this investigation and their formation and decay schemes are

$$^{116}\text{Sn}(n, \gamma)\,^{117m}\text{Sn} \xrightarrow[t_{1/2} = 14\text{ d}]{\text{I.T.}} {}^{117}\text{Sn}(\gamma 0.16\text{ MeV}) \tag{4.18a}$$

$$^{31}\text{P}(n, \gamma)\,^{32}\text{P} \xrightarrow[t_{1/2} = 14.3\text{ d}]{\beta^-} {}^{32}\text{S}(\beta 1.71\text{ MeV}). \tag{4.18b}$$

Since there are many stable isotopes of tin, there will be a correspondingly large number of radioisotopes formed during the irradiation (^{113}Sn, ^{119m}Sn, ^{121}Sn, ^{123}Sn, ^{125m}Sn, ^{125}Sn). However, these were found not to interfere with the ^{117m}Sn detection.

Also, the silicon matrix is activated as

$$^{30}\mathrm{Si}(n, \gamma)\,^{31}\mathrm{Si} \xrightarrow[t_{1/2} = 2.6\ \mathrm{h}]{\beta^-} {}^{31}\mathrm{P}. \tag{4.18c}$$

The Si activity can be neglected if a decay time of about three days is permitted before counting the samples. Otherwise interference may be produced by the activation of the ^{31}P (generated in equation 4.18c). However, for the irradiation times and flux used in the present investigation, the contribution is negligible.

After irradiation, the wafers were cleaned in ~10% HF for 2 min followed by a water and acetone rinse. Each wafer was then mounted on the end of a Teflon stirring rod using Apiezon W-100 wax (stirring speed: 25 rpm). The surface of the wafers was etched by immersing the mounted wafer in 2 ml of mercury etch solution (proprietary formula) contained in a small polyethylene cup. The etching time was either 2 or 3 min depending on the section thickness desired (etch rate ~170 Å min). An average thickness of each etch step (section) was determined from the total weight lost by the wafer during sectioning. This sectioning procedure was complicated by the fact that the tin in solution had a tendency to plate out on the wafer surface. Therefore, an aqua regia rinse (2 ml for 2 min) and an alcohol rinse (2 ml for 1 min) were inserted before each etch step. The mercury-etch, aqua regia, and alcohol rinse for each step were combined to form a 'cocktail' and placed in a 25 ml polyvial and saved for counting. Standard solutions were prepared by dissolving the solid standards in the appropriate solvents (aqua regia for Sn; 3 : 1 : 2 HNO_3 : HF : HAc for P) and diluting to 100 ml with water. The counting standards were prepared by taking 100 μl of the solutions above and adding to a blank mercury-etch 'cocktail'.

All the Sn samples were counted using a 3 x 3 in. NaI(Tl) scintillation crystal-photomultiplier assembly (Isotopes, Inc. model S-1212-I with Dumont 6363 photomultiplier). Gamma spectra were recorded by a TMC model 404C, 400 channel pulse-height analyzer and integrator-resolver. The recorded counts were integrated over the 0.16 MeV peak from 0.13 to 0.19 MeV. With ^{32}P activity present, there is a bremsstrahlung contribution to the Sn gamma spectrum from the 1.71 MeV β^- particle. In these cases, the net Sn counts were obtained by the triangulation method. After the Sn determinations, those samples requiring P analysis were diluted to 22 ml with de-ionized water and recounted using a Nuclear-Chicago model 6812 liquid scintillation system. The Cerenkov radiation produced by the 1.71 MeV β^- from the ^{32}P was measured to determine the activity present. The Sn contribution to each P sample was removed from the total counts by measuring the contribution of a standard tin sample and subtracting a proportionate amount from each sample count for the amount of Sn known to be present.

Radiotracer Techniques

For the radiotracer technique, the diffusion is carried out using a radioisotope-labeled dopant which should be free of radiochemical impurities. This method is conveniently adaptable to closed-system processes such as capsule diffusion. For open-tube diffusions, suitable precautions must be taken to entrap the relatively large amounts of radioactive material not incorporated into the diffused structure. Successive sectioning and counting of the removed layers is performed as in the neutron activation technique. Because of the absence of interfering radioactive species, this techique permits the use of many more radioassay procedures than the neutron activation technique. A specific activity measurement of the labeled starting material is also necessary for standardization in the case of chemical diffusion of an impurity.

An alternative procedure which avoids the need to collect and prepare the removed section for radioassay is to count the residual activity of the sample following removal of a section. If R is the count rate from a crystal sample after a thickness x has been removed then [18a]

$$R = K \int_x^\infty c(z) \exp(-\mu(z-x))\, dz \tag{4.19a}$$

where $c(z)$ is the concentration of radiotracer at depth z measured from the initial surface ($x = 0$), μ is the linear absorption coefficient for the tracer radiation in the crystal matrix and K is a proportionality factor. Differentiating equation 4.19a with respect to x gives

$$c(x) = \left(\mu R - \frac{\partial R}{\partial x}\right)/K \tag{4.19b}$$

Equation 4.19b is quite general and imposes no conditions on $c(x)$. Two limiting situations of considerable experimental value arise for very strongly absorbed radiation (μ very big) and for very weakly absorbed radiation (μ very small). In the former case equation 4.19b reduces to

$$c(x) = \mu R/K \tag{4.19c}$$

and in the latter situation

$$c(x) = -\frac{\partial R}{\partial x}/K \tag{4.19d}$$

The situations represented by equations 4.19c and 4.19d enable relative concentration profiles to be obtained without knowing μ. If however the profile is an erfc or Gaussian one then [18a]

$$\ln\left(\frac{\partial^2 R}{\partial x^2} - \mu\frac{\partial R}{\partial x}\right) = \text{constant} - \frac{x^2}{4\mathrm{D}t} \tag{4.20a}$$

for the erfc dependence and

$$\ln\left(\frac{\partial R}{\partial x} - \mu R\right) = \text{constant} - \frac{x^2}{4\mathrm{D}t} \tag{4.20b}$$

for the Gaussian profile. A practical disadvantage arises if neither of equations 4.19c or 4.19d apply as then μ must be known in order to evaluate the R data using equations 4.19b, 4.20a or 4.20b. It is common practice in experimental diffusions to use thin plates of the semiconductor crystal (which however are thick enough not to violate the validity of equations 4.1 or 4.2) and to diffuse from both principal faces. Equation 4.19b remains valid but, unless μ is very large, neither of equations 4.20a and 4.20b are, even for erfc or Gaussian profiles, because R will include a component from the back face profile. As a consequence the residual activity method is generally used with samples which have the radiotracer plated on one face only. The technique is then known as the Gruzin method. It is obviously important to ensure that none of the radiotracer travels to the back face by surface diffusion. In practice this may be difficult so that it is wiser to remove a layer of appropriate thickness from the back face prior to analysis of the diffused face. A recent and improved version of the Gruzin method has been described by Cramer and Crow [18b].

Autoradiographic methods can also be used. Their real value however lies in ascertaining whether a radiotracer is uniformly distributed across a face rather than in obtaining quantitative diffusion data.

Problems in Layer-removing Techniques

The accuracy of radiochemical analysis for determining the diffusion coefficients of the impurities is subject to many possible errors, such as may arise during a radiochemical manipulation, standard preparation, counting statistics, and layer-removing procedures. Perhaps the last item, layer-removing procedures, is the most uncertain. Many of the early studies of impurity diffusion in germanium or silicon by the radiotracer technique used mechanical means [19] to remove a parallel layer of material from the sample. The error involved for such profiling techniques can be rather high. The combined method of successive anodization, chemical etching, and liquid scintillation radioassay certainly presents a much improved method over the mechanical means of removing a thin layer of diffused material. The error of this new technique [20] for establishing an impurity profile can be limited to about 10%. Errors due to sectioning procedures are discussed further in section 4.4.

4.2.3 Surface Concentration Measurements

As discussed in section 4.2.1 on the *p-n* junction method, the surface concentration is one of two measured parameters needed to calculate the diffusion coefficient of a particular impurity at a given temperature and time. Four techniques available for silicon are discussed below, with emphasis on the range of application, accuracy, and limitations of each method.

Sheet Resistivity and Junction Depth

The determination of surface concentration was first worked out in detail by Backenstoss [21] and then revised by Irvin [22] through the use of the sheet resistivity and junction depth measurements. The method required that the impurity distribution, the majority carrier mobilities, and the resistivities of the material (in this case, silicon) be known. The sheet resistivity of a diffused layer is generally measured by a four-point probe, as shown in figure 4.8. The apparatus uses four sharply pointed and equally spaced points. A given current is passed through the outer pair of points and the resulting potential differences between the inner pair is then recorded. A diffused layer of a depth x_j can be considered as a thin layer having an insulating boundary with respect to the bottom material.

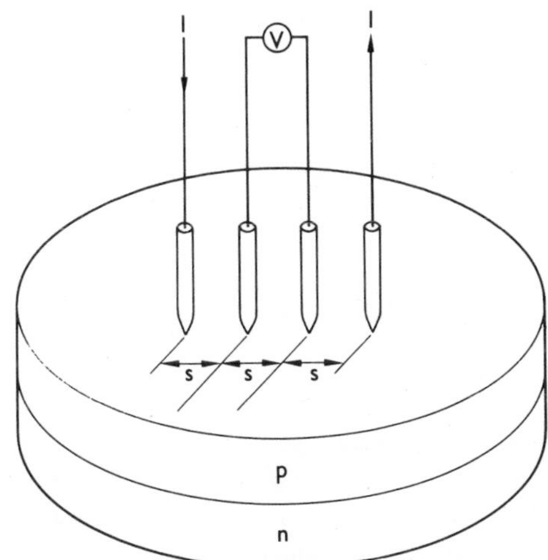

Figure 4.8. Four-point probe resistivity measurement setup.

The sheet resistivity is given by

$$\rho_s = \frac{V}{I} C\left(\frac{a}{d}; \frac{d}{s}\right) \tag{4.21}$$

according to Smits [23]. Table 4.3 gives this factor C for various geometries, where ρ_s is in ohm-cm, V in volts, and I in amps.

TABLE 4.3

Correction factor C for the measurement of sheet resistivities with the four-point probe (Smits [23])

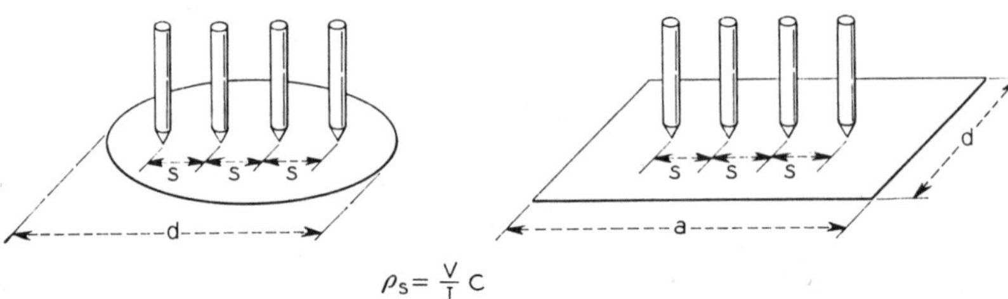

d/s	circle diam d/s	$a/d = 1$	$a/d = 2$	$a/d = 3$	$a/d \geqslant 4$
1.0				0.9988	0.9994
1.25				1.2467	1.2248
1.5			1.4788	1.4893	1.4893
1.75			1.7196	1.7238	1.7238
2.0			1.9454	1.9475	1.9475
2.5			2.3532	2.3541	2.3541
3.0	2.2662	2.4575	2.7000	2.7005	2.7005
4.0	2.9289	3.1137	3.2246	3.2248	3.2248
5.0	3.3625	3.5098	3.5749	3.5750	3.5750
7.5	3.9273	4.0095	4.0361	4.0362	4.0362
10.0	4.1716	4.2209	4.2357	4.2357	4.2357
15.0	4.3646	4.3882	4.3947	4.3947	4.3947
20.0	4.4364	4.4516	4.4553	4.4553	4.4553
40.0	4.5076	4.5120	4.5129	4.5129	4.5129
∞	4.5324	4.5324	4.5324	4.5325	4.5324

Backenstoss [21] first related the surface concentration c_0, to $1/\rho_s x_j$ in the form of curves for a given background concentration, c_B, and a known impurity distribution, such as a complementary error function or Gaussian distribution. Irvin [22] then revised these calculations based on new and more extensive resistivity data. Similarly to Backenstoss, Irvin defined an average conductivity of a diffused layer by the expression:

$$\bar{\sigma} = \frac{1}{\rho_s x_j} = \frac{1}{x_j} \int_0^{x_j} e\mu c\,dx \qquad (4.22)$$

where e = electronic charge,

μ = carrier mobility typical of a total ionized impurity density of $c(x) + c_B$,

c = the density of carriers, equal to $r(c(x) - c_B)$,

r = the fraction of uncompensated diffused impurity atoms that are ionized,

$c(x)$ = the total concentration of diffused impurity atoms at depth x,

c_B = the concentration of acceptors or donors originally in the material, i.e., $c(x_j) = c_B$.

Excerpts† from Irvin's curves are presented in figures 4.9-4.12, which cover both n- and p-type diffused layers (for c_B of 10^{14}, 10^{15}, and 10^{16} cm^{-3}) and both complementary error function and Gaussian distributions.

This method fails, however, if the impurity distribution departs from one of the simple mathematical distributions mentioned. Such failures occur if the diffusion coefficient is concentration-dependent [17], which unfortunately is apt to be the case for surface concentrations over 3×10^{19} cm^{-3}.

Plasma Resonance

Ever since the original work of Spitzer and Fan [24], it has been known that when infrared radiation is reflected from a degenerate semiconductor, a minimum occurs in the reflectivity spectrum. This minimum occurs at the plasma frequency ω_p, which is a function of the free carrier concentration, effective mass of the carriers, and the dielectric constant of the semiconductor.

Recently Gardner *et al.* [25] have shown that infrared plasma reflection can be used to determine the surface concentration of a diffused silicon wafer. Since the absorption coefficient for infrared radiation in the wavelength range 2.5-40 μm is very high for degenerate semiconductors, the depth of penetration is consequently quite low. Thus the plasma minimum is due primarily to free carriers in surface layers. In theory the

† For general use in the silicon device area, Irvin has plotted the average conductivity of a sub-surface layer against the concentration at the surface, i.e., plotting $\bar{\sigma} = 1/\rho_s(x_j - x)$ versus c_o, where x at some point lies between the surface and the junction. He has taken the ratio x/x_j as a parameter. In this chapter we are interested only in a single diffused layer and its surface concentration; therefore we plot $\bar{\sigma} = 1/\rho_s x_j$ versus c_o.

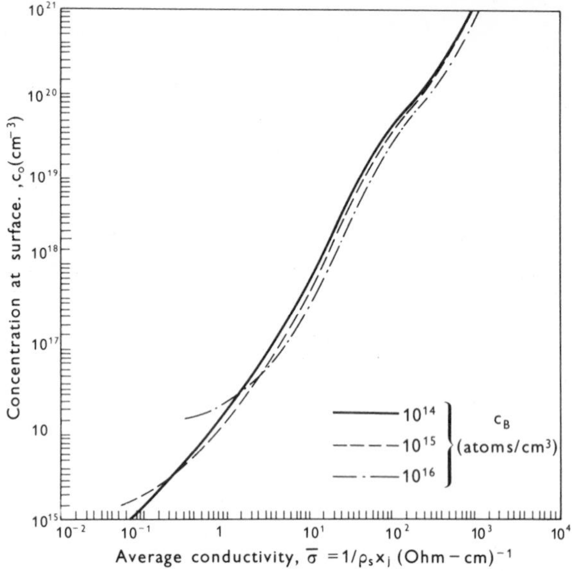

Figure 4.9. Average conductivity of *n*-type complementary error function layers in Si. (After Irvin [22].)

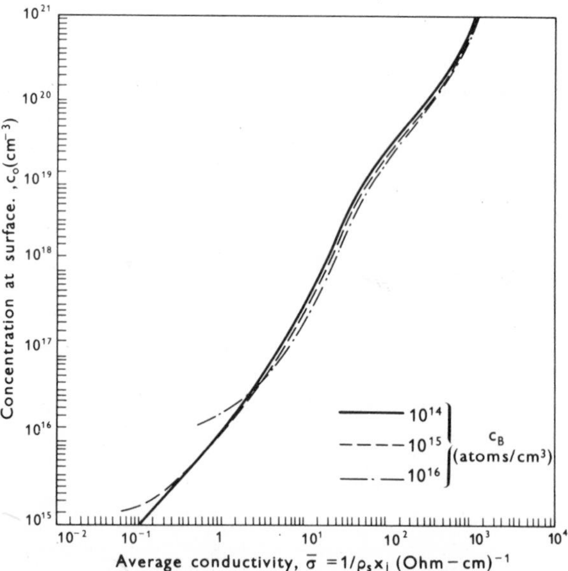

Figure 4.10. Average conductivity of *n*-type Gaussian layers in Si. (After Irvin [22].)

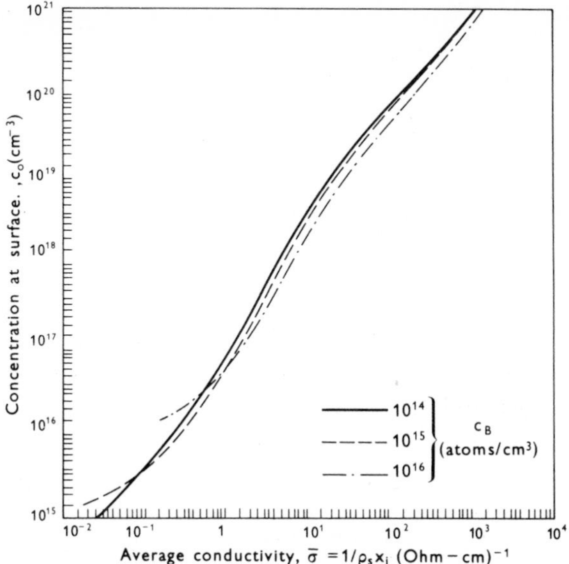

Figure 4.11. Average conductivity of p-type complementary error function layers in Si. (After Irvin [22].)

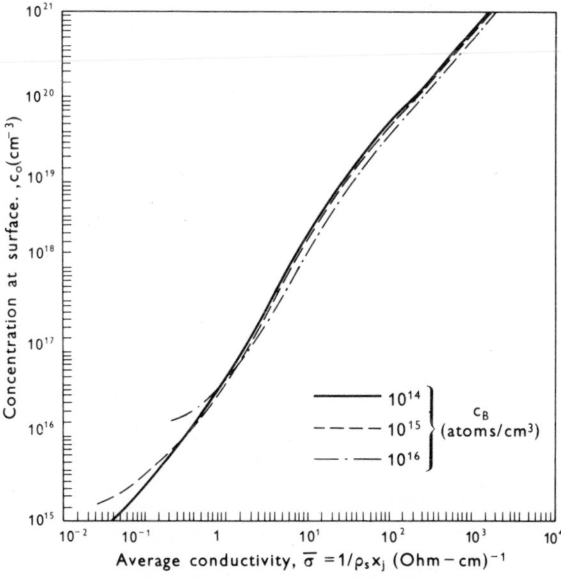

Figure 4.12. Average conductivity of p-type Gaussian layers in Si. (After Irvin [22].)

free carrier concentration, N, can be determined from the relationship [26]:

$$N = 4\pi^2 m^* c^2 \epsilon_0 (\epsilon - 1)/e^2 \lambda_p^2 \tag{4.23}$$

where m^* = effective mass of the free carrier,
c = velocity of light,
ϵ_0 = permittivity of free space,
ϵ = relative dielectric constant of the intrinsic semiconductor, and
λ_p = wavelength corresponding to the plasma minimum.

In practice, it is more convenient to measure the plasma minimum for a number of semiconductor crystals of known carrier concentration, i.e., to establish a calibration curve. This has been done for both n- and p-type silicon by Gardner *et al.* [25] and the results are shown in figure 4.13. The carrier concentrations of these silicon wafers were determined from four-point probe measurements and Irvin's [22] curves of resistivity versus concentration. The two equations shown in the figure for n- and p-type

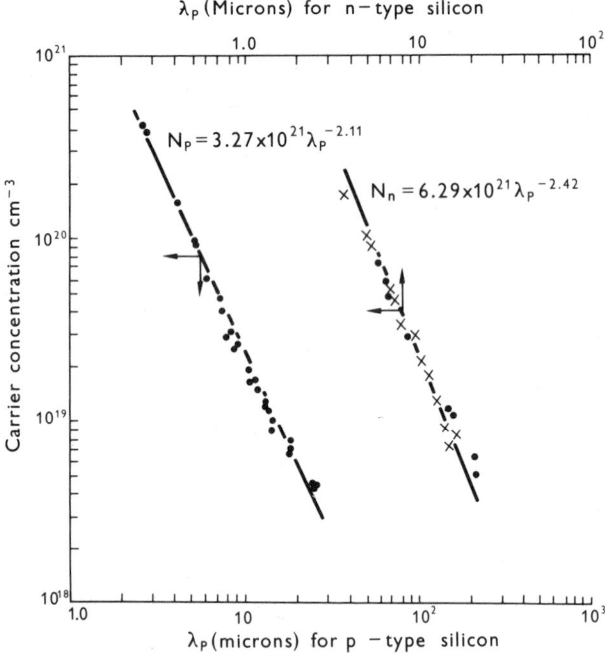

Figure 4.13. Carrier concentration versus plasma resonance minima for n- and p-type Si. N_p and N_n are the free carrier concentrations of p-type and n-type Si respectively. (After Gardner *et al.* [25].)

silicon give the free carrier concentration per cm^3 when the plasma minimum wavelength is expressed in micrometers. Plasma minima, λ_p, were measured using a Perkin-Elmer 521 infrared spectrometer. A typical example is shown in figure 4.14. It shows the reflectance as a function of wave number for an *n*-type silicon specimen with a carrier concentration of 2.7×10^{20} cm^{-3}.

Diffused samples of silicon can first be measured for their plasma minima and then the curves shown in figure 4.13 can be used to obtain their surface concentration, c_o. In Table 4.4, the c_o's determined by

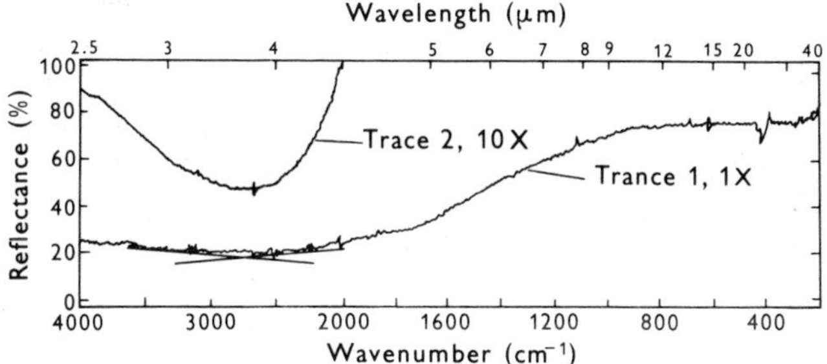

Figure 4.14. Reflectance as a function of wavenumber for an *n*-type silicon specimen with a carrier concentration of 2.7×10^{20} cm^{-3}.

TABLE 4.4

Comparison of c_o's determined by plasma resonance and other techniques (Gardner *et al.* [25])
Reprinted by permission of the American Institute of Physics

Sample No.	$c_o\,(cm^{-3})$ (Plasma resonance)	$c_o\,(cm^{-3})$ (Other technique)
1042-1	5.4×10^{19}	5.7×10^{19}
1041-3	7.9×10^{19}	9.5×10^{19}
1047-2	2.5×10^{19}	1.9×10^{19}
1045-1	2.05×10^{19}	2.9×10^{19}
1050	7.7×10^{18}	6.8×10^{18}
1054-B-1	4.4×10^{18}	1.7×10^{18}
1057-2	5.9×10^{19}	6.3×10^{19}
1061-4	3.0×10^{19}	3.5×10^{19}
1060-B-2	4.0×10^{18}	4.9×10^{19}
1046-6	8.1×10^{19}	9.5×10^{19} [a]
1046-7	7.8×10^{19}	9.5×10^{19} [a]
1046-3	1.95×10^{20}	2.9×10^{20} [a]
1046-4	1.95×10^{20}	2.9×10^{20} [a]

[a] The c_o's of these wafers were determined by neutron activation analysis and the others by the differential conductance technique.

plasma resonance are compared with those found by neutron activation analysis or the differential resistivity technique. There is good agreement between the c_o's determined by the plasma resonance and differential resistivity techniques. It is observed that four c_o's evaluated by neutron activation analysis are higher than the corresponding plasma resonance values. Since the neutron activation analysis is a measure of the total impurity concentration and the plasma resonance technique a function of the free carrier concentration, it is expected that the two values will differ at high concentrations.

This method is also applicable to *n*-type germanium. The concentration range is from 3×10^{18} to 10^{21} atoms per cm^3 for *n*- and *p*-type silicon and *n*-type germanium. Since this method relies on the photon penetration, which certainly occurs within a finite distance from the surface, a correction is required which not only depends on concentration, but also assumes to some extent what kind of profile is involved. Its resolution fails at concentrations below approximately 5×10^{18} cm^{-3} and fundamental reflection properties inhibit its use above approximately 5×10^{20} cm^{-3}.

Spreading Resistance

The three-point probe spreading resistance technique can determine the entire profile along a beveled surface of a diffused layer that includes the surface concentration in question. A correction factor must be applied to the measured value in order to get the actual concentration. The detail of this technique is discussed in section 4.2.4 of this chapter.

Radiochemical

Either neutron activation or the radiotracer technique described in section 4.2.2 of this chapter can be used to determine the surface concentration of the impurity in the diffused layer. Standardization of the radioisotope needs particular attention for accurate determinations.

4.2.4 Impurity Profiles by Electrical Methods

The surface concentration is usually determined by the measurement of junction depth and sheet resistance measurements, utilizing published curves [22]. The surface concentration and junction depth values are then used to determine the diffusion coefficient. Since the curves were established on the assumption that the profile of the diffused impurity obeys either a complementary error function or a Gaussian distribution, the experimentally determined diffusion coefficients are certainly subject to question if the profile departs from the assumed distribution. At present, two techniques can be used to establish the impurity profile of the diffused layer, as described below.

Differential Resistivity Technique

The differential resistivity technique was first described by Tannenbaum [17] in 1961 for establishing the detailed concentration profiles of phosphorus-diffused layers in *p*-type silicon. It consists of measuring repeatedly the sheet resistivity of the diffused layer by the four-point probe after successively removing a thin layer of diffused material from the surface until the desired depth is reached.

Thin layers of silicon were removed parallel to the junction by anodic oxidation. The sample can be anodized in a solution of 0.1 M boric acid, sodium tetraborate [17], or some other chemicals [27], which allows an oxide to grow on the surface. The thickness of the oxide grown in this way can be determined from calibration charts of color versus thickness. The thickness of the silicon oxidized is recognized as a fraction of the SiO_2 thickness and has been calculated from the molecular weights and densities of Si and SiO_2. The impurity in the anodized layer is removed by dissolving the silicon dioxide in HF solution. The success of this technique quite obviously depends upon the ability to accurately measure the sheet resistivity of the sample and to know accurately how much silicon is removed by each anodic oxidation.

The experimental data are plotted in the form of ρ_s^{-1} as a function of x, where ρ_s is the sheet resistance and x is the depth from the original surface. The slope $[d(\rho_s^{-1})/dx]$ at any point on the curve is a measure of the conductance in a plane parallel to the junction passing through the point x. The conductance is related to the concentration by the expression:

$$\frac{d(\rho_s^{-1})}{dx} = \sigma_x = N_x \mu_x' e \tag{4.24}$$

where N_x = the concentration of majority carriers,
e = the electronic charge, and
μ_x' = the effective mobility for the majority carrier.

Recently, Donovan [28] described a simple way to convert the incremental sheet resistivities into impurity profiles as follows:

(1) Plot $\log \rho_s(x)$ versus x, where $\rho_s(x)$ is the sheet resistivity of the remaining diffused layer, and x is the distance from the original surface to the new surface; draw a curve through the experimental points.

(2) Calculate the slope of this curve $(d \log \rho_s)/dx$, and plot it versus x; draw a curve through the points.

(3) Calculate $\rho(x)$ from the following equations:
Since

$$\sigma_s(x) = \int_x^{x_j} \sigma(x)\, dx, \tag{4.25}$$

$$\sigma(x) = -\frac{d\sigma_s(x)}{dx} = -\sigma_s(x)\frac{d \ln \sigma_s(x)}{dx}; \tag{4.26}$$

$$= \frac{1}{\rho_s(x)} \frac{d \ln \rho_s(x)}{dx}; \text{ and}$$

$$= \frac{1}{\rho_s(x) \log e} \frac{d \log \rho_s(x)}{dx};$$

then

$$\rho(x) = \frac{\rho_s(x) \log e}{d \log \rho_s(x)/dx} \tag{4.27}$$

(4) From a curve of resistivity ρ versus $|N_D - N_A|$ where N_D and N_A are the donor and acceptor impurity concentrations such as given by Irvin [22], find $|N_D - N_A|$ for each $\rho(x)$. Plot $\log |N_D - N_A|$ versus x and draw a curve through the points. This is the desired impurity profile for the diffused sample.

An example of both the sheet resistivity and the corresponding impurity concentration for a typical boron diffusion is shown in figure 4.15. The slope of $\log \rho_s(x)$ is given in figure 4.16, and is used to

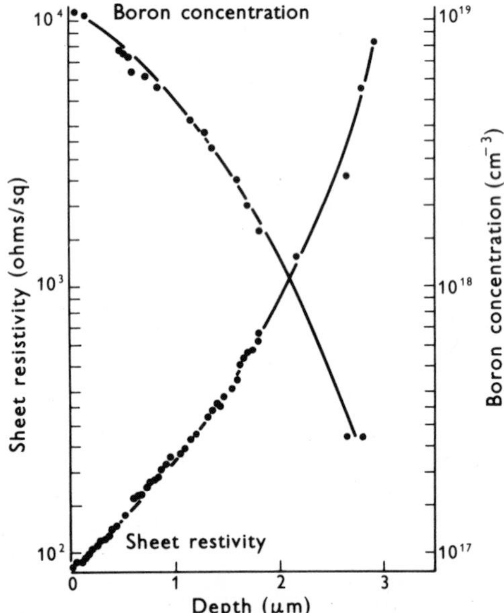

Figure 4.15. Boron distribution as determined by incremental sheet resistivity measurements. (After Donovan [28].)

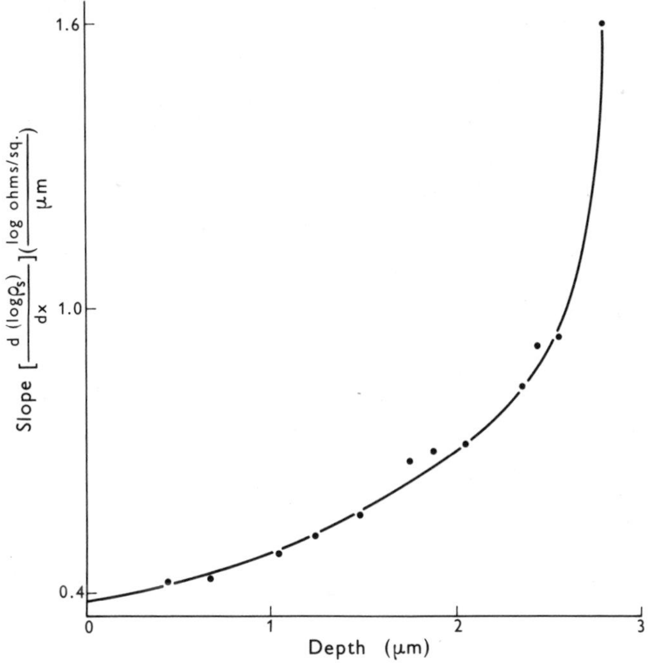

Figure 4.16. Slope of log resistivity $\left[\frac{(d \log \rho_S)}{dx}\right]$ as determined from the data of figure 4.15. (After Donovan [28].)

calculate $\rho(x)$. These $\rho(x)$ values are then used to establish the impurity profile of the sample according to Irvin's curve [22].

Ever since Tannenbaum's disclosure [17] of the differential resistivity technique to establish the impurity profiles in the diffused sample, it has been quite clear that many experimentally determined impurity profiles are very much different from the assumed complementary error function or Gaussian distributions. For example, figure 4.17 gives a representative sampling of the experimental data by Tannenbaum [17]. It is obvious that all the data above 900°C form straight lines over a considerable portion of each curve. Since the slope is a constant, then the concentration is a constant, assuming that the mobility is a constant over the pertinent resistivity range. Using the effective mobility values compiled by Irvin [22] for concentration below about $5 \times 10^{19}\ cm^{-3}$ (assumed to be a constant value of $75\ cm^2\ V^{-1}\ s^{-1}$ above that concentration), concentration profiles are obtained as shown in figure 4.18 and 4.19 on logarithmic and linear scales, respectively. The commonly assumed complementary error

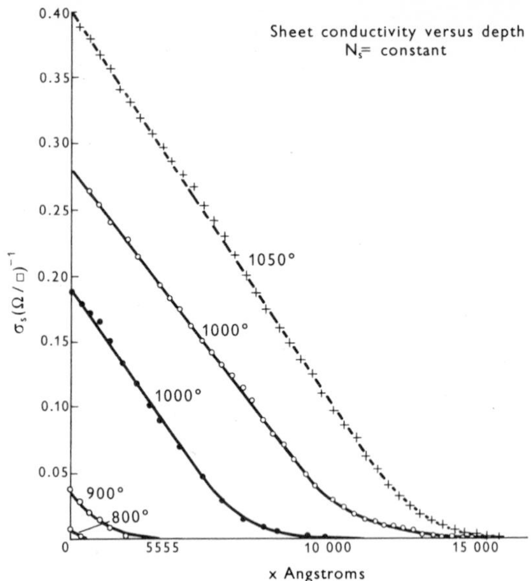

Figure 4.17. Sheet conductance as a function of depth for a series of samples diffused under conditions of constant surface concentration. (After Tannenbaum [17].)

function was calculated for comparison, using Mackintosh's [29] value of the diffusion coefficient and the experimentally determined junction depth. It can be seen from figure 4.19 that although the values of

$$\int_0^\infty N_x dx$$

for the experimental curve and the erfc curve agree within reproducibility limits, the surface concentration of electrically active impurities (calculated assuming an erfc distribution) is far higher than that found experimentally.

Spreading Resistance Probe

The spreading resistance probe was proposed some time ago by Mazur and Dickey [30] and Gardner, Schumann, and Gorey [31] for determining either the thickness of diffused or epitaxial layers of silicon or establishing the impurity profiles for various multilayered structures. Gupta *et al.* [31a] have recently investigated the effect of surface preparation on spreading resistance measurements. They find that mechanical surface damage can reduce the measured resistance by as much as a factor five depending on the bulk resistivity. Probe features and factors affecting reproducibility have been discussed by Severin [31b].

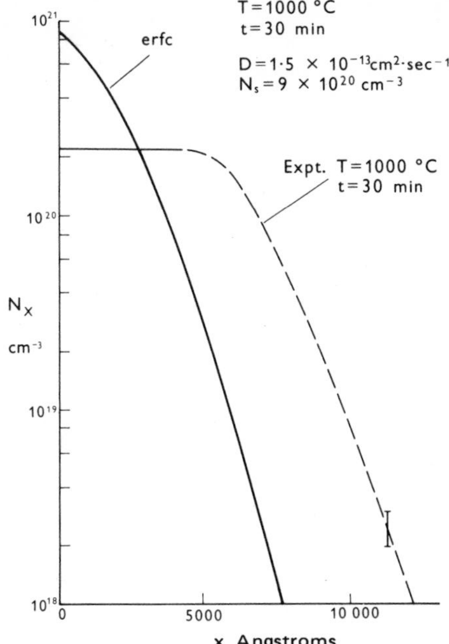

Figure 4.18. Logarithm of the calculated concentration as a function of depth. The vertical line indicates the maximum deviation of the calculated points from the dashed curve. The complementary error function, predicted by simple theory, is shown for comparison. (After Tannenbaum [17].) .)

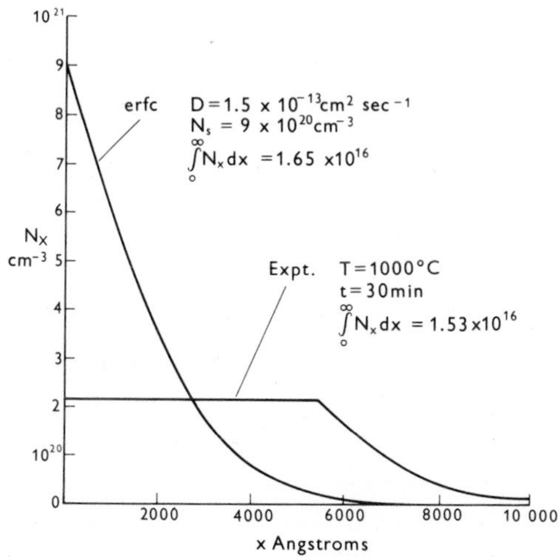

Figure 4.19. Linear plot of the calculated concentration as a function of depth. The data are the same as in figure 4.18. Note that although the distributions differ widely, the integrals of the theoretical and experimental curves are very nearly the same. (After Tannenbaum [17].)

The spreading resistance probe can be used to establish resistivity (or impurity) profiles and the thickness of multilayered silicon structures because it measures the localized resistivity on the sample in an extremely small volume, in the order of 10^{-10} cm^{-3}. If a flat circular voltage probe of radius r_0 makes contact to a semi-infinite conducting material [32], the potential distribution in the vicinity of the probe is as shown in figure 4.20. It is observed that practically all the potential drop occurs

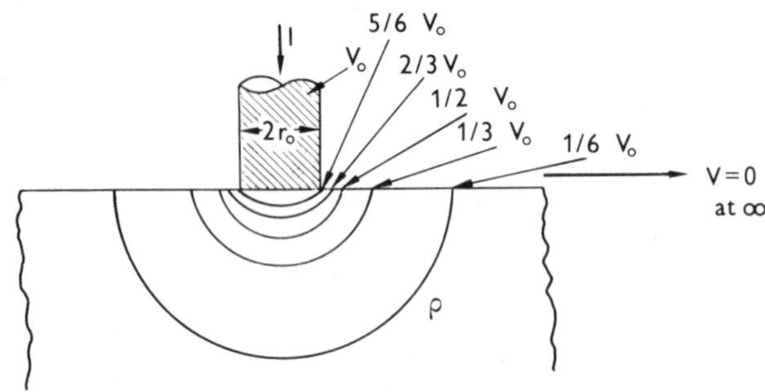

Figure 4.20. Potential distribution of flat circular probe on semi-infinite medium. (After Gardner *et al.* [31].)

within a distance of a few probe radii of the probe center. For an ideal metal-semiconductor contact having no barrier resistance, the potential of the contact is directly related to the semiconductor resistivity ρ by

$$V_0 = \frac{\rho I}{4r_0}. \tag{4.28}$$

Gardner, Schumann, and Gorey [31] have found it convenient to use three probes: two are connected to a current source, and the potential difference is measured between one of these and the third probe as shown in figure 4.21. For this experimental arrangement, the probe connected to both the current source and the voltmeter is the critical probe. A second configuration is to place the left-hand current connection of figure 4.21 on the center probe, in which case the center probe becomes the critical probe. The chief consideration in the design of the apparatus is the control of the common probe's contact parameters: contact area, depth of penetration, contact pattern, and surface damage. To regulate these parameters, the probe tip loading, velocity of contact, and deformation must be controlled precisely. Schumann *et al.*, have discussed these points in detail [33].

If the semiconductor material is semi-infinite in extent and the probe spacing s is large as compared with the contact radius r_0, then the

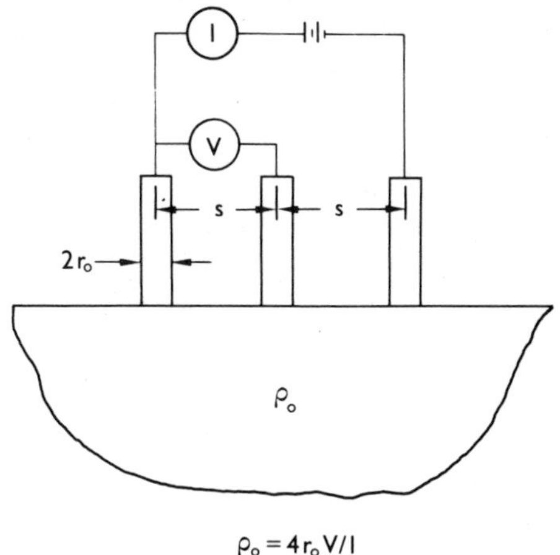

Figure 4.21. Spreading resistance technique on semi-infinite medium. (After Gardner *et al.* [31].)

potential is the same as given by equation 4.27. Thus the measured spreading resistance R_{SR} is related to the resistivity of the material and the effective radius of contact of the probe as follows:

$$R_{SR} = \frac{V}{I} = \frac{\rho_0}{4r_0} \tag{4.29}$$

For a diffused structure, the semiconductor material is not semi-infinite in extent but a layered structure, as shown in figure 4.22. The spreading resistance R_{SR} is given by the expression [34]

$$R_{SR} = \frac{\Delta V}{I} = \frac{\rho_1}{4r_0}\left\{\frac{4}{\pi}\int_0^\infty \left[\frac{1+K_1 \exp(-2H_1X)}{1-K_1 \exp(-2H_1X)}\right] \cdot \left[\frac{J_1(X)}{X^2} - \frac{J_0(SX/2)}{X} + \frac{J_0(SX)}{2X}\right] \sin X \, dX\right\} \tag{4.30}$$

where R_{SR} = spreading resistance,
$K_1 = (\rho_2 - \rho_1)/(\rho_2 + \rho_1)$,
$S = 2s/r_0$,
$H_1 = h_1/r_0$,
s = probe spacing,
r_0 = the effective radius of contact of the probe,
$h_1 = x_0 - x_1$, the layer thickness,
X = dimensionless integration parameter, and
J_1, J_0 = Bessel functions.

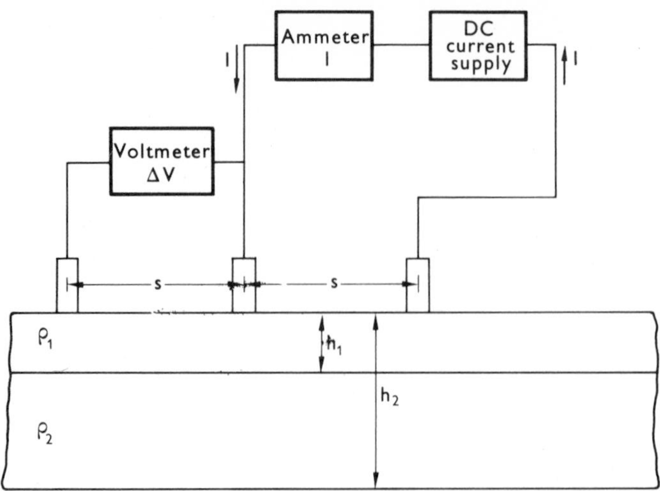

Figure 4.22. Spreading resistance technique on a two-layered structure.

The terms within the curly bracket of equation 4.30 are generally labeled as a correction factor, CF. The corrected resistivity value is then given by the following expression:

$$\rho_{1\,\mathrm{corrected}} = \frac{\rho_1}{CF}. \tag{4.31}$$

The calculation of the correction factors based on the assumption of the unilayer step-junction theory has been discussed in earlier publications [31, 35]. To determine an impurity profile, one can use a small test chip and bevel at a small angle as shown in figure 4.23. The probes are aligned as shown and are moved down the bevel in regular increments. The thickness t needed for the correction factor is shown in the same figure as the distance from the interface of the two layers.

For an n/p silicon structure, the measured spreading resistance values are converted to resistivity values from the calibration curve [31] and plotted against distance as shown in figure 4.24. The calculation of correction factors is based on a distinct structure, n/p, according to equation 4.30, and is considered to be a two-layer problem as shown in figure 4.22.

The first measured resistivity to be corrected is labeled ρ_1, which actually is the last measurement in the n/p structure, and this resistivity is assumed to be uniform throughout the layer thickness of $x_0 - x_1 = h_1$. The resistivity change at the interface n/p is abrupt, and an infinite resistivity can be assumed for ρ_1. Then the K_1 value in equation 4.30 for this problem is equal to $(\rho_0 - \rho_1)/(\rho_0 + \rho_1)$. Upon determining the

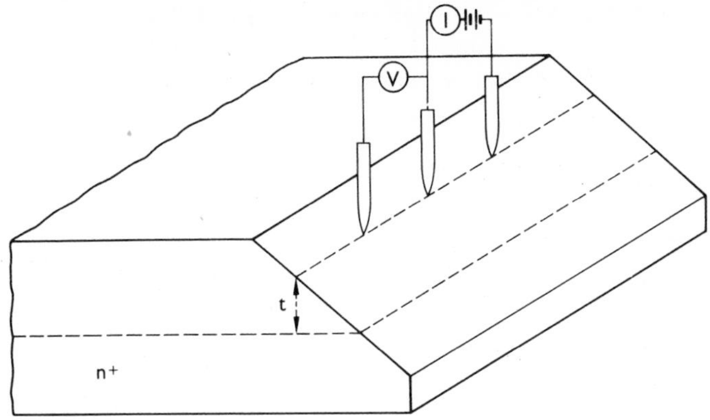

Figure 4.23. Determination of resistivity profiles by moving probes down a beveled test chip.

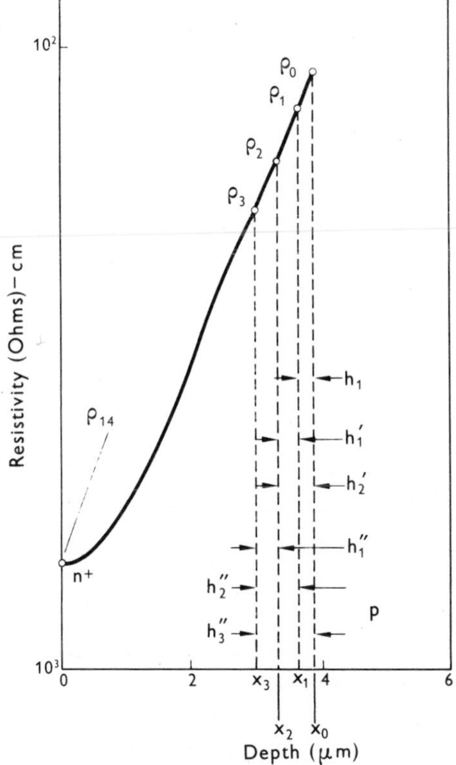

Figure 4.24. Initial resistivity value versus depth for *n/p* structure.

correction factor, the resistivity is corrected using equation 4.31. Then this corrected resistivity value $\rho_{1_{\text{corrected}}}$ may be used to find a new effective radius of contact r_0 from the calibration curve [31], a new correction factor, and a new corrected resistivity. This procedure is repeated to get an exact value of resistivity (i.e., until either the r_0 value or the correction factor does not change), because the r_0 value obtained initially from the calibration curve fro the first correction-factor calculation is not necessarily exact.

Because of the unilayer step-junction assumption, neither the corrected resistivity $\rho_{1_{\text{corrected}}}$ nor its initial resistivity ρ_1 are utilized when the second measured resistivity ρ_2 is corrected. For correcting ρ_2, it is assumed that the resistivity ρ_2 is homogeneous throughout its thickness $x_0 - x_2 = h_1$ even though we know that within its thickness one of the layers has a different resistivity value, $\rho_{1_{\text{corrected}}}$. The calculation of the correction factor for ρ_2 is again evaluated by equation 4.30 using the K_1 value of $(\rho_0 - \rho_2)/(\rho_0 + \rho_2)$.

Undoubtedly, some accuracy of the correction factor is sacrificed for each of the subsequent resistivity values, such as $\rho_3, \rho_4 \ldots \rho_{14}$ for the *n/p* structure because of the unilayer step-junction assumption. This is especially true for the corrected value of ρ_{14}, because in the calculation of its correction factor, this resistivity, ρ_{14}, is assumed to be uniform throughout the total thickness of $x_0 - x_{14}$, even though there are 13 layers of corrected resistivity values, $\rho_{1_{\text{corrected}}} \cdots \rho_{13_{\text{corrected}}}$, below ρ_{14}.

Obviously, if one uses the corrected resistivity values to obtain a correction factor for any one of the measured values, e.g., utilizing $\rho_{1_{\text{corrected}}}$ and $\rho_{2_{\text{corrected}}}$ to correct ρ_3, then the value of $\rho_{3_{\text{corrected}}}$ should be more accurate than the one without the consideration of those values of $\rho_{1_{\text{corrected}}}$ and $\rho_{2_{\text{corrected}}}$.

However, if one wants to use the preceding corrected resistivities for the calculation of the correction factor for the subsequent initial resistivity, one must use the multilayer step-junction theory to solve the problem. The geometry is shown in figure 4.25. The solution to Laplace's equation in cylindrical coordinates for this problem is:

$$R_n = \frac{V_n(R, Z)}{I} = \frac{\rho_n}{2\pi r_0}\left\{\int_0^\infty \frac{\exp(-XZ)\sin X\, J_0(RX)}{X}\,dX + \int_0^\infty \frac{N\theta_n(X/r_0)\exp(-XZ)\sin X\, J_0(RX)}{X}\,dX + \int_0^\infty \frac{N\psi_n(X/r_0)\exp(XZ)\sin X\, J_0(RX)}{X}\,dX\right\}$$

$$n = 1, 2, 3 \ldots N \tag{4.32}$$

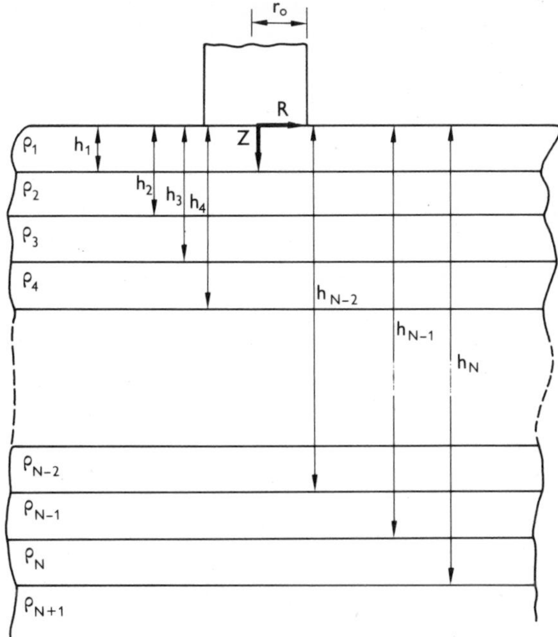

Figure 4.25. Spreading resistance technique on a multilayered structure.

The functions ${}_N\theta_n(X/r_0)$ and ${}_N\psi_n(X/r_0)$ are used in matching boundary conditions between regions. The boundary conditions for this type of problem are relatively standard, with the exception of the current distribution under the contact.

The application of the boundary conditions [34-36] results in a series of linear equations, equations 4.33-4.48:

$$ {}_N\theta_1\left(\frac{X}{r_0}\right) - {}_N\psi_1\left(\frac{X}{r_0}\right) = 0; \tag{4.33}$$

$$\exp(-2H_nX)\cdot {}_N\theta_n\left(\frac{X}{r_0}\right) + {}_N\psi_n\left(\frac{X}{r_0}\right) - \exp(-2H_nX)$$
$$ {}_N\theta_{n+1}\left(\frac{X}{r_0}\right) - {}_N\psi_{n+1}\left(\frac{X}{r_0}\right) = 0; \tag{4.34}$$

$$-\exp(-2H_nX)\cdot {}_N\theta_n\left(\frac{X}{r_0}\right) + {}_N\psi_n\left(\frac{X}{r_0}\right)\beta_n \exp(-2H_nX)$$
$$ {}_N\theta_{n+1}\left(\frac{X}{r_0}\right) - \beta_{nN}\psi_{n+1}\left(\frac{X}{r_0}\right) = (1-\beta_n)\exp(-2H_nX); \tag{4.35}$$

$$ {}_N\psi_{n+1}\left(\frac{X}{r_0}\right) = 0; \tag{4.36}$$

where N = number of layers of different resistivity and finite thickness,

$$n = \text{index, from 1 to } N,$$

$$\beta_n = \frac{\rho_n}{\rho_{n+1}},$$

$$H_n = \frac{h_n}{r_0},$$

and region $N+1$ is infinitely thick.

Using the simplified notation

$$ {}_N\theta_n\left(\frac{X}{r_0}\right) = {}_N\theta_n, \tag{4.37}$$

$$\exp(-2H_nX) = E_n, \tag{4.38}$$

the equations above become:

$$ {}_N\theta_1 - {}_N\psi_1 = 0; \tag{4.39}$$

$$E_n \cdot {}_N\theta_n + {}_N\psi_n - E_n \cdot {}_N\theta_{n+1} - {}_N\psi_{n+1} = 0; \tag{4.40}$$

$$-E_n \cdot {}_N\theta_n + {}_N\psi_n + \beta_n E_n \cdot {}_N\theta_{n+1} - \beta_n \cdot {}_N\psi_{n+1} = (1-\beta_n)E_n; \tag{4.41}$$

$$ {}_N\psi_{n+1} = 0. \tag{4.42}$$

For example, if $N = 2$, the set of linear equations becomes:

$$ {}_2\theta_1 - {}_2\psi_1 = 0; \tag{4.43}$$

$$E_1 \cdot {}_2\theta_1 + {}_2\psi_1 - E_1 \cdot {}_2\theta_2 - {}_2\psi_2 = 0; \tag{4.44}$$

$$-E_1 \cdot {}_2\theta_1 + {}_2\psi_1 + \beta_1 E_1 \cdot {}_2\theta_2 - \beta_1 \cdot {}_2\psi_2 = (1-\beta_1)E_1; \tag{4.45}$$

$$E_2 \cdot {}_2\theta_2 + {}_2\psi_2 - E_2 \cdot {}_2\theta_3 - {}_2\psi_3 = 0; \tag{4.46}$$

$$-E_2 \cdot {}_2\theta_2 + {}_2\psi_2 + \beta_2 E_2 \cdot {}_2\theta_3 - \beta_2 \cdot {}_2\psi_3 = (1-\beta_2)E_2; \tag{4.47}$$

$$ {}_2\psi_3 = 0. \tag{4.48}$$

There are six equations and six unknowns. A matrix can be written with the columns given as below, such that determinants may then be formed to solve for any of the θ's and ψ's. In general, the matrix will be $2N + 3$ by $2N + 2$. The determinants will be $2N + 2$ by $2N + 2$.

${}_2\theta_1$	${}_2\psi_1$	${}_2\theta_2$	${}_2\psi_2$	${}_2\theta_3$	${}_2\psi_3$	Constant
1	-1	0	0	0	0	0
E_1	1	$-E_1$	-1	0	0	0
$-E_1$	1	$\beta_1 E_1$	$-\beta_1$	0	0	$(1-\beta_1)E_1$
0	0	E_2	1	$-E_2$	-1	0
0	0	$-E_2$	1	$\beta_2 E_2$	$-\beta_2$	$(1-\beta_2)E_2$
0	0	0	0	0	1	0

For a sample (*n/p* structure) having 14 layers of resistivity to be corrected by the multilayer theory, one must solve an ever-increasing number of determinants, starting with a 4×4 determinant and then steadily expanding to a 30×30 determinant. If this is done by hand, it is an arduous task. With the aid of an IBM System-360 Model 65 computer, however, 14 corrected resistivities were obtained in 7 min. These values were then converted to a net impurity concentration according to Irvin's curve [22] and plotted against the depth as shown in figure 4.26. In the same figure, the uncorrected impurity profile is also included for comparison.

Two electrical methods have been described above for establishing the impurity profile in a diffused layer. However both methods depend critically on a knowledge of the carrier mobility in a specimen with a certain impurity concentration gradient, or on the use of published curves [22] relating resistivity and impurity concentration. Most mobility data, including the curves mentioned above, represent values applicable only to homogeneous material (i.e. material uniformly doped with impurity) and are not strictly applicable to diffused structures with impurity gradients. In addition there is the lack of actual mobility data in silicon and germanium for impurity concentrations above $5 \times 10^{19} \text{cm}^{-3}$. Recently, therefore, particular attention has been given to radiochemical methods of establishing the *total* impurity concentration profile of a diffused structure (section 4.2.2). Radiochemical analysis measures the total impurity present (i.e. ionized and un-ionized) and is free from the difficulties arising from uncertainty in the mobility data. Electrical resistivity measurements relate to the net ionized impurity concentration so that compensation effects can lead to difficulties additional to those due to uncertainty of mobility data.

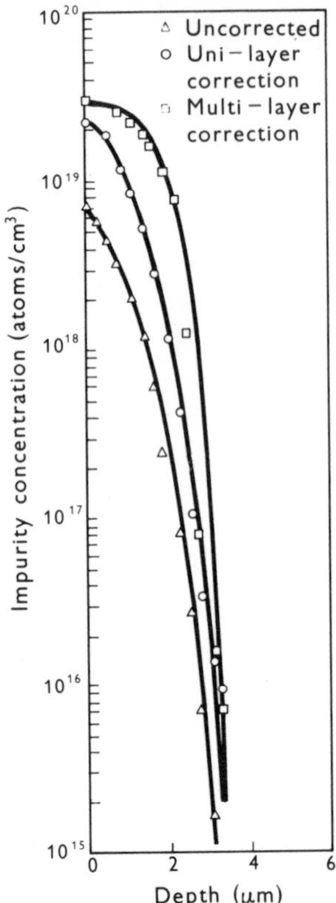

Figure 4.26. Impurity concentration as a function of depth for *n/p* structure.

4.2.5 Diode Capacitance

In most diffusion studies, the impurity profile can be approximated by either abrupt or linearly graded junctions, as shown in figures 4.27a and 4.27b, respectively. The abrupt approximation provides an adequate description for alloyed junctions and for shallowly diffused junctions. The linearly graded approximation is reasonable for deeply diffused junctions. For the purpose of the chapter, only the linearly graded junction and its capacitance measurements, which can be used to determine the diffusion coefficient of the impurity of interest, will be discussed.

The electrical capacitance of a *p-n* junction is closely related to the concentration of donor (N_D) and acceptor (N_A) impurities in the transition region. Consider the diode shown in figure 4.28, with a linearly graded impurity concentration of

$$N_D - N_A = ax \tag{4.49}$$

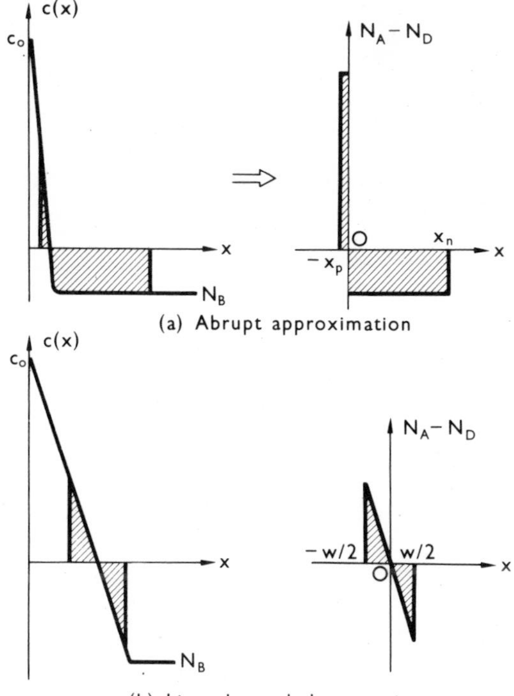

Figure 4.27. Approximate doping profiles. (After Sze [37].)
(a) Abrupt junction.
(b) Linearly graded junction.

where a is the net impurity gradient in cm^{-4}. It is assumed that the diode is of known cross-sectional area and that all the impurities are ionized. All these impurities will be neutralized by the mobile free carriers except in the transition or space charge region.

The Poisson equation for the space charge region of a linearly graded junction is:

$$-\frac{\partial^2 V}{\partial x^2} = \frac{\partial E}{\partial x} = \frac{e}{\epsilon}\,[p - n + ax] \simeq \frac{e}{\epsilon}\,ax \tag{4.50}$$

for

$$-\frac{w}{2} \leqslant x \leqslant \frac{w}{2}$$

where E = the electric field in volt-cm^{-1},
e = the electronic charge in Coulombs,
ϵ = the semiconductor permittivity in farad-cm^{-1}, and
w = the width of the transition region in cm.

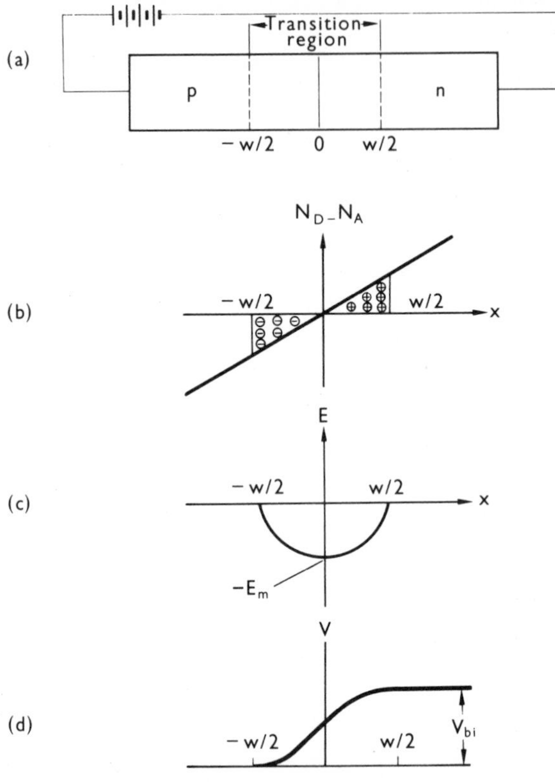

Figure 4.28. Analysis of capacitance method for measuring diffusion coefficient.
(a) Diode under reverse biased condition.
(b) Impurity distribution.†
(c) Field distribution.†
(d) Potential variation with distance.†
† For linearly graded junction in thermal equilibrium.

The electric field distribution as shown in figure 4.28d is obtained by integration of equation 4.50:

$$E(x) = -\frac{ea}{2\epsilon}\left[\left(\frac{w}{2}\right)^2 - x^2\right], \tag{4.51}$$

which has a maximum value of

$$E_m = -\frac{eaw^2}{8\epsilon} \text{ at } x = 0. \tag{4.52}$$

The built-in potential V_{bi} across the transition region can be obtained by integration of equation 4.51:

$$V_{bi} = \frac{eaw^3}{12\epsilon}. \tag{4.53}$$

The values of the built-in potential V_{bi} for Ge, Si, and GaAs are shown in figure 4.29.

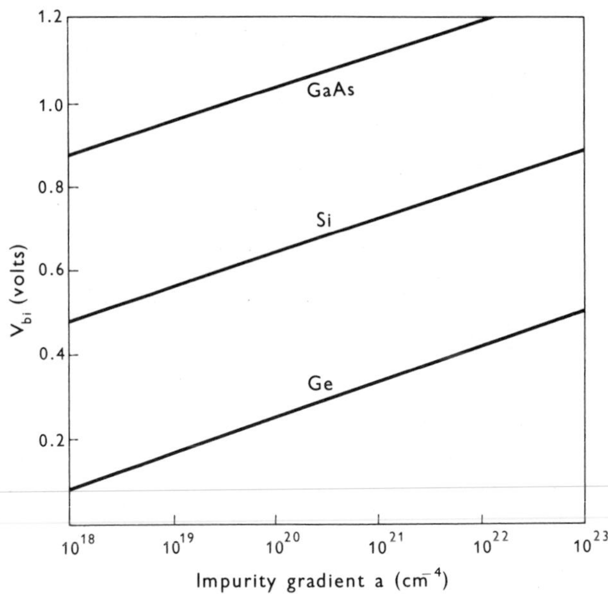

Figure 4.29. Built-in potential V_{bi} for Ge, Si, and GaAs. (After Sze [37].)

The unneutralized charge in the transition region gives rise to a capacitance referred to as the depletion-layer capacitance. An increase in the applied voltage dV across the depletion layer causes an increase in the charge density per unit area dQ. The depletion-layer capacitance per unit area is defined as:

$$C \equiv \frac{dQ}{dV}. \tag{4.54}$$

The charge per unit area in one half of the depletion region is:

$$Q = e \int_0^{w/2} ax\,dx = \frac{eaw^2}{8}; \tag{4.55}$$

thus

$$C = \frac{d(eaw^2/8)}{d(eaw^3/12\epsilon)} = \frac{\epsilon}{w}. \tag{4.56}$$

From equations 4.53 and 4.56,

$$C = \left[\frac{ea\epsilon^2}{12(V_{bi} \pm V)}\right]^{1/3} \tag{4.57}$$

where $V_{bi} + V$ is the total electrostatic potential variation across the junction for the reverse-biased condition, and $V_{bi} - V$ is the total potential drop for the forward-biased condition. The value of the impurity gradient can be obtained easily by measuring the capacitance of the junction at different applied voltages. It is expected from equation 4.57 that by plotting $1/C^3$ against V, a straight line should result with a slope of $12/ea\epsilon^2$, from which the impurity gradient a can easily be calculated. Typical values of the depletion-layer capacitance C in pf-cm^{-2} versus the impurity gradient a for the silicon junction have been measured by Sze [37] and are shown in figure 4.30. These results can also be used for GaAs since both Si and GaAs have approximately the same static dielectric constants. To obtain the depeletion-layer width for Ge, the silicon data should be multiplied by the factor $[\epsilon(\mathrm{Ge})/\epsilon(\mathrm{Si})]^{1/3} = 1.1$.

The value of a, the impurity gradient in cm^{-4}, which is obtained by measuring the junction capacitance before and after a certain diffusion process step can be used together with a knowledge of the concentration of donors and acceptors in the diode well away from the transition region, to calculate the diffusion coefficient **D** of the impurity.

Consider the case where one impurity, N_1, is uniformly distributed and a second impurity, N, is being diffused from a region where its

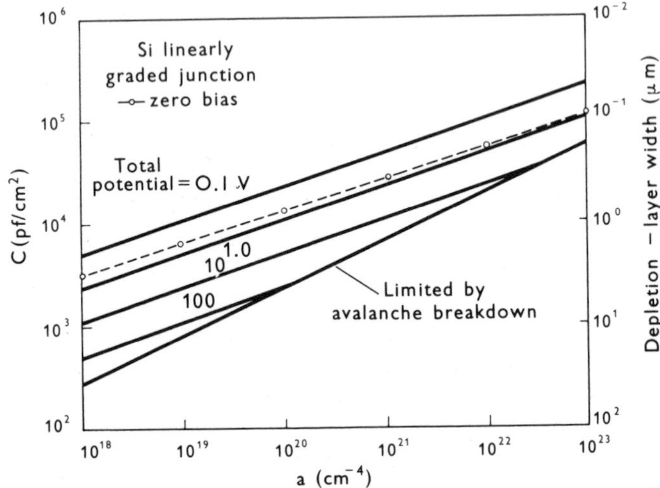

Figure 4.30. Depletion-layer capacitance versus impurity gradient. (After Sze [37].)

concentration is N_2 into a region where its concentration is zero. Assuming the impurity distribution after diffusion is given by the complementary error function:

$$N(x, t) = N_2\left(1 - \text{erf}\ \frac{x}{2\sqrt{\mathbf{D}t}}\right) \tag{4.58}$$

Since at the junction x_1, $N(x_1, t) = N_1$, then

$$N(x_1, t) = N_1 = N_2\left(1 - \text{erf}\ \frac{x_1}{2\sqrt{\mathbf{D}t}}\right) \tag{4.59}$$

or

$$\text{erf}\ \frac{x_1}{2\sqrt{\mathbf{D}t}} = 1 - \frac{N_1}{N_2}. \tag{4.60}$$

Knowledge of N_1 and N_2 yields the argument $x_1/2\sqrt{\mathbf{D}t}$ of the error function directly. If the value of a, the impurity gradient in cm^{-4}, is known by the capacitance and voltage measurement, it can be related to the impurity concentration and diffusion coefficient of the impurity of interest by the following equations:

$$a = \left.\frac{\partial N(x, t)}{\partial x}\right|_{x=x_1} = \frac{N_2}{\sqrt{\pi \mathbf{D}t}} \exp -\left(\frac{x_1^2}{4\mathbf{D}t}\right), \tag{4.61}$$

and the diffusion coefficient $\mathbf{D}$ can be expressed as

$$\mathbf{D} = \frac{N_2^2}{\pi a^2 t} \exp(-2s^2) \tag{4.62}$$

where s is the numerical value of $x_1/2\sqrt{\mathbf{D}t}$ obtained from equation 4.60.

The diode capacitance method, perhaps because of the need to measure many precise but small increments in capacitance with applied voltage and to know very accurately the diode area, has not been used extensively for the determination of diffusion coefficients. If a significant concentration of deep levels is also present the small signal junction capacitance becomes frequency dependent because with increasing signal frequency it becomes increasingly more difficult for the charge states of the deep levels to follow the applied signal. This feature has been studied most in abrupt junctions [38a-d]. For graded junctions the problem is more complex and no solution appears to be available. Equation 4.57 requires a negligible deep level concentration.

In abrupt junctions the doping level in the low doped side of the junction can be derived from the differential capacitance [39]. With Schottky barriers, work in GaAs [40] has shown that surface contamination and bulk trapping centres influence the junction capacitance so causing interpretative problems in the analysis of experimental data.

4.3 METHODS FOR COMPOUND SEMICONDUCTORS

The methods that have been described for Ge and Si can also in principle be used for compound semiconductors. Although the *p-n* junction method has been widely used in compound semiconductors electrical methods are at a disadvantage because of compensation effects which can be severe. Compensation of a diffusing impurity may occur because of: (a) uncontrolled background impurities as the purity levels are not comparable to either Ge or Si; (b) ionized native defects whose concentrations are variable, depend on the degree of non-stoichiometry and are not known in absolute terms with any certainty. There is also now considerable evidence that the chemical diffusion of impurities in the III-V compounds cannot be described by either of equations 4.1 or 4.2 [41-44]. Tracer methods are therefore to be preferred for studying impurity diffusions as they give unequivocal concentration profiles.

It must be emphasized that to obtain meaningful and reproducible results the neutral native defect concentrations in the host lattice must be clearly defined. With elemental semiconductors temperature alone suffices to determine the neutral defect concentration. A binary compound requires temperature and the activity (chemical) of one of its components to be specified (e.g. the partial vapour pressure of As over GaAs). Similarly a ternary system requires two component activities to be controlled. Many previous impurity diffusion experiments in compounds have unfortunately neglected to define or control the degree of non-stoichiometry.

Self and isoconcentration diffusions require an isotope method. Generally a radiotracer is used whenever a suitable one exists and is available or alternatively can be produced by neutron activation. If, however, this is not the case an alternative possibility is to use a stable isotope obtained by enrichment of the appropriate element in what is known as the *isotope exchange* method.† This method has been widely used in studies of self-diffusion of oxygen in oxides. Essentially the oxide sample is immersed in an oxygen atmosphere enriched with O^{18} at time $t = 0$. After time t the total number M, of say, O^{16} atoms lost by exchange with O^{18} atoms is given by [45]

$$M = 2Ac_o(\mathbf{D}t/\pi)^{1/2} \tag{4.63}$$

A and c_0 are the total surface area of the sample and the initial uniform concentration of O^{16} in the crystal. The change in isotope composition of the gas phase is measured using a mass spectrometer.

† This is not a very happy choice of name as the use of a tracer in self and isoconcentration diffusions is strictly also an isotopic exchange system.

Changes in non-stoichiometry (i.e. host, or solvent, chemical diffusion) cannot be followed using radiotracers as the defect diffusion coefficients are very much larger than the tracer coefficients of the host atoms, so that the defects have equilibrated in the tracer diffusion region (see section 1.4.4). Solvent chemical diffusion can be studied by following changes of mass or of electrical conductivity in the sample or by the movement of *p-n* junctions if the crystal exhibits *n*- or *p*-type conductivity according to the degree of non-stoichiometry. If the diffusion samples are quenched to room temperature following the diffusion anneal it is obviously of importance to ensure that the annealing situation is 'frozen in'. Ideally the quench is fast enough to prevent precipitation, evaporation or out-diffusion. If the room temperature measurements are to be used to infer with confidence what processes are operative at the diffusion temperature then, with regard to electrical measurements, the electroneutrality conditions at room temperature and at the diffusion temperature must be the same. Otherwise a detailed knowledge of the temperature variation of the electroneutrality condition is needed. Broderson *et al.* [46] have described the use of the *p-n* junction method in PbSe. In some cases it is possible to measure the chemical diffusion coefficient describing a change in non-stoichiometry by monitoring electrical conductivity versus time at the diffusion temperature. Zanio [47] has described such experiments in CdTe. This type of experiment is to be preferred because it eliminates difficulties arising from the quench. Procedures for measuring solvent chemical diffusion in oxides are discussed in section 8.5.6.

Use of a radiotracer method generally involves a sectioning procedure. Probably because of the difficulty of finding suitable etchants, mechanical lapping methods have found most favour for semiconductor compounds. Buhsmer [48] has recently described a development, of Goldstein's [49] mechanical method, capable of removing sections of 0.1 μm thickness. The vibratory polishing technique can also be used for removal of very thin layers. Whitton [50] has reported removal rates of 20-50 Å/min by this method for silicon and various refractory oxides. A slurry containing 0.05 μm alumina powder was employed. Argon ion sputtering utilizing a radio-frequency glow discharge [51] or an ion gun [51a] is a new sectioning technique that has been described. Removal rates of the order of 200 Å/min can be obtained. The sputtering also creates surface damage which may affect the profile obtained [51a].

Finally it should be pointed out that it is important to carry out both self and chemical diffusion experiments for the solvent species. There is evidence, for example in CdTe (see section 7.4.4), that the dominant electrically active native defect is not necessarily involved in the tracer diffusion process. It is possible that the self and chemical diffusion experiments involve quite different mechanisms.

4.4 THE ASSESSMENT AND EVALUATION OF EXPERIMENTAL DIFFUSION DATA

4.4.1 Electrical versus Radiochemical Methods

Electrical resistance methods yield carrier concentrations and tracer techniques measure total atomic concentrations. For impurity chemical diffusion where both types of method can be used the two concentrations will only be equal if there is no compensation and *all* the impurity is electrically active i.e. no neutral states or precipitation. The *p-n* junction method is of course based on compensation. Obtaining carrier concentrations from sheet and spreading resistance measurements utilizes the published curves [22] relating resistivity and impurity concentration. These data apply to bulk material with a uniform impurity concentration and the validity of their application to a diffused layer with an impurity concentration gradient is not obvious. Reliable carrier mobility data for impurity concentrations $>5 \times 10^{19}$ cm^{-3} is not available even for bulk material: it is customary to take the mobility at 300°K as constant for concentrations $>5 \times 10^{19}$ cm^{-3} in Si and Ge. It is evident that for an impurity diffusion electrical and radiochemical methods measure different parameters so that identical concentration profiles are not generally to be expected. A difference in profiles clearly implies different diffusion coefficients. The tracer diffusion coefficient should be of greater physical significance in relation to the impurity diffusion mechanism. The electrical diffusion coefficient has a more empirical nature but is obviously of greater importance in characterizing the electrical properties of a diffused layer; it is unlikely to provide much insight into the diffusion mechanism.

It has already been remarked in section 4.3 that solvent self and chemical diffusion in semiconductor compounds may involve different mechanisms. In such cases tracer and electrical methods have equal importance with respect to identifying a diffusion mechanism. A solvent chemical diffusion coefficient is also valuable in that it gives the rate of equilibration of the crystal compound under prescribed annealing conditions.

4.4.2 Serial Sectioning Methods

Confining attention to planar diffusion with an infinite source or thin film boundary condition the diffusant concentration is represented by equations 4.1 or 4.2 respectively if **D** is independent of position. In the latter case a plot of $\ln c(x)$ versus x^2 yields a straight line of slope

$-(4\mathbf{D}t)^{-1}$ from which **D** is readily found. With an erfc profile a method due to Hall [52] can be used to find **D** in a very simple way. Arithmetical probability paper is used and the quantity

$$50\left(2-\frac{c(x)}{c_o}\right)$$

is plotted on the probability axis against x on the linear abscissa axis. The ordinate scale is actually linear in $(x/2\sqrt{\mathbf{D}t})$ so that a plot of

$$50\left(2-\frac{c(x)}{c_o}\right)$$

versus x yields a straight line of gradient $(2\sqrt{\mathbf{D}t})^{-1}$ if equation 4.1 is satisfied. This is illustrated in figure 4.31. The extraction of **D** from measurements of R in the residual activity method is more complex even

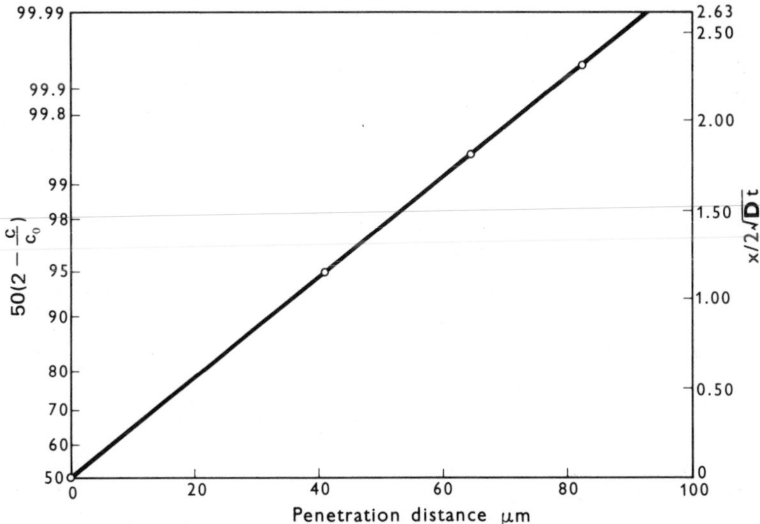

Figure 4.31. A plot of the As isoconcentration profile in Si shown in figure 5.10a using a probability ordinate scale. The profile gives a straight line which intercepts the upper abscissa scale at 92.5 μm. The other coordinate read from the right hand scale is 2.63 so that the gradient of the profile line, $(2\sqrt{\mathbf{D}t})^{-1} = 2.85 \times 10^{6}\ \text{cm}^{-1}$. Hence $\mathbf{D}t = 3.10 \times 10^{-10}\ \text{cm}^2$.

when the profiles are described by equations 4.1 or 4.2. A relative concentration profile using equation 4.19b could be obtained and **D** obtained by either of the methods described above. Alternatively **D** may be obtained from equations 4.20a or 4.20b by plotting the ln term versus x^2. A further alternative is to integrate equation 4.19a and obtain

$$\frac{R}{R(0)} = \left\{ \operatorname{erfc}\left(\frac{x}{2\sqrt{\mathbf{D}t}}\right) - \exp(\mu x + \mu^2 \mathbf{D}t) \operatorname{erfc}\left(\frac{x}{2\sqrt{\mathbf{D}t}} + \mu\sqrt{\mathbf{D}t}\right)\right\} \Big/ \left\{1 - \exp(\mu^2 \mathbf{D}t) \operatorname{erfc}(\mu\sqrt{\mathbf{D}t})\right\} \quad (4.63a)$$

for an erfc profile and

$$\frac{R}{R(0)} = \exp(\mu x) \operatorname{erfc}\left(\frac{x}{2\sqrt{\mathbf{D}t}} + \mu\sqrt{\mathbf{D}t}\right) \Big/ \operatorname{erfc}(\mu\sqrt{\mathbf{D}t}) \quad (4.63b)$$

for a Gaussian profile. $R(0)$ is the value of R at $x = 0$. This is the procedure generally followed in the Gruzin method. In order to obtain $\mathbf{D}$ from equation 4.63b Cramer and Crow [18b] computed master curves of $R/R(0)$ versus x for a wide range of $\mathbf{D}t$ and a given value of μ. $\mathbf{D}$ is then obtained by matching an experimental profile to the set of master curves. Equation 4.63a can be treated in a similar way. When $\mu \gg x/2\mathbf{D}t$, corresponding to strongly absorbed radiation, equations 4.63a and 4.63b simplify to

$$\frac{R}{R(0)} = \operatorname{erfc}(x/2\sqrt{\mathbf{D}t})$$

and

$$\frac{R}{R(0)} = \exp(-x^2/4\mathbf{D}t)$$

respectively. At the other extreme when $\mu x \ll 1$ and $\mu \ll x/2\mathbf{D}t$, corresponding to strongly penetrating radiation, the equations 4.63a and 4.63b now reduce to

$$\frac{R}{R(0)} = \exp(-x^2/4\mathbf{D}t) - \sqrt{\pi}\,\frac{x}{2\sqrt{\mathbf{D}t}} \operatorname{erfc}(x/2\sqrt{\mathbf{D}t})$$

and $R/R(0) = \operatorname{erfc}(x/2\sqrt{\mathbf{D}t})$, respectively. The great advantage of these simplifications is that μ does not need to be known in order to obtain $\mathbf{D}$.

While the above methods are satisfactory provided $\mathbf{D}$ is constant over the profile, such as in self and isoconcentration diffusion, they are not adequate if $\mathbf{D}$ varies with $c(x)$ as may occur in chemical diffusion. In the event of the latter the dependence of $\mathbf{D}$ upon $c(x)$ can be obtained from a penetration profile using the Matano [53] analysis which gives

$$\mathbf{D} = -\frac{1}{2t}\int_0^{c} x\,dc \Big/ \left(\frac{\partial c}{\partial x}\right)_{c} \quad (4.64)$$

Note that equation 4.64 requires the diffusion equation to be

$$\frac{\partial c}{\partial t} = \frac{\partial}{\partial x}\left(\mathbf{D}\,\frac{\partial c}{\partial x}\right) \tag{4.65}$$

so that **D** must be solely a function of c (i.e. of $x/\sqrt{t}$). It also assumes the use of the infinite source boundary condition. A consequence of equation 4.65 is that the depth of a given concentration is $\propto \sqrt{t}$. Such a proportionality if observed experimentally establishes the applicability of equation 4.64, it does not show that **D** is independent of c. No analysis appears to be available for the thin film case when **D** depends upon c.

It is also possible for **D** to vary along the profile but not to be solely a function of c. No analysis exists to extract **D** and its variation for this case. Instead theoretical profiles calculated from an appropriate model need to be matched to the experimental profiles.

Two sources of error are inherent to the sectioning method and arise from: (a) plotting the mean concentration in a section at the mid-point of the section and (b) the removal of non-parallel layers. With reference to figure 4.32 Malkovich [18a] has shown that to a first approximation these two sources of error lead to

$$\frac{\mathbf{D}\,(\text{measured})}{\mathbf{D}} = 1 + \frac{2}{3}\,(\beta^2 + \gamma^2) \tag{4.66}$$

where **D** is the true diffusion coefficient and β, $\gamma = \Delta/2\sqrt{\mathbf{D}t}$, $h/2\sqrt{\mathbf{D}t}$ respectively. Equation 4.66 applies only to Gaussian or erfc profiles. In practice β and γ would be calculated using **D** (measured) and an iterative process. Clearly $\mathbf{D} < \mathbf{D}$ (measured) always. If **D** varies with c the error in **D** (measured) is difficult to quantify. It is probable that the largest error is contributed through the Matano analysis.

Diffusion by short circuit paths (e.g. dislocations) is present even in single crystal samples and is parallel to the volume process. In any crystal section the diffusant concentration is therefore the result of combined volume and short circuit path diffusion. Generally the former is dominant near the surface and the latter process appears as a faster diffusing tail further in from the surface. The separation of the two components may be difficult especially when the volume coefficient, **D**, is small. It is generally advisable to keep section thicknesses $<\sqrt{\mathbf{D}t}$ to minimize short circuit effects [54]. Nebauer [55] has given a discussion of the problems involved in the analysis of a concentration profile due to diffusion via two diffusion paths. It is of interest to note that if, in the isotope exchange method,

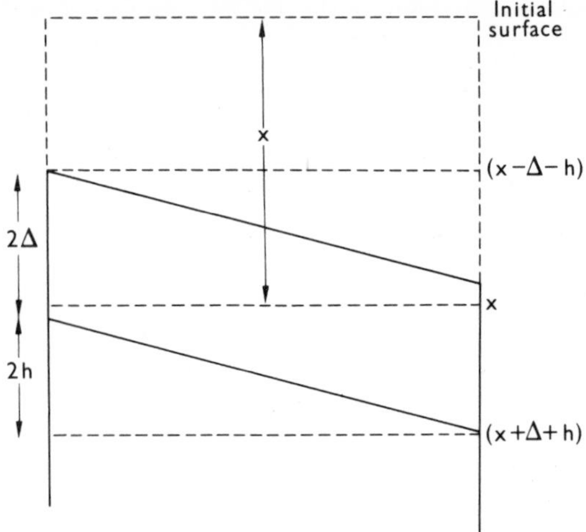

Figure 4.32. The situation considered for calculating the error in **D** due to constructing a concentration profile in which the average concentration in a section is plotted at the mid-point, x, of the section. Parallel layers of thickness 2 b inclined at angle $\tan^{-1} 2b$ to the original surface are removed.

short circuit path diffusion dominates over volume diffusion then [45] $M \propto t^{3/4}$, thereby allowing a ready distinction from volume diffusion for which $M \propto t^{1/2}$ (see equation 4.63).

Malkovich [18a] has drawn attention to the fact that equation 4.2 assumes that all of the diffusant put on the sample surface does in fact diffuse into the crystal. If the diffusant has a low solubility in the crystal matrix care needs to be taken in keeping W of equation 4.2 small enough to ensure that at all times the surface concentration does not exceed the solubility c_o. If W is such, however, that the surface concentration in the crystal is maintained at the solubility limit c_o for a period $\tau <$ the total diffusion time t, then†

$$\mathbf{D} = \mathbf{D}\,(\text{measured})\left(\frac{3b - 2\tan^{-1} b}{2b - \tan^{-1} b}\right) \tag{4.67}$$

where

$$b = \{\tau/(t-\tau)\}^{1/2} = \tan\left(\frac{\pi}{2} \cdot \frac{c(0,t)}{c_o}\right)$$

† A. D. LeClaire, private communication. Equation 4.67 is a better approximation than a previously published expression [56],

$$\mathbf{D} = \mathbf{D}\,(\text{measured})\left(2 - \frac{1}{b}\tan^{-1} b\right)$$

The equation given for b above is exact.

and **D** (measured) is obtained by fitting the experimental profile to equation 4.2. $c(0, t)$ is the surface concentration at time t. Obviously when $t = \tau$ equation 4.1 applies. A solution of the diffusion equation in the more general case of a continuous depletion of the source is given in Appendix A1, of Chapter 1.

4.4.3 The *p-n* Junction Method

This method requires the use of the infinite source boundary condition and if **D** is independent of c its value can be got as described in section 4.2.1. If **D** varies with c the procedures of section 4.2.1 can still be used but now yield an apparent junction diffusion coefficient $\mathbf{D}_j$ which can be used to predict the variation of junction depth x_j with time for given doping conditions. Chang [57] has shown that if **D** varies with c for $c > c_r$ and is constant for $c < c_r$ then for $c < c_r$

$$c = c_o^* \operatorname{erfc}\left(\frac{x}{2\sqrt{\mathbf{D}t}}\right) \tag{4.68}$$

where c_o^* is a fictitious surface concentration and **D** is the constant value appropriate to $c < c_r$. Fuller's method [1] then allows **D** to be obtained provided the impurity concentration at the junction c_j is less than c_r. A further discussion of the problems arising in the interpretation of results from the *p-n* junction and sheet resistivity method is given in section 5.5.1.

The accuracy of **D** determined from junction depth measurements is only as good as the accuracy to which the junction depth x_j can be measured. The typical difficulties have already been described (section 4.2.1). In addition variations in x_j can occur according to the technique of revealing the junction. The scanning light spot method gives the true electrical junction (i.e. ionized acceptor concentration = ionized donor concentration) whereas chemical staining or electrochemical plating reveals only the so-called 'metallurgical junction' which is displaced from the true electrical junction by a distance of the order of a carrier diffusion length.

4.4.4 The Effective Time of a Diffusion Anneal

In practice a sample always takes a finite time to reach and leave the diffusion temperature so that the real period of diffusion t is never just the time at the diffusion temperature. If $\mathbf{D} = D_o \exp(-Q/kT)$ then [58]

$$t = \int_0^{t'} \frac{\exp(-Q/kT(\tau))d\tau}{\exp(-Q/kT)} \tag{4.69}$$

where T is the diffusion temperature, t' is the whole duration of the diffusion anneal including the heating and cooling periods and $T(\tau)$ is the time dependent temperature of the sample with τ being the lapsed time.

4.4.5 The Role of Surface Effects

A sample surface can be prepared for diffusion either by mechanical lapping and polishing or by chemical polishing; a combination of both methods is also possible. If the surface isn't flat then errors will obviously arise in using planar solutions (equations 4.1 and 4.2) to describe the diffusion. It may also happen that the surface, although initially flat, becomes uneven or develops etch pits as a result of the diffusion anneal. Quantitative effects on the penetration profile are difficult to estimate. Borg and Lai [59] suggest that if the ratio (mean depth of undulations)/(root mean square penetration distance) is < 0.05 the error in **D** due to a non-planar surface is probably insignificant in a sectioning experiment. They also point out that if the surface is undulating then chemical sectioning is preferable as this will follow the surface contours.

A further difficulty which can occur is evaporation of the sample surface during the diffusion anneal. Analytical solutions are available for profiles generated using thin film or infinite source boundary conditions (see Appendix A2, Chapter 1) and also for the isotope exchange method [60], when the matrix is evaporating.

Mechanical lapping and polishing will generally produce a damaged surface layer on the semiconductor sample. The damage comprises cracks, high dislocation densities and lattice strain [61, 62, 63, 64]. Quantitative assessment of the damage is difficult as several factors are involved such as particle size of the abrasive, the mode of lapping (e.g. random or unidirectional), the load applied to the specimen during lapping, the hardness of the semiconductor and also the criteria for identifying damage. Most of the available evidence [61, 62] shows that the depth of damage is within a factor 2 of the mean particle size of the abrasive. It is evident that a damaged surface consisting of cracks will no longer provide a planar surface, equally a high dislocation density will give an enhanced short circuit path component to the overall diffusion. The presence of strain is probably of less significance as it appears to anneal out very readily [63]. It is therefore desirable to remove the damaged layer completely prior to a diffusion anneal, such as by chemical polishing or possibly by argon ion sputtering [51, 51a, 63]. If this is not feasible then $\sqrt{\mathbf{D}t}$ should be as large as possible relative to the thickness of the damaged layer, **D** being the volume diffusion coefficient. It should however be noted that Whitton [50] has reported that the vibratory polishing technique, although mechanical, produces no deformation in a silicon surface.

In using equations 4.1, 4.2 and 4.64 to analyze experimental data it is important to ensure that the experimental conditions conform to the boundary conditions appropriate to these solutions. Obvious departures occur if the surface is evaporating or if the source is depleted during the diffusion anneal. Remedies for both situations have been described above. A less obvious departure may arise if the source is in the form of a compound of the diffusant. In such a case the diffusion in the crystal may be controlled by a surface chemical reaction between the matrix and the diffusant source. An ensuing consequence is that the semiconductor can become 'screened' from the diffusant source by a layer of the reaction products.† In thin film experiments although a diffusant may be applied in elemental form, say by evaporation, it is quite possible that by the time the sample is at the diffusion temperature the diffusant is in the form of the oxide. The rate determining step may then be the reduction of the oxide by the matrix. This problem has been discussed by Reimann and Stark [65]. Haul and Just [66] have analyzed the effect of different surface reaction models on volume diffusion in the isotope exchange method.

4.5 CONCLUSIONS

The flow of a diffusant can be followed by using the change in any one of a variety of different parameters. The space and time variation of the particular parameter can then be utilized to obtain a diffusion coefficient associated with that parameter. If two different parameters are used to describe the diffusion of a particular species there is no reason to expect the two resulting diffusion coefficients to be identical: they will only be identical if the parameters have a one to one correspondence. Thus sheet resistivity methods measure primarily the diffusion coefficient of a resistivity change, a tracer sectioning method gives the average tracer atom (or ion) concentration in the section and the corresponding diffusion coefficient.

To be physically meaningful it must be possible to relate the diffusion coefficient, deduced from whatever parameter, to that of the atomic mechanism whose operation is measured by the particular parameter. If this cannot be done, and some of the difficulties in relating electrical

† The general effect will be a reduced c_0 or W in equations 4.1 and 4.2. These values may also be time dependent. The penetration of the diffusant will be reduced but if the source boundary conditions are known and can be handled analytically **D** may still be obtained.

parameters to defect diffusivities have been discussed in earlier sections, the diffusion coefficient can only be regarded as of empirical value. Even when a reliable concentration profile has been obtained the evaluation of the diffusion coefficient of the particular species must take into account the native defect concentrations, neutral and ionized, and the background impurity doping levels [67-72] (see also sections 1.2.4, 1.4.5). These concentrations must be controlled before reproducible and meaningful results can be expected. Further factors which require consideration in impurity chemical diffusion are (a) the role of the internal electrical field (see sections 1.4.2, 1.4.3) and (b) the generation of misfit dislocations because of the strain gradient set up by the impurity gradient at high concentrations [73-79] (see also section 1.4.5).

Most diffusion experiments employ either the thin film or infinite source boundary conditions in the solution of the planar diffusion equation. It is however quite feasible that the boundary conditions that apply during the actual experiment are different from those planned. More importantly the change may not be recognized so that the analysis used is not appropriate to the actual situation. Quite misleading results can therefore be obtained, especially if the diffusion coefficient is concentration dependent. Departures from simple planar volume diffusion with an infinite or thin film source can arise because:

(a) There is a significant short circuit path component e.g. due to a high density of dislocations created by a diffusing impurity [73-79] (see also section 1.4.5).
(b) There is a reaction between an in-diffusing impurity and a different impurity, already present in the crystal, to form a second phase inclusion.
(c) The equilibrium at the interface between the semiconductor and source in the infinite source case is not instantaneous.
(d) There is a chemical reaction between the source and semiconductor.
(e) The high temperature structure is not retained during cool down to room temperature. Precipitation [80-83] may occur as also out-diffusion from, and evaporation of, the sample. A more subtle and less obvious effect is a marked redistribution of an impurity due to strong vacancy fluxes created by the cooling process [84]. This could be important in shallow diffusion layers.

As a final comment one can claim a measure of reassurance, if Gaussian or erfc concentration profiles are found, that the experiment is observing what was planned. If more complex profiles are found considerable care is needed to ensure that the reality has not diverged from the expectations.

ACKNOWLEDGMENTS

The author would like to thank many colleagues in IBM for their discussions, particularly Dr. A. H. Tong, Dr. W. A. Keenan, and Mr. P. A. Schumann, Jr., for their critical comments on the scanning light-spot, spreading resistance, and diode capacitance technique, respectively.

REFERENCES

1. C. S. Fuller, *Phys. Rev.*, **86**, 136 (1952)
2. M. B. Prince, ibid., **93**, 1204 (1954)
3. C. S. Fuller and J. A. Ditzenberger, *J. Appl. Phys.*, **27**, 544 (1956)
4. R. J. Archer, *Phys. Chem. Solids*, **14**, 104 (1960)
5. P. J. Whoriskey, *J. Appl. Phys.*, **29**, 867 (1958)
6. M. O. Thurston and J. Tsai, Ohio State University Research Foundation Sixth Quarterly Report—1233-6Q, April 1, 1962, to June 30, 1962
7. D. R. Turner, *J. Electrochem. Soc.*, **106**, 701 (1959)
8. S. J. Silverman and D. R. Benn, ibid., **105**, 171 (1958)
9. P. A. Iles and P. J. Coppen, *J. Appl. Phys.*, **29**, 1514 (1958)
10. A. H. Tong, P. A. Schumann, Jr., and A. Dupnock, 'Semiconductor Silicon' (Edited by R. R. Haberecht and E. L. Kern), The Electrochem. Soc. (1969)
11. W. J. Pietenpol, *U.S. Patent 2,* **790**, 952 (April 30, 1957)
12. F. S. Goucher *et al.*, *Phys. Rev.*, **81**, 637 (1951)
13. J. Orshnik and A. Many, *J. Electrochem. Soc.*, **106**, 360 (1959)
14. J. N. Shive, *Proc. Inst. Radio Engrs.*, **40**, 1410 (1952)
15. C. A. Hogarth, *Proc. Phys. Soc.*, **B69**, 791 (1956)
16. B. Jansen, *Solid-St. Electron.*, **2**, 14 (1961)
17. E. Tannenbaum, ibid., **2** 123 (1961)
18. T. H. Yeh, S. M. Hu and R. H. Kastl, *J. Appl. Phys.*, **39**, 4266 (1968)
18a. R. S. Malkovich, *Soviet Phys. solid St.*, **1**, 548 (1959)
18b. K. R. Cramer and W. B. Crow, *phys. stat. sol. (a)*, **1**, 81 (1970)
19. W. C. Dunlap, Jr., *Phys. Rev.*, **94**, 1531 (1954)
20. B. J. Masters, Symposium on Silicon Device Processing, June 2-3, 1970, National Bureau of Standards, Gaithersburg, Md.
21. G. Backenstoss, *Bell System Tech. J.*, **37**, 699 (1958)
22. J. C. Irvin, ibid., **41**, 387 (1962)
23. F. M. Smits, ibid., **37**, 711 (1958)
24. W. G. Spitzer and H. Y. Fan, *Phys. Rev.*, **106**, 882 (1957)
25. E. E. Gardner, W. Kappalo and C. R. Gordon, *J. Appl. Phys. Lett.*, **9**, 432 (1966)
26. H. A. Hyden, *Phys. Rev.*, **134**, A1106 (1964)
27. H. Rupprecht, *Bull. Amer. Phys. Soc.*, **8**, 228 (1963)
28. R. P. Donovan, Integrated Silicon Device Tech., Vol. 12, Measurements Techniques ASD-TDR-63-316, or AD643610 (1966); Symposium on Silicon Device Processing, June 2-3, 1970, National Bureau of Standards, Gaithersburg, Md.
29. I. M. Mackintosh, *J. Electrochem. Soc.*, **109**, 392 (1962)
30. R. G. Mazur and D. H. Dickey, *J. Electrochem. Soc.*, **113**, 255 (1966)
31. E. E. Gardner, P. A. Schumann, Jr., and E. F. Gorey, Electrochem. Soc. Symposium Proceedings, 'Measurement Techniques for Thin Films', April 1967
31a. D. C. Gupta and J. Y. Chan, *Rev. Sci. Instruments,* **41**, 176 (1970). D. C. Gupta, J. Y. Chan and P. Wang, *Rev. Sci. Instruments,* **41**, 1681 (1970)
31b. P. J. Severin, *Solid-St. Electron.*, **14**, 247 (1971)
32. R. Holm, Electrical Contacts Handbook. (Springer-Verlag, Berlin, 1958)
33. P. A. Schumann, Jr., J. M. Adley, M. R. Poponiak, C. P. Schneider and A. H. Tong, *J. Electrochem. Soc.*, **116**, 150C (1969)

34. P. A. Schumann, Jr., and E. E. Gardner, *J. Electrochem. Soc.*, **116**, 88 (1968)
35. P. A. Schumann, Jr., and E. E. Gardner, *Solid-St. Electron.*, **12**, 371 (1969)
36. T. H. Yeh and K. H. Khokhani, *J. Electrochem. Soc.*, **116**, 1461 (1969)
37. S. M. Sze, Physics of Semiconductor Devices. (John Wiley and Sons, New York, 1969)
38a. E. G. Schibli and A. G. Milnes, *Solid St. Electron.*, **11**, 323 (1968)
38b. E. G. Schibli, *Solid St. Electron.*, **13**, 392 (1970)
38c. C. R. Crowell and K. Nakano, *Solid St. Electron.*, **15**, 605 (1972)
38d. G. I. Roberts and C. R. Crowell, *J. Appl. Phys.*, **41**, 1767 (1970)
39. W. E. Carter, H. K. Gummel and B. R. Chawla, *Solid St. Electron.*, **15**, 195 (1972)
40. B. L. Smith and E. H. Rhoderick, *Brit. J. App. Phys.*, **2** (Series 2), 465 (1969)
41. F. A. Cunnell and C. A. Gooch, *Solid-St. Electron.*, **15**, 127 (1960)
42. B. Goldstein and C. Dobin, *Solid-St. Electron.*, **5**, 411 (1962)
43. D. L. Kendall, Diffusion in III-V Compounds with Particular Reference to Self-diffusion in In-Sb, Report No. 65-29, Department of Material Science, Stanford University, Stanford, Calif. (August 1965)
44. C. S. Fuller and K. B. Wolfstirn, *J. Appl. Phys.*, **33**, 2507 (1962)
45. A. B. Lidiard and K. Tharmalingam, Disc. Faraday Soc., No. 28, p. 64 (1959)
46. R. W. Broderson, J. N. Walpole and A. R. Calawa, *J. Appl. Phys.*, **41**, 1484 (1970)
47. K. Zanio, *J. Appl. Phys.*, **41**, 1935 (1970)
48. C. P. Buhsmer, *J. Mater. Sci.*, **5**, 1015 (1970)
49. B. Goldstein, *Rev. Sci. Instru.*, **28**, 289 (1957)
50. J. L. Whitton, *J. App. Phys.*, **36**, 3917 (1965)
51. D. Gupta and R. T. C. Tsui, *App. Phys. Lett.*, **17**, 294 (1970)
51a. S. M. Davidson, *J. Phys. E.*, **5**, 23 (1970); *J. Mater. Sci.*, **7**, 473 (1972)
52. L. D. Hall, *J. chem. Phys.*, **21**, 87 (1953)
53. P. G. Shewmon, Diffusion in Solids, p. 28. (McGraw-Hill, New York, 1963)
54. R. E. Pawel and T. S. Lundy, *Acta metall.*, **17**, 979 (1969)
55. E. Nebauer, *phys. stat. sol.*, **36**, K63 (1969)
56. F. A. Smith and L. W. Barr, *Phil. Mag.*, **21**, 633 (1970)
57. J. J. Chang, *IEEE Trans.*, **ED-10**, 357 (1963)
58. D. Y. F. Lai, Diffusion in Body-Centred Cubic Metals, p. 269. (American Society for Metals, 1965)
59. R. J. Borg and D. Y. F. Lai, *J. Appl. Phys.*, **39**, 2738 (1968)
60. D. K. Dawson, L. W. Barr and R. A. Pitt-Pladdy, *Br. J. Appl. Phys.*, **17**, 657 (1966)
61. E. N. Pugh and L. E. Samuels, *J. electrochem. Soc.*, **108**, 1043 (1961); **109**, 409 (1962); **111**, 1429 (1964); *J. Appl. Phys.*, **35**, 1966 (1964)
62. H. C. Gatos, M. C. Lavine and E. P. Warekois, *J. Appl. Phys.*, **31**, 1302 (1960); *J. electrochem. Soc.*, **108**, 645 (1961)
63. A. Taloni and D. Haneman, *Surface Sci.*, **8**, 323 (1967)
64. B. G. Cohen and M. W. Focht, *Solid-St. Electron.*, **13**, 105 (1970)
65. D. K. Reimann and J. P. Stark, *Acta metall.*, **18**, 63 (1970)
66. R. Haul and D. Just, *J. Appl. Phys.*, **33**, 487 (1962)
67. R. L. Longini and R. F. Greene, *Phys. Rev.*, **102**, 992 (1956)
68. M. W. Valenta and C. Ramasastry, ibid., **106**, 73 (1957)
69. R. A. Swalin, *J. Appl. Phys.*, **29**, 670 (1958)
70. P. Baruch, C. Constantin, J. C. Pfister and R. Saintesprit, *Discussions Faraday Soc.*, **31**, 76 (1961)
71. M. F. Millea, *J. Phys. Chem. Solids*, **27**, 309, 315 (1966)
72. G. D. Watkins and J. W. Corbett, *Phys. Rev.*, **134A**, 1359 (1964)
73. Y. Sâto and H. Arata, *J. Appl. Phys. Japan*, **3**, 511 (1964)
74. J. E. Lawrence, *J. Electrochem. Soc.*, **113**, 819 (1966)
75. M. L. Joshi and F. Wilhelm, *J. Electrochem. Soc.*, **112**, 185 (1965)

76. R. C. McDonald, G. G. Ehlenberger and T. R. Huffman, *Solid-St. Electron.*, **9**, 807 (1966)
77. W. Czaja, *J. Appl. Phys.*, **37**, 3441 (1966)
78. E. Levine, J. Washburn and G. Thomas, ibid., **38**, 81 (1967)
79. T. H. Yeh and M. L. Joshi, *J. Electrochem. Soc.*, **116**, 73 (1969)
80. G. Thomas, *Trans. AIME*, **233**, 1608 (1965)
81. M. L. Joshi, B. J. Masters and S. Dash, *Appl. Phys. Lett.*, **7**, 306 (1965)
82. M. L. Joshi, *J. Electrochem. Soc.*, **113**, 45 (1966)
83. M. L. Joshi and S. Dash, *IBM J. Res. Develop.*, **11**, 271 (1967)
84. R. E. Hanneman and T. R. Anthony, *Acta metall.*, **17**, 1133 (1969)

DIFFUSION IN SILICON AND GERMANIUM

5

S. M. Hu

CONTENTS

5.1 INTRODUCTION

Silicon, and to a lesser extent germanium, provides the most important building foundation for modern electronic devices. Its preeminence derives both from the availabilitv of the purest and most perfect crystals known, and from our ability to create electric potential profile structures and carrier concentrations to specifications in this material by controlled doping with impurity atoms. One of the most important methods of controlled doping is that of solid state diffusion. The high density and the high speed of devices in modern integrated circuits often require a vertical dimension of 1-3000 Å and a lateral dimension of several microns. The complexity of vertical potential structures often requires several sequential diffusion steps. The net effect of various diffusion steps on an electric potential structure may be subtractive, and there may be interactions between sequential diffusion steps. The current state of the art of diffusion processes in the fabrication of integrated circuits relies on an empirical approach with trial and error. The difficulty of such an approach will increase exponentially with the increase of the number of diffusion steps, and a semiempirical approach incorporating an understanding of the details of diffusion processes may become essential. Presently there is an intensive effort in the industry in the development of computer-aided design of integrated circuits incorporating optimization of device performance, which is intimately coupled to the exact details of the impurity distributions and their attainability. Modeling and computer simulation of impurity profiles can often provide details in shallow, complex structures which are usually beyond the feasibility of accurate experimental determination. At least in the simple system of arsenic diffusion in silicon, simulation of complex structure from multistep diffusion processes has met considerable success. This has undoubtedly helped promote a common sense for the need of an in-depth understanding of the various diffusion processes in semiconductors.

The intensive technological exploitation of silicon and germanium devices has produced a great wealth of experimental data in the literature on diffusion in these materials. However, interpretations of the data are not often definitive and consistent with all aspects. It is not the purpose of the present review to provide answers to the various problems, which would not only be presumptuous, but unattainable. For, to quote a well-worn phrase, more experimental information will be needed. Nor is it to give a comprehensive compilation and summary of published works on the subject. Rather, what is to be presented is the author's personal perspective of the subject. In that sense, the present chapter is not meant to complement or update those previous reviews [1-5]. The emphasis will be placed on the fundamental aspects of diffusion processes, which is what is problematic. The discussions will not be concerned with the practical implementations of diffusion processes, which are basically uncontroversial, and are characterized by relative merits rather than physical validity.

5.2 LATTICE DEFECTS IN SILICON AND GERMANIUM

5.2.1 Lattice Defects and Diffusion in Silicon and Germanium

Before going into detailed discussions of lattice defects and diffusion mechanisms, it will be useful to cast an overview of the pertinent relationships between lattice defects and diffusions in silicon and germanium. Although it remains rather uncertain in regard to definitive mechanisms for various diffusion processes in silicon and germanium, it is certain that diffusion takes place through the motion of lattice defects in one form or another. The ring mechanism of Zener [6] has not been observed in metals, and has been considered improbable by Huntington and Seitz [7-9] on account of energetics from theoretical calculations. It should be still more improbable in diamond lattices where a smallest ring consists of six atoms and the motion would involve a breaking of 12 covalent bonds. The direct interchange of two neighboring atoms in a diamond lattice, though more probable than in metals in comparison with the ring mechanism because of the openness of the diamond structure, is still highly improbable because it requires a simultaneous breaking of six covalent bonds. A different version of the interchange mechanism was proposed by Millea [10] to account for the observation of enhanced diffusivity of indium in *p*-type silicon (a more detailed discussion will be given in section 5.4.3). The version should be called indirect interchange because in this model a lattice atom will first leave its regular substitutional site and become an interstitial, and then one of its nearest neighboring substitutional atoms moves into the vacancy left behind by the first atom. Then the first atom, now at the interstitial position, moves into the vacancy left behind the second atom, thus completing the cycle. We shall argue that this mechanism is also unlikely on account of energetics. The first step in this process involves the energy of a Frenkel pair. The enthalpy of the formation of a Frenkel pair is of the order of the sum of the enthalpies of formation of an interstitial and a vacancy. The activation enthalpy of the generation of a Frenkel pair will be ΔE more than the enthalpy of its formation, ΔE being the energy barrier to the recombination of the Frenkel pair. One then sees that if ΔE is of the order of the activation energy of vacancy motion, the generation of a Frenkel pair from a given lattice atom will be much less frequent than the occurrence of the event that a vacancy happens to be neighboring to that atom and exchanges sites with it. In other words, such an interchange mechanism would be less likely than a vacancy mechanism. If ΔE is smaller than the activation energy of vacancy motion, then by great odds the Frenkel pair will recombine before the second atom moves into the vacancy. The Frenkel defect need not be a close pair, if the energy barrier to the dissociation of the pair is low e.g., a barrier characteristic of interstitial migration in diamond lattice. This is then the dissociative mechanism, as originally proposed by Frank and Turnbull [11]. Thus one sees that the Frenkel pair need not be short-lived for a small ΔE in this process. For those elements which have energies of substitutional solution comparable to or higher than interstitial solution (e.g., the transition elements), the probability that a substitutional atom of such elements dissociates into a Frenkel pair should be comparable to or even greater than the probability of having a lattice vacancy as one of its nearest neighbors assuming the Coulombic interaction is compensated by the exchange of one solute-solvent bond for a solvent-solvent bond. In diamond type lattices the migration energy of an interstitial is generally very small in contrast to that in metals, as we shall see later (e.g., ~0.4 eV in silicon). Then one can expect the dissociative process to be comparatively more efficient, on the assumption that the migration energy of a vacancy is much higher. Subsequently we shall discuss the finding by Watkins [11] of the surprisingly low activation energy of migration of a simple vacancy in silicon (~0.33 eV). This has posed many difficult problems, including the virtual existence of a dissociative diffusion mechanism, for the simple vacancy mechanism and the indirect interchange mechanism would then be expected to be more efficient, at least from the energetics point of view.

The interstitial diffusion mechanism is not the counterpart of the vacancy mechanism. The latter finds its counterpart in the so-called 'interstitialcy' mechanism proposed by Seitz [9]. In this mechanism it is supposed that the overlap repulsive potential at the saddle point in a normal interstitial path is very high, so that the interstitial atom chooses to move by pushing one of its nearest neighbors into another interstitial site and itself takes up the substitutional site; the same process is repeated by the new interstitial. A closer look reveals however, that this is not an exact counterpart of the vacancy mechanism. To show this, two points can be raised. First, the interstitial can be labeled (for example, it can be a tracer or an impurity atom) while a vacancy cannot (in the sense that we are considering the vacancy in a dilute solution having an elemental solvent as opposed to an alloy); second, in the interstitialcy mechanism the correlation effect of the movement of the labeled atom occurs only alternatively, i.e., the correlation effect exists only when the labeled atom is at the substitutional site. However, the antisymmetry between the interstitialcy and the vacancy mechanisms can be restored quite simply by viewing two consecutive steps as a single step, thereby insuring the interstitial to be always an unlabeled solvent atom and restoring the correlated movement of the labeled atom. Another interesting consequence of this concept is that the relationship between the diagonal and the off-diagonal coefficients of the phenomenological equation for a vacancy mechanism as derived by Hu [13] also can find its counterpart in the interstitialcy mechanism by replacing a vacancy with an interstitial and a negative sign. Thus, the conservation of lattice sites can be stated as

$$[A] + [B] - [I] = [S], \tag{5.1}$$

where $[A]$, $[B]$, $[I]$, and $[S]$ are respectively the local concentrations of solute atoms, solvent atoms, interstitials, and lattice sites. The flux relationship also holds:

$$J(A) + J(B) - J(I) = 0 \tag{5.2}$$

which consequently introduces the counterpart of the chemical pump effect, interstitial-impurity complexes and related aspects. For a fuller discussion of these topics relevant to a vacancy mechanism, the reader is referred to reference 13. The reader may easily extend these analyses to an interstitialcy mechanism. One can also define an interstitial-impurity complex. It is worth pointing out that the hypothetical interstitial-impurity complex can migrate in a diamond lattice without dissociation, in contradistinction from a vacancy-impurity complex.

One question naturally arises. How can one decide which one of the two mechanisms, the interstitial and the interstitialcy will dominate in silicon and germanium? In metals whose lattices are densely packed and the overlap repulsive potential is very high at the saddle point in a normal interstitial path, one can predict that the interstitialcy mechanism should be preferred [7, 8]. On the other hand the diamond lattice has a very open structure, and the overlap repulsive potential should be very small. We shall discuss other types of (mostly Coulombic) interaction in detail in a later section. The energy barrier at the normal interstitial saddle point should be a fraction of an eV. It appears that no theoretical estimation has been available on the formation of an interstitialcy in a diamond type lattice. In fact, the structure of an interstitialcy configuration in the diamond lattice has seldom been described in the literature. Corbett [14] mentioned one such possible configuration, sometimes called a split-interstitial, or a semi-interstitial pair (figure 5.1). Such a configuration appears to have an unclear bonding nature. The symmetry point group of such a configuration is hence unclear; but it should be lower than that of a tetrahedral point group. This is unlikely to be a stable configuration, at least for interstitial impurity ions such as Li and Al, for optical and EPR spectra have shown Li [15, 16] and Al [17, 18] to have tetrahedral symmetry in silicon. Even in the case of boron-lithium pairs in silicon where one might think a lower symmetry is preferred, internal friction studies have revealed that lithium is situated at the tetrahedral site [19]. Although the symmetry environment of boron has been lowered from T_d to C_{3v}, and that it is possible that both boron and lithium could be slightly displaced from their respective T_d sites, evidence from isotope

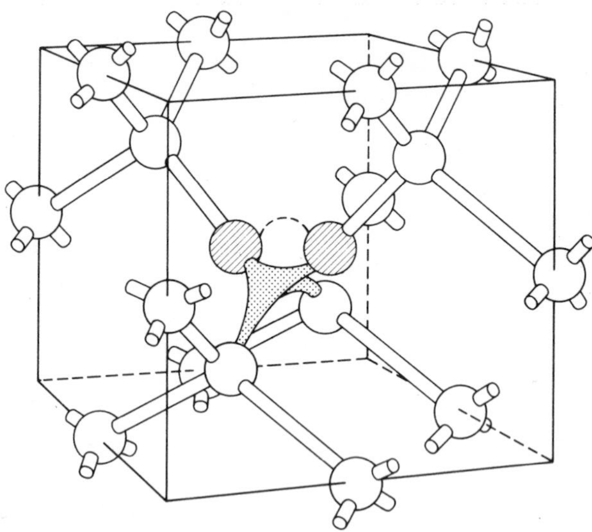

Figure 5.1. A possible split-interstitial configuration in the diamond-type lattice. After Corbett [14].

replacement showed that the motion of either ion of the B-Li is independent of each other [20]. This may be interpreted to mean that a split-interstitial of B-Li of the type depicted in figure 5.1 was not involved. Then, one may conclude that if an interstitialcy mechanism occurs in a diamond type lattice, the split-interstitial should be the saddle point rather than the equilibrium configuration. At this point, it may be noted briefly that most known experimental evidence seems to indicate that lithium, and even doubly positive aluminium are located at the tetrahedral site, instead of the hexagonal site predicted by Weiser [21]. Detailed discussion of this topic and some theoretical considerations will be given in a later section.

The view of a lattice defect simply as a missing atom from the regular lattice site or an extra atom at a regular lattice interstice is merely modeling of a zeroth order approximation; lattice relaxation, however small and in whatever manner, will occur in the vicinity of the defect. A diamond type lattice is unlikely to accommodate a crowdion type interstitial as proposed by Paneth [22] for alkali metals (*bcc*) in which a compressed region due to the presence of an interstitial is linearly extended over several lattice spacings. Seeger *et al.* [3, 23] have proposed an 'extended interstitial' in silicon and germanium, in which a disordered region about the interstitial is extended isotropically to about the second coordination atoms. (There are four first coordination and six second coordination atoms about a tetrahedral interstitial site in a diamond lattice. For the reason mentioned earlier, we will not be concerned with an 'extended interstitial' derived from an interstitial at a hexagonal site which has six first coordination and eight second coordination atoms, the latter consists of two sets of distinct symmetry, 2 and 6 respectively.) Such a model of the relaxed interstitial is a counterpart of a relaxed vacancy model called the 'relaxion' proposed by Nachtrieb and Handler [24] for metals. Seeger and Swanson [23] supported their argument by citing the fact that the density of silicon and germanium increases on melting by ~10% for silicon and ~5% for germanium. This is of course not evidence that an interstitial in silicon or germanium is extended, or conversely that an extended defect in silicon or germanium is an interstitial. Theoretical calculations by Bennemann [25] showed that in an undistorted lattice the formation energy of an interstitial is 1.09 eV in Si and 0.93 eV in Ge, while the formation energy of a vacancy is 2.13 eV in Si and 1.91 eV in Ge. We shall defer comments on Bennemann's analysis for the moment. But if his results are not too far off, one would anticipate the formation of an interstitial to cause less extended distortion of the lattice surrounding the defect than the formation of a vacancy. The formation energies cited above would imply that the interstitial concentration should be much higher than the vacancy concentration in Si and Ge. However, if one assumes the vacancy to have an extended disordered region, and takes Seeger and Swanson's value for an entropy of formation of ~10 k, and comparing the resulting free energy of formation with that of an unrelaxed interstitial, one could then expect the vacancy concentration to be comparable to the interstitial concentration. Seeger and Swanson's model of an extended interstitial is a local molten region, or a 'droplet' consisting of ~10 atoms. They estimated from the entropy of fusion an enthalpy of formation of such a droplet to be ~4.5 eV (=10 $S_m T_m$, where S_m is the entropy of fusion, and T_m is the melting point), which is in close agreement with estimates from the heat of fusion. A question one may raise against this model is why should the 'liquid droplet' interstitial be preferred over the simple interstitial, considering the much higher free energy of formation for the former than for the latter? Since a silicon or a germanium interstitial (as against an aluminium or a lithium interstitial) has not yet been detected in EPR work, one has no knowledge about the symmetry property of the interstitial. However, even if a future finding should prove it to belong to a T_d symmetry, it is itself no evidence against the liquid drop model which is presumed to exist only at high temperatures.

There seems to be less attention paid to the structure of a 'simple vacancy'. It is often assumed that a normal simple vacancy is one lattice site that is unoccupied, plus, perhaps, necessary lattice distortion around it. The 'split vacancy' (to coin a counterpart of the 'split interstitial' in the interstitialcy mechanism of interstitial migration) is usually regarded as the saddle point in the path of vacancy motion. Published results of theoretical calculations tend to corroborate this view [25-27]. Recently, an interesting view has been expressed by Masters [28] who considers the split vacancy (or in his nomenclature 'semi-vacancy pair', which also finds a counterpart in interstitial nomenclature) to be the equilibrium configuration, with the 'normal' configuration as the metastable configuration. The saddle is somewhere between these two positions. While contradicting theoretical calculations, this model attempts to reconcile the results of Watkins and the commonly known self-diffusion data in silicon. We will return to this topic later.

5.2.2 Properties of Point Defects in Silicon and Germanium

Among the properties of lattice defects that are relevant to diffusion phenomena, we will be in greater part concerned with the energy of formation and of migration. Other properties that are also relevant to diffusion phenomena include electron energy levels of the defect states, the interactions and the binding energy between defects and between defects and impurity atoms, the entropy of formation and of migration, and the defect structure and its multiplicity. The latter two properties, i.e., the entropy and the structure, are the least known both theoretically and experimentally. In a sense, a knowledge of the structure of a defect is a prerequisite to the theoretical calculation and, to a lesser degree, the experimental interpretation of the other properties. Practical limitations often compel one to assume a defect structure of the highest symmetry, which in the diamond lattice is the point group T_d. Understandably the entropies of formation and migration which are strongly structure dependent, are most uncertain. Experimentally, the lot of entropy is not much better. While the enthalpy of formation and of activation is easily obtainable from the ubiquitous Arrhenius plot, there is no simple way of determining entropy unequivocally, and the entropy tends to receive a lumped leftover from the estimation of enthalpy. The electron energy level of the defect states is important for the reason that both the defect jump frequency and the defect concentration depend on the defect charge state, which is determined by the energy level of the defect states relative to the Fermi level of the semiconductor bulk.

Properties of Point Defects from Theoretical Calculations

Methods of theoretical calculation of the energies of formation and activation of point defects have been discussed in detail in Chapter 2 of this book. In this section we shall only summarize some of the results of theoretical calculations apropos silicon and germanium. There are two categories of theoretical calculations that have been published. One is semi-empirical, employing a method of calculation that is classical in nature. The energy of the formation of a defect in the lattice is considered to consist of a repulsive potential arising from orbital overlap, and Coulombic interactions of screened point charges and of lattice polarization, with various numbers of atoms allowed to relax around the defect whereby the sum of the above interaction energies over these atoms (considered to approximate the lattice energy due to the defect) is minimized. The second category is the quantum mechanical first principles calculation which takes two different and rather opposite approaches, one being a derivative of the one-electron energy band approach and the other of the molecular orbital approach. The first approach in essence presumes a weak defect potential in the spirit of pseudopotential for the perfect lattice; the second approach presumes a tight binding of localized states. The pros and cons of these two approximations are not easily assessed [29], and the trend of following these two separate approaches is expected to continue.

Table 5.1 summarizes various theoretical results for vacancies and interstitials in silicon and germanium. A discussion of these results follows. First one notices that the tabulated theoretical results from various sources agree with each other rather remarkably, even though the methods employed and the types of interactions emphasized are so different. It is of course, more difficult to assess the accuracy of the semiempirical approach, which, if anything, is more in the spirit of molecular orbital. Yet Swalin's results are very close to those of Bennemann's which are based on free-electron-like pseudopotential scattering theoretic analysis. If one does not attribute this agreement to fortuity, one will have the great difficulty of proving the equivalence of these two apparently different physical models. Bennemann's use of free-electron wave functions in zero order has been criticized by Callaway and Hughes [32]. Bennemann constructed a lattice potential which is a lattice sum of Ge^{4+} (or Si^{4+}) at lattice sites immersed in a uniform electron gas. While this has no effect on the Coulombic interaction of a charged defect at a lattice site, we find this approximation excessively overestimates the interstitial potential in comparison with the opposite model of assuming the majority of the electrons to localize between atoms in the spirit of sp^3 orbitals [33]. Dipole and multipole interactions as well as lattice distortions around the defect were neglected. We feel that the lattice distortion could significantly affect the scattering properties of the model potential of the defect. Bennemann's analysis includes a term (his ΔE_3) which arises from

TABLE 5.1

Theoretical enthalpies of formation and migration of simple point defects in silicon and germanium

Enthalpy[a]	Ge (eV)	Si (eV)	Author(s)	Method
H_V^f	2.07	2.32	Swalin [26]	1
	1.97	2.35	Scholz and Seeger [30]	1
	1.91	2.13	Bennemann [25]	2a
	2.21 ± 0.18		Huang and Watt [31]	2b
H_V^m	0.95	1.06	Swalin [26]	1
	0.98	1.09	Bennemann [25]	2a
H_I^f	0.93	1.09	Bennemann [25]	2a
H_I^m	0.44	0.51	Bennemann [25]	2a
	0-0.25	0-0.22	Hasiguti [37]	1

[a] H_V^f, H_V^m, H_I^f, and H_I^m denote enthalpies of formation and of migration of a vacancy and of an interstitial respectively.

1 Semi-empirical method of calculation.

2a First principle calculation, one-electron band approach.

2b First principle calculation, molecular orbital approach.

the change of the kinetic energy of a uniform electron gas with the compression (due to the formation of interstitials) or the expansion (due to the formation of vacancies) of the crystal volume occupied by the electron gas. This term has a value of 15.7 eV, negative for vacancies and positive for interstitials. While the compression effect on the kinetic energy of an electron gas has been considered important in metals [34], it is difficult to assess the extension of this model to valence crystals like silicon and germanium. Seeger and Chik [3] commented on the ΔE_3 in Bennemann's treatment for his neglect of the lattice distortion and thus overestimating the 'vacancy volume' as well as ΔE_3. Scholz and Seeger [30] estimated the 'vacancy volume' in germanium to be ~0.7 atomic volume. Seeger and Chik [3] then suggested that the overestimated 0.3 atomic volume of the vacancy could contribute an error of ~4.4 eV to ΔE_3. However, it should be noted that the 'vacancy volume' is not necessarily the same as the change in crystal volume, which determines the kinetic energy of the electron gas. Another serious question of Bennemann's analysis is that it involves summation and subtraction of terms of large values of the order of ~15 to ~40 eV (each of which is independently model-dependent) to yield a small value of the order of ~1-2 eV.

A criticism of the empirical approach of Swalin [26] and Scholz and Seeger [30] was given by Huang and Watt [31]. Huang and Watt themselves used a first principles approach of molecular orbitals. This is based on the concept of a defect molecule first introduced by Coulson and Kearsley [35] for diamond and further extended by Yamaguchi [36]. Such an approach involves smaller computational efforts than the energy band approach. Their result was dependent on the cohesive energy chosen and the spread is 0.47 eV (1.91 to 2.38 ± 0.18 eV).

Beside Bennemann, Hasiguti [37] also calculated the migration energy of a self-interstitial in germanium and silicon, giving 0-0.25 eV for Ge and 0-0.22 eV for Si. We feel his method of calculation is of questionable validity. He followed Weiser's [21] model of diffusion of interstitial monovalent ions in silicon and germanium by assuming that a self-interstitial in Ge and Si is also singly ionized. We noted earlier that Weiser's theory is in contradiction with known experimental results in regard to the interstitial's equilibrium position, i.e., the hexagonal site (S_6) versus the tetrahedral site (T_d). Theoretically, one of the most important factors, the non-vanishing lattice potential at both the hexagonal and the tetrahedral site has been neglected in Weiser's model. Depending on the model used, this lattice potential could range from a fraction of an eV to ~20 eV, repulsive to a positively charged ion. Presumably this is also the reason why Weiser's theory predicted a negative and rather large enthalpy of formation of an interstitial in Ge and Si, which in his model is dominated by the energy of lattice polarization. There are also questionable aspects regarding the use of a pair-wise Born-Mayer potential. The use of the covalent radii of Ge and Si for the pair-wise interaction between a host lattice atom and an interstitial ion is questionable because the valence electron orbitals have an sp^3 character, rather than being spherical though it may be empirically justifiable for the pair-wise interaction between two host atoms. On the opposite extreme, Millea [10] suggested the use of the closed shell core radii. The Born-Mayer repulsive potential has the form

$$U(r) = A \exp[(r_0 - r)/\rho], \tag{5.3}$$

where A and ρ are constants characteristic of the pair of atoms; r_0 is the 'equilibrium distance' between the pair, and r is the actual distance between the pair. In its application to an ion at an interstitial site, it is customary to take r as the distance from the interstitial site to the lattice site of the other atom in the pair, and to take $r = r_L + r_I$, where r_I is the radius of the interstitial ion, and r_L is the radius of the lattice atom. The ambiguity as to what is the correct r_L in the direction of the interstitial leads to uncertainty as to what is after all the equilibrium distance between the interstitial and the lattice atom. Then, partly as a consequence of this, the constants A and ρ derived from lattice compressibility data are also questionable when applied to interstitial-lattice atom interaction.

The above discussions of the present state of theoretical calculations of defect properties, if somewhat discouraging, serve to caution the unjudicial use of the theoretical results simply because they tended to agree with each other. However, since the one electron energy band and molecular orbital approaches are fundamentally sound, we can anticipate further efforts on these lines to become of quantitative value.

A theoretical analysis of defect levels in germanium was first made by James and Lark-Horovitz [37a] using a hydrogenic model which is akin to the later effective mass approximation model in the analysis of shallow impurity states in semiconductors. According to this model, the vacancy is a double acceptor with the first ionization level at $\sim(E_v + 0.05)$ eV, and the interstitial is a double donor with a first ionization level at $\sim(E_c - 0.05)$ eV. Blount [37b] disputed the James and Lark-Horovitz model, and he himself put forward some qualitative arguments concerning possible atomic and molecular orbitals of the vacancy and the interstitial. He used these arguments to reassign the four experimentally observed levels in irradiated germanium to the vacancy and the interstitial; but no quantitative calculations were made. The first quantitative analysis of the defect levels using a molecular orbital approach was developed by Coulson and Kearsley [35] for diamond, and later extended by Yamaguchi [36]. We would like to see such calculations performed for defects in silicon and germanium. The only known calculation using the molecular orbital method is that of Hwang and Watt [31] for the vacancy formation energy in germanium, but not for electronic energy levels. Since such energy levels are quantum mechanical eigenvalues, they cannot be obtained with semiempirical methods such as those used for the calculations of the formation and migration energies of point defects in silicon and germanium. More recently, Callaway and Hughes [32] have made some analysis of the energy levels of the monovacancy and the divacancy in silicon, using an energy band approach. However, an energy band method that gives definitive quantitative results remains to be developed.

Experimental Studies of Point Defects in Silicon and Germanium

One difficulty in the experimental study of the properties of point defects in germanium and silicon is the low defect concentrations at thermal equilibrium up to the melting point. Extrinsic defects often dominate the electrical and optical characteristics of Ge and Si. Two categories of the study of defect properties will be discussed. One is the direct study of defect properties using electrical conductivity, photoconductivity, infrared absorption spectroscopy, and electron paramagnetic resonance. The other is the indirect study of such measurable quantities as the rate of dislocation climb, the size of dislocation loop, precipitation, and minority carrier life time through the interaction of point defects with dislocations, impurity atoms, and carriers. A self-diffusion study is unique as it is indirect in the sense that one measures the tracer atoms with which the defects interact, yet it is direct in the sense that it involves nothing other than the defect movement. Unfortunately, self-diffusion, alone, is not only unable to furnish information separately for the energies of defect formation and defect migration, but is also unable to identify the defect involved. On the other hand, it is possible to distinguish the defect interacting with dislocations whether it is a vacancy, or an interstitial by the transmission electron microscopy (TEM) of the dislocation loops [38]. The TEM image contrasts are however, unable to inform on whether the vacancies are simple vacancies or divacancies (the di-interstitials in Ge and Si are presumably unfavored). Seeger and Chik [3] pointed out that in diamond lattices the elementary cell contains two atoms and a jog formed in a dislocation line by the intersection with another dislocation line extends from one $\{111\}$ glide plane to the next nearest parallel $\{111\}$ plane rather than the nearest one. They suggested that the non-conservative motion of such jogs will thus create divacancies rather than simple vacancies (or di-interstitials rather than simple interstitials). However, we do not share this view. At high temperatures ($>T_m^{2/3}$ where T_m is the melting point) where plastic deformation takes place, the binding factor exp (E_b/kT) is $<10^7$, taking E_b the binding energy of the divacancy to be 1.6 eV. Then, unless the monovacancy concentration is $>10^{16}$ cm^{-3} (which is unlikely in silicon and germanium), thermal equilibrium should favor the formation of monovacancies, and the divacancies will dissociate rapidly after their creation. From another point of view, we will be more interested in dislocation climb rather than dislocation glide inasmuch as quenching, diffusion induced dislocations, and precipitations are concerned. Assuming the rate of climb to be controlled by the rate of dissipation (or absorption) of vacancies, it is apparent that monovacancies rather than divacancies are involved for the reason that the activation energy of diffusion of monovacancies is much lower than that of divacancies in germanium and silicon (the converse is true in metals). The climbing plane, consisting of atoms on two parallel planes at only 1/3 of an interatomic distance apart, is hence physically one single plane, though it advances in a zig-zag manner. One finds such a picture consistent with an interpretation of the emitter push effect to be discussed in section 5.5.5.

The EPR experiment can provide information about the symmetry of the defect through the so-called *g* and *D* tensors. Further, hyperfine interactions may help the identification of the impurity atom or the host ^{29}Si through the nuclear magnetic moment. The EPR experiment has been extensively exploited by Watkins in the study of radiation defects in silicon. The results of Watkins and his associates' investigations are the largest single source of information about defect properties in silicon, particularly in structural aspects, as well as kinetics and energetics. His series of studies has combined various effects such as annealing, uniaxial stressing, optical excitation, control of Fermi level by impurity doping, and electron irradiation variables to obtain as much information as possible from EPR. Nevertheless, one should keep in mind that there is no guarantee of the uniqueness of the model derivable from such EPR information. A review of EPR studies in irradiated silicon up to 1963 has been given by Watkins [17], and the reader is referred to it for a perspective of what we know about defects in silicon. Post-review publications include studies of divacancies [39, 40], pairing between vacancies and impurity atoms of Group III acceptors [41], Group V donors [42, 43] in silicon, and the extension of EPR studies to defects in germanium containing oxygen [44], in intrinsic and gallium doped Ge [45], and in arsenic doped Ge [46]. However, these EPR studies in germanium are rather incomplete and indefinitive, and will not be included here. An EPR study of irradiated silicon was first reported by the Purdue group [47]. An anisotropic EPR spectrum, first reported in 1958 by Bemski *et al.* [48] and later labeled the Si-A center by Watkins *et al.* [49] has been identified as due to an oxygen-vacancy complex [49-52]. As the various EPR spectra were observed, they were labeled alphabetically. The Si-E center has been proposed as due to a vacancy-phosphorus atom complex [49, 53]; the Si-C and Si-J centers were assigned as divacancies with three and one trapped electrons respectively [53], and were later re-assigned as singly negative and singly positive divacancies respectively [17, 54]. The flourishing number of EPR spectra observed prompted Watkins to propose a new nomenclature for the designation of Si EPR spectra. In his new nomenclature, a letter and a number is used to designate a spectrum, with the letter signifying the discoverer's affiliation. Thus, G means G. E., B means Bell Telephone Laboratories, and P means Purdue.

Electron states at the defect centers can be measured with various methods, including optical excitation (or bleaching) of EPR spectra as done by Watkins and his associates, infrared absorption and photoconductivity measurements, and, less accurately, by controlled doping and by variation of irradiation dose until saturation with defects. More generally, the ionization levels of the defects are studied by measuring the change of carrier concentration with temperature, usually by measuring the sample electrical conductivity and Hall coefficient, the carrier concentration being related to the defect concentration through Fermi statistics. In contrast, the measurement of minority carrier life time is a determination of the defect concentration through recombination kinetics, generally in accordance with the theory worked out by Hall [64] and Shockley and Read [65]. A review of the theory of recombination and the recombination properties of defects in irradiated germanium and silicon, up to 1959, was given by Wertheim [66]. When the defect concentrations are below 10^{13} cm^{-3}, and other methods of measurement become unsuitable because of sensitivity, the minority carrier life time measurement provides the only means for the determination of the defect concentrations. It thus has been extensively exploited in the study of defects in irradiated germanium [67-74] and silicon [56, 57, 59, 60, 75]. Most studies showed that the electron and hole capture cross-section of these defects are in the order of $\sim 10^{-15}$-10^{-16} cm^2, and only weakly temperature dependent [66], whereas a simple Coulomb well model of the capturing centers gives a cross-section of the order of 10^{-13} cm^2 at 78°K to 10^{-15} cm^2 at 300°K. A larger capture cross-section has been observed experimentally in vacancy-impurity complexes (*E*-center and *E*-center-like complexes) [75], of the order of 10^{-13} cm^2. It is needless to say that the method of minority carrier life time measurement is very susceptible to the contamination of the sample with minute quantities of impurities, such as the ubiquitous copper, having deep energy level in the neighborhood of the midgap.

The various methods mentioned above are only suitable to carry out at low temperatures, where the defect concentrations at thermal equilibrium are probably in the order of 10^{-20} atomic fraction or less in silicon and germanium. Hence, it is necessary to study these defects at their non-equilibrium concentrations. One method of providing defects in excess of equilibrium concentration is by quenching the material from a very high temperature. Another method is the generation of lattice defects in the material at low temperatures by irradiation with energetic particles, preferably electrons in the neighborhood of ~1 MeV, and with γ-rays. The thermal quenching method has the disadvantage that even at a quench rate in the order of 10^3 °C/s, it is not possible to freeze in a significant fraction of the simple defects, since in the initial period of quenching, the time constant for the clustering of simple defects should be commonly of the order of 10^{-3} s. The irradiation method has the disadvantage that the defects so generated are very complex, and can be unrelated to those thermal defects actually present at high temperatures. Such a possibility has been raised by Seeger and his co-workers [3, 23]. Therefore, in taking and applying the properties of defects from the irradiation studies, one should keep such a possibility in mind. We shall not go into a detailed discussion of the properties of radiation defects here, since a very comprehensive and excellent review has been given by Corbett [76]. We will summarize the pertinent results of these studies, including some of the more recent works. Table 5.2 lists some properties of defects in silicon. One may note that the electronic energy levels of simple defects such as mono-vacancies have not been definitely determined. This is partly because these simple defects are short-lived. Watkins [17, 55] has concluded from a series of experiments that the mono-vacancies in silicon are mobile at temperatures as low as ~70-80°K, and have an activation energy of migration of ~0.18 ± 0.02 eV for the doubly negative vacancy, and ~0.33 ± 0.03 eV for the neutral vacancy. Ramdas and Fan [62] also suggested a very low activation energy of vacancy migration, probably <0.24 eV. These experimental observations are somewhat surprising against our intuitive picture of covalent crystals, and the values of activation energies are only ~1/5 to ~1/3 of the theoretical calculations given in Table 5.1. It is also seemingly very difficult to reconcile these results with the results of self-diffusion in silicon, which we will discuss in detail later.

There have been attempts to resolve the seeming contradiction between the self-diffusion data and the high mobility of irradiation induced defects in silicon. We have already described the simple-defect/extended-defect transition model of Seeger and Chik. A second attempt is the suggestion of a vacancy structure by Masters [28]. He argued that a mono-vacancy pictured as a missing atom from a regular lattice site is in fact not a thermal equilibrium structure, but rather a metastable structure. The equilibrium structure is actually one in which one of the four nearest neighbors of the 'metastable' vacancy has moved into a position half way between its original lattice site and the original vacant site, a position conventionally thought of as a saddle point. He called his model a semi-vacancy pair, and, as one may recall, his model is actually a counterpart of the semi-interstitial (or split-interstitial) model. He suggested that the 'metastable' structure (i.e., a conventional vacancy structure) can be produced only in transients by collisions by energetic particles. It will relax into the stable semivacancy pair (or 'split-vacancy') with an activation energy of 0.18 eV for the doubly negative state and ~0.33 eV for the neutral state. The 1.3 eV which was assigned by Watkins and Corbett [39] to the activation energy of migration of a divacancy was

TABLE 5.2

Properties of radiation defects in silicon

Defect	EPR label	Electronic state: Transition[a]	Electronic state: Level (eV)	H_m (eV)	E_b (eV)	Reference
$V^{\cdot}$	Si-G1	$V^{\times}-e$	$<E_v+0.05$	?		17
$V^{\times}$				0.33 ± 0.03		17
				<0.24		62
V'	Si-G2		?	?		17
V''			?	0.18 ± 0.02		55
$V_2^{\cdot}$	Si-G6 (J-center)	$V_2^{\times}-e$	$\sim E_v+0.25$	~ 1.3	$\gtrsim 1.6$	17, 39
			$\sim E_v+0.27$			56
V_2'	Si-G7 (C-center)		?	~ 1.3	$\gtrsim 1.6$	39
V_2''		$V_2'+e$	$\sim E_c-0.4$			17
$(V+O_i)'$	Si-B1 (A-center)	$(V+O_i)^{\times}+e$	$E_c-0.17$			48-50
						56, 57
					$\sim 0.7-0.8$	57a
				~ 1.3		58
$(V+P)^{\times}$	Si-G8 (E-center)	$(V+P)+e$	$E_c-0.4$	0.93 ± 0.05		42, 49, 53
			$E_c-0.4$	0.93	1.04	59, 60, 63
			$E_c-0.47$			61
$(V+As)^{\times}$	Si-G23	$(V+As)^{\cdot}+e$	?	1.07 ± 0.08		43
			$\sim E_c-0.4$	$\begin{Bmatrix}1.07\\1.27\end{Bmatrix}$	1.23	59, 60
			$E_c-0.46$			61
$(V+Sb)^{\times}$	Si-G24	$(V+Sb)^{\cdot}+e$	?	1.29 ± 0.1		43
			$\sim E_c-0.4$	$\begin{Bmatrix}1.28\\1.41\end{Bmatrix}$	1.44	59, 60
			$E_c-0.46$			61
$(V+B_i)^{\times}$		$(V+B_i)^{\cdot}+e$	$\sim E_c-0.4$	$\begin{Bmatrix}1.46\\2.22\end{Bmatrix}$	1.64(?)	59, 60
$(V+Al_s)'$	SiG9		?	?		41

[a] Seeger and Chik [3] have assigned the acceptor level $\sim(E_c-0.4)$ eV for all the complexes of vacancy and group V donors to the transition, e.g., $(V+P)^{\times}+e \rightarrow (V+P)'$. We feel this is incorrect. For at the impurity concentration ($\sim 10^{15}-10^{16}$ cm^{-3}) and the temperature at which EPR was carried out ($\sim 20°$ K), the Fermi level would be $>E_c$, and, if Seeger and Chik's proposed transitions are followed, we would have all complexes in the singly negative state, which has an even number of electrons and does not yield an EPR spectrum.

re-assigned in the semivacancy pair model as the migration energy of the monovacancy, i.e., the barrier height between the equilibrium position and the saddle point of semivacancy pair. We note that this leads to an energy difference of $\sim$1.0 eV between the metastable state and the equilibrium state, instead of the $\sim$1.6 eV which Watkins and Corbett assigned as the binding energy of the divacancy. (If Watkins and Corbett had assigned a divacancy migration energy of $\sim$1.6 eV and a binding energy of $\sim$1.3 eV instead of the other way around, then both models would be equivocal and consistent, not considering self-diffusion). In the undistorted configurations both a divacancy and a semivacancy pair belong to the symmetry point group D_{3d}. It was reasoned [28] that both structures can be similarly distorted to a symmetry of C_{2h}, which is the symmetry of the J-center observed in EPR by Watkins and Corbett. We may note that the relaxation of the metastable state into the equilibrium state of the semivacancy pair should obey a first order kinetics, and is hence consistent with the observation of the exponential decay of the EPR spectrum. There are however, some questions one can raise about the semivacancy pair model. First of all, Watkins and Corbett observed the production rate of the A-center (V-O complex) to be ~5 times that of the J-center. It is relatively easy to explain this as due to the presence of a higher oxygen concentration ($>10^{16}$ cm^{-3}) than the monovacancy concentration. By the mass action law, the formation rate of the V-O complex should then be higher than the formation rate of the divacancy, both rates being limited by the diffusion of monovacancies—the first order decay simply implies an inexhaustible supply of oxygen. On the other hand, if the appearance of a J-center is due to the atomic relaxation of the metastable state into the equilibrium state of the semivacancy pair, one should expect this process to be much faster than the long range migration of the semivacancy pairs on their way to forming V-O complexes. A third explanation was suggested Kiv *et al.* [76a, b]. They proposed a quasimolecular model of the vacancy in which the chemical bonds in the neighborhood of a vacancy are in an excited state, i.e., one of the two electrons of a bond is excited in an antibonding state. Such excitation can be due to the localization of an exciton at the vacancy [76c]. The concentration of excitons in irradiated crystals at low temperature has been found to be very high [76d]. They calculated the difference between the bond energies of the excited state and the ground state, and subtracted it from the value of H_V^m of Swalin (Table 5.1), and obtained a value of migration energy of ~0.2 eV for a vacancy next to an exciton. Very little is known about the interstitials in silicon and germanium. Holding the common assumption that the self-interstitial in silicon and germanium is extremely mobile, investigators have been reluctant to assign certain observed centers as due to interstitials. In principle, the silicon self-interstitial should be observable in the ^{29}Si hyperfine interaction in EPR. So far, it has escaped observation. Watkins has advanced an interesting explanation. In EPR studies of electron irradiated p-type silicon doped with aluminium, he observed a spectrum (G-18) attributable to doubly positive interstitial aluminium ion $Al_i^{\cdot\cdot}$, at an irradiation temperature of 4.2°K. Furthermore, he observed an approximately one-to-one correlation between the production rate of $Al_i^{\cdot\cdot}$

and the concentration of isolated vacancies. Since, one expects one interstitial to be produced for each vacancy produced by electron collision, and since no silicon interstitial was identified in these studies, Watkins was led to the conclusion that the silicon interstitial must be mobile at a temperature as low as 4.2°K. In his model, an interstitial produced from irradiation wanders through the lattice freely, until it runs into a substitutional aluminium atom. Then it ejects the aluminium atom from its substitutional site, and itself takes up the substitutional site [17]. There is also evidence that interstitial boron and gallium can be produced in the same manner [55].

The strong temperature dependence (approximately exponential) of the production rate of defects in irradiated *n*-type silicon has led to a model of close pair Frenkel defects production, with different barriers for the dissociation and recombination of the pair depending on charge state. This subject is beyond the scope of the present chapter.

Generally, we are more confident about the interpretations of the properties of defect-impurity complexes. For one reason, these complexes are generally sufficiently long-lived. For example, the activation energy of migration of E-center type complexes ranges from ~0.93 to 2.2 eV. An A-center is presumably even less mobile. For another reason, one can control the production of these complexes by controlled doping with respective impurities, thus correlating a particular complex with a particular impurity. The question whether the complex is an impurity-interstitial pair or an impurity-vacancy pair is not automatically resolved. The working hypothesis based on the conjecture of Coulombic interactions is that vacancies tend to associate with group V impurities and interstitials tend to associate with group III impurities. This conjecture is based on the hydrogenic model or effective mass model that interstitials are shallow donors and vacancies are shallow acceptors. If a vacancy has also a deep donor level, as suggested by Watkins for silicon [17] ($<E_v + 0.05$ eV), an association of a vacancy with a group III impurity atom is also conceivable. Watkins has assigned an EPR spectrum Si-G9 to a $(V + Al_s)'$ complex. Although both an impurity-vacancy complex and an impurity-interstitial complex have the symmetry point group C_{3h} (not considering Jahn-Teller distortion), it is assumed that if an interstitial is involved in the complex, it should be easily detected through the hyperfine interaction of the isotope ^{29}Si (4.7% abundant). This has not been observed. Watkins and his associates also corroborated their interpretations with LCAO calculations. Later we will discuss the question whether these observations of predominantly impurity-vacancy complexes imply that groups III and V impurities diffuse in silicon via a vacancy mechanism rather than an interstitialcy mechanism.

The properties of defects in germanium are not as well established as in silicon due, in large part, to less successful EPR investigations in germanium. For this reason, we shall not tabulate the properties of defects in Ge. For purpose of reference, we will just quote some values. Figure 5.2 shows the two different interpretations of the four dominant defect levels observed in germanium according to the Blount and the James and Lark-Horovitz models. Note that according to the James and

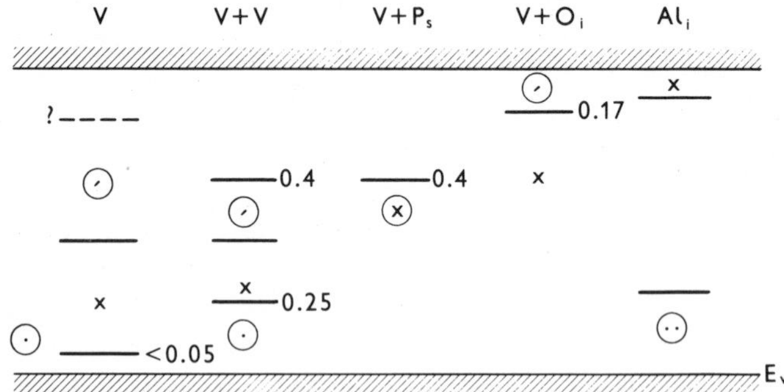

Figure 5.2. Electronic levels associated with some of the defects. The circled charge states are those observed by EPR. Where known, the level position is indicated, the value (in eV) being given to the nearest band edge. After Watkins [17]. For more details on these and other defects in silicon, see Table 5.2.

Lark-Horovitz model, the interstitial is a double donor and the vacancy is a double acceptor, with the second ionization levels deeper; according to the Blount model, the interstitial is both a donor and an acceptor, and so is also the vacancy. Note also that we have taken the averages of the respective energy levels observed by three different experimental methods: neutron, deuteron, and electron irradiations.

Even more uncertain is the activation energy of migration. There is an abundance of published studies on thermal acceptors in germanium in the early 1950's, several years prior to serious interest in silicon. The interpretations of these results have remained ambiguous. It is beyond the present scope to discuss this subject in any detail. A review of these works prior to 1955 was given by Letaw [80]. For more recent works, the reader is referred to references 81-90. If any conclusion can be drawn from these investigations at all, we will say that for Ge, the enthalpy of migration is probably ~1.2 eV, and the enthalpy of formation ~1.9 eV (values taken from Hiraki [89]). However, according to the infrared study of irradiated Ge, Whan [91] gave a migration energy of ~0.2 eV. This is compared with the vacancy migration enthalpy of ~0.18-0.33 eV from Watkins [17]. If this turns out to be true, the interpreted electron energy levels in figure 5.2 can hardly be assigned to either the vacancy or the interstitial. More likely, they could be the levels of less mobile complexes. It appears that the similarity between the migration energy of the defect in Ge observed by Hiraki [89] and the migration energy of the divacancy in Si observed by Watkins and Corbett [39] somewhat parallels the similarity between the migration energy of monovacancies in Ge by Whan [91] and that in Si by Watkins [17, 55]. A plausibility hence exists that the defect in Ge studied by Hiraki is in fact the divacancy rather than the monovacancy.

Thermal defects from quenching in Si are equally uncertain for the same reason, i.e., the enormous mobility of point defects. Mayburg [92] concluded from his quenching experiments that the equilibrium solubility at 1100°C of vacancies or interstitials in Si should be less than 5×10^{12} cm^{-3}. This conclusion appears unwarranted for the simple reason that simple vacancies and interstitials cannot be effectively frozen in. Then, curiously, he also concluded that silicon interstitials of $\sim 10^{16}$ cm^{-3} could have been frozen in. Furthermore, he advanced a mechanism to explain why no frozen-in defects were detected in Ge. He argued that the presence of Frenkel defects aided the diffusion of the excess defects out of the crystal. We find this mechanism rather implausible. First of all, it is questionable that Frenkel pairs could exist as a significant fraction of the interstitials on account of the large difference between the formation energy of interstitials and the vacancies. Secondly, it appears that the presence of Frenkel pairs can facilitate the recombination of extra defects. A closer examination shows that the assumption of the generation of an 'equilibrium' concentration of Frenkel defects has raised the total concentration of respective point defects, instead of decreasing them, in any given time span. Elstner and Kamprath [93] investigated quenched p-type Si and observed donor centers at $E_v + 0.37$ eV after annealing. They attribute these to vacancy clusters, each containing four to six single vacancies. They estimated the formation energy of the quenched-in defects to be $\sim$2.5 eV. This formation energy is probably that of monovacancies which coalesce into the observed vacancy clusters during quenching. Bemski and Dias had earlier also investigated quenched-in defects in Si, and observed donor centers with an energy level of $\sim E_v + 0.4$ eV. They observed an activation energy of motion of the defect to be $\sim$0.3 eV and a diffusion coefficient of 1.2×10^{-7} cm^2 s^{-1} at room temperature, and they suggested that the defects were silicon self-interstitials. Their results are in contradiction with Elstner and Kamprath in that the results of the latter authors did not reveal any room temperature movement of these defects (presumably they are identical defects because of the closeness of the donor levels estimated in the two cases). The suggestion that silicon self-interstitials were observed is of course also in contradiction with Watkins' series of investigations. The only other reported observation of frozen-in Si self-interstitials is by Mayburg [92] ($\sim 10^{16}$ cm^{-3}). More recently, Swanson [94] also studied the quenching of p-type Si from 800-1150°C into liquid nitrogen or water. He also observed $\sim 10^{15}$ cm^{-3} of donor centers created by the quenching, with the energy level at $\sim E_v + 0.4$ eV. However, the migration energy of the defect was given as 0.81 ± 0.04 eV. Swanson suggested that the defects were introduced either by impurity diffusion or quenching strain, but did not identify the defect.

Boltaks and Budarina [95] investigated the quenched-in defects in both n-type and p-type Si (from 950-1200°C into vacuum oil at room temperature, with a quenching rate given as 10^3 °C/s) by the measurements of sample density changes before and after heating-quenching cycle. They reported surprisingly high concentrations of vacancies in the order of 10^{17}-10^{19} cm^{-3} (higher in p-type than in n-type), at least four orders of magnitude higher than what is commonly believed. These results would also imply, *a priori,* either that during quenching the interstitials have quickly diffused out of the crystal, leaving behind them the vacancies despite the strong Coulombic interactions between the vacancy cloud and the fleeing interstitials; or that the point defects in Si at high temperatures are predominantly vacancies rather than interstitials, for quenched-in interstitials should result in an increase rather than a decrease of sample density. This is also in contradiction to the common belief that the formation energy of a self-interstitial in silicon is much smaller than the formation energy of a vacancy (see Table 5.1. The extended interstitial of Seeger and Swanson [23] is a different thing, which has been contrived to possess an extremely high formation energy of ~ 4.5 eV, so that it may reconcile self-diffusion data). Boltaks and Budarina also gave the vacancy formation energy as a function of impurity concentration (they did not distinguish between the p- and the n-type impurities), and extrapolated to give an intrinsic formation energy of 2.8 eV. (We note that there was an error in extrapolation, i.e., they should have extrapolated back to $\log(n/n_i) = 0$ instead of 1, where n is the carrier concentration instead of the impurity concentration, and n_i is the intrinsic carrier concentration. In this way, they should obtain an intrinsic formation energy of ~ 3.1 eV.) We may note that although the measurement of lattice dilation has been employed in the study of defects in such covalent crystals as Si and Ge (for example, see references 96-98), it is questionable whether any conclusive results have been obtained. There are at least three difficulties with this method, i.e., the low precision of the method; the simultaneous presence of various defects such as vacancies, interstitials, vacancy-impurity complexes, interstitial complexes and precipitations, etc., which a volume change alone can never unscramble; and thirdly, the uncertain contribution of the electronic volume effect [96, 98-100].

We mentioned earlier that interactions of point defects with dislocations can be observed through TEM. If silicon or germanium is quenched from T_1 to T_2, both being high temperatures, two types of interactions of the excess point defects (at T_2) with dislocations can occur. In one, edge dislocation segments will climb by absorbing the excess point defects. Both the magnitude of climb and the sign of the dislocation can be determined, thus yielding information of the concentration of the point defects and the nature of the point defects (vacancies or interstitials). Milevskii [101] used transmission polarized infrared microscopy to study such interactions, and concluded that at the melting point, the equilibrium vacancy concentration in silicon is $\sim 10^{15}$ cm^{-3}.

In the second type, excess vacancies in the course of quenching condense to form dislocation helices [102-104]. That the equilibrium shape of a dislocation line in the absence of external mechanical stress and in the presence of a vacancy (or interstitial) supersaturation or undersaturation is a helix has been proved by Weertman [103]. The radius of the helix or the number of turns n per unit length is determined by the equilibrium between the excess vacancy chemical potential $kT \ln([V]/[V]_e)$ and the dislocation line tension. According to the theory worked out by Weertman [103], one has

$$\mathrm{n} = (kT/2\pi G b^4) \ln([V]/[V]_e), \tag{5.4}$$

where G is the modulus of rigidity and b is the magnitude of the Burgers vector. Thus, by the measurement of n of the helices formed in quenching from T_1 to T_2, one can calculate the ratio $[V]_{e,\,T_1}/[V]_{e,\,T_2} \simeq ([V]/[V]_e)_{T_2}$, and obtain the enthalpy of formation of vacancies (or interstitials). Helical dislocations in silicon have been reported by Joshi and Dash [105]. Unfortunately, the above information cannot be obtained from their experiments for the reason that their experimental conditions were not appropriate for this purpose. First, they employed a normal cooling scheme instead of quenching from T_1 to T_2, both at sufficiently high temperatures. Second, the precipitations of the impurities (As, B, P) that they looked for and observed have complicated the whole picture. For now one cannot be sure whether the transition to a supersaturation (of vacancies) is a result of cooling; or to a decrease of electrically active impurities due to precipitation, and hence a change in the Fermi level of the system that affects the concentrations of the charged vacancies; or to the generation of vacancies when substitutional impurity atoms leave their lattice sites to form interstitial type precipitates and/or climbing dislocation planes. This hopeless situation can be avoided by using silicon of low impurity content with the quenching scheme mentioned.

There is another method that has been used to determine the thermal equilibrium vacancy concentrations in Si and Ge. This is based on the interactions of certain interstitial impurities, such as Cu, Ni and Li, with vacancies which become excess on quenching, or with vacancy clusters, or vacancy-impurity complexes. The increase in solubility due to this type of interaction can be measured by various means, and the increase in solubility is then interpreted in terms of defect concentrations and energy of interactions. Fuller and Wolfstirn have attempted such measurements [90, 106]. The results have not been conclusive. Weltzin and Swalin [88] employed lithium as the interacting impurity. The interaction involves oxygen-vacancy complexes acting as nucleation sites for the precipitation of Li. Their results showed a binding energy of the O–V pair of 0.6 eV (this is to be compared with the value of ~0.7-0.8 eV for silicon [57a]). They also observed that the concentration of the O–V complex can be expressed as $[\mathrm{O{-}V}] = 2.6 \times 10^4\ [\mathrm{O}]\ \exp\ (-1.4/kT)$. We interpret this result to mean that the enthalpy of vacancy formation is 2.0 eV (1.4 + 0.6 eV). This is in good agreement with the value of ~1.9 eV from Hiraki [89].

Quenching is the only method that can provide information regarding the concentrations of vacancies and interstitials at thermal equilibrium. The above review has shown us a very foggy picture about these thermal defects in silicon and germanium. It is not possible to decide on the values of formation energies of vacancies in Si and Ge from the literature cited above. We can only note that the formation energies of vacancies in Ge of 1.9 eV [89], and in Si of 2.5 eV [93] are in close agreement with theoretical values (Table 5.1). The self-interstitials in Si and Ge are then not known.

5.3 SELF-DIFFUSION

If self-diffusion is effected through a vacancy mechanism, the tracer self-diffusivity is related to the vacancy diffusivity through the relationship

$$D_S[S] = D(V)[V], \tag{5.5}$$

where D_S and $D(V)$ are the self-diffusivity and the vacancy diffusivity respectively; and $[S]$ and $[V]$ are the concentrations of lattice sites and of vacancies respectively. If self-diffusion is effected through an interstitialcy mechanism, a similar relationship holds,

$$D_S[S] = D_I[I], \tag{5.6}$$

where D_I and $[I]$ are the diffusivity and the concentration of the self-interstitial. If the self-diffusion is effected through a direct interstitial mechanism, meaning that it is far easier for the interstitial to move along the paths of normal interstices than to move by displacing a lattice atom, then clearly the lattice site-interstitial interchange must be the rate controlling mechanism, e.g.,

$$(\widetilde{\mathrm{Si}})_s + (\mathrm{Si})_i \rightleftharpoons (\widetilde{\mathrm{Si}})_i + (\mathrm{Si})_s, \tag{5.7}$$

where $(\widetilde{\mathrm{Si}})_s$ and $(\widetilde{\mathrm{Si}})_i$ are the labeled silicon atoms at a lattice site and at an interstice respectively. Self-diffusion via a divacancy mechanism is similarly expressed by equation 5.5. From equations 5.5 or 5.6, it is clear that the activation energy of self-diffusion is equal to the sum of the energy of formation and the energy of migration of the vacancy (for a vacancy mechanism) or the interstitial (for an interstitialcy mechanism). Now, if we take the theoretical values of formation and migration energies of the silicon vacancy (Table 5.1), we obtain an activation energy for self-diffusion in silicon of 3.2-3.4 eV for the vacancy mechanism. For either an interstitial or an interstitialcy mechanism, it is clear that the reaction of the type described by equation 5.7 must be the rate determining step for self-diffusion. Then the activation energy of self-diffusion is *not* the sum of the formation and the migration energies of the self-interstitial; but it is simply equal to the formation energy of a split interstitial. Unfortunately, this point has not been recognized by most investigators in their theoretical calculations, so that the formation energy of a split-interstitial has not been estimated.

To discuss the mechanisms of self-diffusion in Ge and Si, we will examine experimental data of self-diffusion in the light of theoretical calculations, and of experimental data for the defect properties both from irradiation and thermal quenching experiments. We summarize the experimental results of self-diffusion in Ge and Si in Table 5.3. Combining data from various sources, these cover a temperature range of 730-930°C for Ge, and 1100-1400°C for Si. In these temperature ranges, the Arrhenius plots of self-diffusivity versus reciprocal temperatures are straight lines within experimental accuracy for both Ge and Si. We note that the activation energy of self-diffusion in Ge is quite consistent among various sources, and has an average value of 3.0 eV. We note also that the theoretical activation energy of self-diffusion in Ge by a vacancy mechanism is also ~3.0 eV (3.02 eV according to Swalin [26] and 2.89 eV according to Bennemann [25]), in remarkable agreement with the experimental value (see Table 5.1). But we should note the disagreement between the theoretical migration energy of a vacancy in Ge and the experimental one later reported by Whan [91] as discussed in the preceding section. Experimental investigations of self-diffusion in Si were reported more recently (1966), partly because of the late development of silicon technology. A comparison of Table 5.3 and Table 5.1 shows that

TABLE 5.3

Self-diffusion in germanium and silicon

Semiconductor	D_O(cm²/s)	Q (eV)	Reference
Ge	7.8 ± 3.4	2.98 ± 0.04	108
	22	3.1	109[a]
	44 ± 41	3.15 ± 0.13	110[b]
	10.8 ± 2.4	3.02 ± 0.02	110[b]
Si	9.0×10^3	5.13	111, 112
	1.8×10^3	4.78	113
	1.2×10^3	4.73	114

[a] These values are the author's estimation from figure 5.2 of reference 109.
[b] Both sets of data are from the same reference using ^{71}Ge radioactive isotope, but using respectively the Steigman technique and the Gruzin technique for measurements.

the activation energy of self-diffusion in silicon does not agree with the theoretical prediction, assuming a vacancy mechanism. The discrepancy is ~5.0 eV versus ~3.4 eV. Using the experimental values of the formation and migration energies of a vacancy in silicon (~2.5 eV and 0.33 eV), one finds a greater discrepancy. We also note the abnormally high value of the pre-exponential factor $D_O(10^3\text{-}10^4 \text{ cm}^2 \text{ s}^{-1})$. This discrepancy led Kendall to suggest [4, 115] a divacancy mechanism of self-diffusion in silicon. The possibility of a divacancy mechanism has also been cited by Peart [113], and later also by Ghoshtagore [116]. The divacancy mechanism argument is essentially based on a mixture of the experimental binding energy of a divacancy (>1.6 eV) and its motional energy (~1.3 eV), both according to Watkins [17, 39], and the theoretical energy of formation of monovacancies. If we take Swalin's value of H_V^f (2.32 eV), this would give an activation energy of $\gtrsim$4.34 eV for the self-diffusion by a divacancy mechanism. However, this proposition is an oversimplification of the problem. One cannot simply discard the monovacancy mechanism because the theoretical estimation of the activation energy of self-diffusion according to this mechanism is much smaller than experimentally observed; for it could be that the theoretical estimation is quite incorrect. Or it could be that the defect properties are quite temperature dependent so that these theoretical calculations and irradiation experiments are not pertinent for high temperature diffusion. More seriously, one must raise the question: why should not the monovacancy contribute predominantly to self-diffusion if one wholly accepts the theoretical activation energy of a mere ~3.4 eV for the monovacancy mechanism versus the ~4.3 eV for the divacancy mechanism? Or, for that matter, versus any other mechanism that one can think of that gives an activation energy of 5.1 eV. For, the occurrence of random walks of monovacancies is necessarily accompanied, in a one-to-one correspondence, by the random walks of the host lattice atoms. Clearly, such reasoning suggests that one should question the available theoretical and experimental values of the energies of formation and of migration of monovacancies about their correctness, or their appropriateness, at high temperatures. Seeger and Chik [3] emphatically rejected the divacancy mechanism with this cogent argument: If the monovacancy contribution to the self-diffusion is to be negligible compared with the divacancy contribution, the relationship $H_V^f + H_V^m > H_{V_2}^f + H_{V_2}^m$ must be satisfied. (This, we note, is not an exact criterion because of the possibility that the pre-exponential factor for a divacancy mechanism could be somewhat larger than that for a monovacancy mechanism, possibly by a factor of 10-100. But this would not materially affect the essential logic of Seeger and Chik's argument.) If we accept Watkin's data, the difference between the migration energy of a divacancy and that of a monovacancy is ~ 1.0 eV. Then by rearranging, we have

$$H_V^f - H_{V_2}^f > H_{V_2}^m - H_V^m \gg 0. \tag{5.8}$$

Equation 5.8 would then require that the formation energy of a monovacancy be larger than that of a divacancy by about 1.0 eV; this is extremely implausible. Even if we do not accept Watkins' values for H_V^m and $H_{V_2}^m$, it is still theoretically anticipated that $H_{V_2}^m \gg H_V^m$ in a diamond type lattice. This would require $H_V^f > H_{V_2}^f$, and is still implausible. In other words, since in silicon, and more generally in any diamond lattice, both $H_{V_2}^f$ and $H_{V_2}^m$ are larger than H_V^f and H_V^m respectively, the divacancy mechanism cannot make any significant contribution to self-diffusion compared to the monovacancy mechanism. Thus, the discrepancy between the activation energy of self-diffusion in silicon, and the sum of the energies of formation and migration that we have hitherto assigned to the silicon monovacancy simply indicates that perhaps we have been wrong about the monovacancy. This must be the inevitable conclusion, because if we were correct, and the vacancy does have a migration energy of 0.33 eV (or Swalin's 1.06 eV), and a formation energy of 2.5 eV (or Swalin's 2.32 eV), the swift random walks of the swarm of monovacancies in the lattice must induce, for their part, a silicon tracer diffusivity of the order of $\sim \exp[-(2.83 \text{ to } 3.38)/kT]$. It then would not make sense that the resultant self-diffusivity should be the much smaller value of $9 \times 10^3 \exp(-5.1/kT)$. Although we can consider the paradox resolved by simply rejecting the values of the formation and the migration energies previously assigned to the silicon monovacancy, we can consider the problem solved only after we can account for these formation and migration energies that have been advanced, and/or the determination of the correct energies. As we have already discussed in some detail in section 5.2.2, the theoretical values of the silicon formation and migration energy may have suffered from insufficiently refined models. To account for the experimental monovacancy migration energy of 0.33 eV by Watkins, we have also discussed a proposition of Masters (see section 5.2.2) of a split-vacancy model. Perhaps a more reasonable explanation is one due to Seeger and Swanson [23]. They suggested that the monovacancy in silicon exists in two forms, one at low and one at higher temperatures. The transformation from the low temperature form (which is what has been studied in irradiation experiments) to the high temperature form takes place at some temperature T_0 with an enthalpy of transformation ΔH_V. Then the entropy of the vacancy at any temperature above T_0 is given by

$$S(T) = \frac{\Delta H_V}{T_0} + S_{\text{vibr}}, \tag{5.9}$$

where S_{vibr} is the vibrational entropy which is only weakly temperature dependent above the Debye temperature ($\theta = 658^\circ$K for Si and 366°K for Ge) due to anharmonicity effects. Of course, less restrictively one can also consider the transformation to occur over a very narrow range centered about T_0, in which case one replaces the term $\Delta H_V/T_0$ in equation 5.9 by an appropriate integral; but this is trivial. The essential point of this model is that it can account for the very large pre-exponential factor observed in silicon self-diffusion (see Table 5.3) by the excess entropy $\Delta H_V/T_0$. As an arbitrary example, they cited a transformation temperature of 500°K at which an enthalpy of transformation of 0.56 eV can give an excess entropy of 12.9 k, sufficient to account for the observed pre-exponential factor. More specifically, they suggested an order-disorder transformation such that the high temperature form has an extended region of disorder over the 16 nearest and second-nearest neighbors of the original vacancy site, somewhat similar to the relaxion model of vacancy of Nachtrieb and Handler [24]. The large number of the atoms involved in the defect gives rise to the large entropy of formation. Seeger and Swanson did not propose a model for the motion of such an extended vacancy. Quite obviously, the activation energy of migration would no longer be the low temperature value of 0.33 eV.

The mechanism by which such a disordered region (or 'liquid droplet') moves in a crystalline lattice is completely unclear. At present it appears that neither the point mechanisms of solid state diffusion (vacancy, interstitial, and interstitialcy) nor the currently available models of liquid state diffusion (the localization of free volume [117], and the energy fluctuation [118]) can be adapted for the diffusion of such a droplet in solid. The lack of a motional model makes the extended vacancy (or extended interstitial) difficult to assess. Seeger and Swanson [23] further argued in favor of an extended interstitial over the extended vacancy in silicon and germanium on the ground that both Si and Ge are denser in the liquid state than in the solid state by ~10% and ~5% respectively (see the discussion in section 5.2.1). Their final conclusion is that silicon self-diffusion probably proceeds via an extended interstitial mechanism, and that germanium self-diffusion proceeds via a simple vacancy mechanism. Our query on the validity of Seeger and Swanson's liquid droplet extended interstitial model has been discussed in section 5.2.1, on the grounds that such an extended interstitial, with a formation enthalpy of ~4.5 eV and an entropy of ~15 k, has a much larger free energy of formation than a simple interstitial. For the extended interstitial (or any other mechanism) to be responsible for silicon self-diffusion, it is required that $H_V^f + H_V^m > 5.1$ eV for the monovacancy (or the extended vacancy) not to be responsible. If we take $H_V^f \sim 2.5$ eV from quenching experiments [93], it leaves $H_V^m > 2.6$ eV. This value of H_V^m is not impossible. To account for the fact that monovacancies could not be quenched in, one need only adopt a low activation energy of migration for the low temperature form. Similarly, a modification of Seeger and Swanson's extended interstitial model can be made by discarding the liquid droplet concept and suitably decreasing the enthalpy of formation with a corresponding increase in the energy of migration. This modification can be made such that the free energy of formation of the extended interstitial becomes smaller than the simple interstitial.

In summary, the extended defect models of Seeger and Swanson are attractive for the explanation of self-diffusion in silicon. The models can be made reasonable by a modification mentioned above. However, at present it is not possible to decide on whether the extended interstitial or the extended vacancy contributes predominantly to the self-diffusion.

The effect of impurity doping on self-diffusion in germanium [109] and in silicon [112, 114] has also been studied. However, most experimental data appear to have considerable scatter. In general, doping with donor impurities increases the self-diffusivities in Si and Ge, indicating the acceptor nature of the defect involved in self-diffusion. This is consistent with the interpretation of the defect as a vacancy. The low temperature form of the vacancy in silicon has been found to be a double acceptor, and probably also a deep donor [17, 55]. The levels have not been accurately determined because of the very short life time of the irradiation generated vacancy. The first acceptor level is probably slightly below the midgap. The acceptor level of the E-center is more accurately defined at $\sim E_c - 0.4$ eV, for the reason of longer life time. In general, it is this center (or other E-center-like complexes) that is of interest to us in the calculation of impurity diffusivity, which is directly related to the concentration of this complex center. No second acceptor level and no donor level have been reported for the E-center, or other E-center-like complexes. One does not expect this level to contribute significantly to self-diffusion at least when the impurity concentration is below 10^{20} cm^{-3}. Much less is known about vacancies in Ge. But according to either the James and Lark-Horovitz model, or the Blount model, a vacancy in Ge should have a shallow acceptor level (figure 5.3). Then, in intrinsic

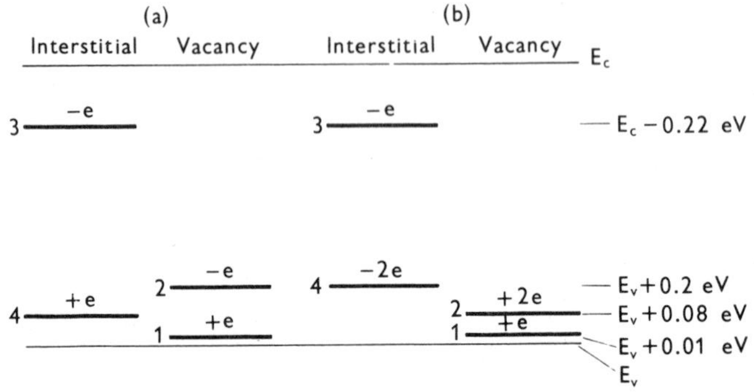

Figure 5.3. Interpretations of defect energy levels in germanium according to (a) the Blount model; (b) the James and Lark-Horovitz model. Note that the numerical values given are averages of three different experimental methods.

or n-type Ge, the vacancies will be predominantly negatively charged. Neglecting the second acceptor level, one can write the vacancy concentration as a function of Fermi level in the approximate expression

$$\frac{[V]}{[V]_i} = \exp\,[(E_F - E_F^i)/kT] = n/n_i, \tag{5.10}$$

where both the subscript and the superscript i denote the intrinsic condition. Valenta and Ramasastry [109] first reported such an effect from the study of self-diffusion in doped germanium, as interpreted from the assumed relation $D_S/D_S^i = [V]/[V]_i$, where D_S and D_S^i are the tracer self-diffusivities in the doped and in the intrinsic semiconductors respectively. They observed that this relation appeared to hold in n-type Ge; in p-type Ge, the ratio D_S/D_S^i was found to be substantially smaller than unity as predicted from an acceptor type vacancy mechanism, but was somewhat larger than given by equation 5.10 (see figure 5.4). They

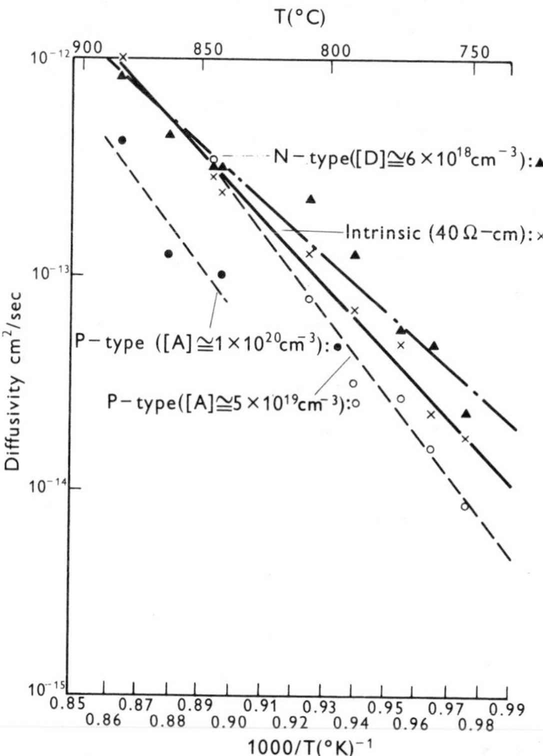

Figure 5.4. Self-diffusion coefficients in germanium. After Valenta and Ramasastry [109].

explained this discrepancy as either caused by the evaporation of gallium (the acceptor dopant) from the surfaces of the p-type specimens, or by the possibility that the acceptor level is not as shallow as depicted in figure 5.3, and could be somewhere near the midgap, making equation 5.10 a poor approximation. For self-diffusion in Si, the diffusivity was also observed to increase in n-type samples [112, 114]. However, conflicting results were reported for self-diffusion in p-type silicon. While Ghoshtagore [114] reported the observation of a decreased diffusivity in p-type silicon, Fairfield and Masters observed increased diffusivities with increased acceptor (boron) concentrations. They reported $D_S/D_S^i = 1.15$ at a boron concentration of 8.0×10^{19} cm^{-3}, and $D_S/D_S^i = 1.75$ at 2.2×10^{20} cm^{-3} at 1090°C. If the vacancy acceptor level is deep, somewhere above the midgap, such an increase in p-type silicon can be explained by the existence of a vacancy donor level. However, if the acceptor level is at or below the midgap (implying the existence of a significant fraction of negative vacancies in intrinsic silicon), or if the donor level is deep ($<E_v + 0.05$ eV as reported by Watkins in low temperature experiments), such an explanation becomes implausible. Fairfield and Masters [112] invoked a model of Hoffman *et al.* [119] to explain this boron doping effect. In this model, self-diffusion is enhanced by the motion of more mobile solute atoms, i.e., a mixing effect, and is given by the expression

$$D_S/D_S^i = 1+\alpha[A]/[S]\,(D_e(A)/D_S^i-1), \tag{5.11}$$

where $[A]$ and $[S]$ are the concentrations of the solute A and the lattice sites respectively, and $D_e(A)$ is the diffusivity of A. α is the effective number of vacancy-solvent atom interchanges involved in the completion of one effective solute atom jump. (We note this is not a very well defined quantity because of the correlation effect.) This model has also been discussed by Reiss [107].

Most of the self-diffusion data in Si are subject to a considerable degree of experimental inaccuracy. To circumvent the short half-life of radioactive ^{31}Si, Ghoshtagore [114] diffused a non-radioactive isotope ^{30}Si into a regular silicon wafer, and followed by a subsequent activation of ^{30}Si to ^{31}Si. He evaporated a 0.1 μm thick layer of 89.1% ^{30}Si isotope onto a standard silicon wafer, and carried out diffusion at 1178°C for 5.6×10^5 s (his figure 5.1). At this temperature, the diffusivity in intrinsic silicon is 4.6×10^{-14} cm^2 s^{-1} (his Table 5.2). The diffusion length $2\sqrt{Dt}$ is hence 3.2×10^{-4} cm. Since the impurity profile is given by $c(x, t) = c_o \exp(-x^2/4Dt)$ as shown in his figure 5.1, the surface concentration of the isotope ^{30}Si can be found from the relationship

$$c_o\int_0^\infty \exp(-x^2/4Dt)\,dx = c_o\sqrt{\pi Dt} = c_s\delta = (0.891)(1 \times 10^{-5}), \tag{5.12}$$

where c_s is the ^{30}Si in the source in atomic fraction (0.891), and δ is the thickness of the source layer in cm (1×10^{-5}). This gives $c_o = 0.0314$ atomic fraction of ^{30}Si at the surface, and the isotope concentration decreases Gaussianly with distance. Since regular silicon contains 3.05% of ^{30}Si (natural abundance), we find that the maximum difference between the total concentration and the background concentration of isotope ^{30}Si is only a factor of 2, which differs substantially from the experimental result.

In Fairfield and Masters' experiments, the main difficulty arises from the very short half-life for ^{31}Si of 2.62 h.

In summary, the available experimental information on silicon self-diffusion is still rather incomplete, and its accuracy is not unquestionable, as demonstrated by the conflicting results both in the activation energy of self-diffusion, and the effect of boron doping. The situation in germanium is somewhat better because of the availability of a rather long-lived isotope ^{71}Ge (half-life 11 d). But the remarkable lack of scatter in the tracer concentration profiles as well as in the Arrhenius plot help to lend confidence in Fairfield and Masters' data [112].

Besides diffusion of radioactive tracers, self-diffusion in silicon and germanium can also be studied indirectly. However, unlike the tracer study of self-diffusion, the validity of the interpretation of the results from these indirect experiments depends critically on the diffusion model assumed. For the purpose of expediency, the simplifying assumption of a rate limiting process is often made, sometimes without sufficient justification. There have been a number of indirect experiments that offered conclusions in reasonable agreement with the results from tracer diffusion studies. These experiments involved the study of diffusion of some transition elements, such as gold [120], copper [83, 123] and nickel [121, 122]. These elements diffuse in silicon and germanium via some complex mechanisms within the general category of the dissociative diffusion mechanism. The model [11] envisages a fast migration of these elements through the interstitial paths, but that the equilibrium concentrations of these elements at the interstitial sites are very low in comparison with their corresponding concentrations at the substitutional sites. As they penetrate the lattice, they combine with the lattice vacancies to become substitutional defects. There are some modifications of this simple model to include the effect of trapping [124] and interactions with dislocations and precipitations [83, 123, 125]. We will return to these topics later. But we will illustrate such an approach with the diffusion of gold in silicon. We make a number of assumptions: (1) there are no dislocation or other types of internal vacancy sources in the silicon; (2) $[\mathrm{Au}_s]_e \gg [V]_e$, which is customarily believed to be true; (3) $D(\mathrm{Au}_i) \gg D(V)$, or at least $D(\mathrm{Au}_i)[\mathrm{Au}_i]_e \gg D(V)[V]_e$. Then we envisage gold atoms to diffuse rapidly through interstitial paths into the silicon and quickly saturate the bulk of the substrate. The interstitial gold atoms Au_i then combine locally with vacancies to form substitutional gold atoms Au_s according to

$$\mathrm{Au}_i + V \rightleftharpoons \mathrm{Au}_s \tag{5.13}$$

Two additional assumptions are then made: (4) the diffusion of Au_i is faster than the reactions of equation 5.13 so that at all times $[\mathrm{Au}_i] \approx [\mathrm{Au}_i]_e$. This assumption is valid only at distance x from the surface if $x < (\pi R_c[V])^{-1/2}$ for the reason that a reaction of the type of equation 5.13 involves the diffusion of Au_i to the location within capture radius R_c of the vacancy. A reasonable value of R_c in silicon is

$\sim 5 \times 10^{-8}$ cm, assuming Coulombic interaction between the negative vacancy and positive Au_i. A reasonable value of the vacancy concentration $[V]$ is $\sim 10^{12}$-10^{13} cm^{-3} at $1100°C$ but is much lower at, e.g., the common gold diffusion temperature of $\sim 800°C$. This gives $x < 10^{-2}$ cm, which is sufficiently deep by semiconductor diffusion standards; (5) the reaction of equation 5.13 is much faster than the diffusion of vacancies. This assumption will be valid only for

$$x > \{\pi R_c [Au_i]_e D(Au_i)/D(V)\}^{-1/2}$$

We take a reasonable value of $[Au_i]_e$ of 10^{15}-10^{16} cm^{-3} at $1000°C$, and a relatively arbitrary value of $D(Au_i)/D(V)$ of ~ 10. We then have $x > 10^{-5}$ cm.

Assumption (5) leads to a local equilibrium at all times given by

$$[Au_i]_e [V] = K[Au_s] \tag{5.14}$$

in the region $\sim 10^{-5}$-10^{-2} cm from the surface. In a very short initial period, an almost uniform $[Au_s]$ exists as a result of the annihilation of Au_i with existing vacancies initially at $[V]_e$, so that initially

$$[Au_s] = [V]_e [Au_i]_e/(K + [Au_i]_e) \tag{5.15}$$

and the vacancy concentration drops to

$$[V] = [V]_e/(1 + [Au_i]_e/K) \tag{5.16}$$

Then, further increase of local $[Au_s]$ in the region 10^{-5}-10^{-2} cm is limited by the diffusion of vacancies inward from the surface, following the relationship of equation 5.14. Now if we neglect the diffusion of Au_s, one has the continuity equation, using equation 5.14

$$\frac{\partial [V]}{\partial t} + \frac{\partial [Au_s]}{\partial t} = D(V)\frac{\partial^2 [V]}{\partial x^2} = D(V)\frac{K}{[Au_i]_e} \cdot \frac{\partial^2 [Au_s]}{\partial x^2} \tag{5.17}$$

Since $[V] \ll [Au_s]$, we omit the first term in equation 5.17. Then the result immediately suggests an effective diffusivity for Au_s of

$$\mathbf{D}(Au_s) = D(V)K/[Au_i]_e = D_S[S]/[Au_s]_e \tag{5.18}$$

where we have used equations 5.5 and 5.14. More generally $\mathbf{D}(Au_s)$ can be obtained from a Matano analysis [126] of the [Au] profile in the region $\sim 10^{-4}$-10^{-2} cm. Wilcox and LaChappelle chose for expediency to fit a segment of the [Au] profile for $x \gtrsim 10^{-2}$ cm by superimposing a log-log plot of a complementary error function. It is felt, however, that a really good fit cannot be obtained this way because the profile could at best be represented by $c_o \operatorname{erfc}(x/2\sqrt{\mathbf{D}t}) + c_1$, where c_o is an appropriate fictitious surface concentration, and c_1 is some constant approximately given by equation 5.15. In any case, after we have obtained $\mathbf{D}(Au_S)$, the self-diffusion coefficient D_S is readily obtained from equation 5.18. Of course, this implies that the self-diffusion proceeds via a vacancy mechanism, whereas in the Si tracer diffusion experiments such an assumption is not necessary.

For comparison, we quote some results of the indirect methods in Table 5.4. These data can be considered to be in close agreement with

TABLE 5.4

Self-diffusion in silicon from the dissociative diffusion of gold and nickel into silicon

Dissociative diffusant	Silicon self-diffusivity D_O (cm^2 s^{-1})	Silicon self-diffusivity Q (eV)	Reference
Au	1.81×10^4	4.87 ± 0.87	120
Ni	30.	4.5 ± 0.4	121
Ni	10^3	4.24	122

NOTE: the accuracy and the interpretation of these results, particularly that of reference [122], should be subject to scrutiny. For a discussion, see section 5.4.2.

those from tracer experiments (Table 5.3), thus lending support to a vacancy mechanism of self-diffusion. Since we have ruled out a divacancy mechanism, only the monovacancy mechanism can be considered appropriate. Whether the vacancy is extended is secondary, but it seems to be a logical model particularly in view of the large pre-exponential factor of the silicon self-diffusivity. The vacancy mechanism is also consistent with the effect of donor and acceptor doping, whereas a reverse effect should be observed if the self-diffusion in silicon proceeds via an interstitial or an interstitialcy mechanism. It should be noted however, that the gold diffusion data of Wilcox and LaChappelle have considerable scatter in the Arrhenius plots which determine a part of the activation energy Q for self-diffusion. Another part which comes through $[Au_S]_e$ (see equation 5.18) is also rather uncertain. Wilcox and LaChappelle had used the diffusivity of Ge in Si in guiding their choice among the scattered values. Dash and Joshi [127] recently reported the TEM observation of interstitial-type dislocation loops in both phosphorus and arsenic diffused silicon samples and concluded that this lends support to Seeger and Chik's interstitial mechanism of self-diffusion in silicon, and probably also for P and As diffusion in silicon. We feel, however, that there is no reason why the presence of interstitial type dislocation loops should require the point defects responsible for diffusion to be interstitials. These interstitial type dislocation loops could have come from climbing dislocations that absorb impurity atoms [128], the dislocation generation and propagation being a result of lattice stress [129]. Or it could simply mean that precipitates nucleate on interstitial type inclusions or edge dislocations because of elastic interactions. Furthermore, for impurity diffusion in Si, Seeger and Chik actually proposed an 'interstitialcy' mechanism, rather than the 'interstitial' mechanism misquoted by Dash and Joshi [127]. In the interstitialcy mechanism, one does not expect the impurity to exist massively as interstitials, as implied by Dash and Joshi.

In concluding this section, the author would like to propose an experiment that could probably settle the question whether self-diffusion in silicon, or any impurity diffusion, proceeds via a vacancy mechanism or an interstitial (or an interstitialcy) mechanism. The experiment involves the study of the pressure effect on self-diffusivity or impurity diffusivity in silicon. The pressure effect on diffusivity comes in through a physical parameter called activation volume. Since the work of Nachtrieb *et al.* [130], there have been some experimental as well as theoretical studies on this subject [131-135]. The activation volume is defined, through the absolute rate theory, by

$$\Delta V_{\text{act}} = -RT\left(\frac{\partial \ln D/D_o}{\partial P}\right)_T, \tag{5.19}$$

where D_o is the preexponential factor and P is the pressure. There have been some theories formulated for the activation volume, but an entirely satisfactory first principle theory is lacking. However, a physical concept of the activation volume is sufficient for the present purpose. We envisage the physical picture of the activation volume to consist of two parts. One part comes from the incremental crystal volume associated with the formation of the defect. Quantitative values aside, one expects the formation of a vacancy to be accompanied by an increase in the crystal volume which is a sizable fraction of the atomic volume. On the other hand, one expects the formation of an interstitial to be accompanied by a decrease in the crystal volume. The second part comes from the increment of crystal volume when the defect moves from its equilibrium position to the strained saddle point. We will just note that this description is not rigorously correct. Nevertheless, it gives us grounds to suppose that the activation volume of self-diffusion or impurity diffusion via an interstitial mechanism in silicon should be close to zero, if not negative, while that for a vacancy mechanism should be a sizable fraction of the atomic volume of silicon (probably 0.5 or more the atomic volume). Keyes' empirical formula [133] might be useful for a rough estimation of the activation volume. If the experimental results show a decrease of self or impurity diffusivity with increasing pressure with an activation volume of a sizable fraction of atomic volume, one may conclude that the diffusion process most probably proceeds via a vacancy mechanism.

5.4 IMPURITY DIFFUSION

5.4.1 General

Solute diffusion in silicon and germanium is frequently referred to as impurity diffusion because the concentrations of the solutes concerned are of the order of 0.01 atomic fraction or lower. For this reason, solute diffusion in Ge and Si is often viewed microscopically as isolated solute atoms independently executing random walks. Theories of dilute solutions can therefore be easily adapted for impurity diffusion in Ge and Si, from the analysis of chemical potentials at finite solute concentrations. Customarily, non-ideal solution behavior of chemical potentials is usually lumped into a so-called 'effective diffusivity', defined in such a way so as to preserve resemblance to the Fick's laws of diffusion, whenever this can be done. This is of course not always possible; then a more general approach using the linear phenomenological equations of irreversible thermodynamics [136] is required. This approach, in principle, can handle situations of any degree of complexity, such as the simultaneous presence of more than one type of solute atom, which can be simultaneously at interstitial sites and substitutional sites, and interacting with a multiplicity of lattice defects, vacancies and self-interstitials. The discussions in this section, however, will be centered at another aspect of the diffusion problem: the mechanism by which the impurity atoms move in the lattices of Ge and Si. Various elements behave differently in Ge and Si, principally according to their chemical properties. It is therefore convenient to discuss them following the grouping in the periodic table. This is also consistent with other types of physical properties, such as the electron energy levels, of various elements in Ge and Si.

The diffusional behavior of various elements are primarily determined by the types of solutions which they form as solutes in the lattices of Si and Ge, i.e., whether they are located at interstitial or substitutional sites, or a mixture of both substitutional and interstitial sites in the Si and Ge lattices. Solutes which predominantly take up interstitial sites in Si and Ge, such as Group I and Group VIII elements, diffuse interstitially without exception. Most transition elements dissolve both interstitially and substitutionally in Si and Ge, the difference in solubility between the two sites being substantial. For example, Wilcox and LaChappelle [120] estimated from the tail region of the gold diffusion concentration profiles that the ratio of substitutional gold solubility to interstitial gold solubility ranged from ~5 at 1200°C to ~10^2 at 700°C. (It should be pointed out that their method of estimating the ratio of substitutional to interstitial is meaningful only if the gold does in fact exist predominantly in the substitutional form; for their method would always predict the ratio to be greater than one because of the nature of the diffusion concentration profile.) Hall and Racette [137] reported the solubility of substitutional copper to be slightly less than an order of magnitude higher than the solubility of interstitial copper in Ge, while in Si, the solubility of substitutional copper is lower than the solubility of interstitial copper by a few orders of magnitude ($[Cu_s]_e/[Cu_i]_e \sim 10^{-4}$ at 700°C). There has been no satisfactory theory for calculating the ratio of substitutional to interstitial solubilities. Weiser's theory [138] of interstitial solubility in silicon and germanium appears to be the only known one; but it overestimates the ratio of substitutional to interstitial solubility for transition elements at least by several orders of magnitude. In any case, the amphoteric nature of these elements tends to suggest the simultaneous existence of the interstitial and the substitutional forms. The interstitial form can diffuse rapidly along the interstitial paths. The substitutional form must diffuse either by a vacancy mechanism, or an interstitialcy mechanism, both being slower than the simple interstitial mechanism by several orders of magnitude. For practical purposes, these substitutional impurities can be simply assumed to be non-diffusing in view of the interstitial-substitutional solubility ratio. The disparity between the mobilities of the interstitial and the substitutional components tends to disproportionate the local concentrations of the two components during diffusion, and requires the establishment of local equilibrium by dissociation of the substitutional or the recombination of the interstitial and the vacancy. This is the dissociative mechanism of diffusion. Group III and Group V elements are located on substitutional sites in Ge and Si at thermal equilibrium. It is generally believed that the equilibrium concentrations of the interstitials should be several orders of magnitude lower than the concentrations of the corresponding substitutionals. At such low concentrations, there is no suitable experimental method for their determination. One might suppose that interstitials of these elements in Ge and Si could possess deep electron levels, and thus could affect minority carrier life time in a measurable way. Such aspects have not been studied even in ion implanted impurities of III and V elements in Ge and Si, where nonequilibrium interstitials exist in considerable concentrations. Theoretical estimation of the interstitial-to-substitutional ratio can be made using an empirical relation due to Weiser [140], originally intended for the estimation of the interstitial-to-substitutional ratio of Zn in GaAs. A better empirical formula than equation 5.3 of reference 140 for the estimation of the interstitial-to-substitutional ratio of boron is given by [141]

$$\frac{[B_i]_{e,i}}{[B_s]_{e,i}} = \frac{[Cu_i]_{e,i}}{[Cu_s]_{e,i}} \frac{k_s(Cu)}{k_s(B)} \exp[-(\text{I.P. of B} - \text{I.P. of Cu})/kT] \tag{5.20}$$

k_s denotes the segregation coefficients of the substitutional species; and I.P. the first ionization potential of the respective elements. Copper is used as a reference on the assumption that reliable $[Cu_i]_{e,i}/[Cu_s]_{e,i}$ values can be obtained from the data of Hall and Racette [137]. At 1110°C, the ratio of interstitial-to-substitutional boron is estimated to be $\sim 3 \times 10^{-6}$ under intrinsic conditions [141]. Needless to say, this is only an order-of-magnitude estimation. The ionization potentials [142] of P and As are higher than that of B by more than 2 eV, and the existence of interstitial phosphorus and arsenic can be excluded according to this model. The ionization potentials [142] of Sb and Bi are comparable to that of B. However, the radii of $Sb^{\cdot}$ and $Bi^{\cdot}$ are much larger than the radius of $Cu^{\cdot}$, and this additional effect should be corrected by incorporating a repulsive potential factor. In view of the above assessment, we can assert that Group III and Group V elements diffuse in Ge and Si by mechanisms involving native point lattice defects, i.e., monovacancy, or lattice self-interstitial (we exclude the divacancy mechanism for the reasons discussed earlier), and exclude the interstitial, as well as the dissociative mechanism.

5.4.2 Interstitial Diffusion in Ge and Si—The Alkali Elements, the Transition Elements, and the Inert Gases

As we mentioned earlier, anything that exists interstitially in Ge and Si will diffuse interstitially. The alkali elements and the inert gases exist exclusively as interstitials in Ge and Si, while portions of each of most known transition element solutes in Si and Ge exist in interstitial form. The diffusivities of these solutes are tabulated in Tables 5.5 and 5.6. Table 5.5 also includes the migration energies of phosphorus and boron which were introduced into Si in nonequilibrium concentrations by ion implantation [143]. From Table 5.5, it is seen that with very few exceptions, the activation energies of interstitial migration in Si fall in the range ~0.4 to ~0.8 eV. These data have been selected according to this author's judgement of accuracy. For example it is believed that the diffusivity of Ni in Si obtained by Bonzel [122] should be the diffusivity of interstitial Ni in Si rather than the 'apparent diffusivity' of the substitution Ni similar to that defined in equation 5.18. The reason is that

TABLE 5.5

Diffusivities of interstitial solutes in silicon

Solute	$[A_i]_e/[A_S]_e$ [a]	D_O(cm²/s)	Q (eV)	Reference
Li	$\sim\infty$	9.4×10^{-3}	0.79	144
		2.3×10^{-3}	0.66	145
		2.2×10^{-3}	0.70	146
		2.5×10^{-3}	0.66	147, 148
		2.65×10^{-3}	0.63	149
Na	$\sim\infty$	1.65×10^{-3}	0.72	150
K	$\sim\infty$	1.1×10^{-3}	0.76	150
H	∞	9.4×10^{-3}	0.48	151
		4.2×10^{-5}	0.56	152
He	∞	0.11	1.26	151
		5.1×10^{-4}	0.58	153
Ag		2.0×10^{-3}	1.60 [g]	154
Au	$\sim 10^{-2} - 10^{-1}$ [b]	1.1×10^{-2}	1.12	155
		2.44×10^{-4}	0.39	120
Cu	$\sim 10^{5}$ [c]	4×10^{-2}	1.00	156
		4.7×10^{-3} [h]	0.43 [h]	137
Fe	$\sim 10^{-2} - 10^{-1}$ [d]	6.2×10^{-3}	0.87	155
			0.78	165
Ni	$\sim 10^{4}$ [e]		$O(E_I^m$ of Si)	157
Zn		0.1	1.4	158
B	Non-equil. [f]		0.62 ± 0.06	159
P	Non-equil. [f]		0.34 ± 0.03	143

[a] The ratio of equilibrium concentrations of interstitial solute A_i to the substitutional solute A_S.

[b] Estimated from figure 5.8, reference [120].

[c] Estimated from figure 5.7, reference [137].

[d] Estimated from figure 5.7, reference [163].

[e] Estimated from a comparison of figure 5.3, reference [121] and figure 5.2, reference [157].

[f] Copious non-equilibrium interstitials were produced by ion implantation.

[g] It is unclear whether this pertains to the interstitial diffusivity or an apparent diffusivity from a dissociative mechanism (cf. Table 5.7).

[h] This is probably the more accurate of the two.

Ni exists in Si predominantly as interstitials (the interstitial-to-substitutional ratio is $\sim 10^4$), and it is undoubtedly this predominant species rather than the minute substitutional Ni that Bonzel measured with the radiochemical technique. This is different from the electrical measurements of Yoshida and Saito [121] which determined just the electrically active substitutional Ni. When the interstitial component of a solute is present as the majority, the total concentration profile of that solute is simply determined by the diffusion of the interstitial solute to a very good approximation, as if the substitutional or the dissociative mechanism were not present. The total impurity profile is determined by the dissociative mechanism only when the substitutional is both much less

TABLE 5.6

Diffusion of interstitial solutes in germanium

Solute	$[A_i]_e/[A_s]_e$ [a]	D_o(cm²/s)	Q (eV)	Reference
Li	$\sim\infty$	1.3×10^{-3}	0.46	144
		2.5×10^{-3}	0.52	145
		9.1×10^{-3}	0.57	149
He	$\sim\infty$	6.1×10^{-3}	0.70	151
		1.8×10^{-3}	0.61 ± 0.08	153
H	$\sim\infty$		0.38	159
Ag		(At 710°C, $D(A_i) = 2 \times 10^{-6}$ cm²/s) [b]		124
Cu	~0.1 – 1	4.0×10^{-3}	0.33	82
Fe		0.13	1.10	160
Au		3.5×10^{-6}	0.63	166

[a] See footnote *a* in Table 5.5.
[b] Interstitial diffusivity. cf. Table 5.7.

TABLE 5.7

Apparent diffusivity (dissociative mechanism) of some transition elements in silicon and germanium

Solute	Lattice	D_o(cm²/s)	Q (eV)	Reference
Ag	Si	2.0×10^{-3}	1.60	154
Ag	Ge	4.4×10^{-2}	1.00	160
		(At 710°C, D = 3.6×10^{-8} cm²/s) [a]		124
Au	Si	1.15×10^{4}	3.11	161
		2.75×10^{-3}	2.06 [b]	120
		1.15×10^{3}	3.12 [c]	120
		1.78×10^{-2}	1.13 [d]	120
Au	Ge	2.25×10^{2}	2.50 [e]	162
Ni	Si	1.3×10^{-2}	1.4 [f]	121
		0.1	1.91 [g]	122
Ni	Ge	0.8	0.91	164
Co	Ge	0.16	1.12	124

[a] Apparent diffusivity, vacancy supply rate limited. cf. Table 5.6.
[b] Interpreted as the diffusivity of substitutional Au by a vacancy mechanism.
[c] Vacancy-controlled apparent diffusivity, equation 5.18.
[d] Unlimited vacancy supply, Au_i diffusion limited apparent diffusivity

$$\mathrm{D(Au)} = D(\mathrm{Au}_i)[\mathrm{Au}_i]/([\mathrm{Au}_i] + [\mathrm{Au}_s]_e).$$

[e] Rate determining mechanism is not clear. It is likely to be temperature, crystal geometry and crystal perfection dependent.

[f] Pertaining to the time constant of the decay of electrically active centers (substitutional Ni) during annealing; the decay process is assumed to be vacancy diffusion controlled, thus yielding an apparent diffusivity similar to equation 5.18.

[g] The validity of this work is questioned. See text.

mobile and the majority component, such as is the case of Au in Si. Ni in Si belongs to the former case; but Bonzel interpreted his results on the basis of a dissociative mechanism, and also calculated from these results the self-diffusivity of silicon. This is probably incorrect. On the other hand, his value of 1.9 eV appears uncommonly high for an activation energy of interstitial migration (cf. Table 5.5). Hence, we did not include Bonzel's data in Table 5.5 for interstitial diffusion.

Oxygen also dissolves interstitially in silicon and germanium. However, covalent bonds Si–O and Ge–O are formed, as evidenced from the infrared bands at 9 μ and 11.9 μ respectively [203-207]. The diffusion mechanism is hence conceived to be quite different from that of the ordinary interstitials, which interact with the host lattice mainly through overlap repulsions and Coulombic attractions. The diffusion of oxygen should hence be discussed in a separate category.

Swalin first pointed out [208] that because of the openness of the diamond structure, many interstitial atoms are smaller than the lattice saddle point constriction, and hence the strain energy contribution to the activation energy should be unimportant. One important factor contributing to the activation energy of interstitial migration is the Coulombic interaction between the impurity and the solvent ions due to imperfect screening. He utilized some semi-empirical observations of Pauling and assumed an 'effective screening coefficient' of 0.4 for valence electrons, and 1.0 (i.e. perfect screening) for core electrons. Thus, he set an effective charge of +2.4 for each lattice atom and +0.6 for the hydrogen atom and +1.2 for the helium atom. The Coulombic interaction energy is then given by

$$\sum_j z_i z_s e^2/\epsilon r_{ij},$$

where z_i and z_s are the effective charge of the interstitial and of the solvent atoms respectively; ϵ is the appropriate dielectric constant and r_{ij} is the distance between the interstitial atom and the jth solvent atom; the sum is over all j solvent atoms. Swalin considered an interstitial to reside in the center of a tetrahedron formed by four solvent atoms; this is the so-called 'tetrahedral site'. On diffusing, the interstitial atom moves into one of its four nearest neighbor tetrahedral interstitial sites by passing through a hexagonal ring of six solvent atoms. The center of the puckered hexagonal ring is considered as the saddle point. In spite of the fact that good agreement was obtained for the calculated activation energies of H, Li$^{\cdot}$, and He in Si and Ge with the experimentally observed values, the use of a screening coefficient of 0.4 is somewhat arbitrary. Luther and Moore [153] used a pairwise interaction potential between unlike inert gases due to Mason [209] for the calculation of the activation energy of He in Si and

TABLE 5.8

Diffusivities of group III and group V elements in silicon

Solute	Representative diffusivity[a] D_0(cm^2/s)	Q (eV)	Reference	References of other data
B	5.1	3.70	210	167-175
Al	8.0	3.47	167	176, 177
Ga	60	3.89	283	167,178, 179
In	16.5	3.91	167	10
Tl	16.5	3.90	167	
P	10.5	3.69	167	180
As	60.0	4.20	201	167, 183-185
Sb	12.9	3.98	186	167
Bi	1.03×10^3	4.63	167	187

[a] The choice of representative diffusivities involves some arbitrariness, and is not intended to imply the best accuracy. We assume these to be appropriate for low concentrations.

TABLE 5.9

Diffusivities of group III and group V elements in germanium

Solute	Representative diffusivity[a] D_0(cm^2/s)	Q (eV)	Reference	References of other data
B	1.1×10^7	4.54	188	189
Al	1.6×10^2	3.24	188	
Ga	40	3.15	189	
In	33	3.03	190	189, 191-193
Tl	1.7×10^3	3.42	194	
P	2.5	2.49	189	
As	10.3	2.51	202	189, 191, 192, 195-197
Sb	3.2	2.42	190	189, 191, 192, 197-200

[a] See footnote in Table 5.8.

Ge, and obtained a poorer fit to experimental data. If we extend Swalin's model to the diffusion of self-interstitials in Si and Ge, both in Ge and Si we should find the relationship $Q(I) = 2Q(\text{He}) = 4Q(\text{H})$ on account of the number of the valence electrons of each interstitial concerned ($Q(I)$ being the activation energy of the self-interstitial). This gives $Q(I) = 1.32$ eV in Ge, and $Q(I) = 1.83$ eV in Si. These values are to be compared with Bennemann's respective theoretical values of 0.44 and 0.51 eV (Table 5.1), and perhaps even smaller experimental values according to Watkins. One incidental point may be noted. The center of the puckered hexagonal ring as the saddle point is an assumption which has not been justified. Depending on the pairwise interaction model, it is conceivable that the saddle point could be at some other site, e.g., the center of the triangle formed by three non-neighboring atoms of the puckered six-membered ring, and not necessarily the center of the six-membered ring itself.

For solutes which normally ionize in Ge and Si, Weiser [21] proposed a theory which considers both the energy due to orbital overlap repulsive potential, and the energy due to the polarization of the lattice by the presence of the solute ions. The lattice polarization energy was neglected in Swalin's theory, and also in Bennemann's first principles treatment even though it has been commonly accepted that self-interstitials in Ge and Si behave as donors (but the levels are probably not very shallow. See figure 5.2). It turns out that the lattice polarization energy in silicon and germanium due to the presence of a singly charged ion is very large, and is dependent on the interstitial site at which the ion resides. For silicon, the polarization energy U_{pol} is -4.67 eV due to an ion at the tetrahedral site, and -5.43 eV due to an ion at the hexagonal site. There are three major approximations made in the calculation of the polarization energy. First, it is assumed that the polarizability is a constant derivable from the macroscopic dielectric constant of either silicon or germanium using the Clausius-Mossotti equation, and is independent of the electric field. The second assumption is that only dipoles are significant and higher multipoles can be neglected. The dipoles can be regarded as geometric points. The third assumption is that dipoles can be grouped in accordance with atomic shells about the ion, and dipoles on each shells are equivalent and radially directed. The third assumption was made primarily because of the magnitude of computational effort involved, rather than mathematical tractability. This assumption, however, can be dropped and at the same time the number of atoms considered can be increased by using group theory, as has recently been done by Hu and Weiser [33]. The results were not very much different from Weiser's earlier calculations however; they obtained U_{pol} of -4.20 eV and -5.16 eV respectively for an ion at the tetrahedral site and at the hexagonal site in Si, and -4.35 eV and -5.02 eV respectively for the tetrahedral and the hexagonal site in Ge. The difference in the lattice polarization energy between the hexagonal site and the tetrahedral site, ΔU_{pol}, tends to favor the ion at the hexagonal site. According to Weiser's theory, there is a competition from the repulsive energy, which tends to favor the ion at the tetrahedral site. Weiser calculated the repulsive energy with an empirical Born-Mayer potential. The activation energy of interstitial migration is then given by

$$H_I^m = |\Delta U_{pol} + \Delta U_{rep}|, \tag{5.21}$$

where ΔU_{rep} is the difference in repulsive energy between the hexagonal and the tetrahedral site. Weiser suggested that if $|\Delta U_{pol}| > |\Delta U_{rep}|$ or $\Delta U_{pol} + \Delta U_{rep} < 0$ the hexagonal site will be the equilibrium interstitial site; and vice versa. Ions of small radii, such as $Li_i^\cdot$ and $Cu_i^\cdot$ for which $|\Delta U_{pol}| > |\Delta U_{rep}|$ or $\Delta U_{pol} + \Delta U_{rep} < 0$ should therefore reside at the hexagonal sites, with an interesting consequence that some values of ionic radii exist for which $H_I^m \sim 0$, and after that, the tetrahedral site becomes the equilibrium interstitial site, accompanied by an increase in H_I^m again. The conclusion that $Li^{\cdot}$ is at the hexagonal site at equilibrium is in disagreement with experimental observations [15, 16, 19]. Bellomonte and Pryce [211] calculated the repulsive overlap energy using molecular orbitals. They concluded that the repulsive energy dominates and the polarization energy is insignificant, and hence the $Li_i^{\cdot}$ ion should reside at the tetrahedral site. However, their estimate of the polarization energy was incorrect by more than an order of magnitude. They calculated the overlap repulsive energy to be +0.375 eV, and the polarization energy to be −0.30 eV. This polarization energy is to be compared with Hu and Weiser's [33] value of ~ -4 to -5 eV. We feel that the repulsive energy in Weiser's calculation could have been overestimated, and the actual repulsive energy for an interstitial in Ge and Si could be much smaller. The reason is that the sp^3 orbitals are quite directional, so that the electron density should be very low in the interstitial space. Consequently, the repulsive energy of an atom at an interstitial tetrahedral site should be much less than that of an atom at a regular lattice site, even though the number of nearest neighbors and the distance to the nearest neighbors are the same in the two cases. Millea [10] took the other extreme approach of calculating the overlap repulsive energy by assuming that it results mainly from the interaction between the interstitial ion and the core electrons of the host atoms, i.e., Ge^{4+} and Si^{4+}, which is equivalent to assuming the valence electrons of Ge and Si are entirely concentrated between the neighboring lattice atoms in the spirit of covalent bonds. This model, in which the interstitial ion interacts repulsively mainly with the core electrons and not the valence electrons, leads to a physical picture that the interstitial space in silicon is larger than in germanium, in agreement with experimental findings. The reverse is true if the interstitial repulsive energy is calculated using covalent radii. One may also note that an empirical pairwise potential such as the Born-Mayer potential (with physical parameters normally derivable from bulk compressibilities [212]), such as used by Hasiguti [3], is of questionable applicability to the interstitial interaction.† Swalin simply neglected the interstitial overlap energy.

† We note that in covalent crystals, such as Si and Ge, the electron density distribution is very directional and localized, as the sp^3 orbitals tend to be concentrated between the two nearest neighboring atoms. Because of this, the overlapping repulsive pair-wise potential, which is assumed to have a spherical symmetry (i.e., the potential is assumed to be a function of scalar r rather than vector $\mathbf{r}$, q.v. equation 5.3), is obviously inadequate. The bulk compressibility results primarily from the overlapping electron orbits concentrated between two neighboring silicon (or Ge) atoms. The repulsive potential $U(r)$ derived from it is then a gross overestimation for that existing between an interstitial ion and a lattice silicon atom, because between them there is a negligible valence electron density.

Inasmuch as the interstitial overlap energy is concerned, we may conclude that the 0.375 eV for $Li_i^{\cdot}$ at a tetrahedral interstitial site in Si from the molecular orbital calculation by Bellmonte and Pryce [211] is probably the best among these available estimations. Unfortunately, similar calculations of overlap energies for $Li_i^{\cdot}$ in a hexagonal site in silicon, and for other ions in both silicon and germanium are unavailable in the literature.

Since with Weiser's calculations of the polarization energy and the overlap repulsive energy, it has already been concluded, in contradiction to experimental observations, that a $Li_i^{\cdot}$ ion resides at the hexagonal interstitial site in silicon, the use of more accurate overlap energy calculation as done by Bellmonte and Pryce could only force us more to that conclusion. Recently, the problem has been dicussed elsewhere [33]. It was suggested that [33] the main fault in Weiser's theory is the neglect of the nonvanishing crystal potential at the interstitial space. It turns out that the crystal potential at the interstitial space is not only a very large contributor to the formation energy of an interstitial, but is also very difficult to evaluate. The reader is referred to reference 33 for a detailed analysis. It suffices to point out here that the evaluation of the crystal potential depends critically on an accurate knowledge of the electron distribution, which is not available. Various approximative models had to be assumed, giving radically different results. In one model potential, the Ge and Si lattices are viewed as Ge^{4+} or Si^{4+} ions located at lattice sites and immersed in a uniform electron gas, as has been done by Bennemann [25]. The potential of each unit cell in this model is long ranging, unlike a spherically screened potential, and can be obtained sufficiently accurately only by carrying out Ewald's sum [213]. Bennemann [25] obtained a crystal potential of +21.4 eV at the hexagonal interstitial site and of +17.5 eV at the tetrahedral site for Si; and of +20.6 eV, at the hexagonal site and of +16.8 eV at the tetrahedral site for Ge. If we use this set of crystal potentials in conjunction with Weiser's classical model, we see that the crystal potential contribution would overwhelm the lattice polarization contribution, for if we denote the difference between the crystal potential at the hexagonal site and that at the tetrahedral site as ΔU_{crys}, we have $\Delta U_{crys} \sim 4$ eV for both Ge and Si. This would of course put $Li_i^{\cdot}$ on the tetrahedral site, in agreement with experimental observations. But then, together with the Born-Mayer repulsive potential, we would predict an activation energy of diffusion of approximately ~3.5 eV, using

$$Q = |\Delta U_{crys} + \Delta U_{pol} + \Delta U_{rep}|. \tag{5.22}$$

This value is of course in radical disagreement with known experimental values. In Bennemann's quantum mechanical treatment, this crystal potential is more than compensated by an even larger kinetic energy term due to the scattering of valence electrons by the interstitial ion. We had questioned the suitability of Bennemann's approach earlier in section 5.2.2. Conceivably, we can set up an opposite model in which the valence electrons are considered to localize at the midpoint between two neighboring lattice atoms. Intuitively, one would predict a larger potential at a hexagonal interstitial site than at a tetrahedral interstitial site because of the larger number of nearest neighbors with slightly shorter distance. Actually, the summation of contributions from 316 atoms in the tetrahedral case and from 310 atoms in the hexagonal case (corresponding to 16 and 18 atomic shells respectively for the two cases) showed the opposite to be true, with a tetrahedral interstitial potential of +5.97 eV and a hexagonal interstitial potential of +5.67 eV [33]. This would lead us back again to the conclusion that the equilibrium position for $Li_i^{\cdot}$ in silicon is the hexagonal interstitial site. A reasonable suggestion is that the actual valence electron distribution is intermediate between these two opposite models. For example, if we simply assume two-thirds of the valence electrons are localized, we would immediately obtain a reasonable activation energy of interstitial diffusion, and at the same time satisfy the requirement of an equilibrium position at the tetrahedral site. It has also been found that similar results can be obtained by using a very simple exponentially screened potential for each atom, which is conceptually plausible, and which affords an adjustable parameter [33]. Since Hasiguti's estimation of the migration energies of self-interstitials in silicon and germanium followed precisely Weiser's original theory, it suffered from the same fault that we mentioned above. It should be noted that an attractive feature of Weiser's original theory of the equilibrium site is its ability to satisfy the experimental observation that the activation energy of $Cu_i^{\cdot}$ is lower than that of $Li_i^{\cdot}$ in both Si and Ge. Now in the new model with the tetrahedral site as the equilibrium position, a transition to the saddle point (the hexagonal site) would require a larger activation energy for a $Cu_i^{\cdot}$ ion than for a $Li_i^{\cdot}$ ion, if one assumes that the electrical interactions in terms of interstitial potential and lattice polarization is independent of the species of the interstitial ion, while the repulsive overlap energy increases with ionic radius. Even though the actual value of the repulsive energy of an interstitial in silicon and germanium could be very small, its trend with the radius is not expected to be reversed in any reasonable overlap model. To explain the above phenomenon without invoking a valence electron energy contribution beside that of polarization, a concept of effective charge appears of value. A useful model would then require the effective charge of a $Cu_i^{\cdot}$ ion to be smaller than that of a $Li_i^{\cdot}$ ion, possibly through screening.

Having discussed the difficulty with Weiser's theory, we now tabluate in Table 5.10 some activation energies calculated by Weiser [21] and by Millea [10] according to this theory, to show its features. Weiser has also calculated the preexponential factor D_o according to the theory of Wert and Zener [214],

$$D_o = (1/6)gd(Q/2m)^{1/2} \exp(\Delta S/R), \tag{5.23}$$

where g is the number of saddle points surrounding the equilibrium site; d is the distance between two nearest equilibrium sites; Q and ΔS are the activation enthalpy and activation entropy of diffusion; and m is the mass of diffusing particle. One sees that $g = 2$ for the hexagonal site and $g = 4$ for the tetrahedral site. The calculated values of D_o generally fall in the range of $\sim 10^{-4}$-10^{-3} cm^2/s. Such a magnitude of D_o appears to be characteristic of interstitial diffusion in Si and Ge, as can be seen from Tables 5.5 and 5.6.

TABLE 5.10

Activation energies of interstitial diffusion in Ge and Si according to Weiser's theory[a]

Solute/Solvent	Weiser [21] ΔU_{rep}	Weiser [21] Q	Site[b]	Millea [10] ΔU_{rep}	Millea [10] Q	Site[b]	Q (expt.)[c]
Li/Ge	0.18	0.57	*H*	0.067	0.68	*H*	0.52
Na/Ge	0.53	0.22	*H*				
Cu/Ge	0.54	0.21	*H*	0.19	0.56	*H*	0.33
Au/Ge	1.83	1.08	*T*	1.24	0.49	*T*	0.63
Li/Si	0.23	0.52	*H*	0.023	0.73	*H*	0.66
Na/Si	0.68	0.07	*H*				0.72
Cu/Si	0.69	0.06	*H*	0.11	0.64	*H*	0.43
Au/Si	2.30	1.55	*T*	0.47	0.28	*H*	0.39

[a] See text for a discussion.
[b] ΔU_{rep} and Q in eV; H = hexagonal site, T = tetrahedral site.
[c] Experimental values of Q taken selectively from Tables 5.5 and 5.6.

A more recent study of interstitial diffusion in silicon and germanium was carried out by Nardelli and Reatto [215], applying Kubo's formalism of the statistical-mechanical theory of irreversible processes [216]. Starting from the Liouville equation for the total system they derived the evolution equation for a particle moving in a periodic potential and coupled to the lattice vibrations taken at thermal equilibrium; they then obtained an approximate solution of the evolution equation, giving the expression of the diffusion coefficient of ionized impurities in silicon and germanium. The expression agrees formally with the absolute rate theory which underlies the Wert-Zener theory of interstitial diffusion in that the preexponential factor contains a factor $(Q/2m)^{1/2}$, predicting a reciprocal square root relationship of D_o with the mass of the diffusing particle. It differs from the D_o of the absolute rate theory in that it contains a linearly temperature dependent factor $\bar{\lambda}(T)$, which may be suitably called a mean free path, while the latter D_o contains an atomic jump distance d which is independent of temperature. Their D_o can be expressed as follows:

$$D_o = (4/3\pi)^{1/2} \frac{J}{J'} \bar{\lambda}(T)(Q/2m)^{1/2}, \tag{5.24}$$

where J and J' are pure numbers given by some definite double integrals, and have the values of 0.499 and 0.525 respectively. The $\bar{\lambda}(T)$ is always smaller than the atomic jump distance d, whilst in the absolute rate theory it is assumed to be equal to d. For Si, $\bar{\lambda}(1300^\circ\text{K}) = 0.54$ Å; for Ge, $\bar{\lambda}(1100^\circ\text{K}) = 0.86$ Å; while $d = 2.352$ Å and 2.449 Å for Si and Ge respectively. The entropy factor $\exp(\Delta S/RT)$ is neglected in equation 5.24 because of its assumed small contribution.

The results of Nardelli and Reatto's calculations of D_o for interstitial diffusion of Li in Si and Ge agree with experimental data very well, and of Cu, Ag and Au in Si and Ge fairly well, presumably because of the much less reliable experimental data of Cu, Ag, and Au than Li in Ge and Si. It should be noted that their calculations are based on the same assumption as made by Swalin [208] that only the electrostatic interaction contributes significantly to the interactions between the interstitial ion and the lattice, and further they used Swalin's 'effective charge' of +2.4 for the lattice atoms Ge and Si.

In addition to Nardelli and Reatto's classical approach to Kubo's formalism, Gosar has studied the interstitial diffusion quantum-mechanically using a temperature dependent Green's function method [217]. We summarize the results of Nardelli and Reatto and of Gosar, together with Weiser's calculations based on the Wert-Zener theory in Table 5.11.

In summary, theoretical calculations of the preexponential factor D_o for interstitial diffusion, whether by the absolute rate theory or by the Kubo formalism, agree quite well with experimental data; and the uniformity of the experimental values of D_o as seen from Tables 5.5 and 5.6 also attests to the general accuracy of the assumed interstitial mechanism. It is remarkable that in the diffusion of these interstitials it is not necessary to invoke the concept of an enlarged region of disorder such as the relaxion or the extended interstitial, even though the diffusion temperature in some cases (e.g., Ag in Si) is as high as 1350° C. Since there has been no theory to guide us to predict how D_o (and indeed Q) might be

TABLE 5.11

Theoretical values of preexponential factor of interstitial diffusion in Si (1300° K) and Ge (1100° K)

	D_o(cm²/s)			
Solute/Solvent	Weiser[a]	Nardelli and Reatto	Gosar	Expt.
Li/Ge	4×10^{-4}	3.63×10^{-3}	0.44×10^{-3}	2.5×10^{-3}
Cu/Ge	3×10^{-4}	0.97×10^{-3}		4.0×10^{-3}
Ag/Ge	9×10^{-4}	1.29×10^{-3}		4.4×10^{-2} (?)
Au/Ge	9×10^{-4}	1.51×10^{-3}		2.25×10^{-2} (?)
Li/Si	2×10^{-4}	2.57×10^{-3}	0.94×10^{-3}	2.5×10^{-3}
Cu/Si	2×10^{-4}	0.69×10^{-3}		4.7×10^{-3}
Ag/Si	11×10^{-4}	1.02×10^{-3}		2.0×10^{-3}
Au/Si	10×10^{-4}	3.70×10^{-4}		2.44×10^{-4}

[a] In the Wert-Zener theory of D_o on which Weiser's calculations are based, there is no explicit relationship between D_o and the diffusion temperature. So it is assumed to be temperature independent here.

significantly altered for the extended interstitial model, we would not hastily rule out the possibility of its existence. When the labeled interstitial atom happens to be an isotope of the host lattice, such as ^{31}Si in ^{28}Si in Fairfield and Masters' experiments, the situation is different. The large preexponential factor presumably can be attributed to the entropy factor for the formation of the defect as suggested by Seeger and Swanson [23]. On the other hand, the available theories on the activation energy of interstitial diffusion have not been satisfactory. Clearly, a better knowledge of the crystal interstitial potential is badly needed.

5.4.3 Interstitial-Substitutional Diffusion in Silicon and Germanium

The Dissociative Diffusion Mechanism

Many solutes dissolve both interstitially and substitutionally in silicon and germanium, although the ratio of the interstitial solubility to the substitutional solubility varies over a range of many orders of magnitude. For many solutes belonging to the transition elements, the ratio generally ranges from 10^{-4} to 10^4. In a spatially homogeneous solution, an equilibrium ratio of $[A_i]_e/[A_s]_e$ can be assumed to hold for a solute A occupying both interstitial and substitutional sites, 'i' and 's' respectively. In a spatially inhomogeneous solution diffusion will take place. Since the interstitial and the substitutional species diffuse at disparate rates, the local ratio $[A_i]/[A_s]$ will be upset in such a manner that in the region of high concentration, (i.e., the region where the divergence of flux is negative), $[A_i]/[A_s] < [A_i]_e/[A_s]_e$; and in the region of low concentration, $[A_i]/[A_s] > [A_i]_e/[A_s]_e$. This is because the interstitial diffusivity is usually much higher than the substitutional diffusivity of any known model. This is then accompanied by a kinetic process to reestablish equilibrium, according to the reaction

$$A_s \underset{k_2}{\overset{k_1}{\rightleftarrows}} A_i + V. \tag{5.25}$$

Clearly, the forward reaction will take place in the region of high concentration, and the diffusion is said to proceed dissociatively [11]. In a region of low concentration, the reverse reaction will take place. The resulting impurity concentration profile is rather complicated. In principle, it is obtainable from a solution of the following set of equations of continuity:

$$\frac{\partial [A_i]}{\partial t} = D(A_i)\nabla^2 [A_i] + k_1 [A_s] - k_2 [A_i] [V], \tag{5.26}$$

$$\frac{\partial [A_s]}{\partial t} = D(A_s)\nabla^2 [A_s] - k_1 [A_s] + k_2 [A_i] [V], \tag{5.27}$$

$$\frac{\partial [V]}{\partial t} = D(V)\nabla^2 [V] + k_1 [A_s] - k_2 [A_i] [V]. \tag{5.28}$$

The list of these simultaneous equations can be expanded and each can be lengthened by considerations of further details such as formation of complexes such as *A-V* pairs, *A-X* pairs where *X* is some second impurity, say oxygen. But the three equations 5.26-5.28 listed above are sufficient for the purpose of the present discussion. Some simplifying assumptions of rate controlling mechanisms are necessary both for mathematical tractability and conceptual insight. Because of the known fact of extremely fast interstitial diffusion, one may start out by making an assumption about the interstitial. One of two assumptions are usually made, that the interstitial concentration is always at quasi-equilibrium condition ($[A_i] \simeq [A_i]_e$), or that it is at a quasi-steady state ($\partial [A_i]/\partial t \simeq 0$). The latter assumption has been made by Wei [124] in his discussion of the 'two-stream mechanism'. In making these assumptions, we have in effect neglected the initial period during which the interstitial diffusion is clearly the control mechanism. The supposedly fleeting existence of this period is of no great interest to us. In the present discussion, we will not make a distinction between the diffusion of a solute inward from the surface and the precipitation from a supersaturated solution. We make the further assumption that $D(A_s)$ is extremely small so that the ∇^2 term in equation 5.27 can be dropped. In the first case, we assume $[A_i] \simeq [A_i]_e$, with the criterion of its justification to be discussed later. Then the problem simplifies to finding solutions for the set

$$\frac{\partial [A_s]}{\partial t} = -k_1 [A_s] + k_2 [A_i]_e [V], \tag{5.29}$$

$$\frac{\partial [V]}{\partial t} = D(V) \nabla^2 [V] + k_1 [A_s] - k_2 [A_i]_e [V]. \tag{5.30}$$

The above set of simultaneous equations is linear. Analytical solutions both for the one-dimensional case [218] and for the three-dimensional case [123] are available. The one-dimensional solution is appropriate when only the crystal surface acts as the sink or the source. When there is a sufficiently high density of dislocations that act as sources or sinks, one must use the three-dimensional formulation, with cylindrical boundary conditions [219]. It is noted that beside the ∇^2 term, terms involving the Coulombic and the elastic interactions between the vacancy and the dislocations [220] have been omitted in equation 5.30; this is justifiable after a short initial period so that a capture radius of the dislocation for the point defect can be defined [221]. The general solution to the problem can be expressed in the form

$$[A_s] = \sum_{n=0}^{\infty} \alpha_n \phi_n (r \text{ or } x) \exp(-\mu_n k_2 [A_i]_e t), \tag{5.31}$$

where $\phi_n(r)$ is some linear combination of Bessel functions for the case of dislocation sinks (sources), and $\phi_n(x)$ is a sine or a cosine function for the case of surface sink (source); α_n is some appropriate constant, and μ_n are the eigenvalues. A similar expression is obtained for $[V]$. For more detailed discussions, the reader is referred to references 123, 218 and 222.

Pseudo Homogeneous Reaction Approximation

If one is interested in the change of total (or average) concentration with time, a pseudo first-order homogeneous reaction approximation is useful. One replaces the diffusion terms by suitable first-order reactions. Thus, equations 5.26 and 5.28 become

$$\frac{\partial [A_i]}{\partial t} = k_I([A_i]_e - [A_i]) + k_1 [A_s] - k_2 [A_i][V], \tag{5.32}$$

$$\frac{\partial [V]}{\partial t} = k_V([V]_e - [V]) + k_1 [A_s] - k_2 [A_i][V], \tag{5.33}$$

where k_I and k_V are pseudo reaction rate constants in a similar sense to k_2.

They define some reactions which are diffusion controlled. The values k_I, k_V and k_2 are hence obtainable as the smallest eigenvalues of the respective diffusion equations [223] with suitable boundary conditions. These can be listed as follows [222-224].

$$k_2 = 4\pi R_V(D(A_i) + D(V)). \tag{5.34}$$

If the sinks and the sources are dislocations, we have for k_I and k_V

$$k_{I \text{ or } V} = 2\pi\rho_d D/\ln(l/R_d). \tag{5.35}$$

where D stands for $D(A_i)$ or $D(V)$ as the case may be. If the sinks and the sources are the surfaces, we have

$$k_{I \text{ or } V} = (\pi/L)^2 D. \tag{5.36}$$

In the above expressions, R_V is the capture radius of the vacancy for the interstitial (and vice versa); R_d is the capture radius of the dislocation for either the interstitial or the vacancy; ρ_d is the dislocation density; l the half mean distance between dislocations, given by $\sim(\pi\rho_d)^{-1/2}$; and L is some average distance of the diffusion region of interest to the surface. (In the case of precipitation, which starts out from a homogeneous solution, L is taken as the thickness of the semiconductor slab.) The dissociation rate constant k_1 can be obtained from the relationship [225].

$$k_1 = k_2[A_i]_e[V]_e/[A_s]_e. \tag{5.37}$$

The solution of the set of equations 5.29 and 5.33 is a combination of two exponential functions of time (there is no spatial distribution) with time constants given as

$$1/\tau = (1/2)\left\{k_1 + k_2[A_i]_e + k_V \pm \sqrt{(k_1 + k_2[A_i]_e + k_V)^2 - 4k_1k_V}\right\}. \tag{5.38}$$

Such an exponential decay type of kinetics is therefore expected for diffusion controlled precipitation which proceeds via a dissociative mechanism. This agrees with the observation of Tweet [81] of Cu in Ge. Even in the case where the diffusion of the interstitial is the rate controlling mechanism, an approximate exponential decay function for $[A_s]$ is still appropriate [125]. Such kinetics were not observed for Cu precipitation in Si [226], where $[A_i]_e \gg [A_s]_e$. Instead, the kinetics were observed to follow a power law $(\text{time})^n$, where $n = 0.687 \pm 0.043$. Cottrell and Bilby [227], and Harper [228] had earlier obtained a theoretical expression of precipitation rate which is proportional to $(\text{time})^{2/3}$ in the model of drift approximation (i.e., by assuming the force field term to dominate). This has been criticized by Ham [221, 229]. It has been pointed out by Bullough and Newman [220, 230, 231] that the boundary condition (at the interface between the host lattice and the line precipitate) may be such that there is a finite velocity of transfer of solute atoms which is a function of the degree of supersaturation. In such a case they showed that [220] kinetics of the form (time)^n may result. The interactions between solutes and dislocations will be discussed further in a later section.

Criteria for the Simplifying Assumptions $[A_i] = [A_i]_e$, *or* $[V] = [V]_e$

The pseudo first order homogeneous reaction approximation is also useful in the discussion of the condition for the suitability of the usual simplifying assumptions made in the dissociative diffusion mechanism, i.e., $[A_i] = [A_i]_e$, or $[V] = [V]_e$. Yoshida [222] has shown that the criterion for the validity of the assumption $[A_i] = [A_i]_e$ is

$$k_1 k_I + k_2 k_I [A_i]_e + k_I k_V > k_1 k_V + k_2 k_V [V]_e; \qquad (5.39)$$

and the criterion for the validity of the assumption $[V] = [V]_e$ is

$$k_1 k_V + k_2 k_V [V]_e + k_I k_V > k_1 k_I + k_2 k_I [A_i]_e. \qquad (5.40)$$

For the case of the in-diffusion or the annealing of Ni in Si in the temperature range 800-1200°C, Yoshida assumed that $D(A_i)$ and $D(V) \sim 10^{-5}$-10^{-6} cm^2 s^{-1}; then $k_2 \sim 10^{-12}$ cm^3 s^{-1}; and if the thickness of the specimen is $\sim 10^{-1}$ cm and the dislocation density is $\sim 10^4$ cm^{-2}, k_I and k_V (qv. equation 5.35 and 5.36) are of the order of 10^{-2} s^{-1}; and for $[A_i]_e > [A_s]_e > [V]_e$, $k_2 [A_i]_e > k_1$ (qv. equation 5.37). Since $[A_i]_e > 10^{15}$ cm^{-3}, he concluded that the term $k_2 k_I [A_i]_e$ dominates and equation 5.39 is satisfied, and the assumption of a quasi-equilibrium for the interstitials is justified. In order of magnitude, the above cited physical parameters are also appropriate for gold in silicon, except that $[A_s]_e > [A_i]_e > [V]_e$. We then see that the term $k_2 k_V [A_i]_e$ still dominates (since $[A_i]_e \sim 10^{15}$-10^{16} cm^{-3}), and equation 5.39 is satisfied. Hence, the assumption of quasi-equilibrium gold interstitials is valid, and it leads to the relationship between the vacancy diffusivity and the apparent gold diffusivity in silicon (qv. equation 5.18). The assumption $[A_i] = [A_i]_e$ is even more well justified for Cu in Si, where $[A_i]_e \gg [A_s]_e > [V]_e$. As a rule of thumb, one may then say that unless $D(A_i)[A_i]_e < D(V)[V]_e$, the dissociative diffusion mechanism will be accompanied by a vacancy non-equilibrium—undersaturation in the case of in-diffusion and supersaturation in the case of annealing-precipitation.

In their original proposal of the dissociative diffusion mechanism for the interpretation of copper precipitation in germanium, Frank and Turnbull [11] had considered two possible cases, which may be stated as follows:

(a) If the dislocation density is high, one can assume $[V] = [V]_e$. Then

$\partial[V]/\partial t$ in equation 5.28 vanishes. By the addition of equation 5.27 to equation 5.26 and by neglecting the $D(A_s)\nabla^2[\mathrm{Au}_s]$ term, one immediately obtains, utilizing

$$[A_i] = ([A_i] + [A_s])[A_i]_e/([A_i]_e + [A_s]_e) \quad \text{at} \quad [V] = [V]_e,$$

$$\partial([A_i] + [A_s])/\partial t = D(A_i)\,\nabla^2[A_i]$$

$$= \{D(A_i)[A_i]_e/([A_i]_e + [A_s]_e)\}\,\nabla^2([A_i] + [A_s])$$

or

$$\partial[A_{\text{tot}}]/\partial t = \mathbf{D}(A_{\text{tot}})\,\nabla^2[A_{\text{tot}}] \qquad (5.41)$$

where the subscript 'tot' denotes the total solute irrespective of lattice position. Hence obtains the apparent diffusivity

$$\mathbf{D}(A_{\text{tot}}) = D(A_i)[A_i]_e/([A_i]_e + [A_s]_e) \qquad (5.42)$$

(b) If the dislocation density is low, the diffusion of vacancies will be rate controlling, and the assumption $[A_i] = [A_i]_e$ can be made. Then one obtains an expression similar to equation 5.18, i.e.

$$\mathbf{D}(A_s) = D(V)[V]_e/([V]_e + [A_s]_e) \simeq D(V)[V]_e/[A_s]_e \qquad (5.43)$$

However, from the criteria set forth, it appears that for Cu in Ge, both for the case of high and the case of low dislocation density, only the assumption $[A_i] = [A_i]_e$ is justified, at least in the case of precipitation from a supersaturated solution. Consequently the validity of equation 5.42 is questionable for case (a).

The Two-Stream Model

Observation of two simultaneously diffusing streams was first reported by Kosenko [232] for Ag, In, Zn, and Te in Ge. He observed these solutes to have a fast diffusing component, generally with a low concentration of $\sim 10^{15}$ cm^{-3}, and a high diffusivity of $\sim 10^{-7}$ cm^2 s^{-1}; and a corresponding slow component of higher concentration ($\sim 10^{18}$ cm^{-3}) and lower diffusivity ($\sim 10^{-12}$ cm^2 s^{-1}) at 800°C. The radioactive tracer profiles from such diffusion exhibited a discontinuity, as shown in figure 5.5. Such a profile would suggest that the two streams diffused almost independently. If, as interpreted by Kosenko [232], the two streams followed an interstitial and a vacancy mechanism respectively, it would imply a very small or negligible rate of interstitial-vacancy recombination. This would mean the dominance of the term $k_I([A_i]_e - [A_i])$ over the term $k_2[A_i][V]$ in equation 5.32 of the homogeneous reaction approximation. Using equations 5.34 and 5.35, one sees that this can happen when $[V]_e < (R_V L^2)^{-1}$. Taking $R_V \sim 10^{-8}$ cm,

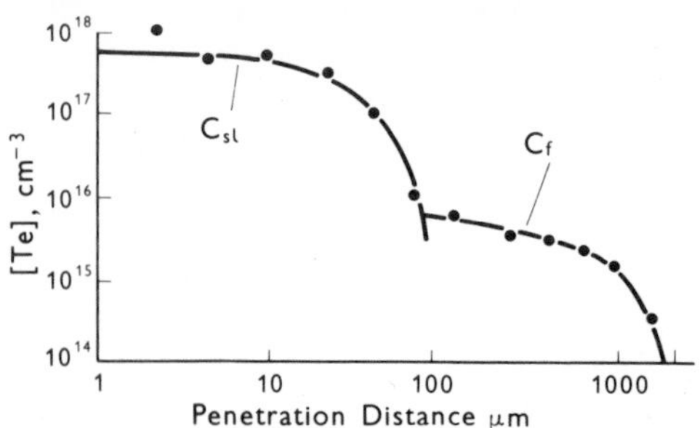

Figure 5.5. Diffusion of tellurium in germanium at 800°C. After Kosenko [232].

$L \sim 10^{-2}$ cm (figure 5.5) this means $[V]_e < 10^{12}$ cm^{-3}, which is not unreasonable (a vacancy formation energy of 2.0 eV would give $[V]_e \sim 10^{13}$ cm^{-3} at 800°C). Nonetheless the possibility of contamination by impurities such as Cu should be further examined. The element indium is interesting in that, though belonging to Group III, it has a rather low ionization potential (5.8 eV), and is hypothetically expected to exist with an appreciable interstitial component following a relationship similar to equation 5.20. This appears to be compatible with the observations of In in Ge by Kosenko [232] and of In in Si by Millea [10]. We have however previously criticized Millea's proposed interchange mechanism on other grounds, particularly the consideration of energetics (see section 5.2.1), and have suggested that it would be less favorable than either a dissociative mechanism or a vacancy mechanism. Boron has also been suspected by Rupprecht and Schwuttke [233] to diffuse in silicon by a two-stream mechanism because of the observation of slightly kinked profiles. In fact, Boltzmann-Matano analyses of boron concentration profiles gave an approximately linear (actually slightly superlinear) relationship between the diffusivity and the concentration [234]. Other studies of boron concentration profiles have been reported elsewhere [171, 174]. We see that such observations are compatible with a vacancy mechanism only if the vacancy has a relatively shallow donor level. But it can be easily explained by either a dissociative mechanism in which the relative concentration of the interstitial component is enhanced by the lowering of the Fermi level; or an interstitialcy mechanism in which the silicon self-interstitial (which functions similarly to a vacancy as a transport vehicle), being a shallow donor, is enhanced by the lowering of the Fermi level with increasing boron concentration.

Two-Stream with Trapping [124]

A finer classification of the interstitial-substitutional diffusion processes may be necessary. According to Wei [124], the diffusion of Ag, Co and Fe is distinctively different from the diffusion of Cu and Ni in Ge. He considered Cu and Ni (and also Li) to be very weakly bonded to Ge atoms, and Ag, Co and Fe weakly bonded to Ge atoms. The activation energy of the diffusion of Cu and Ni in Ge is ~0.5 eV, and that of Ag, Co and Fe in Ge is ~1 eV. The apparent diffusivity of the latter group is about two orders of magnitude lower than the former group (10^{-6}-10^{-7} versus 10^{-4}-10^{-5} cm^2 s^{-1} at 800° C). The maximum solubility of the latter group is 10^{-8}-10^{-9} atomic fraction, against that of the former group of 10^{-4}-10^{-7} atomic fraction. Now in the limit of low dislocation density, we have already shown the validity of the vacancy diffusion controlled dissociative mechanism for the diffusion of Cu, Ni, and Au in Si. The simple relationship equation 5.43 says that the apparent diffusivity of the substitutional component should be simply given by the quotient of the solvent self-diffusivity to the equilibrium substitutional solute concentration in atomic fraction. The observation of the smaller equilibrium concentrations of the group Ag, Co and Fe together with their lower diffusivities appears in discord with the relationship equation 5.43. The physical meaning for this is that for Ag, Co and Fe in Si, $D(A_i)[A_i]_e$ could be smaller than $D(V)[V]_e$. Such a condition could satisfy the criterion of equation 5.40. Hence, in place of the assumption $[A_i] = [A_i]_e$, one has $[V] = [V]_e$. The diffusion of these impurities is then described by the set of equations 5.26 and 5.27, with $[V] = [V]_e$. By replacing suitable constants K_1 and K_2 for the constants k_1 and $k_2[V]_e$ in equations 5.26 and 5.27, the trapping centers will then not necessarily be limited to vacancies. To interpret his data, Wei [124] conceived that vacancies may not be a determining factor in Fe diffusion in Ge. Instead, some other kind of trap, possibly oxygen may play an important role. If the role of vacancies becomes insignificant, the dislocation density would have little effect on the concentration and diffusivity of the solute. The impurity traps may have a higher concentration than the equilibrium concentration of vacancies and may produce pockets in the crystals. A high rate of trapping would give the non-Fickian diffusion that he observed. More specifically, he observed the concentration profile to follow a simple exp $(-x/L)$ law instead of a complementary error function. We rewrite equations 5.26 and 5.27 to conform with the concept that the impurity traps may not necessarily be vacancies (and accordingly the trapped impurities may not necessarily become substitutionals)

$$\frac{\partial [A_i]}{\partial t} = D(A_i)\frac{\partial^2 [A_i]}{\partial x^2} + K_1[A_{tr}] - K_2[A_i], \tag{5.44}$$

$$\frac{\partial [A_{tr}]}{\partial t} = D(A_{tr})\frac{\partial^2 [A_{tr}]}{\partial x^2} - K_1[A_{tr}] + K_2[A_i], \tag{5.45}$$

where the subscript tr refers to the trapped species. Instead of the simplifying assumption of a quasi-equilibrium concentration for $[A_i]$ which has been made previously for the dissociative mechanism, we must allow $[A_i]$ to be a variable in the present case because of the lower $[A_i]$ and lower $D(A_i)$. A general solution of this problem has been given by Kucher [218]. Although he assumed the traps to be vacancies, the generality of equations 5.44 and 5.45 and their solution by Kucher should make no distinction between vacancies and impurities as traps. We give a sample case in figure 5.6 due to Kucher, for $D(A_{tr})/(D(A_i) = 10^{-3}$,

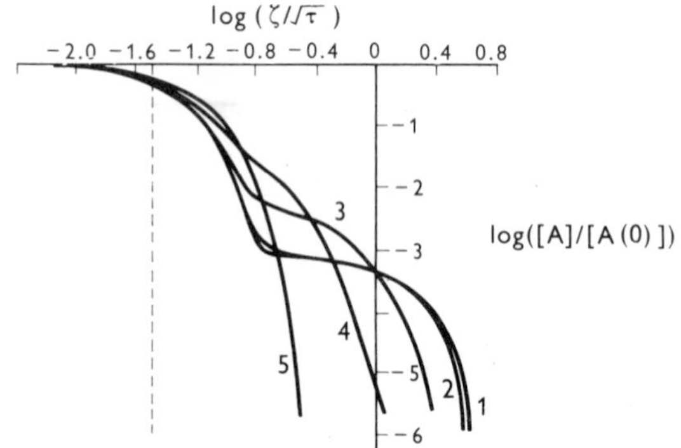

Figure 5.6. The distribution of total impurity concentration ($[A] = [A_i] + [A_{tr}]$) from a two-stream mechanism of diffusion. $D(A_{tr})/D(A_i) = 10^{-3}$; $[A_i]_e/[A_{tr}]_e = 10^{-3}$; values of $\tau = K_2 t$ are (1), 0.05; (2) 0.1; (3) 10; (4) 10^2; (5) 10^4. $\zeta = (K_2/D(A_i))^{1/2}$. Note that because of the use of the variable $\log(\zeta/\sqrt{\tau})$ each curve has been displaced by an amount $\log\sqrt{\tau}$ to the left in the log-log plot. Thus, the spatial extent of the curves should increase with τ. After Kucher [218].

$[A_i]_e/[A_{tr}]_e = 10^{-3}$ ($[A_{tr}]_e$ is simply $[A_s]_e$ for vacancy type traps) and for five values of dimensionless time τ. The reduced variables are: $\tau = K_2 t$; $\zeta = x(K_2/D(A_i))^{1/2}$. Note that the use of the variable $\log(\zeta/\sqrt{\tau})$ has effectively displaced curves to the left in the log-log plot by an amount $\log\sqrt{\tau}$, a scheme for space saving. For approximate solutions, Wei made the simplifying assumption of a quasi-steady state for the interstitial, i.e. $\partial[A_i]/\partial t = 0$. He considered two cases:

(a) Dead trapping, $K_1 = 0$. The solutions are

$$[A_i] = [A_i(0)] \exp(-x/L)$$

$$[A_{tr}] = K_2 t [A_i(0)] \exp(-x/L) \tag{5.46}$$

where $L = (D(A_i)/K_2)^{1/2}$. It may be noted that the linearization that leads to equations 5.44 and 5.45 implies a much higher concentration of traps than the concentration of trapped species.

Otherwise, K_2 should be replaced by $k_2([X] - [A_{tr}])$, where $[X]$ is the initial trap concentration, thus making the diffusion equations nonlinear. If the traps are vacancies, however, a high $[V]_e$ is not implied, since vacancies are highly mobile species, and are always kept at $[V]_e$ in the system under consideration.

(b) Shallow traps, $K_1 \neq 0$. Approximate solutions are

$$[A_i] = [A_i(0)] \exp(-x/L);$$

$$[A_{tr}] = K_2 K_1^{-1} [A_i(0)] [1 - \exp(-K_1 t)] \exp(-x/L), \qquad (5.47)$$

where $L = (D(A_i)/K_2)^{1/2} \exp(K_1 t/2)$. It is very difficult to assess the suitability of the quasi-steady state approximation. Presumably, there exists some time period, probably in the short to intermediate time period and some space range where this assumption may be reasonable. In the long time period, one expects $K_1[A_{tr}] - K_2[A_i]$ to vanish, and equation 5.44 reduces to Fick's second law. Clearly, some interesting features of the profile as shown in figure 5.6 have been lost in the quasi-steady state approximation.

Recently, Boctor and Watt [322] have given a more complete analysis of the double-stream diffusion mechanism with trapping by relaxing the quasi-steady state assumption. However, unlike Kucher's analysis [218], they assumed the diffusivity of one diffusant to be zero.

Summary Notes on the Interstitial-Substitutional Diffusion Mechanism

We have, in the above discussion, distinguished three categories of interstitial-substitutional diffusion mechanisms. For solutes whose solubility (particularly the interstitial component) is low, and whose interstitial diffusivity is also low (with activation energy $\gtrsim$1 eV), a diffusion trapping mechanism is suitable. The interstitial component $[A_i]$ is below its equilibrium value, but the vacancies can be assumed to attain equilibrium during the diffusion process. The process is typified by Ag, Co, and Fe diffusion in Ge. For solutes whose solubility is intermediate, and whose interstitial diffusivity is very high (with an activation energy of ~0.5 eV), the dissociative diffusion mechanism is appropriate with the rate controlled by vacancy diffusion, and the interstitial at equilibrium. The third category includes those solutes with high solubility and low interstitial diffusivity, such as Zn and Te in Ge, and possibly In in Ge and Si. A two-stream mechanism has been proposed. B in Si has also been suspected to follow this mechanism. We should note the discrepancy in the results of diffusion of Ag in Ge from Wei's [124] and from Kosenko's [232] observations. The kinked concentration profiles of the solute/solvent systems listed in Table 5.12 appear to be more convincing evidence

TABLE 5.12

Two-stream diffusion mechanism in Ge after Kosenko [232]

Solute	D_{sl}(cm^2/s)	D_f(cm^2/s)	D_f/D_{sl}	c_{sl} (cm^{-3})	c_f (cm^{-3})	c_{sl}/c_f
Ag	2.0×10^{-12}	9.0×10^{-7}	4.5×10^5	4.0×10^{18}	3.0×10^{14}	1.3×10^4
In	1.3×10^{-12}	6.9×10^{-8}	5.3×10^4	4.0×10^{18}	3.4×10^{15}	1.0×10^3
Zn	6.5×10^{-13}	2.0×10^{-8}	3.0×10^4	2.5×10^{18}	2.6×10^{14}	9.6×10^3
Te	3.2×10^{-11}	5.0×10^{-7}	1.6×10^4	6.0×10^{17}	2.5×10^{15}	2.4×10^2

of a two-stream mechanism than the kinked boron and phosphorus concentration profiles in silicon. Further investigations in this area are clearly badly needed. We also note an interesting phenomenon in the *two-step diffusion* of Zn in GaAs and in InSb [235, 236]: The first step was carried out in the presence of a zinc source; Zn diffused into n-type GaAs dissociatively. In the second step, which is equivalent to the well known drive-in step in silicon and germanium technology, the source was removed and the sample surface was cleaned, and the heat treatment was carried out. Kendall found that there was little or no movement of the p-n junction. He suggested that in the absence of an infinite zinc source, the diffusion of the ionized interstitial Zn donor was retarded by the internal electrostatic field due to the immobile substitutional Zn acceptor profile. He also suggested that [237] as the same phenomenon has not been observed in the two-step diffusion of boron or phosphorus in silicon this would exclude an interstitial-substitutional mechanism.

Dissociative Diffusion Under Extrinsic Conditions

At the substitutional sites in Si and Ge, the transition elements generally possess deep acceptor levels, while at the interstitial sites they possess deep donor levels. For example, Collins *et al.* [217a] have reported for gold in silicon a donor level of 0.35 eV above the valence band edge and an acceptor level of 0.54 eV below the conduction band edge. Consequently, the activities and the solubilities of interstitial and substitutional species of each such element are dependent on the Fermi level of Si or Ge. Such a dependency has been propounded by Reiss [217b] and Reiss *et al.* [217c]. A practical example of the dissociative diffusion under extrinsic conditions is the diffusion of gold in Si (to provide recombination centers for minority carriers) in the presence of phosphorus, arsenic or boron. The most general approach to such a problem is to evaluate the activities of all species concerned by evaluating the change of the system free energy with respect to the concentration of each species. Such an approach is frequently difficult. The concept of mass action is then useful. We rewrite equation 5.25 as

$$A_s' - e' \underset{k_2'}{\overset{k_1'}{\rightleftharpoons}} A_s^{\times} \underset{k_2}{\overset{k_1}{\rightleftharpoons}} A_i^{\cdot} + V' \underset{k_2''}{\overset{k_1''}{\rightleftharpoons}} A_i^{\cdot} + V^{\times} + e' \tag{5.47a}$$

The reactions between $A_s^{\times}$ and e', and between $V^{\times}$ and e' can be regarded as instantaneous. Hence,

$$[A_s^{\times}] = (k_1'/k_2')[A_s']/n \tag{5.47b}$$

and

$$[V'] = (k_2''/k_1'')[V^{\times}]n \tag{5.47c}$$

Both factors (k_1'/k_2') and (k_2''/k_1'') depend on the acceptor levels of A_s and V, while n depends on the Fermi level, and the ratios $[A_s']/([A_s^{\times}] + [A_s'])$ and $[V']/([V^{\times}] + [V'])$ are governed by Fermi statistics. We may for simplicity assume that ionized interstitials are by far more mobile than either the uncharged interstitials or the substitutionals. We may then rewrite equations 5.26-5.28 as

$$\frac{\partial [A_i]}{\partial t} = \nabla \cdot \left\{ D(A_i)(\nabla [A_i^{\cdot}] - e[A_i^{\cdot}]\mathbf{E}/kT) \right\} + k_1[A_s^{\times}] - k_2[A_i^{\cdot}][V'] \tag{5.47d}$$

$$(1 + nk_2'/k_1') \frac{\partial [A_s^{\times}]}{\partial t} = -k_1[A_s^{\times}] + k_2[A_i^{\cdot}][V'] \tag{5.47e}$$

$$(1 + k_1''/k_2''n) \frac{\partial [V']}{\partial t} = \nabla \cdot \left\{ D(V)(\nabla (1 + k_1''/k_2''n)[V'] - e[V']\mathbf{E}/kT) \right\} + k_1[A_s^{\times}] - k_2[A_i^{\cdot}][V'] \tag{5.47f}$$

We may assume that $n(x)$ is time invariant during the diffusion of the transition element, an assumption that is reasonable when $n(x)$ is dominated by the shallow level impurities of Groups III or V, whose diffusivity is generally several orders of magnitude lower. In the above expression, **E** is the internal field due to the presence of shallow level impurities.

Because of its amphoteric nature (i.e., the possession of a donor level as an interstitial and an acceptor level as a substitutional) the solubility of gold is increased in both heavily p- and n-doped silicon. However, the apparent diffusivity of gold will be decreased in heavily n-doped silicon, while increased in the p-doped silicon. Badalov and Shuman [217d] reported experimental results of the effect of n-type dopants (with donor concentrations of $<n_i$) on the diffusion of gold in silicon. No simple interpretation appears available for their results.

5.4.4 Diffusion of Group III and Group V Impurities

Group III and Group V elements as solutes in silicon and germanium are distinguished from elements of other groups in their ability to form strong covalent bonds with the lattice atoms. This results in their high solubilities in Si and Ge [238], and their existence in the lattices of Si and Ge almost entirely in the substitutional form. Two important consequences are that they diffuse in Si and Ge predominantly by either a vacancy mechanism or an interstitialcy mechanism, and that their high concentrations under typical conditions make the solid solutions nonideal even at diffusion temperatures, mainly but not solely through electrical interactions. From presently available information, it seems not sufficiently warranted to make a definite statement concerning the suitability of either the vacancy or the interstitialcy mechanism. Some arguments in favor of an interstitialcy mechanism for impurity diffusion in Si and of a vacancy mechanism in Ge can be found in the excellent review article by Seeger and Chik [3]. In particular, such an interpretation is able to explain why acceptor impurities diffuse faster than donor impurities in Si, while the reverse is true in Ge (see Tables 5.8 and 5.9). On the other hand, an interstitialcy mechanism for boron diffusion does not appear to be compatible with the current model of the emitter push effect [128] and with radiation enhanced diffusion [239, 240]. (Silicon self-interstitials created by irradiation are not expected to exist in appreciable supersaturations due to their very short life time. This argument is in accord with the fact that silicon self-interstitials have not been detected in low temperature irradiation experiments [17].) An unequivocal answer to this question may probably come from the pressure effect experiment suggested in section 5.3. With the above comments in mind, we shall discuss the diffusion of Group III and Group V elements in Si and Ge on the assumption of a vacancy mechanism, making reference to the interstitialcy mechanism where needed.

From Tables 5.8 and 5.9, we see that both the preexponential factor  D_0 and the activation energy Q of impurity diffusion are lower than those for self-diffusion in silicon and germanium respectively. The same fact has been reported for diffusion in dilute solid metallic solutions by Stiegman, Shockley and Nix (SSN), who first proposed the vacancy mechanism of diffusion [241]. SSN observed that Q for impurity diffusion is sometimes as low as half the Q for self-diffusion. To explain this phenomenon, Johnson [242] suggested a model in which solute atoms and vacancies form pairs of a solute-vacancy complex. Pairing occurs when the binding between a solute atom and its neighboring solvent atoms is less tight than the binding between neighboring solvent atoms. He visualized a solute-vacancy complex to migrate in a lattice without dissociation by a series of cyclic process of reorientation and inversion. In a diamond type lattice, partial dissociation of the pair is necessary to implement a reorientation process [13, 243, 244]. He showed that if the vacancy and the impurity atom do not diffuse as a pair, the activation energy of the impurity diffusion should be exactly equal to that of solvent self-diffusion regardless of the impurity vacancy interaction, because then the vacancy has to disappear into the bulk of the lattice, and this becomes the rate controlling mechanism which is identical to that of self-diffusion. At a time before the concept of the correlation effect was introduced [245], he was able to argue qualitatively that the preexponential factor D_o for impurity diffusion should also be lower than that for self-diffusion [242], because of the greater number of vacancy jumps required to complete an effective impurity displacement cycle. It would however, be incorrect to assume that the correlation factor, such as that given by Hu [244], would completely go into the preexponential factor, because according to Hu's model, the correlation factor would be temperature dependent. Later Hu [246] found his earlier model of the interaction potential [244] to be an oversimplification, and can yield incorrect D_o and Q values for impurity diffusion in relation to self-diffusion. Let us consider the relationship between the impurity diffusivity and the self-diffusivity. A general expression for the impurity diffusivity via a vacancy mechanism has been shown to be [13]

$$D(A) = \frac{1}{6} r^2 \omega_A f_A [AV] / Z[A], \tag{5.48}$$

where r is the nearest neighbor distance and ω_A is the vacancy-impurity exchange frequency, f_A is the correlation factor for an effective displacement of the impurity atom; $[AV]$ is the concentration of the vacancy-impurity complex; and Z is the number of the nearest neighbors. In the limit of infinite dilution, equation 5.48 reduces to

$$D(A) = \frac{1}{6} r^2 \omega_A f_A ([V]/[S]) \exp(E_b/kT), \tag{5.49}$$

where E_b is the vacancy-impurity binding energy. Also at infinite dilution, the self-diffusivity is given by

$$D_S = \frac{1}{6} r^2 \omega_V f_V [V]/[S], \tag{5.50}$$

where ω_V is the vacancy jump frequency in the bulk, and f_V is the correlation factor, which is 1/2 for tracer self-diffusion in the diamond lattice [247]. Therefore,

$$D(A) = D_S(\omega_A f_A / \omega_V f_V) \exp(E_b/kT) \tag{5.51}$$

Swalin had earlier suggested an expression [248] which is somewhat similar to equation 5.51, but with the factor $(\omega_A f_A/\omega_V f_V)$ missing. Neglecting f_A is not well justified, because f_A is very much smaller than unity, and is a complicated function of the two parameters, (ω_A/ω_V) and (E_b/kT). According to Hu's model of the interacting potential f_A turns out to be approximately proportional to ω_V/ω_A (see figure 5.3 in reference 244). For a qualitative discussion, we may write $f_A \sim f'_A(E_b/kT)\omega_V/\omega_A$, where f'_A is some function of (E_b/kT) only. Taking f_V as unity one has

$$D(A) \simeq D_S f'_A(E_b/kT). \tag{5.52}$$

Now if one assumes a vacancy-impurity interaction potential that terminates at the second nearest site (which Hu did for the sole reason of simplicity [244]), one obtains a very approximate relation $f'_A \sim \exp(-E_b/kT)$, which will practically cancel out the association factor $\exp(E_b/kT)$. Then one would be led to an obviously incorrect result of $D(A) \simeq D_S$. More specifically, $D_o(A) = D_o(S)$, and $Q(A) = Q(S)$ where S denotes the solvent species. This unrealistic result does not come from a faulty concept of correlation effect, which has been shown to be physically sound. Rather one must conclude that a simplified model interaction potential which terminates at the second nearest site is unrealistic. Two modifications of the model potential are hence necessary. First, the interaction potential should have a rather long range, perhaps reaching several coordination sites rather than terminating at the second coordination site (see figure 5.7). It perhaps grossly resembles, in profile rather than in detail, the interaction potential for the vacancy-phosphorus pair pictured by Watkins and Corbett [42]. The reason this picture is important is that in the diamond lattice, unlike the *bcc* or *fcc* lattices, the 'Johnson cyclic process' cannot be realized. To achieve an effective impurity displacement cycle, the vacancy must first partially dissociate from the impurity atom, and go at least as far as the third coordination site to close a path around the impurity atom [244]. The probability of the vacancy making such a path is of the order of $\exp(-\Delta E/kT)$, where ΔE is the difference between the potential energy of the vacancy at the third coordination site and at the first coordination site. (For such a qualitative argument, we have omitted the weighting factor of the number of sites in respective stages of transition. A quantitative analysis, however, can be readily obtained using a suitable transition matrix in place of the one used in [244].) Since the correlation factor is primarily determined by the probability of the vacancy taking paths to the other side of the impurity atom, one sees that to a 0th order, $f'_A \sim \exp(-\Delta E/kT)$. In the case of a long range interaction potential between the vacancy and the

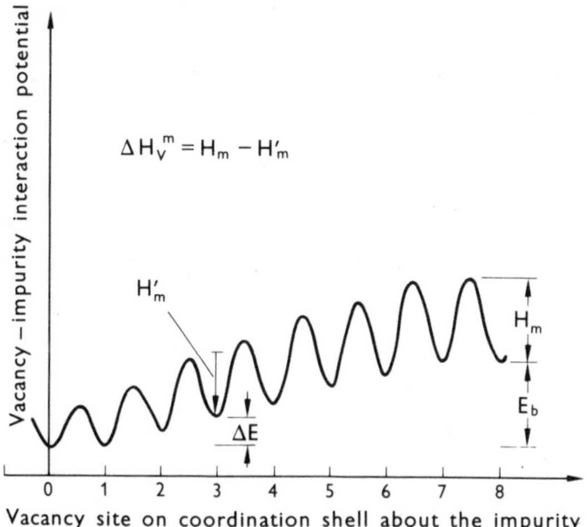

Figure 5.7. Schematic diagram showing a long-range vacancy-impurity interaction potential that is required to account for the lower activation energy for impurity diffusion than for self-diffusion in diamond-type lattices (see text). This replaces the model potential given earlier in [244].

impurity atom, it is clear that ΔE will be only a small fraction of E_b. Now, instead of 5.52, we may write as a 0th order approximation

$$D(A) \sim D_S\theta \exp[(E_b - \Delta E)/kT], \tag{5.53}$$

where θ is some constant which is appreciably smaller than unity, and is dependent upon the model potential. The second modification is to change the model potential from one which is a function only of the coordination site of the vacancy from the impurity atom to one which is a function of the position vector of the vacancy. This modification is necessary because of the presence of a Coulombic interaction between the vacancy and the impurity atom. We will not discuss the effect of this modification here. Equation 5.53 enables us to explain the observed decrease in both D_o and Q for impurity diffusion relative to self-diffusion. According to Hu's earlier model potential, we would then have $\Delta E = E_b$, so that $D(A) \sim D_S\theta$. In the long range potential model the activation energy of impurity diffusion $Q(A)$ is hence related to the activation energy of self-diffusion $Q(S)$ as $Q(A) = Q(S) - E_b + \Delta E$.

In the above discussion which led to equation 5.53, we did not distinguish between the contributions to diffusion by different species of vacancies. We turn now to this problem. There are two main factors that determine the relative contributions of different species of vacancies. The first is a dynamical factor that concerns the vacancy jump frequencies, ω_A, ω_V, etc. For example, Watkins has reported a lower activation energy of migration for a doubly negative vacancy than for a neutral vacancy in silicon (see Table 5.2). The second is a statistical factor that concerns the relative concentrations of the various vacancy species as well as the pairs of these with impurity atoms. The binding energy of the A-V pair may also vary with vacancy species. In fact, we must also consider the various charge species of the impurity atom. In the diffusion of Groups III and V elements in Ge and Si, we are often concerned with surface concentrations in the range of $\sim 10^{18}$-10^{21} cm^{-3}. At such high concentrations of electrically active diffusants these semiconductors are extrinsic even at diffusion temperatures. This effect renders the system nonideal, and thus distinguishes this category of diffusants from those previously considered. The effect of this nonideal behavior on the chemical potential of electrons (i.e., the Fermi level), and of the impurity has been treated in a classical thermodynamical framework in combination with a one-electron energy band model [13, 249-255].

It has been recognized that the presence of an impurity concentration profile in the semiconductor during the diffusion process will establish an internal electric field. Hence, various works have been devoted to the analysis of the inclusion of a field drift term for the diffusion of the ionized impurities [256-264]. Experimental data have also been analyzed in this way [265, 266]. Though simple and intuitively appealing, such an approach may be viewed as of a patchwork nature, lacking completeness in scope and susceptible to inconsistency. For example, if indeed an internal field exists in a certain case of diffusion it must mean that there exists a significant spatial variation in $(E_c - E_F)$, the difference between the conduction band edge and the Fermi level. Consequently, the equilibrium concentrations of defects such as vacancies and interstitials which possess electronic levels within or in the neighborhood of the semiconductor band gap, will be dependent on the impurity concentration [13, 250, 267, 268]. Since the diffusion of Groups III and V elements (or any other substitutional impurities), according to our previous discussion, is probably effected through the intermediary of vacancies or lattice self-interstitials, the addition of an ion drift term in the internal field, alone, is an incomplete analysis. One must also consider the variation of diffusivity with impurity concentration [13, 269]. The simple field drift expression is also subject to a self-consistency problem regarding charge shielding, charged carrier population, and the effect of the formation of a band tail. Therefore, the most natural and foolproof way to treat the problem is to analyze the entire system thermodynamically [13]. In this approach, we analyze the chemical potentials of various components. Since the chemical potential is defined as the partial derivative of the system free energy with respect to the concentration of the component concerned, it is inclusive of entropy of configuration, of vibration entropy, of chemical binding, electrostatic potential, etc. In principle, any desired degree of self-consistency can be achieved by suitable minimization of the system free energy. In practice, this involves great difficulty in a complex system, and certain simplifying assumptions are often made. For example, the local charge neutrality condition is often imposed as an additional constraint. Since the vacancy (or interstitial) concentration is negligible insofar as its influence, at diffusion temperature, on the Fermi level is concerned, the Fermi level of the system can therefore be calculated first, using a straightforward conventional method [270]. However, when the impurity concentration is high, the problem becomes more complicated. It is not enough just to replace the Boltzmann statistics by Fermi-Dirac statistics; one must also consider the phenomenon of energy band tailing. Various theories of the formation of band tail have been developed [271-275], and have been employed in the calculation of the Fermi level in GaAs [276, 277], and Si [13]. If an unperturbed band structure is assumed, then approximate expressions for the activity coefficients of the impurity and of the vacancy are given respectively by

$$\gamma_A \simeq (1+\zeta_i)/(1+\zeta), \tag{5.54}$$

$$\gamma_V \simeq (1+\xi_i)/(1+\xi), \tag{5.55}$$

where

$$\zeta = g_A^{-1} \exp\,[(E_A - E_F)/kT]; \quad \xi = g_V \exp\,[(E_F - E_1)/kT]. \tag{5.56}$$

In 5.56, g_A and g_V are the degeneracy factors for the impurity and the vacancy respectively, and E_1 is the first vacancy level (see figure 5.2). The subscript i denotes the intrinsic condition. One can easily include higher vacancy acceptor levels following the method of Shockley and Last [267]. Examples of activity coefficients for a donor and the vacancy in silicon are given in figures 5.8 and 5.9. Equation 5.55 becomes a poor approximation for the vacancy activity coefficient at high impurity concentration in the presence of a strong vacancy-impurity association. It should then be replaced by [13]

$$\gamma_V = \frac{1+\xi_i}{1+\xi}\left(1 + \frac{[A]}{[S]}\left\{\frac{\alpha}{1+\xi}\left(1 + \frac{1+\beta\zeta}{1+\zeta}\,\xi\right) - (Z+1)\right\}\right), \tag{5.57}$$

where Z is the number of the first coordination sites, and

$$\alpha = Z \exp\,(E_\alpha/kT), \quad \beta = \exp\,(E_\beta/kT). \tag{5.58}$$

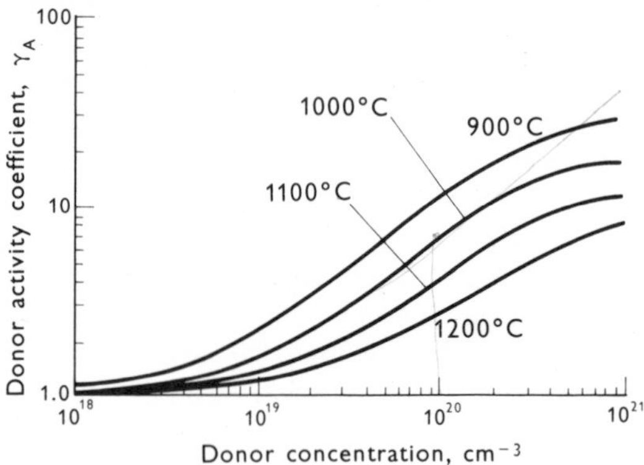

Figure 5.8. Donor activity coefficient in Si calculated using Fermi statistics, on the assumption of constant donor level and constant energy-band structure. After Hu [13].

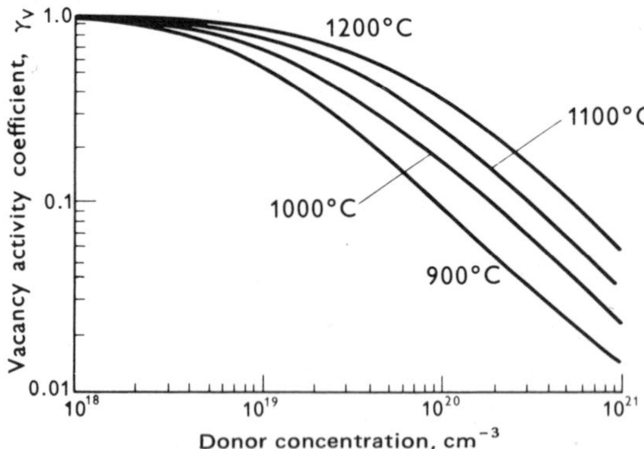

Figure 5.9. Vacancy activity coefficient in Si calculated using Fermi statistics, on the assumption of constant donor and vacancy levels and constant energy-band structure. After Hu [13].

E_β is the Coulombic component of the vacancy-impurity binding energy, and E_α is the non-Coulombic component; $E_\alpha + E_\beta = E_b$. Equation 5.57 is derived from the assumption that E_b terminates at the second coordination site. This, as we have mentioned, is not correct (see figure 5.7). But this simplifying assumption is necessary to render this part of the problem tractable. From equation 5.57 one can easily calculate the equilibrium concentration of vacancies from the relation $[V]_e\gamma_V = ([V]_e\gamma_V)_i = [V]_{e,i}$, and obtain $[V']$ and $[V^\times]$ from Fermi statistics. But from the standpoint of the diffusivity, the required information is the concentrations of the vacancy-impurity complexes of the four species $A^\times V^\times$, $A^\cdot V'$, $A^\times V'$, and $A^\cdot V^\times$ (assuming the diffusant is a donor). Of these, one can probably neglect $A^\times\ V^\times$ in favor of $A^\cdot V'$. The concentrations of these are given by [13]

$$[A^\cdot V^\times] = [V]\left\{\frac{[A]}{[S]}\ \frac{\alpha\zeta}{1+\zeta}\left(\frac{1+\beta\xi}{1+\xi}\right)\right\}\delta, \tag{5.59}$$

$$[A^\times V'] = [V]\left\{\frac{[A]}{[S]}\ \frac{\alpha\xi}{1+\xi}\left(\frac{1+\beta\zeta}{1+\zeta}\right)\right\}\delta, \tag{5.60}$$

$$[AV] = [V]\left\{\frac{[A]}{[S]}\ \frac{\alpha}{1+\xi}\left(1+\ \frac{1+\beta\zeta}{1+\zeta}\ \xi\right)\right\}\delta, \tag{5.61}$$

$$\delta = \left\{1+\ \frac{[A]}{[S]}\left[\frac{\alpha}{1+\xi}\left(1+\ \frac{1+\beta\zeta}{1+\zeta}\ \xi\right)-(Z+1)\right]\right\}. \tag{5.62}$$

where

$$[AV] = [A^\times V^\times]+[A^\cdot V']+[A^\cdot V^\times]+[A^\times V'].$$

The remaining task is to find suitable model potentials for these A-V pairs, so that their respective ω_A, ω_V and f_A can be determined. Summing over these species their respective contributions to $D(A)$ in the form of equation 5.53, one can then relate the impurity diffusivity to the semiconductor self-diffusivity. This is

$$D(A) = D_S\{\theta(A^\cdot V')[A^\cdot V'] + \theta(A^\cdot V^\times)[A^\cdot V^\times] + \theta(A^\times V')[A^\times V']\} \Big/ \{1+\xi\omega_{V'}/\omega_{V^\times}\} \tag{5.63}$$

This differs from Swalin's expression [248] which neglects the significant contribution of the impurity diffusion correlation factor, as well as the difference in $\omega_{V'}$ and $\omega_{V^\times}$. The exact application of equation 5.63 is complicated, and is probably not essential. It is useful to consider the limiting case (which is also quite typical) where $[A^\cdot V^\times]$ is completely negligible due to a sufficient dopant concentration, and where $[A^\cdot V']$ overwhelms $[A^\times V']$ due to the Coulomb binding energy. In that case, the extrinsic impurity diffusivity $D(A)$ can be related to the intrinsic impurity diffusivity $D_i(A)$ as follows [13]:

$$D(A) = D^i(A)[V]/[V]_{e,\,i} \tag{5.64}$$

We define the intrinsic impurity diffusivity as the impurity diffusivity at infinite dilution of impurity concentrations, and when the vacancy (or self-interstitial) concentration is at thermal equilibrium. In cases where the cross terms in the phenomenological equations of diffusion are negligible, one can also define an apparent diffusivity as

$$\mathbf{D}(A) = D^i(A)(1+\partial\ln\gamma_A/\partial\ln[A])[V]/[V]_{e,\,i} \tag{5.65}$$

It has been shown that [13] in a tight-binding model, or one such as depicted in figure 5.7, there will be a negligible chemical pump effect [9, 278]. Using this model, it can be shown from an inequality estimation that [279] the maximum vacancy undersaturation is of the order of

$$|([V]-[V]_e)/[V]_e|_{\max} \gtrsim 8D(A(0))/D(V), \tag{5.66}$$

where $D(A(0))$ is the impurity diffusivity at the surface, presumably a maximum value. This is entirely negligible. One may then replace $[V]$ by $[V]_e$ in equation 5.65, or $[V]/[V]_{e,i}$ by γ_V^{-1},

$$\mathbf{D}(A) = D^i(A)\gamma_V^{-1}(1+\partial \ln \gamma_A/\partial \ln [A]). \tag{5.67}$$

The second term in the parentheses in equation 5.67 is often replaced by a field drift term. The field drift concept is unnecessarily restrictive. It does not allow inclusively the consideration of such effects as strain interaction, clustering and ion pairing, etc., which would otherwise be all included in γ_A. Equation 5.67 can be further simplified in the limiting case where the impurity concentration is sufficiently dilute so that clustering and complex formation are negligible, and the system becomes non-degenerate. Then, since the vacancy is a shallow acceptor, one has

$$\gamma_V^{-1} = \frac{1+\xi}{1+\xi_i} \cong \xi/\xi_i = \exp[(E_F - E_F^i)/kT] = n/n_i$$

$$= [A]/2n_i + (1+[A]^2/4n_i^2)^{1/2}; \tag{5.68}$$

$$\partial \ln \gamma_A/\partial \ln [A] \cong [A]/([A]^2+4n_i^2)^{1/2}. \tag{5.69}$$

An apparent diffusivity such as given in equation 5.67, together with the expressions of equations 5.68 and 5.69, has been observed to describe quite well the diffusion of arsenic in silicon up to a concentration of $\sim 3 \times 10^{20}$ cm^{-3}. Profiles generated by diffusion from a constant surface concentration into a semiinfinite body using $\mathbf{D}(A)$ in equation 5.67 are shown in figure 5.10, in which $\rho = [A(0)]/2n_i$ is a parameter expressing the degree of extrinsicity, $[A(0)]$ being the surface concentration of A. The agreement of arsenic diffusion profiles in silicon with that depicted in figure 5.10 lends strong support to a vacancy mechanism.

An alternative experimental method for the study of concentration-dependent diffusivity is the so-called 'isoconcentration diffusion'. In this method the semiconductor is homogeneously doped with the impurity of interest at various concentration levels; a radioactive isotope of this impurity is then diffused into the semiconductor under chemical equilibrium conditions. The process is that of a mixing of atoms of the labeled and the unlabeled species. The absence of a chemical concentration gradient offers a number of desirable features: there is no internal field;

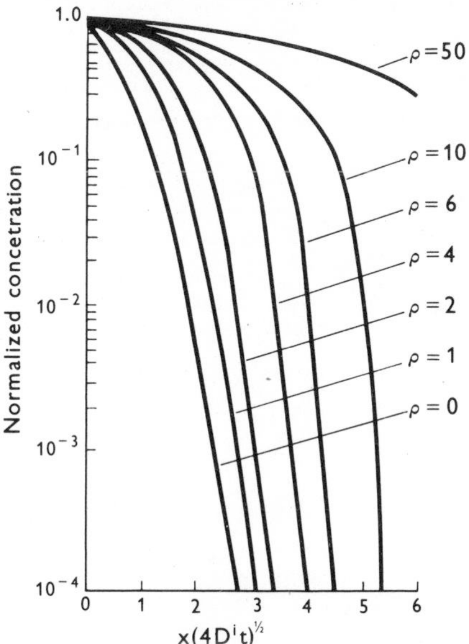

Figure 5.10. Impurity concentration profiles produced by diffusion from a constant surface concentration $[A(0)]$, and an apparent diffusivity given by equation 5.67. The parameter $\rho = [A(0)]/2n_i$, is a measure of extrinsicity. After Hu and Schmidt [269].

the local charge neutrality assumption is rigorously valid; and there will be no departure from the equilibrium condition for the vacancy concentration and the complex concentrations. This permits us to examine the diffusion process in its elementary form. Masters and Fairfield [201] have studied arsenic diffusion in silicon using the isoconcentration method. The arsenic profiles in their work are illustrated in figure 5.10a. Note that the arsenic profile from isoconcentration diffusion is well described by an erfc distribution. Masters and Fairfield observed the relationship $D_e(\widetilde{\mathrm{As}}) \simeq D^i(\widetilde{\mathrm{As}})n/n_i$.

Intrinsic carrier concentrations, required for the calculation of activity coefficients, have been given by Morin and Maita. For germanium [280],

$$n_i^2 = 3.10 \times 10^{32}\ \mathrm{T}^3 \exp(-\epsilon_g/kT)\ \mathrm{cm}^{-3} \tag{5.70}$$

where the energy gap in eV is

$$\epsilon_g = 0.785 - 4.61 \times 10^{-10}\ n_i^{1/2}\ \mathrm{T}^{-1/2}. \tag{5.71}$$

For silicon [281],

$$n_i^2 = 1.5 \times 10^{33}\ \mathrm{T}^3 \exp(-\epsilon_g/kT)\ \mathrm{cm}^{-3}, \tag{5.72}$$

where

$$\epsilon_g = 1.21 - 7.1 \times 10^{-10}\ n_i^{1/2}\ \mathrm{T}^{-1/2}\ \mathrm{eV}. \tag{5.73}$$

The results are also given in figures 5.11a and b.

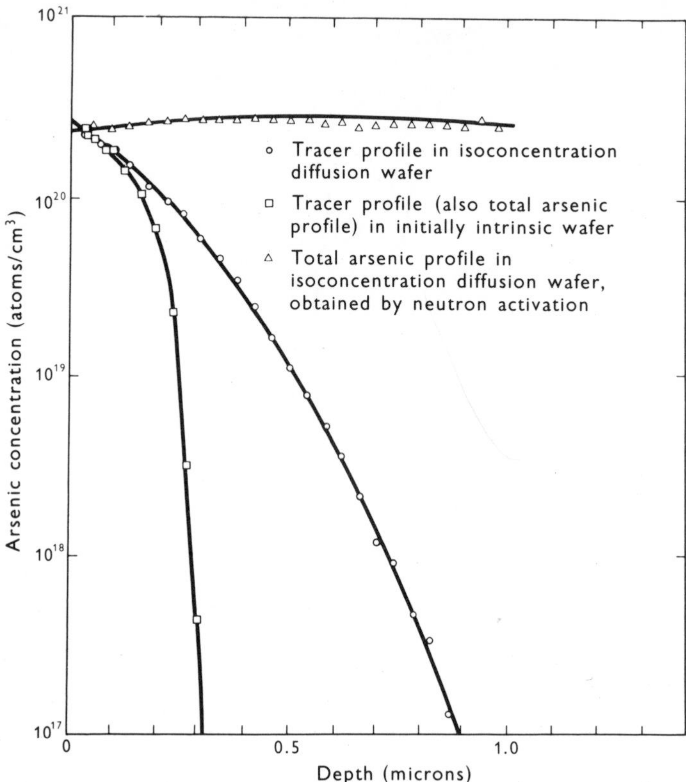

Figure 5.10a. Arsenic concentration profiles in silicon. After Masters and Fairfield [201].

It should be noted that the description of arsenic diffusion in silicon by this model is good only up to an arsenic concentration of $\sim 3 \times 10^{20}$ cm^{-3}. A trend of an anomalous decrease of apparent diffusivity sets in above this concentration. This subject will be discussed in more detail in section 5.5.2. Historically, a detailed analysis of an impurity profile in silicon was first carried out by Tannenbaum [182]. She analyzed the phosphorus concentration profiles by the Boltzmann-Matano method, making the assumption that the departure from a complementary error function distribution is the result of a concentration dependence of the diffusivity. She found an anomalous increase of diffusivity as the concentration went above $\sim 10^{20}$ cm^{-3} in a superlinear manner. Since then, a considerable number of investigators have published more detailed analyses of phosphorus diffusion profiles in silicon, revealing a complex nature. The diffusion of phosphorus deserves a separate discussion.

The diffusion of Group III acceptors in silicon is interesting. Mikhailova [174] first reported boron concentration profiles in silicon showing large

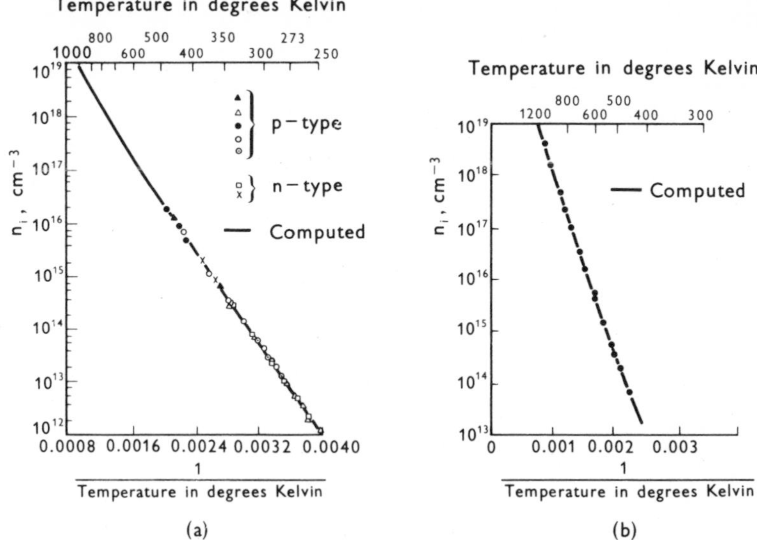

Figure 5.11a. Intrinsic carrier concentration in germanium. After Morin and Maita [280].

Figure 5.11b. Intrinsic carrier concentration in silicon. After Morin and Maita [281].

departure from erfc profiles. She did not offer an explanation; but the boron profiles she reported exhibit similarity with those in figure 5.10. However, she did point out that the internal electric field would be too small to account for the large departure from erfc profiles. Maekawa and Oshida [171] later reported results of boron diffusion profiles in silicon both from conductivity measurements and from proton activation analysis. Their profiles generally confirmed the results of Mikhailova [174]. Unfortunately, their results appeared to be of a qualitative nature, since no data points were given accompanying the profile curves. The inaccuracy of their profiles apparently made it impossible for them to carry out a Boltzmann-Matano analysis [126]. Instead they obtained the diffusivity from the *p-n* junction method. They suggested the cause of non-erfc profiles to be an increased diffusivity with concentration due to solute-induced lattice strain. It is difficult to estimate the effect of lattice strain on diffusivity for lack of a model of such an effect. Nevertheless, it is believed that the effect should be small because of the short range of elastic interactions. From the consideration of a discrete model, Hall [282] has shown the elastic displacement due to a point defect to decay more rapidly (and in an oscillatory manner) than the inverse square law.

More recently, details of boron profiles have permitted Boltzmann-Matano analyses. It appears that the boron diffusivity in silicon is almost linear with concentration, or slightly superlinear, for concentrations of $\sim 10^{20}$ cm^{-3} and above [234]. A sublinear relationship has also been reported [210]. The linear relationship between the apparent diffusivity and the boron concentration can be explained by three different mechanisms: the interstitialcy mechanism in which the concentration of the silicon self-interstitial donor increases with the boron concentration; the vacancy mechanism with the assumption of the existence of certain vacancy donor levels; and the two-stream mechanism in which the boron interstitial behaves as a donor and its relative concentration increases with the total boron concentration. We noted earlier that an interstitialcy mechanism for boron in silicon appears incompatible with the current model of the emitter-push effect [128].

Millea [10] observed an approximate proportionality between the diffusivity of indium in silicon and the concentration of holes. He suggested an 'interchange mechanism' which we have criticized earlier in section 5.2.1. Recent experimental investigations of gallium diffusion in silicon also showed a similar effect [283]. Thus, one is led to the conclusion that this effect is of a general nature common to Group III elements in silicon. One immediate implication is that in a sequential diffusion process, e.g., the sequential diffusion of boron and arsenic, the pile-up and the retardation effect will be substantially less than that calculated previously by Hu and Schmidt [269]. This is because that when the driving force due to the internal field is larger than that due to the concentration gradient, the rate of pile-up, or 'up-hill' migration, will decrease with the decrease of diffusivity. The analysis in [269] was based on the assumption that the diffusivity of boron in intrinsic or in n-type silicon is nearly constant, an assumption that is justified if there exists no vacancy donor, or one which is very deep. Fairfield and Masters [112] found that the silicon self-diffusivity was increased only to 1.75 times its intrinsic value at a boron doping level of 2.2×10^{20} cm^{-3} at 1090°C. This concentration corresponds to approximately $p/n_i \simeq 10$. Therefore, if a vacancy donor level exists, one of these two possibilities would be inferred: the vacancy donor level must be very deep; or that negatively charged vacancies are much more mobile at diffusion temperatures. Presumably only the positively charged vacancies and perhaps neutral vacancies contribute to boron (or other Group III elements) diffusion because of Coulombic interaction, which is absent in silicon self-diffusion.

Little information is available concerning the concentration dependent diffusivities of Group III and Group V elements in germanium. This is perhaps not surprising since the technological interest had already shifted from germanium to silicon when Tannenbaum made her first detailed analysis of the phosphorus diffusion profile. Recent experimental study of arsenic diffusion in germanium [284] showed a linear dependency of diffusivity on arsenic concentration from $\sim 2 \times 10^{18}$ cm^{-3} to

$\sim 2 \times 10^{19}$ cm^{-3} at 750°C. Arsenic profiles in Ge are qualitatively similar to those in figure 5.10 [285]. Information on the concentration-diffusivity relation of Group III elements in germanium would be very interesting since, presumably, an interstitialcy mechanism is less favored in germanium than in silicon. Unfortunately, such information is still unavailable.

Finally, we note two outstanding questions. The first concerns the phenomenon that Group III elements diffuse generally faster than Group V elements in silicon, by about a factor of $\sim$10-10^2, while the opposite is true in germanium, under the condition of light doping (near the intrinsic condition). Swalin [248] suggested that this could be explained by the difference between the relative contributions from Coulombic interaction and from the elastic interaction (solute atomic size effect) in these two semiconductors. He reasoned that in Ge, since the Fermi level is usually higher than the vacancy acceptor level, a major fraction of the vacancies are charged. The Coulombic interaction between the charged vacancies and the charged solute ions outweighs the contribution due to the size difference, and hence the attractive Coulomb interaction between the vacancies and the Group V ions leads to a faster diffusion rate. In silicon, on the other hand, the Fermi level is lower than the vacancy level so that a large fraction of the vacancies are uncharged, and the Coulombic interaction between the vacancies and the solute ions makes a smaller contribution to the diffusivity than the size effect. In general, Group III solutes introduce more lattice strain than Group V elements. It should be noted, however, that Swalin had incorrectly assigned a value of $(E_c - 0.16)$ eV to the silicon vacancy acceptor level. Sometime after Swalin's work, the level at $(E_c - 0.16)$ eV has been accurately assigned to the silicon A-center, i.e., the vacancy-oxygen complex [49] (qv. Table 5.2). The acceptor level for the silicon monovacancy has never been accurately determined. But the acceptor levels for *V-A* complexes, where *A* is a Group V atom, have been quite well defined, and all are located at about $(E_c - 0.4)$ eV. Since we are primarily interested in the concentration of *V-A* complexes, this level should be the correct one to use. Using this value, one is led to the conclusion that the Coulombic interaction would make an overwhelming contribution to diffusion in silicon as well. This fact suggests a necessity to seek a different diffusion model for Group III elements in silicon. The recent experimental information on the concentration dependent diffusivity of Group III elements in silicon and the three possible mechanisms mentioned earlier in this section are compatible with the fact just mentioned.

The second question is that in silicon, the diffusivity of Group V elements decreases with the increase of atom size. In germanium, the trend is reversed (see figure 5.12). No explanation has been offered in the literature for this phenomenon. If we accept Seeger and Chik's model of impurity diffusion in silicon, i.e., the interstitialcy mechanism (and at the same time assume a vacancy mechanism for diffusion in germanium) this paradox can be resolved readily merely by postulating that the elastic

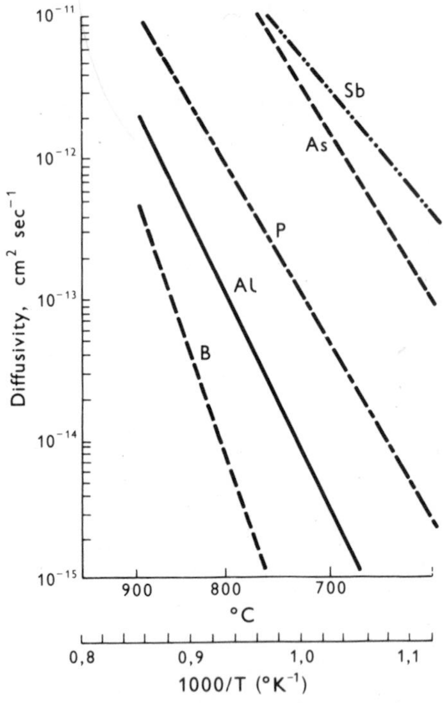

Figure 5.12. Diffusivities of some solutes in germanium. After Meer and Pommerrenig [188].

interaction is such that a silicon self-interstitial tends to be more strongly attracted to a solute atom of smaller size, and a vacancy to a solute atom of larger size. An interstitialcy mechanism for the diffusion of Group V elements in silicon requires that the silicon self-interstitial is amphoteric, i.e., it also behaves as an acceptor in *n*-doped silicon, both in order to have some degree of association between the self-interstitial and the solute atom, and to account for the increase of the diffusivity of Group V elements with concentration.

5.4.5 Diffusion of Oxygen and Carbon

Oxygen is often present in Si and Ge crystals grown by pulling from quartz crucibles. Silicon may contain as much as 10^{18} atoms cm^{-3} of oxygen.

The solid solubility of oxygen in Si at the melting point is 1.7×10^{18} cm^{-3} [286, 287], with a heat of solution of 2.3 ± 0.3 eV. Oxygen can also be introduced into silicon and germanium during solid state diffusion in an oxygen environment. Although at typical diffusion temperatures the oxygen solubility is lower, an oxygen concentration of the order of $\sim 10^{17}$ cm^{-3} in silicon can be reached. Oxygen can precipitate in the form of SiO_2 particles during subsequent heat treatment at lower temperatures, such as epitaxy. In silicon and in germanium, the dissolved oxygen is believed to exist in a so-called 'bonded interstitial' state, in which the oxygen resides at an interstitial site without replacing any lattice atom; but is simultaneously bonded to two neighboring lattice atoms replacing the Si–Si bond by Si–O–Si bond [204, 205, 288]. At lower temperatures, the dissolved oxygen can be converted into oxygen complexes which act as donors. This is known as thermal conversion, which can convert a lightly *p*-doped material into *n*-type material. The formation of the donor complex, which is thought [289] to be SiO_4, is a clustering process with the diffusion of bonded interstitial oxygen as the rate limiting mechanism at low temperatures. It probably has no sharp transition temperature; but there is some optimum temperature range for complex formation above which the complexes are thermodynamically unstable, and below which the rate of oxygen diffusion is too small for complex formation. A temperature range of 450-500°C for maximum conversion has been reported. The kinetics of the formation of these donors has been studied by Kaiser *et al.* [289]. Such kinetics in principle can furnish information on the oxygen diffusivity (or vice versa). Accurate results on the oxygen diffusivity in Si and Ge were obtained by Haas [290] from analyses of the internal friction data of Southgate [291, 292], and by Corbett *et al.* [293] from data on the annealing of the dichroism of the 9 μm band in silicon and the 11.7 μm band in germanium. The results by Haas are: 0 in Si, $D_o = 0.21$ cm^2 s^{-1}, $Q = 2.55$ eV; in Ge, $D_o = 0.17$ cm^2 s^{-1}, $Q = 2.02$ eV. The results by Corbett *et al.*, are, in Si, $D_o = 0.23$ cm^2 s^{-1}, $Q = 2.561 \pm 0.005$ eV; in Ge, $D_o = 0.40$ cm^2 s^{-1}, $Q = 2.076 \pm 0.010$ eV. Oxygen free ($\leqslant 10^{14}$ cm^{-3}) crystals can be grown by the float zone technique.

Carbon, like oxygen, also gives rise to a localized mode of vibration (16.5 μm band in Si [294]). But unlike oxygen, it is normally present in both Czochralski grown as well as in float zone silicon. Also, unlike oxygen, carbon has been shown to occupy substitutional sites in silicon [295, 296]. Reports on the carbon content of grown silicon are not in general agreement. While Dash [297] reported a solid solubility of $\sim 10^{18}$ cm^{-3}, Newman and Wakefield [295] reported crystals containing carbon up to $\sim 10^{19}$ cm^{-3}. On the other hand, Baker *et al.* [296] reported a carbon content of 10^{16}-10^{17} cm^{-3}, and in Czochralski crystals, it is possible to have a carbon content of $<10^{16}$ cm^{-3}. Precipitation of carbon in the form of β-SiC (12.6 μm absorption band) has been observed, and, like oxygen complex formation, presumably is also rate-limited by carbon diffusion. Like SiO_2 precipitates, SiC precipitates can act as nucleation sites for stacking faults in epitaxial growth. The diffusivity of carbon in silicon is reported [295] to be $D_o = 0.33$ cm^2 s^{-1}, $Q = 2.92 \pm 0.25$ eV.

5.4.6 Diffusion in an External and Internal Electric Field

The drift of ionized impurities in an externally applied electric field was first utilized by Fuller and Severiens [145] in the study of lithium diffusivity in silicon and germanium. The Einstein relationship was found to hold with the lithium ion possessing a full positive charge. Gallagher [298] measured the drift of Cu, Fe, and Au in Si, and found that more than 60% of the rapidly diffusing copper is positively charged, while Fe and Au do not show the field effect.

Konstantinov and Badenko [299] reported an interesting effect in the drift of indium and antimony in germanium. Below 600° C, indium moved towards the anode and antimony moved towards the cathode as expected of their respective charge states. *At higher temperatures, however, the direction of transport was reversed*, and above 800° C, indium moved

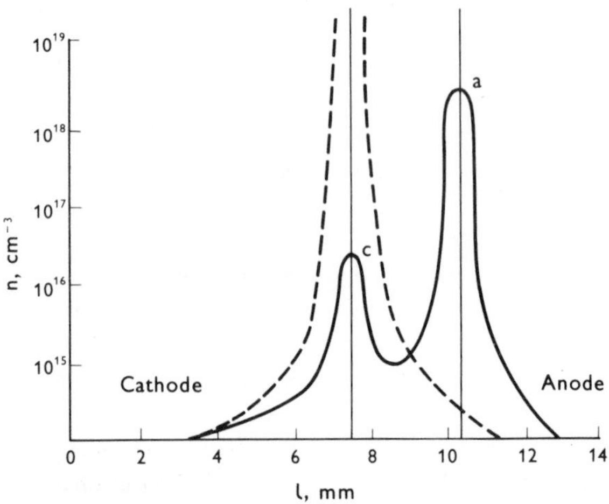

Figure 5.13a. Drift in an external field. The distribution of antimony in a germanium sample. The broken curve represents the initial distribution; the continuous curve gives the distribution after passage of a current at 800° C. Electric field = 0.67 V/cm. After Konstantinov and Badenko [299].

towards the cathode and antimony moved towards the anode, and at the same time the respective mobilities were increased by approximately an order of magnitude. Their results are shown in figures 5.13a and b. In any case, the mobilities were observed to be several orders of magnitude greater than the values from the Einstein relationship [300]. Their experiments were carried out at a current density of 150-350 A/cm^2, giving a field of 0.4-1 V/cm. It appears that *the entrainment of the ions by*

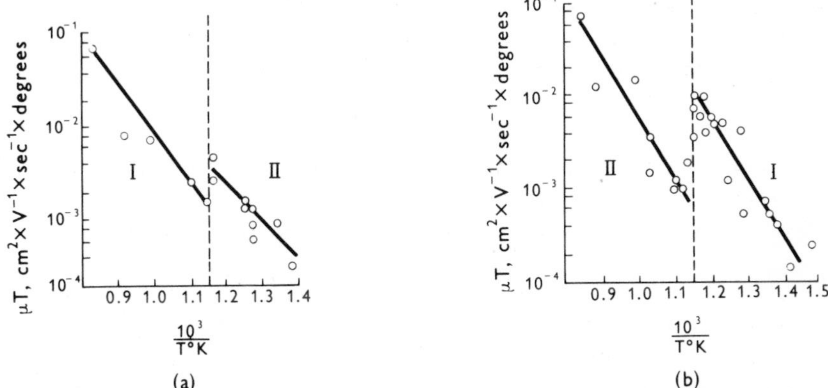

Figure 5.13b. Mobility of ions in an external field in Ge. Dependence of $\log \mu T = f(1/T)$ for tests with antimony (a) and indium (b) impurities in germanium. (I) Transport toward the anode; (II) toward the cathode. After Badenko [300].

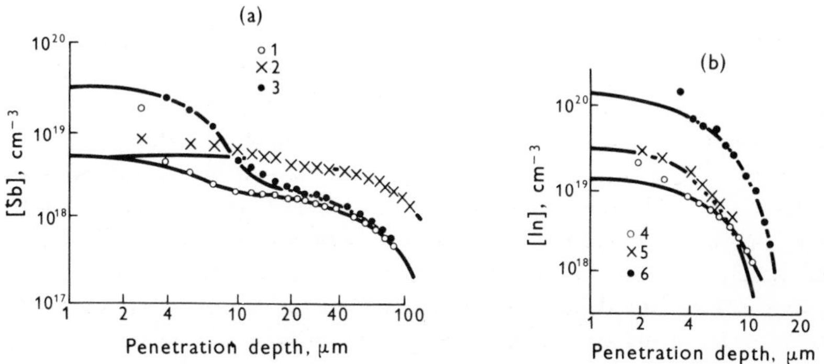

Figure 5.13c. Distribution of antimony (a) and indium (b) in germanium double diffusion (1, 3, 4, 6) and in the control samples (2, 5). The solid curves are based on theory. (a) N_0, cm^{-3}: (1) $5.1 \cdot 10^{18}$; (2) $4.9 \cdot 10^{18}$; (3) $3.8 \cdot 10^{19}$; P_0, cm^{-3}: (1) $1.6 \cdot 10^{19}$; (2) 0; (3) $1.8 \cdot 10^{20}$. (b) P_0, cm^{-3}: (4) $1.6 \cdot 10^{19}$; (5) $4 \cdot 10^{19}$; (6) $1.8 \cdot 10^{20}$; N_0, cm^{-3}: (4) $5.1 \cdot 10^{18}$; (5) 0; (6) $3.8 \cdot 10^{19}$; N_0 = [Sb(0)]; P_0 = [In(0)]. After Vas'kin *et al.* [264].

the electrons could have caused the phenomenon of the reversal of ion transport direction. Fiks [301] has considered the ion drift theoretically taking into account the entrainment by electron flux. He showed that, using a free electron approximation, the force F acting on the ion by the entraining electrons is given by

$$F = zel\sigma nE, \tag{5.74}$$

where E is the intensity of the electric field; ze is the absolute charge of the ion; n is the free electron concentration; l is the mean free path of electrons in the semiconductor (assumed to be energy independent), and σ is the average electron scattering cross section of the ion. Hence, the effective mobility of the ion in an external field can be written as

$$\mu_{\mathrm{eff}} = \mu_0(1 - nl\sigma). \tag{5.75}$$

When $nl\sigma > 1$, the effective mobility of positive ions changes its sign, and the entrainment effect overwhelms the drift of the ions. The population of electrons increases exponentially with temperature. He estimated that a reversal in ion transport direction will take place at $\sim$700°C in Ge and at $\sim$1300°C in Si (assuming the electrical conductivity of the semiconductors to be dominated by the intrinsic carriers. For experimental intrinsic carrier concentrations, see figures 5.11a and b.) In metals, the electron entrainment (or the 'electron wind' effect) overwhelms because of the inherently high electron concentration.

The effect of an internal electric field, due to a given electrically active concentration profile, on the diffusion of another impurity of opposite type and of lower local concentration has been demonstrated conclusively in various experiments. These include those in which the field determining impurity 1 is diffused first, followed by the diffusion of impurity 2 of an opposite electrical character [266, 302], as well as those in which impurities 1 and 2 are diffused simultaneously [264, 303, 304]. In these experiments, a significant amount of retardation of the diffusion of the impurity No. 2 was observed. Figure 5.13c shows an example of the simultaneous diffusion of antimony and indium in germanium at 795°C by Vas'kin *et al.* [264], with the Sb and the In profiles displayed separately in a and b. It is interesting to note that for the theoretical curves shown, the diffusivity was assumed to be constant. The results of these experiments are, however, subject to different interpretations. The investigators of these experiments themselves appear to hold the unanimous view that the retardation so observed is due to the effect of the internal electrostatic field. In the concept of thermodynamics, this is equivalent to saying that there is a chemical potential gradient of impurity

No. 2 due to the spatial inhomogeneity of the overwhelming impurity No. 1, and impurity No. 2 will go to where its activity coefficient is lowest. Beside this interpretation, it is also reasonable to anticipate that other effects may have contributed to the retardation phenomenon. Two of the more important effects are the ion pairing due to Coulombic interaction between the oppositely charged impurities [305]; and the change in the population of vacancies of various charge states. It is important to note that the acceptance of the concept of a variation in the equilibrium vacancy concentrations of various charge state must be concomitant with the acceptance of existence of an internal electrostatic field, for it is precisely this potential inhomogeneity for which thermodynamics predicts a change in vacancy activity coefficient, and hence in the vacancy equilibrium concentration. In other words, one must either accept that both there is no internal field and the vacancy concentration is unchanged, or there exists an internal field and the vacancy concentration may or may not change (depending on the depth of the vacancy electron levels). For one diffusant system, one must take both Figures 5.8 and 5.9, or none of these. Here one can appreciate the advantage of the self-consistency in the thermodynamical approach over the fragmentized consideration of the internal field and other effects. An experiment which demonstrates unequivocally the important contribution of the retardation phenomenon was reported by Boltaks and Dzhafarov [306]. They investigated the effect of the local electric field created by the concentration gradient of indium or gallium on the diffusion of antimony in germanium. They observed that if the concentration gradients of the donor (Sb) and of the acceptors (In) are co-directional, the field retards the diffusion of Sb. *However, if these concentration gradients are in opposite directions, the electric field accelerates the displacement of the diffusing Sb ions.* The concentration gradients of In or Ga which is opposite to that of the in-diffusing Sb, was obtained by the evaporation of In or Ga from the pre-doped samples. One sees clearly that *if ion pairing and/or the decrease of the vacancy concentration were the principal cause of retardation, one would have observed the retardation of Sb diffusion regardless of the direction of the concentration gradient of the indium (or the gallium).* Thus, one can conclude that the internal field is the dominant factor in the retardation phenomenon. Boltaks *et al.*, however, suggested that a fraction of the retardation could have come from ion pairing [307]. Recently Kennedy [308] has questioned the existence of an internal electric field, with the implication that the enhanced diffusivity in a single diffusant system as well as the retardation phenomenon in the 'ambipolar diffusion' system are caused by an alteration in the diffusivity due to the change in vacancy concentration. We mentioned earlier that such a concept is thermodynamically inconsistent. And it is also in disagreement with the experimental results of Boltaks *et al.* [306, 307] mentioned above. The arguments given in [308] are that at high impurity concentrations the Debye length is comparable to or smaller than the separation between the impurity atoms, and the impurity ions will thus be neutralized by mobile charges within the 'structure' (the 'structure' presumably means the average unit cell for each impurity ion), and consequently, according to Gauss' theorem, there should be no internal field. These arguments basically arise from an incorrect application of the Debye screened potential. The electric potential due to a point charge j in an electrolyte is given by

$$V_j = \frac{z_j e}{\epsilon r_j} \exp(-r_j/r_s) \tag{5.76}$$

where ϵ is the macroscopic dielectric constant of the electrolyte, r_j is the distance between the test charge and the jth ion having absolute charge $z_j e$, and r_s is an appropriate screening length. In nondegenerate as well as in degenerate semiconductors, r_s has been given by Dingle [309]. It is given by the Debye length L_D in nondegenerate electrolytes [310]. We should note that the important assumption in the derivation of equation 5.76 is that the electrolyte is homogeneous, and is macroscopically electrically neutral, i.e., that the total number of the screening ions of one charge type is equal to the total number of the ions of the opposite charge type. In the presence of spatial inhomogeneity and the attendant electric field, the charge shielding no longer has a spherical symmetry. Furthermore, because of the presence of space charge, equation 5.76 no longer describes the potential due to a point charge, even if we could neglect the nonspherical symmetry aspect of the problem. A self-consistent treatment of this problem is not available. But if we focus our attention on a region which contains the ion j and which is of the dimensions of the average inter-impurity separation, we could make the approximative assumption that the region is spatially homogeneous and contains a space charge due to the depletion of the screening carriers to some other region. Let f be the fraction of the ionic charged that is unbalanced. Then, in a very crude approximation, one may write

$$V_j = \frac{z_j e f}{\epsilon r_j} + \frac{z_j e(1-f)}{\epsilon r_j} \exp(-r_j/r_s') \tag{5.77}$$

in place of equation 5.76. In equation 5.77, r_s' is some screening length which is only slightly different from r_s because of the slightly smaller concentration of the screening carriers in the present case than in the case of charge neutrality. Summing all potentials (given by equation 5.77) at the test charge site due to all the ions, one obtains a potential which is equivalent to one built from the net unscreened ionic charges according to Gauss' theorem, since the screened portion of potentials vanishes in the limit $r_s < r_j$. That the net charge is unscreened follows from definition. We see that it is precisely this misapplication of Debye's formula (equation 5.76) that led to the conclusion in [308]. The space charge fraction f is built to self-consistency for a given impurity profile and its accompanying internal field by simultaneously satisfying Boltzmann's statistics (or Fermi-Dirac statistics)

$$\sinh u = (1-f)[A]/2n_i, \tag{5.78}$$

and Poisson's equation

$$\nabla^2 u = f[A]/2n_i L_D^2, \tag{5.79}$$

where $[A]$ is the concentration of the impurity ions, and u is the normalized electrostatic potential $(E_c - E_F)/kT$. Rigorously speaking, equations 5.78 and 5.79 are to be solved simultaneously. But if f is very small compared to unity, u can be readily obtained from equation 5.78 alone by neglecting f,

$$u = \sinh^{-1}([A]/2n_i). \tag{5.80}$$

This is, of course, the well known local charge neutrality approximation. On the other hand it is obviously incorrect to suggest that one can also obtain u from equation 5.79 by setting $f = 0$. This mistake is equivalent to the above case in which one applies Gauss' theorem to obtain the macroscopic field and uses Debye's charge neutrality screening formula for a point charge potential.

Having cleared up the questions raised against the existence of an internal field, we now should note that the actual internal field may be appreciably less than obtained from equation 5.80 (with $E = (kT/e)du/dx$) for a number of reasons, among which are the local charge neutrality approximation, the complete ionization approximation, the non-degeneracy approximation, and the problem of the formation of the impurity band tail discussed earlier. The term 'ionization' in heavily doped semiconductors is here used in a special sense to denote the transition of electrons from the impurity band to the conduction band. We shall herein call the electrons in the impurity band 'un-ionized', even though they are delocalized.

In the normal sequence of a transistor fabrication process, the base impurity, having a lower concentration, is diffused first, and the emitter impurity, having a high concentration and hence field determining is diffused next. In such a sequence, the influence of the internal field due to the second impurity on the diffusion of the first impurity is much smaller than in the reverse sequence. Hence, the retardation phenomenon in the normal transistor diffusion sequence is very small and difficult to observe. (Nevertheless, depletion of the impurity concentration in the base region may still be substantial. The depletion phenomenon was recently confirmed by Ziegler *et al.* [391] who used a new boron profiling technique based on atomic mutation.) It has been shown [269] that, unlike in the reverse sequence, the retardation effect will be observable in the normal sequence only when a number of requirements are satisfied. Among these requirements, the following are emphasized: the base diffusivity should be higher than the emitter diffusivity; the base width should be sufficiently narrow, and should be a small fraction of the initial base diffusion length; and the initial base diffusion length should be only a small fraction ($<\frac{1}{2}$) of the final diffusion length. An example of the retardation phenomenon is shown in figure 5.14, which was experimentally exaggerated by heat

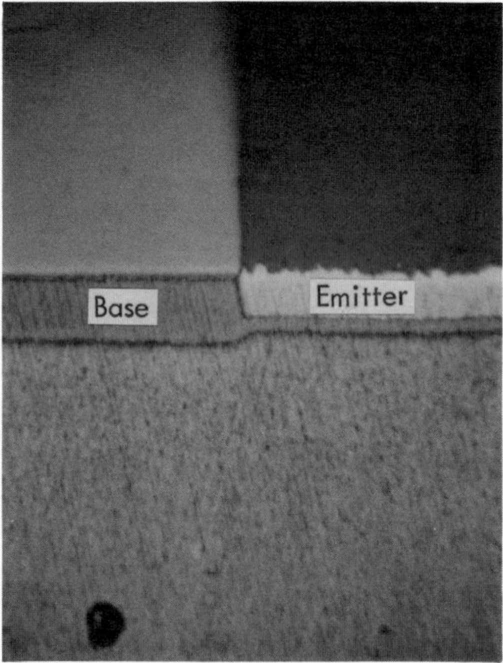

Figure 5.14. The retardation phenomenon in a boron-arsenic sequential diffusion in silicon. After the emitter diffusion at 1000° C for 2 h 45 min, the sample was subject to heat treatment in argon at 900° C for 24 h to enhance the retardation phenomenon.

treatment, in an argon atmosphere, at 900°C for 24 h. The low temperature heat treatment was intended to enhance the effect of the internal field, since the extrinsicity factor $[A]/2n_j$ increases with the decrease of temperature. The observed retardation was considerably smaller than what would be predicted by the model of Hu and Schmidt [269]. There are two possible causes for this discrepancy: first, in the model of Hu and Schmidt, the boron diffusivity was assumed to be unaffected by the presence of n-type impurities at very high concentration, an assumption that comes from the model that boron diffuses via a vacancy mechanism, and that the vacancy concerned has only an acceptor level. Recently, Makris and Masters [283] observed that the diffusivity of gallium in silicon decreases with arsenic doping. It may be reasonable to predict that this effect also exists for boron diffusion in silicon. Consequently, the drift velocity of boron in the internal field due to an n-type impurity profile should be smaller than predicted in the model of Hu and Schmidt, resulting in little surface pile-up and a decreased retardation of junction movement. The second cause is the probable overestimation of the Fermi energy due to a number of reasons, among which are the chemical clustering of impurity atoms (section 5.5.2), the formation of a conduction band tail, and the contribution of multiple-band density of states (see Appendix 3), all of which occur at high impurity concentrations. A less well settled problem (at least in the present author's opinion) is what happens to the impurity band when the impurity concentration exceeds the critical concentration of metal-nonmetal transition ($\sim 2 \times 10^{19}$ cm^{-3} in Si and $\sim 10^{18}$ cm^{-3} in Ge, as determined from NMR Knight shift [311a]). A prevalent concept is that the impurity band which begins to form after the first critical concentration (3×10^{18} cm^{-3} in Si) will disappear after the second critical concentration (2×10^{19} cm^{-3} in Si) because of complete screening of the impurity ions [275, 311b]. There seems to be lack of self-consistency here, since the 'complete screening' must imply the existence of impurity states, at least in the sense of virtual states. Recently, Kaplan [311c] has shown that when charge neutrality is taken into account, the virtual states exist regardless of doping, and these states are found to peak at the Fermi level. If one accepts this view, one must then include these virtual states in the statistics of carrier distribution, and consider the ionization problem (i.e., electrons in these virtual states, though delocalized, can be considered to be un-ionized, and behave differently from free carriers in the conduction band states). We do not wish to complicate our present discussion with this yet unsettled problem. However, the concept of virtual states and incomplete ionization may prove relevant in the discussion of impurity clustering (section 5.5.2).

5.5 SOME 'ANOMALOUS' DIFFUSION PHENOMENA

5.5.1 The Fictitious Background Doping Effect

Williams [170] first reported that the diffusivity of boron in silicon increases with the *n*-type background impurity doping concentration, with a factor of 2 difference in the range observed. Maekawa and Oshida [171] later reported an even greater dependency of the boron diffusivity in silicon on the *n*-type background doping concentration—almost a three-fold increase from a background concentration of 0.9×10^{15} to 1.4×10^{19} cm^{-3}. Mackintosh [180] reported a large increase in phosphorous diffusivity in silicon with *p*-type background doping. Since this effect was observed for background concentrations with a wide range below the intrinsic carrier concentrations at the reported diffusion temperatures it is implausible that this effect could arise from electronic interactions. The background concentrations were also too low to cause appreciable ionic pairing at the diffusion temperatures. Further, these effects, if any, would tend to retard the diffusion rather than enhance it. In this section we will show that this effect is fictitious, and will present an analysis [311] of this phenomenon.

The results of the above mentioned phenomenon have all been obtained with experimental methods that involved the measurements of junction depth and sheet conductivity, with the assumption that the impurity profile is given by an erfc distribution. In reality, the impurity profiles should closely resemble those in figure 5.10, at least for boron and for arsenic single diffusion. The detail of these profiles is a function of the extrinsicity factor $\rho = [A(0)]/2n_i$. In the manner of the conventional method of characterization of a diffusion profile by junction depth and sheet conductivity measurements, one defines a fictitious impurity profile by

$$[A(\xi^*)]^* = [A(0)]^* \ \operatorname{erfc}(\xi^*), \tag{5.81}$$

where $[A(0)]^*$ is some fictitious surface concentration of the impurity A, and ξ^* is the fictitious normalized distance. $\xi^* = x/2\sqrt{D^*t}$, where D^* is a fictitious diffusivity. Equation 5.81 can be rewritten as

$$[A(\xi)]^* = [A(0)]^* \ \operatorname{erfc}(\xi\sqrt{D^i/D^*}), \tag{5.82}$$

where D_i is the real intrinsic diffusivity, and ξ is the real normalized distance. The fictitious surface concentration and the fictitious diffusivity are determined such that the fictitious erfc profile gives a sheet conductivity and a junction depth that are identical with those of the real profile $[A(\xi, \rho)]$. The equal sheet conductivity condition is

$$[A(0)]^* \int_0^{\xi_j} \mu[A]^* \operatorname{erfc}(\xi\sqrt{D^i/D^*})\,d\xi = \int_0^{\xi_j} \mu[A]\,[A(\xi, \rho)]\,d\xi, \tag{5.83}$$

where μ is the carrier mobility, and ξ_j is the real junction depth (measured). The requirement of equal junction depth gives

$$[A(0)]^* \operatorname{erfc}(\xi_j\sqrt{D^i/D^*}) = [B], \tag{5.84}$$

where $[B]$ is the background impurity concentration. To simplify the presentation of the results of numerical computation, we neglect the concentration dependence of μ. This should be an adequate approximation because the concentration dependence of the carrier mobility generally levels off at impurity concentrations $\gtrsim 10^{19}$ cm^{-3}, and it is primarily the high concentration region of a diffused layer that determines the sheet conductivity. The junction depth ξ_j is given implicitly by the relation

$$[A(\xi_j, \rho)] = [B] \tag{5.85}$$

After first integrating the left-hand side of equation 5.83, and the substitution of equation 5.84 into equation 5.83, the resulting equation is solved numerically together with equation 5.85. The principle underlying the analysis is shown in figure 5.15, and the results of numerical

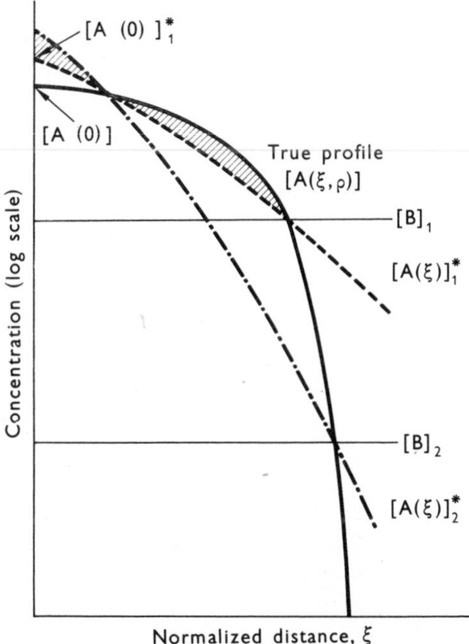

Figure 5.15. Schematic diagram showing two fictitious erfc profiles $[A(\xi)]^*_1$ and $[A(\xi)]^*_2$, for two background concentrations $[B]_1$ and $[B]_2$ respectively, that give the same junction depth and sheet conductance as the true profile $[A(\xi, \rho)]$. The shaded areas are equal in size. $\rho = [A(0)]/2n_i$.

calculation for the fictitious diffusivity and fictitious surface concentration are given in figures 5.16 and 5.17. From these figures, it is seen that the fictitious diffusivity for a fictitious erfc profile can change by as much as a factor of 10 for different background doping concentrations. Also, one sees that the fictitious surface concentration can be more than three times the real surface concentration. Figures 5.16 and 5.17 are of practical utility in estimating the true surface concentrations from the method of junction-depth and sheet-conductivity, which due to its simplicity is still the most widely used for diffusion process characterization.

Petro *et al.* [312] reported an anomalous phenomenon in the diffusion of antimony in silicon with aluminium background doping, which was observed with the junction depth and sheet conductivity techniques. This is shown in figure 5.18, in which the diffusivity of Sb is seen to first increase, reach a maximum, and then decrease with the alumunium doping concentration. In the light of the above analysis, we can conceive two explanations for this phenomenon. (Of course, an experimental re-examination of this phenomenon is desirable.) The reason for the increase of the fictitious diffusivity (which we assume it is) with background doping concentration has already been discussed above. The

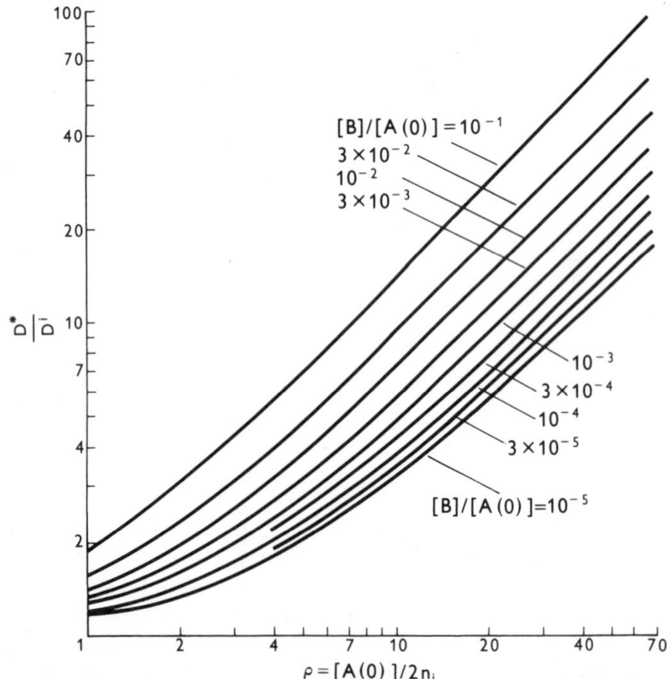

Figure 5.16. D^*/D^i versus ρ for various $[B]/[A(0)]$. After Hu [311].

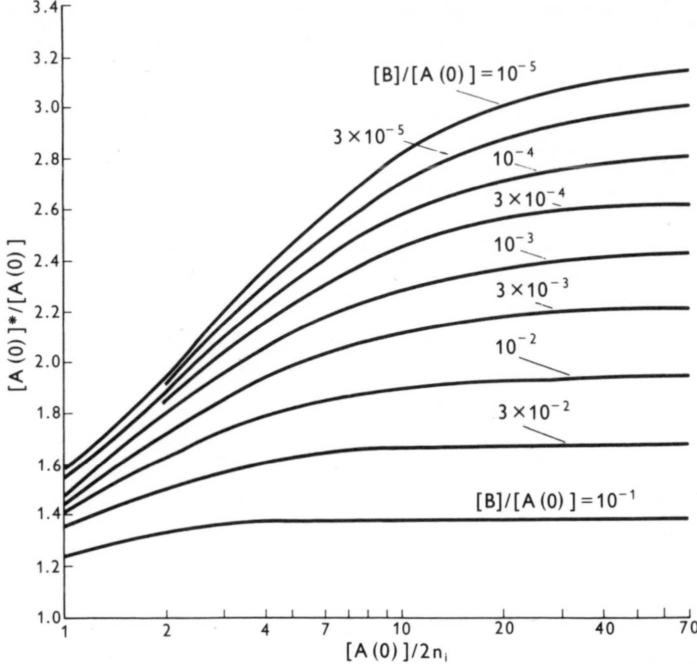

Figure 5.17. $[A(0)]^*/[A(0)]$ versus ρ for various $[B]/[A(0)]$. After Hu [311].

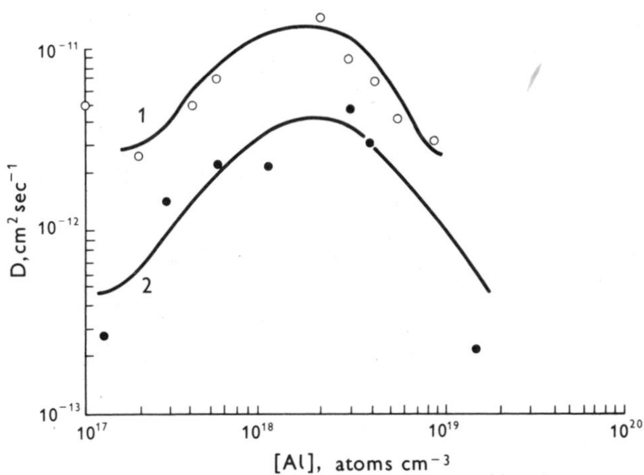

Figure 5.18. Diffusion isotherms for antimony in silicon samples variously doped with aluminium: (1) $T = 1300°$ C; (2) $T = 1215°$ C. After Petro *et al.* [312].

decrease could either be caused by a more complex Sb profile, such as the existence of a kink, or by some real process of ion pairing, etc.

5.5.2 Phenomenon at High Concentrations—an Example of Arsenic Clustering in Silicon

Arsenic diffusion in silicon has been shown [201, 269] to follow a model discussed earlier that embodies equation 5.64. However, considerable discrepancy between the model and the actual diffusivity develops at arsenic concentrations $\gtrsim 3 \times 10^{20}$ cm^{-3} [313a], and there is a decrease of diffusivity with arsenic concentration at higher concentrations [313b]. Such a phenomenon is explainable in terms of the cluster model [314], which is largely based on the vapor pressure data of Sandhu *et al.* [315]. The cluster is envisaged to consist of four arsenic atoms forming a tetrahedron either with a normal interstitial site or with a silicon atom at its center. A conjecture of the first structure is shown in figure 5.19. Note that in this structure the arsenic atoms are not nearest neighbors to each other.

The formation of the two As–As covalent bonds makes the complex electrically inactive. However, there exists a possibility that it may act as a deep donor. The latter structure can be written as ($SiAs_4$). Either will be simply referred to as the tetratomic arsenic complex. The number of arsenic atoms in the complex was arrived at by analyses of various models from As_2 to As_5, and finding the one that gave best fit to the vapor

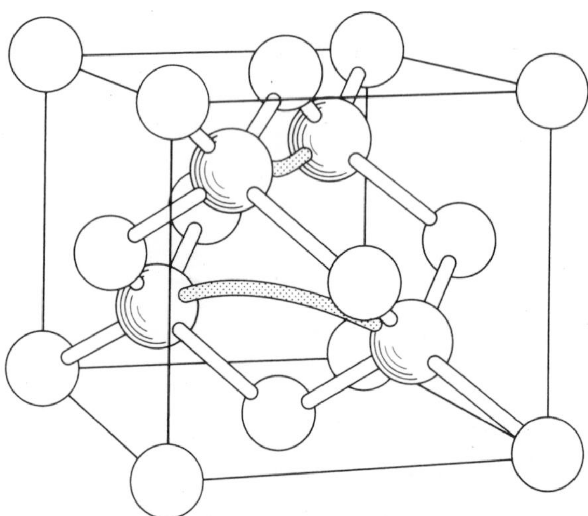

Figure 5.19. A conjecture of the structure of the arsenic complex in silicon. Open circles = silicon atoms; shaded circles = arsenic atoms. The four arsenic atoms form a tetrahedron with a normal interstitial site as its center.

pressure data. The basis of such analysis is the assumption that, for chemical reasons, only one complex structure dominates in a certain temperature range.

Let us define γ' as the activity coefficient of arsenic in the absence of clustering. γ' is identical with γ_A given by equation 5.54. If one defines α as the ionization factor, assuming the existence of virtual states in the impurity band (see last paragraph of section 5.4.6) (p. 301), one has $(1-\alpha) = 1/((1+\zeta)$, where $\zeta = g_A^{-1} \exp[(E_A - E_F)/kT]$. One can then rewrite equation 5.54 as

$$\gamma' = (1-\alpha)/(1-\alpha_i), \tag{5.86}$$

where α_i is the ionization factor in intrinsic silicon. Note that α_i does not approach unity at infinite dilution, and hence $(1-\alpha_i)$ does not vanish in equation 5.86. This is because the Fermi energy E_F is never given by negative infinity, but by some finite value. By extending the analysis in [13] to include the formation of the tetratomic arsenic complex, one obtains the activity coefficient of arsenic in silicon as

$$\gamma = \gamma'\gamma'', \tag{5.87}$$

where γ'' is given by

$$\gamma'' = 1/\{1 + 4MK(1-\alpha)^4 [A_1]^3/[S]^3\}, \tag{5.88}$$

where $[A_1]$ is the concentration of the monatomic arsenic ionized as well as un-ionized, and is related to the total arsenic concentration $[A]$ through the relationship

$$[A] = [A_1] + 4MK(1-\alpha)^4[A_1]^4/[S]^3 = [A_1]/\gamma''. \tag{5.89}$$

$[S]$, as before, denotes the concentration of lattice sites. M is the configurational multiplicity, which for the conjectured structure in figure 5.19, is 24. $K = \exp(\Delta H/RT)$, where ΔH is the energy of formation of the complex. Hence, γ'' is simply the fraction of arsenic atoms that is in monatomic form, and γ' is the fraction of monatomic arsenic that is un-ionized at the concentration concerned relative to that at infinite dilution (equation 5.86). The physical meaning of equation 5.87 is then that the activity of arsenic in the solid solution is given by the concentration of its neutral monatomic species. We compute the activity of arsenic in silicon according to equations 5.86-5.89 by iteration, using Fermi-Dirac statistic in the evaluation of E_F. The value K is obtained by best fit to the vapor pressure data of arsenic over arsenic-silicon solid solution. An example is given for 1000° C, with $K = 1.2 \times 10^5$, corresponding to a ΔH of 1.3 eV. The calculated activity coefficient γ is plotted against the total arsenic concentration in figure 5.20. For

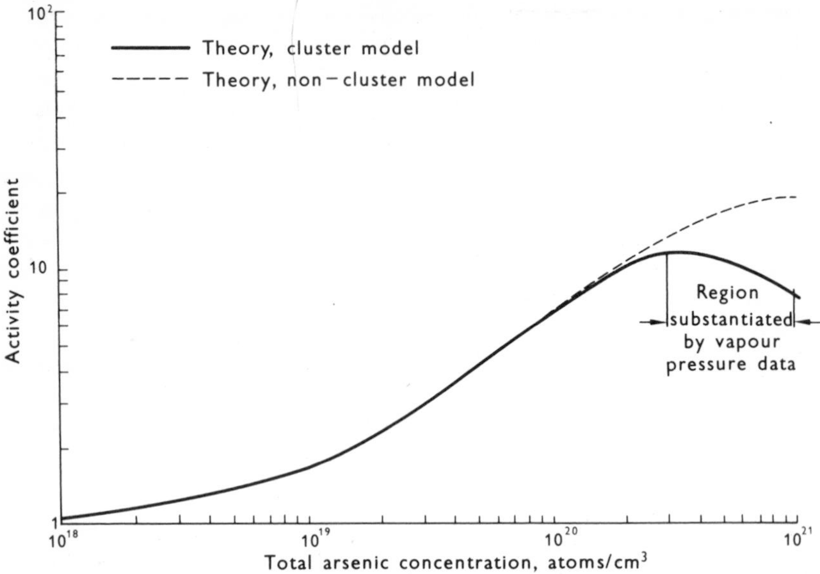

Figure 5.20. Activity coefficient of arsenic in silicon at 1000° C—a comparison of the cluster theory, non-cluster theory, and experimental vapor pressure data. The extrapolated vapor pressure data are believed to be inaccurate. After Hu [314].

comparison, we have calculated the activity coefficient of arsenic from the vapor pressure data of Sandhu *et al.* [315]. Note that, following convention, our theoretical activity coefficients have been normalized to infinite dilution (i.e. $\gamma = 1$ at infinite dilution). From the experimental data point of view, however, a normalization with respect to infinition dilution is not practicable since the vapor pressure at very low concentration cannot be measured with a reasonable degree of accuracy. To circumvent this difficulty, we have adopted a normalization by equating the activity coefficient calculated from the vapor pressure data at the highest concentration point to the theoretically calculated activity coefficient at the same concentration. This is done based on the assumption that the vapor pressure measurement at a higher concentration is subject to less error.

The assumption underlying the above analysis of the chemical potential of arsenic in silicon is that equilibrium (or quasi-equilibrium) exists in the ensemble of a chosen small region surrounding the complex. The rate of formation of arsenic clusters in this small region is diffusion controlled, but is much faster than the macroscopic diffusion. This condition prevails when the dimension of the small region, of the order of $[A]^{1/3}$, is much smaller than the diffusion length. If this is not the case, it becomes necessary to solve the diffusion problem by including, in the equations of continuity, terms representing the kinetics of clustering. Because of some degree of similarity between our conjectured arsenic cluster and the better known oxygen complex (SiO_4), we may adopt the simplifying assumptions of Kaiser *et al.* [289] concerning the inequality relations among various rate constants in the detailed reaction scheme, and reduce the clustering process to an effective reaction of

$$d[A_4]/dt = k_f[A_1]^4 - k_b[A_4], \tag{5.90}$$

where k_f and k_b are the forward and the backward rate constants. (However we have assumed the absence of further polymerization reactions beyond A_4.) The continuity equation for A_1 is then

$$d[A_1]/dt = -\partial J(A_1)/\partial x - k_f[A_1]^4 + k_b[A_4]. \tag{5.91}$$

We derive an expression for the apparent diffusivity for the case where the quasi-equilibrium condition prevails for the arsenic complex during the diffusion process. It is natural to assume that the complex is immobile before it decomposes. Hence, one has

$$J(A) = J(A_1) = -D(A_1)(1 + \partial \ln \gamma'/\partial \ln [A_1])\partial \ln [A_1]/\partial x. \tag{5.92}$$

Utilizing the relationship $[A_1] = \gamma''[A]$ (equation 5.89), one can show that equation 5.92 is equivalent to

$$J(A) = -D(A_1)\gamma''(1 + \partial \ln \gamma'/\partial \ln [A_1])(1 - \partial \ln \gamma''/\partial \ln [A_1])\partial [A]/\partial x. \tag{5.93}$$

If we define an apparent diffusivity $\mathbf{D}(A)$ by the relation

$$J(A) = -\mathbf{D}(A)\partial[A]/\partial x, \tag{5.94}$$

and utilize the relation $D(A_1) = D^i(A_1)[V']/[V']_i$, we obtain

$$\mathbf{D}(A) = D^i(A_1)([V']/[V']_i)\gamma''(1 + \partial \ln \gamma'/\partial \ln [A_1])/(1 - \partial \ln \gamma''/\partial \ln [A_1]). \tag{5.95}$$

The apparent diffusivity given in equation 5.95 is just what one would obtain from the Boltzmann-Matano analysis of a concentration profile, and is easily calculated using equations 5.86 and 5.88. It may be desirable to express the apparent diffusivity as a function of the readily measurable total concentration $[A]$. This cannot be accomplished explicitly, but can be done numerically. An example of the theoretical apparent diffusivity according to the cluster model is given in figure 5.21. For comparison the experimental apparent diffusivity, obtained from Boltzmann-Matano analyses of arsenic diffusion profiles, is also plotted in the same figure. It should be noted, however, that, unlike the theoretical apparent diffusivity, it is difficult to obtain an exact ratio of the extrinsic apparent diffusivity to the intrinsic diffusivity. For it is often not possible to obtain the intrinsic diffusivity from a high concentration diffusion profile because of

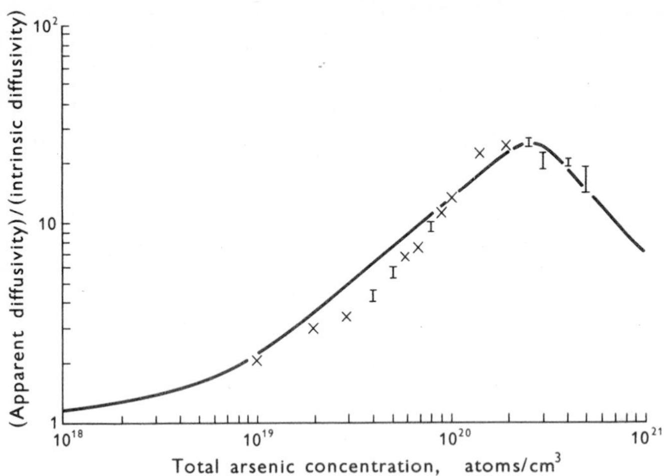

Figure 5.21. Diffusivity of arsenic in silicon at 1000° C—a comparison of theory (solid curve) based on the tetratomic cluster mode and the experimental results of Boltzmann-Matano analysis of arsenic diffusion profiles from activation analysis.

the uncertainty in the gradient in the low concentration region. Nor can it guarantee sufficient accuracy for the present purpose to compare the extrinsic diffusivity from a high concentration diffusion profile with the intrinsic diffusivity of a separate low concentration diffusion profile because of inevitable variations of diffusion conditions. Therefore, we choose the maximum experimental diffusivity in the diffusivity versus concentration plot as a reference, and set it equal to the maximum theoretical diffusivity. The experimental diffusivity at any given concentration is then normalized to the maximum diffusivity. In view of the inherent inaccuracy and the large scatter in the diffusivity obtained by Matano analysis, we may consider the experimental data and the theoretical prediction to be in good agreement. The convincing point of the tetratomic cluster model is corroborated by its good fit with the vapor pressure data. One should of course, expect the theory to agree better with the experiment from vapor pressure measurement because the vapor pressure is an equilibrium thermodynamic property, while the diffusion is a non-equilibrium process. In the thermodynamic activity calculation, there is no problem of the presence of space charge due to an internal field, nor is there a problem of a possible departure from quasi-equilibrium for the arsenic complex. This thermodynamic property, presumably, also has experimental simplicity. One consequence of clustering is that, even without precipitation, the concentration of electrically active centers will be considerably lower than the total arsenic concentration for concentrations $\gtrsim 2 \times 10^{20}$ atoms/cm^3.

5.5.3 Phenomena at High Impurity Concentrations—Lattice, Strain, Generation of Dislocations and Precipitation During Diffusion

Mismatch of atomic (or ionic) size between the solute and the host lattice introduces strain surrounding each solute atom. At high impurity concentrations, the accumulation of these local strains develops into a macroscopic strain in the lattice. The macroscopic lattice strain has been shown to be given empirically by the linear relation (Vegard's law)

$$\epsilon = \beta[A] \tag{5.96}$$

where ϵ is the macroscopic lattice strain, and β is a coefficient of lattice contraction. The experimental values of β have been determined for boron and phosphorus in silicon by a number of investigators [316-319]. The values given by Pearson and Bardeen [316] are $\beta[S] = -0.11$ for boron, and $\beta[S] = -0.05$ for phosphorus. $[S]$, as before, denotes the concentration of lattice sites (5.5×10^{22} cm^{-3} in silicon). The coefficient of lattice contraction β can also be estimated approximately:

$$\epsilon = (1 + \{(r_A/r_S)^3 - 1\}\,[A]/[S])^{1/3} - 1. \tag{5.97}$$

After binomial expansion of equation 5.97 and neglecting the second and the higher order terms, one obtains, by a comparison with (5.96),

$$\beta = \{(r_A/r_S)^3 - 1\}/3[S], \tag{5.98}$$

where r_A and r_S are the atomic (or ionic) radii of the solute and the host crystal respectively. Note that the above expression differs from a formula described by Lawrence [334] by a factor of 3. It should be noted that deviation from Vegard's law is often appreciable in many solid solutions [319a-319c]. The free energy associated with the deformation of the lattice is given by $(9/2)K(\beta[S])^2[A]$, where K is the bulk compressibility of the solvent. Then, if none other than the configurational entropy and the strain energy contribute to the chemical potential of the impurity, one may write, following Lyubov [334a],

$$\mu_A = kT \ln([A]/[S]) + \frac{2Y}{1-\nu}\beta^2[A], \tag{5.98a}$$

where Y is the Young's modulus, and ν is the Poisson's ratio. One may then write the chemical potential gradient as

$$\frac{\partial \mu_A}{\partial x} = \left(\frac{kT}{[A]} + \frac{2Y\beta^2}{1-\nu}\right)\frac{\partial [A]}{\partial x} \tag{5.98b}$$

In the above expression, the lattice strain has been assumed to be localized, an assumption which is analogous in spirit to the local charge neutrality assumption for the expression of an internal field, or local potential (equation 5.80). To give the reader a perspective of the relative contributions to the thermodynamical force of transport, we compare the usual diffusive term (the first term) with the strain term (the second term) in the parentheses in equation 5.98b. Typically, $kT \sim 0.1$ eV $\sim 1 \times 10^{-13}$ erg, $Y \sim 1.5 \times 10^{12}$ dynes/cm^2 for silicon (average) [334b], $\nu \sim 1/3$. And for phosphorus in silicon, $\beta \sim 0.05/[S]$, where $[S] = 5.5 \times 10^{22}$ cm^{-3}. After substitution, we find that the ratio of the second term to the first term is $\sim 2.1[A]/[S]$. Since $[A]/[S] \gtrsim 0.02$ (solubility limit), one sees that the pure strain contribution to the enhancement of diffusion is $\gtrsim$4%. For boron in silicon, $\beta \sim 0.11/[S]$, and $[A]/[S] \gtrsim 0.01$, the enhancement due to pure macroscopic strain is $\gtrsim$10%. In addition to its effect on the chemical potential of the diffusant, the macroscopic stress arising from misfitting solute atoms may also affect the diffusivity according to equation 5.19. The diffusivity in an interstitialcy mechanism may increase with compressive stresses (such as due to oversized solute atoms), while in a vacancy mechanism it increases with tensile stresses (such as due to undersized solute atoms). The macroscopic stresses are truncated to an upper limit equal to O_c, the critical yield stress of the substrate, by the generation of misfit dislocations. Take silicon, $O_c \simeq 0.5 \times 10^9$ dynes cm^{-2}; $\Delta V_{\text{act}} \simeq +0.5$ atomic volume (approximate estimation for a vacancy mechanism). At the upper limit due to tensile stress, $-P\Delta V_{\text{act}} \simeq 0.005$ eV. Thus, at 1000°C, the change in D/D_o is $\sim$5%. Finally, one may note that, if the macroscopic stress is isotropic (e.g. a hydrostatic pressure) there will be no misfit dislocations generated, and hence no upper limit of the macroscopic stress.

The macroscopic lattice strain so introduced can be relieved by the generation of dislocations. The experimental observation of this phenomenon was first reported by Queisser [320]. He found that when boron is diffused into silicon with a total amount of $W = 5 \times 10^{16}$ boron atoms/cm^2, a dense slip pattern network developed. The dislocations in the network have Burgers vectors parallel to the surface plane of the silicon wafers [325], and are similar in nature to the misfit dislocations in the epitaxial films grown on substrates of disparate lattice parameters. The original theory of misfit dislocations of van der Werwe [320a], has been later extended to include the consideration of a diffusion zone in which the lattice parameter varies linearly with distance, by Vermaak and van der Werwe [320d]. The van der Werwe theory, and its derivatives, predicts equilibrium dislocation density by minimizing the energy sum of 'misfit' energy (which implies dislocation self energy) and lattice strain energy. Application of this theory to a diffusion zone is not straightforward. Dislocation-dislocation interactions, impurity-dislocation interactions, etc., are not taken into account. A more sophisticated but preliminary analysis has been given by Pavlov and Pashkov [320e], which considers the structure of the dislocation network as well, i.e., finding a dislocation density tensor, and passing to discrete structure by detailed enumeration. The dislocation network is found by minimization of an energy functional. Two simpler approximate theories have been given by Shockley [321] and Prussin [129] respectively.

In Shockley's theory [321], the generation and penetration of a dislocation is considered as a process that minimizes the free energy of the system. Due to an image force, the energy released in the material with a uniform strain by the presence of an edge dislocation at a distance x from the surface is given by

$$E_s = \sqrt{2} b x \epsilon G(1+\nu)/(1-\nu), \tag{5.99}$$

where G is the modulus of rigidity, ν is Poisson's ratio, and b is the Burgers vector of the dislocation. The energy of the formation of a dislocation is given, approximately, by

$$E_d = [b^2 G/4\pi(1-\nu)] \ln (2^{3/2} x/r_0), \tag{5.100}$$

where r_0 is the effective dislocation core radius. In a diffused layer, where the strain is not uniform, an average value of $x\epsilon$ is used in equation 5.99

$$\langle x\epsilon\rangle = \beta \int [A]\, dx = \beta W. \tag{5.101}$$

Equation 5.101 also serves to define W. The generation of dislocations is favored if $E_s > E_d$, which leads to a criterion for the condition

$$W_c \geqslant [b/\beta 4\pi\sqrt{2}(1+\nu)] \ln (2\sqrt{2x}/r_0) \tag{5.102}$$

for the generation of dislocations.

In Prussin's model [129], the self-energy of the dislocation has been neglected, perhaps unjustifiably. The model simply assumes that the number of extra planes m introduced by a dislocation density ρ_d is just enough to compensate for the lattice contraction due to the impurity concentration $[A]$ above a critical concentration $[A]_c$:

$$m = \int_x^{x_c} \rho_d dx = \alpha^{-1} \beta([A] - [A]_c), \tag{5.103}$$

where α is the component of the Burgers vector in a direction perpendicular to x. x_c is the distance where $[A] = [A]_c$. The critical concentration $[A]_c$ is defined through the expression

$$\sigma_c = \beta [A]_c Y/(1-\nu), \tag{5.104}$$

where σ_c is the critical stress for plastic deformation at the diffusion temperature and Y is Young's modulus. The Prussin model predicts the generation of dislocations when the impurity concentration exceeds $[A]_c$.

In contrast, the Shockley model predicts the generation of dislocations only after the total amount of impurity diffused into the crystal has exceeded W_c. The Prussin model appears to disagree with most experimental observations [127]. Unfortunately, the Shockley model can only serve to give a criterion of W_c, and is incapable of predicting the density distribution of dislocations. Another model, due to Czaja [323] is merely a mixture of the Shockley model and the Prussin model. Finally, we may note that an equilibrium structure of dislocation network may not obtain in actual cases of diffusion. A dynamical model of dislocation generation is not yet available.

Since Queisser's pioneering work [320], a number of reports of experimental observations of diffusion induced dislocations have appeared [319, 324-330]. However, most published works have been rather cursory, and did no more than attest to the existence of misfit dislocations. Obviously, work beyond the mere observation of dislocations is badly required. In particular, we would like to see some quantitative correlations between the diffusion induced stress and dislocation density distribution, the kinetics of dislocation generation, and the interactions between dislocations and impurity atoms.

The effect of dislocations on impurity diffusion has not been well established. While moving dislocations, particularly climbing dislocations are instrumental in the phenomenon of enhanced base diffusion known as the emitter-push effect, the diffusion of the dislocation-generating impurity itself seems to be retarded. This, for example, has been observed in phosphorus diffusion in silicon [331, 332]. Calhoun and Heldt [332] reported that plastic deformation has no effect on the diffusion of antimony in germanium, and suggested that the moving dislocations in Ge produced divacancies which are less mobile than monovacancies. This hypothesis is at least questionable: one cannot ignore the possibility that the dislocations were conservative in their experiments, or they could even have caused precipitation. Lawrence [334, 335] has studied the effect of plastic deformation due to external stress (pin loading) on the diffusion of some impurities in silicon. He found that phosphorus diffusion was retarded, whereas the diffusivities of antimony, gallium and boron were enhanced. He interpreted the enhancement of the diffusion of B, Ga and Sb as due to the generation of excess vacancies by the moving dislocations, and the retardation of the diffusion of P as due to its precipitation at the moving dislocations. According to Lawrence's results, it appears that the Coulombic interaction between the acceptor dislocations and donor phosphorus is not a prime contributor to the precipitation of phosphorus, for antimony is also a donor, but an opposite effect was observed. We must note however that, unfortunately, Lawrence did not report the respective concentrations of the various solutes he studied. The observation by Widmer [336] of an enhanced self-diffusion in germanium which had been plastically deformed is understandable, for one should not expect the germanium tracer to preferentially precipitate out at the dislocation over its isotopic host lattice atoms. This is in contrast to antimony diffusion in Ge by Calhoun and Heldt [332].

Similarly, most published works on the phenomenon of precipitation during diffusion have not gone much beyond reporting observations of the precipitations. They have, however, been of considerable value in our conceptual interpretation of some anomalous diffusion phenomena. Precipitates were first observed for phosphorus in silicon by Kooi [337] and Schmidt and Stickler [338]. Kooi suggested that the phosphorus precipitates, probably SiP, are the cause of the discrepancy between the electrically active profile and the total phosphorus profile as discussed by Tannenbaum. He observed the precipitates to occur mostly at the oxide-silicon interface. Schmidt, Stickler, and O'Keefe [338, 339] have reported on the observations of rod-shaped SiP precipitates, up to 1 μm in (110) wafers, and have established the crystallographic orientation relationship between the precipitates and the silicon matrix. The crystallographic identification of the precipitates as SiP was further confirmed by Beck and Stickler [340] by X-ray diffraction and comparison with SiP and SiAs crystals from vapor growth. They also suggested that since in the Si-P phase diagram, SiP is stable only at temperatures below 1131°C, only at higher temperatures can phosphorus be diffused into silicon in sufficient concentrations to cause the formation of a diffusion-induced dislocation network. At high temperature, according to Kooi [337], a phosphorus rich molten layer can be formed at the SiO_2-Si interface, with the precipitation of P on cooling. Work in the area of the observation of precipitates has been pursued by a number of investigators [105, 330, 341-343]. Joshi [342] reported the observation of incoherent phosphorus precipitates, randomly oriented in the silicon matrix. Later, Joshi and Dash [105] reported the observation of vacancy-type coherent precipitates of B and P in Si, which they suggested, from contrast observation, to be the result of the smaller atom size of B and P relative to silicon. Jaccodine [343] has however identified the phosphorus precipitates as of the interstitial type. The suggestion of interstitial type precipitate of B and P appears to be logical, and is consistent with the expectation of macroscopic strain relief. Similar to precipitation at climbing dislocations, these looped precipitates are expected to generate vacancies (perhaps both types can be regarded as interstitials). In fact, helices (caused by condensation of excess vacancies [103, 104]) have been observed by Joshi and Dash [105]. However, they

TABLE 5.13

Size effect interaction for various impurities in silicon at a temperature of 1200°C

Impurity element	Interstitial (i) Substitutional (s)	Radius r_i of impurity atom (Å)	$\epsilon = \dfrac{r_i - r_0}{r_0}$	E_1 (max) eV[a]	Capture radius[c] in Angstroms
Boron	s	0.88	−0.25	0.75	24
Boron[b]	s	–	−0.28	0.80	25
Aluminium	s	1.26	0.077	0.23	7
Gallium	s	1.26	0.077	0.23	7
Indium	s	1.44	0.23	0.70	21
Phosphorus	s	1.10	−0.060	0.18	5
Arsenic	s	1.18	0.009	0.03	8
Antimony	s	1.36	0.16	0.49	15
Carbon	s	0.77	−0.34	1.0	31
Germanium	s	1.22	0.086	0.26	9
Copper	s	1.35	0.15	0.47	14
Gold	s	1.5	0.28	0.85	26
Oxygen[b]	i	–	0.19	0.57	17
Copper	i	1.28	0.09	0.28	9
Gold, Silver	i	1.44	0.23	0.70	21
Nickel	i	1.24	0.06	0.18	5
Iron	i	1.26	0.08	0.23	7

[a] Interaction energy.

[b] Estimates of ϵ from X-ray data.

[c] If one makes the assumption that the impurity atom is a sphere, then the elastic interaction between the impurity atom and the edge dislocation can be represented by

$$E_1 = (4/3)\frac{(1+\nu)}{(1-\nu)} \cdot \frac{Gb(r_i - r_0)r_0^3 \sin\theta}{r}$$

where (r, θ) are the cylindrical-polar coordinates, r_i and r_0 are the radii of the impurity atom and the host lattice atom, and ν, G are the Poisson's ratio and the shear modulus respectively. A capture radius R_c is defined as the distance from the dislocation line where the potential well depth is 1 kT, and within which the interaction energy is greater than the thermal energy. Since the interaction energy as given in the above equation is not cylindrically symmetrical, we define R_c at $\theta = \pm\,\pi/2$.

suggested a number of mechanisms, other than the interstitial-type precipitates, as possible causes for the generation of excess vacancies. More recently, they have revised their view [127], and regarded the precipitates to be of the interstitial type.

There are two thoughts concerning the mechanism of precipitation on dislocations. In the first, dislocations are merely viewed as nucleation centers, and precipitation is considered as a normal phase transition process when the solid solution becomes supersaturated upon cooling. The second thought is that precipitation actually occurs during the isothermal diffusion process. The driving force for precipitation can be the stress field (elastic as well as Coulombic) of the dislocations. A thorough treatment of the subject of interactions between solute atoms and dislocations in Si and

TABLE 5.14

Size effect interaction for various impurities in germanium at a temperature of 700°C

Impurity element	Interstitial (i) Substitutional (s)	Radius r_i of impurity atom (Å)	$\epsilon = \frac{r_i - r_0}{r_0}$	E_1 (max) eV[a]	Capture radius[c] in Angstroms
Boron	s	0.88	−0.28	0.79	38
Aluminium	s	1.26	0.03	0.09	4
Gallium	s	1.26	0.03	0.09	4
Indium	s	1.44	0.18	0.51	24
Phosphorus	s	1.10	−0.10	0.28	13
Arsenic	s	1.18	0.03	0.09	4
Antimony	s	1.36	0.12	0.33	16
Silicon	s	1.17	0.04	0.12	6
Copper	s	1.35	0.11	0.30	14
Gold	s	1.5	0.23	0.65	31
Oxygen	i	1.39[b]	0.14	0.40	19
Copper	i	1.28	0.05	0.14	7
Gold, Silver	i	1.44	0.18	0.51	24
Nickel	i	1.24	0.02	0.05	2
Iron	i	1.26	0.03	0.09	4

[a] Interaction energy.
[b] Estimated from X-ray data for oxygen in silicon.
[c] See footnote in Table 5.13.

Ge has been given by Bullough and Newman [231]. Tables 5.13 and 5.14, due to Bullough and Newman, list some pertinent parameters for the elastic interaction between some solute atoms and dislocations. The dislocation will not become saturated by precipitating solute atoms if these precipitating atoms form the climbing interstitial plane of the dislocation. Thermodynamically, we view this process as the increase of the impurity activity coefficient in the bulk due to a decrease of macroscopic strain ϵ as a result of the development of the dislocation network, and at the same time, there is a decrease of activity coefficient in the region surrounding newly created dislocations.

From the foregoing discussions, we may hypothesize the following general rule of the effect of diffusion-induced dislocations on the impurity diffusion process: For an impurity at very high concentrations, the dislocations cause precipitation, immobilize a significant fraction of the impurity atoms, and result in a retardation of diffusion. For impurity at low concentrations, very much below the solid solubility limit, non-conservatively moving dislocations will enhance the impurity diffusion due to the generation of vacancies. Note that in the latter case, the dislocations can only be generated by the diffusion of another impurity of higher concentration ($>[A]_c$), or by an externally induced plastic deformation. We shall see that this general rule can explain consistently the phosphorus diffusion phenomenon (the first case) in the next section, as well as the emitter-push effect (the second case) in section 5.5.5.

5.5.4 Diffusion of Phosphorus into Silicon

Tannenbaum [182] first obtained the detail of a phosphorus diffusion profile in silicon which showed a departure from an erfc distribution. Also, she demonstrated the discrepancy between the phosphorus concentration profiles as determined by differential sheet conductivity measurement and by the radioactive tracer technique. Her profiles, obtained from diffusion with P_2O_5 source, were relatively smooth and simple other than a deviation from erfc profile. Using Boltzmann-Matano analysis, she obtained a concentration dependent diffusivity as shown in figure 5.22. Recently, Bakeman and Borrego [344] proposed a theory to explain Tannenbaum's diffusivity-concentration relationship. Based on the assumption that the atomic jump probability is proportional to the number of available lattice sites which in turn is proportional to the difference between the saturation concentration and the local concentration, they arrived at the following expression for the diffusivity

$$\frac{D(A)}{D^i(A)} = \frac{[A]_{max} + [A]}{[A]_{max} - [A]}, \tag{5.105}$$

where $[A]_{max}$ is the saturation concentration of A. This expression appears unreasonable. It would predict an infinite diffusivity in saturated

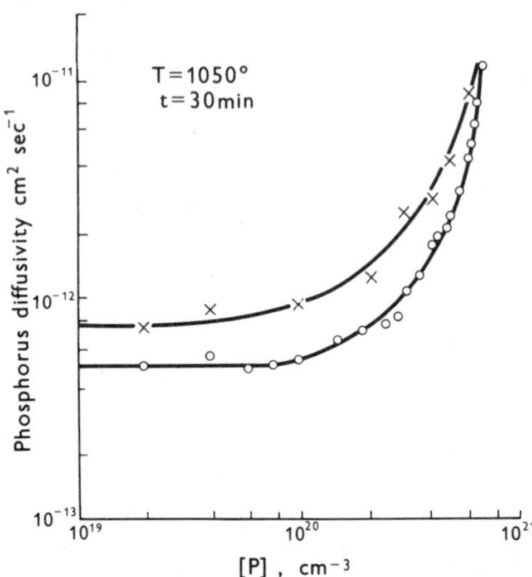

Figure 5.22. Diffusion coefficient of phosphorus in silicon at 1050° C obtained from Boltzmann-Matano analysis of radioactive-tracer data. Crosses and circles represent two different samples diffused under the same conditions. After Tannenbaum [182].

solutions, and a negative diffusivity (which cannot exist physically because it would imply the direction of irreversibility is to decrease entropy) in supersaturated systems. They also arrived at an activity coefficient of $\gamma_A = (1 - [A]/[A]_{max})^{-2}$, which is again unreasonable. The solubility limit does not imply an infinite chemical potential; rather the coexistence of two phases results from minimization of the free energy of the system as a whole. The implication that the concentration function of a chemical activity is only dependent on solubility and not on any physical model of impurity interaction is physically unreal and results from the premise made which contradicts the principle of microscopic reversibility.

Shaw [344a] has also criticized the Bakeman-Borrego model for a number of reasons. In particular, he pointed out that implicit in the Bakeman-Borrego model, $D_i(A)$ is only temperature dependent, and is independent of concentration; yet two values of $D_i(A)$ (4.8×10^{-13} and 7.4×10^{-13} cm^2/s) were required to fit Tannenbaum's results on two samples diffused under identical conditions. Shaw [344a] also questioned the validity of Tannenbaum's results, on which the Bakeman-Borrego model is based. It should also be remembered that the application of the Matano analysis of diffusivity from a diffusion profile is valid only when the diffusivity is a unique function of local concentration, and is independent of x and t (other than through $[A(x, t)]$). Furthermore, the validity of the Matano analysis also requires that the diffusion process obeys Fick's second law, from which the Boltzmann-Matano analysis is derived. Hence, the Matano analysis is invalid for any diffusion processes which involve kinetic terms and other types of interactions (such as precipitation, generation of non-equilibrium vacancy concentrations by climbing dislocations, dissociative diffusion, trapping and complex formations that are far from equilibrium, etc.). Since high concentration phosphorus diffusion in Si is often accompanied by the generation of dislocations and precipitation, it is not describable by the simple Fick's second law, and the Matano analysis (which Tannenbaum employed) is inapplicable.

After Tannenbaum, a large number of investigators [181, 345-349] have studied the detailed feature of phosphorus diffusion profiles. Invariably they used some form of phosphorus-oxygen compound, like P_2O_5 and $POCl_3$. These studies showed the phosphorus profiles to be much more complicated than observed by Tannenbaum. A typical example [331] is given in figure 5.23. Note in particular the kink in each profile. This phenomenon has not been completely understood. None of the models that have been suggested so far is satisfactory. One model [349a] is that a thin barrier to diffusion might have often existed on the surface of silicon, and diffusion through the thin film is a rate controlling process in the initial period. A simple analysis will show that, unless a

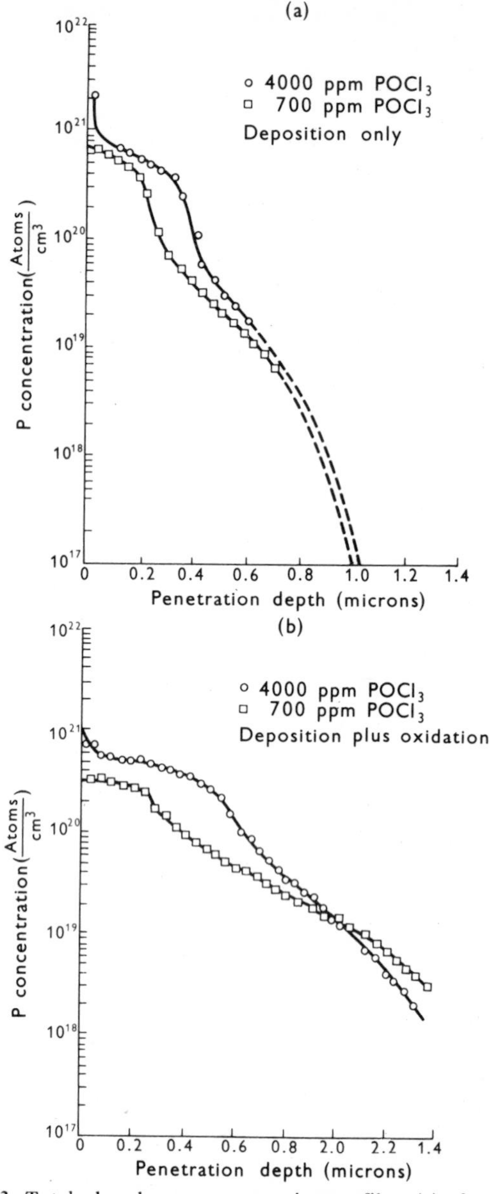

Figure 5.23. Total phosphorus concentration profiles: (a) after deposition only, (b) after both deposition and drive-in 970° C. After Duffy *et al.* [331].

sudden change in the barrier some time during the diffusion occurs, this model will not produce a kink in the profile. It can readily be shown that this model will predict a simple smooth profile both for diffusion of short duration (total diffusion time shorter than time-to-rupture of the barrier film) as well as for long duration (time-to-rupture of barrier film becomes a negligible fraction of the total diffusion time). This is in contradiction with known experimental results that the kinked characteristics become more pronounced with diffusion time, and the 'tail' part becomes proportionally longer (see figures 5.23 and 5.24). To accommodate various profiles, Makris, Ferris-Prabhu and Joshi [349a] assumed various times-to-rupture of the barrier film. It seems physically unreasonable that the barrier film could have anticipated the total diffusion time and took a time-to-rupture accordingly (e.g., an 8-min time-to-rupture for a 10-min diffusion and a 16-min time-to-rupture for a 20-min diffusion). The low surface concentration before rupture of the barrier film ($\sim 5 \times 10^{-3}$ of the

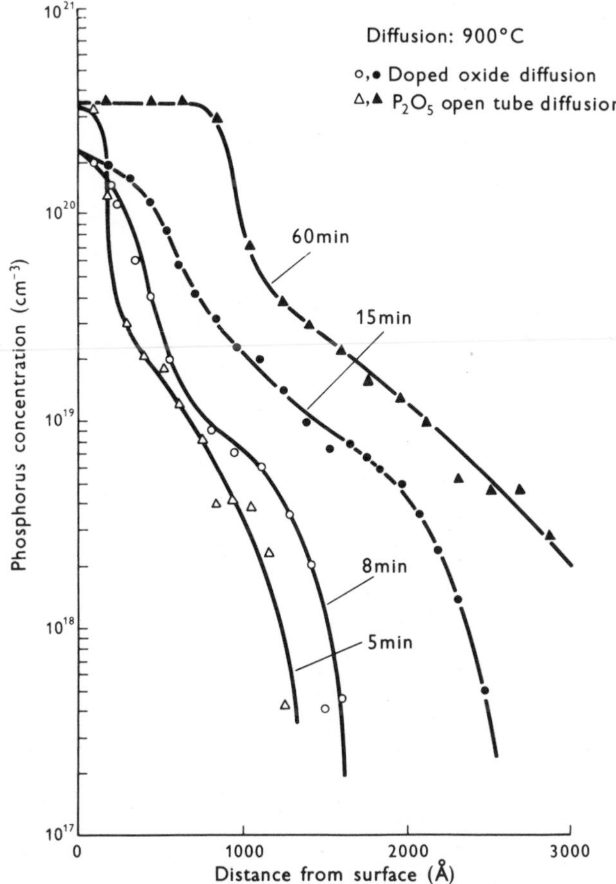

Figure 5.24. Low temperature and shallow diffusion of phosphorus into silicon at 900° C. After Sato *et al.* [351].

source concentration) as required by the model has not been observed. The second model assumes a two-stream diffusion mechanism. A characteristic feature of this mechanism is that it produces a kinked diffusion profile, as shown in figure 5.6. But this introduces another question: what are the two species of phosphorus that partake in the two-stream diffusion process? *A priori* one is inclined to rule out the existence of phosphorus interstitials. It is also implausible for phosphorus complexes such as the hypothetical SiP_4 to play the role of the slow diffusing species, because such a complex will be completely immobile. Perhaps this argument is somewhat narrow minded, because it is possible to conjecture a model with an appropriate combination of a suitable complex structure and a suitable migration mechanism so that the complex will migrate in the diamond lattice without dissocation. One such model that we can conceive is a P_2 pair which migrates by an interstitialcy mechanism, i.e., through the intermediary of a silicon self-interstitial. The third explanation invokes a phenomenon suggested elsewhere by Casey *et al.* [350, 351] in their model of the diffusion of Zn in GaAs. Their model suggests that the formation of an impurity band causes the hole activity coefficient to display a marked dependency on Zn concentration. Calculations [13] using Kane's model of the impurity band, however, did not reveal this phenomenon in Si. Results of lower temperature diffusion (900°C) by Sato *et al.* [349] appear to accentuate the kinked feature (figure 5.24). One could not attribute the apparent flattened region to the existence of a molten phase because the eutectic temperature of the Si-P system is 1131°C [353]. If we accept the results of Sato *et al.*, the flattened region cannot be attributed to precipitation either, because these profiles were obtained from differential sheet conductivity measurements. The profiles of Duffy *et al.* [331] were measured by neutron activation analysis, and exhibit similar features. Further, the more prominently kinked profiles obtained after only the deposition step showed no evidence of dislocations or precipitation [331]. This argues against the suggestion that dislocations and precipitation are the causes of the kink in the profiles. Dislocations and accompanying precipitation occur only after the drive-in cycle [331]. This observation has raised the question about the suitability of the Prussin model of misfit dislocations (section 5.3), which predicts the generation of the dislocations to be primarily governed by the solute surface concentration, and to take place immediately at the incipient diffusion. Rather, this observation is qualitatively compatible with the Shockley model, which suggests a critical total amount of solute atoms incorporated into the diffused layer as a prerequisite (section 5.5.3). The generation of dislocations and the accompanying precipitation cause a retardation in the diffusion. This is shown in figure 5.25a and 5.25b, where the diffused junction depth is plotted versus the source

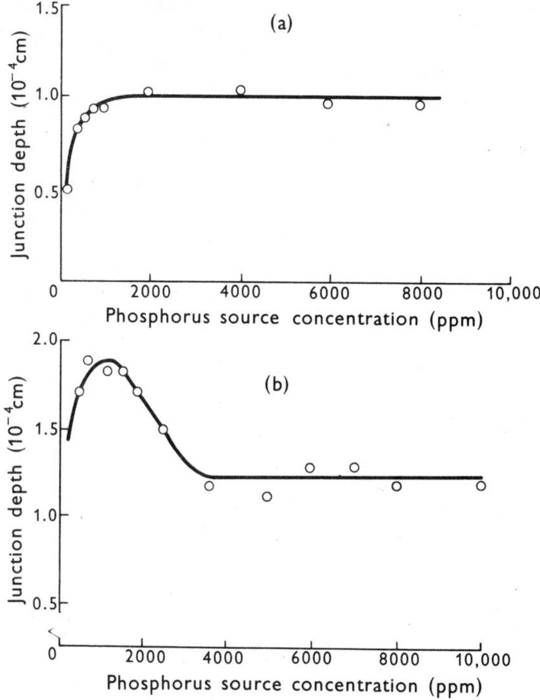

Figure 5.25a. Junction depth versus phosphorus concentration after the deposition cycle only (35 min).

Figure 5.25b. Junction depth versus phosphorus concentration after both the deposition and drive-in cycle (drive-in cycle totalled 75 min). After Duffy *et al.* [331].

concentration with otherwise similar conditions. The phosphorus diffusion is seen to increase with source concentration, probably as a result of an enhanced vacancy concentration. Then it reaches a turning point and starts to decrease with source concentration. It is in this retardation region where the phosphorus precipitation plays an important role. This retardation appears consistent with Lawrence's pin-loading experiment (section 5.5.3). It remains unexplained, however, why the junction depth levels off at higher source concentrations. Dash and Joshi [127] reported that, for shallow junction depths (~2000 Å) the phosphorus diffusion profiles are given by an erfc distribution even at a surface concentration as high as 1.2×10^{21} atoms/cm^3. Their results are seen to contradict those of Sato *et al.* [351] (figure 5.24), and are presently not understood. It is generally agreed that one cannot obtain accurate experimental profiles for shallow diffusion of the order of 0.2 μm. Dash and Joshi [127], and also Joshi and Howard [349b], maintained that their shallow diffusion profiles are erfc distributions; hence the concentration dependence of diffusivity is fictitious, and that the unmistakable deviation from the erfc distribution for deeper diffusion which they also observed is the result of dislocation generation. Such a view is inconsistent with, among others, the results of Masters and Fairfield [201].

Tsai [348] has added some further experimental observations on the kink phenomenon of phosphorus diffusion in silicon. He also gave a mathematical analysis of the phenomenon. He assumed the phosphorus diffusion profile to consist of three regions: (1) a surface region; (2) a transition region; and (3) a 'normal' diffusion region. He considered the surface region to be a mixture of silicon saturated with phosphorus and SiP (this is incorrect because SiP is a separate phase; it is possible that he actually meant some form of phosphorus complexes). A phase boundary separates the surface region and the transition region. (This assumption again is questionable unless the surface region consists entirely of the SiP phase or is a molten layer.) The phase boundary is assumed to move at a constant rate. The transition region contains two distinguishable diffusing species, for which he suggested a slow diffusant of some silicon-phosphorus compound such as silicon phosphide, and a fast diffusant of normal phosphorus atoms. The suggestion of SiP as a slow diffusant is questionable, for a diatomic SiP does not exist as a compound in the silicon matrix, and the polymeric SiP has zero diffusivity. In any case, he considered the two species to diffuse independently of each other and with constant diffusivities. This naturally yields a kink in the total concentration profile. In fact, the postulation of a separate phase in the surface region is unnecessary for the explanation of the plateau feature of the profile in the surface region. Such a feature can also be obtained by adopting a concentration dependent diffusivity for the SiP complex.

Recently, Thai [356, 356a] suggested that the ratio between the extrinsic and the intrinsic diffusivity, $D(A)/D^i(A)$, can be expressed as a product of two enhancement factors F_p and F_f, arising respectively from plastic deformation, and from the internal field. He neglected one important factor, i.e. the Fermi-level dependency of the equilibrium vacancy concentration. He followed Prussin's model of lattice stress due to a diffused layer of misfitting solute atoms, and applied Saada's theory of plasticity for the calculation of the number of vacancies generated by plastic deformation. He then gave the local vacancy concentration after plastic deformation as

$$[V] = \beta^2 Y[A]^2/6(1-\nu)Gb^3 + [V]_e, \tag{5.106}$$

where G is the shear modulus and the other terms have already been defined. The first term in the right-hand side of equation 5.106 is the total number of vacancies generated by plastic deformation. This is incorrect. For it must imply an infinite vacancy life-time. It has been shown previously by the present author [128, 376] that the excess vacancies typically have a life time of $\sim 10^{-2}$ s, compared to a typical impurity diffusion time of 10^3 s. The majority of the vacancies generated should have escaped to the crystal surface. A more efficient (and probably also more appropriate) mechanism for the generation of vacancies is dislocation climb, which will be discussed in the next section. Therefore, the enhancement factor given by Thai

$$F_p = ([A]/[A]^*)^2 + 1, \tag{5.107}$$

where $[A]^*$ is a characteristic concentration is erroneous.

5.5.5 The Emitter-Push and the Emitter-Pull Effect

In the transistor fabrication process, it has been frequently observed that the diffusion of the boron (or gallium) base in the area immediately beneath the phosphorus emitter is enhanced relative to base diffusion in the peripheral area. This results in a dip of the base-collector junction. An example is shown in the low angle leveled section view in the photomicrograph (figure 5.26). This phenomenon has been variously

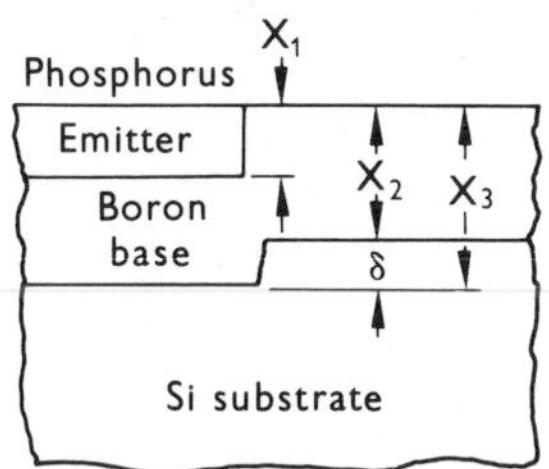

Figure 5.26. Bevel sectioned photomicrograph of a phosphorus-boron transistor structure showing the emitter-push effect. $x_1 = 1.6$ μm; $x_2 = 2.88$ μm; $x_3 = 3.20$ μm.

referred to as the emitter-dip effect [353, 354], the emitter-push effect [128], cooperative diffusion [334], the push-out effect [355], anomalous diffusion [356], run-on [357], etc. The phenomenon was first mentioned in the literature by Miller [358]. The suggestion by Sato and Arata [359] that this effect is caused by the enhanced diffusion through the dense dislocations that have penetrated deep through the collector junction is certainly incorrect. For, the emitter-push effect has been observed in most cases in samples where dislocations have not penetrated up to the emitter diffusion front. In fact, most published results show that [233, 322, 326, 360] the density of diffusion induced dislocations may be as high as 10^9 cm^{-2} in the region near the surface, but becomes immeasurably low near the emitter junction and beyond. Misfit dislocations have been observed to penetrate beyond the phosphorus diffusion front only in substrates with mechanically damaged surfaces [330]. A concise review on the subject of the emitter-push effect has been given by Willoughby [355], covering published literature up to 1967. There have been numerous speculations on the mechanism of the emitter-push effect; but few of these can be regarded as plausible. Miller [358] suggested that it might be caused by the enhanced equilibrium vacancy concentrations as a result of a raised Fermi level in the emitter region. This is obviously incorrect because the Fermi level effect is localized, and cannot extend from the emitter region to the base region. Further, only the concentration of the negatively charged vacancies, but not the concentration of neutral vacancies is enhanced by a raised Fermi level [268], and negative vacancies are not effective vehicles for boron diffusion. In fact, recent experiments by Masters and Makris [283] have shown a retardation of gallium diffusion in silicon homogeneously and heavily doped with arsenic. Similarly, the suggestion by Moore [361] that macroscopic elastic strain caused by the emitter solute might have increased the vacancy concentration in the base region is unreasonable because the elastic strain is also very much localized. It should be noted that to account for the enhanced diffusion of the base underneath the emitter relative to the base in the periphery, an increase by a factor of 10 or more in the base diffusivity is required. Gereth *et al.* [353] studied the emitter-push effect in more detail, and reported the following observations:

(1) the effect was observed only in *npn,* but not in *pnp* structures; (2) the effect was observable only when the phosphorus surface concentration exceeded the intrinsic carrier concentration, but above this, the phosphorus surface concentration had no effect on the magnitude of the push-out; (3) the effect did not depend on the emitter diffusion time; and (4) the push-out was increased by slow cooling, and by successive steps of annealing-cooling cycles. All these observations of Gereth *et al.,* disagree with subsequent observations of various other investigators. Lawrence [334], Adler *et al.* [362, 363], and ,Prahl [364] have reported the observation of the emitter-push effect in *pnp* structures, though it is not as pronounced as in *npn* structures.

Hu and Yeh [128], and Allen [357] have observed the push-out to increase with emitter surface concentration (see figure 5.27), in disagreement with the second point of Gereth *et al.* In disagreement with Gereth's third point, Hu and Yeh [128], and the Adler group [362-364] have observed the magnitude of the push-out to increase with the emitter diffusion time (see figure 5.27). Finally, the Adler group [362-364] has established by convincing arguments that the emitter-push effect does not take place to a significant extent during the cooling period. The evidence so far, particularly from the observations of Hu and Yeh [128], Allen [357], and Adler *et al.* [362-364], appears to consolidate the idea that the emitter-push effect is caused by a local excess of vacancies generated by

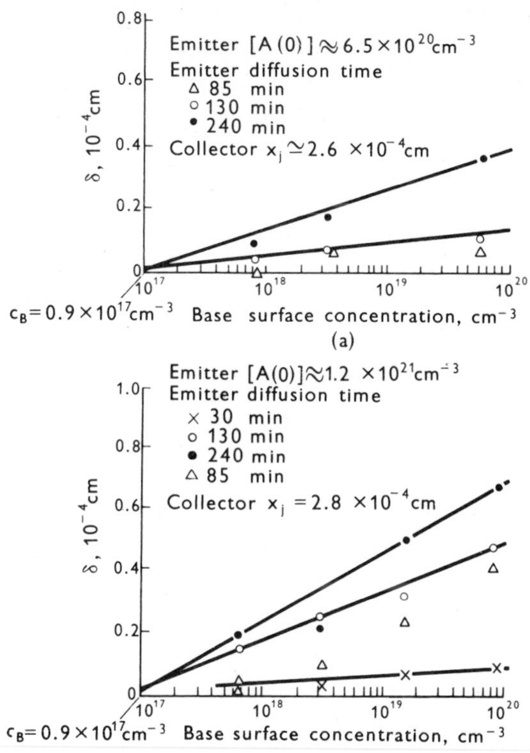

(b)

Figure 5.27a and b. Experimental observations of emitter push-out in relation to the base surface concentration, emitter surface concentration, and emitter diffusion time. The push-out δ is defined as the difference in base-collector junction depths between the area under the emitter and the peripheral area. $[A(0)]$ and c_B are the emitter surface and background doping concentrations respectively. After Hu and Yeh [128].

non-conservatively moving dislocations during the emitter diffusion. It recently has been reported by Yoshida and Kanamori [365] that the emitter-push effect was observed even if dislocations were not induced by phosphorus diffusion. They did not give experimental details.) Baruch *et al.* [239] had earlier shown that proton irradiation produced an enhanced diffusion of gallium, similar to the emitter-push effect, and suggested the cause to be a local supersaturation of vacancies produced by proton irradiation. Balluffi and Ruoff [366] invoked such an idea to explain the phenomenon of enhanced diffusion in plastically strained metals. Lawrence [334] showed that indeed plastic deformation caused by an externally applied stress can produce enhanced diffusion. Nevertheless, these observations are not an identification of the mechanism of the emitter-push effect; but they show the plausibility and consistency of the idea.

The possible role of vacancy reflux in the emitter-push effect has recently been discussed by Prussin [367]. His theory is based on the phenomenological equation of diffusion, in which fluxes of various species interfere through the off-diagonal terms. Seitz [9, 278] has shown that in a system where the diffusivity of the solute differs from that of the solvent, a vacancy flux given by $(D(A) - D_S)\,\partial[A]/\partial X$ will be induced by the solute fluxes (in addition to any that may result from the vacancy concentration gradient). Seitz called this a *chemical pump* effect. According to this model, the influx of a fast diffusing impurity such as phosphorus or boron in silicon will cause an efflux of lattice vacancies. Extending this line of reasoning, one then expects the local outward flux of vacancies, due to the in-diffusion of the emitter impurity, to also enhance the local inward flux of an existing solute such as the base impurity. In particular, the off-diagonal coefficients have usually been obtained by a reasoning that neglects the correlation effect and the vacancy-solute association [9, 278, 368-371]. This model has been one of the thoughts on the cause of the emitter-push effect for some time; but Prussin's represents the first attempt to put this idea on a formal, quantitative basis. While the approach using the phenomenological equations is fundamentally sound, a meaningful application of the phenomenological equations, however, depends on whether one can achieve a self-consistent set of phenomenological coefficients. The first attempt to achieve self-consistency in the phenomenological coefficients was made by Hu [13]. He has established a self-consistency relationship between the diagonal and the off-diagonal phenomenological coefficients for the ternary system solute-lattice-vacancy. He also showed that [13] these phenomenological coefficients, particularly the relationship among the diagonal elements, are dependent on an atomistic model of the random walk process. For a model which envisages a strong vacancy-impurity binding, the induced local outward flux of vacancies is negligibly small [279], in contradistinction to the result predicted from Seitz's model. Evidence for the existence of a strong binding between a vacancy and some Group V impurities has been reported by Watkins *et al.* [42, 43] and Hirata *et al.* [59, 60], as has been discussed previously (see Table 5.2).

Experimentally, an unequivocal answer to this problem has been provided by the recent work of Adler *et al.* [362, 363]. They grew epitaxial homojunctions on a high resistivity silicon substrate. The film doping was achieved by controlled amounts of phosphine and diborane in the gas stream. The first 20 μm of the film were heavily doped with impurity of one conductivity type (say, phosphorus). Then the heavy doping source was abruptly shut off, and approximately two more microns of epitaxial silicon were grown of lightly doped opposite conductivity (say with boron dopant), thus forming a 'base-collector' junction. Two sets of such junctions, np^+ and pn^+, were grown. The heavy doping side had a doping level of $\sim 10^{19}$ cm^{-3}, and the light doping side $\sim 10^{15}$ cm^{-3}. Then an 'emitter' of either phosphorus (2×10^{20}-1×10^{21} cm^{-3} for $[A(0)]$) or boron ($\sim 5 \times 10^{20}$ cm^{-3} for $[A(0)]$) was diffused into the structure to a junction depth of 1.0 μm (P) or 1.3 μm (B), forming a combination of four structures: n^+pn^+, p^+np^+, n^+np^+, p^+pn^+. In each of these four cases, they observed a 'pull' effect instead of a 'push' effect exerted on the collector junction by the emitter. This indicates an enhanced diffusion of the collector impurity down its own concentration gradient. Therefore, the emitter-push effect in the conventional transistor structure as well as the 'emitter-pull' effect in the unconventional Adler's structure must be interpreted in terms of an enhanced diffusivity of the base impurity (the first case) or of the 'collector' impurity (the second case). Since the direction of the vacancy flux induced by the emitter diffusion is the same (i.e., outward) in both cases, one can only conclude that the vacancy reflux (or the chemical pump effect) at most plays an insignificant role. The emitter-pull effect, as reported by Adler *et al.* is not as pronounced as the conventional emitter-push effect (see figures 5.28 and 5.29). This probably arises from a decay of vacancy supersaturation with increasing distance away from the emitter. Such a decay is also suggested to have significant effect in the case of the emitter-push [372, 373]. A more pronounced emitter pull was recently reported by Lee and Willoughby [373a], in which a p-type buried layer exhibited an enhanced diffusion even at a separation of ~ 8 μm from the emitter junction. This indicates the long range nature of the effect, consistent with the nature of vacancy dynamics.

An attempt was first made by Parker to give some quantitative arguments for the mechanism that the emitter-push effect is caused by the generation of excess vacancies by the nonconservative moving dislocations. In spite of some questionable points in his analysis [376, 377], his work has made a significant contribution to some quantitative aspects of the rate of vacancy generation, both by the dislocation glide [374] and by the dislocation climb [375]. Later, Hu [376] suggested a criterion for the condition of occurrence of the emitter-push effect.

$$\tau \langle d[V]/dt \rangle \gtrsim [V]_e, \tag{5.108}$$

where τ is the average vacancy life-time, given by $\sim x_c^2/2D(V)$ (for a definition of the penetration depth x_c, see equation 5.103 and the text); $\langle d[V]/dt \rangle$ is the average rate of vacancy generation by the non-conservatively moving dislocation. Hu and Yeh [128] conceived the dislocations to propagate mainly by climbing. The reason for favoring the climb over the glide mode of dislocation propagation is as follows: The tensile stresses induced in the diffused layer lie in the plane of the surface.

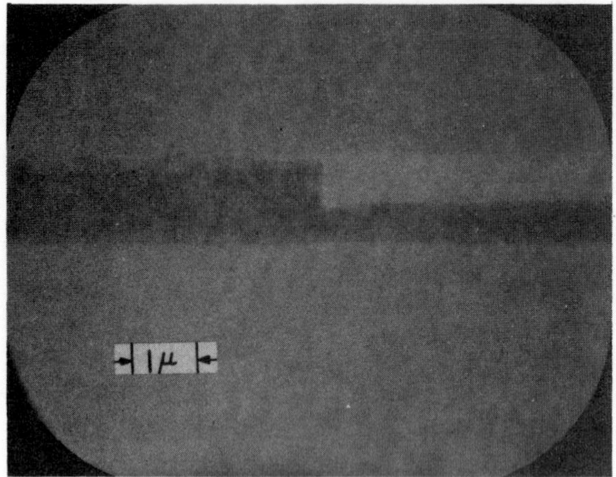

Figure 5.28. Bevel-sectioned photomicrograph of an n^+pn^+ structure (see text for explanation) showing the 'emitter-pull' effect. After Adler *et al.* [362].

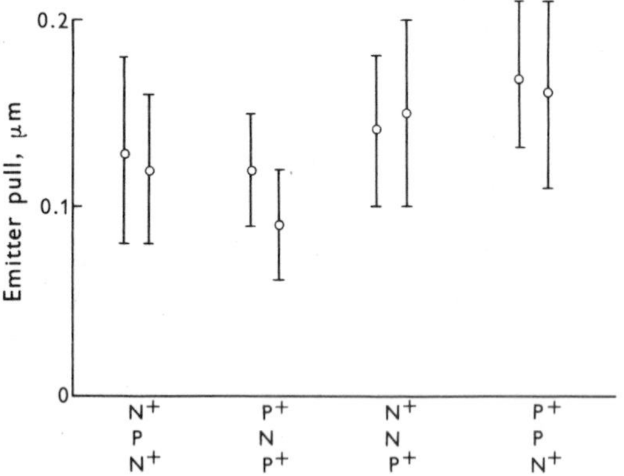

Figure 5.29. The 'emitter-pull' is relatively insensitive to particular structures (four are shown), in constradistinction to the 'emitter-push' effect. After Adler *et al.* [362].

To maximize the stress relief, dislocations with Burgers vectors lying in the plane of the surface will be preferentially induced if conditions permit. It is easily seen that the Burgers vectors $a/2\langle 1\bar{1}0\rangle$ are compatible with the surface planes $\{111\}$, $\{110\}$, and $\{100\}$. Since Burgers vectors lying in the surface plane have zero component in the normal to the plane, such dislocations cannot glide down from the surface into the interior of crystal; they can only propagate inward by climb. In fact, this has been observed experimentally [326], at least inasmuch as $\langle 111\rangle$ and $\langle 110\rangle$ substrates are concerned. There have been suggestions [326, 328] that the Burgers vector in the surface plane $\{001\}$ can split up according to

$a/2[110] \rightarrow a/2[101] + a/2[01\bar{1}]$. The latter two are not in the surface plane, and can hence glide inward. We feel such evidence is not sufficiently strong to suggest glide as the predominant mode. The energy required for splitting has not been estimated; but we feel it is not likely to be small, and is probably of the order of the energy of an additional dislocation. Dense dislocation networks and high concentrations may tend to enhance dislocation climb, for the strong Coulombic and elastic interactions between the impurity atoms and the dislocations enhance the climbing dislocations. For the climb mode of dislocation movement, Parker has given [375]:

$$d[V]/dt \approx \rho v/b^2, \tag{5.109}$$

when v is the velocity of climb. Adopting Prussin's model of dislocation distribution, Hu and Yeh have shown that [128, 373]

$$\langle d[V]/dt \rangle \cong 8a_o^{-3}\beta t^{-1}\pi^{-1/2}[A(0)]\left\{1-\pi^{1/2}\ \mathrm{ierfc}(x_c/2\sqrt{D(A)t}\right.$$
$$\left. -\pi^{1/2}[A]_c x_c/[A(0)]\,2\sqrt{D(A)t}\right\}, \tag{5.110}$$

where a_o is the lattice parameter. They also obtained, from a 0th order approximation, the expression for the amount of emitter-push δ (δ is defined as the difference between the base-collector junction depths under the emitter and in the periphery) for a base impurity initially having a Gaussian distribution:

$$\delta = (32\beta/\pi^{1/2}0.73^2\, a_o^3)\ln([B(0)]/c_B)f\,\frac{[A(0)]\,D(A)D(B)t}{[S]\,D_S^i x_j} \tag{5.111}$$

where f is the quantity in the curly brackets of equation 5.110, and is between zero and unity. $[B(0)]$ and c_B are the surface base impurity and the background doping concentrations respectively, and x_j is the junction depth.

The effect of crystal orientation has been studied by Allen [357] and by Adler *et al.* [362, 363]. They all found the emitter-push effect to be less pronounced in the $\{100\}$ materials than in the $\{111\}$ materials. Since the component α of the Burgers vector does not appear in equation 5.110, a plausible explanation of the crystal orientation effect is that a small fraction of the dislocations do indeed propagate in the glide mode by a split mechanism, as suggested by Queisser [328], and Levine *et al.* [326]. Glide, through the formation of jogs, can also generate vacancies (or interstitials). Detailed discussions on the subject can be found in the work of Parker [374, 375]. However, glide is not as efficient as climb in the generation of vacancies.

To conclude this section let us examine the question of whether lattice strain in itself rather than or in addition to the strain induced nonconservative motion of dislocations has a role in the emitter-push effect. We replace equation 5.98b by

$$\frac{\partial \mu_B}{\partial x} = \frac{kT}{[B]} \frac{\partial [B]}{\partial x} + \frac{2Y}{1-\nu} \beta^2 \frac{\partial [P]}{\partial x} \tag{5.111a}$$

where P and B denote phosphorus and boron, respectively. To a good approximation, $(1/[\mathrm{B}])\partial[\mathrm{B}]/\partial x \sim (1/[\mathrm{P}])\partial[\mathrm{P}]/\partial x$. We are led to the conclusion that the enhancement of the base diffusion by lattice strain due to phosphorus is no more than 4%, which is far too small to account for the enhancement by a factor of 10 or more as observed experimentally.

5.5.6 Phenomena Associated with Boundary Conditions

The rate of diffusion of boron into silicon has been found to vary with crystal orientation [357, 378, 379, 379a-e]. Such a crystal orientation dependence of diffusion has been reported to be insignificant for phosphorus [379a, 379b] and for arsenic and antimony [379b] in silicon, while it has also been reported [379c] that similar crystal orientation dependence was observed for phosphorus in silicon. For crystals of cubic symmetry such as Si and Ge, theory predicts the diffusivity to be isotropic. The results of Kovalev *et al.*, on the crystal orientation effect are given in Table 5.15. Their results are compared to those of Wills in figure 5.30. Note that in both Wills' and Kovalev *et al.*'s experiments, the B_2O_3 glass source had been removed prior to drive-in. It appears that thermal oxidation with a moving boundary is responsible for this effect, according to the results in Table 5.15. Analyses of impurity redistribution

TABLE 5.15

Crystal orientation effect on the diffusion of boron into silicon. After Kovalev *et al.* [379]

Sample	Orientation	Diffusion conditions			Penetration depth of the *p-n* junctions, μm
		T, °C	t, h	atmosphere	
1	{111}				2.65
2	{110}	1150	1.75	Dry oxygen	2.95
3	{100}				3.40
4	{111}				2.85
5	{110}	1150	1.75	Dry oxygen	3.00
6	{100}				3.41
7	{111}				12.10
8	{110}	1200	10	Argon	12.00
9	{100}				12.07

Note. A pyrolytic layer of SiO_2 was deposited before diffusion on samples 4-9.

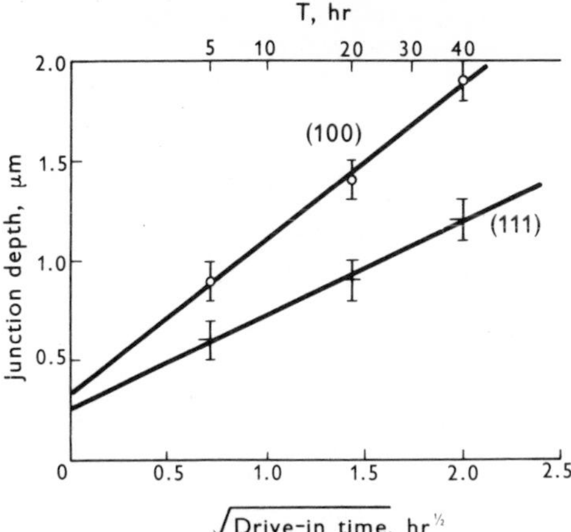

Figure 5.30. The crystal orientation dependence of diffusion of boron into silicon. Drive-in at 1050° C in oxygen. B_2O_3 glass layer had been removed prior to drive-in. After Wills [378].

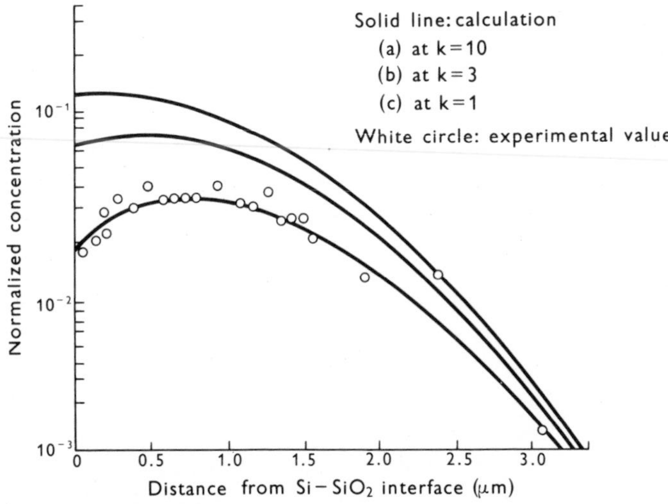

Figure 5.31. Redistribution of boron in silicon during thermal oxidation (k = segregation coefficient). Calculated concentration profile for

$$D_0 t_j = 3.6 \times 10^{-10} \text{ cm}^2$$
$$D = 2.0 \times 10^{-12} \text{ cm}^2/\text{s},$$
$$t = 3100 \text{ s and}$$
$$K = 1.02 \times 10^{-13} \text{ cm}^2/\text{s},$$

and comparison with experimental values. D_0 and D are the boron diffusivities during the 'deposition' and oxidation periods respectively; t_j is the duration of the deposition period; K is the rate constant for oxide growth. After Kato and Nishi [172].

during thermal oxidation are available in the literature [172, 380]. The phenomenon was first discussed by Atalla and Tannenbaum [381], and a summary of this phenomenon can be found in the book by Grove [382]. For impurities with segregation coefficients $>\alpha$, such as P in Si (where α is the molar volume ratio of SiO_2 to Si), the impurity will be continuously rejected into the adjacent region from the region of the semiconductor being converted into oxide. This is the so-called 'snow plow' phenomenon, which results in an enhanced movement of the impurity profile. On the other hand, for impurities having a segregation coefficient of $<\alpha$, such as B in Si, a surface depletion phenomenon will be produced, resulting in a retarded movement of the impurity profile. A typical boron profile in silicon after thermal oxidation is shown in figure 5.31.

However, the model of impurity segregation between the oxide phase and the semiconductor phase during thermal oxidation with a moving boundary, as analyzed by the above cited investigators, cannot explain the crystal orientation effect on diffusion. Thermodynamics requires that the segregation coefficient (or partition coefficient) is independent of crystal orientation for any crystal system. On the other hand, most surface and interface rate processes are dependent on the structure of the surface or the interface. Well known examples are the markedly disparate rates of oxidation, etching, and crystal growth among various crystallographic planes of a given crystal. In the simple model of Kato and Nishi [172], or of Grove *et al.* [380], the interface process, that is the impurity atoms leaving the crystalline interface plane of one phase and rearranging themselves into the interface plane of another phase, is considered or implied to occur instantaneously. There may exist a potential barrier at the interface, which regulates the impurity flux across the interface. Such an interface potential will, of course, be dependent on the crystalline orientation of the interface. Such an interface potential, for example, has been observed by Miller and Smits [383]. They found that an antimony atom in the gas phase incident on a germanium surface must surmount an energy barrier of 1.2 eV in order to enter the germanium crystalline lattice. This has been argued as a rate limiting process in the out-diffusion of Sb from Ge [199, 383]. It appears likely that a similar rate limiting interface process exists in the thermal oxidation, so that the impurity concentrations at the interface in the semiconductor and the oxide phases are not in equilibrium as given by the segregation coefficient. This will explain the crystal orientation effect.

We can treat the problem of an interface rate limiting process in a simple way as follows: We assume that the number of impurity atoms per unit surface area of a phase boundary is proportional to its bulk concentration, an assumption, for example, that is valid for a Langmuir isotherm in the limit of infinite dilution. Then we can calculate the relative concentration of impurity A in the two phases at the phase boundary by the continuity condition;

$$J(A)_{x=0} = -D(A)(\partial[A]/\partial x)_{x=0} = C([A]_{\mathrm{Si}} - k[A]_{\mathrm{ox}})_{x=0},$$

where k is the segregation coefficient, and C is some constant which is an exponential function of the interface barrier, and depends on crystal orientation. The subscripts Si and ox denote silicon and oxide phases respectively. The evaporation kinetics and its effect on the concentration profile of arsenic in germanium has also been discussed by Miller and Smits [199]. The evaporation of Sb from Ge has been discussed by Makimoto [384]. The evaporation velocity of arsenic in silicon has been given by Arai and Terunuma [385]. In all cases, the surface concentration was found to be nonzero, and in many cases, it is about ~0.3-0.5 of the bulk concentration. These results appear to be good evidence for the existence of a rate limiting process at the surface.

An alternative explanation is that excess vacancies are involved in this phenomenon of crystal orientation dependence. Katz [379d] suggested that excess vacancies are generated by nonconservative motion of dislocations which vary for different crystalline directions. He has, however, adopted the questionable work hardening model of Thai [356, 356a] in his rather strenuous analysis. Higuchi *et al.* [379e] suggested that the excess vacancies come just as a result of oxidation from an unknown mechanism and differ in various crystalline directions. He further observed that this effect is long ranging, reaching 7×10^{-4} cm (of the epitaxial layer thickness), which is compatible with the dynamical nature of vacancies.

ACKNOWLEDGEMENTS

The author wishes to thank Dr. B. J. Masters for valuable discussions on a number of important topics, and Dr. K. Weiser for acquainting him with problem areas in the theory of interstitial diffusion in Si and Ge.

The major portion of this work was written while the author was on a visiting assignment at the IBM Thomas J. Watson Research Center.

He also wishes to thank the typists for their sincere efforts.

APPENDIX 1

SOLUBILITIES OF SOME IMPURITIES IN SILICON AND GERMANIUM (figures 5.32 and 5.33)

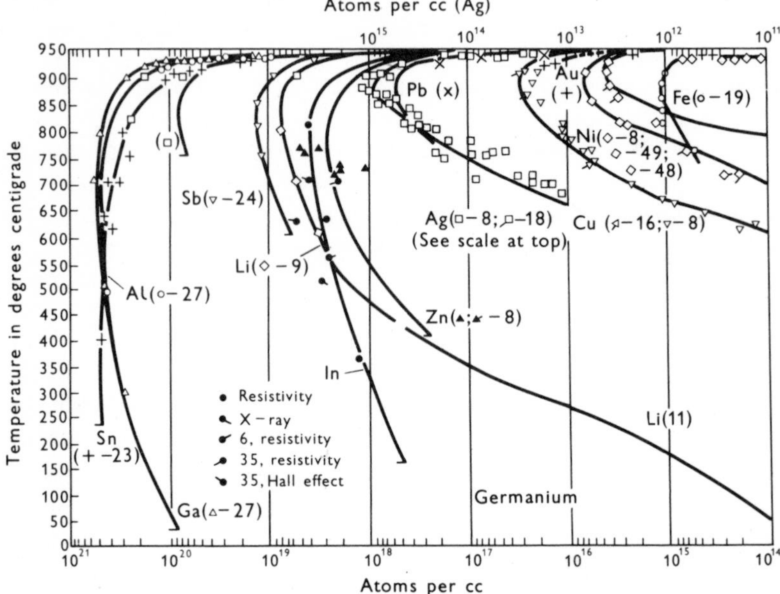

Figure 5.32. Solubility of some impurities in Ge. After Trumbore [238].

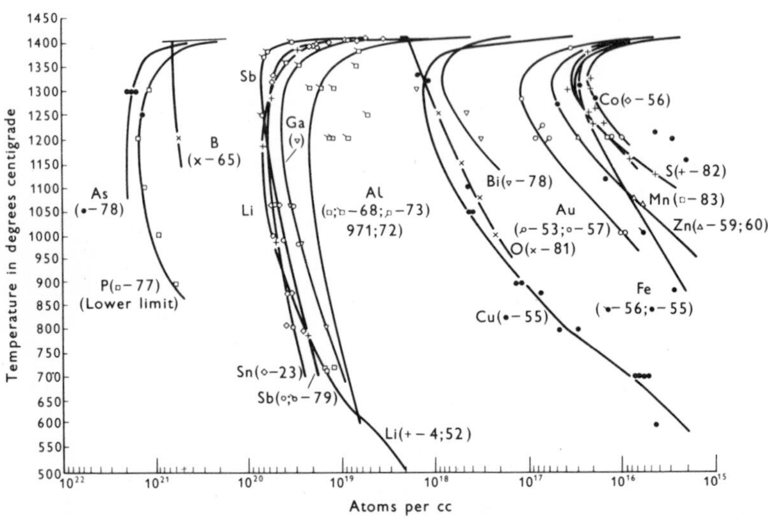

Figure 5.33. Solubility of some impurities in Si. After Trumbore [238].

APPENDIX 2

THE VALIDITY OF THE ELECTRONEUTRALITY CONDITION

In the derivation of the effective diffusivity **D**(A) as given in equation 5.67, from the phenomenological equation, an assumption has been made that the activity coefficient γ_A is a function of local intensive variables, which in the case concerned is the concentration $[A]$. That is, we have made the assumption of

$$\frac{\partial \ln \gamma_A}{\partial x} = \frac{\partial \ln \gamma_A}{\partial [A]} \frac{\partial [A]}{\partial x} \tag{A2.1}$$

(from the expression $\partial \mu_A / \partial x = \partial (kT \ln \gamma_A [A] / \partial x)$.

Relation A2.1 is valid only under the condition that $[A]$ is a well-defined function of x and γ_A is a unique function of $[A]$. In reality, n/n_i, and hence γ_A is a nonlocal function of $[A]$. Equation 5.67 is valid under the local charge neutrality assumption from which $n(x)/n_i$ is a unique function of $[A(x)]$. We now examine how good the neutrality approximation is with respect to both $n(x)/n_i$ and $\partial \ln \gamma_A / \partial \ln [A]$. Note that the term $\partial \ln \gamma_A / \partial \ln [A]$, as given by equation 5.69 in the nondegeneracy and local charge neutrality approximation, approaches unity for high $[A]$. The possibility of $\partial \ln \gamma_A / \partial \ln [A] > 1$ under degeneracy conditions has been pointed out by Shockley [386] assuming, at the same time, complete ionization.

We write the Poisson-Boltzmann equation as follows

$$\frac{d^2 u}{d\xi^2} = \left(\frac{\lambda}{L_D}\right)^2 \left(\sinh u + \frac{[A]}{2n_i}\right), \tag{A2.2}$$

where u is the normalized potential $(E_c - E_F)/kT$; $\xi = x/\lambda$, where λ is the impurity diffusion length, and L_D is the intrinsic Debye length. Assuming a semi-infinite body, and a charge balance in the entire system, one has the boundary condition

$$du/d\xi = 0 \quad \text{at} \quad x = 0 \quad \text{and} \quad x = \infty. \tag{A2.3}$$

We consider the case $[A] = [A(0)] \operatorname{erfc}(x/\lambda)$, with $[A(0)]/2n_i = 5$, as an example. The results of the numerical solution to A2.2-A2.3 are given in figures 5.34 and 5.35. It is seen from figures 5.34 and 5.35 that the local-charge-neutrality approximation is reasonably good for $\lambda/L_D \lesssim 6$, except for the region very close to the surface. It should be noted that in diffusion in Si at, e.g., 1000°C, $n_i \cong 10^{19}$ cm^{-3}; $L_D = 1.87 \times 10^{-7}$ cm. This means that the neutrality approximation is reasonable for $\lambda \lesssim 1.1 \times 10^{-6}$ cm, which prevails for typical diffusion processes except for the very beginning period.

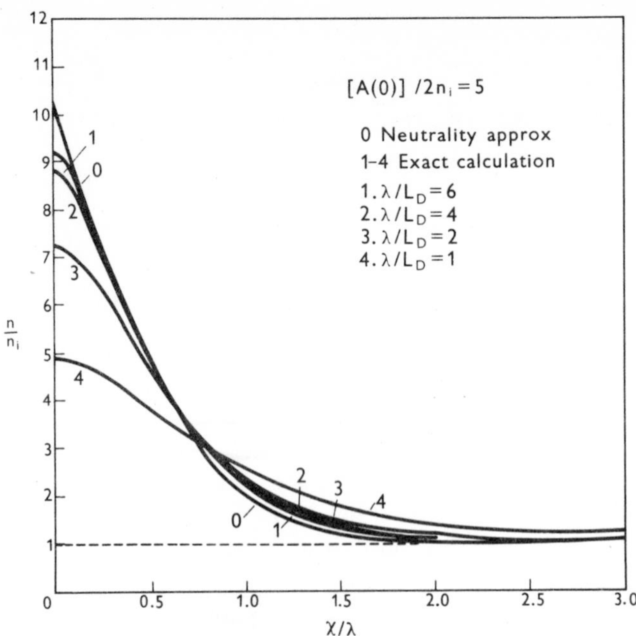

Figure 5.34. Local electron concentration $n(x)$ for a sample erfc profile at various values for the ratio of diffusion length to Debye length.

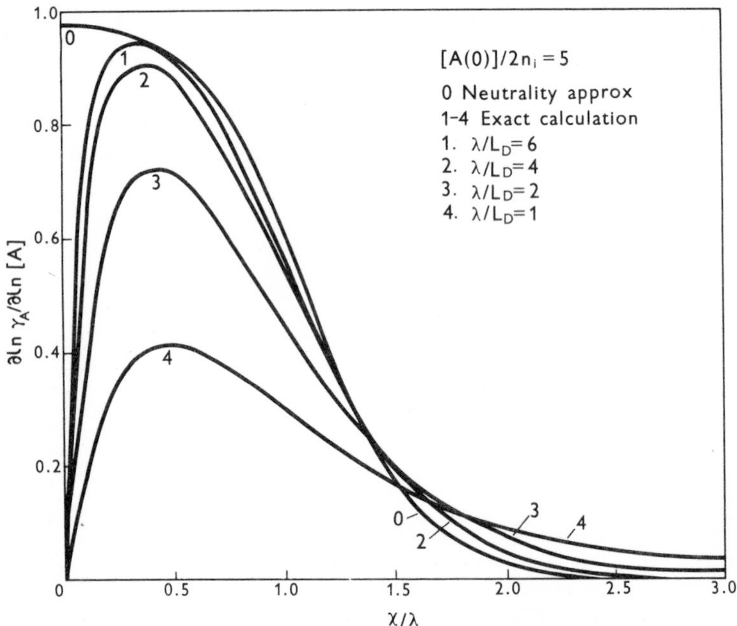

Figure 5.35. The internal field enhancement factor $\partial \ln \gamma_A/\partial \ln [A]$ for a sample erfc impurity profile for various values of the ratio of diffusion length to Debye length.

APPENDIX 3

THE POSITION OF E_F IN DEGENERATE MATERIAL

When dealing with degenerate Si, with doping in the range of 10^{20}-10^{21} cm^{-3}, a one-band approximation and the assumption of complete ionization may give a Fermi energy E_F as high as 0.35 eV above the conduction band edge. However, the one-band approximation is no longer valid for Si at $E > (E_c + 0.13)$ eV, and for Ge at $E > (E_c + 0.15)$ eV. In Ge, at $E \geqslant (E_c + 0.15)$ eV the Γ_2' band suddenly begins to contribute considerably to the total density of states, and at $E > (E_c + 0.2)$ eV the constant energy surface ellipsoid about Δ_1^m begins to contribute to the total density of states. Herman *et al.* [387], have computed the density of states in Ge using the Gilat-Raubenheimer k-space sampling method, and their results dramatically demonstrate this sudden increase of the total density of states. (Herman *et al.*, did not indicate the incipient contribution due to the energy ellipsoid about Δ_1^m; however, we believe they have actually included that in their calculations.)

There have also been some calculations of the total density of states in silicon [388], but these have not been worked out in as much detail. But from the large number of calculations of the energy bands of Si available, it appears that the one-band density of states would fail for $E > (E_c + 0.13)$/eV, where bands about $X_1(\Delta_2', \Sigma_1, \Sigma_3)$ begin to make a contribution.

(Dresselhaus and Dresselhaus [389] regard the X_1 point not to be a critical point in the joint density of states because, with spin-orbit interactions, all bands approach the X_1 point with non-zero slopes. Nonetheless, the calculations of Herman *et al.* [387] indicate the contributions of bands beginning at X_1.)

Let us now consider a two-band approximation for germanium in which the conduction band density of states is made up of states on the ellipsoidal energy surfaces about L_1 and Δ_1^m. The curvatures of various bands at the X_1 point cannot be described by simple effective masses [388], and we shall neglect the presence of the X_1 point, which is approximately 0.2 eV above the Δ_1^m point. The contribution of states in the neighborhood of the Γ_2' point is also neglected as the associated effective mass is over an order of magnitude smaller than the effective masses at L_1 and Δ_1^m [389]. Assuming the annihilation of the impurity states because of strong screening, and neglecting the formation of conduction band-tails, we may write for a monovalent donor D

$$[D^{\cdot}] = 2\left(\frac{2\pi kT}{h^2}\right)^{3/2}\left\{(m_n)_{L_1}^{3/2} F_{1/2}(\eta_c/kT) + (m_n)_{\Delta_1^m}^{3/2} F_{1/2}\left(\frac{\eta_c - \Delta\epsilon}{kT}\right)\right\}$$

where $(m_n)_{L_1}$ and $(m_n)_{\Delta_1^m}$ are the density-of-states effective masses at L_1

and Δ_1^m respectively, $\eta_c = E_F - E_c$, $\Delta\epsilon$ is the energy difference between L_1 and Δ_1^m (which we take as 0.2 eV [390]) and $F_{1/2}$ is the Fermi-Dirac integral. The density-of-states effective masses, calculated using the longitudinal and transverse effective masses both at L_1 and Δ_1^m from reference 389, have the values $(m_n)_{L_1} = 0.55\ m_e$, $(m_n)_{\Delta_1^m} = 0.75\ m_e$.† Figure 5.36 shows the results of a sample calculation of the Fermi energy

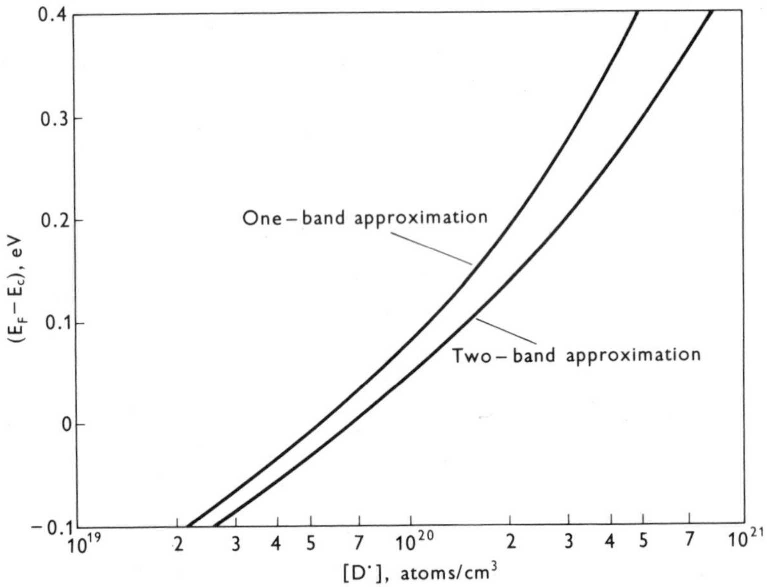

Figure 5.36. The Fermi energy in heavily doped germanium at 800° C from one-band and two-band approximations.

versus donor concentrations, in both the one-band and the two-band approximations. Note that our two-parabolic-band approximation apparently underestimates the total density of states in comparison with the results of Herman *et al.* [387]. It appears that, inasmuch as the Fermi energy is concerned, using the one-band approximation may incur more error than neglecting the conduction band-tail.

† The density of states effective masses were calculated from the formula $m_n = (N^2 m_{/\!/} m_\perp^2)^{1/3}$ where N is the number of equivalent symmetry points of concern in the Brillouin zone. $m_{/\!/}$ and $m_\perp$ are the longitudinal and transverse effective masses respectively.

REFERENCES

1. H. Reiss and C. S. Fuller, Semiconductors, p. 222 (N. B. Hannay, ed.). (Reinhold Publ. Corp., New York, 1959)
2. B. I. Boltaks, Diffusion in Semiconductors. (Academic Press, 1963)
3. A. Seeger and K. P. Chik, *Phys. Stat. sol.,* **29**, 455 (1968)
4. D. L. Kendall and D. B. DeVries, Electrochem. Soc. Meeting, New York, May (1969)
5. A. M. Smith, Fundamentals of Silicon Integrated Device Technology, Vol. 1, p. 181 ff. (R. M. Burger and R. P. Donovan, eds.). (Prentice-Hall, Inc., Englewood Cliffs, New Jersey, 1967)
6. C. M. Zener, *Acta Cryst.,* **3**, 346 (1950)
7. H. B. Huntington and F. Seitz, *Phys. Rev.,* **61**, 315 (1942)
8. H. B. Huntington and F. Seitz, *Phys. Rev.,* **76**, 1728 (1949)
9. F. Seitz, *Acta Cryst.,* **3**, 355 (1950)
10. M. F. Millea, *J. Phys. Chem. Solids,* **27**, 315 (1966)
11. F. C. Frank and D. Turnbull, *Phys. Rev.,* **104**, 617 (1956)
12. G. D. Watkins, *J. Phys. Soc. Japan,* **18**, 22 (1963)
13. S. M. Hu, *Phys. Rev.,* **180**, 773-784 (1969)
14. J. W. Corbett, Radiation Effects in Semiconductors, p. 3 (F. L. Vook, ed.). (Plenum Press, New York, 1968)
15. R. L. Aggarwal, P. Fisher, V. Mourzine and A. K. Ramda, *Phys. Rev.,* **138**, A882 (1965)
16. G. D. Watkins and F. S. Ham, *Phys. Rev.,* **B1**, 4071 (1970)
17. G. D. Watkins, Radiation Damage in Semiconductors, p. 97. (Dunod, Paris, 1965)
18. K. L. Brower, *Phys. Rev.,* **B1**, 1908 (1970)
19. B. Berry, *J. Phys. Chem. Solids,* **31**, 1827 (1970)
20. M. Balkanski and W. Nazarewica, *J. Phys. Chem. Solids,* **27**, 671 (1966)
21. K. Weiser, *Phys. Rev.,* **126**, 1427 (1962)
22. H. Paneth, *Phys. Rev.,* **80**, 708 (1950)
23. A. Seeger and M. L. Swanson, Lattice Defects in Semiconductors, p. 93 (R. R. Hasiguti, ed.). (University of Tokyo Press, Tokyo, 1968)
24. N. H. Nachtrieb and G. S. Handler, *Acta metall.,* **2**, 797 (1954)
25. K. H. Bennemann, *Phys. Rev.,* **137**, A1497 (1965)
26. R. A. Swalin, *J. Phys. Chem. Solids,* **18**, 290 (1961)
27. R. A. Swalin, see chapter 2 of this book.
28. B. J. Masters, *Solid State Comm.,* **9**, 283 (1971)
29. For some arguments on this subject, see F. Herman, J. Callaway and A. B. Lidiard, Radiation Effects in Semiconductors, p. 82 ff. (F. L. Vook, ed.). (Plenum Press, New York, 1968)
30. A. Scholz and A. Seeger, *Phys. Stat. Sol.,* **3**, 1480 (1963)
31. C. J. Huang and L. A. K. Watt, *Phys. Rev.,* **171**, 958 (1968)
32. J. Callaway and A. J. Hughes, *Phys. Rev.,* **156**, 860 (1967)
33. S. M. Hu and K. Weiser (to be published)
34. For example, see F. G. Fumi, *Phil. Mag.,* **46**, 1007 (1955)
35. C. A. Coulson and M. J. Kearsley, *Proc. Roy. Soc. (London),* **A241**, 433 (1957)
36. T. Yamaguchi, *J. Phys. Soc. Japan,* **17**, 1359 (1962); ibid., **18**, 923 (1963)
37. R. R. Hasiguti, *J. Phys. Soc. Japan,* **21**, 1927 (1966)

37a. M. James and K. Lark-Horovitz, *Z. Physik. Chem.,* **198**, 107 (1951)

37b. E. I. Blount, *J. Appl. Phys.,* **30**, 1218 (1959)

38. For example, see a discussion by A. Howie, in Direct Observation of Imperfections in Crystals, p. 269 ff. (J. B. Newkirk and J. H. Wernick, eds.). (Interscience Publ., New York, 1962)
39. G. D. Watkins and J. W. Corbett, *Phys. Rev.,* **138**, A543 (1965)
40. J. W. Corbett and G. D. Watkins, *Phys. Rev.,* **138**, A555 (1965)

41. G. D. Watkins, *Phys. Rev.*, **155**, 802 (1967)
42. G. D. Watkins and J. W. Corbett, *Phys. Rev.*, **134**, A1359 (1964)
43. E. L. Elkin and G. D. Watkins, *Phys. Rev.*, **174**, 881 (1968)
44. J. A. Baldwin, Jr., *J. Appl. Phys.*, **36**, 793 (1965); ibid., **36**, 2079 (1965)
45. D. Trueblood, *Phys. Rev.*, **161**, 828 (1967)
46. R. R. Hasiguti and K. Tanaka, Radiation Effects in Semiconductors, p. 89 (F. L. Vook, ed.). (Plenum Press, New York, 1968)
47. E. Schultz-Dubois, M. Nisenoff, H. Y. Fan and K. Lark-Horowitz, *Phys. Rev.*, **98**, 1561 (1955)
48. G. Bemski, G. Feher and E. Gere, *Bull. Am. Phys. Soc. Ser. II*, **3**, 135 (1958)
49. G. D. Watkins, J. W. Corbett and R. M. Walker, *J. Appl. Phys.*, **30**, 1198 (1959)
50. G. Bemski, *J. Appl. Phys.*, **30**, 1195 (1959)
51. G. D. Watkins and J. W. Corbett, *Phys. Rev.*, **121**, 1001 (1961)
52. J. W. Corbett, G. D. Watkins, R. M. Chrenko and R. S. McDonald, *Phys. Rev.*, **121**, 1015 (1961)
53. G. D. Watkins and J. W. Corbett, *Disc. Faraday Soc.*, **31**, 86 (1961)
54. J. W. Corbett and G. D. Watkins, *Phys. Rev. Ltrs.*, **7**, 314 (1961)
55. G. D. Watkins, Radiation Effects in Semiconductors, p. 67 (F. L. Vook, ed.). (Plenum Press, New York, 1968)
56. G. K. Wertheim, *Phys. Rev..*, **110**, 1272 (1958)
57. G. K. Wertheim, *Phys. Rev.*, **105**, 1730 (1957)
57a. Y. Inuishi, *J. Phys. Soc. Japan*, **18**, Suppl. III, 141 (1963)
58. J. W. Corbett, G. D. Watkins and R. S. McDonald, *Phys. Rev.*, **135**, A1381 (1964)
59. M. Hirata, M. Hirata, H. Saito and J. H. Crawford, Jr., *J. Appl. Phys.*, **38**, 2433 (1967)
60. M. Hirata, M. Hirata and H. Saito, *J. Appl. Phys. Japan*, **27**, 405 (1969)
61. E. Sonder and L. C. Tempelton, *J. Appl. Phys.*, **38**, 3295 (1963)
62. A. K. Ramdas and H. Y. Fan, *J. Phys. Soc. Japan*, **18**, Suppl. II, 33 (1963)
63. H. Saito, M. Hirata and T. Horiuchi, *J. Phys. Soc. Japan*, **18**, Suppl. III, 240 (1963)
64. R. N. Hall, *Phys. Rev.*, **87**, 387 (1952)
65. W. Shockley and W. T. Read, *Phys. Rev.*, **87**, 835 (1952)
66. G. K. Wertheim, *J. Appl. Phys.*, **30**, 1166 (1959)
67. R. G. Shulman, *Phys. Rev.*, **102**, 1451 (1956)
68. O. L. Curtis, Jr., J. W. Cleland, J. H. Crawford, Jr. and J. C. Pigg, *J. Appl. Phys.*, **28**, 1161 (1957)
69. O. L. Curtis, Jr., J. W. Cleland and J. H. Crawford, Jr., *J. Appl. Phys.*, **29**, 1722 (1958)
70. P. Baruch, *J. Phys. Chem. Solids*, **8**, 153 (1959)
71. L. S. Smirnov and V. S. Vavilov, *Sov. Phys.-Tech. Phys.*, **2**, 387 (1957)
72. V. S. Vavilov, A. V. Spitsyn, L. S. Smirnov and M. W. Chukichev, *Sov. Phys.-JETP*, **5**, 579 (1957)
73. O. L. Curtis, Jr., *J. Appl. Phys.*, **30**, 1174 (1959)
74. O. L. Curtis, Jr. and J. W. Cleland, *J. Appl. Phys.*, **31**, 423 (1960)
75. M. Hirata, M. Hirata and H. Saito, *J. Appl. Phys.*, **37**, 1867 (1966)
76. J. W. Corbett, Electron Radiation Damage in Semiconductors and Metals. Solid State Phys. Suppl. 7. (Academic Press, New York, 1966)
76a. A. E. Kiv and F. T. Umarova, *Sov. Phys. Semicon.*, **4**, 474 (1970)
76b. A. E. Kiv, F. T. Umarova and Z. A. Iskanderova, *Sov. Phys. Semicon.*, **4**, 1543 (1971)
76c. L. B. Redei, *Proc. Roy. Soc. London*, **A270**, 383 (1962)
76d. G. Faraci, I. F. Quercia, M. Spadoni and E. Turrisi, *Nuovo Cimento*, **B60**, 228 (1969)
77. G. K. Wertheim, *Phys. Rev.*, **115**, 568 (1959)
78. J. W. Mackay and E. E. Klontz, *J. Appl. Phys.*, **30**, 1269 (1959)

79. F. L. Vook and H. J. Stein, Radiation Effects in Semiconductors, p. 99 (F. L. Vook, ed.). (Plenum Press, New York, 1968)
80. H. Letaw, Jr., *J. Phys. Chem. Solids,* **1**, 100 (1956)
81. A. G. Tweet, *Phys. Rev.,* **106**, 221 (1957)
82. H. H. Woodbury and W. W. Tyler, *Phys. Rev.,* **105**, 84 (1957)
83. P. Penning, *Philips Res. Rept.,* **13**, 17 (1958)
84. A. G. Tweet and W. W. Tyler, *J. Appl. Phys.,* **29**, 1578 (1958)
85. A. G. Tweet, *Phys. Rev.,* **111**, 57, 67 (1958)
86. V. A. Zhidkov, *Soviet Phys. solid St.,* **3**, 335, 339 (1961)
87. A. Hiraki and T. Suita, *J. Phys. Soc. Japan,* **18**, Suppl. III, 254 (1963)
88. R. Weltzin and R. A. Swalin, *J. Phys. Soc. Japan,* **18**, Suppl. III, 136 (1963)
89. A. Hiraki, *J. Phys. Soc. Japan,* **21**, 34 (1966)
90. C. S. Fuller and K. B. Wolfstirn, *J. Phys. Chem. Solids,* **27**, 1431 (1966)
91. R. E. Whan, *Phys. Rev.,* **140**, A690 (1965)
92. S. Mayburg, *Acta metall.,* **4**, 52 (1956)
93. L. Elstner and W. Kamprath, *phys. stat. sol.,* **22**, 541 (1967)
94. M. L. Swanson, *phys. stat. sol.,* **33**, 721 (1969)
95. B. I. Boltaks and S. I. Budarina, *Soviet Phys. solid St.* (Eng. transl.), **11**, 330 (1969)
96. J. S. Prener and F. E. Williams, *J. Chem Phys.,* **35**, 1803 (1961)
97. F. L. Vook, *Phys. Rev.,* **125**, 855 (1962)
98. J. C. North and R. C. Buschert, *Phys. Rev.,* **143**, 609 (1966)
99. R. W. Keyes, *IBM J. Res. Develop.,* **4**, 266 (1961)
100. T. Figielski, *phys. stat. sol.,* **1**, 306 (1961)
101. L. S. Milevskii, *Soviet Phys. solid St.* (Eng. transl.), 4, 1376 (1963)
102. F. Seitz, *Advances in Phys.,* **1**, 43 (1952)
103. J. Weertman, *Phys. Rev.,* **107**, 1259 (1957)
104. J. W. Mitchell, *J. Appl. Phys.,* **33**, 406 (1962)
105. M. L. Joshi and S. Dash, *IBM J. Res. Develop.,* **11**, 271 (1967)
106. C. S. Fuller and K. B. Wolfstirn, *J. Phys. Chem. Solids,* **26**, 1463 (1965)
107. H. Reiss, *Phys. Rev.,* **113**, 1445 (1959)
108. H. Letaw, Jr., W. M. Portnoy and L. Slifkin, *Phys. Rev.,* **102**, 636 (1956)
109. M. W. Valenta and C. Ramasastry, *Phys. Rev.,* **106**, 73 (1957)
110. H. Widmer and G. R. Gunther-Mohr, *Helv. phys. Acta.,* **34**, 635 (1961)
111. B. J. Masters and J. M. Fairfield, *Appl. Phys. Ltrs.,* **8**, 280 (1966)
112. J. M. Fairfield and B. J. Masters, *J. Appl. Phys.,* **38**, 3148 (1967)
113. R. F. Peart, *Phys. Stat. Sol.,* **15**, K119 (1966)
114. R. N. Ghoshtagore, *Phys. Rev. Letters,* **16**, 890 (1966)
115. D. L. Kendall, Ph.D. Dissertation, Stanford University, 1965
116. R. N. Ghoshtagore, *Phys. Stat. Sol.,* **20**, K89 (1967)
117. M. H. Cohen and D. Turnbull, *J. Chem. Phys.,* **31**, 1164 (1959)
118. R. Swalin, *Acta metall.,* **7**, 736 (1959); **9**, 379 (1961)
119. R. E. Hoffman, D. Turnbull and E. W. Hart, *Acta metall.,* **3**, 417 (1958)
120. W. R. Wilcox and T. J. LaChappelle, *J. Appl. Phys.,* **35**, 240 (1964)
121. M. Yoshida and K. Saito, *J. Appl. Phys. Japan,* **6**, 573 (1967)
122. H. P. Bonzel, *Phys. Stat. Sol.,* **20**, 493 (1967)
123. K. P. Chik and A. Seeger, *Helv. Phys. Acta,* **41**, 742 (1968)
124. L. Y. Wei, *J. Phys. Chem. Solids,* **18**, 162 (1961)
125. M. D. Sturge, *Proc. Phys. Soc. Lond.,* **73**, 297 (1959)
126. J. Crank, The Mathematics of Diffusion, p. 232 ff. (Oxford University Press, London, 1956)
127. S. Dash and M. L. Joshi, AIME Conf. Defects in Electronic Materials for Devices. Boston (Aug. 24-27, 1969)
128. S. M. Hu and T. H. Yeh, *J. Appl. Phys.,* **40**, 4615 (1969)
129. S. Prussin, *J. Appl. Phys.,* **32**, 1876 (1961)
130. N. H. Nachtrieb, J. A. Weil, E. Catalano and A. W. Lawson, *J. Chem. Phys.,* **20**, 1189 (1952)

131. N. H. Nachtrieb, H. A. Resing and S. A. Rice, *J. Chem. Phys.*, **31**, 135 (1959)
132. S. A. Rice and N. H. Nachtrieb, *J. Chem. Phys.*, **31**, 139 (1959)
133. R. W. Keyes, *J. Chem. Phys.*, **29**, 467 (1958)
134. R. M. Emrick, *Phys. Rev.*, **122**, 1720 (1961)
135. J. B. Hudson and R. E. Hoffman, *Trans. AIME*, **221**, 761 (1961)
136. For example, see S. R. DeGroot and P. Mazur, Non-Equilibrium Thermodynamics. (North-Holland Publishing Co., Amsterdam, 1962)
137. R. N. Hall and J. H. Racette, *J. Appl. Phys.*, **35**, 379 (1964)
138. K. Weiser, *J. Phys. Chem. Solids*, **17**, 149 (1960)
139. For example, see L. Eriksson, J. A. Davis, N. G. E. Johansson and J. W. Mayer, *J. Appl. Phys.*, **40**, 842 (1969); K. Bjorkqvist, B. Domeij, L. Eriksson, G. Fladda, A. Fontell and J. W. Mayer, *Appl. Phys. Ltrs.*, **13**, 379 (1968)
140. K. Weiser, *J. Appl. Phys.*, **34**, 3387 (1963)
141. K. Weiser (private communication). The author is indebted to Dr. Weiser for suggesting the formula and for calculating the interstitial-to-substitutional ratio of boron in silicon.
142. For example, see Handbook of Physics and Chemistry. (Chemical Rubber Publishing Co., Cleveland)
143. V. M. Gusev and V. V. Titov, *Sov. Phys. Semicon.*, **3**, 1 (1969)
144. C. S. Fuller and J. A. Ditzenberger, *Phys. Rev.*, **91**, 193 (1953)
145. C. S. Fuller and J. C. Severiens, *Phys. Rev.*, **96**, 21 (1954)
146. T. M. Shashkov and I. P. Akimchenko, *Dokl. Akad. Nauk. SSSR*, **128**, 937 (1959)
147. E. M. Pell, *J. Phys. Chem. Solids*, **3**, 77 (1957)
148. E. M. Pell, *Phys. Rev.*, **119**, 1014, 1222 (1960)
149. B. Pratt and F. Friedman, *J. Appl. Phys.*, **37**, 1893 (1966)
150. L. Svob, *Solid-St. Electron.*, **10**, 991 (1967)
151. A. Van Wieringen and N. Warmoltiz, *Physica*, **22**, 849 (1956)
152. T. Ichimiya and A. Furuichi, *Internat. J. Appl. Radiation Isotopes (GB)*, **19**, 573 (1968)
153. L. C. Luther and W. J. Moore, *J. Chem. Phys.*, **41**, 1018 (1964)
154. B. I. Boltaks and S.-Y. Hsueh, *Soviet Phys. solid St.*, **2**, 2383 (1961)
155. J. D. Struthers, *J. Appl. Phys.*, **27**, 1560 (1956); **28**, 516 (1957)
156. B. I. Boltaks and I. Sozinov, *Zhur. Tekh. Fiz.*, **28**, 3 (1958)
157. M. Yoshida and K. Furusho, *J. Appl. Phys. Japan*, **3**, 521 (1964)
158. R. Sh. Malkovich and N. A. Alimbarashvili, *Soviet Phys. solid St.*, **4**, 1725 (1963)
159. R. C. Frank and J. E. Thomas, *Bull Am. Phys. Soc. (Ser. 2)*, **4**, 411 (1959)
160. A. A. Bugai, V. E. Kosenko and E. G. Miselyuk, *Sov. Phys. Tech. Phys.*, **2**, 183 (1957)
161. B. I. Boltaks, G. S. Kulikov and R. Sh. Malkovich, *Soviet Phys. solid St.*, **2**, 2134 (1961)
162. W. C. Dunlap, Jr., *Phys. Rev.*, **97**, 614 (1955)
163. C. B. Collins and R. O. Carlson, *Phys. Rev.*, **108**, 1409 (1957)
164. F. Van Der Maesen and J. A. Brenkman, *Philips Res. Rept.*, **9**, 225 (1954)
165. W. H. Shepherd and J. A. Turner, *J. Phys. Chem. Solids*, **23**, 1697 (1962)
166. O. N. Gromova and K. M. Khodunova, *Fiz. Khim. Obr. Mater.*, **No. 5**, 150-154 (1968)
167. C. S. Fuller and J. A. Ditzenberger, *J. Appl. Phys.*, **27**, 544 (1956)
168. A. D. Kurtz and R. Yee, *J. Appl. Phys.*, **31**, 303 (1960)
169. J. Yamaguchi, S. Horiuchi and K. Matsumura, *J. Phys. Soc. Japan*, **15**, 1541 (1960)
170. E. L. Williams, *J. Electrochem. Soc.*, **108**, 795 (1961)
171. S. Maekawa and T. Oshida, *J. Phys. Soc. Japan*, **19**, 253 (1964)
172. T. Kato and Y. Nishi, *J. Appl. Phys. Japan*, **3**, 377 (1964)
173. T. Nagano, S. Iwauchi and T. Tanaka, *J. Appl. Phys. Japan*, **7**, 1361 (1968)
174. D. N. Mikhailova, *Fiz. Tver. Tela*, **4**, 2992 (1962)

175. M. Okamura, *J. Appl. Phys. Japan,* **8**, 1440 (1969)
176. R. C. Miller and S. Savage, *J. Appl. Phys.,* **27**, 1430 (1956)
177. Y. C. Kao, *Electrochem. Tech.,* **5**, 90 (1967)
178. A. D. Kurtz and C. L. Gravel, *J. Appl. Phys.,* **29**, 1456 (1958)
179. B. I. Boltaks and T. D. Dzhafarov, *Soviet Phys. solid St.,* **5**, 2649 (1963)
180. I. M. Mackintosh, *J. Electrochem. Soc.,* **109**, 392 (1962)
181. S. Maekawa, *J. Phys. Soc. Japan,* **17**, 1592 (1962)
182. E. Tannenbaum, *Solid-St. Electron.,* **2**, 123 (1961)
183. W. J. Armstrong, *J. Electrochem. Soc.,* **109**, 1065 (1962)
184. P. S. Raju, N. E. K. Rao and E. V. K. Rao, *Indian J. Pure Appl. Phys.,* **2**, 353 (1964)
185. Y. W. Hsueh, *Electrochem. Tech.,* **6**, 361 (1968)
186. J. J. Rohan, N. E. Pickering and J. Kennedy, *J. Electrochem. Soc.,* **106**, 705 (1959)
187. D. Pommerrenig, *Acta Phys. Austriaca,* **20**, 338 (1965)
188. W. Meer and D. Pommerrenig, *Z. angew. Phys.,* **23**, 370 (1967)
189. W. C. Dunlap, Jr., *Phys. Rev.,* **94**, 1531 (1954)
190. P. V. Pavlov, *Soviet Phys. solid St.,* **8**, 2377 (1967)
191. C. S. Fuller, *Phys. Rev.,* **86**, 136 (1952)
192. W. Bösenberg, *Z. Naturf.,* **10A**, 285 (1955)
193. A. V. Sandulova, M. I. Dronyuk and V. M. Rybak, *Soviet Phys. solid St.,* **3**, 2128 (1962)
194. V. I. Tagirov and A. A. Kuliev, *Soviet Phys. solid St.,* **4**, 196 (1962)
195. W. Albers, *Solid-St. Electron.,* **2**, 85 (1961)
196. R. Wölfle and H. Dorendorf, *Solid-St. Electron.,* **5**, 98 (1962)
197. A. R. H. Niedermeyer, *Phys. Sta. Sol.,* **6**, 741 (1964)
198. W. E. Kosenko, *Izv. Akad. Nauk. SSSR,* **20**, 1526 (1956)
199. F. M. Smits and R. C. Miller, *Phys. Rev.,* **104**, 1242 (1956)
200. B. I. Boltaks and T. D. Dzhafarov, *Soviet Phys. solid St.,* **5**, 2061 (1964)
201. B. J. Masters and J. M. Fairfield, *J. Appl. Phys.,* **40**, 2390 (1969)
202. N. Isawa, *J. Appl. Phys. Japan,* **7**, 81 (1968)
203. W. Kaiser, P. H. Keck and C. F. Lange, *Phys. Rev.,* **101**, 1264 (1956)
204. W. Kaiser, *Phys. Rev.,* **105**, 1751 (1957)
205. H. J. Hrostowski and R. H. Kaiser, *Phys. Rev.,* **107**, 966 (1957)
206. W. L. Bond and W. Kaiser, *J. Phys. Chem. Solids,* **16**, 44 (1960)
207. J. W. Corbett, R. S. McDonald and G. D. Watkins, *J. Phys. Chem. Solids,* **25**, 873 (1964)
208. R. Swalin, *J. Phys. Chem. Solids,* **23**, 154 (1962)
209. E. A. Mason, *J. Chem. Phys.,* **23**, 49 (1955)
210. M. Okamura, *J. Appl. Phys. Japan,* **8**, 1440 (1969)
211. L. Bellmonte and M. H. L. Pryce, *Proc. Phys. Soc. (London),* **89**, 967, 973 (1966)
212. G. Leibfried, Encyclopedia of Physics, Vol. VII, pt. 1, p. 104 (S. Flugge, ed.). (Springer Verlag, Berlin, 1955)
213. For example, see B. R. A. Nijboer and F. W. DeWette, *Physica,* **23**, 309 (1957)
214. C. A. Wert and C. Zener, *Phys. Rev.,* **76**, 1169 (1949)
215. G. F. Nardelli and L. Reatto, *Physica,* **31**, 541 (1965)
216. R. Kubo, *J. Phys. Soc. Japan,* **12**, 570, 1203 (1957)
217. P. Gosar, *Nuovo Cimento,* **31**, 781 (1964)
217a. C. Ḅ. Collins, R. O. Carlson and C. J. Gallagher, *Phys. Rev.,* **105**, 1168 (1957)
217b. H. Reiss, *J. Chem. Phys.,* **21**, 1209 (1953)
217c. H. Reiss, C. S. Fuller and F. J. Morin, *Bell Syst. Tech. J.,* **35**, 535 (1956)
217d. A. Z. Badalov and V. B. Shuman, *Sov. Phys. Semicon.,* **3**, 1137 (1970)
218. T. I. Kucher, *Soviet Phys. solid St.,* **6**, 623 (1964)
219. P. Penning, *Philips Res. Repts.,* **14**, 337 (1959)
220. For example, see R. Bullough and R. C. Newman, *Proc. Roy. Soc.,* **266**, 198, 209 (1962)

221. F. S. Ham, *J. Appl. Phys.*, **30**, 915 (1959)
222. M. Yoshida, *J. Appl. Phys. Japan*, **8**, 1211 (1969)
223. A. C. Damask and G. J. Dienes, Point Defects in Metals, p. 81. (Gordon and Breach, New York, 1963)
224. T. R. Waite, *Phys. Rev.*, **107**, 463 (1957)
225. P. Penning, *Phys. Rev.*, **110**, 586 (1958)
226. R. C. Dorward and J. S. Kirkaldy, *Phil. Mag.*, **17**, 929 (1968)
227. A. H. Cottrell and B. A. Bilby, *Proc. Phys. Soc.*, **A62**, 49 (1949)
228. S. Harper, *Phys. Rev.*, **83**, 709 (1951)
229. F. S. Ham, *J. Phys. Chem. Solids*, **6**, 335 (1959)
230. R. Bullough and R. C. Newman, *Phil. Mag.*, **6**, 403 (1961)
231. R. Bullough and R. C. Newman, Progress in Semiconductors, Vol. 7, p. 99 ff. (A. F. Gibson and R. E. Burgess, eds.). (John Wiley, New York, 1963)
232. V. E. Kosenko, *Soviet Phys. solid St.*, **3**, 1526 (1962); [*Fiz. Tver. Tela*, **3**, 2102 (1961)]
233. H. Ruppecht and G. H. Schwuttke, *J. Appl. Phys.*, **37**, 2862 (1966)
234. H. Ghosh, S. M. Hu and A. Aldridge (unpublished)
235. D. L. Kendall, Semiconductors and Semimetals, Vol. 4, pp. 223, 243 (R. K. Willardson and A. C. Beer, eds.). (Academic Press, Inc., New York, 1968)
236. D. L. Kendall, J. A. Kanz and B. S. Reed (to be published)
237. D. L. Kendall (private communication)
238. For example, see F. A. Trumbore, *Bell Syst. Tech. J.*, **39**, 205 (1960)
239. P. Baruch, C. Constantin, J. C. Pfister and R. Saintesprit, *Disc. Faraday Soc.*, **31**, 76 (1962)
240. J. C. Pfister and P. Baruch, *J. Phys. Soc. Japan*, **18**, Suppl. 3, 251 (1963)
241. J. Stiegman, W. Shockley and F. C. Nix, *Phys. Rev.*, **56**, 13 (1939)
242. R. P. Johnson, *Phys. Rev.*, **56**, 814 (1939)
243. J. R. Manning, *Phys. Rev.*, **116**, 819 (1959)
244. S. M. Hu, *Phys. Rev.*, **177**, 1334 (1969)
245. J. Bardeen and C. Herring, Atomic Movements, p. 51. (Symposium of American Society of Metals, Cleveland, 1951)
246. S. M. Hu (unpublished)
247. K. Compaan and Y. Haven, *Disc. Faraday Soc.*, **23**, 105 (1957)
248. R. A. Swalin, *J. Appl. Phys.*, **29**, 670 (1958)
249. H. Reiss, *J. Chem. Phys.*, **21**, 1209 (1953)
250. R. L. Longini and R. F. Greene, *Phys. Rev.*, **102**, 992 (1956)
251. A. J. Rosenberg, *J. Chem. Phys.*, **33**, 665 (1960)
252. W. W. Harvey, *Phys. Rev.*, **123**, 1666 (1961)
253. W. W. Harvey, *J. Phys. Chem. Solids*, **23**, 1545 (1962)
254. R. F. Brebrick, *J. Appl. Phys.*, **33**, 422 (1962)
255. W. W. Harvey, *J. Phys. Chem. Solids*, **24**, 701 (1963)
256. S. Zaromb, *IBM J. Res. Devel.*, **1**, 57 (1957)
257. F. M. Smits, *IRE Proc.*, **46**, 1049 (1958)
258. W. Shockley, *J. Appl. Phys.*, **32**, 1402 (1961)
259. K. Lehovec and A. Slobodskoy, *Solid-St. Electron.*, **3**, 45 (1961)
260. V. V. Vas'kin, V. A. Uskov and M. Ya. Shirobokov, *Soviet Phys. solid St.*, **7**, 2703 (1966)
261. D. Shaw and A. L. J. Wells, *Brit. J. Appl. Phys.*, **17**, 999 (1966)
262. T. Klein and J. R. A. Beale, *Solid-St. Electron.*, **9**, 59 (1966)
263. N. W. Bordina, A. M. Vasilev and D. A. Popov, *Soviet Phys. solid St.*, **8**, 1791 (1967)
264. V. V. Vas'kin, V. S. Metrikin, V. A. Uskov and M. Ya. Shirobokov, *Soviet Phys. solid St.*, **8**, 2779 (1967)
265. A. D. Kurts and R. Yee, *J. Appl. Phys.*, **31**, 303 (1960)
266. C. Fa and R. Zuleeg, *Solid-St., Electron.*, **3**, 18 (1961)
267. W. Shockley and J. T. Last, *Phys. Rev.*, **107**, 392 (1957)

268. W. Shockley and J. L. Moll, *Phys. Rev.*, **119**, 1480 (1960)
269. S. M. Hu and S. Schmidt, *J. Appl. Phys.*, **39**, 4272 (1968)
270. For example, see J. S. Blakemore, Semiconduct Statistics, p. 106 ff. (Pergamon Press, Inc., New York, 1962)
271. F. Stern and R. M. Talley, *Phys. Rev.*, **100**, 1638 (1955)
272. E. O. Kane, *Phys. Rev.*, **131**, 79 (1963)
273. T. N. Morgan, *Phys. Rev.*, **139**, A343 (1965)
274. B. I. Halperin and M. Lax, *Phys. Rev.*, **148**, 722 (1966); **153**, 802 (1967)
275. V. L. Bonch-Bruyevich, The Electronic Theory of Heavily Doped Semiconductors. (Elsevier Publishing Co., Inc., New York, 1966)
276. M. B. Panish and H. C. Casey, Jr., *J. Phys. Chem. Solids*, **28**, 1673 (1967)
277. C. J. Hwang, *J. Appl. Phys.*, **41**, 2668 (1970)
278. F. Seits, *Phys. Rev.*, **74**, 1513 (1948); *Acta metall.*, **1**, 355 (1953); *J. Phys. Soc. Japan*, **10**, 679 (1955)
279. S. M. Hu and M. S. Mock, *Phys. Rev.*, **B1**, 2582 (1970)
280. F. J. Morin and J. P. Maita, *Phys. Rev.*, **94**, 1525 (1954)
281. F. J. Morin and J. P. Maita, *Phys. Rev.*, **96**, 28 (1954)
282. G. L. Hall, *J. Phys. Chem. Solids*, **3**, 210 (1957)
283. J. Makriş and B. J. Masters, *J. Appl. Phys.*, **42**, 3750 (1971)
284. C. S. Chang (private communication)
285. G. F. Foxhall and L. E. Miller, *J. Electrochem. Soc.*, **113**, 698 (1966)
286. W. Kaiser and P. H. Keck, *J. Appl. Phys.*, **28**, 882 (1957)
287. R. A. Logan and A. J. Peters, *J. Appl. Phys.*, **30**, 1627 (1959)
288. T. Arai, *J. Phys. Soc. Japan*, **18**, Suppl. II, 43 (1963)
289. W. Kaiser, H. L. Frisch and H. Reiss, *Phys. Rev.*, **112**, 1546 (1958)
289a. C. S. Fuller and R. A. Logan, *J. Appl. Phys.*, **28**, 1427 (1958)
290. C. Haas, *J. Phys. Chem. Solids*, **15**, 108 (1960)
291. P. D. Southgate, *Proc. Phys. Soc. Lond.*, **B70**, 800 (1957)
292. P. D. Southgate, *Phys. Rev.*, **110**, 855 (1958)
293. J. W. Corbett, R. S. McDonald and G. D. Watkins, *J. Phys. Chem. Solids*, **25**, 873 (1964)
294. A. R. Bean, R. C. Newman and R. S. Smith, *J. Phys. Chem. Solids*, **31**, 739 (1970)
295. R. C. Newman and J. Wakefield, *J. Phys. Chem. Solids*, **19**, 230 (1961)
296. J. A. Baker, T. N. Tucker, N. E. Moyer and R. C. Buschert, *J. Appl. Phys.*, **39**, 4365 (1968)
297. W. C. Dash, *J. Appl. Phys.*, **30**, 459 (1959)
298. C. J. Gallager, *J. Phys. Chem. Solids*, **3**, 82 (1957)
299. B. P. Konstantinov and L. A. Badenko, *Soviet Phys. solid St.*, **2**, 2400 (1961)
300. L. A. Badenko, *Soviet Phys. solid St.*, **6**, 762 (1964)
301. V. B. Fiks, *Soviet Phys. solid St.*, **1**, 1212 (1959)
302. R. J. Jaccodine, *Appl. Phys. Ltrs.*, **11**, 370 (1967)
303. M. Okamura and T. Ogawa, *J. Appl. Phys. Japan*, **4**, 823 (1965)
304. M. Okamura, *J. Appl. Phys. Japan*, **7**, 1067, 1231 (1968)
305. H. Reiss, C. S. Fuller and F. J. Morin, *Bell Syst. Tech. J.*, **35**, 535 (1956)
306. B. I. Boltaks and T. D. Dzhafarov, *Soviet Phys. solid St.*, **5**, 2061, 2649 (1964)
307. B. I. Boltaks, V. P. Grabchak and T. D. Dzhafarov, *Soviet Phys. solid St.*, **6**, 2542 (1965)
308. D. P. Kennedy, *Proc. IEEE (Ltrs)*, **57**, 1202 (1969)
309. R. B. Dingle, *Phil. Mag.*, **46**, 831 (1955)
310. For example, see H. S. Harned and B. B. Owen, The Physical Chemistry of Electrolytic Solutions, 3rd ed. (Reinhold Publ. Corp., New York, 1958)
311. S. M. Hu, *J. Appl. Phys.*, **42**, 4102 (1971)
311a. M. N. Alexander and D. F. Holcomb, *Rev. Mod. Phys.*, **40**, 815 (1968)
311b. N. Mikoshiba, *Rev. Mod. Phys.*, **40**, 833 (1968)
311c. D. Kaplan (to be published)

312. D. Petro *et al.*, IMET Rept. on 'Impurity Diffusion Processes in Semiconductor', pp. 214, 215. (Quoted in Ref. 2, 1958)
313a. S. M. Hu, IBM Internal Memo, September 13, 1968 (unpublished)
313b. D. P. Kennedy and P. C. Murley, *Proc. IEEE,* **59**, 335 (1971)
314. S. M. Hu (unpublished)
315. J. S. Sandhu and V. Lyons (unpublished); J. S. Sandhu and J. L. Reuter, *IBM J. Res. Dev.*, **15**, 464 (1971)
316. G. L. Pearson and J. Bardeen, *Phys. Rev.*, **75**, 865 (1949)
317. F. H. Horn, *Phys. Rev.*, **97**, 1521 (1955)
318. B. G. Cohen, *Solid-St. Electron.*, **10**, 33 (1967)
319. T. H. Yeh and M. L. Joshi, *J. Electrochem. Soc.*, **116**, 73 (1969)
319a. J. Friedel, *Advan. Phys.*, **3**, 446 (1954)
319b. J. Friedel, *Phil. Mag.*, **46**, 514 (1955)
319c. R. Munoz, *Phil. Mag.*, **14**, 1105 (1966)
320. H. J. Queisser, *J. Appl. Phys.*, **32**, 1776 (1961)
320a. J. H. van der Merwe, *J. Appl. Phys.*, **34**, 117, 123 (1963)
320b. W. A. Jesser, *J. Appl. Phys.*, **41**, 39 (1970)
320c. J. H. van der Merwe, *J. Appl. Phys.*, **41**, 4725 (1970)
320d. J. S. Vermaak and J. H. van der Merwe, *Phil. Mag.*, **10**, 785 (1964); **12**, 453 (1965)
320e. P. V. Pavlov and V. I. Pashkov, *Sov. Phys. Solid state,* **11**, 2501 (1970)
321. W. Shockley (quoted in Ref. 320)
322. S. A. Boctor and L. A. K. Watt, *J. Appl. Phys.*, **41**, 2844 (1970)
323. W. Czaja, *J. Appl. Phys.*, **37**, 3441 (1966)
324. G. H. Schwuttke and H. J. Queisser, *J. Appl. Phys.*, **33**, 1540 (1962)
325. J. Washburn, G. Thomas and H. J. Queisser, *J. Appl. Phys.*, **35**, 1906 (1964)
326. E. Levine, J. Washburn and G. Thomas, *J. Appl. Phys.*, **38**, 81, 87 (1967)
327. M. L. Joshi and F. Wilhelm, *J. Electrochem.*, **112**, 185 (1965)
328. H. J. Queisser, *Disc. Faraday Soc.*, **38**, 305 (1964)
329. J. E. Lawrence, *J. Electrochem. Soc.*, **113**, 819 (1966)
330. Y. Yukimoto, *J. Appl. Phys. Japan,* **8**, 568 (1969)
331. M. C. Duffy, F. Barson, J. M. Fairfield and G. H. Schwuttke, *J. Electrochem. Soc.*, **115**, 84, 1291 (1968)
332. T. J. Parker, *J. Electrochem. Soc.*, **115**, 1290 (1968)
333. C. D. Calhoun and L. A. Heldt, *Acta metall.*, **13**, 932 (1965)
334. J. E. Lawrence, *J. Appl. Phys.*, **37**, 4106 (1966)
334a. B. Ya. Lyubov, Mobility of Atoms in Crystal Lattices, p. 44 ff. (V. N. Svechnikov, ed.). (Academy of Sciences of the Ukrainian SSR, Kiev, 1965; translated by Israel Program for Scientific Translations, Jerusalem, 1970)
334b. J. J. Wortman and R. A. Evans, *J. Appl. Phys.*, **36**, 153 (1965)
335. J. E. Lawrence, *Brit. J. Appl. Phys.*, **18**, 405 (1967)
336. H. Widmer, *Phys. Rev.*, **125**, 30 (1962)
337. E. Kooi, *J. Electrochem. Soc.*, **111**, 1383 (1964)
338. P. F. Schmidt and R. Stickler, *J. Electrochem. Soc.*, **111**, 1188 (1964)
339. T. W. O'Keefe, P. F. Schmidt and R. Stickler, *J. Electrochem. Soc.*, **112**, 818 (1965)
340. C. G. Beck and R. Stickler, *J. Appl. Phys.*, **37**, 4683 (1966)
341. V. M. Al'tshuller and V. I. Pil'don, *Soc. Phys. Solid State,* **9**, 648 (1967)
342. M. Joshi, *J. Electrochem.. Soc.*, **113**, 45 (1966)
343. R. J. Jaccodine, *J. Appl. Phys.*, **39**, 3105 (1968)
344. P. E. Bakeman, Jr. and J. M. Borrego, *J. Electrochem. Soc.*, **117**, 688 (1970)
344a. D. Shaw, *J. Electrochem. Soc.*, **117**, 1587 (1970)
345. V. K. Subashiev, A. P. Landsman and A. A. Kukharskii, *Soviet Phys. solid St.*, **2**, 2406 (1961)
346. P. A. Iles and Leibenhaut, *Solid-St. Electron.*, **5**, 331 (1962)
347. B. I. Boltaks and N. N. Matveeva, *Soviet Phys. solid St.*, **4**, 444 (1962)

348. J. C. C. Tsai, *Proc. IEEE,* **57**, 1499 (1969)
349. K. Sato, Y. Araki and T. Abe (to be published)
349a. J. Makris, A. Ferris-Prabhu and M. L. Joshi, *IBM J. Res. & Dev.,* **15**, 132 (1971)
349b. M. L. Joshi and J. K. Howard, Silicon Device Processing, p. 313. (Nat'l Bur. Stand., Spec. Publ. 337, 1970)
350. H. C. Casey, Jr., M. B. Panish and L. L. Chang, *Phys. Rev.,* **162**, 660 (1967)
351. K. Sato, A. Miyazaki and T. Abe (to be published)
352. B. Giessen and R. Vogel, *Z. Metallkunde,* **5**, 174 (1959)
353. R. Gereth, P. G. G. Van Loon and W. Williams, *J. Electrochem. Soc.,* **112**, 323 (1965)
354. R. Gereth and G. H. Schwuttke, *Appl. Phys. Lett.,* **8**, 55 (1966)
355. A. F. W. Willoughby, *J. Mater. Sci.,* **3**, 89 (1968)
356. N. D. Thai, *Solid-St. Electron.,* **13**, 165 (1970)
356a. N. D. Thai, *J. Appl. Phys.,* **41**, 2859 (1970)
357. W. G. Allen (The author is grateful to Mr. Allen for communicating his results prior to publication)
358. L. E. Miller, Properties of Elemental and Compound Semiconductors, p. 303 (H. Gates, ed.). (Interscience Publishers, Inc., New York, 1960)
359. Y. Sato and H. Arata, *J. Appl. Phys. Japan,* **3**, 511 (1964)
360. M. L. Joshi, C. H. Ma and J. Makris, *J. Appl. Phys.,* **38**, 725 (1967)
361. G. E. Moore, Microelectronics, p. 282 (E. Keonjian, ed.). (McGraw-Hill, New York, 1963)
362. R. B. Adler, J. S. Brownson and H. Pearce (to be published). The author is grateful to Professor Adler for communicating his results prior to publication.
363. R. B. Adler, H. Pearce, J. Brownson and E. Prahl (to be published)
364. E. L. Prahl, M.S. Thesis, Massachusetts Institute of Technology (1970)
365. M. Yoshida and S. Kanamori, *J. Appl. Phys. Japan,* **9**, 338 (1970)
366. R. W. Balluffi and A. L. Ruoff, *J. Appl. Phys.,* **34**, 1634 (1963)
367. S. Prussin, Electrochemical Soc. Meeting, Los Angeles, May 10-15, 1970. (Extended Abstr. No. 120, p. 312)
368. F. Seitz, *J. Phys. Soc. Japan,* **10**, 679 (1955)
369. A. D. LeClaire, *Phil. Mag.,* **3**, 921 (1958)
370. J. P. Borel, *Phys. Stat. Sol.,* **13**, 3 (1966)
371. S. Prussin, *Scripta Met.,* **3**, 827 (1969)
372. D. B. Lee and A. F. W. Willoughby, *J. Appl. Phys.,* **42**, 2576 (1971)
373. S. M. Hu and T. H. Yeh, *J. Appl. Phys.,* **42**, 2153 (1971)
373a. D. B. Lee and A. F. W. Willoughby, *J. Appl. Phys.,* **43**, 245 (1972)
374. T. J. Parker, *J. Appl. Phys.,* **38**, 3471, 3475 (1967)
375. T. J. Parker, *J. Appl. Phys.,* **39**, 2043 (1968)
376. S. M. Hu, *J. Appl. Phys.,* **40**, 4684 (1969)
377. T. J. Parker, *J. Appl. Phys.,* **41**, 424 (1970)
378. G. N. Wills, *Solid-St. Electron.,* **12**, 133 (1969)
379. R. A. Kovalev, V. B. Bernikov, Yu. I. Pashintsev and V. A. Marasanov, *Soviet Phys. solid St.,* **11**, 1571 (1970)
379a. K. E. Bean and P. S. Gleim, *Proc. IEEE,* **57**, 1469 (1969)
379b. T. C. Chan and C. C. Mai, *Proc. IEEE,* **58**, 588 (1970)
379c. M. Okamura, *J. Appl. Phys. Japan,* **9**, 849 (1970)
379d. L. E. Katz, Silicon Device Processing, p. 192. (Natl. Bur. Stand. Spec. Publ. 337, 1970)
379e. H. Higuchi, M. Maki and Y. Takano, Electrochem. Soc. Meeting, Washington, D.C., May 9-13, 1971. (Extended Abstr. No. 78)
380. A. S. Grove, O. Leistiko and C. T. Sah, *J. Appl. Phys.,* **35**, 2695 (1964)
381. M. M. Atalla and E. Tannenbaum, *Bell Syst. Tech. J.,* **39**, 933 (1960)
382. A. S. Grove, Physics and Technology of Semiconductor Devices, pp. 69-75. (John Wiley, Inc., New York, 1967)

383. R. C. Miller and F. M. Smits, *Phys. Rev.*, **107**, 65 (1957)
384. T. Makimoto, *J. Appl. Phys. Japan*, **4**, 487 (1967)
385. E. Arai and Y. Terunuma, *J. Appl. Phys. Japan*, **9**, 410 (1970)
386. W. Shockley, *J. Appl. Phys.*, **32**, 1402 (1961)
387. F. Herman, R. L. Kortum, C. D. Kuglin and J. P. VanDyke, Methods in Computational Physics, Vol. 8, p. 204 (S. Fernbach and M. Rotenberg, eds.). (Academic Press, Inc., New York, 1968)
388. For example, see E. O. Kane, *Phys. Rev.*, **146**, 558 (1966); D. Brust, *Phys. Rev.*, **139**, A489 (1965)
389. G. Dresselhaus and M. S. Dresselhaus, *Phys. Rev.*, **160**, 649 (1967)
390. R. Braustein, A. R. Moor and F. Herman, *Phys. Rev.*, **109**, 695 (1958)

Note added in proof: (p. 279) More recent analyses (1, 2) of the correlation factor of impurity diffusion in diamond lattice improve over that of Hu [244].

1. H. Bakker and H. V. M. Mirani (to be published in *Z. Naturforsch*, 1972)
2. J. R. Manning (to be published in *Phys. Rev.*, 1972)

DIFFUSION IN THE III-V COMPOUND SEMICONDUCTORS 6

H. C. CASEY, Jr.

CONTENTS

6.1 INTRODUCTION

6.1.1 The III-V Compound Semiconductors and Their Applications

Of the various compound semiconductors, the III-V compounds have properties most similar to the group IV elemental semiconductors. Like Si and Ge, the III-V compounds may readily be doped as *n*- or *p*-type to form *p-n* junctions, the most useful and widespread application of semiconductors. They are 1-to-1 chemical compounds of the group III elements B, Al, Ga, and In with the group V elements N, P, As, and Sb. The III-V compounds are tetrahedrally coordinated, and the majority crystallize in the zinc-blende structure illustrated in figure 6.1 for GaAs.

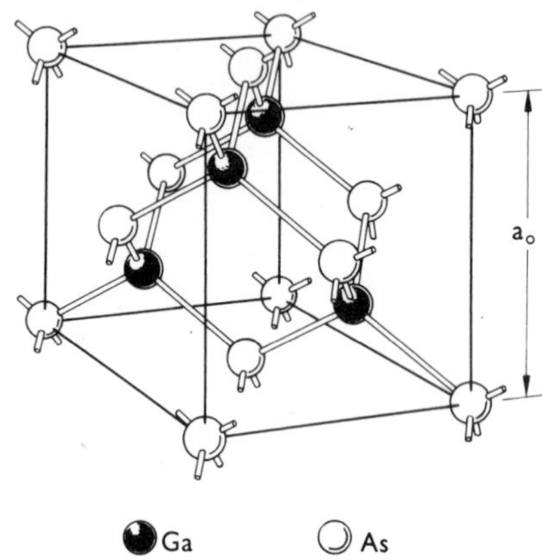

Figure 6.1. The zinc-blende structure for GaAs. The lattice constant is a_o.

The zinc-blende structure is the diamond lattice of Si or Ge, but with group III and V atoms occupying adjacent lattice sites. Although the diamond and zinc-blende structures are similar, the differences in lattice constant, the presence or absence of *d*-shell electrons, and the ionicity of the III-V compounds result in significant differences in the band structure [1]. The varied band structures and large range of energy gaps possible with the III-V compounds have led to many potential applications. The crystal structure, lattice constant, type of energy gap, and the room temperature energy gap are summarized in Table 6.1 for the elemental semiconductors Si and Ge and the III-V compound semiconductors.

The greatest impact of the III-V compound semiconductors has been not in the areas dominated by Si and Ge, such as the bipolar transistor, but in applications that depend on unique properties of these compounds. One example is the bulk-effect device, called the Gunn oscillator, which generates microwave oscillations when the applied d.c. field exceeds a critical threshold. This class of device does not require a *p-n* junction, but depends on the field-induced transfer of electrons from the high-mobility direct conduction-band minimum to the low-mobility indirect minimum such as that found in GaAs and the other direct-gap semiconductors. Most

TABLE 6.1

Crystal structure and energy gaps of the elemental semiconductors Si and Ge and the III-V compound semiconductors[a]

Crystal	Type	Lattice constant, Å	Type of energy gap	300°K energy gap, eV
Si	Diamond	5.430951 ± 0.000005	indirect	1.120
Ge	Diamond	5.646133 ± 0.000010	indirect	0.663
BN	Zinc blende	3.615 ± 0.001	indirect	>5?
BP	Zinc blende	4.538	indirect	2.0
BAs	Zinc blende	4.7778	indirect	0.85 cal.
BSb	No compound	–	–	–
AlN	Wurtzite	$a = 3.111$ $c = 4.978$	indirect	5.9
AlP	Zinc blende	5.451	indirect	2.45
AlAs	Zinc blende	5.6622	indirect	2.16
AlSb	Zinc blende	6.1355 ± 0.0001	indirect	1.5
GaN	Wurtzite	$a = 3.189$ $c = 5.185$	direct	3.7
GaP	Zinc blende	5.45117	indirect	2.261
GaAs	Zinc blende	5.65321 ± 0.0003	direct	1.435
GaSb	Zinc blende	6.09593 ± 0.00004	direct	0.72
InN	Wurtzite	$a = 3.533$ $c = 5.693$	direct	2.4?
InP	Zinc blende	5.86875 ± 0.0001	direct	1.351
InAs	Zinc blende	6.0584 ± 0.0001	direct	0.35
InSb	Zinc blende	6.47937 ± 0.00003	direct	0.180

[a] These values were compiled by H. C. Casey, Jr. and F. A. Trumbore in a review paper published in the *Journal of Material Science and Engineering*, 6, 69 (1970). Further description and the original references are given in that paper.

? These experimental values are uncertain at the present time.

cal. Calculated value; no experimental value is presently available.

of the other commercial applications utilize some form of a *p-n* junction. The impact-ionization avalanche transit-time (IMPATT) diode is a reverse-biased *p-n* junction or Schottky barrier that generates microwave oscillations. Although most work has been on Si and Ge, Zn-diffused GaAs *p-n* junctions have shown considerable promise [2]. Both the Gunn oscillator and the IMPATT diode use lightly doped *n*-type layers grown epitaxially on heavily doped substrates, and the outdiffusion of impurities into the layers during growth is of considerable concern [3-5].

One rapidly evolving technology where Si, Ge, and other semiconductors are unable to compete with the III-V compounds is electroluminescence. Electroluminescence is the emission (visible and near infrared) associated with the application of a small d.c. voltage to a *p-n* junction. In 1962, the observation of high quantum efficiencies in the infrared for GaAs [6, 7], and in the visible for GaP [8], and the achievement of laser action in GaAs [9-11] resulted in considerable research and development activity in the III-V compounds. In the preparation of *p-n* junctions, whether by diffusion from the vapor phase, liquid- or vapor-phase epitaxy, or growth from solution, it is necessary to consider the behavior of impurities in the lattice, especially the movement in the solid by diffusion. As discussed in section 6.4.2, the diffusion rate of an impurity in the solid may have a strong influence upon both the surface concentration for vapor-phase diffusion and the amount of impurity incorporated into the solid during crystal growth.

6.1.2 Diffusion Behavior of Impurities in the III-V Compound Semiconductors

Interstitial-Substitutional Diffusion

Attention was focused on the noncomplementary error function shape of the Zn diffusion profiles in GaAs by Cunnell and Gooch in 1960 [12]. That paper gave a detailed account of the Zn-concentration-versus-distance curves described in a brief note by Allen and Cunnell [13]. Determination of the diffusion coefficient from these diffusion profiles by the Boltzmann-Matano method [14] revealed a concentration-dependent diffusion coefficient. Most explanations of the Zn diffusion in GaAs have been based on Longini's [15] suggestion of an interstitial-substitutional model in which the more rapidly diffusing interstitial donor dominates the diffusion process at high-substitutional concentrations. Thus, a diffusion model that can be associated with the III-V compound semiconductors is the diffusion of an impurity by the interstitial mode, and equilibration of the mobile interstitial impurity with vacancies to form an immobile substitutional impurity. There are cases other than that of Zn in GaAs where the interstitial-substitutional model can be shown to be applicable. Unfortunately, insufficient information prevents analysis of most of the experimental data on impurity diffusion in the III-V compounds. Other than for Zn in various III-V compounds, the diffusion results are generally given in terms of the usual Arrhenius expression, $D = D_0 \exp(-Q/kT)$. Whether diffusion in these cases also proceeds by the interstitial-substitutional mode, by substitutional diffusion through divacancies, or by some other mechanism is presently not clear. No matter what diffusion mechanism dominates, it is necessary to consider the diffusion in terms of the impurity-Ga-As ternary system.

Ternary Considerations

An unpublished report by Allen and Pearson [16] emphasized that an understanding of diffusion requires relating the diffusion conditions and system compositions to the ternary phase diagram. The relationship of the ternary phase diagram to the diffusion analysis was illustrated for Zn in GaAs by Casey and Panish [17]. If a diffusion system is to be interpreted in terms of a fundamental diffusion model, the diffusion source must reach an equilibrium composition as defined by the components in the system and the temperature. For example, in the Zn-GaAs system the *starting* source material for Zn diffusion in sealed ampoules has generally been either elemental Zn, dilute solutions of Zn in Ga, or various combinations of Zn and As. The *actual* diffusion source in any ternary system must either consist of an equilibrium ternary mixture or be approaching this mixture in composition. The composition of the ternary mixture is determined by the total amount of each element present, the temperature, and the volume of the system. The actual diffusion source composition will determine the partial pressures of the impurity, Ga, and As. These partial pressures control the impurity surface concentration and the vacancy concentrations in the III-V compound semiconductor. Any diffusion mechanism for impurities that reside substitutionally in the lattice depends upon the vacancy concentration and hence depends on the actual source composition. Empirical behavior can be obtained by simply introducing a known amount of the elemental impurity in the diffusion system, but the assignment of the actual source conditions and composition is usually not possible. Thus, neglect of ternary considerations in diffusion experiments has prevented the development of fundamental explanations for the diffusion of most impurities in the III-V compound semiconductors.

6.1.3 The Scope and Content of This Chapter

This chapter will first consider lattice vacancies and self-diffusion. Treatment of vacancies requires consideration of the binary phase diagrams and the dissociation pressures. Most of the discussion on impurity diffusion in the III-V compound semiconductors will be on the diffusion of Zn in GaAs. Consideration of Zn in GaAs permits a fundamental and quantitative presentation of the interstitial-substitutional diffusion mechanism, which can be substantiated by comparison with experimental results. The results derived for Zn in GaAs are directly applicable to Zn in GaP, and Zn diffusion in the other III-V compounds also shows the strong concentration-dependence characteristic of interstitial-substitutional diffusion. Zinc in GaAs is certainly not the only system of interest; however, an in-depth development of this system permits presentation of the many concepts that are necessary for an understanding of diffusion in the III-V compound semiconductors.

After the development of the interstitial-substitutional diffusion model for Zn in GaAs, Zn diffusion in the other III-V binary compounds and the ternary compounds $GaAs_{1-x}P_x$ and $InAs_{1-x}P_x$ will be considered. The empirical results of donor diffusion will also be presented. To demonstrate impurity diffusion in a commercial application, the techniques used in preparation of GaAs and $GaAs_{1-x}P_x$ electroluminescent devices will be briefly described. And finally, for completeness, the experimental diffusion coefficients for impurity and self-diffusion in the III-V compounds are compiled in section 6.6. It should be noted that diffusion coefficients for the III-V compound semiconductors were summarized by Kendall [18] and Yarbrough [19] in 1968.

6.2 LATTICE VACANCIES AND SELF-DIFFUSION

6.2.1 General Comments

Although the formation and the properties of vacancies can be demonstrated to have a pronounced effect on the incorporation and diffusion of impurities in the III-V compound semiconductors, very little data on vacancies have been reported. The related problem of self-diffusion also has not received much attention. Many questions arise concerning vacancies and self-diffusion. Not only is it difficult to estimate the concentration of these atomic defects, but it is not clear whether or not vacancies can result in energy levels within the energy gap and behave as ionized donors or acceptors. It would also be useful to know whether the group III or group V vacancies tend to associate and form vacancy pairs on adjacent lattice sites (divacancies), vacancy clusters, or even more complex associations. However, it is clear that deviations from stoichiometry are small for these compounds.

It is difficult to speculate on the mechanism of self-diffusion when so little is known about the vacancy behavior. Unless an atom can reside in a vacant nearest-neighbor site as an antistructure defect (see figure 6.1), self-diffusion must occur through divacancies or by the substitutional atom becoming interstitial and moving to a vacancy on its own sublattice. Also, self-diffusion by a vacancy mechanism in its own sublattice, such as that demonstrated for Ag in AgCl (reference 20), is possible. In this case, the atom simply moves through an interstitial site in reaching the adjacent sublattice site. From the self-diffusion data that have been reported, one fact has been demonstrated: Self-diffusion in the III-V compound semiconductors is extremely slow. Some results are briefly described below in section 6.2.3.

Because there is so much uncertainty about the nature of vacancies in the III-V compound semiconductors, the discussion in the next section will concentrate on the effect of the partial pressure of the volatile group V elements on the relative vacancy concentration. In addition, the expressions utilized for vacancies in the discussion of Zn diffusion will be presented. Gallium arsenide is taken as the representative III-V compound semiconductor.

6.2.2 Vacancy Equilibria

Solid-Liquid-Vapor Equilibria

Control of the vacancy concentration requires consideration of the liquidus curve and the equilibrium pressures of the various vapor species along the liquidus curve. The relationships among the nonstoichiometric compound, the liquid, and the vapor are illustrated in figures 6.2 and 6.3 for GaAs. The GaAs liquidus curve in figure 6.2 was taken from Thurmond [21] whose analysis was based on the data of Hall [22] and Köster and Thoma [23]. The melting point was taken as 1238°C [23]. Partial pressures of the vapor species for GaAs that are shown in figure 6.3 were taken from Arthur [24]. These results are based on Arthur's mass spectrometric measurements at low pressures, the total high-pressure data of Richman [25], and the analysis of Thurmond [21]. The total pressure

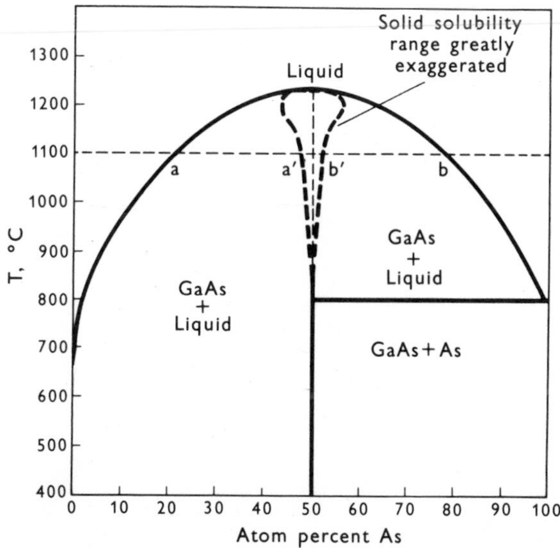

Figure 6.2. The GaAs liquidus curve was adapted from reference 21. The points *a*, *a'* represent the equilibrium on the Ga-rich side while *b*, *b'* represent the As-rich side. The corresponding pressures of the vapor species are given in figure 6.3.

at the melting point is 0.976 atm (reference 24). For the partial pressures of the GaP vapor species, there are considerably fewer experimental data. The partial pressures for GaP shown in figure 6.4 were obtained by Thurmond [21] on the basis of Richman's [25] high-temperature measurements and Johnston's [26] low-temperature decomposition pressures. The results of Nygren *et al.* [27] gave a total pressure of 32 atm at the melting point of GaP. Panish and Arthur [28] determined the phosphorus partial pressure for InP along the In-rich side as shown in figure 6.5. A summary of the total vapor pressures and phase diagrams of several III-V systems has been given by Sirota [29]. Table 6.2 lists the melting points and total pressures at those temperatures for the III-V compounds. The compounds with unreported melting points are estimated to have very high melting points, and have been prepared by chemical reaction at temperatures below their melting points.

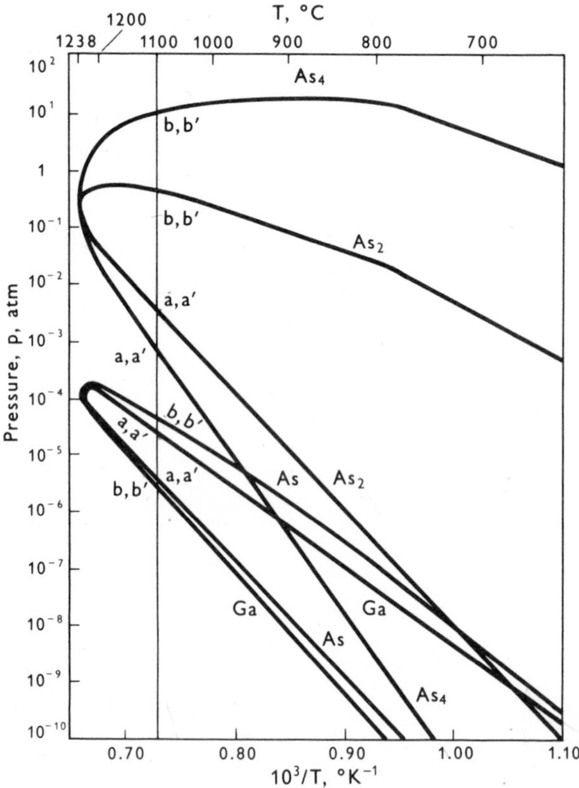

Figure 6.3. Partial pressures of the Ga and As vapor species in equilibrium with the liquidus of figure 6.2 (reference 24).

TABLE 6.2

Melting points and total pressures of the III-V compounds

Compound	Melting point, °C	References	Total pressure, atm	References
BN	≈3000	a	*	a
BP	≈3000	b	†	c
BAs	Unknown		‡	c
BSb	No compound			
AlN	>2400	d	§	d
AlP	2530 ± 50	e		
AlAs	1740 ± 20	f	1.0	f
AlSb	1060	g	‖	
GaN	1500	d	¶	d
GaP	1465	h	32	i
GaAs	1238	h	0.976	j
GaSb	712	k	‖	
InN	1200	d	**	d
InP	1068 ± 2	l	18 ± 5	l
InAs	943 ± 3	m	0.33	m
InSb	525	n	‖	

* Begins to dissociate in vacuum at 2700° C.[a]

† Above 1100° C, BP loses phosphorus and B_6P is formed.[c]

‡ At a temperature of about 1100° C, BAs goes to a more stable lower arsenide with an orthorhombic structure.[c]

§ Begins to dissociate in vacuum at 1750° C.[d]

‖ Although the pressures for the Sb compounds are not reported, they are known to be low.

¶ Begins to dissociate in vacuum at 1050° C.[d]

** Begins to dissociate in vacuum at 620° C.[d]

a. I. E. Campbell, C. F. Powell, D. H. Nowicki and B. W. Gonser, *J. Electrochem. Soc.*, **96**, 318 (1949)
b. F. V. Williams, Compound Semiconductors: Preparation of III-V Compounds, Vol. 1, p. 171. (R. K. Willardson and H. L. Goering, eds.). (Reinhold Publishing Corp., New York, 1962)
c. F. V. Williams and R. A. Ruehrwein, *J. Am. Chem. Soc.*, **82**, 1330 (1960)
d. T. Renner, *Z. Anorg. Chem.*, **298**, 22 (1959)
e. W. Kischio, *J. Inorg. Nucl. Chem.*, **27**, 750 (1965)
f. W. Kischio, *Z. Anorg. Allgem. Chem.*, **328**, 187 (1964)
g. H. Welker, *Z. Naturforsch.*, **8a**, 248 (1953)
h. C. D. Thurmond, *J. Phys. Chem. Solids*, **26**, 785 (1965)
i. S. F. Nygren, C. M. Ringel and H. W. Verleur, *J. Electrochem. Soc.*, **118**, 306 (1971)
j. J. R. Arthur, *J. Phys. Chem. Solids*, **28**, 2257 (1967)
k. N. N. Sirota, Semiconductors and Semimentals, Vol. 4, p. 60, Physics of III-V Compounds (R. K. Willardson and A. C. Beer, eds.). (Academic Press, New York, 1968)
l. M. B. Panish and J. R. Arthur, *J. Chem. Thermo.*, **2**, 299 (1970)
m. J. van den Boomgaard and K. Schol, *Philips Res. Repts.*, **12**, 127 (1957)
n. T. S. Liu and E. A. Peretti, *Trans. Am Soc. Metals*, **44**, 539 (1952)

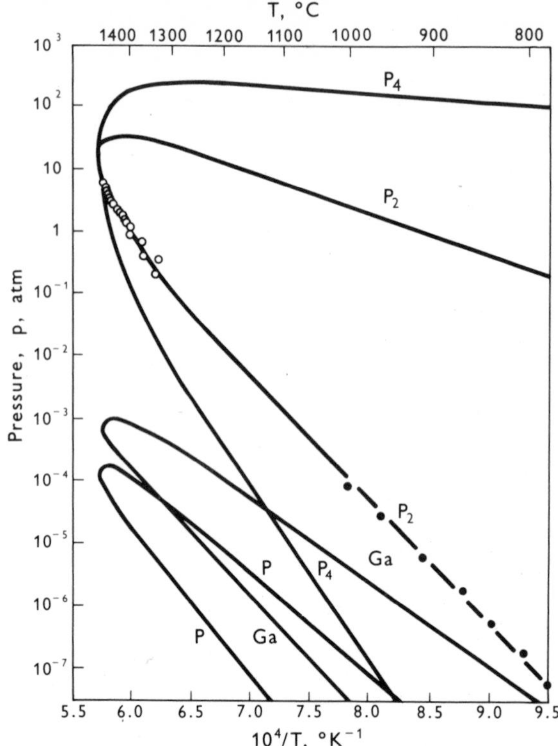

Figure 6.4. Partial pressures of the Ga and P vapor species in equilibrium with the Ga-P liquidus (reference 21).

The partial pressure curves for GaAs represent the partial pressures of each vapor species in equilibrium with the liquidus and solidus in figure 6.2. The 1100° C equilibrium points are shown in figures 6.2 and 6.3 for GaAs, Ga-As liquid and the vapor species Ga_G, As_G, $(As_2)_G$, and $(As_4)_G$.

Figure 6.3 shows that along the Ga-rich portion of the liquidus curve As_2 and As_4 partial pressures are generally greater than the partial pressure of Ga. Therefore, GaAs will decompose when heated to give a Ga-rich liquid phase and As_2 and As_4 in the vapor. However, at lower temperatures $p_{Ga} = 2p_{As_2} + 4p_{As_4}$ and evaporation will be congruent and the vapor will have the same composition as the solid phase. This temperature, called the decomposition temperature, is 637° C for GaAs [24]. Now that the solid-liquid-vapor equilibria have been illustrated, the relationship between the partial pressure and the vacancy concentration will be presented.

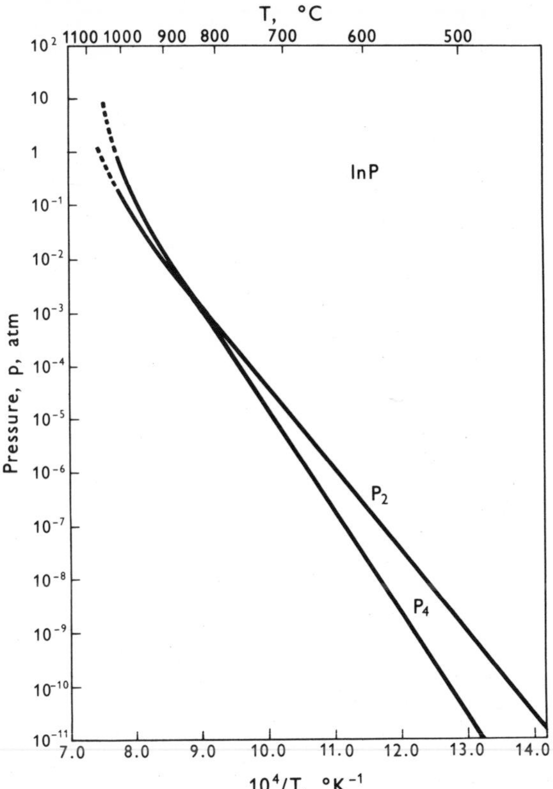

Figure 6.5. The phosphorus partial pressure in equilibrium with the liquidus in the In-P system on the In-rich side (reference 28).

Vacancy Concentration

Vacancies in the crystal lattice constitute defects in the pure crystal. Whether or not the vacancies in GaAs are ionized has not been demonstrated. Studies of the incorporation of Zn in GaAs [30] suggested that vacancy ionization may be neglected. With this assumption, the role of arsenic pressure in the control of the vacancy concentration may readily be demonstrated. From the decomposition reaction,

$$\mathrm{GaAs}_S \rightleftharpoons \mathrm{Ga}_L + \tfrac{1}{2}(\mathrm{As}_2)_G, \tag{6.1}$$

it follows that

$$\{\mathrm{Ga}_L\}\, p_{\mathrm{As}_2}^{1/2} = K_1(T), \tag{6.2}$$

where $\{\mathrm{Ga}_L\}$ is the Ga activity and $K_1(T)$ is a temperature-dependent equilibrium constant. From the reactions for the formation of As and Ga

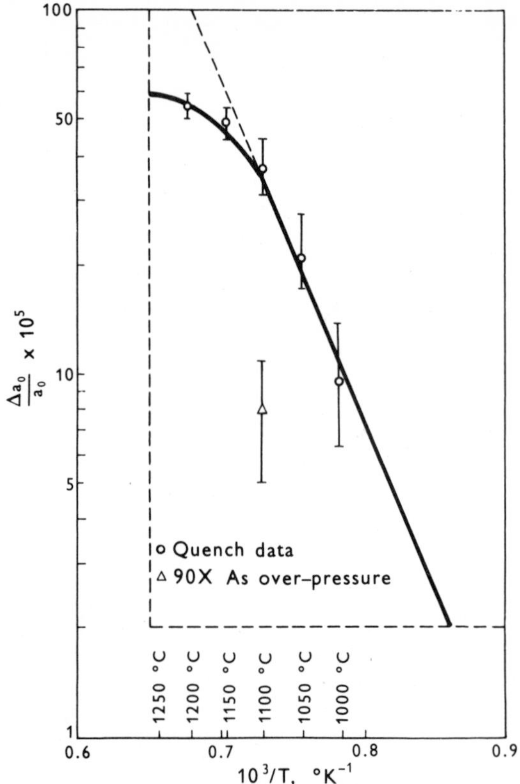

Figure 6.6. Comparison of $\Delta a_0/a_0$ from samples quenched in excess arsenic pressure with data from samples quenched in equilibrium As partial pressure over GaAs (reference 31).

vacancies,

$$\mathrm{As_{As}} \rightleftharpoons \tfrac{1}{2}(\mathrm{As_2})_G + V_{\mathrm{As}} \tag{6.3}$$

and

$$\mathrm{Ga_{Ga}} \rightleftharpoons \mathrm{Ga}_L + V_{\mathrm{Ga}}, \tag{6.4}$$

it follows that the vacancy concentrations may be expressed as

$$X(V_{\mathrm{As}}) = K_2(T)/p_{\mathrm{As_2}}^{1/2} \tag{6.5}$$

and

$$X(V_{\mathrm{Ga}}) = K_3(T)/\{\mathrm{Ga}_L\} = [K_3(T)/K_1(T)]p_{\mathrm{As_2}}^{1/2}, \tag{6.6}$$

where $X(V_{\mathrm{Ga}})$ and $X(V_{\mathrm{As}})$ are the atom fractions of Ga and As vacancies.

When equations 6.5 and 6.6 are combined, the product of the Ga and As vacancy concentrations can be shown to be a function of temperature only:

$$X(V_{Ga})X(V_{As}) = [K_2(T)K_3(T)/K_1(T)] = K_V(T). \tag{6.7}$$

Expressions similar to equations 6.5 and 6.6, which relate the As and Ga vacancy concentrations to the arsenic partial pressure, may be written with a one-quarter power dependence for the As_4 partial pressure.

The lattice constant measurements by Potts and Pearson [31] on heat-treated and quenched crystals permit identification of the primary defect species as As monovacancies, and the assignment of an absolute concentration for these As vacancies. They heated GaAs crystals in evacuated fused-silica ampoules, along with metallic As when desired, at temperatures between 1000° and 1200° C for 24 h. These samples were quenched to 0° C and the change in the bulk lattice constant a_o was determined. The results are shown in figure 6.6. It should be noted that the lattice constant becomes larger as the heat-treatment temperature increases. Also, the change in a_o at 1100° C was less if As was added to the ampoule to increase the arsenic partial pressure. To interpret these data, it is first necessary to establish whether the expansion of the lattice is due to the creation or annihilation of vacancies.

The present results indicate that for GaAs the lattice expands in the vicinity of a vacancy [31-33] rather than contracting as observed for the covalent semiconductors Si and Ge. Vook found that high-energy electron irradiation, which creates lattice vacancies and interstitials, increased the lattice constant [32]. As pointed out by Pearson *et al.* [33], it is difficult to assign the lattice strain to interstitial atoms because the zinc-blende structure is such an open lattice. In addition, Swalin's vacancy formation model [34] predicts expansion about the vacancy for materials with partial ionic bonding such as GaAs [33]. Therefore, it appears reasonable to relate an increased lattice constant to an increase in vacancy concentration. When a single defect species is dominant, the lattice constant change Δa_o may be related to vacancy concentrations as [31]

$$\Delta a_o/a_o = \alpha X(V) = X(V), \tag{6.8}$$

where a_o is the lattice constant before heat treatment, and α is the proportionality constant. Potts and Pearson estimated that an appropriate proportionality constant for the Kossel-line method of lattice constant measurement is unity, which gives the equality between $\Delta a_o/a_o$ and $X(V)$ [31]. This expression assumes that $X(V)$ before heat treatment is sufficiently small compared with the high-temperature value to permit assignment of the change in $X(V)$ as the high-temperature value.

The assignment of the dominant defect as the As monovacancy was based on the arsenic pressure dependence of $\Delta a_o/a_o$. The total vacancy concentration is given by

$$X(V_{\text{total}}) = X(V_{\text{As}}) + X(V_{\text{Ga}}) = K_2(T)/p_{\text{As}_2}^{1/2} + [K_3(T)/K_1(T)]\, p_{\text{As}_2}^{1/2}. \quad (6.9)$$

Because $\Delta a_o/a_o$ decreased for heat treatment in the presence of added As, and the change was that represented by equation 6.5 for As vacancies, the As vacancy was designated as the dominant vacancy [31]. This assignment of As as the dominant lattice vacancy is in agreement with the independent lattice constant measurements of Straumanis and Kim [35]. They measured samples prepared at 1000° C. For the Ga-rich sample, a_o = 5.65326 Å and for the As-rich sample, a_o = 5.65298 Å. Going from the Ga-rich to the As-rich sample, which is equivalent to increasing the arsenic pressure in Potts and Pearson's work, leads to a smaller lattice constant. These results by Straumanis and Kim corroborate the vacancy assignment by Potts and Pearson.

The usual thermodynamic relationships apply to the equilibrium expressed by equation 6.5, and the equilibrium constant may be written as

$$K_2(T) = \exp(\Delta S_2/k) \exp(-\Delta H_2/kT), \quad (6.10)$$

where ΔS_2 and ΔH_2 are entropy and enthalpy changes for the equilibrium relation of equation 6.3. Thus, the concentration of As vacancies can be written as

$$X(V_{\text{As}}) = \exp(\Delta S_2/k) \exp(-\Delta H_2/kT)/p_{\text{As}_2}^{1/2}. \quad (6.11)$$

Equation 6.11 represents the vacancy concentration at a given temperature when the component partial pressure controls the non-stoichiometry.

The ΔS_2 and ΔH_2 in equation 6.11 may be evaluated from figure 6.6 by the assignment of $X(V) = \Delta a_o/a_o$, and with the As_2 pressure given in figure 6.3. The As vacancy concentration (atom fraction) is then given by

$$X(V_{\text{As}}) = 1.13 \times 10^{10} \exp(-4.0/kT)/p_{\text{As}_2}^{1/2}, \quad (6.12)$$

and is plotted in figure 6.7. It should be noted that for a given temperature the As vacancy concentration can have values within the boundaries represented by the equilibrium partial pressure of As_2 on the Ga-rich and As-rich sides of the liquidus line. For example, at 800° C the As vacancy concentration can be decreased by a factor of 300 by going from the Ga-rich to the As-rich side of the liquidus line. It is difficult to assess the correctness of the As vacancy magnitudes in figure 6.7. However, the assumptions in the interpretation of Potts and Pearson's measurements suggest an uncertainty of less than a factor of 10 for the high-temperature values of $X(V_{\text{As}})$. As discussed below, the assignment of the Ga-vacancy concentration is a more formidable problem.

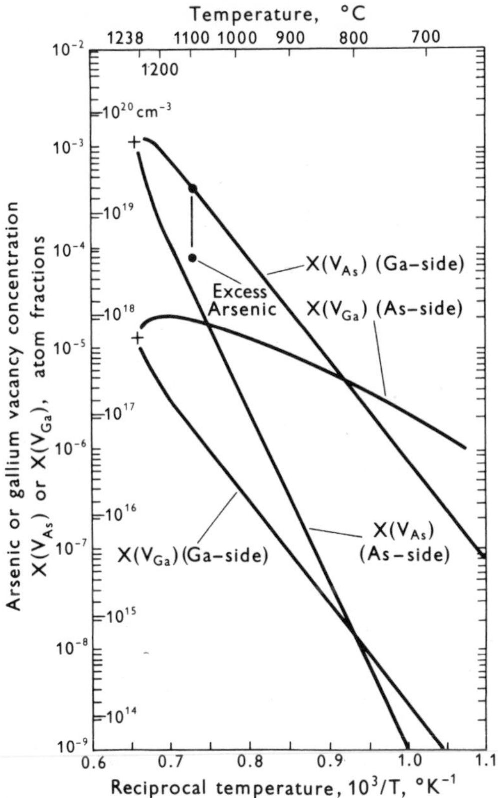

Figure 6.7. The Ga and As vacancy concentration as a function of temperature along the Ga-As liquidus. The melting point is indicated by a cross (+). The absolute value of $X(V_{As})$ is assigned from the data of Potts and Pearson [31], while the absolute value of $X(V_{Ga})$ has arbitrarily been assigned as discussed in conjunction with equation 6.13.

With no experimental data that can be related to the Ga-vacancy concentration, it is difficult to select a representative model for the Ga vacancy. One approach is to consider the Ga-vacancy, As-vacancy concentration product. Jordan [36] estimated the enthalpy of reaction for the defect equilibria represented in equation 6.7 in order to calculate the enthalpy change in the Zn incorporation in GaAs. The resultant value of 4.0 eV represents $(\Delta H_2 + \Delta H_3 - \Delta H_1)$ for the equilibrium constants of the Ga-vacancy, As-vacancy concentration product. From Potts and Pearson's data, as given by equation 6.12, ΔH_2 is 4.0 eV. Therefore, ΔH_1 and ΔH_3 may be considered equal and equation 6.6 becomes

$$X(V_{Ga}) = \exp\left[(\Delta S_3 - \Delta S_1)/k\right] p_{As_2}^{1/2}. \qquad (6.13)$$

The Ga-vacancy concentration calculated from equation 6.13 is also shown in figure 6.7. The entropy term $(\Delta S_3 - \Delta S_1)/k$ has arbitrarily been assigned a value of 2.8×10^{-5} to insure that $X(V_{Ga}) < X(V_{As})$ at 1100° C where the excess arsenic pressure lattice constant measurements were made by Potts and Pearson [31]. The shape of the Ga-vacancy curve should be representative of the Ga vacancy temperature dependence, but the absence of experimental data makes the absolute value arbitrary at the present time.

Another question to consider is the charge state of the lattice vacancies. As previously mentioned, studies of the incorporation of Zn in GaAs [30] suggested that vacancy ionization may be neglected. In addition, the dominant shallow levels in GaAs and many deep levels have been related to the presence of a chemical impurity. The only commonly observed level that has not been assigned to a known chemical impurity is located 0.17 eV below the conduction band [37, 38]. The concentrations assigned to this level by Basinski for pulled crystals were between 10^{15} and 10^{16} cm^{-3}. The III-V compound GaSb, when highly purified, has been found to be p-type with a room-temperature hole concentration near 10^{17} cm^{-3} [39, 40]. This hole concentration is generally ascribed to charged vacancies, but because of chemical analysis limitations at this concentration, residual impurities are difficult to rule out. The effects of charged vacancies and vacancy associations have been presented by Kendall [18]. Internal friction measurements on heat treated GaAs samples by Chakraverty and Dreyfus [41] were interpreted as nearest-neighbor pairs of Ga vacancies. However, this assignment of Ga divacancies is not particularly rigorous or unambiguous. Therefore, with no existing data that clearly demonstrate a charged or associated nature of vacancies, only neutral, unassociated vacancies will be considered in the discussion that follows.

6.2.3 Self-diffusion

There are many difficulties in obtaining self-diffusion data for the III-V compounds. In materials such as GaAs, the volatile As is readily lost from the sample and condenses on cooler spots in the ampoule. The sample loss not only requires analysis of the diffusion data in terms of an evaporating surface, but also results in surface pits [42] which make interpretation difficult. It is clear, however, that self-diffusion proceeds very slowly. The self-diffusion coefficient $D_e(\widetilde{H})$ is generally expressed by

$$D_e(\widetilde{H}) = D_o \exp(-Q/kT), \tag{6.14}$$

where D_o is a pre-exponential factor, and Q is the activation energy. From data by Harper [43] and Kendall [44], Kendall assigned values of $D_o = 0.7$ and $Q = 3.2$ eV for As self-diffusion in GaAs. This gives $D_e(\widetilde{\text{As}})$ as $\approx 10^{-17}$ cm^2 s^{-1} at 700°C and $\approx 10^{-13}$ cm^2 s^{-1} at 1000°C. Other data on self-diffusion for AlSb, GaAs, GaSb, InP, and InSb have been compiled and evaluated by Kendall [18].

In a compound it is difficult to visualize a substitutional self-diffusion mechanism, because this model requires significant As atoms on Ga sites and vice versa. Therefore, a self-diffusion mechanism through the lattice would imply a significant concentration of antistructure defects, but presently no results suggest their presence. However, as mentioned in section 6.2.1, self-diffusion by movement through an interstitial site in reaching an adjacent vacancy on its own sublattice [20] is a more reasonable vacancy mechanism. Self-diffusion through divacancies is another possibility. Kendall and Huggins [45] proposed the In : Sb divacancy as the mechanism for In and Sb self-diffusion in InSb. Theoretical calculations were given to demonstrate the plausibility of In and Sb self-diffusion through divacancies. Several interstitial mechanisms can be suggested for self-diffusion. In the interstitialcy mechanism, lattice vacancies are not required. The interstitial atom moves into a normal lattice site and the atom which originally occupied the lattice site is pushed into a neighboring interstitial site. Because interstitial concentrations are expected to be very small [46], slow self-diffusion would result if an atom became interstitial to move to a nonadjacent vacancy on its own sublattice. The interstitial-substitutional self-diffusion is described below to illustrate the manner in which this process can be treated. It should be noted that considerably more experimental self-diffusion studies in the III-V compounds are necessary before a particular mechanism can be assigned as the dominant process.

The radioactive flux for either group III or V atoms can be written as

$$J = -D_e(\widetilde{H}_i)\partial\widetilde{c}_i/\partial x, \tag{6.15}$$

where $D_e(\widetilde{H}_i)$ is the interstitial diffusion coefficient, and $\widetilde{c}_i$ is the interstitial radioactive tracer concentration. Equation 6.15 represents a concentration gradient in the radioactive tracer concentration and not the total concentration, which is constant with distance. This equation may be written

$$J = -[D_e(\widetilde{H}_i)\partial\widetilde{c}_i/\partial\widetilde{c}]\partial\widetilde{c}/\partial x, \tag{6.16}$$

where $\widetilde{c} = \widetilde{c}_l + \widetilde{c}_i$ is the total radioactive tracer concentration in lattice sites ($\widetilde{c}_l$) and interstitial sites ($\widetilde{c}_i$). By analogy with equation 6.15, the tracer self-diffusion coefficient may be written as

$$D_e(\widetilde{H}) \equiv D_e(\widetilde{H}_i)\partial\widetilde{c}_i/\partial\widetilde{c}. \tag{6.17}$$

It should be noted that for self-diffusion with a radioactive tracer, the total concentration of atoms c is the sum of radioactive atoms $\tilde{c}$ and nonradioactive atoms $\bar{c}$, $c = \bar{c} + \tilde{c}$. The nonradioactive atoms are the sum of the lattice site and interstitial atoms, $\bar{c} = \bar{c}_l + \bar{c}_i$. The same fraction of the radioactive and nonradioactive atoms is in interstitial sites, $\tilde{c}_i/\tilde{c} = \bar{c}_i/\bar{c}$, and $\bar{c}_i > \tilde{c}_i$ and $\bar{c} > \tilde{c}$. Then with $\tilde{c}_i/\tilde{c} = c_i/c$, it is useful to write equation 6.17 as

$$D_e(\tilde{H}) = D_e(\tilde{H}_i)c_i/c, \tag{6.18}$$

or since the number of lattice vacancies is small and c in units of atom fraction is nearly unity,

$$D_e(\tilde{H}) = D_e(\tilde{H}_i)\mathrm{c}_i. \tag{6.19}$$

Thus, even if $D_e(\tilde{H}_i)$ is 10^{-5} cm^2 s^{-1}, c_i in atom fraction is probably 10^{-7} or less, which gives $D_e(\tilde{H}) < 10^{-12}$ cm^2 s^{-1} for an interstitial-substitutional self-diffusion mechanism. The activation energy of $D_e(\tilde{H})$ may be obtained by representing the interstitial diffusion coefficient by the usual Arrhenius expression,

$$D_e(\tilde{H}_i) = D_o(\tilde{H}_i) \exp(-Q_i/kT), \tag{6.20}$$

and the interstitial concentration can be represented by [46]

$$\mathrm{c}_i = A \exp(-\Delta H_i/kT). \tag{6.21}$$

These equations give the effective diffusion coefficient for self-diffusion as

$$D_e(\tilde{H}) = AD_o(\tilde{H}_i) \exp[-(Q_i + \Delta H_i)/kT], \tag{6.22}$$

which by comparison with equation 6.14 gives the activation energy for self-diffusion by the interstitial-substitutional process as $(Q_i + \Delta H_i)$. For Cu in GaAs (reference 47), Q_i and ΔH_i are of the order of 1 eV, which suggests an activation energy of about 2 eV for self-diffusion in GaAs by the interstitial-substitutional process. This value is somewhat less than the 3.2 obtained by Kendall [44] for As self-diffusion. Additional work must be done before a definitive description is possible.

The vacancy diffusion coefficient is related to self-diffusion by [48]

$$D_e(\tilde{H}) = 0.78\, D_e(V_\mathrm{H})X(V_\mathrm{H}), \tag{6.23}$$

where the factor 0.78 is the vacancy correlation factor for the f.c.c. lattice. Thus, for small vacancy concentrations, $D_e(V)$ can be comparatively rapid even though $D_e(\tilde{H})$ is very small. For example, at 700°C $D_e(\tilde{H})$ for As is $\approx 10^{-17}$ cm^2 s^{-1}, and from figure 6.7 $X(V_\mathrm{As})$ is $\approx 5 \times 10^{-7}$. These values in equation 6.23 give the As vacancy diffusion coefficient as $\approx 2 \times 10^{-11}$ cm^2 s^{-1}.

At the present time, there is an almost total absence of definitive studies on vacancies and self-diffusion in the III-V compounds. Much fundamental work is obviously needed. The concepts presented here illustrate the expected behavior and introduce the concepts that will be utilized in the discussion of Zn diffusion in GaAs. Although little is known about Ga vacancies, such concepts as the one that states that the product $X(V_{Ga})X(V_{Ga})$ is constant at a given temperature are still extremely useful in interpreting and controlling impurity diffusion.

6.3 ACCEPTOR DIFFUSION IN GaAs

6.3.1 Selection of Zn in GaAs to Illustrate Acceptor Diffusion

Elements that are known to be acceptors in GaAs are listed in Table 6.3 together with their ionization energy levels. For completeness, the donors are also given. Of all the acceptors, Zn is the most widely used, and therefore has been the most extensively studied. Copper has also received considerable attention, and several studies have been reported for Cd. Only limited information is available on the diffusion behavior of the other acceptors. There appear to be more publications on Zn in GaAs than on the other donors and acceptors combined.

Because of the extensive studies of Zn in GaAs, there are sufficient data available to determine the basic diffusion mechanism. It is now well established that Zn diffuses interstitially as an ionized donor, reacts with a Ga vacancy, and becomes an immobile, substitutional, ionized acceptor. *Qualitatively, the fact that the immobile, negatively charged, substitutional Zn enhances the concentration of the highly mobile and much more dilute, positively charged, interstitial Zn results in a concentration-dependent diffusion coefficient.* The following discussion of Zn in GaAs illustrates the manner in which the properties of the ternary system influence and control the diffusion behavior. It must be emphasized that it is necessary to know the partial pressures of the ternary components in order to quantitatively represent the diffusion behavior of an impurity in a III-V compound.

6.3.2 The Ga-As-Zn Ternary Phase Diagram

The first step in establishing the diffusion conditions is to determine the *actual* diffusion source composition from the *starting* diffusion source composition. The discussion presented here of the relationship between the ternary phase diagram and the properties of the diffusion source is extracted, in part, from the description by Casey and Panish [17]. Only closed diffusion systems are considered because open or flowing vapor systems require maintaining constant vapor pressures in the flowing gas and represent control problems outside the scope of this presentation.

TABLE 6.3

Impurity ionization[a] energies in GaAs

Element	Type impurity	Level (eV)
S	donor	0.00610
Se	donor	0.00589
Te	donor	0.0058
Si	donor	0.00581
Ge	donor	0.00608
Sn	donor	0.006
C	donor	Similar to Sn
O	donor	0.75
O	donor	0.4
Unknown	donor	0.17
Zn	acceptor	0.029
Cd	acceptor	0.030
Ge	acceptor	0.038
Cu	acceptor	0.15
Cu	acceptor	0.47
Li	acceptor	0.023, 0.05
C	acceptor	0.019[b]
Si	acceptor	0.030[b]
Cr	acceptor	0.79
Mn	acceptor	0.10
Fe	acceptor	0.52
Co	acceptor	0.16
Ni	acceptor	0.21
Be	acceptor	0.030
Mg	acceptor	0.030
Au	acceptor	0.090
Ag	acceptor	0.11

[a] Donor ionization energies are measured from the conduction band, and the acceptor ionization energies are measured from the valence band. These values were also compiled by H. C. Casey, Jr. and F. A. Trumbore in the *Journal of Material Science and Engineering,* 6, 69 (1970).

[b] The uncertainty of these values is much greater than for the other ionization energies, and further studies are required to establish these ionization energies.

For Zn diffusion into GaAs, the *starting* source material in sealed ampoules has generally been either elemental Zn, dilute solutions of Zn in Ga, small amounts of $ZnAs_2$, or various combinations of Zn and As. The *actual* diffusion source in any ternary system must be approaching or must be an equilibrium ternary mixture. The composition of the ternary mixture is determined by the total amount of each element present, the temperature, and the volume of the system. As will be shown below, the diffusion source composition controls the Zn surface concentration, the Ga and As vacancy concentrations, and hence, the resulting diffusion.

Figure 6.8 shows the projected ternary diagram of reference 49. The position of the boundaries in the high-As portion of the diagram is not known, and the dashed parts simply represent the situation schematically. The possible diffusion sources and diffusion behavior depend greatly on the temperature. Above 1015° C, the only solid is GaAs doped with Zn which is in equilibrium with the liquid composition represented by the liquidus line. Between 1015° and 950° C, both GaAs doped with Zn and Zn_3As_2 doped with Ga are present. A representative isothermal section between 1015° and 950° C is shown in figure 6.9. Below 950° C the ternary isotherms become more complicated. The liquidus curves *ab* and *cd* shown in figure 6.9 for the two primary fields meet, and new regions occur in which more than one solid phase is in equilibrium with the liquid

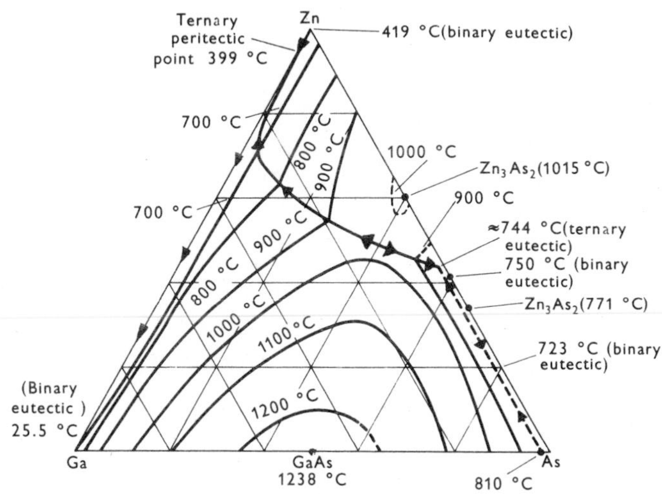

Figure 6.8. The Ga-As-Zn ternary phase diagram (reference 49).

and vapor. The isothermal sections continue to change as the temperature is lowered and are rather complex (see reference 17). The system becomes simpler again at lower temperatures because there is no liquid below 723° C in the As-Zn binary except at the Zn-rich end. The isothermal section at 700° C is shown in figure 6.10.

The relationship between the *starting* source composition and the *actual* source composition is illustrated by considering figures 6.9 and 6.10. In figure 6.9, any overall diffusion source composition within the liquidus curves *ab* and *cd* represented as regions *B* and *C* will involve a

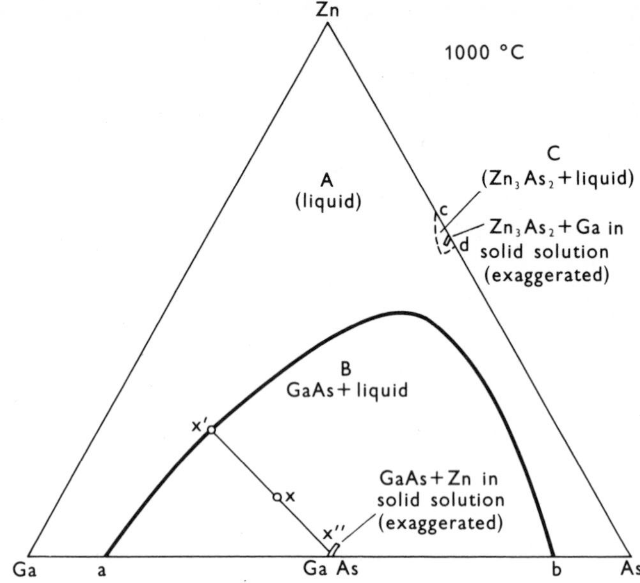

Figure 6.9. The Ga-As-Zn ternary phase diagram at 1000° C (reference 17).

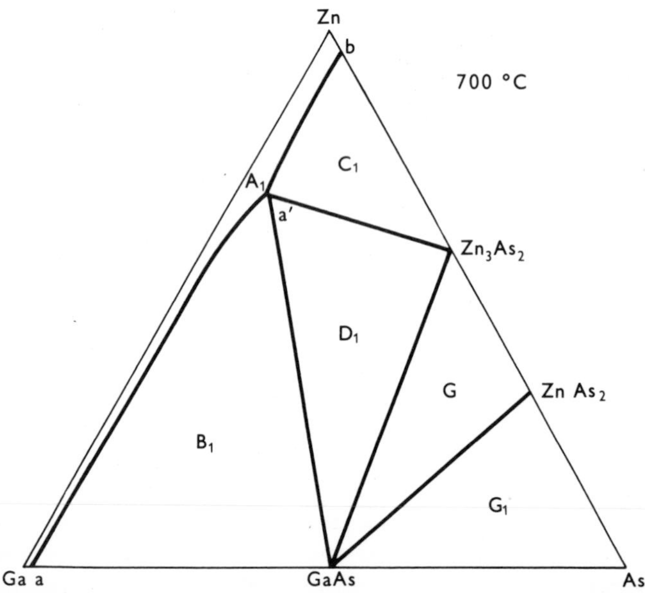

Figure 6.10. The Ga-As-Zn ternary phase diagram at 700° C (reference 17).

liquid in equilibrium with one solid. The solid phase in region B will be GaAs doped with Zn, while Zn_3As_2 doped with Ga is the solid phase in equilibrium with the liquid region C. The composition of the liquid phase will be somewhere on curve ab or cd; for example, a source of overall composition X in region B of figure 6.9 leads to a composition X' in the liquid and X'' in the solid. For an isothermal system in contact with the vapor, the number of degrees of freedom, F, allowed in various source composition ranges is given by the phase rule [50] :

$$F = 3 - P, \tag{6.24}$$

where P is the number of condensed phases. Therefore, the phase rule gives a single degree of freedom which is a univariant region. The one degree of freedom means that the composition of the liquid can be varied along curves ab or cd by varying the overall composition while still maintaining the desired solid phase. This liquid composition then determines the solid solubility of the solid phase.

Because no solid phase exists within region A, starting compositions within this region are not desirable as diffusion sources. With elemental Zn or Ga-Zn sources the system starts with a liquid somewhere in region A. To establish equilibrium the source must react with GaAs until the solid is gone or the boundary ab is reached by the liquid composition. For an elemental source the liquid can form only on the sample surface since the vapor pressure of Zn is much greater than the Ga vapor pressure. With the Ga-Zn source the ternary liquid will form both in the source and on the sample surface. Arsenic must vaporize from the sample and transport to the source, and the As must distribute itself to maintain the same concentration in both the liquid in the source and the new liquid on the sample surface. For both the elemental and Ga-Zn source there are equilibrium As, Zn, and Ga pressures within the ampoule that are supplied from the phases present and completely determined by the liquid composition at equilibrium.

Below about 950°C the liquidus curves ab and cd of the two primary fields meet, and there are regions where more than one solid phase is in equilibrium with the liquid and vapor. There is no liquid below 723°C in the As-Zn binary except at the Zn-rich end. The phase diagram for 700°C appears in figure 6.10. The GaAs, Zn_3As_2, and $ZnAs_2$ solid-solution regions are not shown. Regions A_1, B_1, and C_1 correspond to the regions A, B, and C in the 1000°C isotherm shown in figure 6.9. The region D_1 is an invariant region with GaAs and Zn_3As_2 in equilibrium with a liquid of composition a'. Below 744°C the invariant region G forms because there is a ternary eutectic at about 744°C at which Zn_3As_2, GaAs, and $ZnAs_2$ are in equilibrium. There is no more liquid in the Zn-As binary below 723°C, and G_1 is an invariant region with $ZnAs_2$, GaAs, and As in equilibrium. Ternary diffusion sources that are used at temperatures and compositions which put them in the *D*- or *G*-type regions should be most convenient to use because any composition within an invariant region gives the same surface concentration and diffusion behavior. Therefore, within those regions, no careful control of the system composition or ampoule volume is necessary. The *G*-type sources are preferable, however, since they contain no liquid that could cause surface damage and result in nonplanar diffusions. Sources in region *G* may be used at temperatures up to about 744°C and region G_1 sources up to 723°C.

For a given *starting* source composition, the *actual* source composition can generally be determined from the ternary phase diagram. The actual source composition then determines the Zn and As partial pressures. The remainder of this section first demonstrates the relationship of these partial pressures to the Zn surface concentration in GaAs, and then presents a detailed discussion of diffusion. The discussion of surface concentration is really the consideration of solid solubility.

6.3.3 Zinc Surface Concentration

In diffusion studies of Zn, the limiting Zn concentration at the diffused surface is often taken as the equilibrium solid solubility, but in general the conditions of the diffusion are not sufficiently defined to permit an unambiguous determination of the equilibrium conditions during diffusion. When the equilibrium conditions can be determined, as was done by Chang and Pearson [51] and Shih, Allen, and Pearson [52], the surface concentrations were in agreement with solubilities obtained by solution growth [30]. It therefore appears that for Zn in GaAs, the limiting Zn concentration at the surface is the equilibrium solid solubility. The relationship between the surface concentration and the diffusion source composition is given by the incorporation reaction and the resulting equilibrium relation. For the incorporation of Zn in the liquid Zn_L or vapor into the lattice as a singly ionized substitutional acceptor Zn'_{Ga} on a Ga site plus a hole $h^{\cdot}$, the incorporation reaction is represented by [30]

$$Zn_L + V^{\times}_{Ga} \rightleftharpoons Zn'_{Ga} + h^{\cdot}. \tag{6.25}$$

Analytical measurements of Zn in the solid give the total Zn, and therefore it is necessary to determine if significant un-ionized Zn is present at the growth temperature.

For dilute acceptor concentrations in semiconductors, the isolated acceptor atoms lead to a localized impurity level located at energy E_A in the forbidden band with a density equal to the acceptor density, as

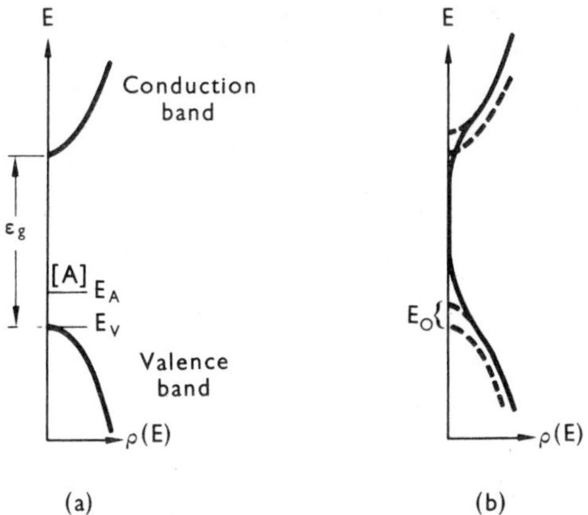

Figure 6.11. The density of states $\rho(E)$ as a function of energy E.
(a) Low impurity concentration, localized acceptor level located at an energy $(E_A - E_v)$ above the valence band with an acceptor density $[A]$.
(b) High impurity concentration, merged impurity band showing the band tails. The band edges are shifted upward for acceptors (downward for electrons) relative to their position in the pure semiconductor by the mean potential E_o.

represented in figure 6.11(a). The neutral acceptor concentration is given by

$$[A^{\times}] = [A]/[1 + (1/2)\exp(E_F - E_A)/kT], \qquad (6.26)$$

with the Fermi level E_F determined from the neutrality condition

$$[A'] + n = p, \qquad (6.27)$$

and with the hole concentration given by

$$p = N_v F_{1/2}[(E_v - E_F)/kT]. \qquad (6.28)$$

In these equations N_v is the effective density of states for the valence band, $F_{1/2}$ is the Fermi-Dirac integral, and E_v is the valence band edge. At high temperatures and high concentrations, these conditions demand the presence of significant amounts of neutral Zn, even for $(E_A - E_v)$ taken as zero [51].

As the impurity concentration is increased from a diminishingly dilute concentration, several effects lead to a decrease in the ionization energy. The first effect is the formation of a band (near the valence band) by the first excited states of the acceptor level [53-55]. Holes are then thermally activated from the acceptor level to this 'excited-state band' at less energy than for activation to the valence band. Also, fluctuations in local potential, which result from randomly distributed impurity ions, lead to broadening of the well-defined low-concentration impurity level [56].

Further increase in impurity concentration results in overlap of the hydrogen-like wave functions of the hole bound to the acceptor center, and enough free carriers are present to make hole-hole and hole-impurity interactions significant. Finally, at very high concentrations the impurity level merges with the valence band and forms tails as shown in figure 6.11(b). Note also that the band edges shift relative to their position in the pure semiconductor by the amount E_o. The density of states when band tails are present may be calculated from the theory of Halperin and Lax [57]. The application of this theory to Zn in GaAs has been described in detail by Panish and Casey [30].

The best criterion for the absence of neutral impurities is the impurity concentration at which metallic impurity conduction occurs. This transition has been predicted [58, 59] to occur when the ratio of the average separation r_A to the radius of the hydrogenic acceptor r_B is about 3:

$$r_A/r_B \approx 3. \tag{6.29}$$

In equation 6.29, r_A is given by $(3/4\pi c)^{1/3}$, where c is the impurity concentration, and r_B is 0.53×10^{-8} $(\epsilon m_n/m_p)$ cm with ϵ the dielectric constant, m_n the free electron mass, and m_p the effective hole mass. For ϵ of 12.5 [60], equation 6.29 predicts the transition to metallic impurity conduction to occur at 4×10^{18} acceptors for a hole effective mass of $0.47m_n$ [61]. For donors with an electron effective mass of $0.067m_n$ [61], the transition is expected to occur at 10^{16} donors/cm^3. Transition to metallic impurity conduction is indeed observed at $\approx 5 \times 10^{18}$ acceptors/cm^3 for Zn-doped GaAs (reference 62) and at $\approx 10^{16}$ donors/cm^3 for GaAs doped with shallow donors [63]. In the following discussion, the assignment of a particular impurity concentration as 'high' refers to concentrations that satisfy equation 6.29, and not to a particular fraction of lattice sites occupied. High concentrations for GaAs are above 10^{16} cm^{-3} for shallow donors, while an increase in impurity concentrations by 500 times is necessary for high acceptor concentrations. When the impurity concentration approaches the value specified by equation 6.29, the expression for neutral impurities, as given for acceptors by equation 6.26, is no longer valid, and the Fermi level must be determined from the effect of the high impurity concentration on the density of states [30]. *The absence of neutral-substitutional Zn at high concentrations also has major significance in the interpretation of Zn diffusion.*

In the absence of neutral Zn, the ionized Zn concentration represents the total Zn in the solid. The equilibrium relation for equation 6.25 can be written as

$$K_4(T) = \frac{\gamma(\mathrm{Zn}'_{\mathrm{Ga}})[\mathrm{Zn}'_{\mathrm{Ga}}]\gamma_p p}{\gamma(\mathrm{Zn}_L)X(\mathrm{Zn}_L)X(V_{\mathrm{Ga}})}, \tag{6.30}$$

where γ_p is the activity coefficient of holes in the solid, p is the hole concentration, $\gamma(\mathrm{Zn}'_{\mathrm{Ga}})$ and $[\mathrm{Zn}'_{\mathrm{Ga}}]$ are the activity coefficient and concentration of ionized Zn on the Ga sites, and $\gamma(\mathrm{Zn}_L)$ and $X(\mathrm{Zn}_L)$ are the activity coefficient and atom fraction of Zn in the liquid. From equation 6.6, the Ga vacancy atom fraction is given as $X(V_{\mathrm{Ga}}) = [K_3(T)/K_1(T)]p_{\mathrm{As}_2}^{1/2}$, and the activity of arsenic is

$$\gamma(\mathrm{As}_L)X(\mathrm{As}_L) = (p_{\mathrm{As}_2}/p^0_{\mathrm{As}_2})^{1/2}, \tag{6.31}$$

where $\gamma(\mathrm{As}_L)$ and $X(\mathrm{As}_L)$ are the activity coefficient and atom fraction of arsenic in the liquid, and $p^0_{\mathrm{As}_2}$ is the As_2 pressure over pure arsenic. Therefore, $X(V_{\mathrm{Ga}})$ is proportional to $\{\mathrm{As}_L\}$, and the equilibrium relation of equation 6.30 may be written as

$$K_5(T) = \frac{\gamma(\mathrm{Zn}'_{\mathrm{Ga}})[\mathrm{Zn}'_{\mathrm{Ga}}]\gamma_p p}{\gamma(\mathrm{Zn}_L)X(\mathrm{Zn}_L)\gamma(\mathrm{As}_L)X(\mathrm{As}_L)} \tag{6.32}$$

For discussion of the solid solubility, it is useful to simplify equation 6.32. Since the solid solutions of Zn in GaAs are dilute with maximum concentrations of about 5×10^{20} cm^{-3} ($\approx$1 atom percent), Henry's law should be obeyed and $\gamma(\mathrm{Zn}'_{\mathrm{Ga}})$ is taken as constant. Thus, equation 6.32 becomes

$$K'_5(T) = \frac{[\mathrm{Zn}'_{\mathrm{Ga}}]\gamma_p p}{\gamma(\mathrm{Zn}_L)X(\mathrm{Zn}_L)\gamma(\mathrm{As}_L)X(\mathrm{As}_L)}. \tag{6.33}$$

With the further restraints of electrical neutrality expressed by equation 6.27, and the hole-electron mass action relation,

$$\gamma_p p n = n_i^2, \tag{6.34}$$

where n is the free electron concentration, and n_i is the intrinsic concentration. The activity coefficient for electrons has been taken as unity because of the low electron concentration in Zn-doped GaAs. As discussed in section 6.3.4, the hole activity coefficient γ_p also goes to unity at low hole concentrations. Equations 6.27, 6.33, and 6.34 may be combined to yield a relation for the Zn solid solubility:

$$[\mathrm{Zn}'_{\mathrm{Ga}}] = \left\{\frac{K'_5(T)\gamma(\mathrm{Zn}_L)X(\mathrm{Zn}_L)\gamma(\mathrm{As}_L)X(\mathrm{As}_L)}{\gamma_p[1 + n_i^2/K'_5(T)\gamma(\mathrm{Zn}_L)X(\mathrm{Zn}_L)\gamma(\mathrm{As}_L)X(\mathrm{As}_L)]}\right\}^{1/2}. \tag{6.35}$$

For extrinsic conditions of $[\mathrm{Zn}'_{\mathrm{Ga}}] > n_i$, the n_i^2 term in the denominator may be neglected. Both solubility and diffusion behavior are strongly

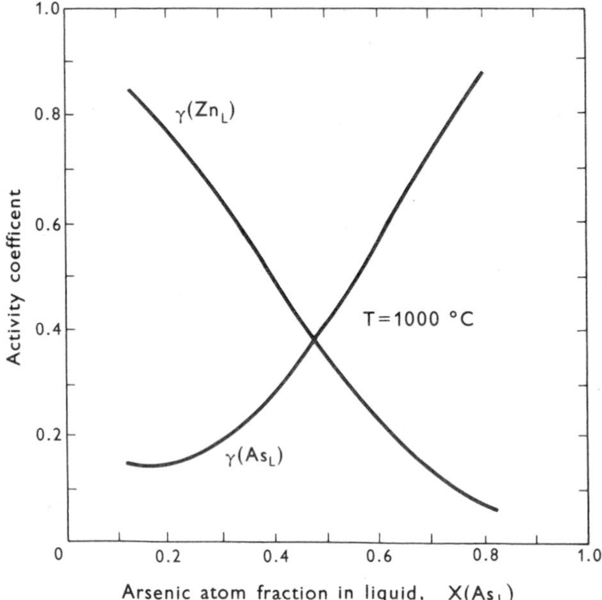

Figure 6.12. The activity coefficients of arsenic and zinc along the Ga-As-Zn liquidus line at 1000° C (reference 65).

dependent upon whether the acceptor or donor concentration exceeds n_i. The values of n_i as a function of temperature for GaAs and the other commonly available III-V compounds are given in the Appendix. Over most of the range of $[\mathrm{Zn}'_{\mathrm{Ga}}]$, the $\gamma(\mathrm{Zn}_L)\gamma(\mathrm{As}_L)$ product changes slowly, γ_p is unity, and the solid solubility may be approximated by

$$[\mathrm{Zn}'_{\mathrm{Ga}}] = [K''_5(T)X(\mathrm{Zn}_L)X(V_{\mathrm{Ga}})]^{1/2} \tag{6.36}$$

when $[\mathrm{Zn}'_{\mathrm{Ga}}] > n_i$. This equation emphasizes that under extrinsic conditions the equilibrium surface concentration for diffusion is proportional to the square root of the Zn concentration in the liquid and the square root of the Ga vacancy concentration. The complete solid solubility over the ternary system may be calculated from equation 6.35 with appropriate values of the activity coefficients and equilibrium constants.

The equilibrium partial pressures of zinc and arsenic for several compositions along the 1000° C liquidus isotherm were determined by Shih, Allen, and Pearson [64]. These measurements and partial pressures of As_4, As_2, and Zn along the liquidus isotherm calculated by Jordan [65] permit the evaluation of $\gamma(\mathrm{Zn}_L)$ and $\gamma(\mathrm{As}_L)$ in equation 6.35. Values for $\gamma(\mathrm{Zn}_L)$ and $\gamma(\mathrm{As}_L)$ at 1000° C are shown in figure 6.12. By a critical

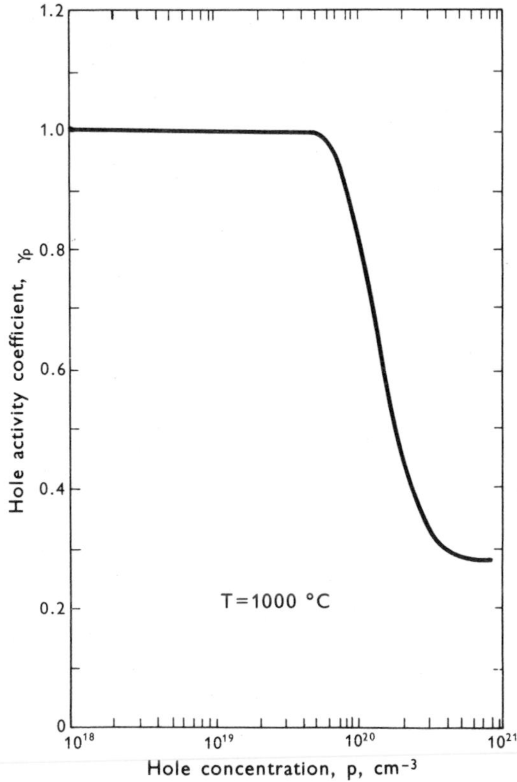

Figure 6.13. Variation of the hole activity coefficient with hole concentration at 1000° C (reference 66).

evaluation of the published Zn solubility data, Jordan [36] utilized the Zn and As activities to evaluate the Zn concentration over a wide temperature range. The concentration dependence of the 1000° C hole activity coefficient obtained by Casey *et al.* [66] is shown in figure 6.13 to illustrate the effect of γ_p in evaluating equation 6.35. The solid solubility along the liquidus isotherms of the Ga-As-Zn ternary system (figure 6.8) is shown in figure 6.14. The dashed curve for 1000° C has been calculated for $\gamma_p = 1$ and demonstrates the necessity of including the $\gamma_p < 1$ at high Zn concentrations. This figure illustrates how the temperature and the composition of the *actual* diffusion source influence the surface concentration for diffusion.

6.3.4 The Hole Activity Coefficient

Comparison of the solid and dashed curves in figure 6.14 demonstrates that the experimental solubility data require a hole activity coefficient

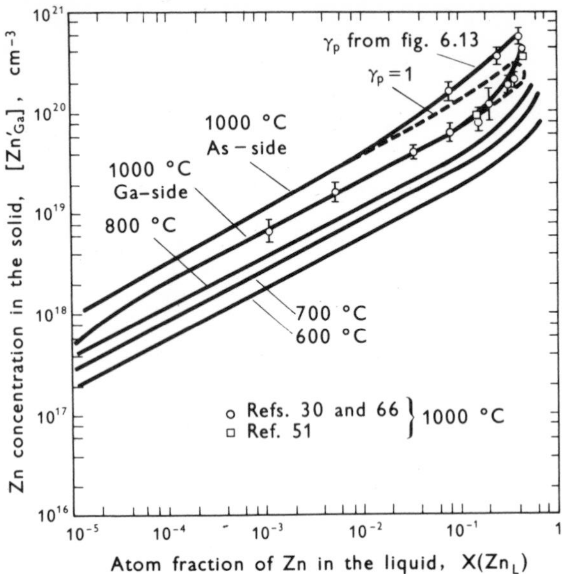

Figure 6.14. The Zn concentration in GaAs versus the atom fraction of Zn in the liquid along the 600°, 700°, 800° and 1000° C liquidus isotherms in the Ga-As-Zn system (reference 36). The isotherms at 800° C and below are terminated at the approximate liquid compositions where Zn_3As_2, a secondary solid phase, appears.

with a significant departure from unity at high Zn (hole) concentrations. The hole activity coefficient values given in figure 6.13 were obtained from the diffusion behavior [66] by the analysis described in section 6.3.6. The hole activity coefficient may be calculated from the position of the Fermi level. The Fermi level can be found by determining the concentration-dependent density of states as given by the theory of Halperin and Lax [57]. The relationships among the Fermi level, the chemical potential, and the activity coefficient at high impurity concentration is briefly considered in the discussion given below. These relationships have been treated in detail by Brews and Hwang [67, 68].

The electrochemical potential $\bar{\mu}$ and the chemical potential μ are related by

$$\bar{\mu} = \mu + e\Phi, \tag{6.37}$$

where Φ is the electrostatic potential. In a heavily doped semiconductor, the electrostatic potential for a hole (or electron) is due to the Coulomb potential of ionized acceptors and other holes. An alternative approach to equation 6.37 is to include the electrostatic potential in the wave-equation Hamiltonian, and find a new density of states. This procedure accounts for the electrostatic potential by inclusion in the Hamiltonian and the resulting density of states, and therefore eliminates the external potential in equation 6.37 so that $\bar{\mu} = \mu$.

The concepts of nonideal solutions relate the activity coefficient to the chemical potential for holes $\mu(h^{\cdot})$ by

$$\mu(h^{\cdot}) = \mu_o + kT \ln (\gamma_p p/N_v), \tag{6.38}$$

where μ_o is the reference potential that is taken as independent of impurity effects, and γ_p includes all the hole interactions due to the high concentration. From the reaction for the creation of a hole-electron pair,

$$(e'h^{\cdot}) \rightleftharpoons e' + h^{\cdot}, \tag{6.39}$$

the sum of the electron chemical potential $\mu(e')$ and the hole chemical potential is zero: $\mu(e') + \mu(h^{\cdot}) = 0$. With statistical mechanics, it may be shown that $\mu(e') = E_F$, and then $\mu(h^{\cdot}) = -E_F$. For dilute hole concentrations, the hole activity coefficient is unity, and $F_{1/2}\,[(E_v - E_F)/kT]$ in equation 6.28 may be taken as $\exp\,[(E_v - E_F)/kT]$ to give

$$p = N_v \exp\,[(E_v - E_F)/kT]. \tag{6.40}$$

Rewriting equation 6.40 as

$$-E_F = -E_v^0 + kT \ln (p/N_v) \tag{6.41}$$

shows that μ_o in equation 6.38 is $-E_v^0$, the valence band edge in the dilute case. At high hole concentrations, γ_p departs from unity and with equations 6.38 and 6.40, it may be seen that N_v is replaced by N_v/γ_p.

Previous treatment of electron (or hole) activity coefficients by Rosenberg [69] introduced the activity coefficient to account for the difference in Boltzmann and Fermi statistics. However, in semiconductors the departure from Boltzmann statistics is accompanied by changes in the density of states as shown in figure 6.11(b). Therefore, Rosenberg's treatment did not include the major effects of departure from ideal behavior. The activity coefficients calculated from equation 6.38 and the density of states model of Halperin and Lax [57] give the limiting γ_p at high concentrations that is shown in figure 6.13. Although this discussion is concerned with holes, it should be noted that the activity coefficient for electrons over an extensive concentration and temperature range has been calculated by Hwang and Brews [68]. In the evaluation of E_F, the density of states for holes is found to be displaced upward from the pure semiconductor case by a mean potential E_o and to have 'band tails' as shown in figure 6.11(b). The Fermi level is found from the condition that

$$p = [\mathrm{Zn}'_{\mathrm{Ga}}] = \int_{-\infty}^{\infty} \rho(E - E_v) f(E)\,dE, \tag{6.42}$$

where $\rho(E - E_v)$ is the concentration-dependent density of states which includes the mean displacement E_o and band tails, and $f(E)$ is the Fermi distribution function: $f(E) = 1/[+ \exp(E - E_F)/kT]$. Thus, E_F must be found from the concentration-dependent density of states as described for holes in reference 30 and for electrons in reference 68. The Fermi level and γ_p are related by equation 6.38 with $\mu_o = -E_v^0$.

6.3.5 Concentration Gradient Diffusion of Zn in GaAs

The noncomplementary error function shape of the Zn-concentration-versus-distance curves in GaAs that were obtained by Cunnell and Gooch [12] initiated extensive studies of Zn diffusion in the III-V compound semiconductors. Their diffusion profiles for 1000°C are shown in figure 6.15. Diffusion profiles of this shape were verified by extensive,

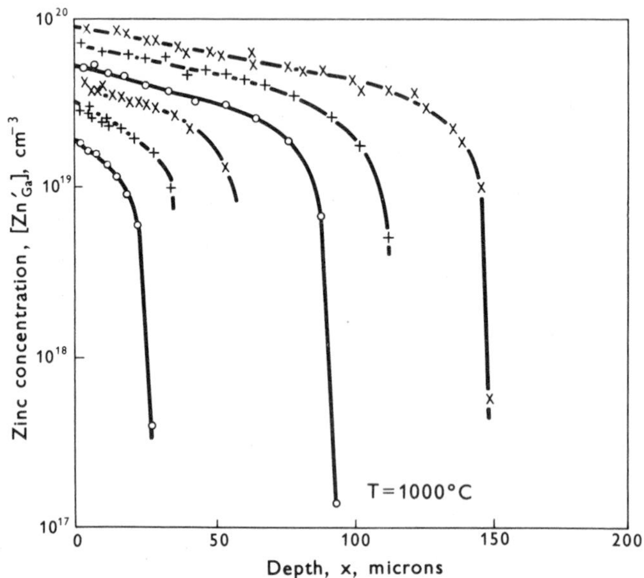

Figure 6.15. Diffusion profiles of Zn in GaAs at 1000° C for diffusion time of 10^4 s. The different surface concentrations were obtained by maintaining the Zn source at temperatures in the range of 600° to 800° C (reference 12).

although unpublished, studies by Kendall and Jones [70]. These profiles are characterized by the very abrupt drop in Zn concentration at the diffusion front.

When the diffusion coefficient **D** is a function of concentration only, the diffusion coefficient may be obtained from the experimental $[Zn'_{Ga}]$-versus-x plot such as in figure 6.15 by the Boltzmann-Matano analysis [14]. This method makes a transformation in the one-dimensional diffusion equation and carries out the first integration. The resulting expression for the diffusion coefficient is

$$\mathbf{D} = -1/(2t)\left[\int_0^{c_x} x dc/(dc/dx)_{c_x}\right]. \qquad (6.43)$$

For a known diffusion time t, the value of the diffusion coefficient **D** at a given concentration, $c = c_x$, is obtained by graphically evaluating the flux from 0 to c_x together with the concentration gradient at c_x. By using the Boltzmann-Matano analysis, Cunnell and Gooch [12] obtained a diffusion coefficient that varied as the square of the Zn concentration. The diffusion coefficient reached a maximum and decreased slightly at the highest concentrations.

Several explanations were initially advanced to explain the concentration-dependent diffusion coefficient. Allen [71] attributed this diffusion behavior to the differing substitutional diffusion coefficients for the neutral and the ionized Zn. Goldstein pointed out that this explanation was not reasonable because neutral Zn was not observed at these concentrations [72]. The criterion for the absence of neutral Zn is given by equation 6.29. Kendall and Jones [70] suggested that the Zn diffused substitutionally and that the Ga vacancies were triply ionized donors. The Ga vacancy concentration is then enhanced by the high Zn concentration and results in more rapid diffusion at high Zn concentration. Ionized vacancy behavior has yet to be demonstrated, and the solubility studies described in section 6.3.3 suggest that ionized Ga vacancies are not present at high Zn concentrations. Longini [15], in an attempt to explain the deterioration of GaAs tunnel diodes, suggested that Zn diffused interstitially as a donor although it is a substitutional acceptor. Only plausibility arguments were given. Careful examination of the radioactive Zn65 diffusion profiles strongly suggests that Zn diffuses by this interstitial-substitutional mechanism. Certain aspects of the model have been disputed by various investigators, but the basic concept of interstitial-substitutional diffusion is now well established. Weisberg and Blanc [73] applied computer solutions to calculation of diffusion profiles for concentration-dependent diffusion coefficients. Very good fit to the data shown in figure 6.15 was obtained for $\mathbf{D} \propto [Zn'_{Ga}]^2$. Present concepts are based on their demonstration of the agreement between the interstitial-substitutional diffusion model and Cunnell and Gooch's 1000°C data.

6.3.6 Interstitial-Substitutional Diffusion

In their diffusion experiments, Cunnell and Gooch [12] used a two-zone heating system to reduce the vapor pressure of the Zn (which also reduces the partial pressure of the arsenic vapor species), and obtained diffusion profiles for surface concentrations only up to 8×10^{19} cm^{-3}. This concentration is much less than the 'pseudobinary' solubility of 3.5-4.0 x 10^{20} cm^{-3} or maximum solubility at 1000°C of 6 x 10^{20} cm^{-3} shown in figure 6.14.

The derivation of the effective diffusion coefficient that is presented here is taken, in part, from the analysis of Casey, Panish, and Chang [66]. The total Zn flux J(Zn) is the sum of the interstitial and substitutional flux, each of which consists of a flux term due to the concentration gradient and a flux term for the built-in field. The one-dimensional expression for the total flux along x is

$$J(\mathrm{Zn}) = -D(\mathrm{Zn}_i)(\partial c_i/\partial x) \pm v(\mathrm{Zn}_i)c_i\mathbf{E} - D(\mathrm{Zn}_s)(\partial c_s/\partial x) \pm v(\mathrm{Zn}_s)c_s\mathbf{E}, \quad (6.44)$$

where $D(\mathrm{Zn}_i)$ is the diffusion coefficient for the interstitial species and c_i its concentration, $D(\mathrm{Zn}_s)$ and c_s the substitutional diffusion coefficient and concentration, and **E** the built-in field. The proper sign for the field term is determined by both the direction of the field and the sign of the ionized impurity. The mobility v is related to the diffusion coefficient by the expression referred to as the Einstein equation, which is defined as

$$v = eD/kT, \quad (6.45)$$

where e is the electronic charge.

For conditions of equilibrium between the holes and electrons, there can be no net flow of electric current. Therefore, in inhomogeneous material a built-in field must exist to counteract the current due to diffusion by the chemical potential gradient. The most mobile of the species present in the largest concentration will dominate the expression for current, i.e., the hole current

$$i_p = -e\mu_p p\left[(1/e)\,\frac{\partial[\mu(h^{\cdot}) - \mu_o]}{\partial x} - \mathbf{E}\right] = 0, \quad (6.46)$$

where μ_p is the hole mobility. Note that in equation 6.46 the hole diffusive and drift currents are equal but flow in opposite directions. The reference potential μ_o is $-E_v^0$ and $\mu(h^{\cdot})$ is $-E_F$ in equation 6.38, which gives $[\mu(h^{\cdot}) - \mu_o]$ as

$$\mu(h^{\cdot}) - \mu_o = E_v^0 - E_F = kT \ln(\gamma_p p/N_v). \quad (6.47)$$

The hole activity coefficient is utilized here because the diffusion expressions are simpler mathematically with γ_p than with E_F. Therefore, the built-in field is

$$\mathbf{E} = \frac{1}{e}\,\frac{\partial[\mu(h^{\cdot}) - \mu_o]}{\partial x} = \frac{kT}{e}\left(\frac{1}{p}\,\frac{\partial p}{\partial x} + \frac{1}{\gamma_p}\,\frac{\partial \gamma_p}{\partial x}\right). \quad (6.48)$$

For a constant γ_p, equation 6.48 may be written in terms of the usual diffusion coefficient times the hole gradient. Whenever a species in a current (flux) equation departs from ideality, as shown by a nonunity activity coefficient, the diffusion term normally written as a concentration gradient times a diffusion coefficient must be written as a chemical potential gradient times a mobility-concentration product [74], as in equation 6.46.

It should be noted that the usual Einstein relation of equation 6.45 is not directly applicable to the flux equation if the potential gradient is taken simply as the usual concentration gradient. Writing the diffusive current term of equation 6.46 with $e^{-1}\partial[\mu(h^{\cdot})-\mu_o]/\partial x$ given by equation 6.48 gives

$$i_p = -\mu_p p \frac{\partial[\mu(h^{\cdot})-\mu_o]}{\partial x} = -eD(p)p\left[\frac{1}{p}\frac{\partial p}{\partial x} + \frac{1}{\gamma_p}\frac{\partial \gamma_p}{\partial x}\right], \tag{6.49}$$

where $D(h^{\cdot}) = \mu_p kT/e$ is the hole diffusivity. Rewriting as

$$i_p = -eD(h^{\cdot})\left[1 + \frac{p}{\gamma_p}\frac{\partial \gamma_p}{\partial p}\right]\frac{\partial p}{\partial x} \tag{6.50}$$

gives an effective hole diffusion coefficient $\mathbf{D}(h^{\cdot})$ as

$$\mathbf{D}(h^{\cdot}) = \frac{i_p}{e\partial p/\partial x} = D(h^{\cdot})\left[1 + \frac{p}{\gamma_p}\frac{\partial \gamma_p}{\partial p}\right], \tag{6.51}$$

or

$$\mathbf{D}(h^{\cdot}) = \frac{\mu_p kT}{e}\left[1 + \frac{p}{\gamma_p}\frac{\partial \gamma_p}{\partial p}\right]. \tag{6.52}$$

Thus, a diffusive flux equation such as equation 6.49 must be written with the hole diffusion coefficient $D(h^{\cdot})$ times the potential gradient or with an effective diffusion coefficient $\mathbf{D}(h^{\cdot})$ as in equation 6.51, times the concentration gradient. In the potential gradient case, the relationship between mobility and diffusivity is given by the usual Einstein relation of equation 6.45, while for the concentration gradient case, they are related by equation 6.52.

The built-in field given by equation 6.48 and the Einstein relation of equation 6.45 permit writing the Zn flux equation 6.44 as

$$J(\mathrm{Zn}) = -D(\mathrm{Zn}_i)\left[\frac{\partial c_i}{\partial c} \pm c_i\left(\frac{1}{p}\frac{\partial p}{\partial c} + \frac{1}{\gamma_p}\frac{\partial \gamma_p}{\partial c}\right)\right]\frac{\partial c}{\partial x}$$

$$-D(\mathrm{Zn}_s)\left[\frac{\partial c_s}{\partial c} \pm c_s\left(\frac{1}{p}\frac{\partial p}{\partial c} + \frac{1}{\gamma_p}\frac{\partial \gamma_p}{\partial c}\right)\right]\frac{\partial c}{\partial x}, \tag{6.53}$$

where the total Zn concentration $c = c_i + c_s = [\mathrm{Zn}'_{\mathrm{Ga}}] = p$, and $(\partial/\partial x)$ has been replaced by $(\partial/\partial c)(\partial c/\partial x)$. An effective diffusion coefficient $\mathbf{D}(\mathrm{Zn}) = -J(\mathrm{Zn})/(\partial c/\partial x)$ may be identified as

$$\mathbf{D}(\mathrm{Zn}) = D(\mathrm{Zn}_i)\,\frac{\partial c_i}{\partial c} \pm D(\mathrm{Zn}_i)c_i\left(\frac{1}{p}\,\frac{\partial p}{\partial c} + \frac{1}{\gamma_p}\,\frac{\partial \gamma_p}{\partial c}\right)$$

$$+D(\mathrm{Zn}_s)\,\frac{\partial c_s}{\partial c} \pm D(\mathrm{Zn}_s)c_s\left(\frac{1}{p}\,\frac{\partial p}{\partial c} + \frac{1}{\gamma_p}\,\frac{\partial \gamma_p}{\partial c}\right) \qquad (6.54)$$

The proper designation of the sign for the interstitial and substitutional field terms depends on the charge state of the interstitial and substitutional species. Ions of charge state identical to that of the dominant free carrier (the hole) will be retarded and have a minus sign in equation 6.54, while oppositely charged ions will be aided and enter equation 6.54 as the plus sign.

To evaluate equation 6.54, the relationship between c_i and c must be obtained from the equilibrium reaction. At equilibrium the interstitial Zn, all of which is assumed to be a singly ionized donor, reacts with a neutral Ga vacancy to form an ionized substitutional Zn acceptor and two holes:

$$\mathrm{Zn}_i^{\cdot} + V_{\mathrm{Ga}}^{\times} \rightleftharpoons \mathrm{Zn}'_{\mathrm{Ga}} + 2h^{\cdot}. \qquad (6.55)$$

For these charge states, the built-in field will retard the interstitial $\mathrm{Zn}_i^{\cdot}$ diffusion and aid the substitutional $\mathrm{Zn}'_{\mathrm{Ga}}$ diffusion. Since by equation 6.6 the equilibrium Ga vacancy concentration in the solid is proportional to the square root of the As_2 pressure, the equilibrium reaction gives c_i as

$$c_i = \frac{c_s(\gamma_p p)^2}{K_6(T)p_{\mathrm{As}_2}^{1/2}}, \qquad (6.56)$$

with the activity coefficients of interstitial and substitutional Zn taken as constants. As described in conjunction with equation 6.29, the hole concentration equals c_s at concentrations above $5 \times 10^{18}\ \mathrm{cm}^{-3}$, and equation 6.56 may be written as

$$c_i = \gamma_p^2 c_s^3 / K_6(T)p_{\mathrm{As}_2}^{1/2}, \qquad (6.57)$$

where $K_6(T)$, a function of temperature only, represents the equilibrium constant and other constants associated with the reaction of equation 6.55.

With the condition $c_i \ll c_s$ [75], so that $c_s = c = p$, and with the elimination of c_i from equation 6.54 by equation 6.57, the effective diffusion coefficient becomes

$$\mathbf{D}(\mathrm{Zn}) = \frac{D(\mathrm{Zn}_i)}{K_6(T)p_{\mathrm{As}_2}^{1/2}}\, c_s^2\gamma_p^2 \left[\left(3 + 2\frac{c_s}{\gamma_p}\frac{d\gamma_p}{dc_s}\right) - \left(1 + \frac{c_s}{\gamma_p}\frac{d\gamma_p}{dc_s}\right)\right]$$

$$+D(\mathrm{Zn}_s)\left[1 + \left(1 + \frac{c_s}{\gamma_p}\frac{d\gamma_p}{dc_s}\right)\right]. \qquad (6.58)$$

Equation 6.58 was written in the above form to emphasize that the term $[1 + (c_s/\gamma_p)(d\gamma_p/dc_s)]$ represents the effect of the built-in field. Several treatments of diffusion have considered the field term for substitutional diffusion [76, 77]. The treatment given here is different because at high concentrations p has been taken as equal to c_s for the electrical neutrality condition rather than $n - p = [A'] - [D^{\cdot}]$. Near the p-n junction where the Zn concentration is only slightly in excess of the background donor concentration, the analysis such as in reference 76 or 77 applies, and in particular van Opdorp has considered Zn in GaAs. The effect of the field near the p-n junction will be considered later.

As seen from the results of Cunnell and Gooch [12] where the Zn concentrations are low enough to assign $\gamma_p = 1$, the effective diffusion coefficient can vary as the square of the Zn concentration only if substitutional diffusion is neglected. Nelgecting $D(\mathrm{Zn}_s)$ in equation 6.58 permits writing the effective diffusion coefficient as

$$\mathbf{D}(\mathrm{Zn}) = \frac{2D(\mathrm{Zn}_i)c_s^2}{K_6(T)p_{\mathrm{As}_2}^{1/2}}\, \gamma_p^2 \left[1 + \frac{c_s}{2\gamma_p}\frac{d\gamma_p}{dc_s}\right]. \qquad (6.59)$$

At low concentrations where $\gamma_p = 1$, $D(\mathrm{Zn}) \propto c_s^2$. For higher concentrations where $\gamma_p < 1$, the derivative $d\gamma_p/dc_s$ is negative and equation 6.59 leads to a decrease in **D**(Zn).

Weisberg and Blanc [73] fit the data of Cunnell and Gooch [12]. For these diffusion profiles, the Zn concentrations are low enough to give $\gamma_p = 1$. Under these conditions equation 6.59 reduces to

$$\mathbf{D}(\mathrm{Zn}) = \frac{2D(\mathrm{Zn}_i)}{K_6(T)p_{\mathrm{As}_2}^{1/2}}\, c_s^2. \qquad (6.60)$$

Their computer-generated diffusion profiles for the data of figure 6.15 are shown in figure 6.16. The slightly greater apparent experimental Zn penetration was ascribed by van Opdorp [77] to the effect of the built-in fields near the junction region. The excellent agreement shown in figure 6.16 strongly supports the interstitial-substitutional mechanism.

Diffusions at high Zn concentrations where $\gamma_p < 1$ were performed by Casey, Panish, and Chang [66]. The approximate liquidus compositions for the diffusion sources are shown in the ternary phase diagram of

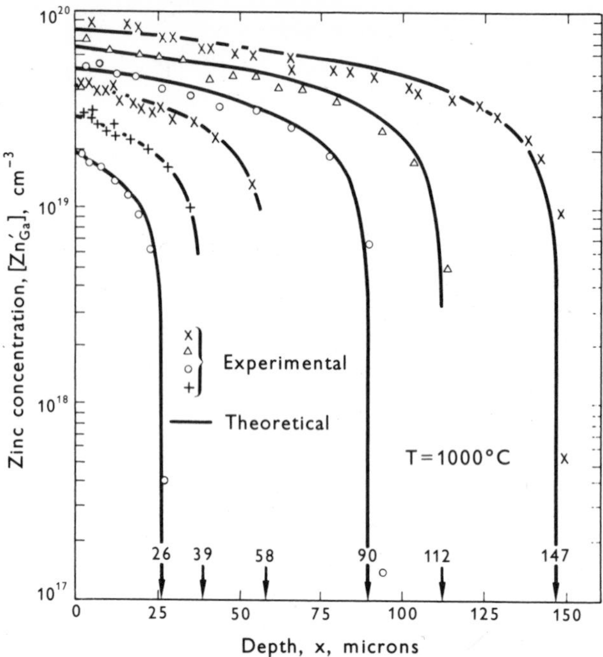

Figure 6.16. Comparison of Cunnell and Gooch's [12] experimental Zn diffusion profiles in GaAs with Weisberg and Blanc's [73] calculated diffusion profiles with $D(Zn) \propto [Zn'_{Ga}]^2$. The arrows indicate the 'effective zero' for each theoretical curve.

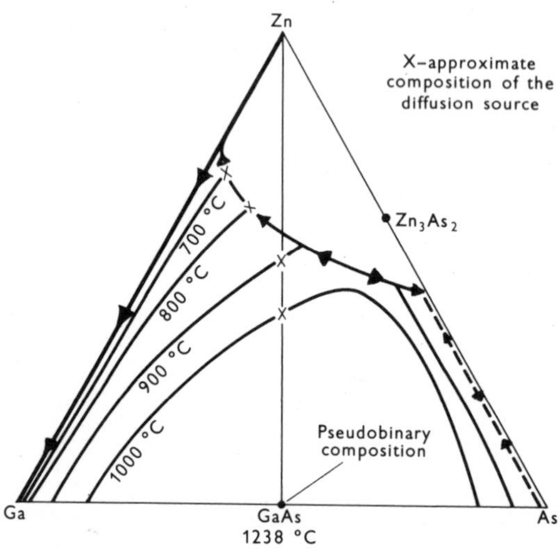

Figure 6.17. The Ga-As-Zn ternary phase diagram illustrating the approximate compositions of the diffusion sources used for the profiles of figure 6.18 (reference 66).

figure 6.17. The diffusion profiles obtained with radioactive Zn65 concentrations taken as the substitutional Zn, $c_S = [\mathrm{Zn}'_{\mathrm{Ga}}]$, are shown in figure 6.18. Diffusion times were 1.5 h for 800°, 900°, and 1000°C profiles and 168 h for the 700°C profile. The 700°C profile was normalized to 1.5 h according to the $x/t^{1/2}$ relation. The detailed experimental procedure for obtaining the diffusion profiles is described in reference 78.

The concentration-dependent diffusion coefficients may be obtained for these diffusion profiles by the Boltzmann-Matano analysis with equation 6.43. The effect of the diffusion profile shape on the accuracy of this type of analysis requires examination of the profile plotted linearly rather than semi-logarithmically. The 1000°C profile of figure 6.18 is plotted linearly in figure 6.19 and shows that the apparently 'flat' region near the surface in the semi-logarithmic plot is indeed a very steep concentration gradient. At low concentrations, the experimental data have their greatest errors, and therefore both the slope and *xdc* in the integral

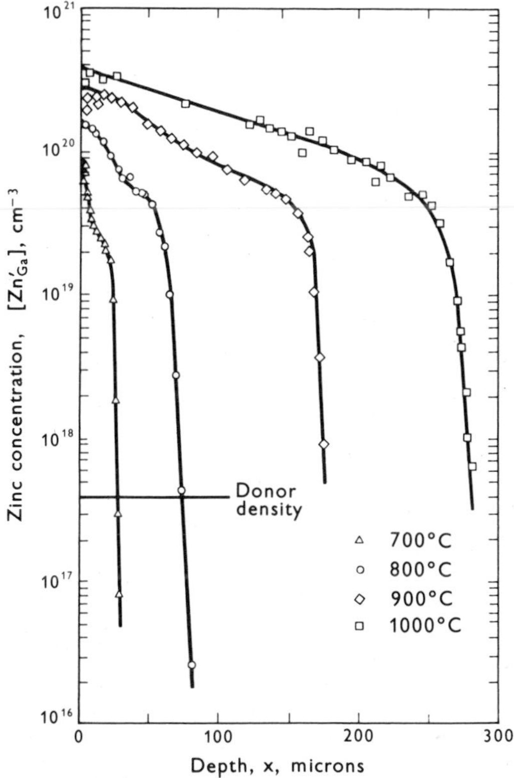

Figure 6.18. Diffusion profiles of Zn in GaAs at 700°, 800°, 900°, and 1000° C. Diffusion time is normalized to 1.5 h (reference 66).

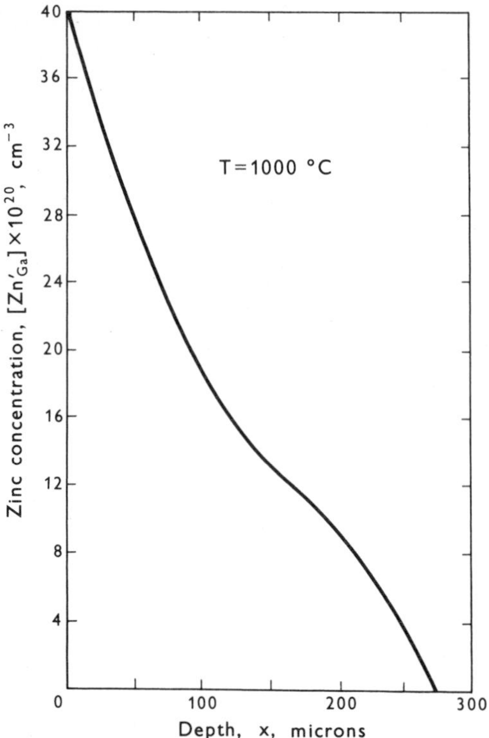

Figure 6.19. A linear plot of the 1000° C diffusion profile of figure 6.18 which illustrates the steep Zn concentration gradient near the surface.

of equation 6.43 are difficult to evaluate. At high concentrations, the slope evaluation is subject to significant errors. Therefore, the greatest errors in the Boltzmann-Matano analysis occur at the lowest and highest concentrations. The concentration-dependent diffusion coefficients obtained in this manner are shown in figure 6.20 and are for the range of $[Zn'_{Ga}]$ where these errors are least significant. Isoconcentration diffusion values at 900° C which were taken from references 44 and 79 are shown for comparison. Section 6.3.7 considers isoconcentration diffusion in detail.

It should be noted in figure 6.20 that for each temperature the D-versus-$[Zn'_{Ga}]$ curve varies as $[Zn'_{Ga}]^2$ at the lower concentration, goes through a peak and valley, and then increases. These features will now be considered in terms of the effective diffusion coefficient given in equation 6.59. Both the charge state of the species in the reaction of the interstitial Zn with a Ga vacancy to form the substitutional Zn and maintenance of equilibrium must be questioned. Also, regardless of the ionization state of the interstitial species for the interstitial-substitutional diffusion, the effective diffusion coefficient at any temperature should

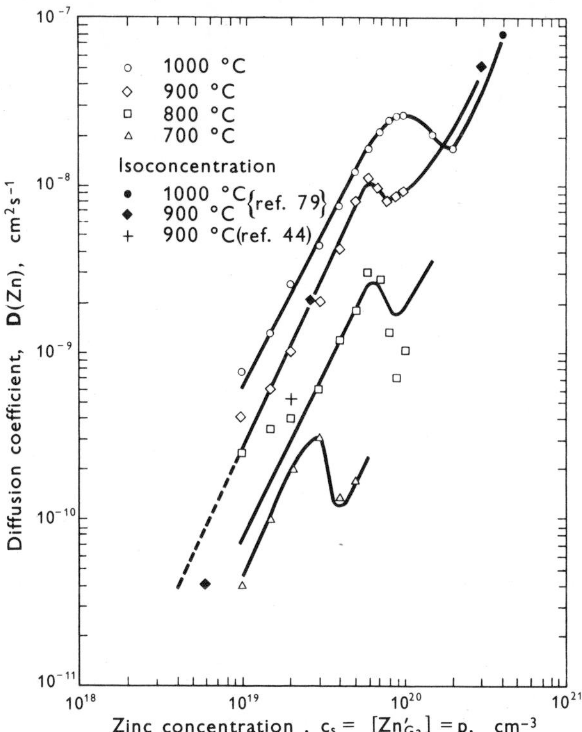

Figure 6.20. Diffusion coefficient of Zn in GaAs versus Zn concentration at specified temperatures as derived from a Boltzmann-Matano analysis (reference 66).

vary inversely with the Ga vacancy concentration, which is represented by the $p_{As_2}^{-1/2}$ dependence of **D**(Zn).

As discussed in the solubility analysis for the surface concentration, the data could only be reasonably fit with a neutral Ga vacancy and a singly ionized substitutional acceptor. The interstitial Zn has previously been considered either as a singly or doubly ionized donor. Longini [15] suggested doubly ionized because Zn normally loses two electrons in chemical bonding, but Weisberg and Blanc [73] pointed out that the second ionization potential (in vacuum) of Zn (17.9 eV) exceeds the first potential of hydrogen (13.5 eV) and is close to the second ionization for Cu (20.3 eV). Since in GaAs hydrogen is neutral, while interstitial Cu is a single donor, Weisberg and Blanc concluded that interstitial Zn is probably a single donor [73]. If the interstitial Zn is taken as doubly ionized, then the effective diffusion coefficient would be

$$\mathbf{D}(\mathrm{Zn}) = \frac{3D(\mathrm{Zn}_i)}{K_6(T)p_{As_2}^{1/2}}\, c_s^3\gamma_p^3 \left[1 + \frac{2}{3}\frac{c_s}{\gamma_p}\frac{d\gamma_p}{dc_s}\right] \tag{6.61}$$

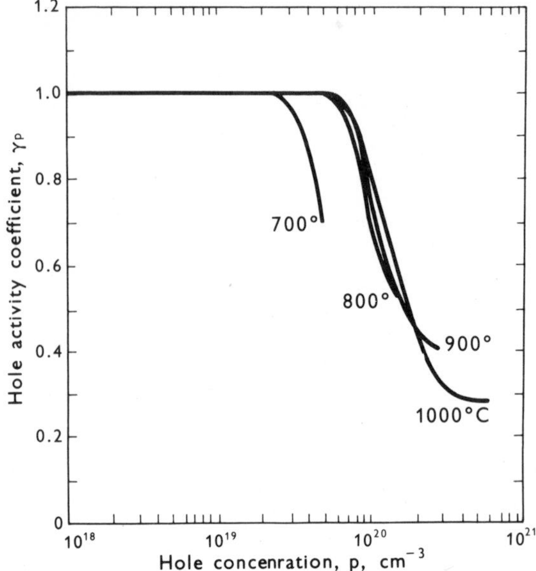

Figure 6.21. Variation of the hole activity coefficient with hole concentration at specified temperatures (reference 66).

and would vary as the concentration cubed at low concentrations. Thus, the data of figure 6.20 are consistent with a singly ionized interstitial Zn.

At high concentrations γ_p is expected to decrease from unity. Since γ_p enters equation 6.59 as $[\gamma_p^2 + (c_s\gamma_p/2)(d\gamma_p/dc_s)]$, an explicit expression for γ_p cannot be obtained, and the values of γ_p as a function of Zn concentration in the solid must be found by trial and error. The possible values of γ_p are further limited by the restriction that the Fermi level is a single-valued, monotonic function of p. This restriction means that in the region where $\gamma_p < 1$, there is a minimum possible value of γ_p at any given c_s for E_F [given by equation 6.42] to decrease smoothly. The γ_p curves shown in figure 6.21 were obtained in this manner from the curves of figure 6.20. For the Zn concentrations considered in this diffusion analysis, the Zn is fully ionized so that $c_s = [Zn'_{Ga}] = p$. The initial departure from the c_s^2 dependence indicates the concentration where γ_p becomes less than unity. The valley in the **D**(Zn)-versus-c_s curve occurs approximately at the inflection point of a linear γ_p-versus-c_s plot. At this point the quantity $(c_s/2\gamma_p)(d\gamma_p/dc_s)$ is negative and approaching unity. The approach again of **D**(Zn) to a c_s^2 dependence indicates γ_p is again approaching a constant value, but for $\gamma_p < 1$. Even if the total shape in the vicinity of the decrease of **D**(Zn) in figure 6.20 is not entirely due to γ_p, solubility results illustrated in figure 6.14 and theoretical considerations [30] require $\gamma_p < 1$. The γ_p-versus-p curves would be essentially the same for D(Zn)-versus-c_s curves that increase monotonically and do not have the minimum at high concentrations.

6.3.7 Isoconcentration Zn Diffusion

For isoconcentration diffusion [70, 79], nonradioactive Zn (preferably a ternary composition) is first diffused into the sample for a sufficiently long time to obtain a uniform Zn concentration throughout the entire sample. The sample is then diffused with radioactive Zn65 from a source of the same composition, and nonradioactive Zn diffuses out while the radioactive Zn diffuses in. The total Zn concentration throughout remains constant at all times. The resulting Zn65 profile is a complementary error function which readily permits determination of $D_e(\tilde{Zn})$ at a fixed Zn concentration. The isoconcentration and concentration-gradient diffusions are compared in figure 6.22 (reference 79) for Zn in GaAs at 900° C.

As pointed out by Shaw and Showan [80], derivation of the effective diffusion coefficient for isoconcentration diffusion differs from the procedure followed for the concentration-gradient diffusion. There is no

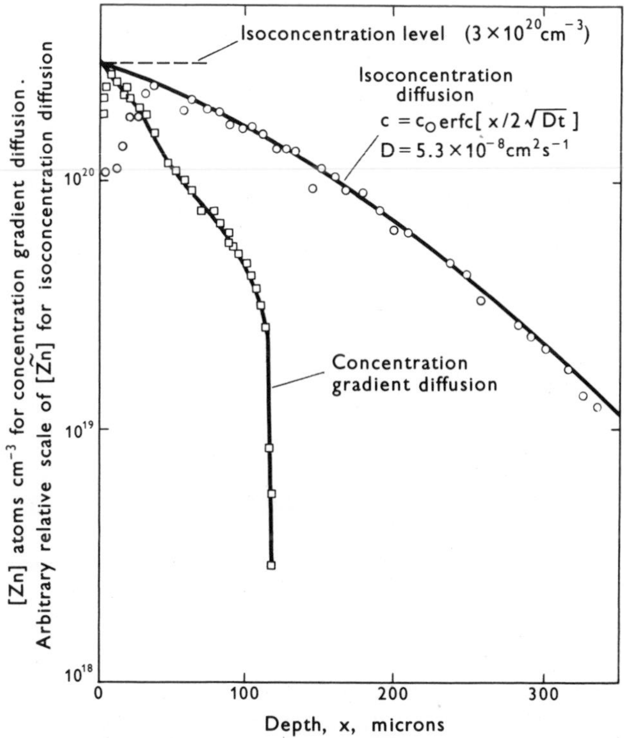

Figure 6.22. Comparison of diffusion profiles of concentration-gradient and isoconcentration diffusions for Zn in GaAs at 900° C for 0.75 h (reference 79).

gradient in the hole concentration or in the total Zn concentration, only in the radioactive Zn concentration. In addition, there is no built-in field. Equation 6.56 for the relation between the total interstitial concentration c_i and total substitutional concentration c_s must be written as the sum of the radioactive Zn, $\tilde{c}_i$ and $\tilde{c}_s$, and nonradioactive Zn, $\bar{c}_i$ and $\bar{c}_s$: $c_i = \bar{c}_i + \tilde{c}_i$ and $c_s = \bar{c}_s + \tilde{c}_s$. Equation 6.56 becomes

$$(\bar{c}_i + \tilde{c}_i) = \frac{(\gamma_p p)^2 (\bar{c}_s + \tilde{c}_s)}{K_6(T) p_{As_2}^{1/2}}, \tag{6.62}$$

and

$$\tilde{c}_i = \frac{(\gamma_p p)^2}{K_6(T) p_{As_2}^{1/2}} \tilde{c}_s \frac{(1 + \bar{c}_s/\tilde{c}_s)}{(1 + \bar{c}_i/\tilde{c}_i)}. \tag{6.63}$$

For the condition that the radioactive species is the same fraction of the interstitial and substitutional Zn, $(\bar{c}_i/\tilde{c}_i) = (\bar{c}_s/\tilde{c}_s)$, equations 6.44 and 6.63 give the $\widetilde{\text{Zn}}$ diffusion coefficient $D_e(\widetilde{\text{Zn}}) = -J(\text{Zn})/(\partial \tilde{c}_s/\partial x)$ as

$$D_e(\widetilde{\text{Zn}}) = \frac{D_e(\text{Zn}_i)(\gamma_p p)^2}{K_6(T) p_{As_2}^{1/2}}, \tag{6.64}$$

and with the neutrality relation, $p = [\text{Zn}'_{\text{Ga}}] \equiv c_s$,

$$D_e(\widetilde{\text{Zn}}) = \frac{D_e(\text{Zn}_i) \gamma_p^2 c_s^2}{K_6(T) p_{As_2}^{1/2}}. \tag{6.65}$$

Equation 6.65 for $D_e(\widetilde{\text{Zn}})$ has a Ga vacancy concentration dependence represented by $p_{As_2}^{1/2}$. As the Zn concentration for an isoconcentration diffusion is varied along the ternary liquidus of figure 6.8, the As_2 pressure decreases by a factor of 10 from the Zn-GaAs pseudobinary to the Ga-As binary at 900°C (reference 65). Thus, for a concentration-gradient diffusion with any given source composition, $X(V_{\text{Ga}})$ is the same at every c_s (under equilibrium conditions), but for the isoconcentration diffusion, equilibrium is established prior to the radioactive tracer diffusion, and $X(V_{\text{Ga}})$ varies for each choice of c_s.

The isoconcentration diffusion coefficients obtained for Zn in GaAs by Kendall [44] and Chang and Pearson [79] are shown in figure 6.23. The data have been fit with three different dependences. Curve *A* is for $D_e(\widetilde{\text{Zn}}) \propto c_s^2$, curve *B* is for $D_e(\widetilde{\text{Zn}}) \propto c_s^2/p_{As_2}^{1/2}$, while curve *C* also includes γ_p taken from figure 6.21 to give $D_e(\widetilde{\text{Zn}}) \propto \gamma_p^2 c_s^2/p_{As_2}^{1/2}$. The values of p_{As_2} used to calculate curves *B* and *C* were taken from the work of Jordan [65]. It should be noted that the $D_e(\widetilde{\text{Zn}})$ at 2.8×10^{19} cm^{-3} seems to be slightly in error and is not fit for any assumed model of fully ionized Zn unless the point at 3×10^{20} cm^{-3} is neglected. These data have previously been fit to suggest a c_s^3 dependence and hence a doubly ionized interstitial donor [79]. The fit shown in figure 6.23 with equation 6.65 shows that the

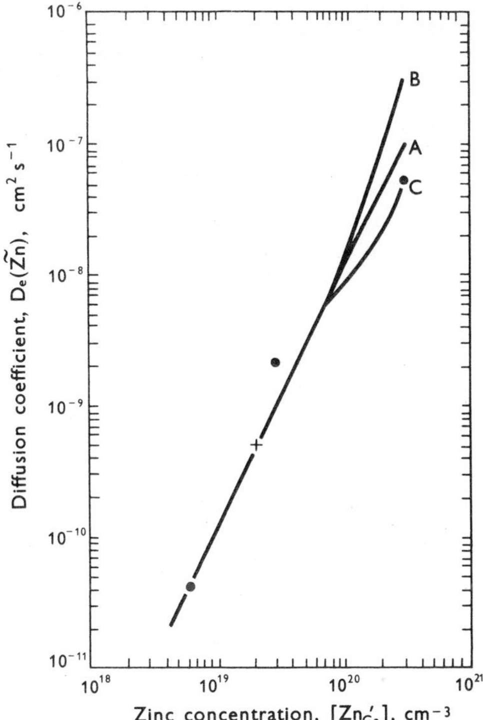

Figure 6.23. Evaluation of the Zn in GaAs isoconcentration diffusion coefficient at 900° C. Curve A, $D_e(\widetilde{\mathrm{Zn}}) \propto c_s^2$. Curve B, $D_e(\widetilde{\mathrm{Zn}}) \propto c_s^2/p_{\mathrm{As}_2}^{1/2}$. Curve C, $D_e(\widetilde{\mathrm{Zn}}) \propto \gamma_p^2 c_s^2/p_{\mathrm{As}_2}^{1/2}$.

donor is singly ionized. Further isoconcentration data in the concentration range from 3 x 10^{19} to 3 x 10^{20} cm^{-3} would be very helpful. Figure 6.23 also illustrates the effect of γ_p^2 on the diffusion coefficient, which for concentration-gradient diffusion has the additional term $[1 + (c_s/2\gamma_p)(d\gamma_p/dc_s)]$. Thus, the hole activity coefficient must contribute at least in part to the D(Zn)-versus-c_s behavior shown in figure 6.20. However, departures from vacancy equilibrium may also contribute to the shape of the diffusion profiles and will be discussed in section 6.3.9 after the effects of arsenic pressure have been considered.

Ting and Pearson [80a] have measured $D_e(\widetilde{\mathrm{Zn}})$ in GaAs at different temperatures between 600 and 1000° C. The Zn concentration was maintained close to 2.0 x 10^{19} cm^{-3} in all the experiments. They found $D_e(\widetilde{\mathrm{Zn}}) = 9.0 \times 10^{-10}\ \mathrm{cm}^2\,\mathrm{s}^{-1}$ (after corrections for deviations in Zn concentration from the value 2.0 x 10^{19} cm^{-3}), with a standard deviation of 1.15, for the temperature range and Zn concentration cited. The results are consistent with the interstitial-substitutional model but require that $c_i D_e(\mathrm{Zn}_i)$ be independent of temperature.

6.3.8 Effects of Arsenic Pressure

The effective diffusion coefficient for interstitial-substitutional diffusion, as expressed by equation 6.59, varies as $p_{\mathrm{As}_2}^{1/2}$ at a fixed temperature and Zn concentration. Numerous authors have used excess arsenic to obtain more planar *p-n* junction interfaces, but most of these studies concern only the empirical behavior. However, the work of Shih, Allen, and Pearson [52, 64] provided a quantitative understanding of the effect of arsenic pressure on the Zn diffusion. In their work, the diffusion source compositions were varied from the pseudobinary composition to the As-rich side of the liquidus isotherm as shown in figure 6.24. In the binary

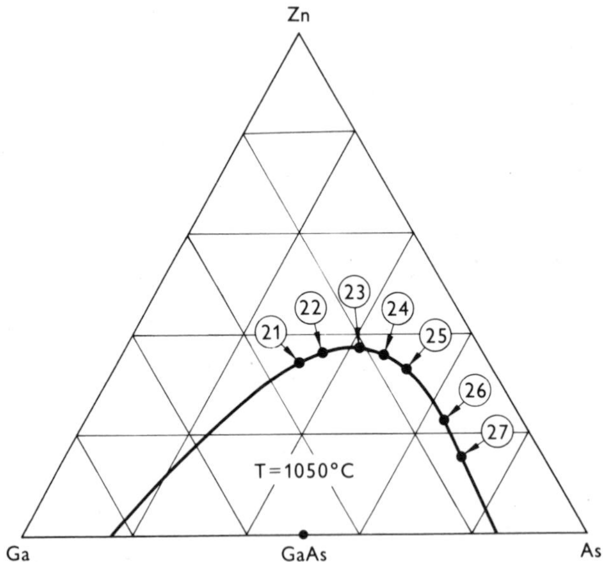

Figure 6.24. The 1050° C Ga-As-Zn ternary isotherm illustrating the diffusion source compositions that were used to obtain the diffusion profiles in figure 6.26 (reference 52).

Ga-As system, As_4 dominates on the As-rich side, and therefore the arsenic pressure measurements of Shih *et al.* [64] on the As-rich side of the liquidus isotherm are expected to measure the As_4 pressure. For this reason, it is more convenient to express equation 6.59 in terms of $p_{\mathrm{As}_4}^{1/4}$ as

$$\mathbf{D}(\mathrm{Zn}) = \frac{2D(\mathrm{Zn}_i)c_s^2}{K_6'(T)p_{\mathrm{As}_4}^{1/4}}\gamma_p^2\left[1 + \frac{c_s}{2\gamma_p}\frac{d\gamma_p}{dc_s}\right]. \tag{6.66}$$

Since p_{As_2} and p_{As_4} are related in a known manner, as shown in figure 6.25, either expression may be used. Shih *et al.* [64] assumed that

As_4 was the dominant species under all conditions of their measurements, and As_2 was neglected. The recent measurements by Arthur [24] showed that As_2 was the dominant species in the Ga-rich region. Jordan [65] recalculated the pressure measurements of Shih *et al.*, to give the proper pressure assignments shown in figure 6.25. Fortunately, As_4 dominates for the diffusion source compositions illustrated in figure 6.24 for 1050° C.

The diffusion profiles at 1050° C for 1 h with the As-rich diffusion sources of figure 6.24 are shown in figure 6.26. With increasing X(As), and hence greater p_{As_4}, the diffusion profiles become progressively shallower.

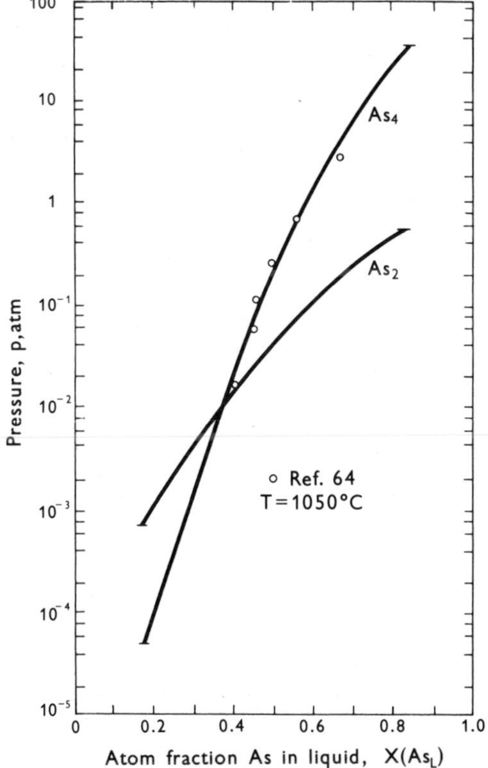

Figure 6.25. The As_2 and As_4 partial pressures versus $X(As_L)$ along the 1050° C liquidus isotherm in the Ga-As-Zn system (reference 65).

The diffusion coefficients obtained from these profiles by the Boltzmann-Matano analysis are plotted in figure 6.27 and show that the diffusion coefficient at a fixed Zn concentration indeed decreases. When these diffusion coefficient values for constant Zn concentrations are plotted as a function of As_4 pressure in figure 6.28, a reasonable $p_{As_4}^{-1/4}$

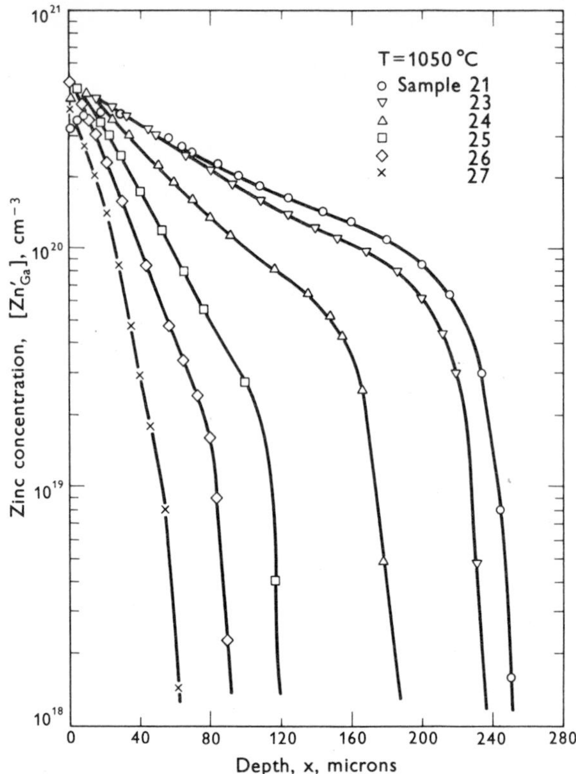

Figure 6.26. Diffusion profiles of Zn in GaAs at 1050° C for 1 h with excess arsenic added (reference 52). The sample numbers refer to the diffusion source compositions indicated in figure 6.24.

dependence is obtained. Data from profiles for 900° C are also included in this figure. As the arsenic pressure increases, the profiles become shallower and accurate diffusion coefficients become more difficult to obtain, and the fit at each concentration to $p_{As_4}^{-1/4}$ is not exact. When the difficulties of obtaining quantitative diffusion profiles and arsenic pressure are considered, the results are quite reasonable.

The results of Shih, Allen, and Pearson demonstrate two very basic concepts. First, the assignment of an interstitial-substitutional diffusion model is verified by the observed dependence of the effective diffusion coefficient on the arsenic pressure. *Second, these results illustrate that in the interpretation of diffusion in the III-V compound semiconductors it is necessary to consider the ternary nature of the system; the results derived from treatment of the compound as one element in a binary system can be very misleading.*

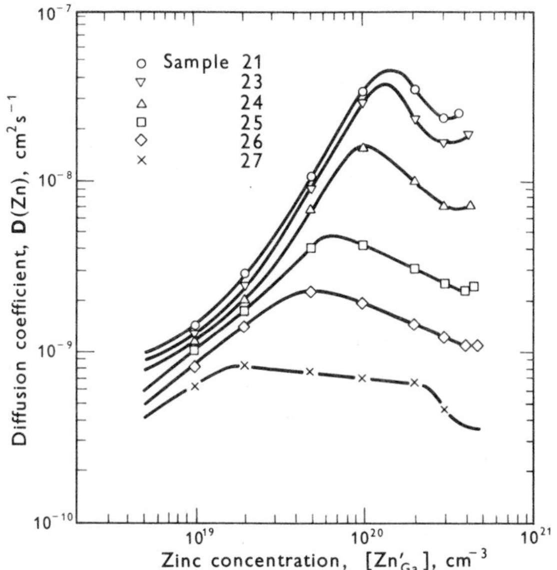

Figure 6.27. Diffusion coefficient of Zn in GaAs versus Zn concentration at 1050° C as derived from the Boltzmann-Matano analysis (reference 52).

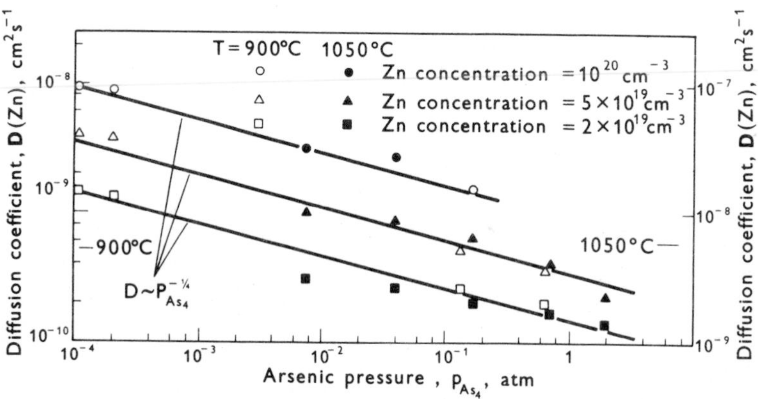

Figure 6.28. Diffusion coefficient at a given Zn concentration as a function of p_{As_4} at 900° and 1050° C (reference 52).

6.3.9 Departure from Equilibrium

The analysis given above in sections 6.3.5 and 6.3.6 for the diffusion of Zn in GaAs has assumed homogeneous concentrations for the native defects which are equal to their equilibrium values and are determined by the temperature and diffusion source. In this section the evidence for and consequences of nonequilibrium diffusion will be considered. Unlike the analyses in the previous sections, analysis for nonequilibrium conditions does not lead to analytical expressions that readily permit comparison with experimental results. Several experimental observations that may be related to nonequilibrium are presented here. One is the irregular junction interface that is frequently observed for low arsenic pressure sources. Another is the observation that if the diffusion depth or time exceeds certain values, the diffusion behavior changes and the diffusion profile no longer varies as the square root of time. These changes in diffusion profile are accompanied by diffusion-induced dislocations. In addition, it is necessary to consider the rate and mechanism of vacancy supply.

Numerous publications have dealt with diffusion-induced dislocations and precipitation, especially for phosphorus and boron in silicon [81, 82]. A complete description of these subjects for Zn in GaAs is not available at the present time. However, certain facts have emerged. The work of Black and Jungbluth [83, 84] and Maruyama [85] indicates that a certain period of time or 'incubation' period is required before diffusion-induced dislocations occur. It should be emphasized that many routine techniques permit preparation of *p-n* junctions by diffusion without diffusion-induced dislocations. Present results indicate that for a given Zn surface concentration, there is an incubation period of diffusion without diffusion-induced dislocations. Excess dislocations in the entire diffused layer are observed for diffusion times in excess of this period. This incubation period is shorter the higher the diffusion temperature. Although there is considerable evidence for diffusion-induced precipitation, there have not been sufficient experimental data to describe the nature of the precipitates in detail. For Zn in GaAs, agreement between the hole concentration and the total Zn concentration was within the accuracy of the Zn analysis technique [83]. This result does not rule out the precipitation of some Zn, but does indicate that the majority of the Zn is still present substitutionally in the lattice.

The work of Ting [86] and Nygren and Pearson [87] shows what may be happening as the diffusion passes through the critical conditions for diffusion-induced dislocations. First of all, the variation of junction depth with the square-root of time has a kink as illustrated in figure 6.29. Similar results have been obtained for Zn in GaP [87]. These studies [86, 87] showed that the shallower diffused junctions had no diffusion-induced defects, while beyond the break in the curves in figure 6.29 the diffused layers contained high dislocation densities. In addition, there is evidence that the shape of the diffusion profile is different in the region before and after the break in the curves in figure 6.29. Beyond the knee in the curves, the diffused layers are no longer transparent, but become opaque

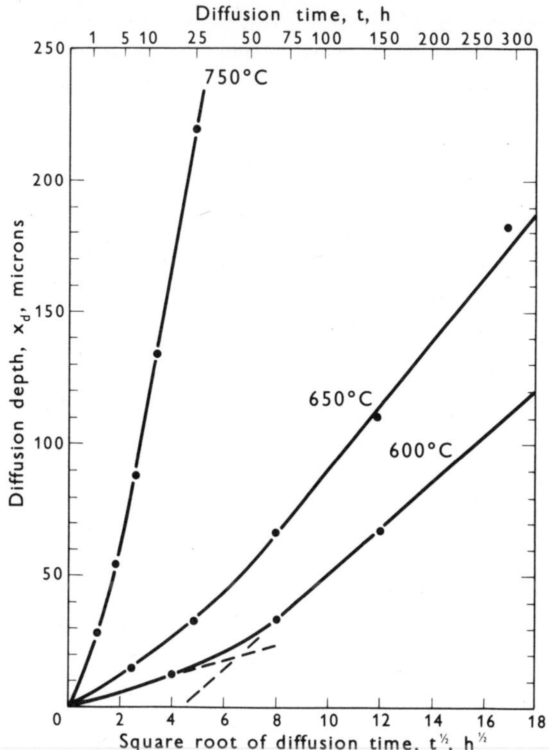

Figure 6.29. Diffusion depth at a concentration of 10^{18} cm^{-3} as a function of the square-root of time for Zn in GaAs at 600°, 650°, and 750° C (reference 86).

[84, 85, 87]. The 'incubation' period discussed by Black and Jungbluth [83, 84] appears to be related to the kink in the junction-depth-versus-square-root of time curve.

The analysis of diffusion-induced dislocations is usually based on stress that is related to the change in lattice constant due to the presence of the impurity. However, measurements by Black and Lublin [88] failed to show changes in the lattice constant for heavily Zn-doped GaAs. A particularly sensitive method of measuring the small strain produced by the high impurity concentration of diffused layers is by use of a double crystal spectrometer [89] in a double beam mode. No strain due to high Zn concentrations could be observed by this method [90]. In the absence of lattice strain data, application of the analysis for diffusion-induced dislocations is not possible.

The application of the Boltzmann-Matano analysis requires that the diffusion profiles vary as $x/t^{1/2}$. Application of this analysis in the vicinity of a pronounced kink in the junction-depth-versus-square-root of time plot violates this condition. If, however, the profile is in the initial region before the kink, or for sufficiently long times that the pre-kink region is an insignificant portion of the overall diffusion, then the analysis should be representative of the diffusion coefficient.

The question of vacancy equilibrium is particularly difficult to resolve. There are really two aspects that must be considered. One is the achievement of the proper stoichiometry required by the ternary partial pressures, and the other is the maintenance of the local vacancy equilibrium as the high substitutional Zn concentration fills the Ga vacancies. Figure 6.7 illustrates the stoichiometry question. Because the As vacancy concentration can be assigned a quantitative value, it is used in figure 6.30 to illustrate the stoichiometry during diffusion. The Ga vacancy concentration is related to $X(V_{As})$ by equation 6.7: $X(V_{Ga})X(V_{As}) = K_V(T)$. For a fixed diffusion temperature, the ternary source composition determines the crystal stoichiometry. At 1000° C, the

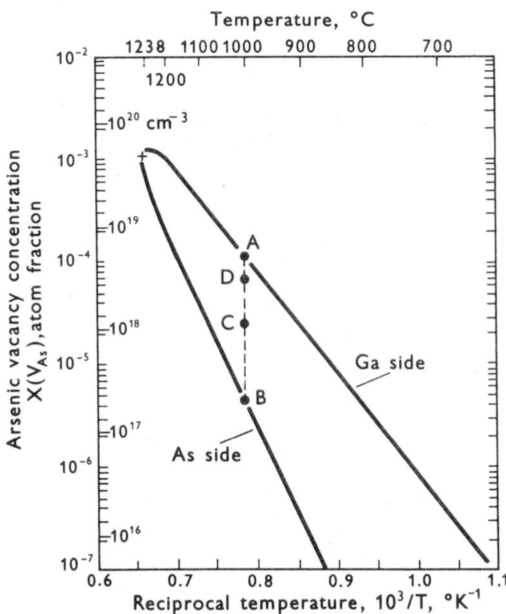

Figure 6.30. The As vacancy concentration as a function of temperature along the Ga-As liquidus (see figure 6.7). The melting point is indicated by a cross (+). Points *A* and *B* are the vacancy concentrations on the Ga- and As-rich sides respectively. Point *C* is taken as a representative $X(V_{As})$ for a crystal pulled from the melt. Point *D* represents the As vacancy concentration for a Ga-rich ternary diffusion source.

vacancy concentration moves from A to B as the ternary source composition is varied from the Ga-rich to the As-rich side as shown in figure 6.30. A crystal pulled from the melt would have a stoichiometry determined by the melt growth conditions and could reasonably be expected to have a vacancy concentration at 1000° C between A and B as represented by C. If a ternary source has an arsenic pressure that would represent point D, vacancy and self-diffusion from the surface are necessary for establishment of the new conditions. Only diffusion from the surface can establish these conditions of a new crystal equilibrium, and this vacancy diffusion must proceed more rapidly than the Zn diffusion if Ga and As vacancy equilibria are to be maintained in the diffused region.

Even if the correct stoichiometry is established, it is not clear that the Ga-vacancy concentration can be maintained as the Zn substitutionally fills the Ga vacancies. These vacancies can be supplied by internal sources such as dislocations and do not have to depend on supply from the surface. Shaw and Showan have considered Zn diffusion when the rate of supply of Ga vacancies dominates the diffusion behavior [80]. However, at the present time, there are rather formidable difficulties in a quantitative approach to this problem.

Figure 6.31 shows, for Zn diffusion in GaP, the irregular diffusion front

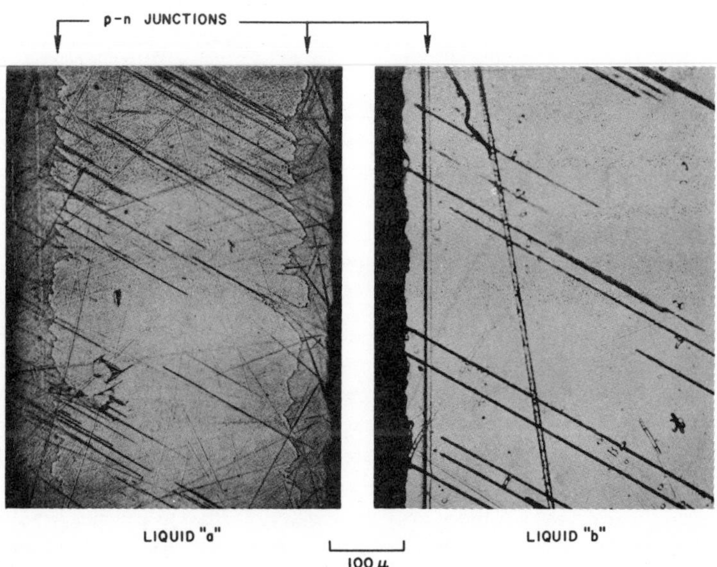

Figure 6.31. Etched cross sections of GaP crystals comparing junction planarity of samples that were Zn diffused to approximately the same depth at 900° C with the Ga-rich liquid 'a' and the P-rich liquid 'b' (reference 87).

often encountered for Ga-rich diffusion sources (labeled liquid 'a'). The behavior is the same in GaAs, but generally not as pronounced. A P-rich diffusion source (labeled liquid 'b') is shown to result in very planar *p-n* junctions. Several As-rich sources for GaAs that have been discussed [17, 91] also result in planar junction interfaces. Because a planar diffusion can be obtained in the same material by just increasing the group V partial pressure, the dislocations cannot inherently act as 'diffusion pipes.' These results may indicate nonequilibrium near the dislocations, and if the diffusion is slowed down by a high phosphorus or arsenic pressure, then the external vapor enforces a uniform vacancy equilibrium upon the crystal ahead of the Zn diffusion front.

The question of nonequilibrium should not detract from the basic concepts of interstitial-substitutional diffusion. These nonequilibrium effects, which are mentioned above, may be contributing factors, but certainly are not the dominant factors in controlling diffusion. These concepts of departure from equilibrium are presented only to give a complete representation of the status of the behavior of Zn in GaAs. Consideration of Zn in GaAs does bring forth in considerable detail many techniques and diverse concepts that must be considered in any impurity diffusion in the III-V compounds.

6.3.10 Zinc Diffusion in GaP, InAs, InP, AlSb, and InSb

Only for the Ga-As-Zn system have both the ternary phase diagram and the partial pressures been obtained experimentally. For Zn in GaP, the ternary phase diagram has been determined by Panish [92], and the partial pressures have been calculated by Jordan [65]. Although detailed analysis is not possible without both partial pressures and the ternary phase diagram, useful diffusion profiles have been obtained also for Zn diffusion into InAs, InP, AlSb, and InSb.

Diffusion profiles for Zn in GaAs (reference 66) and GaP (reference 78) are compared in figure 6.32(a). For comparable diffusion source compositions, the profiles have similar shape, but the surface concentration and total depth of diffusion are less for GaP. For InP (reference 93) and InAs (reference 94) shown in figure 6.32(b), the steep diffusion front in the interior also occurs as for GaAs and GaP. However, for InP and InAs with diffusion source compositions similar to those for GaAs and GaP in figure 6.32(a), the surface concentration is greater, and the profiles drop very rapidly near the surface. Diffusion profiles for Zn in InSb (reference 44) and Zn in AlSb (reference 95) are shown in figure 6.33(a) and (b) respectively. The profiles for InSb are relatively constant in concentration, and then decrease abruptly at the diffusion front. The profiles for AlSb shown in part (b) of figure 6.33 are similar to those of GaAs and GaP.

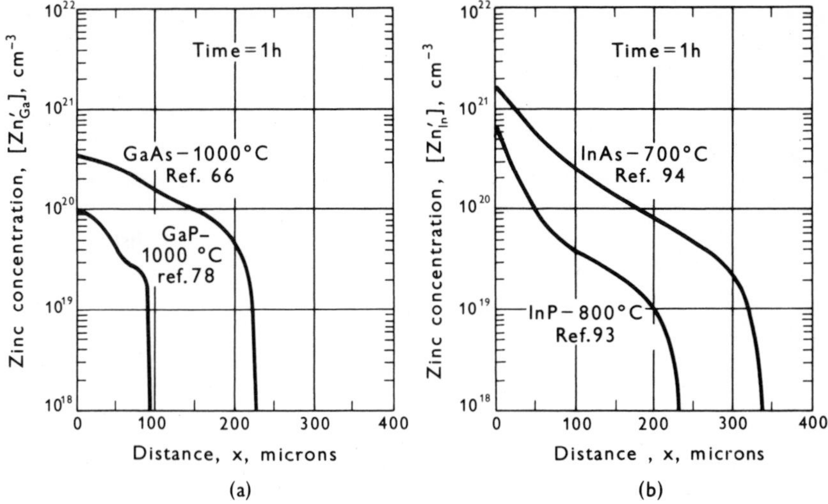

Figure 6.32. (a) Diffusion profile of Zn in GaAs and GaP, and (b) diffusion profile of Zn in InAs and InP.

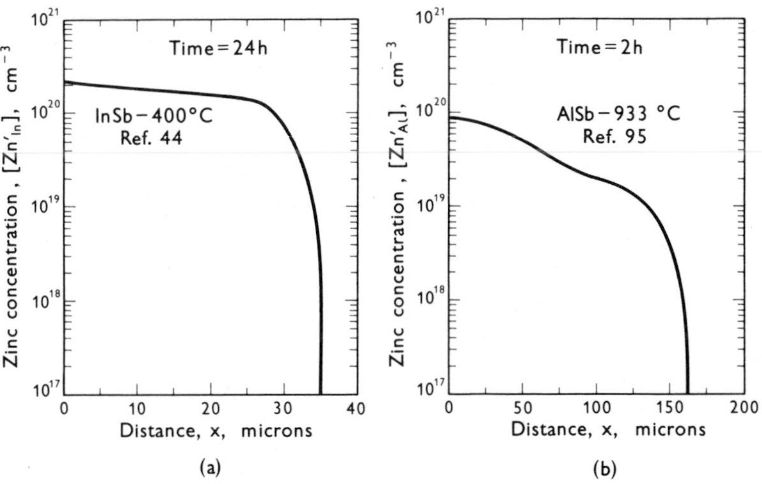

Figure 6.33. (a) Diffusion profile of Zn in InSb, and (b) diffusion profile of Zn in AlSb.

The concentration-dependent diffusion coefficient for the Zn diffusion profiles of figure 6.32(a) and (b), as determined by the Boltzmann-Matano analysis, are compared in figure 6.34. All these concentration-dependent diffusion coefficients tend to have a concentration-squared dependence near 10^{19} Zn atoms cm^{-3}, and then decrease at higher Zn concentrations. Therefore, the concepts developed for the interstitial-substitutional diffusion mechanism of GaAs should also be applicable to these III-V

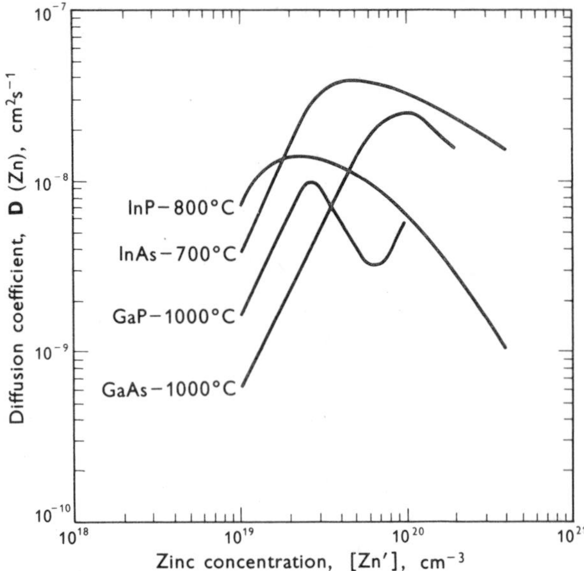

Figure 6.34. Diffusion coefficient of Zn in InP, InAs, GaP, and GaAs versus Zn concentration at specified temperatures as derived from a Boltzmann-Matano analysis.

binary semiconductors. Assignment of a unique cause for the decrease in the effective diffusion coefficient at the high Zn concentrations does not appear possible at the present time. As mentioned previously, the departure of γ_p from unity, vacancy nonequilibrium, diffusion-induced dislocations, and precipitation may all play roles in the departure from the concentration-squared dependence of the diffusion coefficient.

6.3.11 Zinc Diffusion in the Ternary III-V Semiconductors

Although Zn diffusion is routinely used in the preparation of $GaAs_{1-x}P_x$ electroluminescent *p-n* junctions, little quantitative data have been presented. Nuese *et al.* [96] measured junction depths for diffusion into $GaAs_{1-x}P_x$ from sources containing 1 mg Zn and 10 mg As in 5 cm^3 ampoules. The crystals were $GaAs_{0.67}P_{0.33}$, and the diffusion temperature and time were varied to obtain the junction-depth-versus-square-root of time curve shown in figure 6.35(a). Then the diffusion time and temperature were held constant for samples of different mole fraction of P. The result shown in figure 6.35(b) indicates a smooth decrease in diffusivity as the ternary compound varies from GaAs to GaP. The comparison of diffusion in GaAs and GaP shown in figure 6.32(a) also suggests a decreased diffusivity as the P mole fraction becomes larger.

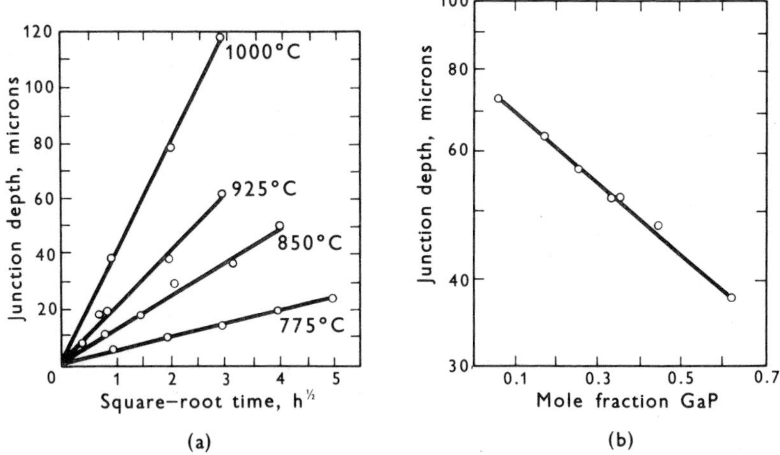

Figure 6.35. Zinc diffusion in $GaAs_{1-x}P_x$ (reference 96).
(a) Junction-depth-versus-square-root of time at the indicated temperatures for $GaAs_{0.67}P_{0.33}$. The background electron concentration is 1.3×10^{18} cm^{-3}.
(b) Junction depth as a function of GaP mole fraction at 925° C and 14 h.

Neuse *et al.*, also found that the junction depth decreased with increased arsenic pressure as was observed for GaAs, and the higher arsenic pressure eliminated nonplanar junction interfaces in the same manner as illustrated for GaP in figure 6.31.

Isoconcentration Zn and Cd diffusion studies have been made by Arseni *et al.* [97] for the ternary systems $InAs_{1-x}P_x$ and $GaAs_{1-x}P_x$. Determination of the variation of the diffusion coefficient $\mathbf{D}_e(\tilde{Zn})$ with composition x is complicated by the accompanying variation in solubility with x for a given diffusion source. Because $\mathbf{D}_e(\tilde{Zn})$ is also concentration dependent, it is difficult to identify which portion of the variation in $\mathbf{D}_e(\tilde{Zn})$ is due to the variation in ternary compound composition. However, it appears that the concepts developed for Zn in GaAs are applicable to the ternary compounds.

6.4 DONOR DIFFUSION IN GaAs

6.4.1 General Behavior

Unlike the great wealth of data for Zn diffusion, only limited data exist for donor diffusion in the III-V compounds. As in the case of Zn, however, the most extensive data are for GaAs. One general feature of donor diffusion in the III-V compounds is a very low diffusivity. Both high diffusion temperatures within 200° to 300°C of the host lattice melting point and long diffusion times are necessary to achieve measurable donor penetration. These conditions result in sample surface deterioration due to transport of the volatile group V element in the presence of small temperature gradients within the ampoule. In addition, for donor diffusion with S, Se, and Te, the surface deterioration is enhanced by both the formation of compounds such as Ga_2S_3 on the surface [98] and increased vapor transport in the presence of these impurities.

The experimental results of Young and Pearson [99] for sulfur in GaAs and GaP are presented in section 6.4.3 to illustrate donor diffusion in the III-V compounds. For an understanding of donor surface concentration, the solid solubility results for Te in GaAs of Casey, Panish, and Wolfstirn [100] are first considered. These Te results limit the possible interpretations of the S diffusion data. Additional experiments are required before a complete and fundamental explanation of donor diffusion can be presented.

6.4.2 Donor Surface Concentration

The impurity incorporation concepts that were presented for Zn in GaAs cannot be used for donors in unmodified form. Examination of existing results for other impurities in GaAs and GaP suggests that it is necessary to include surface band bending in the interpretation of donor incorporation in III-V compounds. Zschauer and Vogel [101] have considered the relationship of surface band bending, crystal growth rate, and diffusivity in the solid to impurity incorporation. The application of surface band bending to Te in GaAs by Casey *et al.* [100] illustrates the basic concepts.

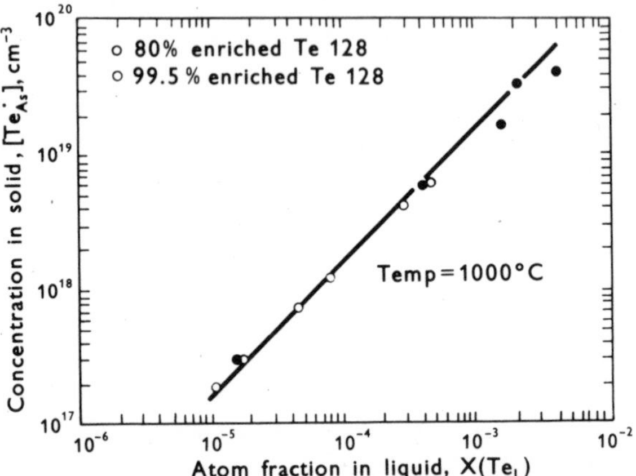

Figure 6.36. A portion of the 1000° C solid-solubility isotherm for Te in GaAs (reference 100).

In that study, the 1000°C solid solubility isotherm for Te in GaAs between 2×10^{17} and 4×10^{19} Te atoms cm^{-3} was found to vary linearly with atom fraction in the liquid as shown in figure 6.36. For extrinsic conditions and dilute liquid solutions, the concepts that result in equation 6.36 for fully ionized impurities lead to a square-root dependence on the impurity concentration in the liquid phase. One condition for the linear relationship between the impurities in the solid and liquid phases is that the electron concentration at the surface of the solid must be independent of the impurity concentration in the solid. Treatment of the liquid-solid interface as a metal-semiconductor Schottky barrier gives an electron concentration at the surface that is a function of temperature only [100].

The reaction for incorporation of Te as a simple fully ionized substitutional donor on an As site during liquid-phase epitaxial growth may be described by

$$\mathrm{Te}_L + V_{\mathrm{As}} \rightleftharpoons \mathrm{Te}^{\cdot}_{\mathrm{As}} + e'. \tag{6.67}$$

It is well established that unassociated substitutional Te is fully ionized for concentrations in excess of 10^{16} cm^{-3} at all temperatures [63], and in the absence of complexes [100, 102], the ionized Te concentration represents the total Te concentration in the solid. By equations 6.2 and 6.5, $X(V_{\mathrm{As}})$ may be shown to be proportional to $\gamma(\mathrm{Ga}_L)X(\mathrm{Ga}_L)$, and the equilibrium relationship for equation 6.67 may be written as

$$K_7(T) = \frac{\gamma(\mathrm{Te}_S)[\mathrm{Te}^{\cdot}_{\mathrm{As}}]\gamma_n n}{\gamma(\mathrm{Te}_L)X(\mathrm{Te}_L)\gamma(\mathrm{Ga}_L)X(\mathrm{Ga}_L)}, \tag{6.68}$$

where γ_n is the activity coefficient of electrons in the solid, n is the electron concentration, and $\gamma(\mathrm{Te}_S)$ and $[\mathrm{Te}^{\cdot}_{\mathrm{As}}]$ are the activity coefficient and concentration of $\mathrm{Te}^{\cdot}_{\mathrm{As}}$ on As sites. The quantities $\gamma(\mathrm{Te}_L)$ and $\gamma(\mathrm{Ga}_L)$ are the activity coefficients for Te and Ga in the liquid. It is convenient to express $[\mathrm{Te}^{\cdot}_{\mathrm{As}}]$ in units of atoms cm^{-3} and the concentrations of the composition in the liquid as atom fractions, $X(\mathrm{Te}_L)$ and $X(\mathrm{Ga}_L)$.

Equation 6.68 may readily be simplified. When the solid solutions are 0.1 atom percent or less, Henry's law should be obeyed and $\gamma(\mathrm{Te}_S)$ is taken as constant. The Te concentrations in the liquid are quite dilute, and $X(\mathrm{Ga}_L)$ is essentially constant at 0.88 atom fraction. Therefore, $\gamma(\mathrm{Te}_L)$ and $\gamma(\mathrm{Ga}_L)X(\mathrm{Ga}_L)$ may also be taken as constant. Treatment of the electron activity coefficient in the same manner as for holes shows that γ_n is relatively constant at a value of approximately 0.4 at 1000°C in the concentration range considered here [68]. Equation 6.68 reduces to the simple expression

$$[\mathrm{Te}^{\cdot}_{\mathrm{As}}] = \frac{K'_7(T)X(\mathrm{Te}_L)}{n}. \tag{6.69}$$

If $[Te_{As}^{\cdot}] = n$, then $[Te_{As}^{\cdot}] = [K_7'(T)X(Te_L)]^{1/2}$, as for Zn in equation 6.36. However, the experimental results shown in figure 6.36 give a linear dependence between $X(Te_L)$ and $[Te_{As}^{\cdot}]$.

The electron concentration is given by the position of the Fermi level E_F and the density of states $\rho(E)$ by [103]

$$n = \int_{-\infty}^{\infty} \frac{\rho(E)dE}{[1 + \exp(E - E_F)/kT]} \tag{6.70}$$

The effects of the high impurity concentration may be included in γ_n, so that equation 6.70 gives n at the surface as

$$n(0) = N_c F_{1/2}(-\Phi_{Bn}/kT), \tag{6.71}$$

where N_c is the effective density of states, Φ_{Bn} is the position of the Fermi level relative to the conduction band at the surface, and $F_{1/2}(-\Phi_{Bn}/kT)$ is the Fermi-Dirac integral.

Treatment of the liquid-solid interface as a metal-semiconductor Schottky barrier results in the energy-band diagram shown in figure 6.37.

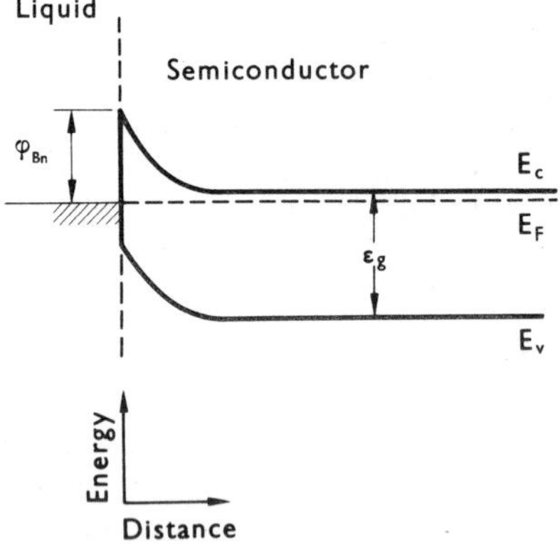

Figure 6.37. Energy-band diagram for a liquid n-type semiconductor interface. The separation between the valence band E_v and the conduction band E_c is the energy gap ϵ_g. The Fermi level is E_F and the barrier height Φ_{Bn} is determined by the position of E_F at the metal-semiconductor interface (reference 100).

Experimental data on Si suggest that the position of the Fermi level at the surface remains at a fixed energy above the valence band as the temperature varies. For GaAs at 1000° C, $\Phi_{Bn} \approx 0.4$ eV (reference 100) for the assumption that GaAs behaves similarly to Si and GaP. Then, for $\Phi_{Bn} > kT$, the Boltzmann approximation may be used for $F_{1/2}(-\Phi_{Bn}/kT)$ and

$$[Te_{As}^{\cdot}] = \frac{K_7'(T)X(Te_L)}{N_c \exp(-\Phi_{Bn}/kT)}, \tag{6.72}$$

and hence $[Te^{\cdot}_{As}]$ varies linearly with $X(Te_L)$ at a given temperature.

For GaAs grown near the melting point, chemical analysis showed that a linear relation results between the amount of Se (references 104 and 105) and Te (reference 104) in the solid and melt. For S (reference 104), the relationship was linear up to concentrations in the solid of $\approx 2 \times 10^{19}$ cm^{-3}; above 2×10^{19} cm^{-3} there was less than linear increase in the solid. The acceptor Ge in GaAs has a linear dependence on the amount in the liquid [106], while Zn varies as the square-root under the same conditions. The distinction between the linear and square-root behavior appears to be related to the diffusivity of the impurity in the solid. Only for a relatively high diffusivity such as for Zn can the liquid achieve equilibrium with the semiconductor bulk. For low diffusivity, the liquid is in equilibrium with the surface layer, and the solubility is determined by the Fermi level at the surface rather than in the bulk.

The surface concentration for donor diffusion from the vapor phase is less readily understood than impurity incorporation during growth from

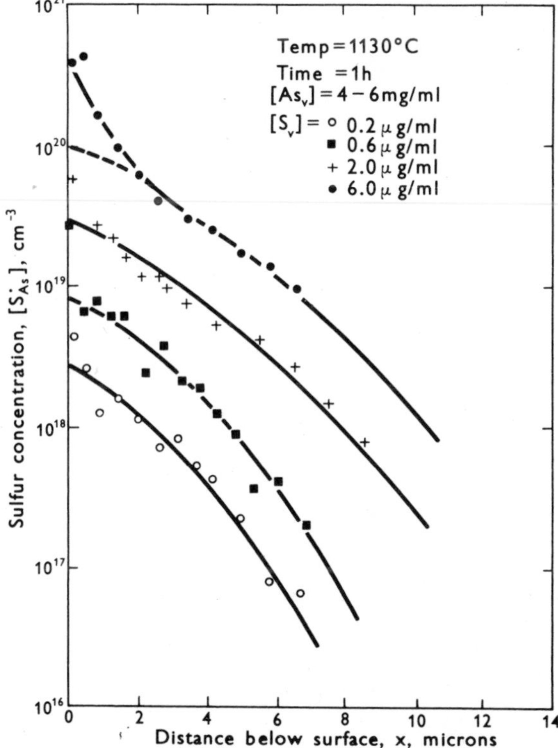

Figure 6.38. Diffusion profiles of sulfur in GaAs at various sulfur pressures (reference 99).

the equilibrium liquid. One problem is the interpretation of vapor-phase diffusion data without knowledge of the dominant impurity vapor species. The diffusion profiles obtained by Young and Pearson [99] by use of radioactive S35 are shown in figure 6.38. To prevent surface deterioration and large sample weight losses, the sample was 'sandwiched' between two fused-silica plates. The loss on each face was ≈0.25 μm. These profiles were obtained for varying amounts of S at constant temperature and arsenic pressure. Several of the profiles have a S-rich layer at the surface which may be due to surface contamination; however, some fundamental phenomena, although presently unknown, might be responsible [99]. Extension of the smooth curve to the surface as shown in figure 6.38 for the 6.0 μg/ml sample has been taken as its surface concentration. In figure 6.39, these surface concentrations have been plotted as a function of the sulfur vapor density $[S_v]$, the S mass per cm^3 of ampoule volume, and show a linear relationship.

The molecular species present in sulfur vapor are octatomic, hexatomic, and diatomic molecules, but near 1000°C the dominant species is S_2(reference 107). However, the dominant sulfur vapor species in the Ga-As-S ternary system is not known. For S_2 as the dominant vapor species in the ternary system,

$$(S_2)_G \rightleftharpoons 2S_L, \tag{6.73}$$

and an analysis following the procedure that led to equation 6.72 gives a

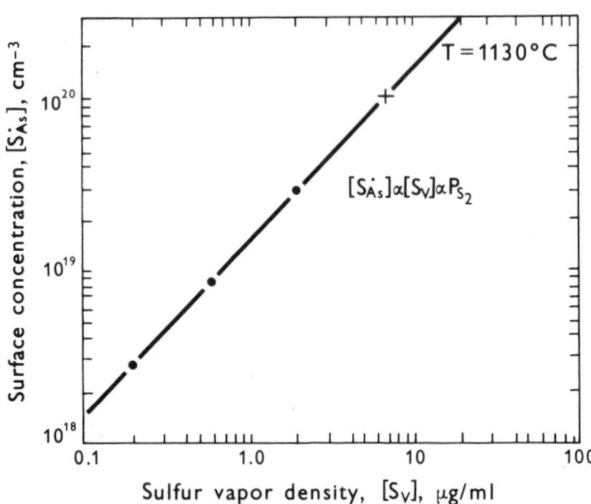

Figure 6.39. Dependence of the sulfur surface concentration (figure 6.38) on the sulfur vapor density with 4-6 mg/ml arsenic. The point indicated by a cross (+) is the erfc extrapolation to the surface as shown by the dashed curve for the 6.0 μg/ml sample in figure 6.38 (reference 99).

square-root dependence on the sulfur vapor density rather than the linear dependence shown in figure 6.39. A linear relationship between the sulfur vapor density and the surface concentration results if the dominant vapor species is Ga_2S. This species is suggested by Arthur's mass spectrograph measurements on the Ga-Te system [108]. Quantitative assignment of the ratios of the various tellurium species could not be made; but Te_2 was an order of magnitude less than Ga_2Te. In addition, Ga_2O is the vapor species observed for Ga-Ga_2O_3 solutions [109]. Therefore, the assumption of a Ga_2S vapor species is not unreasonable. The incorporation reaction equivalent to equation 6.67 is

$$(Ga_2S)_G + V_{As} \rightleftharpoons S^{\cdot}_{As} + 2(Ga)_G + e'. \tag{6.74}$$

For the pressure of Ga_2S proportional to the sulfur vapor density, then the equation similar to equation 6.72 is

$$[S^{\cdot}_{As}] = \frac{K'_8(T)[S_v]}{N_c \exp(-\Phi_{Bn}/kT)}. \tag{6.75}$$

The validity of this expression depends upon the assignment of Ga_2S as the dominant vapor species; however, this model is consistent with the solid solubility analysis presented above for the incorporation of Te from the liquid phase.

6.4.3 Diffusion of S in GaAs

Young and Pearson's [99] diffusion coefficient of S in GaAs as a function of temperature is shown in figure 6.40 together with the results obtained in other studies. The reason for the differences is not understood. However, Goldstein's [110] low values have been suggested to be due to vapor etching of the sample during diffusion [111]. Kendall [111], Vieland [112], and Frieser [113] obtained **D**(S) from the *p-n* junction depth and not from radioactive S diffusion profiles. In addition, Frieser used Al_2S_3 as the S diffusion source. As illustrated by figure 6.41, **D**(S) varies with arsenic pressure. Therefore, these data in figure 6.40 probably represent different diffusion conditions with different arsenic pressures for the various experimental techniques.

The arsenic pressure given in figure 6.41 was calculated from the amount of As added to the diffusion ampoule [99]. In terms of the Ga-As-S ternary system it is not possible to determine the actual diffusion source composition and actual arsenic pressures. Therefore, it is difficult to judge how well the pressure given in figure 6.41 represents the actual arsenic pressure. There is, however, clearly an increase in **D**(S) with As added to the diffusion ampoule. From the results of Vieland [112] for the variation of junction depth with arsenic pressure for S, Sn, and Si

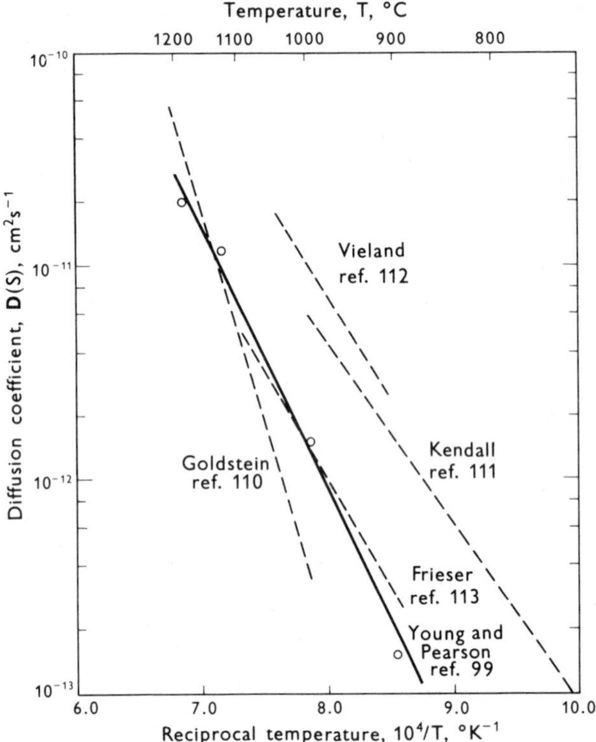

Figure 6.40. Comparison of the temperature dependence of the sulfur diffusion coefficients in GaAs as reported in indicated references.

diffusion, the behavior illustrated by figure 6.41 appears to be a fundamental property of donor diffusion. Without a detailed knowledge of the ternary behavior, it is difficult to interpret the knee in the **D**(S)-versus-p_{As_4} pressure curve. The data below the knee are fit best by a square-root dependence on the As_4 pressure, while at the highest As_4 pressures, **D**(S) becomes constant. One explanation suggested by Young and Pearson [99] is diffusion via the Ga divacancy ($V_{Ga}V_{Ga}$). The S on an As site is assumed immobile, but by reaction with the Ga divacancy, it becomes mobile. However, the results of the Zn solid solubility analysis indicate that a significant concentration of Ga divacancies is not observed [30]. Although unique assignment of a donor diffusion mechanism is not presently possible, Young and Pearson's work clearly demonstrates that donor diffusion is very slow. From their arsenic pressure dependence of S diffusion, it is again shown that ternary considerations are crucial in interpretation of the diffusion behavior and the assignment of diffusion

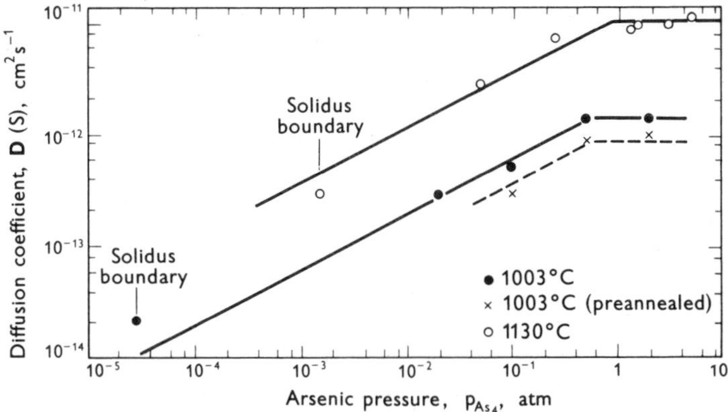

Figure 6.41. Dependence of the diffusion coefficient of sulfur in GaAs on the arsenic pressure with a sulfur vapor density $[S_v]$ of 0.2 μg/ml (reference 99).

coefficients in the III-V compounds. It should be noted that S diffusion in GaP was also investigated in reference 99.

6.5 JUNCTION DEVICE PREPARATION BY DIFFUSION

Application of Zn diffusion to the preparation of electroluminescent *p-n* junctions in GaAs and $GaAs_{1-x}P_x$ will be briefly described to illustrate how the diffusion behavior influences the processing of III-V semiconductor devices. Because the emitted radiation is absorbed by the *p*-layer at high Zn concentrations, it has been found desirable to use Zn surface concentrations between 3×10^{18} and 10^{19} cm^{-3} (reference 114). Diffused junction depths are typically ≈10 μm for GaAs and 1 to 2 μm for $GaAs_{1-x}P_x$ (reference 114). The factors that are considered in the selection of the junction depth for optimum light output have been presented by Archer and Kerps [115]. The properties of electroluminescence in $GaAs_{1-x}P_x$ have been presented by Pilkuhn and Rupprecht [116] and Herzog, Groves, and Craford [117].

Two basic diffusion sources have been used. One is Ga-Zn, or preferably the Ga-As-Zn source. Reference to figure 6.14 permits selection of the necessary Zn composition for a ternary Ga-As-Zn ternary source to give a desired concentration at a given temperature. The other commonly used diffusion source is $ZnAs_2$ [118]. For small amounts of $ZnAs_2$ (of the order of milligrams per cm^3 of ampoule volume) there may be no condensed phase for the source. That is, all the Zn and As is in the vapor. Such sources yield planar junction interfaces as was discussed in section 6.3.9 for high pressure arsenic and phosphorus sources. The surface concentration and junction depth are controlled by the source-weight-to-ampoule-volume ratio, and reproducible diffusion is obtained if this ratio is held constant. The diffusion temperature is typically 650°C for the $ZnAs_2$ source, and the n-type material generally has a donor concentration between 10^{17} and 10^{18} cm^{-3} (reference 114). Diffusion times are a few hours.

Other techniques have been used to diffuse Zn into GaAs such as diffusion through SiO_2 films [119] and diffusion from Zn-doped SiO_2 films [120]. Films of SiO_2 and Si_3N_4 may also be used to prevent Zn diffusion into the GaAs, and thus selected area diffusion may readily be achieved. The SiO_2 film must generally have a thin phosphate glass layer to mask Zn diffusion. Only closed diffusion systems have been considered. Open or flowing vapor systems generally do not result in constant vapor pressures; therefore, they present problems of interpretation outside the scope of this discussion.

6.6 COMPILATION OF DIFFUSION COEFFICIENTS IN THE III-V COMPOUND SEMICONDUCTORS

The reported self- and impurity-diffusion coefficients are compiled in Tables 6.4 through 6.10. Although self-diffusion data have been reported for several of the III-V compounds, it appears that only Kendall and Huggins [45] working with InSb have attempted to deal with the difficult experimental problems encountered in self-diffusion studies. Surface

TABLE 6.4

Self- and impurity-diffusion coefficients in AlSb; $D = D_0 \exp(-Q/kT)$

Diffusant	D_0 (cm^2 s^{-1})	Q (eV)	References	Comments
Al	7.3×10^{-1}	1.8	a	X-ray measurements on the growth of AlSb in sandwich specimens of Al and Sb layers.
Sb	2.9×10^{-1}	1.6	a	
Zn	Concentration dependent		b	See figure 6.33 (b)
Cd	Concentration dependent		c	Radioactive profiles
Cu	3.5×10^{-3}	0.36	d	Radioactive profiles

a. B. Ya. Pines and E. F. Chaikovskii, *Soviet Phys. solid St.*, **1**, 864 (1959); translated from *Fiz. Tver. Tela.*, **1**, 946 (1959)
b. S. R. Showan and D. Shaw, *Phys. Stat. Sol.*, **32**, 97 (1969)
c. D. Shaw and S. R. Showan, *Phys. Stat. Sol.*, **34**, 475 (1969)
d. R. H. Wieber, J. C. Gorton and C. S. Peet, *J. Appl. Phys.*, **31**, 608 (1960)

TABLE 6.5

Self- and impurity-diffusion coefficients in GaAs; $D = D_O \exp(-Q/kT)$

Diffusant	$D_O(\text{cm}^2\ \text{s}^{-1})$	Q (eV)	References	Comments
As	7×10^{-1}	3.2	a	Radioactive As
Ga	1×10^{7}	5.6	b	Radioactive Ga
Ag	2.5×10^{1}	2.27	c	Radioactive profiles
Au	2.9×10^{1}	2.64	c	Radioactive profiles
Be	7.3×10^{-6}	1.2	d	Incremental sheet resistivities
Cd	Concentration dependent		e	Radioactive profiles
Cr	4.3×10^{3}	3.4	c	Radioactive profiles, data scatters
Cu	3×10^{-2}	0.53	f	Radioactive profiles
Hg	$D(1000°\text{C}) = 5 \times 10^{-14}$		g	Radioactive profiles
In	$D(1000°\text{C}) = 7 \times 10^{-11}$		h	Radioactive profiles
Li	5.3×10^{-1}	1.0	i	Chemical analysis
Mg	2.6×10^{-2}	2.7	j	Sheet resistance and p-n junction measurements
Mn	6.5×10^{-1}	2.49	k	Radioactive profiles with excess arsenic pressure
O	2×10^{-3}	1.1	l	Estimated from mass-spectrometric data on O out-diffusion
S	1.85×10^{-2}	2.6	m	Radioactive profiles at high arsenic pressure
Se	3.0×10^{3}	4.16	b	Radioactive profiles, surface alloying
Si	Junction depth at 900° and 1000°C only		n	Junction depth measurement
Sn	3.8×10^{-2}	2.7	o	Radioactive profiles and junction depth
Te	$D(1000°\text{C}) = 10^{-13}$ $D(1100°\text{C}) = 2 \times 10^{-12}$		p	Radioactive profiles, diffusion through SiO_2 films
Tm	2.3×10^{-16}	(−)1.0	q	Radioactive profiles
Zn	Concentration dependent			See section 6.3.6 and figures 6.18, 6.20, 6.22, and 6.23

a. D. L. Kendall, unpublished
b. B. Goldstein, *Phys. Rev.*, **121**, 1305 (1961)
c. K. B. Wolfstirn, unpublished
d. E. A. Poltoratskii and V. M. Stuchebnikov, *Soviet Phys. solid St.*, **8**, 770 (1966); translated from *Fiz. Tver. Tela*, **8**, 963 (1966)
e. S. R. Showan and D. Shaw, *Phys. Stat. Sol.*, **35**, K79 (1969)
f. R. N. Hall and J. H. Racette, *J. Appl. Phys.*, **35**, 379 (1964)
g. J. A. Kanz, unpublished
h. D. L. Kendall, *Appl. Phys. Lett.*, **4**, 67 (1964)
i. C. S. Fuller and K. B. Wolfstirn, *J. Appl. Phys.*, **33**, 2507 (1962)
j. R. G. Moore, Jr., M. Belasco and H. Strack, *Bull. Am. Phys. Soc.*, **10**, 731 (1965)
k. M. S. Seltzer, *J. Phys. Chem. Solids*, **26**, 243 (1965)
l. J. Rachmann and R. Biermann, *Solid State Comm.*, **7**, 1771 (1969)
m. A. B. Y. Young and G. L. Pearson, *J. Phys. Chem. Solids*, **31**, 517 (1970)
n. G. R. Antell, *Solid-St. Electron.*, **8**, 943 (1965)
o. R. W. Fane and A. J. Goss, *Solid-St. Electron.*, **6**, 383 (1963)
p. J. F. Osborne, K. G. Heinen and H. Riser, unpublished
q. H. C. Casey, Jr. and G. L. Pearson, *J. Appl. Phys.*, **35**, 3401 (1964)

deterioration is an especially difficult problem and can result in considerable error in the assignment of diffusion coefficient values for shallow profiles. Table 6.5 for GaAs is presented in graphical form as D-versus-$1/T$ in figure 6.42. The data are expressed by the usual Arrhenius relation $D = D_0 \exp(-Q/kT)$. These diffusion data for impurities in the III-V compounds, except for a few cases, have been obtained by introducing a small amount of elemental impurity into the diffusion ampoule. As demonstrated by the detailed treatment of Zn in GaAs, it is difficult to determine the *actual* ternary diffusion conditions for the reported data. Different group V partial pressures can lead to significantly different diffusion behavior. Also, to correctly determine diffusion behavior, radiotracer diffusion profiles should be obtained by sectioning. Often *p-n* junction depth measurements are used with an assumed erfc profile to determine the diffusion coefficient.

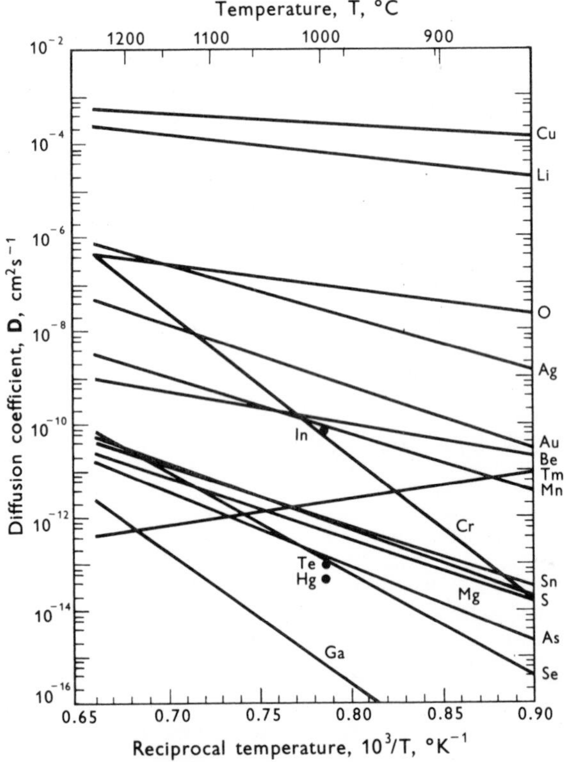

Figure 6.42. Summary of the self- and impurity-diffusion coefficients in GaAs. See Table 6.5 for the literature references.

TABLE 6.6

Impurity-diffusion coefficients in GaP; $D = D_O \exp(-Q/kT)$

Diffusant	D_O($cm^2\ s^{-1}$)	Q (eV)	Ref.	Comments
Zn	Concentration dependent		a	See figures 6.32(a) and 6.34
S	3.2×10^3	4.7	b	Radioactive profiles
Be	Concentration dependent		c	Atomic absorption profiles

a. L. L. Chang and G. L. Pearson, *J. Appl. Phys.*, **35**, 374 (1964)
b. A. B. Y. Young and G. L. Pearson, *J. Phys. Chem. Solids*, **31**, 517 (1970)
c. M. Ilegems and W. C. O'Mara, *J. Appl. Phys.*, **43**, 1190 (1972).

TABLE 6.7

Self- and impurity-diffusion coefficients in GaSb; $D = D_O \exp(-Q/kT)$

Diffusant	D_O($cm^2\ s^{-1}$)	Q (eV)	Reference	Comments
Ga	3.2×10^3	3.15	a	Radioactive profiles
Sb	3.4×10^4	3.45	a	Radioactive profiles
Cd	1.5×10^{-6}	0.72	b	Junction depth measurements
In	1.2×10^{-7}	0.53	c	Radioactive profiles
Sn	2.4×10^{-5}	0.80	c	Radioactive profiles
Te	3.8×10^{-4}	1.20	c	Radioactive profiles

a. F. H. Eisen and C. E. Birchenall, *Acta metall.*, **5**, 265 (1957)
b. J. Bougnot, L. Szepessy and S. F. Da Cunha, *Phys. Stat. Sol.*, **26**, K127 (1968)
c. B. I. Boltaks and Ya. A. Gutorov, *Soviet Phys. solid St.*, **1**, 930 (1960), translated from *Fiz. Tver. Tela*, **1**, 1015 (1960)

No discussion of these diffusion data will be given here. It should be emphasized that these data have been compiled only for completeness. Kendall [18] and Yarbrough [19] have thoroughly summarized and discussed the experimental diffusion data of the III-V compounds. Tables 6.4 through 6.10 include data obtained since the publication of references 18 and 19 in 1968. Instead of a listing of all known references for a particular diffusant, which would require extensive discussion where conflicting results have been reported, only a representative reference has been given.

TABLE 6.8

Self- and impurity-diffusion coefficients in InAs; $D = D_0 \exp(-Q/kT)$

Diffusant	D_0(cm^2 s^{-1})	Q (eV)	Reference	Comments
As	3×10^7	4.45	a	Radioactive profiles
In	6.0×10^5	4.0	a	Radioactive profiles
Ag	7.3×10^{-4}	0.26	b	Radioactive profiles
Au	5.8×10^{-4}	0.65	c	Radioactive profiles
Cd	7.4×10^{-4}	1.15	d	Radioactive profiles
Cu	3.6×10^{-3}	0.52	e	Radioactive profiles
Ge	3.74×10^{-6}	1.17	f	Measurement of p-n junction depth
Mg	1.98×10^{-6}	1.17	f	Measurement of p-n junction depth
S	6.78	2.20	f	Measurement of p-n junction depth
Sn	1.49×10^{-6}	1.17	f	Measurement of p-n junction depth
Se	12.6	2.20	f	Measurement of p-n junction depth
Te	3.43×10^{-5}	1.28	f	Measurement of p-n junction depth
Zn	Concentration dependent			See figures 6.32(b) and 6.34
Hg	1.45×10^{-5}	1.32	g	Radioactive profiles

a. H. Kato, M. Yokozawa, R. Kohara, Y. Okabayashi, S. Takayanagi, *Solid-St. Electron.*, **12**, 137 (1969)
b. B. I. Boltaks, S. I. Rembeza and B. L. Sharma, *Sov. Phys.-Semicond.*, **1**, 196 (1967); translated from *Fiz. Tekh. Poluprov.*, **1**, 247 (1967)
c. S. I. Rembeza, *Sov. Phys.-Semicond.*, **1**, 516 (1967); translated from *Fiz. Tekh. Poluprov.*, **1**, 615 (1967)
d. K. A. Arseni, B. I. Boltaks and S. I. Rembeza, *Soviet Phys. solid St.*, **8**, 2248 (1967); translated from *Fiz. Tver. Tela*, **8**, 2809 (1966)
e. C. S. Fuller and K. B. Wolfstirn, *J. Electrochem. Soc.*, **114**, 856 (1967)
f. E. Schillman, Compound Semiconductors: Preparation of III-V Compounds, Vol. 1, p. 358. (R. K. Willardson and H. L. Goering, eds.). (Reinhold Publishing Corp., New York, 1962)
g. B. L. Sharma, R. K. Purohit and S. N. Mukerjee, *J. Phys. Chem. Solids*, **32**, 1397 (1971)

TABLE 6.9

Self- and impurity-diffusion coefficients in InP; $D = D_0 \exp(-Q/kT)$

Diffusant	D_0(cm^2 s^{-1})	Q (eV)	Reference	Comments
In	1×10^5	3.85	a	Radioactive profiles
P	7×10^{10}	5.65	a	Radioactive profiles
Ag	3.6×10^{-4}	0.59	b	Radioactive profiles
Au	1.32×10^{-5}	0.48	c	Radioactive profiles
Cd	1.8	1.9	d	Radioactive profiles
Cu	3.8×10^{-3}	0.69	e	Radioactive profiles
Zn	Concentration dependent			See figures 6.32(b) and 6.34

a. B. Goldstein, *Phys. Rev.*, **121**, 1305 (1961)
b. K. A. Arseni and B. I. Boltaks, *Soviet Phys. solid St.*, **10**, 2190 (1969); translated from *Fiz. Tver. Tela*, **10**, 2783 (1968)
c. S. I. Rembeza, *Sov. Phys.-Semicond.*, **3**, 519 (1969); translated from *Fizz. Tekh. Poluprov.*, **3**, 612 (1969)
d. K. A. Arseni, B. I. Boltaks, V. L. Gordin and Ya. A. Ugai, *Inorg. Mat.*, **3**, 1465 (1967); translated from *Izv. Akad. Nauk. SSSR, Neorg. Mater.*, **3**, 1679 (1967)
e. K. A. Arseni, *Soviet Phys. solid St.*, **10**, 2263 (1969); translated from *Fiz. Tverd. Tela*, **10**, 2864 (1968)

TABLE 6.10

Self- and impurity-diffusion coefficients in InSb; $D = D_O \exp(-Q/kT)$

Diffusant	D_O(cm² s⁻¹)	Q (eV)	Reference	Comments
In	1.76×10^{13}	4.3	a	Radioactive profiles
Sb	3.1×10^{13}	4.3	a	Radioactive profiles
Ag	$\sim 1.0 \times 10^{-7}$	~0.25	b	Radioactive profiles
Au	7.0×10^{-4}	0.32	c	Radioactive profiles
Co	$\sim 1.0 \times 10^{-7}$	~0.25	b	Radioactive profiles
Cd	1.3×10^{-4}	1.2	d	Measurement of *p-n* junction depth
Cu	9.0×10^{-4}	1.08	e	Radioactive profiles
Fe	$\sim 1.0 \times 10^{-7}$	~0.25	b	Radioactive profiles
Hg	4.0×10^{-6}	1.17	f	Radioactive profiles
Li	7.0×10^{-4}	0.28	g	Electrical measurements
Se	1.6	1.87	h	Measurement of profile by capacitance-voltage
Sn	5.5×10^{-8}	0.75	i	Radioactive profiles
Te	1.7×10^{-7}	0.57	j	Radioactive profiles
Zn	Concentration dependent			See figure 6.33(a)

a. D. L. Kendall and R. A. Huggins, *J. Appl. Phys.*, **40**, 2750 (1969)
b. L. A. K. Watt and W. S. Chen, *Bull. Am. Phys. Soc.*, **7**, 89 (1962)
c. B. I. Boltaks and V. I. Sokolov, *Soviet Phys. solid St.*, **6**, 600 (1964); translated from *Fiz. Tver. Tela,* **6**, 771 (1964)
d. R. B. Wilson and E. L. Heasell, *Proc. Phys. Soc. (London),* **79**, 403 (1962). For diffusion profiles at 505° C, see reference 18, p. 252
e. H. J. Stocker, *Phys. Rev.*, **130**, 2160 (1963)
f. I. A. Gusev and A. N. Murin, *Soviet Phys. solid St.*, **6**, 1229 (1964); translated from *Fiz. Tver. Tela,* **6**, 1563 (1964)
g. T. Takabatake, H. Ikari and Y. Uyeda, *J. Appl. Phys. Japan*, **5**, 839 (1966)
h. G. I. Rekalova, A. A. Shakov and V. V. Gavrushko, *Sov. Phys.-Semicond.*, **2**, 1452 (1969); translated from *Fiz. Tekh. Poluprov.*, **2**, 1744 (1969)
i. S. M. Sze and L. Y. Wei, *Phys. Rev.*, **124**, 84 (1961)
j. B. I. Boltaks and G. S. Kulikov, *Sov. Phys.-Tech. Phys.*, **2**, 67 (1957); translated from *Zh. Tekh. Fiz.*, **27**, 82 (1957)

ACKNOWLEDGMENTS

The vacancy analysis for GaAs is an outgrowth of extensive discussions with M. B. Panish, A. S. Jordan, and C. D. Thurmond. F. Ermanis assisted with the calculation of the intrinsic carrier concentrations given in the Appendix. F. A. Trumbore contributed many suggestions which greatly facilitated the preparation of the manuscript.

6.A APPENDIX

6.A.1 Intrinsic Carrier Concentration

The carrier concentration at a given temperature in a crystal that has no impurity levels is called the intrinsic carrier concentration n_i. The intrinsic carrier concentration is a significant quantity to consider because the diffusion behavior can be drastically different for intrinsic conditions, $n = p = n_i$, and for extrinsic conditions, $n > n_i > p$ or $p > n_i > n$. For intrinsic conditions the substitutional Zn concentration will not enhance the interstitial Zn concentration, as expressed by equation 6.56, and therefore the diffusion coefficient will not be concentration dependent.

The intrinsic carrier concentration n_i is given by [121]

$$n_i = \sqrt{N_c N_v} \exp(-\epsilon_g/2kT). \tag{6.A.1}$$

The effective density of states for the valence band is

$$N_v = 2(2\pi m_p kT/h^2)^{3/2}, \tag{6.A.2}$$

and the effective density of states for the conduction band is

$$N_c = 2M(2\pi m_{n_1} kT/h^2)^{3/2}\,[1 + (m_{n_2}/m_{n_1})^{3/2} \exp(-\Delta\epsilon/kT)]. \tag{6.A.3}$$

In these equations, ϵ_g is the value of the separation between the valence band and the lowest lying conduction band at the absolute temperature T, m_p is the effective density of states mass for holes, m_{n_1} is the effective density of states mass for the lowest lying conduction band with M as the number of equivalent conduction band minima, m_{n_2} is the effective density of states mass for the next lowest lying conduction band, and $\Delta\epsilon$ is the separation in energy of the lowest lying and next lowest lying conduction band. The quantity M is unity for direct energy gap materials, and for indirect energy gap materials there are generally three, four, or six equivalent minima. Often, the value given for m_{n_1} includes the value of M; i.e., M is brought inside the parenthesis of equation 6.A.3 as $3^{2/3}$, $4^{2/3}$, or $6^{2/3}$. In most cases, $\Delta\epsilon$ is sufficiently large so that the term in the brackets goes to unity and N_c may be written as the usual expression

$$N_c = 2M(2\pi m_{n_1} kT/h^2)^{3/2}. \tag{6.A.4}$$

Evaluation of the constants permits writing equations 6.A.2 and 6.A.3 as

$$N_v = 2.5 \times 10^{19} (m_p/m_e)^{3/2} (T/300)^{3/2}, \tag{6.A.5}$$

and

$$N_c = 2.5 \times 10^{19} M (m_{n_1}/m_e)^{3/2} (T/300)^{3/2} \times [1 + (m_{n_1}/m_{n_2})^{3/2} \exp(-\Delta\epsilon/kT)]. \tag{6.A.6}$$

Thus, to evaluate n_i it is necessary to know the temperature dependence of the energy gap ϵ_g, the effective density of states mass for holes and electrons, and also $\Delta\epsilon$ and m_{n_2} if applicable. The evaluation of n_i for the seven III-V compounds whose diffusion coefficients are summarized in Tables 6.4 through 6.10 is described below.

6.A.2 Evaluation of the Intrinsic Carrier Concentration for GaAs

The calculation of the intrinsic carrier concentration will be illustrated for GaAs. Not only is GaAs one of the most commonly encountered III-V compounds and is utilized extensively in the diffusion analysis, but also the evaluation of n_i for GaAs demonstrates a case where $\Delta\epsilon$ must be considered. A simplified band diagram of GaAs is shown in figure 6.A.1.

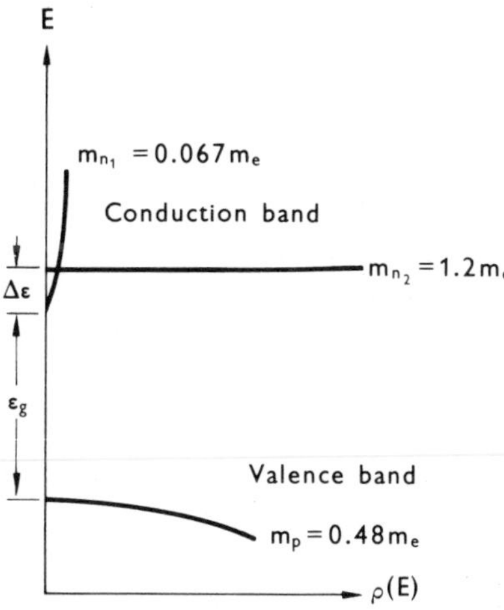

Figure 6.A.1. Schematic representation of the density of states $\rho(E)$ as a function of energy E for GaAs. The direct conduction band is designated by m_{n_1}, and the indirect conduction band is designated by m_{n_2}. The direct energy gap is ϵ_g, and the separation between the direct and indirect energy gap is $\Delta\epsilon$. Because $m_{n_2} \gg m_{n_1}$, the indirect conduction band has a large influence on the value of n_i at high temperatures.

The direct energy gap is the smallest separation between the valence and conduction bands. However, the separation $\Delta\epsilon$ is small enough to greatly influence the value of n_i at high temperatures.

The temperature dependence of the energy gap ϵ_g was measured by Panish and Casey [122] and can be represented by

$$\epsilon_g = 1.522 - 5.8 \times 10^{-4}\, T^2/(T+300)\ \text{eV}. \qquad (6.A.7)$$

In cases where the temperature dependence of the energy gap is not known, the general form of the energy-gap temperature dependence given

TABLE 6.A.1

Electron and hole effective density of states masses and the temperature-dependent energy gaps of the III-V compounds

Compound	Hole mass m_p	Ref.	Electron mass m_{n_1}	Ref.	m_{n_2}	Ref.	Energy gap ϵ_g, eV	Ref.	$\Delta\epsilon$, eV	Ref.
AlSb	$0.9m_e$	a	$0.39m_e$	b			$1.6\text{-}6.0 \times 10^{-4} T^2/(T+292)$	c		
GaAs	$0.48m_e$	d	$0.067m_e$	d	$1.2m_e$	e	$1.522\text{-}5.8 \times 10^{-4} T^2/(T+300)$	f	0.36	e
GaP	$1.0m_e$	g	$0.40m_e$	h			$2.338\text{-}6.2 \times 10^{-4} T^2/(T+460)$	f		
GaSb	$0.34m_e$	i	$0.042m_e$	j	$0.57m_e$	k	$0.813\text{-}6.0 \times 10^{-4} T^2/(T+265)$	l	$0.078\text{-}2 \times 10^{-5} T$	k
InAs	$0.43m_e$	m	$0.023m_e$	n			$0.426\text{-}3.2 \times 10^{-4} T^2/(T+93)$	o		
InP	$1.0m_e$	p	$0.069m_e$	q			$1.421\text{-}4.9 \times 10^{-4} T^2/(T+327)$	o		
InSb	$0.43m_e$	r	$0.0145m_e$	r			$n_i = 5.76 \times 10^{14} T^{3/2} \times \exp(-0.129/kT)$	r, s		

by equations 6.A.7 is

$$\epsilon_g = \epsilon_g(0) - \alpha T^2/(T + \beta), \quad (6.A.8)$$

where $\epsilon_g(0)$ is the energy gap at 0° K, β is approximately the 0° K Debye temperature $\theta_D(0)$, and α is an empirical constant that is approximately 6×10^{-4} [122]. Values of $\theta_D(0)$ for the III-V compounds have been given

---

a. D. N. Nasledov and S. V. Slobodchikov, *Sov. Phys.-Tech. Phys.*, **3**, 669 (1958): translated from *Zh. Tekh. Fiz.*, **28**, 715 (1958)
b. T. S. Moss, A. K. Walton and B. Ellis, Proc. of the Inter. Conf. on Semiconductor Phys., p. 295, Exeter. (Inst. of Phys. and Phys. Soc., London, 1962)
c. The 0° K value of ϵ_g is taken from R. F. Blunt, H. P. R. Frederikse, J. H. Becker and W. R. Hosler, *Phys. Rev.*, **96**, 578 (1954), and the value at 300° K is given by C. A. Mead and W. G. Spitzer, *Phys. Rev. Lett.*, **11**, 358 (1963). The temperature dependence of ϵ_g is based on the general expression of equation 6.A.8 with $\beta = \theta_D(0)$ from reference 123
d. Q. H. F. Vrehen, *J. Phys. Chem. Solids*, **29**, 129 (1968)
e. H. Ehrenreich, *Phys. Rev.*, **120**, 1951 (1960)
f. M. B. Panish and H. C. Casey, Jr., *J. Appl. Phys.*, **40**, 163 (1969)
g. No experimental values presently exist for m_p. Theoretical calculations by R. A. Faulker suggest a value of $1.0m_e$ is reasonable
h. A. Onton, *Phys. Rev.*, **186**, 786 (1969) has given the transverse and parallel conduction band masses as $m\perp = 0.191m_e$ and $m\| = 1.7m_e$ to give $m_{n_1} = (m_\| m_\perp^2)^{1/3} = 0.40m_e$. Considerable uncertainty exists as to whether there are three or six conduction band minima. For this calculation, M has been taken as six. The value of n_i for $M = 3$ would be $\sqrt{3/6}$ or 0.707 times smaller
i. A. K. Walton and U. K. Mishra, *Proc. Phys. Soc.*, **90**, 1111 (1967)
j. T. O. Yep and W. M. Becker, *Phys. Rev.*, **144**, 741 (1966)
k. E. H. Van Tongerloo and J. C. Woolley, *Can. J. Phys.*, **47**, 241 (1969). The density of states mass is given as $(m_\| m_\perp^2)^{1/3} = 0.226m_e$, and for four minima this gives m_{n_2} as $M^{2/3}(m_\| m_\perp^2)^{1/3} = 0.57m_e$
l. The 0° K value of ϵ_g is taken from E. J. Johnson and H. Y. Fan, *Phys. Rev.*, **139**, A1991 (1965), and the value at 300° K is given by E. J. Johnson, I. Filinski and H. Y. Fan, Proc. of the Inter. Conf. on Semiconductor Phys., p. 375, Exeter. (Inst. of Phys. and Phys. Soc., London, 1962.) The temperature dependence of ϵ_g is based on the general expression of equation 6.A.8 with $\beta = \theta_D(0)$ from reference 123
m. Following the suggestion of C. R. Pidgeon, D. L. Mitchell and R. N. Brown, *Phys. Rev.*, **154**, 737 (1967), m_p for InAs has been taken as the InSb value
n. C. W. Litton, R. B. Dennis and S. D. Smith, *J. Phys. C. (Solid St. Phys.)*, **2**, 2146 (1969)
o. Y. P. Varshni, *Physica*, **34**, 149 (1967)
p. No experimental values presently exist for m_p. Thermoelectric power measurements on *p*-type InP by V. V. Galavanov, S. G. Metreveli and S. P. Starosel'tseva, *Sov. Phys.-Semicond.*, **3**, 1159 (1970): translated from *Fiz. Tekh. Poluprov.*, **3**, 1391 (1969) indicate a much larger value than for GaAs and about the same value as for GaP. For this reason, m_p has been taken as $1.0m_e$.
q. F. P. Kesamanly, D. N. Nasledov, A. Ya. Nashel'skii and V. A. Skripkin, *Sov. Phys.-Semicond.*, **2**, 1221 (1969): translated from *Fiz. Tekh. Poluprov.*, **2**, 1463 (1968).
r. R. W. Cunningham and J. B. Gruber, *J. Appl. Phys.*, **41**, 1804 (1970)
s. In the absence of data on the temperature dependence of ϵ_g, the expression for n_i given in reference r has been used.

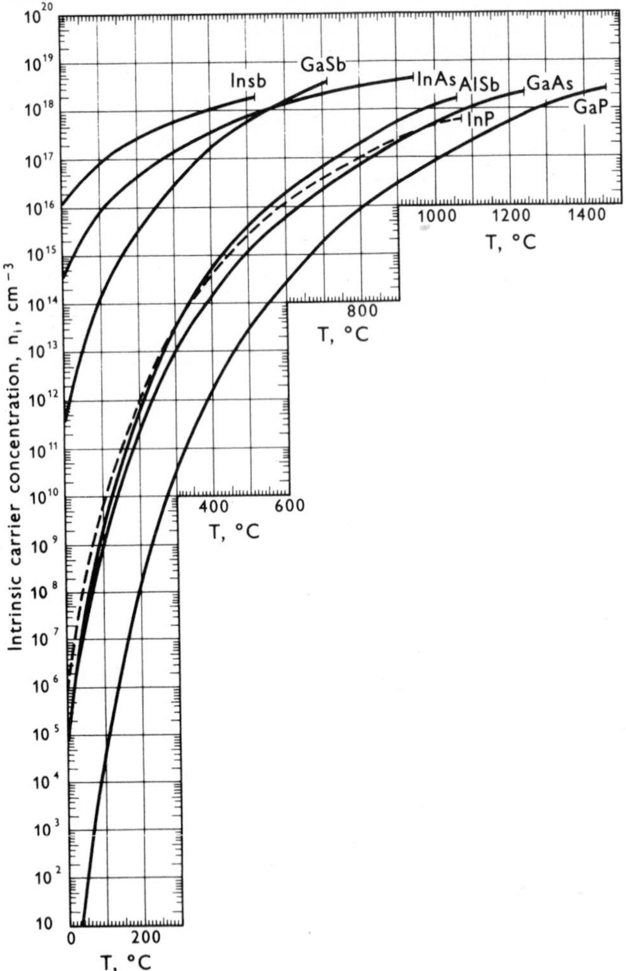

Figure 6.A.2. The intrinsic carrier concentration as calculated from the mass and energy gap data in Table 6.A.1. The values of n_i are given from 0° C to the compound melting point.

by Steigmeier [123]. This general form of the ϵ_g temperature dependence is discussed by Panish and Casey [122] and Varshni [124].

The density of states effective mass for the valence band m_p is $0.48m_e$ (reference 61). The density of states mass for the lowest conduction band m_{n_1} is $0.067m_e$ (reference 61). The assignment of $\Delta\epsilon$ and the density of states mass m_{n_2} is somewhat uncertain. Comparison of numerous publications shows that $\Delta\epsilon$ is on the order of 0.36 eV and m_{n_2} is approximately m_e. Recent high pressure Hall measurements gave

$\Delta\epsilon = 0.38 \pm 0.01$ eV, and $m_{n_2} = 0.85 \pm 0.10 m_e$ (reference 125). Analysis of Aukerman and Willardson's [126] high temperature Hall data led Ehrenreich [127] to assign $\Delta\epsilon = 0.36$ eV and $m_{n_2} = 1.2 m_e$. For the calculation of n_i given here, Ehrenreich's values will be used. The values used in the evaluation of equations 6.A.1, 6.A.5, and 6.A.6 are given in Table 6.A.1, and the resulting calculated values of n_i as a function of temperature are given in figure 6.A.2.

REFERENCES

1. J. A. Van Vechten, *Phys. Rev.*, **187**, 1007 (1969)
2. J. J. Baranowski, V. J. Higgins, C. K. Kim and L. D. Armstrong, *Microwave J.*, **12**, 71 (1969)
3. K. L. Lawley, *J. Electrochem. Soc.*, **113**, 240 (1966)
4. A. S. Grove, A. Roder and C. T. Sah, *J. Appl. Phys.*, **36**, 802 (1965)
5. S. M. Hu, *J. Appl. Phys.*, **39**, 3844 (1968)
6. J. I. Pankove and M. Massoulie, Electron Div. Abstr., Spring Meeting Electrochem. Soc., Los Angeles, p. 71 (1962)
7. R. J. Keyes and T. M. Quist, *Proc. IRE*, **50**, 1822 (1962)
8. M. Gershenzon and R. M. Mikulyak, *IRE Trans. Electron Devices*, **ED9**, 503 (1962)
9. D. N. Nasledov, A. A. Rogachev, S. M. Ryvkin and B. V. Tsarenkov, *Soviet Phys. solid St.*, **4**, 782 (1962): translated from *Fiz. Tverd. Tela*, **4**, 1062 (1962)
10. M. I. Nathan, W. P. Dumke, G. Burns, F. H. Dill, Jr. and G. Lasher, *Appl. Phys. Lett.*, **1**, 62 (1962)
11. R. N. Hall, G. E. Fenner, J. D. Kingsley, T. J. Soltys and R. O. Carlson, *Phys. Rev. Lett.*, **9**, 366 (1962)
12. F. A. Cunnell and C. H. Gooch, *J. Phys. Chem. Solids*, **15**, 127 (1960)
13. J. W. Allen and F. A. Cunnell, *Nature*, **182**, 1158 (1958)
14. P. G. Shewmon, Diffusion in Solids, p. 28. (McGraw-Hill Book Company, Inc., New York, 1963)
15. R. L. Longini, *Solid-St. Electron.*, **5**, 127 (1962)
16. J. W. Allen and G. L. Pearson, NASA Cr-438, Stanford University, Stanford, Calif., April 1966
17. H. C. Casey, Jr. and M. B. Panish, *Trans. Met. Soc. AIME*, **242**, 406 (1968)
18. D. L. Kendall, Semiconductors and Semimetals, Vol. 4, p. 163. Physics of III-V Compounds. (R. K. Willardson and A. C. Beer, eds.). (Academic Press, New York, 1968)
19. D. W. Yarbrough, Res. and Dev. Tech. Report ECOM-2942, U.S. Army Electronics Command, Fort Monmouth, N.J., March 1968
20. J. P. Gracey and R. J. Friauf, *J. Phys. Chem. Solids*, **30**, 421 (1969)
21. C. D. Thurmond, *J. Phys. Chem. Solids*, **26**, 785 (1965)
22. R. N. Hall, *J. Electrochem. Soc.*, **110**, 385 (1963)
23. W. Köster and B. Thoma, *Z. Metall.*, **46**, 291 (1955)
24. J. R. Arthur, *J. Phys. Chem. Solids*, **28**, 2257 (1967)
25. D. Richman, *J. Phys. Chem. Solids*, **24**, 1131 (1963)
26. W. D. Johnston, *J. Electrochem. Soc.*, **110**, 117 (1963)
27. S. F. Nygren, C. M. Ringel and H. W. Verleur, *J. Electrochem. Soc.*, **118**, 306 (1971)
28. M. B. Panish and J. R. Arthur, *J. Chem. Thermo.*, **2**, 299 (1970)
29. N. N. Sirota, Semiconductors and Semimetals, Vol. 4, p. 35. Physics of III-V Compounds (R. K. Willardson and A. C. Beer, eds.). (Academic Press, New York, 1968)

30. M. B. Panish and H. C. Casey, Jr., *J. Phys. Chem. Solids,* **28,** 1673 (1967)
31. H. R. Potts and G. L. Pearson, *J. Appl. Phys.,* **37,** 2098 (1966)
32. F. L. Vook, *J. Phys. Soc. Japan,* **18,** Suppl. II, 190 (1963)
33. G. L. Pearson, H. R. Potts and V. G. Macres, Proc. 7th Inter. Conf. on the Phys. of Semiconductors, Radiation Damage in Semiconductors, p. 197. (Dunod, Cie., Paris, 1965)
34. R. A. Swalin, *J. Phys. Chem. Solids,* **18,** 290 (1961)
35. M. E. Straumanis and C. D. Kim, *Acta Cryst.,* **19,** 256 (1965)
36. A. S. Jordan, *J. Electrochem. Soc.,* **118,** 781 (1971)
37. J. Basinski, *Can. J. Phys.,* **44,** 941 (1966)
38. R. J. Sladek, *Phys. Rev.,* **140,** A1345 (1965)
39. F. J. Reid, R. D. Baxter and S. E. Miller, *J. Electrochem. Soc.,* **113,** 713 (1966)
40. Y. J. Van Der Meulen, *J. Phys. Chem. Solids,* **28,** 25 (1967)
41. B. K. Chakraverty and R. W. Dreyfus, *J. Appl. Phys.,* **37,** 631 (1966)
42. D. L. Kendall, Ph.D. Dissertation, Stanford University (1965)
43. J. A. Harper, unpublished
44. D. L. Kendall, unpublished
45. D. L. Kendall and R. A. Huggins, *J. Appl. Phys.,* **40,** 2750 (1969)
46. K. Weiser, *J. Phys. Chem. Solids,* **17,** 149 (1960)
47. R. N. Hall and J. H. Racette, *J. Appl. Phys.,* **35,** 379 (1964)
48. J. R. Manning, Diffusion Kinetics for Atoms in Crystals, pp. 95, 166. (D. van Nostrand Co., Princeton, N.J., 1968)
49. M. B. Panish, *J. Electrochem. Soc.,* **113,** 861 (1966)
50. J. E. Ricci, The Phase Rule and Heterogeneous Equilibrium, p. 1. (D. van Nostrand Co., Princeton, N.J., 1951)
51. L. L. Chang and G. L. Pearson, *J. Phys. Chem. Solids,* **25,** 23 (1964)
52. K. K. Shih, J. W. Allen and G. L. Pearson, *J. Phys. Chem. Solids,* **29,** 379 (1968)
53. H. Fritzsche, *J. Phys. Chem. Solids,* **6,** 69 (1958)
54. H. Nishimura, *Phys. Rev.,* **138,** A815 (1965)
55. E. A. Davis and W. D. Compton, *Phys. Rev.,* **140,** A2183 (1965)
56. T. N. Morgan, *Phys. Rev.,* **139,** A343 (1965)
57. B. I. Halperin and M. Lax, *Phys. Rev.,* **148,** 722 (1966)
58. N. F. Mott, *Phil. Mag.,* **6,** 287 (1961)
59. N. F. Mott and W. D. Twose, *Advan. Phys.,* **10,** 107 (1961)
60. K. G. Hambleton, C. Hilsum and B. R. Holeman, *Proc. Phys. Soc. (London),* **77,** 1147 (1961)
61. Q. H. F. Vrehen, *J. Phys. Chem. Solids,* **29,** 129 (1968)
62. F. Ermanis and K. B. Wolfstirn, *J. Appl. Phys.,* **37,** 1963 (1966)
63. J. Basinski and R. Olivier, *Can. J. Phys.,* **45,** 119 (1967)
64. K. K. Shih, J. W. Allen and G. L. Pearson, *J. Phys. Chem. Solids,* **29,** 367 (1968)
65. A. S. Jordan, *Metal. Trans.,* **2,** 1965 (1971)
66. H. C. Casey, Jr., M. B. Panish and L. L. Chang, *Phys. Rev.,* **162,** 660 (1967)
67. J. R. Brews and C. J. Hwang, *J. Chem. Phys.,* **54,** 3263 (1971)
68. C. J. Hwang and J. R. Brews, *J. Phys. Chem. Solids,* **32,** 837 (1971)
69. A. J. Rosenberg, *J. Chem. Phys.,* **33,** 665 (1960)
70. D. L. Kendall and M. E. Jones, AIEE-IRE Device Research Conference, Stanford (1961)
71. J. W. Allen, *J. Phys. Chem. Solids,* **15,** 134 (1960)
72. B. Goldstein, *Phys. Rev.,* **118,** 1024 (1960)
73. L. R. Weisberg and J. Blanc, *Phys. Rev.,* **131,** 1548 (1963)
74. Reference 14, pp. 23, 122, and 140
75. K. Weiser, *J. Appl. Phys.,* **34,** 3387 (1963)
76. D. Shaw and A. L. J. Wells, *Brit. J. Appl. Phys.,* **17,** 999 (1966)
77. C. Van Opdorp, *J. Appl. Phys.,* **38,** 5411 (1967)

78. L. L. Chang and G. L. Pearson, *J. Appl. Phys.*, **35**, 374 (1964)
79. L. L. Chang and G. L. Pearson, *J. Appl. Phys.*, **35**, 1960 (1964)
80. D. Shaw and S. R. Showan, *Phys. Stat. Sol.*, **32**, 109 (1969)
80a. C. H. Ting and G. L. Pearson, *J. Appl. Phys.*, **42**, 2247 (1971)
81. S. Prussin, *J. Appl. Phys.*, **32**, 1876 (1961)
82. G. H. Schwuttke and H. J. Queisser, *J. Appl. Phys.*, **33**, 1540 (1962)
83. J. F. Black and E. D. Jungbluth, *J. Electrochem. Soc.*, **114**, 181 (1967)
84. J. F. Black and E. D. Jungbluth, *J. Electrochem. Soc.*, **114**, 188 (1967)
85. M. Maruyama, *J. Appl. Phys. Japan*, **7**, 476 (1968)
86. C. H. Ting and G. L. Pearson, *J. Electrochem. Soc.*, **118**, 1454 (1971)
87. S. F. Nygren and G. L. Pearson, *J. Electrochem. Soc.*, **116**, 648 (1969)
88. J. Black and P. Lublin, *J. Appl. Phys.*, **35**, 2462 (1964)
89. R. W. Janes, The Optical Principles of the Diffraction of X-rays, pp. 306-316. (Cornell Univ. Press, Ithaca, N.Y., 1965)
90. B. G. Cohen, private communication
91. H. Rupprecht and C. Z. Lemay, *J. Appl. Phys.*, **35**, 1970 (1964)
92. M. B. Panish, *J. Electrochem. Soc.*, **113**, 224 (1966)
93. L. L. Chang and H. C. Casey, Jr., *Solid-St. Electron.*, **7**, 481 (1964)
94. M. G. Buehler and G. L. Pearson, unpublished
95. S. R. Showan and D. Shaw, *Phys. Stat. Sol.*, **32**, 97 (1969)
96. C. J. Nuese, G. E. Stillman, M. D. Sirkis and N. Holonyak, Jr., *Solid-St. Electron.*, **9**, 735 (1966)
97. K. A. Arseni, B. I. Boltaks and T. D. Dzhafarov, *Phys. Stat. Sol.*, **35**, 1053 (1969)
98. T. H. Yeh, *J. Electrochem. Soc.*, **111**, 253 (1964)
99. A. B. Y. Young and G. L. Pearson, *J. Phys. Chem. Solids*, **31**, 517 (1970)
100. H. C. Casey, Jr., M. B. Panish and K. B. Wolfstirn, *J. Phys. Chem. Solids*, **32**, 571 (1971)
101. K. H. Zschauer and A. Vogel, 'GaAs: 1970 Symp. Proc.', p. 100. (Inst. of Phys., London, 1971)
102. G. Schottky, *J. Phys. Chem. Solids*, **27**, 1721 (1966)
103. J. S. Blakemore, Semiconductor Statistics, p. 77. (Pergamon Press, New York, 1962)
104. M. G. Mil'vidskii and O. V. Pelevin, *Inorg. Mater.*, **3**, 1024 (1967); translated from *Izv. Akad. Nauk. SSSR, Neorg. Mater.*, **3**, 1159 (1967)
105. L. J. Vieland and I. Kudman, *J. Phys. Chem. Solids*, **24**, 437 (1963)
106. F. E. Rosztoczy, F. Ermanis, I. Hayashi and B. Schwartz, *J. Appl. Phys.*, **41**, 264 (1970)
107. J. R. West, *Indust. and Eng. Chem.*, **42**, 713 (1950)
108. J. R. Arthur, private communication
109. S. Antkiw and V. H. Dibeler, *J. Chem. Phys.*, **21**, 1890 (1953)
110. B. Goldstein, *Phys. Rev.*, **121**, 1305 (1961)
111. Reference 18, p. 199
112. L. J. Vieland, *J. Phys. Chem. Solids*, **21**, 318 (1961)
113. R. G. Frieser, *J. Electrochem. Soc.*, **112**, 697 (1965)
114. R. A. Burmeister, Jr., private communication
115. R. J. Archer and D. Kerps, GaAs: 1966 Symp. Proc., p. 103. (Inst. of Phys. and Phys. Soc., London, 1967)
116. M. Pilkuhn and H. Rupprecht, *J. Appl. Phys.*, **36**, 684 (1965)
117. A. H. Herzog, W. O. Groves and M. G. Craford, *J. Appl. Phys.*, **40**, 1830 (1969)
118. M. H. Pilkuhn and H. Rupprecht, *Trans. Met. Soc. AIME*, **230**, 296 (1964)
119. S. R. Shortes, J. A. Kanz and E. C. Wurst, Jr., *Trans. Met. Soc. AIME*, **230**, 300 (1964)
120. H. Becke, D. Flatley, W. Kern and D. Stolnitz, *Trans. Met. Soc. AIME*, **230**, 307 (1964)
121. Reference 103, p. 96

122. M. B. Panish and H. C. Casey, Jr., *J. Appl. Phys.*, **40**, 163 (1969)
123. E. F. Steigmeier, *Appl. Phys. Lett.*, **3**, 6 (1963)
124. Y. P. Varshni, *physica,* **34**, 149 (1967)
125. G. D. Pitt and J. Lees, *Solid-State Comm.*, **8**, 491 (1970)
126. L. W. Aukerman and R. K. Willardson, *J. Appl. Phys.*, **31**, 939 (1960)
127. H. Ehrenreich, *Phys. Rev.*, **120**, 1951 (1960)

DIFFUSION IN THE CHALCOGENIDES OF Zn, Cd and Pb 7

D. A. STEVENSON

CONTENTS

7.1 INTRODUCTION

7.1.1 General Comments on the Chalcogenides

The chalcogenides of Zn, Cd and Pb comprise a group of binary semiconducting compounds that have been the subject of many fundamental investigations and device applications. Until recent years, property studies and applications of crystals of these compounds have been made with relatively little regard to the preparative history of the crystals. Recently, however, there have been a number of careful studies showing the relationship between the crystal preparation, the point defect structure and the properties of these semiconductors.

These compounds may be roughly classified into three groups. The *II-VI compounds* are the one-to-one compounds formed between Zn, Cd and S, Se and Te plus ZnO. These compounds have large band gaps, varying from ~1.5 to 3.7 eV, and possess attractive opto-electronic properties. *The oxides,* CdO and PbO†, have band gaps of 2.2 and 2.3 eV between Pb and S, Se and Te and, having band gaps between 0.29 and 0.42 eV, are of primary interest for their infrared and thermoelectric properties. *The oxides* CdO and PbO†, have band gaps of 2.2 and 2.3 eV respectively and possess properties which are distinctly different from the other groups. The chalcogenides will be given the general designation *MX* in the ensuing discussion.

Some important generalizations may be made contrasting the properties of these compounds. The larger band gap materials are expected to exhibit

† The higher oxides of Pb, viz., PbO_2, Pb_2O_3 and Pb_3O_4, will not be considered in the present discussion.

atomic disorder, as discussed below in section 7.2.4. In addition, there is a tendency for one-carrier type to predominate; all of the II-VI compounds except CdTe are one-carrier type in their pure state, irrespective of stoichiometry. Of particular significance is the fact that the carrier type is not readily changed by doping; only high resistivity crystals of the opposite carrier type may be obtained. This behavior has been explained by the tendency of crystals with larger band gaps to self-compensate electrically active impurities (Mandel, 1963) [1] and has limited the use of such materials for many device applications.

7.1.2 Device Applications of the Chalcogenides

The photosensitivity of the larger band gap compounds, CdS and CdSe, finds application for photoconductive radiation detectors for the range from the near infrared to gamma radiation, whereas the smaller band gap IV-VI compounds, PbS, PbSe and PbTe, find application as photoconductive infrared detectors. The luminescence properties of the II-VI compounds, particularly ZnS and ZnO, find application in cathode ray tube screens and lighting panels. Injection electroluminescence applications in the pure crystals are limited by the difficulty in producing low resistivity *n-p* junctions; however, limited success has been achieved in CdTe and in mixed crystals, such as $ZnSe_yTe_{1-y}$ (Aven, 1965) [2].

The strong piezoelectric coupling of ZnO, CdS, and CdSe is utilized in their applications in acousto-electric devices, such as ultrasonic amplifiers (Hutson, McFee and White, 1961; McFee, 1963; Midford, 1964) [3-5]. Both CdTe and CdS have attracted interest as materials for photovoltaic energy converters (Cusano, 1963; Grimmeis and Memming, 1962) [6, 7], with the former material being particularly attractive because of its optimum band gap for the solar spectrum.

High purity CdTe finds application as an infrared window, with the particular advantage that it is stable with respect to changes in moisture. Cadmium telluride is also used as a material for high temperature gamma ray particle detectors (Zanio, 1969) [8], room temperature spectrometers (Akutagawa and Zanio, 1969) [9] and electro-optic modulators (Kieffer and Yariu, 1969) [10]. *N*-type CdTe has exhibited the Gunn effect (Foyt and McWhorter, 1966) [11]; however, the critical voltage is rather high for practical use in microwave generation. Zinc oxide is the key material in the Electrofax system of electrophotography (Young and Grieg, 1954) [12], and it also finds extensive use as a catalyst for reactions in the gas phase and in aqueous solution (Heiland, Mollwo and Stockmann, 1959; Hauffe, Martinez, Range and Schmidt, 1968) [13, 14]. The oxide, PbO, finds application as a photoconducting pickup tube for television (Vidicon tube). In addition, there is interest in its application as a solid state electrolyte because of its high oxide electrolytic conductivity (Heyne, Beekmans and deBeer, 1972) [15].

Both PbTe and PbSe have a favorable thermoelectric figure of merit, with the former compound considered one of the most promising *n*-type materials for thermoelectric energy conversion.

7.1.3 Relevance of Diffusion Studies

In general, the above device applications depend upon properties of these compounds that are sensitive to native defect and impurity concentrations which, in turn, are influenced by crystal growth as well as post-growth annealing processes.

Rather impressive changes in properties are produced by slight changes in stoichiometry—often less than 10^{-5} atomic fraction. Some of these compounds, for example the IV-VI compounds and CdTe, change carrier type with a slight shift in stoichiometry. Others, CdSe for example, change from semiconducting crystals to insulating crystals with slight shifts in stoichiometry. In a practical sense, the rate of change of crystal composition upon annealing is relevant to the processing of those materials for device application. Changes in composition are controlled by the interdiffusion coefficient, $\underset{\sim}{D}$, which is related to the self-diffusion coefficients and the thermodynamic factor (section 7.1.4). The self-diffusion coefficients are in turn related to the individual defect concentrations and mobilities. Thus, the dependence of self-diffusion coefficients upon component pressure provides insight into the dominant native defects existing in the respective compounds.†

This chapter will consider component diffusion in the chalcogenides, with particular emphasis on the dependence of the self-diffusion coefficient upon component pressure and impurity content. This emphasis is motivated by the relative abundance of recent work on this topic and its fundamental significance in terms of defect models for those compounds. Self-diffusion measurements have the advantage that the measured radiotracer profile is established at elevated temperatures where equilibrium may be attained, and the profile is easily preserved upon cooling to room temperature where the profile is measured. This contrasts with the difficulty in quenching electrically and optically active defects for measurement at room temperature and below. In addition, it is difficult to distinguish between an interstitial defect of one component and a vacancy of the other by electrical and optical measurements alone. In this

† The term 'native defect' is used to describe defects involving the binary components M and X such as vacancies and interstitials: V_M, V_X, M_i and X_i. These defects may arise from a deviation from the ideal stoichiometric ratio.

chapter, comparison will be made with recent investigations of the high temperature electrical properties, since in many cases such a comparison shows the difficulty in explaining both types of information with the existing theory of defect solids. The diffusion of impurities in the chalcogenides is not included, because of the necessity to limit the length of this chapter. Two recent reviews of diffusion in the II-VI compounds (Woodbury, 1967, 1968) [16, 17] discuss impurity diffusion in some detail for these systems. Furthermore, virtually all studies of impurity diffusion in the chalcogenides are somewhat limited in their interpretation, because sufficient information on the thermodynamics in the relevant ternary systems is not available. One is referred to the work in the Ga-As-Zn system, discussed in Chapter 6, for an example of the extensive amount of work necessary to properly treat diffusion in ternary systems.

7.1.4 Fundamental Diffusion Equations†

There are two principal diffusion quantities of interest in discussing component diffusion, the self-diffusion coefficients, $D_e(\tilde{M})$ and $D_e(\tilde{X})$ and the interdiffusion coefficient, $\underset{\sim}{D}(MX)$, often referred to as the isoconcentration diffusion and chemical diffusion coefficients respectively. Since there are no concentration differences in true self-diffusion, Fick's law should be obeyed and an experimental profile should be 'normal' in the sense that it may be analytically represented by the appropriate solution to Fick's law. The interdiffusion coefficient, by contrast, may vary with position since interdiffusion occurs in the presence of concentration gradients. The interdiffusion coefficient is related to the self-diffusion coefficients by the Darken equation (Darken, 1948) [21],

$$\underset{\sim}{D}(MX) = [D_e(\tilde{M})[\underset{\sim}{X}] + [D_e(\tilde{X})[\underset{\sim}{M}]] \frac{\partial \ln \{\underset{\sim}{M}\}}{\partial \ln [\underset{\sim}{M}]}, \tag{7.1}$$

where $[\underset{\sim}{M}]$, $[\underset{\sim}{X}]$ and $\{\underset{\sim}{M}\}$ are the mole fractions of M and X and the activity of M, respectively, and the quantity $\partial \ln \{\underset{\sim}{M}\}/\partial \ln [\underset{\sim}{M}]$ is called the thermodynamic factor, T.F. The dependence of $\underset{\sim}{D}(MX)$ upon composition may result from the dependence of both $D_e(\tilde{M})$, $D_e(\tilde{X})$ and the T.F. upon composition.‡ The value of the $\underset{\sim}{D}(MX)$ for the chalcogenides is expected

† For further details on the diffusion equations, the reader is referred to any standard text on diffusion, such as Shewmon (1963) [18], Jost (1960) [19] or Manning (1968) [20].

‡ The self-diffusion coefficients are related to the mobility, B_i, by the Einstein relations, e.g., $D_e(\tilde{M}) = B(M)/kT$ and to the intrinsic diffusion coefficient, $\underset{\sim}{D}(M)$, by the relation $\underset{\sim}{D}(M) = D_e(\tilde{M})(\text{T.F.})$

to be much larger than the largest of the self-diffusion coefficients because of the large values expected for the T.F. in these systems—c.f. section 7.2.3.

An important equation in the interpretation of diffusion results is the relation between the self-diffusion coefficient and the individual concentration-mobility products of the mobile defects. Considering, for example, the M sublattice, with defects $V_M^{\times} V_M', M_i^{\times}, M_i^{\cdot}$,

$$D_e(\widetilde{M}) = D_e(M_i^{\times})[\underset{\sim}{M}_i^{\times}] + D_e(M_i^{\cdot})[\underset{\sim}{M}_i^{\cdot}] + fD_e(V_M^{\times})[\underset{\sim}{V}_M^{\times}] + fD_e(V_M')[\underset{\sim}{V}_M'] \quad (7.2)$$

where f is the correlation factor.† It is expected that one term will dominate, usually the term corresponding to the defect with the highest concentration; for example,

$$D_e(\widetilde{M}) = D_e(M_i^{\times})[\underset{\sim}{M}_i^{\times}], \quad (7.3)$$

if neutral interstitials are the dominant mobile defects.

7.2 PHYSICAL CHEMISTRY OF THE CHALCOGENIDE SEMICONDUCTORS

7.2.1 Significance of the Component Pressure as an Experimental Variable

Any meaningful diffusion experiment must be performed on materials that are in a precisely defined thermodynamic state, in the sense of the Gibbs phase rule. Component diffusion experiments in pure crystals of Zn, Cd and Pb chalcogenides correspond to binary systems, with $F = C + 2 - \phi = 4 - \phi$ independently variable intensive properties, where C is the number of components and ϕ is the number of phases. If diffusion is studied in a closed ampoule or in an open system with only solid and vapor phases in equilibrium, it is necessary to specify two intensive variables to define the system. The choice of temperature, T, and composition in mole fraction, x_i, are commonly selected for solid phases exhibiting a broad region of existence (x_i is the generic term for mole fraction of the ith component). For the limited region of existence of the chalcogenide compounds, it is appropriate to select one of the component pressures, p_i, as an intensive variable in place of the composition; there is an extremely large change in the component pressure, usually several orders of magnitude for a very small change in composition over the solid

† This relation assumes that interstitial-substitutional exchange is sufficiently rapid to establish a uniform distribution of tracer between interstitial and substitutional sites. The symbols $^{\cdot}$ and $'$ refer to effective charges of + and −, in accord with the normal convention for atomic defect notation.

stability range—typically 10^{-2} to 10^{-6} in x_i. It should be noted that there are dramatic changes in properties produced by these slight changes in composition. For example, compounds, such as CdTe and PbTe, change from n- to p-type conductivity, and other compounds, such as CdSe, change from semiconducting to insulating crystals with this slight shift in composition. Furthermore, the large relative change in component pressure with composition underlies the large values of the thermodynamic factor in these systems, as discussed below.

In diffusion experiments involving solid and vapor phases only, the two intensive variables T and p_i are appropriate selections to characterize the system. The pressure is typically controlled by one of two techniques: a two temperature zone system with one of the pure components in the lower temperature zone (de Nobel, 1956; Lorenz, 1962; Van Doorn, 1962) [22-24]; or use of the ideal gas law in conjunction with the knowledge of the weight of excess component and the volume of a capsule.

There are three states of the solid compound that are easiest to establish experimentally: $MX(S)$ in equilibrium with M-rich liquid and vapor, $MX(S)]_{M\ \mathrm{SAT}}$; $MX(S)$ in equilibrium with X-rich liquid and vapor, $MX(S)]_{X\ \mathrm{SAT}}$; and the congruent subliming composition or minimum total vapor pressure, $MX(S)]P_{\mathrm{min}}$. The first two states are achieved by equilibrating excess M or X liquid with the solid MX. There are three phases coexisting, hence a variance of one ($F = C + 2 - \phi = 1$), and a specification of T is sufficient to define the equilibrium properties of the solid. The congruent subliming composition is obtained by placing the solid compound in either a flowing inert atmosphere or in a dynamic vacuum at a well-defined temperature. Any excess M or X will selectively vaporize until eventually the composition of the vapor is identical to that of the solid phase and, hence, congruent sublimation occurs. In accord with the Gibbs-Konovalov theorem (Prigogine and Defay, 1962) [25], this corresponds to the minimum total pressure, P_{min}, and the system may be treated as a one-component two-phase system. As a consequence, $F = 1 + 2 - 2 = 1$, and a specification of the temperature is sufficient to define the state of the system.

In summary, p_i and T emerge as the two operationally significant experimental variables for sharp binary compounds. There are three cases where T alone defines the state of the crystal: M_{SAT}, X_{SAT}, and P_{min}. For other cases, the component pressure must be controlled and specified. The following discussion of the thermodynamics of binary compounds provides further information regarding this topic. The CdTe system will be used as an illustrative example.

7.2.2 Solid-Vapor Equilibria; Vaporization Constants for Binary Compounds

Chalcogenide crystals vaporize at elevated temperatures according to the reaction†:

$$MX(S) = M(G) + \tfrac{1}{2}X_2(G); \qquad K_D = p_M p_{X_2}^{1/2}. \tag{7.4}$$

The corresponding equilibrium constraint, the decomposition constant, gives the explicit interdependence of the two component pressures:

$$p_M = K_D/p_{X_2}^{1/2} \quad \text{or } p_{X_2} = K_D^2/p_M^2. \tag{7.5}$$

This equilibrium constant, K_D, may be evaluated by direct measurement of the partial pressure product over the compound, or from the standard enthalpy and entropy of formation from the elements using the relation:

$$-RT \ln K_D = \Delta G^\circ = \Delta H^\circ - T\Delta S^\circ. \tag{7.6}$$

The temperature dependence is given by the van't Hoff equation

$$\partial \ln K_D/\partial(1/T) = -\Delta H^\circ/R. \tag{7.7}$$

A summary of these data for many chalcogenides is given by Lorenz (1967) [27].

In addition to interrelating the two component partial pressures, the vaporization constant provides information on the minimum total vapor pressure, P_{min}. Using equation 7.5, the total pressure over $MX(S)$ is expressed as

$$P = p_M + p_{X_2} = p_M + K_D^2/p_M^2. \tag{7.8}$$

Minimizing the total pressure by setting $\partial P/\partial p_M = 0$ and using equation 7.5 leads to the condition

$$p_M = 2p_{X_2} = 2^{1/3} K_D^{2/3}, \tag{7.9}$$

which establishes that P_{min} occurs at the congruent subliming composition.‡ The total pressure, P is related to K_D by the equation,

† Experimental results show that $MX(G)$ species are minor species (Goldfinger and Jeunehomme, 1963) [26]. In some instances, e.g., with sulfides and selenides, other polymeric species such as X_4, X_6, and X_8 must be considered.

‡ The analysis may be applied to the general compound $M_a X_b$, where a and b may be non-integral values, such as 1.0001, characteristic of solids with a slight deviation from an ideal stoichiometric ratio. The minimization of P still leads to the condition of congruent vaporization, where the composition now is off stoichiometry. In general, the congruent subliming composition may be off stoichiometry. For the chalcogenides under consideration, the phase field is sufficiently narrow so that deviation from the ideal ratio is usually not detected.

It is also interesting to reflect on the conditions necessary for congruent vaporization; the vapor pressures of both components in the compound must be

$$P_{min} = (3/2)2^{1/3} K_D^{2/3}. \tag{7.10}$$

In summary, the vaporization constant interrelates the two component vapor pressures so that only one is independent. In addition, the decomposition constant may be related to P_{min}, which corresponds to congruent vaporization.

7.2.3 Phase Diagrams: *T-x*, *P-x*, *P-T* Diagrams†

The interrelation of the various phase diagrams provides important background in discussing the thermodynamic state of binary compounds.

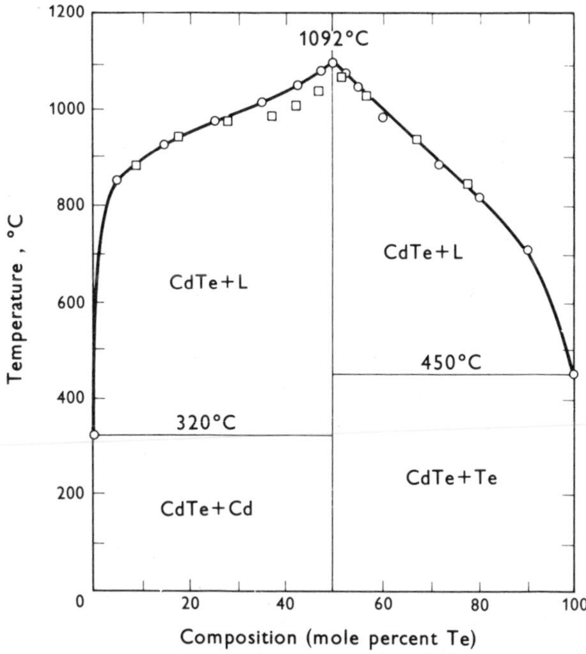

Figure 7.1. Temperature-composition diagram for the cadmium-tellurium system: ○ Lorenz (1962) [23]; □ deNobel (1959) [22]. After Borsenberger (1967) [40].

The normal *T-x* phase diagrams of the chalcogenides, when represented on a normal-unexpanded composition scale, show a typical 'line compound', as seen in figure 7.1 for CdTe. There is, however, a small but finite width to the phase field, typically 10^{-4}-10^{-6} atomic fraction. In figure 7.2 the *T-x* diagram is indicated schematically, with the *MX* phase field

---

comparable, otherwise the vapor is always richer in one component than the solid phase and selective loss of that component occurs. This situation exists in many of the III-V compounds, except at low temperatures, cf. Thurmond (1965) [28].

† The symbol x will be used as a generic symbol for mole fraction, i.e. $x_M \equiv [\underset{\sim}{M}]$.

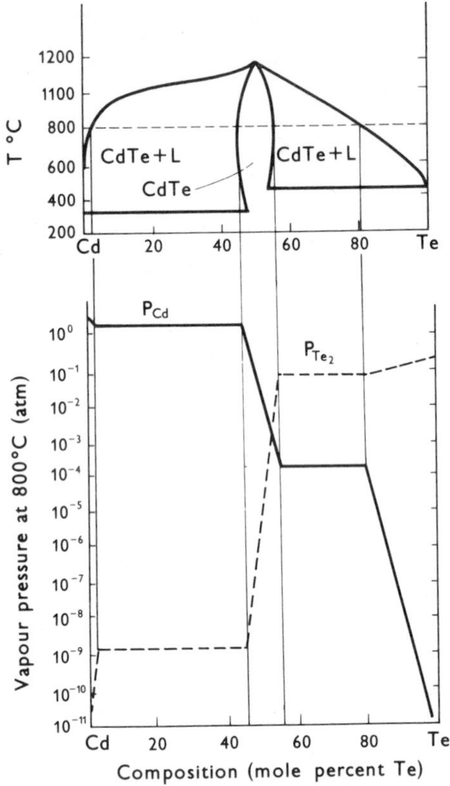

Figure 7.2. Pressure-composition isotherm for the cadmium-tellurium system at 800°C. (Width of CdTe phase field exaggerated for clarity.) After Borsenberger (1967) [40].

exaggerated for clarity. A matching *P-x* diagram is also given, which shows the variation of p_i across the diagram for both components. The extrema in p_i corresponds to the univariant states where there is the coexistence of three phases: $MX(S)$, M-rich liquid, vapor; and $MX(S)$, X-rich liquid, vapor. These states are designated respectively: M_{SAT}, $p_M)_{\max}$ or $p_{Te_2})_{\min}$; X_{SAT}, $p_M)_{\min}$ or $p_{Te_2})_{\max}$. Pressure isotherms may be estimated in the following manner: assuming Raoult's law for M in the M-rich liquid and X in the X-rich liquid, calculate, respectively, p_M and p_{x_2} from the vapor pressure of the pure elements and the two respective liquidus compositions; calculate p_{X_2} and p_M at the two respective points using K_D and assume a linear variation of p_i across the phase field. The assumption of Raoult's law is particularly good at lower temperatures where the corresponding liquidus is, essentially pure M and X, respectively. The assumption of a linear variation of p_i may only show the average trend, as discussed below.

One particularly significant consequence of the large change in component pressure with a small change in composition is the large value of the thermodynamic factor that is to be expected in the chalcogenides. This is the basis of the large observed differences between the interdiffusion coefficient, $\underset{\sim}{D}(MX)$, and the self-diffusion coefficients, $D_e(\widetilde{M})$ and $D_e(\widetilde{X})$ in these systems. Considering a typical case where the metal pressure changes from 1 atm to 10^{-4} atm as the composition changes by 0.0001 in terms of $[\underset{\sim}{M}]$,† an average value of the T.F. can be estimated, cf. for example, Wagner (1953) [29]. Assuming ideal vapor and $p_M = 1$ atm and $[\underset{\sim}{M}] = 0.50005$ at M_{SAT} and $p_M = 10^{-4}$ atm and $[\underset{\sim}{M}] = 0.49995$ at X_{SAT}, then

$$\text{T.F.} = \frac{\partial \ln \{\underset{\sim}{M}\}}{\partial \ln [\underset{\sim}{M}]} \cong \frac{\Delta \ln \{\underset{\sim}{M}\}}{\Delta \ln [\underset{\sim}{M}]} \cong [\underset{\sim}{M}] \frac{\Delta \ln (p_M/\overset{\circ}{p}_M)}{\Delta [\underset{\sim}{M}]}$$

$$= \{[M] \ln [(p_M)_{M\,SAT}/(p_M)_{X\,SAT}]\}/[([M])_{M\,SAT} - ([M])_{X\,SAT}]$$

$$= [0.5 \ln (1/10^{-4})]/10^{-4} \cong 4 \times 10^4.$$

In accord with the Darken Equation, equation 7.1, the interdiffusion coefficient is expected to exceed the larger of the component self-diffusion coefficients by approximately four orders of magnitude.

The assumption of linear variation of p_i across the phase field may not be realistic, in the light of reported variations of T.F. with p_i. For example, Zanio (1970) [30] reported T.F. values in CdTe at 700°C which varied from 4×10^5 to 3×10^6. He measured the carrier density as a function of p_{Cd} and interrelated the carrier density and composition by assuming a particular defect model.

A concise representation of the pressure limits for the stable solid phase is given in a *P-T* diagram, a plot of ln (p_i) versus $1/T$, such as shown in figure 7.3 for CdTe. Since only one pressure is independent, specification of p_{Cd} determines p_{Te_2} through K_D. The region within the blade corresponds to $S + G$ and outside to $L + G$ and the line of demarcation corresponds to the liquidus line in the *T-x* diagram where *S*, *L* and *G* are in equilibrium. Because the liquidus line in the *T-x* diagram approaches pure Cd and Te, respectively, at the two ends of the diagram as the temperature is lowered, the line corresponding to the coexistence of three phases on the *P-T* diagram approaches the vapor pressure of pure Cd, $\overset{\circ}{p}_{Cd}$, and the

† This is consistent with a maximum concentration of native defects as high as 10^{18}-10^{19} cm^{-3}. For a lower concentration, the value of the T.F. would be larger than for the example selected.

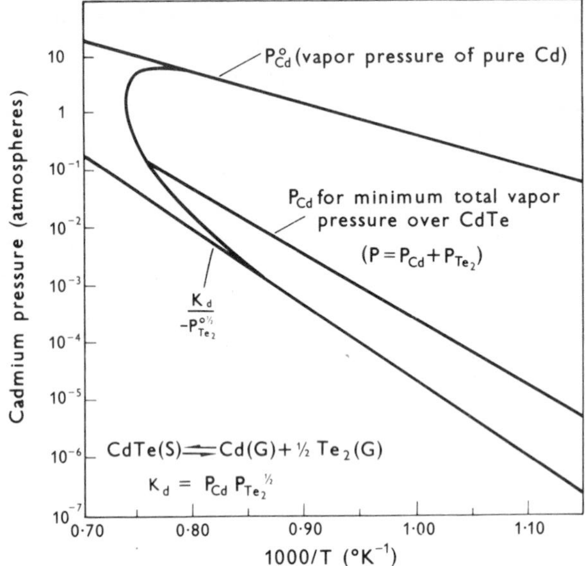

Figure 7.3. Pressure-temperature diagram for the cadmium-tellurium system. After Borsenberger (1967) [40].

value of p_{Cd} over CdTe in equilibrium with pure Te, $K_D/p^{o\frac{1}{2}}_{Te_2}$. These *P-T* diagrams are convenient for obtaining the possible extrema in p_i at a specific temperature.

In summary, the component pressures in the chalcogenides typically vary by several orders of magnitude for a small change in composition. These pressures can be estimated from the vapor pressure of the pure elements, the *T-x* diagram, and the value of K_D, and are concisely represented in a *P-T* diagram. A large value of the T.F. is expected for such systems with the consequence that $\underset{\sim}{\mathbf{D}} \gg D_e$.

7.2.4 Dependence of the Point Defect Concentration on the Component Pressure; Defect Concentration Isotherms

The theory of defect chemistry explains the deviation from the ideal stoichiometric ratio in compounds in terms of native point defects, such as vacancies, interstitials or misplaced component atoms. In sharp binary compounds, the component pressure replaces composition as an operational independent variable and a theoretical framework has been established to relate native defect concentrations to the component pressure. This framework was developed by Wagner and Schottky (1931) [31] and refined and formalized by Brouwer (1954) [32], Kröger and Vink (1956) [33], Vink (1963) [34] and Kröger (1964) [35]. The theory invokes mass action and electrical neutrality constraints on the equilibrium concentrations of native defects. The underlying assumption is the presumed validity of Henry's law for the native defects, which are normally present in low concentrations. This type of analysis is central to the interpretation of diffusion in the chalcogenides and a few simple examples will be given for illustrative purposes.

Consider, as a postulated defect model, a singly ionized Frenkel defect on the M sublattice. Using the established atomic defect notation, the following independent equilibria and equilibrium constraints may be written:

$$M_M \rightleftharpoons V_M^{\times} + M_i^{\times}; \qquad K_F = [V_M^{\times}]\,[M_i^{\times}] \tag{7.11}$$

$$M(G) \rightleftharpoons M_i^{\times}; \qquad K_R = [M_i^{\times}]/p_M \tag{7.12}$$

$$M_i^{\times} \rightleftharpoons M_i^{\cdot} + e'; \qquad K_1 = [M_i^{\cdot}]\,n/[M_i^{\times}] \tag{7.13}$$

$$V_M^{\times} \rightleftharpoons V_M' + h^{\cdot}; \qquad K_2 = [V_M']p/[V_M^{\times}] \tag{7.14}$$

$$0 \rightleftharpoons e' + h^{\cdot}; \qquad K_i = np, \tag{7.15}$$

and in addition, the electrical neutrality condition,

$$n + [V_M'] = p + [M_i^{\cdot}]. \tag{7.16}$$

There are six unknown concentrations, $[M_i^{\times}]$, $[M_i^{\cdot}]$, $[V_M^{\times}]$, $[V_M']$, n and p, and six independent constraints; therefore the problem is solvable, and each of the six concentrations can be expressed as a function of p_M if the values of the constants are known. The solution is conveniently represented on a graph of the logarithm of the various defect concentrations plotted against the logarithm of one of the component pressures; such a graph will be designated a defect concentration isotherm.

The solution of equations 7.11 through 7.16 is simplified by considering regions where the charge neutrality condition involves two defect species only. In the present example the following possibilities are: Region I, highest p_M, $n = [M_i^{\cdot}]$; Region II, intermediate p_M either $[M_i^{\cdot}] = [V_M']$ (dominant atomic disorder) or $n = p$ (dominant electronic disorder); and Region III, lowest p_M, $p = [V_M']$. Diagrams for these two cases are given in figures 7.4 and 7.5. Important reference points on the diagram are given by the values of mass action constants; for example, the value of n and p is $K_i^{1/2}$ at the point where $n = p$ and the value of $[V_M']$ and $[M_i^{\cdot}]$ is $K_{F'}^{1/2}$ when $[V_M'] = [M_i^{\cdot}]$, where $K_{F'} = [V_M']\,[M_i^{\cdot}] = K_1 K_2 K_F/K_i$ by virtue of equations 7.11, 7.13, 7.14 and 7.15. The type of behavior in Region II depends upon the relative magnitudes of K_i and $K_{F'}$. The value of K_i may be calculated from g_c, g_v,

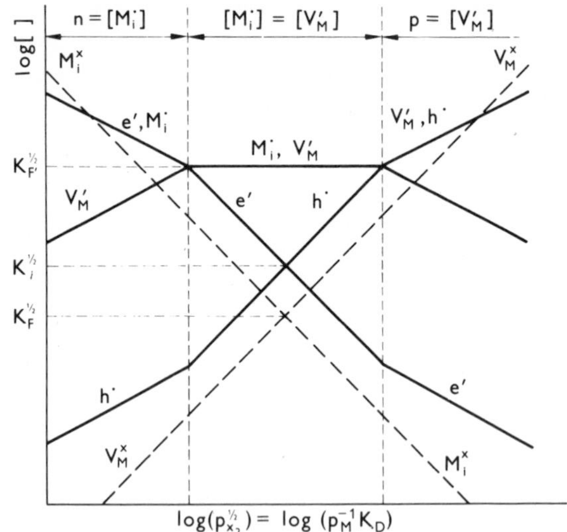

Figure 7.4. Defect equilibrium isotherm; dominant singly ionized Frenkel disorder on the M sublattice, with $K_{F'} > K_i > K_F$.

N_c, and N_v, the degeneracies and densities of states for the conduction and valence bands, respectively, and ϵ_g, the band gap,†

$$K_i = g_c g_v N_c N_v \exp(-\epsilon_g/kT). \tag{7.17}$$

In general, K_i will be large for the smaller band gap materials, such as the lead chalcogenides, and one would expect electronic disorder to dominate in Region II, whereas atomic disorder would be expected in this region for large band gap compounds. Unfortunately, the value of $K_{F'}$ cannot be calculated from fundamental properties, as is the case for K_i, although estimates have been made—Mandel (1963) [1].

The relevance of these diagrams arises from their relationship to the variation of the electrical conductivity, σ, and the self-diffusion coefficients, $D_e(\widetilde{M})$ and $D_e(\widetilde{X})$, with component pressure. Referring to the well established relationships of these two transport coefficients to concentration-mobility products:

$$\sigma_{\text{electronic}} = ne\mu_n + pe\mu_p \tag{7.18}$$

$$\sigma_{\text{ionic}} = \left\{eD_e(M_i^{\cdot})[\underset{\sim}{M}_i^{\cdot}] + eD_e(V_M')[\underset{\sim}{V}_M']\right\} kT \tag{7.19}$$

† A compilation of these parameters for the chalcogenides is given by D. Long (1968) [36].

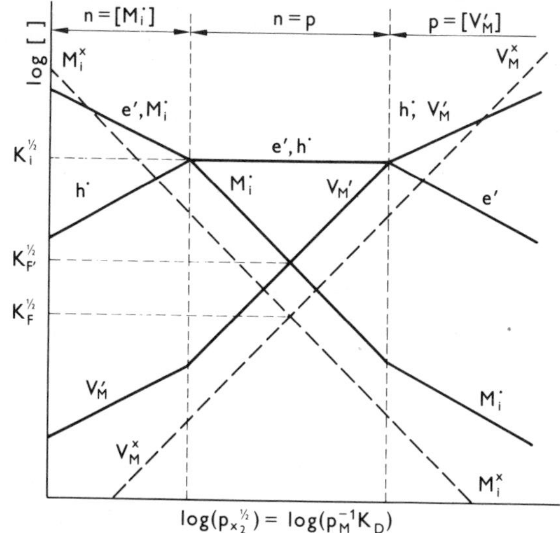

Figure 7.5. Defect equilibrium isotherm; dominant electronic disorder, with $K_i > K_{F'} > K_F$.

$$D_e(\widetilde{M}) = D_e(M_i^{\times})[\underset{\sim}{M}_i^{\times}] + D_e(M_i^{\cdot})[\underset{\sim}{M}_i^{\cdot}] + fD_e(V_M^{\times})[\underset{\sim}{V}_M^{\times}] + fD_e(V_M')[\underset{\sim}{V}_M'], \quad (7.20)$$

where μ_n, μ_p and D_e are the mobilities and diffusion coefficients of the respective species.

If one assumes that one concentration-mobility product dominates and that the mobility or defect diffusivity is independent of p_i, then these properties will show a dependence upon p_i given by p_i^{γ}, where γ, taken from the defect concentration isotherm, is typically 0, ±1/2, or ±1. Since the mobility of one defect might be higher than the others, that defect might dominate in the transport process even if it is in lower concentration than the others. Assuming, for illustrative purposes, an identical mobility for all species, the variations in $\sigma_{\text{electronic}}$ and $D_e(\widetilde{M})$ and σ_{ionic} are shown schematically in figure 7.6 for the disorder described in figure 7.4.

In the analysis of real systems, the type of defect model is implied from the experimentally observed component pressure dependence of measured properties, such as $\sigma_{\text{electronic}}$ (or n or p if the Hall coefficient is measured), $D_e(\widetilde{M})$ or $D_e(\widetilde{X})$. Defect concentration isotherms are devised, based on hypothetical models, and models accepted which lead to predictions consistent with observed behavior.

It must be emphasized that these diagrams correspond to equilibrium conditions at the temperature in question and one must be cognizant of

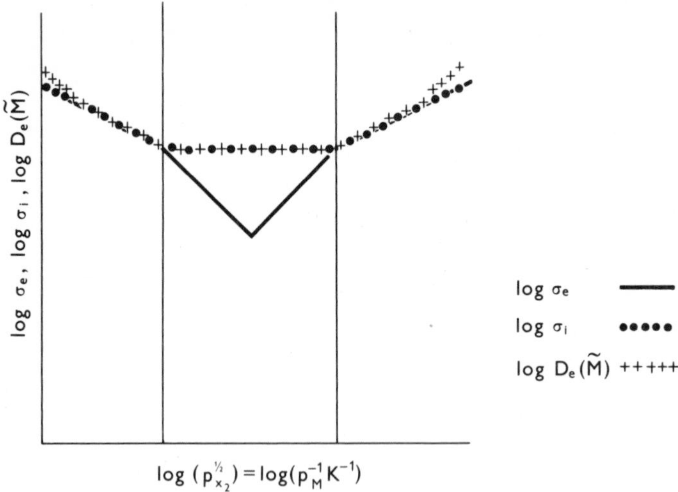

Figure 7.6. Variation of σ_e, σ_i and $D_e(\widetilde{M})$ for disorder described in figure 7.4 for a hypothetical case in which all electronic and atomic defects have identical mobility.

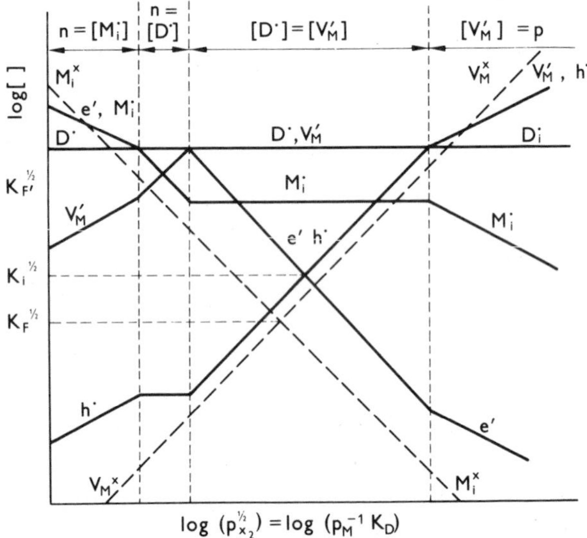

Figure 7.7. Influence of a shallow donor upon the defect equilibrium isotherm. Modification of figure 7.4, upon addition of a shallow donor, D, with $[D^{\cdot}] > K_{F'}^{1/2}$.

changes that occur in the crystal upon cooling from the equilibrium temperature. In addition, the actual stability range for the compound MX may encompass only one region, such as I, II or III in figures 7.4 and 7.5.

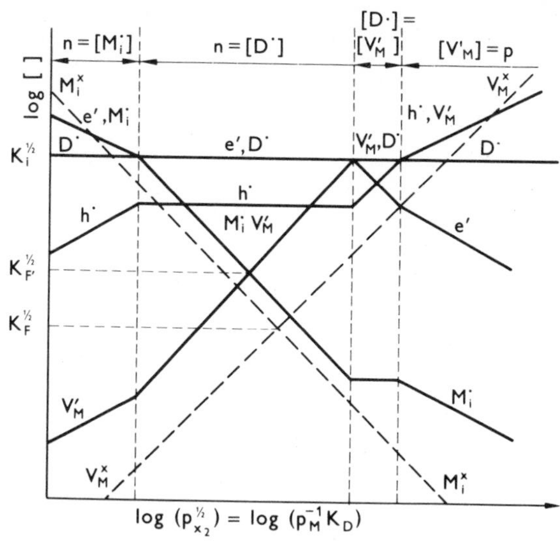

Figure 7.8. Influence of a shallow donor upon the defect equilibrium isotherm. Modification of 7.5 upon addition of a shallow donor D, with $[D^\cdot] > K_i^{1/2}$.

The influence of electrically active impurities on the electrical properties and the self-diffusion coefficients provides additional information concerning the nature of the native disorder in binary compounds. To produce a change in the defect structure of the compound, the impurity concentration must exceed either $K_{F'}^{1/2}$, for the case of atomic disorder, or $K_i^{1/2}$ for electronic disorder. Considering the example of a shallow completely-ionized donor impurity, $D^\times \rightarrow D^\cdot + e'$, the resulting defect concentration isotherms are shown in figures 7.7 and 7.8.

It is enlightening to consider the influence of a donor impurity in Region II for a compound displaying dominant atomic disorder. The addition of a donor impurity is compensated by the native acceptor, V'_M, so that the change in electron density is far less than the added impurity concentration. This behavior, called self-compensation, underlies the difficulty in changing the carrier type of most of the II-VI compounds which in turn limits their device applications—Mandel (1963) [1].

The relationship of the impurity content to the transport properties is obvious. If, for example, the self-diffusion of M proceeds via a V_M^1 defect, the value of $D_e(\tilde{M})$ would be enhanced by donor concentrations that exceed $K_{F'}^{1/2}$. A measurement of the minimum value of $[D^\cdot]$ to produce enhancement of $D_e(\tilde{M})$ may be used to evaluate $K_{F'}^{1/2}$. For a given impurity concentration, $[D^{\cdot}]$, independent of temperature, one may note intrinsic and extrinsic regions for $D_e(\widetilde{M})$, for temperatures where $[D^{\cdot}]$ is less than or greater than $K_{F'}^{1/2}$ respectively.

It should be noted that in the above analysis of impurity effects, it is assumed that the impurity concentration is frozen-in and that the system may still be regarded as a binary one. If this is not the case, there is another degree of freedom and an additional intensive variable must be specified to define the system.

7.2.5 Comments on Prediffusion Annealing and Anomalous Diffusion Profiles

There are a few salient experimental features that deserve special mention. As previously discussed, the defect structure of binary compounds is expected to vary with the imposed component pressure. Since self-diffusion measurements must conform to the condition of uniform defect composition, the crystals should, in a strict sense, be homogenized by prediffusion annealing under precisely the same conditions of component pressure and temperature as in the diffusion anneal.

Many investigations reported in the literature have carried out this preannealing procedure.† There are often instances, however, where identical results are obtained, irrespective of whether or not the samples are preannealed. This may be the case when the self-diffusion coefficients are independent of component pressure, for the case for dominant ionized atomic disorder. It is also true that prediffusion annealing is essentially unnecessary in instances where the compositional range of existence is very small. In these cases, the amount of tracer uptake by the crystal to change crystal composition is negligible compared to the tracer exchange necessary to establish a measurable profile. Furthermore, the establishment of uniform concentration is rapid since it is governed by the interdiffusion coefficient which is typically several orders of magnitude larger than the self-diffusion coefficient. In summary, there are many compound systems which are homogenized rapidly at temperature and the homogenization involves a negligible uptake of component compared to the radio-tracer exchange used to establish the diffusion profile.

There are several reports of anomalous profiles for self-diffusion studies in the chalcogenides, with three possible causes for this behavior: lack of isoconcentration conditions in the sample; a diffusion barrier at the sample

† It is assumed that any disruption in the state of the crystal introduced by quenching to room temperature after the prediffusion anneal and prior to the introduction of the tracer source will be removed rapidly at the diffusion temperature. These disruptions may be the formation of precipitates due to retrograde solubility; however, it is presumed that the precipitates redissolve rapidly, to establish uniform composition.

source interface; and two or more diffusion paths, a slow one and a fast one. A discussion of these profiles and their analytical representation is given by Nebauer (1969) [37]. Evidence for enhanced grain boundary diffusion in the chalcogenides based on autoradiographic evidence has been reported by Reynolds and Stevenson (1969) [38] and by Roberts and Wheeler (1957) [39]—and it is clear that single crystals of coarse grained samples must be used to obtain reliable volume diffusion coefficients. It was pointed out by Borsenberger (1967) [40] that anomalous profiles may also be established by entrainment during sectioning or by edge effects. In summary, profiles that do not conform to normal Fick's law behavior should be tested by autoradiography to establish the significance of diffusion short circuits, edge effects or entrainment.

7.3 DIFFUSION IN THE ZINC CHALCOGENIDES†

7.3.1 Diffusion in ZnO

Diffusion, electrical and optical data relating to the defect structure of ZnO have been reviewed by Garrett (1961) [41]. Kröger (1964) [35], Heiland *et al.* (1959) [13], Boltaks (1963) [42], Harrop (1968) [43], and O'Keeffe (1965) [44]. Although there have been numerous studies on ZnO, descriptions of the principal defects in ZnO appear inconsistent. Contributing causes to this dilemma are the variety of experimental methods and crystal samples used to study diffusion in ZnO.

Diffusion of Zn in ZnO

An early study by Miller (1941) [45] established a value of 4×10^{-11} cm^2 s^{-1} for $D_e(\widetilde{\mathrm{Zn}})$ at 1000°C using a Zn-65 film source, mechanical sectioning technique and a sintered pellet sample.

Lindner (1952) [46] investigated the diffusion of Zn-65 in sintered-compressed pellets of ZnO in air. Two experimental techniques were employed: the activity decrease of the positron emission from a thin-film Zn-65 source was followed as a function of the time of the diffusion anneal, and a radioactive $Zn^{65}O$ pellet was placed in contact with a non-radioactive pellet and the tracer exchange was measured as a function of the time of diffusion annealing. Three different sources of ZnO powder were employed to make pressed tablets that were subsequently sintered in air at 1300°C (densities were 5.67, 5.60 and 5.63, respectively, versus a theoretical density of 5.70). The results of the thin film method showed a dependence of D on the density of the compact,

† A tabular summary of the literature discussed is given in Table 7.1.

TABLE

Summary of component diffusion in

Diffusion system	D_o(cm^2 s^{-1})	Q(eV)	Experimental method	Type of sample; and preannealing treatment
$D_e(\widetilde{Zn}$ in ZnO)			Mechanical sectioning; Zn-65 source	Pellet sintered in air at 1000°C for 12 h
	1.3	3.2	Thin film source; surface decrease	Compressed and sintered pellets from three different sources; sintered in air at 1300°C
	3.5	3.0	Tracer exchange; $Zn^{65}O$ pellet source	Compressed and sintered pellets from three different sources; sintered in air at 1300°C
			Profile by etching and counting	Single crystals 0.3 mm diam. x 10 mm long
	4.8	3.17	Tracer exchange; $Zn^{65}O$ sample	Acicular single crystals circa 0.01 mm diam. x 5.0 mm long; as grown
			Thin film Zn-65 source; surface decrease	Compressed pellets sintered in air 100 h at 1150°C
			Thin film source, Zn-65; sectioning and measuring	Dense polycrystals

[a] Symbols

$D_e(\widetilde{M}$ in $MX)$	Self-diffusion coefficient of M tracer in compound MX
$\underset{\sim}{D}(MX)$	Interdiffusion coefficient in MX
D_{gb}	Grain boundary diffusion coefficient
δ	Grain boundary width
θ	Preannealing temperature in °K
P_{min}	Minimum total pressure, i.e., congruent subliming composition

Diffusion equations correspond to units of cm^2 s^{-1}, with component pressure expressed in atmosphere, unless otherwise specified.

7.1

the Zn, Cd and Pb chalcogenides†

Diffusion temperature (°C) and atmosphere	Remarks	Reference
1000	$D_e(\widetilde{Zn}) = 4 \times 10^{-11}$ at 1000°C	Miller (1941) [45]
950-1380; in air	A dependence of $D_e(\widetilde{Zn})$ upon impurities and compact density was observed	Lindner (1952) [46]
655-1370; in air		Lindner (1952) [46]
850-1100; Zn-65 in closed capsule	The observed diffusion coefficient was 10-100 times larger than Lindner's values and depended upon concentration	Münnich (1955) [47]
900-1026; 1 atm Zn	An isotherm at 1000°C (4 points: 0.2 to 2 atm zinc pressure) showed $D_e(\widetilde{Zn}) \propto p_{Zn}^{0.65}$	Secco and Moore (1957) [48]
1000-1250; air	A moderate amount of scatter; nonuniform coloration of samples; however, $D_e(\widetilde{Zn})$ values were not appreciably different from corresponding values for crystals exposed to Zn vapor	Secco and Moore (1957) [48]
800-1300; oxygen and argon	Autoradiographic evidence of grain boundary diffusion with the grain boundary diffusion coefficient, D_{gb}, estimated to be $D_{gb} = 10^{3 \pm 2} \exp[(-3.0 \pm 0.5)/kT]$ cm² s⁻¹	Roberts and Wheeler (1957) [39]

Table 7.1–*(continued)*

Diffusion system	D_o(cm^2 s^{-1})	Q(eV)	Experimental method	Type of sample; and preannealing treatment
			Thin film source, Zn-65; sectioning and measuring	Dense polycrystals
	1.7×10^2	3.3	Etching and counting the residual Zn-65	Crystals
	1.25×10^{-5}	1.87 ± 0.50	Thin film source mechanical sectioning and counting the material removed	Single crystals 3-5 mm long x 0.9-1.4 mm thick preannealed in oxygen either 1 or 6 h at 700°C
	10^2 (from 720°-780°C) 10 (from 800°-840°C)	3.17 ± 0.21	Zn-65 exchange; static system	Powder circa 3×10^{-5} cm diam.
	0.40	3.3	Zn-65 exchange; static system	Powder circa 3×10^{-5} cm diam. Preannealing at 760°C, 5 h with $p_{Zn} = 0.15$ atm for selected samples
$D_e(\widetilde{O}$ in ZnO)	6.52×10^{11}	7.16	0-18 exchange; mass spectrometry	Small single crystal spheres (~0.4 g)
	1.05×10^3	4.1	0-18 exchange; mass spectrometry	Single crystal needles grown by H_2 transport broken into approximate spheres with radii 0.4-0.8 mm; identical temperature and p_{O_2} as the diffusion anneal for at least the length of time of the anneal
$\underset{\sim}{D}$(ZnO)	5.3×10^{-4}	0.55	Relaxation of conductivity	Needle-like single crystals 0.013-0.026 cm in diam.

Diffusion temperature (°C) and atmosphere	Remarks	Reference
800-1400; oxygen argon and mixtures thereof	For 1 atm O_2 $D^{1/2}/\delta D_{gb} = 10^{4.0 \pm 1.5} \exp[(+1.04 \pm 0.03)/kT]$ $s^{1/2}$ cm^{-2}. Values decreased with decreasing p_{O_2} and were sensitive to sample history	Roberts and Wheeler (1960) [56]
800-1190; zinc vapor		Lamatsch (1958) [57]
1000-1265; oxygen	Established that diffusion perpendicular and parallel to the c axis is essentially the same	Moore and Williams (1959) [49]
720-840; zinc vapor 8.5×10^{-2}-2.5×10^{-1} atm	Exchange rate was independent of p_{Zn}; initial first order rate process observed; discontinuity of D_O at 800°C	Secco (1960) [61]
760 and 800; zinc vapor	Preannealing in Zn vapor caused a large decrease in exchange rate	Secco (1961) [62]
1100-1300; isotherm at 1275°C 0.15 atm $\leqslant p_{O_2} \leqslant$ 0.85 atm	Isotherm at 1275°C yielded $D_e(\tilde{O} \text{ in ZnO}) \propto p_{O_2}^{1/2}$	Moore and Williams (1959) [49]
1150-1400; $p_{O_2} \sim 200$ torr Isotherms at 1200, 1300, 1350, 1400	$D_e(\tilde{O} \text{ in ZnO}) \propto p_{O_2}^{-1.5}$ at 1200°C; $D_e(\tilde{O} \text{ in ZnO}) \propto p_{O_2}^{-0.5}$ at 1300-1400°C	Hoffman and Lauder (1970) [66]
180-350; air		Thomas (1957) [51]

Table 7.1–*(continued)*

Diffusion system	D_0(cm^2 s^{-1})	Q(eV)	Experimental method	Type of sample; and preannealing treatment
$\underset{\sim}{D}$(ZnO)	1.6×10^{-2}	1.7	Relaxation of conductivity	ZnO crystals; samples preannealed in Zn vapor at temperature θ = 700, 800, 900, 1150 and 1330
$D_e(\widetilde{Zn}$ in ZnS)	3.2×10^{7}	4.2	Radiotracer exchange-Zn-65; static system	Small single crystal pellets 0.10-0.50 mm thick x 1 cm x 0.5 cm; vapor grown; wurtzite modification as grown
	4.5	2.7	Radiotracer exchange-Zn-65; static system	I: Reagent grade II: crushed single crystals; both samples ~6.1×10^{-5} cm radius; wurtzite modification
	0.1	2.9 ± 0.2	Kinetics of transformation	Reagent grade powder, 20% sphalerite 80% wurtzite; preannealed in vacuum, 0.1 atm, sulfur 0.1 atm, zinc
$D_e(\widetilde{S}$ in ZnS)	8×10^{-5}	2.2	Autoradiographic determination of profile S-35	Single crystal slices perpendicular to the (111); 1.5 x 5 x 7 mm zinc blende as grown
	7×10^{5}	3.4	S-35 profile, mechanical sectioning	Single crystal slices, 45° from *c* axis; hexagonal modification preannealed 2 days in saturated zinc at the diffusion temperature
$\underset{\sim}{D}$(ZnS)	1.2×10^{-2}	1.39	Rate of film growth	Zinc single crystal and Zn liquid
$D_e(\widetilde{Zn}$ in ZnSe)	10^{3}	3.45 ± 0.22	Zn-65 tracer exchange	Powder, particle radius ~8.7×10^{-4} cm

Diffusion temperature (°C) and atmosphere	Remarks	Reference
480-1050; air	$\underset{\sim}{D}$(ZnO) depends upon preannealing temperature, $\tilde{\theta}$, in accord with the relation $\underset{\sim}{D}(\text{ZnO}) = 1.6 \times 10^{-2} \exp\left(\frac{1.0}{k\theta} - \frac{1.7}{kT}\right)$	Pohl (1959) [65]
925-1075; 1 atm Zn vapor; isotherm at 1025; $p_{Zn} = 0.25$, 1, 2 atm	$D_e(\widetilde{\text{Zn}}$ in ZnS) varied with sample thickness; three separate temperature regions were proposed; isotherm at 1025°C was represented as $D_e(\widetilde{\text{Zn}}$ in ZnS) $\propto p_{Zn}^{1.5}$; initial first order rate process was observed	Secco (1958) [75]
720-960; 0.11 atm zinc vapor	Markedly different results were obtained for sample II, viz. $Q = 0.87 \pm 0.13$ eV	Secco (1964) [77]
800-900; zinc vapor		Bansagi, Secco, Srivastava and Martin (1968) [78]
740-110; S_2 vapor 0.5 atm	$D_e(\widetilde{\text{S}}$ in ZnS) increases with increasing sulfur pressure	Gobrecht, Nelkowski, Baars and Weigt (1967) [79]
700-891; S_2 vapor, 0.5 atm	Preannealing and diffusion annealing may have caused transformation to zinc blende modification	Blount, Marlor and Bube (1967) [80]
320-600; S-rich atmosphere		Sirota and Koren (1962) [81]
720-800; zinc vapor 0.1 atm	No initial first order rate process was observed, in contrast to ZnO and ZnS	Secco and Su (1968) [83]

Table 7.1—*(continued)*

Diffusion system	$D_o(cm^2\ s^{-1})$	Q(eV)	Experimental method	Type of sample; and preannealing treatment
	9.8	3.03	Zn-65 profile; mechanical sectioning	Single crystal slices, melt and vapor grown ~1 cm^2 x 0.15 cm thick preannealed under identical conditions of the diffusion anneal
$D_e(\widetilde{Se}$ in ZnSe)	0.13 (Se saturated crystals)	2.57	Se-75 profile; mechanical sectioning	Single crystal slices melt and vapor grown (~1 cm^2 x 0.15 cm thick) preannealed under identical conditions to the diffusion anneal
$\underset{\sim}{D}$(ZnSe)	8.3 x 10^{-4}	1.13	Rate of film growth	Single crystals of zinc and zinc liquid
$D_e(\widetilde{Zn}$ in ZnTe)	14	2.69 ± 0.08	Zn-65 profile; mechanical sectioning	Single crystal slices (2 mm thick x 8-12 mm diam.); preannealed under identical conditions to the diffusion anneal
	2.34 ± 0.37	2.04 ± 0.22	Zn-65 exchange	Single crystal slab
$D_e(\widetilde{Te}$ in ZnTe)	2 x 10^4 (Te saturated crystals)	3.8 ± 0.4	Te-123 profile; mechanical sectioning	Single crystal slices, (2 mm thick x 8-12 mm diam.) preannealed under identical conditions to the diffusion anneal
D(ZnTe)	1.1 x 10^{-4}	1.08	Rate of film growth	Zn single crystals and Zn liquid
$D_e(\widetilde{O}$ in CdO)	3.8 x 10^6	4.00 ± 0.17	0-18 exchange; mass spectrometry	Small crystals 40-70 μm and pressed pellets; preannealing was shown to have no influence

Diffusion temperature (°C) and atmosphere	Remarks	Reference
1000-1150; p_{Se_2} saturation with intermediate points	Two segments of the diffusion profile were observed; $D_e(\widetilde{Se}$ in ZnSe$) \propto p_{Se_2}^{\gamma}$; $0.4 \lesssim \gamma \lesssim 0.5$, in the Se-rich portion of the phase field	Woodbury and Hall (1967) [90]
860-1020; p_{Se_2} saturation isotherm at 1021	Normal diffusion profiles; $D_e(\widetilde{Se}$ in ZnSe$) \propto p_{Se_2}^{0.35}$ on the Se side of the phase field and ~ independent of p_{Se_2} on the Zn side of the phase field	Henneberg and Stevenson (1971) [85]
300-577; Se-rich atmosphere		Sirota and Koren (1962) [91]
693-1031; Zn and Te saturated crystals and congruent subliming composition	$D_e(\widetilde{Zn}$ in ZnTe) was independent of p_{Zn} and was enhanced by Al doping	Reynolds and Stevenson (1969) [38]
760-860; 0.1 atm Zn vapor		Secco and Yeo (1971) [93]
750-970; Zn and Te saturated crystals	$D_e(\widetilde{Te}$ in ZnTe) decreased with decreasing p_{Te_2}. Values at Te-saturation were too low to reliably measure	Reynolds and Stevenson (1969) [38]
320-520; Te vapor		Sirota and Koren (1963) [94]
630-855; isotherm at 756 from p_{O_2} = 70 torr to 700 torr	$D_e(\widetilde{O})$ increases with decreasing p_{O_2} and with incorporation of the univalent cation Li	Haul and Just (1962) [95]

Table 7.1–*(continued)*

Diffusion system	$D_o(\text{cm}^2\ \text{s}^{-1})$	$Q(\text{eV})$	Experimental method	Type of sample; and preannealing treatment
$D_e(\tilde{\text{Cd}}$ in CdS)	3.4 (Cd)	2.0	Cd-109 or Cd-105 profile; etching and counting	Single crystal cubes ~3 mm on edge
			Cd-115m profiles; mechanical sectioning	Single crystal slices 0.3-0.5 cm^2 x 0.1 cm thick; diffusion parallel to *c* axis
$D_e(\tilde{\text{Cd}}$ in CdS)			Cd-109 profile; mechanical sectioning	CdS single crystal wafers 5 x 5 x 1 mm^3 pure and indium doped preannealed under the diffusion conditions for $T < 600°\text{C}$ and for pressures near P_{min}
$D_e(\tilde{\text{S}}$ in CdS)	2.58 x 10^{-6}	1.9	Cd-109 profile; mechanical sectioning	CdS single crystal wafers 5 x 5 x 1 mm^3, pure and indium doped, preannealed under the diffusion conditions for $T < 600°\text{C}$ and for pressures near P_{min}
$\underset{\sim}{D}$(CdS)	2.1 x 10^{-5} (p_{Cd} = 4 atm)	0.6	Relaxation of conductivity for a step change in component pressure	Single crystals
	8 x 10^{-3}	0.68	Relaxation of conductivity for a step change in component pressure	Single crystals
$D_e(\tilde{\text{Cd}}$ in CdSe)	9.2 x 10^{-2} (Cd saturated crystals) 0.33 (Se saturated crystals)	1.68 2.37	Cd-109 profile; mechanical sectioning	Single crystal wafers sectioned perpendicular to the *c* axis; ~7 x 7 x 2 mm^3; prediffusion annealed for 7 days under conditions identical to the diffusion anneal

Diffusion temperature (°C) and atmosphere	Remarks	Reference
700-1150; Cd and S saturated crystals and some intermediate pressures	$D_e(\widetilde{\text{Cd}})$ increases with increasing p_{Cd} and is enhanced by donor solutes in solutes in sulfur saturated crystals	Woodbury (1964) [99]
850 isotherm from a sulfur pressure of 1.7×10^3 torr to a cadmium pressure of 1.8×10^3 torr	$D_e(\widetilde{\text{Cd}})$ increases with p_{Cd}, with three distinct regions	Shaw and Whelan (1969) [102]
700-900; isotherm for $p_{\text{Cd}} \sim 0.01$ atm to Cd SAT	$D_e(\widetilde{\text{Cd}}) = 7.29 \times 10^{-5} \exp(-1.26/kT) p_{\text{Cd}}^{2/3}$ (for Cd-rich crystals); indium doped crystals show three distinct regions	Kumar and Kröger (1971) [108]
700-900; Isotherm for $p_{\text{Cd}} \sim 0.01$ atm to Cd SAT	$D_e(\widetilde{\text{S}}) \propto p_{\text{Cd}}^{1/3}$ for Cd-rich crystals, and $D_e(\widetilde{\text{S}}) = 0.0111 \exp(-2.08/kT) p_{\text{S}_2}^{1/2}$ for S-rich crystals	Kumar and Kröger (1971) [108]
660-950; Cd-rich portion of the phase field		Boyn, Goode and Kuschnerus (1965) [103]
660-760; Cd-rich portion of the phase field	$\underset{\sim}{D}(\text{CdS}) = \dfrac{3}{[V_{\text{S}}^{\cdot\cdot}]} D_e(\widetilde{\text{Cd}})$	Kumar and Kröger (1971) [109]
545-939; Cd SAT, Se SAT, isotherm at 798°C	$D_e(\widetilde{\text{Cd}}) = 6.2 \times 10^{-5} \exp(-0.99/kT) p_{\text{Cd}}^{0.46}$	Borsenberger, Stevenson and Burmeister (1967) [111]

Table 7.1—*(continued)*

Diffusion system	D_O(cm^2 s^{-1})	Q(eV)	Experimental method	Type of sample; and preannealing treatment
$D_e(\widetilde{Se}$ in CdSe)	2.6×10^{-3} (Se saturated crystals)	1.55	Se-75 profile; etching and counting	Single crystal cubes
	2.2×10^{-2} (P_{min})	2.2		
$\underset{\sim}{D}$(CdSe)	3×10^6	4.0	Electrical conductivity profile using mechanical sectioning	Single crystal bars $2 \times 2 \times 13$ mm^3
$D_e(\widetilde{Cd}$ in CdTe)	1.26 (+1.67 −0.72)	2.07 ± 0.08	Cd-115m profile; mechanical sectioning	Single crystals; it was established that preannealing had no influence
	16	2.4	Cd-109 profile; mechanical sectioning	Single crystal wafers; preannealed under conditions identical to the diffusion anneal
$D_e(\widetilde{Te}$ in CdTe)			Te-123 profile; etching and counting	Single crystal cubes ~3 mm on an edge
	1.66×10^{-4} (Te saturated crystals)	1.38	Te-123 profile; mechanical sectioning	Single crystal slices
	8.54×10^{-7} (Cd saturated crystals)	1.42		
$\underset{\sim}{D}$(CdTe)	4	1.15 ± 0.1	Relaxation of conductivity for a step change in component pressure	Single crystal bar
$D_e(\widetilde{Pb}$ in PbO)	10^5	2.86	Pb-212 uptake; contact method	Pressed and sintered tablets
		2.86 $T > 488°$C; 0.58 $T < 488°$C	Pb-212 α recoil method	Oxidized Pb wafers; as oxidized
		2.78	Pb-210 surface decrease using Bi-210 daughter	

Diffusion temperature (°C) and atmosphere	Remarks	Reference
700-1000; Cd SAT, Se SAT, and intermediate pressures	$D_e(\widetilde{Se}) = 1.3 \times 10^5 \exp(-4.43/kT) p_{Cd}^{-1}$ $D_e(\widetilde{Se})$ showed a different pressure dependence as Cd SAT was approached. $D_e(\widetilde{Cd})$ was independent of donor concentration	Woodbury and Hall (1967) [90]
900-1100; Cd + Ar atmospheres Cd-rich		Shiozawa and Jost (1965) [115]
700-1000; isotherms at 800 and 900°C	$D_e(\widetilde{Cd})$ was independent of p_{Cd}; diffusion profiles consisted of two distinct regions	Whelan and Shaw (1967) [116]
510-920; Cd SAT, Te SAT, isotherm at 775	$D_e(\widetilde{Cd})$ was independent of p_{Cd} and was enhanced by both Al and Au dopants	Borsenberger and Stevenson (1968) [117]
Isotherm at 800	$D_e(\widetilde{Te}) = 1.6 \times 10^{-14} p_{Cd}^{-1}$	Woodbury and Hall (1967) [90]
510-930	$D_e(\widetilde{Te})$ increases with p_{Te_2} at all temperatures studied	Borsenberger and Stevenson (1968) [117]
600-800		Zanio (1970) [30]
400-600; nitrogen or vacuum		Lindner (1952) [124]
200-580	Kinetics of oxidation were more rapid than predictions based on $D_e(\widetilde{Pb})$	Lindner and Terem (1954) [125]
600-670; 3 points		Das Gupta, Sitharamarao and Palkar (1965) [127]

Table 7.1–*(continued)*

Diffusion system	D_O(cm² s⁻¹)	Q(eV)	Experimental method	Type of sample; and preannealing treatment
$D_e(\tilde{O}$ in PbO)	5.39×10^{-5}	0.97	O-18 isotope exchange; mass spectrometry	Small spheres of PbO, 90-160 μm; yellow modification; annealed in air for 24 h at the same temperature as the diffusion anneal
			O-18 isotope exchange exchange using mass spectrometry CO_2 isotropic source	Pellets (3 mm thick by 8 mm diam.) pressed at 30 kilobar and single crystal plates and spheres
$D_e(\tilde{Pb}$ in PbS)	1.3	1.83	Pb-210 profile; mechanical sectioning	Pressed powder samples
	8.6×10^{-5} (for P_{min})	1.52	Pb-210 profile; mechanical sectioning and some autoradiography	Single crystal wafers; identical p_{S_2} as for diffusion anneal; no prediffusion anneal for inert gas annealed
			Pb-210 profile mechanical sectioning	Single crystal wafers prediffusion annealing under conditions identical to the diffusion anneal
			Pb-210 and Bi-210 profiles; mechanical sectioning	Single crystal wafers prediffusion annealing under conditions identical to the diffusion anneal
$\underset{\sim}{D}$(PbS)			*p-n* junction method	Natural PbS single crystals; excess Pb, *n*-type
$D_e(\tilde{S}$ in PbS)	4.56×10^{-5} (for S excess)	1.22	S-35 profile; mechanical sectioning	Single crystal wafers; identical to diffusion anneal
$D_e(\tilde{Pb}$ in PbSe)	4.98×10^{-6}	0.83	Pb-210 profile; mechanical sectioning; residual activity	Single crystals; undoped and doped with Bi or Ag
$D_e(\tilde{Se}$ in PbSe)	2.1×10^{-5}	1.2	Se-75 profile	

Diffusion temperature (°C) and atmosphere	Remarks	Reference
500-650	Parabolic oxidation constants, based on values of $D_e(\widetilde{O})$, were in reasonable agreement with experimental values	Thompson and Strong (1963) [128]
514; p_{O_2} = 10 torr pure CO_2	$D_e(\widetilde{O}) \cong 4 \times 10^{-11}$; $D_e(\widetilde{O})$ independent of p_{O_2} and enhanced with Bi doping	Heyne, Beekmans and de Beer (1972) [15]
460-770; vacuum and sulfur vapor	$D_e(\widetilde{Pb})$ increases with increasing sulfur pressure	Anderson and Richards (1946) [129]
500-800; inert gas and varying p_{S_2} from ~10 to 10^2 atm	$D_e(\widetilde{Pb})$ is enhanced with donor and acceptor additions and by excess Pb or S in pure samples	Simkovich and Wagner (1963) [130]
Isotherm at 700; varying p_{S_2} from ~1 to ~10^{-8} atm	Log $D_e(\widetilde{Pb})$ versus log p_{S_2} shows four distinct zones	Seltzer and Wagner (1963) [132]
Isotherm at 700; varying p_{S_2} from ~1 to ~10^{-8} atm	Diffusion rates of Pb and Bi were comparable for all values of p_{S_2}	Zanio and Wagner (1968) [133]
550; excess sulfur from a reservoir at 123°C	$\underset{\sim}{D}(PbS) = 2 \times 10^{-6}$	Brebrick and Scanlon (1954) [135]
500-750; varying p_{S_2} between ~1 and ~10^{-8} atm	$D_e(\widetilde{S})$ increases with p_{S_2} and slightly enhanced with Bi doping	Seltzer and Wagner (1965) [136]
400-800; inert atmosphere (to develop P_{min})	$D_e(\widetilde{Pb})$ was enhanced by Ag doping and and depressed by Bi doping for $T < 700°C$	Seltzer and Wagner (1962) [137]
650-850		Boltaks and Mokhov (1959) [138, 139]

Table 7.1—*(continued)*

Diffusion system	D_o(cm^2 s^{-1})	Q(eV)	Experimental method	Type of sample; and preannealing treatment
	8.74×10^{-7} (P_{min})	0.63 ± 0.10	Se-75 profile; mechanical sectioning	Single crystals; pure and doped with Bi or Ag; preannealed for 2 days under conditions identical to the diffusion anneal
$\tilde{D}$(PbSe)			*p-n* junction method	Single crystals, *n* or *p* type
			p-n junction method	Single crystals, *n* or *p* type
D_e($\widetilde{Pb}$ in PbTe)			Pb-210 profile; residual activity method	Single crystal wafers; 100 h at 700° C under conditions identical to the diffusion anneal
	3.1×10^{-6} (Pb-rich crystals) 2.5×10^{3} (Te-rich crystals)	1.04 2.65	Pb-210 profile; mechanical sectioning	Single crystal wafers 4 x 4 x 1.5 mm homogenized 120 h at 400° C in closed ampoules
D_e($\widetilde{Te}$ in PbTe)	2.7×10^{-6}	0.75	Te-127 profile; mechanical sectioning	Single crystal wafers
	2.4×10^{-6} (Te-rich crystals) 0.65×10^{-6} (Pb-rich crystals)	1.04 1.04	Te-127 profile; mechanical sectioning	Single crystal wafers of different composition, homogenized 120 h at 400° C in closed ampoules
$\tilde{D}$(PbTe)			*p-n* junction penetration kinetics	Single crystal; *p*-type
$\tilde{D}$(PbTe)	2.9×10^{-5}	0.6	*p-n* junction penetration kinetics	Single crystal; *p*-type preannealed 100 h in vacuum ampoules
			p-n junction penetration kinetics	Single crystals; *p*-type

Diffusion temperature (°C) and atmosphere	Remarks	Reference
350-850; inert atmosphere (to develop P_{min})	$D_e(\widetilde{\mathrm{Se}})$ increases with increasing p_{Se_2}	Ban and Wagner (1970) [140]
650; diffusion atmosphere comprised of a 2 phase Pb- or Se-rich ingot	$\tilde{D}(\mathrm{PbSe})_{n\rightarrow p} = 9 \times 10^{-9}$ $\tilde{D}(\mathrm{PbSe})_{p\rightarrow n} = 4 \times 10^{-8}$	Butler (1964) [141]
650; diffusion atmosphere comprised of a 2 phase Pb- or Se-rich ingot	$\tilde{D}(\mathrm{PbSe})_{n\rightarrow p} \cong 3 \times 10^{-11}$ $\tilde{D}(\mathrm{PbSe})_{p\rightarrow n} \cong 10^{-9}$	Brodersen, Walpole and Calawa (1971) [142]
700; Pb SAT, Te SAT and congruent subliming conditions	$D_e(\widetilde{\mathrm{Pb}})$ shows a minimum value at P_{min}; $D_e(\widetilde{\mathrm{Pb}})_{\mathrm{Pb\ SAT}} = 3.6 \times 10^{-11}$ $D_e(\widetilde{\mathrm{Pb}})_{\mathrm{Te\ SAT}} = 3.2 \times 10^{-10}$	George and Wagner (1971) [144]
550-700; enclosed evacuated ampoules (small free volume)	$D_e(\widetilde{\mathrm{Pb}})$ shows a minimum value at intermediate compositions in the PbTe phase field	Gomez, Stevenson and Huggins (1971) [145]
		Boltaks and Mokhov (1958) [138, 139]
560-720; enclosed evacuated ampoules	$D_e(\widetilde{\mathrm{Te}})$ increases with increasing Te concentration	Gomez, Stevenson and Huggins (1971) [145]
500; Pb saturated vapor	$\tilde{D}(\mathrm{PbTe}) = 5.6 - 9.1 \times 10^{-8}$	Brady (1954) [148]
250-500; Pb-rich vapor		Boltaks and Mokhov (1956, 1957) [149, 150]
50; Pb saturated vapor	$\tilde{D}(\mathrm{PbTe}) = 6 \times 10^{-7}$	Butler (1964) [141]

with the material of lower density giving higher D values by as much as one order of magnitude. A selection of the best results was represented as:

$$D_e(\widetilde{Zn}) = 1.3 \exp[-3.2/kT] \text{ cm}^2 \text{ s}^{-1}, \tag{7.21}$$

for the temperature range 950-1380° C. The pellet contact method gave reproducible values over the temperature range 655-1370° C, described by $D_e(\widetilde{Zn}) = 3.5 \exp[-3.0/kT] \text{ cm}^2 \text{ s}^{-1}$, with these results being about an order of magnitude higher than those from the thin film source method. Lindner attributed this difference to an enhancement of the diffusion coefficient caused by the γ radiation of the $Zn^{65}O$ tablet source as compared to a smaller amount of radiation from the thin film source.

Münnich (1955) [47] studied the diffusion of Zn-65 in ZnO by following the exchange of radiotracer vapor with the non-active Zn in ZnO single crystals 0.3 mm diameter x 10 mm long. A tracer profile was established by measuring the activity of small layers of the crystal removed by etching. The values of the diffusion coefficient were 10-100 times larger than those of Lindner and the subsequent studies of Secco and Moore (1957) [48] and Moore and Williams (1959) [49]. The exposure times may have been short compared to times required to saturate with zinc, suggesting that the observed diffusion behavior would be a combination of interdiffusion and self-diffusion. Münnich's profiles indicated a concentration dependence for the diffusion coefficient with values near the surface being as much as 30 times smaller than for the inside of the sample. For example, at 1000° C the diffusivity varies between 4×10^{-11} and 2×10^{-10} $\text{cm}^2 \text{ s}^{-1}$ from the surface to the interior.

Secco and Moore (1955) [50] and (1957) [48] studied the exchange of zinc between $Zn^{65}O$ crystals and Zn vapor. Small acicular single crystals (circa 0.01 mm in diameter and 5.0 mm long) were prepared by reacting radioactive zinc with atmospheric oxygen. Approximately 1000 such needles were suspended in a quartz vessel where the tracer zinc in the ZnO crystals exchanged with the non-radioactive zinc in the vapor in accord with the reaction:

$$Zn^{65}O(S) + Zn(G) \rightleftharpoons ZnO(S) + Zn^{65}(G) \tag{7.22}$$

The equilibrium zinc pressure was computed from the weight of inactive zinc, assuming ideal behavior of the zinc vapor. The crystals were periodically withdrawn and their residual activity determined. The exchange reaction appeared to be diffusion controlled, except for the initial stages. The diffusion coefficient under 1 atm pressure was represented as

$$D_e(\widetilde{Zn}) = 48 \exp[-3.17/kT] \text{ cm}^2 \text{ s}^{-1} \tag{7.23}$$

between 900° and 1025° C. The dependence of $D_e(\widetilde{Zn})$ on the zinc pressure was determined at 1000° C for pressures ranging from 0.2 to 2.0 atm and showed a dependence, $D_e(\widetilde{Zn}) \propto p_{Zn}^{0.66}$. This is close to a value of 0.5 which would be expected if singly ionized zinc interstitials were the diffusing species and the charge neutrality condition was given by $[Zn^{\cdot}] = n$, i.e.,

$$Zn(G) \rightleftharpoons Zn_i^{\cdot} + e'; \qquad K = [Zn_i^{\cdot}]\, n/p_{Zn}; \tag{7.24}$$

$$[Zn_i^{\cdot}] = n = K^{1/2} p_{Zn}^{1/2}. \tag{7.25}$$

Secco and Moore (1957) [48] also studied diffusion in compressed ZnO pellets exposed to air. Radioactive $Zn(OH)_2$ was precipitated on one of the faces of a pellet and decomposed to ZnO prior to the diffusion anneal. The diffusion was followed by the decrease in the soft positron activity of the Zn-65, with the hard radiation being monitored to insure against evaporation losses. The diffusion coefficients were found to be time dependent in these pellets and showed an activation energy ~3.9 eV. Although there was more scatter than in the exchange data, the results agreed fairly well with those of Lindner (1952) [46], performed in the same atomosphere of air. Although the diffusivities in zinc-rich atmospheres were somewhat larger than those in oxygen-rich atmospheres, the difference is by no means as large as one would expect if an interstitial mechanism were observed over the entire phase field, since the interstitial zinc concentration should be orders of magnitude different in zinc-rich or oxygen-rich atmospheres. The diffusivities were only about an order of magnitude different. As pointed out by Thomas (1957) [51], one may estimate the expected difference, assuming an interstitial mechanism, i.e.,

$$D_e(\widetilde{Zn}) = [Zn_i^{\cdot}]\, D_e(Zn_i^{\cdot}) \tag{7.26}$$

The decomposition constant for ZnO at 1000° C, calculated from Wicks and Block (1963) [52], is

$$ZnO(S) = Zn(G) + \tfrac{1}{2}O_2(G); \qquad K_D = p_{Zn} p_{O_2}^{1/2} = 2.6 \times 10^{-9}\ \text{atm}^{3/2}. \tag{7.27}$$

For air, $p_{O_2} = 0.2$ and $p_{Zn} = 2.6 \times 10^{-9}\ (0.2)^{-1/2} = 5.8 \times 10^{-8}$ atm. Using the postulated defect equilibrium, equation 7.25, one calculates

$$[Zn_i^{\cdot}]_{air}/[Zn_i^{\cdot}]_{1\ atm\ Zn} = \left\{\frac{5.8 \times 10^{-8}}{1}\right\}^{1/2} = 2.4 \times 10^{-4}.$$

The same variation in $D_e(\widetilde{Zn})$ would be expected for samples annealed in air and zinc, respectively. The experimental observations on the pellets, however, suggested a decrease of no more than a factor of 10. This discrepancy could arise because impurities are important in determining the charge neutrality condition or because the predominant defects at low and high p_{Zn} are different. For example, at p_{Zn} much less than 1 atm, the charge neutrality condition could be $[Zn_i^{\cdot}] = [V_{Zn}'] = \sqrt{K_F'}$, which depends only upon temperature.

Roberts and Wheeler (1957) [39] measured the diffusion of Zn-65 in dense polycrystalline samples between 800-1300°C in oxygen or argon atmospheres by sectioning and measuring the residual γ activity. They concluded that the observed diffusion proceeded predominantly via grain boundaries. Their evidence consisted of autoradiographs, which showed a concentration of activity of the diffusing tracer at the grain boundaries, as well as an analytical fit of the tracer profile to the grain boundary diffusion theory of Fisher (1951) [53]. For diffusion in 1 atm oxygen, their results gave

$$D^{1/2}/\delta D_{gb} = 10^{4\pm2} \exp\left[(1.04 \pm 0.05)/kT\right] \text{ s}^{1/2} \text{ cm}^2 \tag{7.28}$$

where D, δ, and D_{gb} are the volume diffusion coefficient, grain boundary width, and grain boundary diffusion coefficient, respectively. By estimating an equation for D from their own work and that of Spicar (1956) [54, 55], they deduced an equation for the grain boundary diffusion coefficient,

$$D_{gb} = 10^{3\pm2} \exp\left[(-3.0 \pm 0.5)/kT\right] \text{ cm}^2 \text{ s}^{-1}. \tag{7.29}$$

In a later study of a similar nature, Roberts and Wheeler (1960) [56] determined the same quantity, $D^{1/2}/\delta D_{gb}$, between 800° and 1400°C for dense sintered ZnO samples in oxygen, argon and oxygen + argon mixtures. In 1 atm of oxygen,

$$D^{1/2}/\delta D_{gb} = 10^{4.0\pm1.5} \exp\left[(1.04 \pm 0.03)/kT\right] \text{ s}^{1/2} \text{ cm}^{-2}, \tag{7.30}$$

with values decreasing with decreasing p_{O_2} and with values sensitive to sample history.

Lamatsch (1958) [57] measured the diffusion of Zn-65 vapor in crystals of ZnO by etching the crystals and measuring the residual γ activity. His results, discussed by Heiland and Mollwo (1959) [58], are represented by

$$D_e(\widetilde{Zn}) = 1.7 \times 10^2 \exp\left[-3.3/kT\right] \text{ cm}^2 \text{ s}^{-1}. \tag{7.31}$$

Moore and Williams (1959) [49] investigated the diffusion of Zn-65 in relatively large single crystals of ZnO (3-5 mm long and 0.9-1.4 mm thick). A thin film source was applied by vaporizing Zn-65 or evaporating $Zn^{65}(NO_3)_2$ on the surface of the ZnO crystal, followed by an anneal at

700° C in oxygen for 6 h or 1 h, respectively. The tracer profile was determined by mechanical sectioning and analysis of the material removed. Measurements made parallel and perpendicular to the c-axis showed no anisotropy, within the limits of the experimental error. Their data were expressed as

$$D_e(\widetilde{Zn}) = 1.25 \times 10^{-5} \exp [(-1.87 \pm 0.5)/kT] \text{ cm}^2 \text{ s}^{-1}. \quad (7.32)$$

The large uncertainty was attributed to evaporation of the crystal during annealing and the possible effects of impurities on the high temperature defect equilibria. Moore and Williams compared the previous high temperature Zn diffusion data for ZnO and concluded that there was a much smaller variation of $D_e(\widetilde{Zn})$ with component pressure than would be expected for the electrical neutrality condition $[Zn_i^{\cdot}] = n$, and that the defects responsible for the diffusion of Zn arise from thermal disorder rather than departures from stoichiometry. Moore and Williams were criticized by Roberts (1959) [59] for comparing their results in oxygen on single crystals with those of Lindner (1952) [46] in air on polycrystalline material, since Roberts and Wheeler (1957) [39] showed the predominance of grain boundary diffusion in polycrystalline samples. It is clear, however, that Moore and Williams' supposition regarding the insensitivity of diffusivity to the component pressure may be substantiated by comparing their results in oxygen atmospheres with those of Secco and Moore (1957) [48] in zinc atmospheres, both on single crystals, and that the main evidence to date supports a thermal disorder model on the Zn sublattice.

There are a number of tracer exchange studies on ZnO by Secco (1959, 1960, 1961, 1969) [60-63] with the objective of studying different rate controlling conditions for the exchange process, for example, phase boundary reaction rate control as opposed to diffusion control. The method employed for the exchange studies was that described by Secco and Moore (1957) [48]. In Secco's 1959 [60] paper, he reports exchange kinetics for Zn-65 vapor at 0.25 atm Zn, and $S = 0.45$ (S is the mole fraction of Zn in the solid) performed on very small particles—circa 3×10^{-5} cm in diameter—from 790° to 820° C. For these conditions, the exchange process followed a first order rate law and was attributed to an 'elementary exchange reaction', rather than a diffusion reaction. Evidence was presented that the reaction was autocatalytic. In a later paper (Secco, 1960) [61], the exchange studies were extended to zinc pressures from 8.5×10^{-2} to 2.5×10^{-1} atm and values of S between 0.4 to 0.6. The exchange rate was found to be independent of both zinc pressure and S for the temperature range studied, 720°-840° C. Two processes were distinguished: an initial rapid first order process with a rate constant

$$k_e = 4.6 \times 10^5 \exp [-1.7/kT] \text{ s}^{-1}, \quad (7.33)$$

and a slower process at later times, which was interpreted as diffusion-controlled. Corresponding diffusivities were evaluated. A discontinuity in $D_e(\widetilde{Zn})$ was observed at 800° C with

$$D_e(\widetilde{Zn}) = 10^2 \exp[-3.17/kT] \text{ cm}^2 \text{ s}^{-1} \tag{7.34}$$

for the temperature range 720-780° C, and the preexponential factor decreasing discontinuously to 10 for temperatures from 800° to 840° C. The activation energy of 1.7 eV for the initial exchange process was interpreted as the energy required to displace a normal lattice zinc atom by an incoming atom of zinc vapor. There are, however, many possible rate limiting processes which could involve either interstitial zinc, oxide vacancies, or other species. The possibility also exists that there is an initial transient period when the solid adjusts composition, and during this time the zinc exchange is determined by interdiffusion rather than self-diffusion. The initial period could involve the same processes as those controlling the resistivity and color changes during atmosphere annealing of ZnO observed, for example, by Thomas (1957) [51], Arneth (1955) [64] or Pohl (1959) [65]. Secco (1960) [61] suggested that the observed activation energy for diffusion, 3.2 eV, be interpreted either as the direct movement of interstitials or as a composite quantity: 1.7 eV for exchange and 1.5 eV for transport components of an interstitial mechanism. Secco (1961) [62] introduced an important additional experimental condition for the exchange studies which previous papers on Zn diffusion in ZnO have failed to mention explicitly, namely, the influence of preannealing on the exchange behavior, particularly in the early stages. Secco preannealed some samples of ZnO in 0.15 atm of Zn at 760° C for 5 h and reported a large decrease in the exchange rate for preannealed samples, in contrast to as-prepared samples, figure 7.9. The preannealed samples gave results that agreed with previous exchange data,

$$D_e(\widetilde{Zn}) = 0.40 \exp[-3.3/kT] \text{ cm}^2 \text{ s}^{-1}. \tag{7.35}$$

For the two temperatures, 760° and 800° C, these observations are possibly due to the large difference that usually exists between the self-diffusion and interdiffusion for phases of limited composition. It would indeed be interesting to follow the exchange process for longer times in the samples that were not preannealed to see if values comparable to the preannealed values are eventually obtained.

In discussing his results, Secco proposes that the diffusion coefficient obtained from the exchange studies is the product of a defect diffusivity, D_i, and the exchange constant, k_e, i.e., $D_e(\widetilde{M}) = D_i k_e$, and that these be interpreted as mobility terms and exchange terms respectively. This is an interesting phenomenological correlation for this system, but it should be

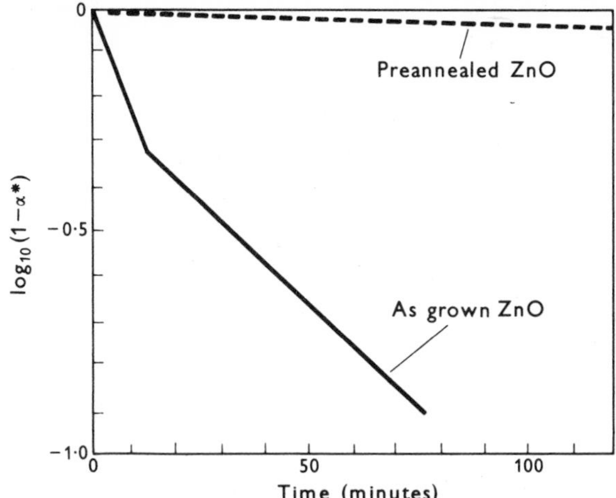

Figure 7.9. Influence of the preannealing of ZnO in Zn vapor upon the exchange rate of Zn-65 in ZnO at 760° C. $(1 - \alpha^*)$ is the unexchanged fraction. -------- Exchange behavior for ZnO, preannealed in zinc vapor ($p_{Zn} \sim 0.15$ atm at 760° C for 5 h). ——— Exchange behavior for ZnO, as-grown. Secco, 1961 [62].

noted that it differs considerably from the established relation that the tracer diffusion coefficient is the defect concentration-defect mobility product, $D_e(\widetilde{M}) = D_e(M_i)[\underset{\sim}{M}_i]$, for an interstitial mechanism, and $D_e(\widetilde{M}) = fD_e(V_M)[\underset{\sim}{V}_M]$ for a vacancy mechanism. In addition, the relation is not dimensionally correct if the quantities are expressed in conventional units. Secco (1961) [62] introduced the relation $D = D_i f_i$, where f_i is defined as the 'fraction of interstices available for occupancy', and it is assumed that f_i varies from 1 for the unsaturated crystal (not preannealed) to 0 for the saturated crystal (preannealed). The trend in the factor f_i is qualitatively related to the chemical potential gradient which may assume values much larger than unity for interdiffusion, but will approach zero as saturation is achieved. It would seem most logical to explain the observed difference in the exchange rate with this phenomenology.

Diffusion of O in ZnO

Moore and Williams (1959) [49] investigated the diffusion of O in ZnO by measuring the O-18 exchange between small ZnO single crystal spheres and an O-18 enriched oxygen atmosphere. For the temperature range studied, 1100-1300° C and oxygen pressures in the range 200-220 torr, the results were expressed as

$$D_e(\widetilde{O}) = 6.52 \times 10^{11} \exp[-7.16/kT] \text{ cm}^2 \text{ s}^{-1}. \qquad (7.36)$$

An isotherm at 1275°C for 0.15 atm $\leqslant p_{O_2} \leqslant$ 0.65 atm showed that $D_e(\tilde{O}) \propto p_{O_2}^{1/2}$. This pressure dependence implies a diffusing specie of some form of atomic oxygen, and the large value of the pre-exponential in the Arrhenius equation suggests an important role for diffusion along dislocations.

Hoffman and Lauder (1970) [66] studied the diffusion of O in ZnO between 1150° and 1400°C using the O-18 gaseous exchange technique. Samples, approximating spheres of radii 0.4-0.8 mm, were broken from single crystal needles which were grown by H_2 transport of ZnO; the samples were always preannealed under the identical temperature and p_{O_2} and for at least the same time as the exchange experiment. The container vessel for the exchange study was made of platinum. The temperature dependence was expressed as

$$D_e(\tilde{O}) = 1.05 \times 10^3 \exp(-4.1/kT) \text{ cm}^2 \text{ s}^{-1} \quad (7.37)$$

for $p_{O_2} \sim 200$ torr in the temperature range 1150-1400°C. Pressure isotherms at 1200°, 1300°, 1350° and 1400°C yielded pressure exponents of −1.5, −0.58, −0.48 and −0.54, respectively. The pressure dependence was interpreted as evidence for diffusion via an oxygen vacancy, with a neutral vacancy consistent with the pressure exponents observed for temperatures of 1300°C and greater. The rather large difference in the results of this study and that of Moore and Williams (1959) [49] was discussed at some length by the authors, who suggested that surface evaporation of the ZnO and background exchange with the ceramic tube may have been a source of error in the earlier study. The authors comment that exchange studies give no information on the diffusion profile in the solid and, hence, cannot evaluate the role of such factors as grain boundary diffusion.

Robin, Solem, Heuer and Cooper (1970) [67] are investigating the oxygen diffusion in ZnO using proton activation techniques. After O-18 exchange between an O-18 enriched environment and a ZnO crystal, the concentration profile of O-18 is obtained by transforming O-18 to N-15 using 0.8 MeV protons and counting the resulting emitted alpha particles on a surface barrier detector. Initial results indicate a possible extrinsic range from 940° to 1075°C, with $D_e(\tilde{O}) = 8 \times 10^{-12} \exp[-0.83/kT]$ cm^2 s^{-1}. Above 1075°C, the activation energy appears to increase. Preliminary studies of the influence of component pressure indicate that decreasing p_{O_2} produces a significant enhancement in $D_e(\tilde{O})$.

Interdiffusion and Defect Diffusion in ZnO

Three types of studies are relevant: relaxation measurements of the conductivity (donor diffusion), relaxation measurements of optical adsorption coefficients, and kinetics of oxidation. Thomas (1957) [51] measured the donor diffusivity in ZnO single crystal needles in the temperature range 180-350°C by measuring the relaxation time for a change in conductivity. Crystals were saturated with zinc at 600°C, quenched to room temperature, and then annealed in air in the temperature range 180-350°C. The rate of change of conductivity was related to the interstitial diffusion coefficient, $D_e(\mathrm{Zn}_i^{\cdot})$, assuming the charge neutrality condition $[\mathrm{Zn}_i^{\cdot}] = n$. The data were represented by

$$2D_e(\mathrm{Zn}_i^{\cdot}) = 5.3 \times 10^{-4} \exp[-0.55/kT] \text{ cm}^2 \text{ s}^{-1}. \qquad (7.38)$$

In the same study, the solubility of the interstitial zinc was determined by annealing ZnO crystals in Zn at various temperatures, quenching to room temperature and measuring the conductivity, with the results for $410°\mathrm{C} < T < 700°\mathrm{C}$ expressed as

$$[\mathrm{Zn}_i^{\cdot}] = n = 3.4 \times 10^{20} \exp[-0.65/kT] \text{ atoms cm}^{-3}. \qquad (7.39)$$

These results were compared with the parabolic rate constant for oxidation, k', using the relation $k' = 2D_e(\mathrm{Zn}_i^{\cdot})[\underset{\sim}{\mathrm{Zn}}_i^{\cdot}]$ cm^2 s^{-1}. Fair agreement was obtained at 400°C with the values of Wagner and Grünewald (1938) [68] and Moore and Lee (1951) [69]. Comparison with the oxidation data of liquid zinc (Cope, 1961) also indicates fair agreement. Thomas discussed a number of inconsistencies in the activation energies and pressure dependence of self-diffusion coefficients, donor concentrations, diffusion coefficients and parabolic oxidation rate constants and suggested that a simple interstitial defect model does not adequately describe the ZnO system.

Pohl (1959) [65] employed essentially the same technique, with the added feature that samples were presaturated with zinc at different temperatures, θ. The out-diffusion of donors was reported to depend upon the temperature of out-diffusion as well as the initial saturation temperature, θ, as shown in figure 7.10. This behavior was represented by the expression:

$$\underset{\sim}{D}(\mathrm{ZnO}) = 1.6 \times 10^{-2} \exp[1.0/k\theta - 1.7/kT] \text{ cm}^2 \text{ s}^{-1}. \qquad (7.40)$$

Arneth (1955) [64] measured the interdiffusion in ZnO by following the increase in optical absorption when ZnO is annealed in Zn vapor in the temperature range 780-1100°C. The results, shown in figure 7.10, are represented by

$$\underset{\sim}{D}(\mathrm{ZnO}) = 1.5 \times 10^{-2} \exp[-1.7/kT] \text{ cm}^2 \text{ s}^{-1}. \qquad (7.41)$$

An excellent summary and discussion of diffusion in ZnO is given by Heiland and Mollwo (1959) [58] who point out the difficulties in

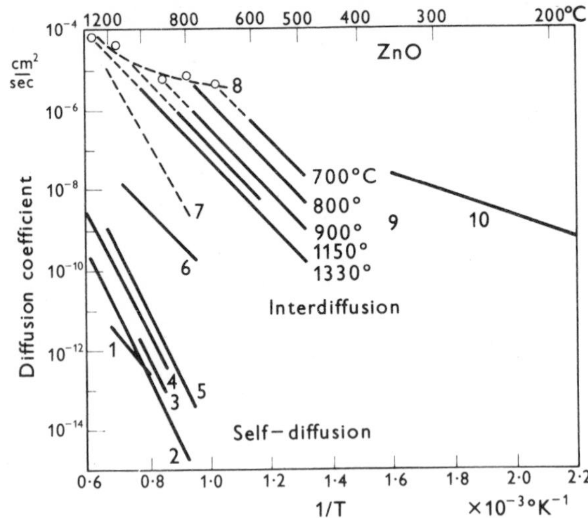

Figure 7.10. A selection of diffusion results in ZnO—compiled by Heiland and Mollwo (1959) [58].

1. Crystals, tracer, in air, after Moore and Williams, 1959 [49].
2. Sintered specimens, tracer, beta-radiation, in air, after Lindner, 1952 [46].
3. Crystals, grown with Zn^{65}, tracer, in zinc vapor of 1 atm, after Secco and Moore, 1957 [48].
4. Sintered specimens, tracer, in air, after Lindner, 1952 [46].
5. Crystals, tracer, in saturated zinc vapor, after Lamatsch, 1958 [57].
6. Crystals, optical absorption after heating in saturated zinc vapor, after Arneth, 1955 [64].
7. Calculated from curve 5 using the mole fraction of interstitial zinc (conductivity).
8. Extrapolated from curves 9 for $T = \theta$.
9. Crystals, decay of conductivity (produced by heating in zinc vapor) during annealing in air at temperature T. Parameter of the different curves is temperature, θ, of pretreatment in zinc vapor, after Pohl, 1959 [65].
10. Crystals, method similar to No. 9, after Thomas, 1957 [51]. After Heiland and Mollwo, 1959 [58].

interrelating and interpreting a selection of data on diffusion in ZnO shown in figure 7.10. There is a rather large dispersion for both self-diffusion and interdiffusion data, not surprising considering the diverse selection of diffusion samples and diffusion conditions. As expected, the self-diffusion values are several orders of magnitude lower; however, attempts to quantitatively relate the difference were not successful. For example, Heiland and Mollwo (1959) [58] show that the following assumptions lead to a calculated value of $D_e(Zn_i^{\cdot})$ lower than observed values—curves 5, 7 and 8 in figure 7.10: Zn diffuses only interstitially, i.e., $D_e(\tilde{Zn}) = D_e(Zn_i^{\cdot})\,[\underset{\sim}{Zn_i^{\cdot}}]$; and $[Zn_i^{\cdot}]$ may be calculated on the premise that only Zn_i provides donors (Pohl, 1955) [65]. They suggest that other donors, such as oxygen vacancies, may be significant. Other attempts to relate the two types of diffusivities simply establish that the general order of magnitude of the difference is reasonable (Moore and Williams, 1959 [49] ; Thomas, 1957 [51]).

Several studies of the oxidation kinetics of Zn to form ZnO have been made and are discussed in a number of references, e.g., Hauffe (1965, 1966) [71, 72] and Kofstad (1966) [73]. Of particular significance is the investigation of the influence of alloying elements on the oxidation of Zn by Hauffe and Gensch (1950) [74]. At 390°C, Zn containing 0.1 and 1 atomic % Al showed a significant decrease in oxidation rate whereas Zn containing 0.4 atomic % Li showed a significant increase in the oxidation rate. Aluminium incorporated on the Zn lattice acts as a donor, $Al^{\cdot}_{Zn}$, suppressing the $Zn_i^{\cdot}$ native donor concentration, whereas Li'_{Zn} has the opposite effect. A reasonably good quantitative correlation with impurity concentration was obtained, using relevant defect equilibria.

7.3.2 Diffusion in ZnS

Diffusion of Zn in ZnS

Secco (1958) [75] studied the exchange rate of Zn-65 with ZnS single crystal platelets of the wurtzite modification which were vapor grown. The experimental technique was essentially the same as Secco and Moore (1957) [48]. The measured exchange was interpreted as diffusion controlled for the temperature range studied, 925-1075°C, and Secco suggested that there are three distinct ranges.

$$D_e(\widetilde{Zn}) = 3 \times 10^{-4} \exp[-1.52/kT] \text{ cm}^2 \text{ s}^{-1}, \quad \text{for } 925^\circ\text{C} < T < 940^\circ\text{C} \tag{7.42}$$

$$D_e(\widetilde{Zn}) = 1.5 \times 10^{4} \exp[-3.25/kT] \text{ cm}^2 \text{ s}^{-1}, \quad \text{for } 940^\circ\text{C} < T < 1030^\circ\text{C} \tag{7.43}$$

and

$$D_e(\widetilde{Zn}) = 1 \times 10^{16} \exp[-6.51/kT] \text{ cm}^2 \text{ s}^{-1}, \quad \text{for } 1030^\circ\text{C} < T < 1075^\circ\text{C}. \tag{7.44}$$

There were only a total of ten data points, and therefore each individual region is not well characterized. Reynolds (1965) [76] suggested that a single equation represents the data over the entire range:

$$D_e(\widetilde{Zn}) \cong 3.2 \times 10^{7} \exp[-4.2/kT] \text{ cm}^2 \text{ s}^{-1}. \tag{7.45}$$

The dependence of $D_e(\widetilde{Zn})$ upon zinc pressure was determined at 1025°C, based on values at 0.25, 1 and 1 atm zinc, and is given as $D_e(\widetilde{Zn}) \propto p_{Zn}^{3/2}$. The data, however, are so limited in number (three points) and the pressure range so limited, that no specific conclusions may be drawn concerning the dominant defect equilibria. Furthermore, it should be noted that the diffusivities increased with increasing sample size, suggesting that the observed values were not observed under isoconcentration conditions. A later study was made by Secco (1964) [77] at lower temperatures, 720-960° C, using the same technique on polycrystalline samples from two sources: (1) reagent grade ZnS and, (2) crushed vapor-grown single crystals. In both cases, the particle radius was circa 6.1×10^{-5} cm. A distinction should be made between small polycrystalline particles, (1), and small single crystals, (2), since grain boundary effects, even within small particles, may be significant (Roberts and Wheeler, 1957) [39]. Secco, in his 1964 paper, reported that X-ray evidence substantiated that only the wurtzite modification of ZnS was present, but a later paper referring to the results (Secco, 1969) [63] states that X-ray analysis shows 80% wurtzite, 20% sphalerite. It is possible that samples measured below the transformation temperature could show different extents of transformation. Two distinct rate processes were observed, the initial rate controlling process being first order while the second was interpreted as diffusion controlled. Different results were obtained for the two samples. For sample (1) the diffusion coefficient was given as

$$D_e(\widetilde{\mathrm{Zn}}) = 4.5 \exp\,[-2.7/kT]\ \mathrm{cm}^2\ \mathrm{s}^{-1}, \tag{7.46}$$

and for sample (2) an activation energy of 0.87 ± 0.13 eV was reported for the diffusion. Sintering of sample (2) during the exchange was proposed to explain the difference in the two samples; however, there were no micrographic data presented to support this hypothesis nor did later sintering studies by Secco support this supposition. The large difference in behavior for the two samples, 1 and 2, was not explained. In addition, there was no discussion or comparison of the results of this study with the previous one (Secco, 1958) [75].

The results of the initial first order exchange process for sample (1), $k_e = 3.3 \times 10^3 \exp\,[-1.4/kT]\ \mathrm{s}^{-1}$, were combined with the diffusion results in accord with the postulate made by Secco (1961) [62] for ZnO and discussed earlier in section 7.3.1, $D_e(\widetilde{\mathrm{Zn}}) = k_e D_i$, so that the respective activation energies show the following relationship: $Q_i = Q_{\mathrm{Zn}} - Q_e \cong 2.7 - 1.4 = 1.3$ eV. Secco assumed that the time for the change from one rate process to another, t_t, is also the time required for saturation. He then determined Q_i by two methods: plotting $\log\,(1/t_t)$ versus $1/T$, and using the relationship $r^2 = 2D_i t_t$, with $D_i = D_i^{\circ} \exp\,[-Q_i/kT]$. Both methods gave an activation energy ≈ 1.2 eV.

If indeed the initial period of the exchange corresponds to the time for saturation, then clearly, chemical diffusion of zinc into the sample is occurring during this time interval. This is expected to be faster than the self-diffusion occurring after saturation is achieved, as observed experimentally. This would lead to a quite different interpretation of the two processes and their corresponding activation energies.

A study was carried out by Bansagi, Secco, Srivastava and Martin (1968) [78] on the kinetics of the wurtzite-sphalerite transformation in the temperature range 800-900° C using X-ray techniques. Reagent grade powder was used, with initially 80% wurtzite, 20% sphalerite and three annealing conditions were imposed: vacuum, sulfur and zinc atmospheres. The kinetic data in the first two cases were analyzed with first and second order rate laws, whereas the latter case showed diffusion control. Analysis of the kinetic data lead to the following expression:

$$D = 0.1 \exp[(-2.9 \pm 0.2)/kT] \text{ cm}^2 \text{ s}^{-1}. \qquad (7.47)$$

Secco (1969) [63] reviewed previous exchange studies in ZnS and presented new data on the influence of a flowing radioactive zinc atmosphere of variable flow rate, on the kinetics of exchange. The initial first order exchange process was lower by a factor of 0.5 when compared to the static measurements, but was independent of the flow rate (40-680 ml/min). Secco interpreted the results as evidence for a difference in the concentration of the adsorbed species for flow versus static method; active sites are occupied completely for all flow rates, but other sites require the longer exposure times prevailing in the static method. This interpretation seems difficult to justify since the effective vapor pressures of zinc around the ZnS sample should be the same for a static or a flow system.

Diffusion of Sulfur in ZnS

The diffusion of S-35 in ZnS was investigated by Gobrecht, Nelkowski, Baars and Weigt (1967) [79] in the temperature range 740-1100° C. Single crystal slices of a boule of 'slightly disordered zincblende structure', 1.5 x 5 x 7 mm^3, were cut perpendicular to the (111) direction and were diffused with a tracer sulfur source under conditions of constant surface composition. The resulting profile was determined by angle lapping the sample followed by autoradiographic determination of the S-35 profile using strippable film. At a pressure of 0.5 atm S_2, assuming ideal vapor behavior,† the diffusion coefficient was found to be

$$D_e(\widetilde{S}) = 8 \times 10^{-5} \exp[-2.2/kT] \text{ cm}^2 \text{ s}^{-1}. \qquad (7.48)$$

Measurements done at 870° C for different values of sulfur pressure showed an increase in diffusivity with increasing pressure, but the specific

† There was no discussion of the equilibria between polymeric species of sulfur.

pressure dependence was not given. In this study, there were no comments on the preannealing conditions; however, the range of compositional change possible in ZnS and the length of the diffusion anneals were such that the sample was probably in equilibrium with the imposed sulfur pressure for essentially the entire diffusion anneal.

Blount, Marlor and Bube (1967) [80] measured the diffusion of S-35 in ZnS in the temperature range 700-891°C. The diffusion samples were sliced at a 45° angle to the *c* axis of 'pure hexagonal' ZnS single crystals. They were pre-annealed for two days at the diffusion temperature in saturated Zn vapor, prior to the diffusion anneal in 0.5 atm sulfur vapor. The diffusion profile was determined by sectioning the sample and counting the activity of each section removed. The results may be represented by

$$D_e(\widetilde{S}) = 7 \times 10^5 \exp[-3.4/kT] \text{ cm}^2 \text{ s}^{-1}. \tag{7.49}$$

Comparison with the results of Gobrecht *et al.* (1967) [79] shows no agreement whatsoever. There are, however, two major differences between the studies. First, Gobrecht *et al.* used slices from a zincblende single crystal, whereas Blount *et al.* used slices from a wurtzite crystal. In the preannealing treatment in zinc vapor, however, it is apparent that the sample in the latter study would transform completely to the zincblende form (Bansagi *et al.*, 1968 [78]). It is not clear what the grain size would be in the transformed crystal. Second, Blount *et al.* preannealed their samples in zinc whereas Gobrecht *et al.* presumably used slices as grown. Although in the former case the initial rate of diffusion would be much higher, due to the chemical gradient, it is probable that the difference in the overall diffusion anneals would be minimal; the amount of tracer sulfur introduced into the ZnS for the change in composition should be small compared to the total exchange. The large dispersion between the two results seems difficult to resolve.

Interdiffusion in ZnS

Interdiffusion in ZnS was studied by Sirota and Koren (1962) [81] in the temperature range 320-600°C by measurement of the rate of parabolic film growth. The resulting diffusion coefficient was expressed as $\underset{\sim}{D}(\text{ZnS}) = 1.2 \times 10^{-2} \exp[-1.39/kT] \text{ cm}^2 \text{ s}^{-1}$. A summary of the interdiffusion coefficients of the zinc chalcogenides was given subsequently by Sirota and Koren (1963) [82] who established a linear relation between the diffusion activation energies (for interdiffusion) and the respective quantities: lattice energy, heat of formation, and the melting temperature.

7.3.3 Diffusion in ZnSe

Diffusion of Zn in ZnSe

An investigation of the diffusion of Zn-65 in ZnSe was made by Secco and Su (1968) [83] in the temperature range 720-800° C using the static exchange method described by Secco and Moore (1957) [48]. The diffusion sample was ZnSe powder, assumed to be spherical particles with radius approximately 8.7×10^{-4} cm and the diffusion atmosphere was 0.10 atm zinc. The authors interpreted the exchange to be diffusion controlled from the initial stages, in contrast to results on ZnO and ZnS where first order processes were observed in the initial stages. Analysis of the fraction exchanged as a function of time, combined with the measurement of the particle radius, gave a diffusion coefficient

$$D_e(\widetilde{Zn}) = 10^3 \exp [(-3.45 \pm 0.22/kT] \text{ cm}^2 \text{ s}^{-1}. \qquad (7.50)$$

The authors implied that there may be a structural basis for the observation that there is no observed initial first order exchange process for ZnSe, in contrast to ZnS and ZnO.

Henneberg (1970) [84] and Henneberg and Stevenson (1971) [85], investigated the self-diffusion of Zn-65 in ZnSe in the temperature range 760-1150° C. Diffusion samples were sliced from single crystal boules from two sources: crystals purchased from the Eagle-Pitcher company, which were grown from the melt, and crystals grown by the investigators using a sublimation technique similar to that described by Piper and Polich (1961) [86]. Samples were preannealed at the exact temperature and component pressure conditions of the diffusion anneal prior to annealing with a Zn-65 source. The tracer profile was determined by mechanically sectioning and weighing. The radioactivity of the sections that were removed was determined by counting with a gamma-ray spectrometer. Periodic autoradiographs were made to ensure homogeneity of the radiotracer distribution. For most of the temperature range studied, two conditions of component pressure were studied, Zn and Se saturation, the coexistence of solid, vapor and Zn- or Se-rich liquid. For either of these conditions, the variance is one and a specification of the temperature defines the system. The component pressures were taken from the pressure-temperature diagram of the ZnSe system. At 878° C component pressures between the extremes were studied by adding precisely weighed amounts of Se or Zn to the capsule and assuming ideal gas behavior; the equilibria between polymeric species of Se were taken into consideration (Illariânov and Lapina, 1957) [87].

The diffusion coefficient was represented by the relation

$$D_e(\widetilde{Zn}) = 9.8 \exp [-3.0/kT] \text{ cm}^2 \text{ s}^{-1} \qquad (7.51)$$

for the temperature range 760-1150°C for both Zn- and Se-saturated crystals. In order to confirm the observed lack of dependence of $D_e(\widetilde{Zn})$ on component pressure over the range of existence of ZnSe, four pressures within the phase field at 878°C were studied, and the results showed diffusivities identical to the saturated values within experimental accuracy. The $D_e(\widetilde{Zn})$ values were found to increase with both acceptor (Cu) and donor (Al or In) doping when the doping concentrations were above critical values. The authors proposed that the defect responsible for the self-diffusion of Zn, and also dominating the charge neutrality condition, may be a singly ionized Frenkel defect, i.e., Zn_i and V'_{Zn} with $K'_F = [Zn_i^{\cdot}][V'_{Zn}]$. This conclusion was shown to be consistent with a recent study of the electron concentration in ZnSe at high temperatures by Smith (1969) [88]. Smith measured the high temperature Hall coefficient of ZnSe as a function of Zn pressure over a limited range of pressures and found the electron concentration to increase as $p_{Zn}^{1/2}$, i.e.,

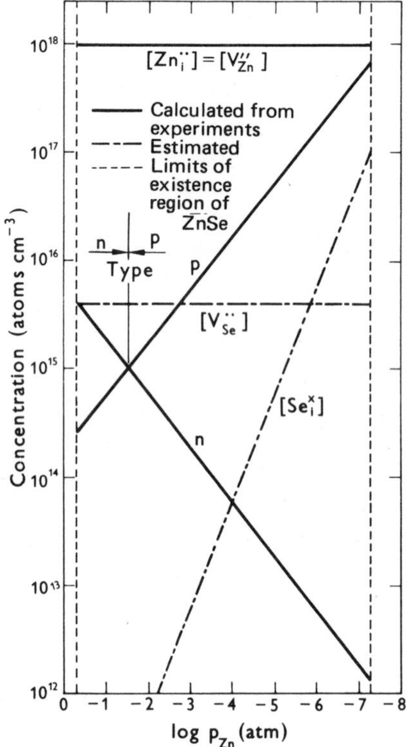

Figure 7.11. Tentative defect isotherm for ZnSe at 850° C. After Henneberg and Stevenson (1971) [58].

$n \propto p_{Zn}^{1/2}$. A tentative defect equilibrium diagram for 850°C was proposed by Henneberg and Stevenson (1971) [85] and is shown in figure 7.11. This diagram is consistent with all observations to date on the observed high temperature electrical and diffusion properties in ZnSe but predicts that p-type conductivity at elevated temperatures occurs except in a small region near the maximum zinc pressure. Although this seems unlikely, considering the observation that ZnSe is n-type at room temperature even after annealing in maximum Se pressure, it may be possible if the relevant acceptor is deep or if retrograde solubility causes precipitation of native acceptors upon cooling.

Diffusion of Selenium in ZnSe

The self-diffusion of Se-75 in ZnSe was investigated by Woodbury (1964) [89] for the temperature range 800-1000°C and later by Woodbury and Hall (1967) [90] at 1150° and 1000°C. The later results are selected as the more recent and significant. The same general technique that was used by Woodbury and coworkers for a number of studies was employed and is described briefly below.

Single crystal samples were cut into small cubes, 2-4 mm on a side, which were preannealed in quartz ampoules under the same thermodynamic conditions as the diffusion anneal. The radioisotopes were usually applied to the diffusion ampoule in aqueous solution form, treated with HNO_3 to remove HCl, and then vacuum dried. Additional amounts of elements or argon were added to fix the desired component pressure and the quartz ampoule was then sealed. After diffusion annealing, the concentration profile was determined by repetitive sequences of weighing the sample, etching, counting the radioactivity of the etchant, and reweighing the sample. The nonuniformity of the etch was checked in some samples with a micrometer, and appropriate corrections made; these corrections never changed the diffusivities by more than a factor of 2. Diffusion profiles were determined for Se and Zn-saturated conditions at 1000°C and 1050°C and intermediate pressures at 1050°C. The diffusion profiles–log (activity) versus x^2–showed two distinct segments with diffusion coefficients assigned to each segment and differing by a factor of approximately 2.5. For Se-saturated crystals, the diffusion coefficient was represented as

$$D_e(\widetilde{\mathrm{Se}}) = 0.23 \exp\left[-2.7/kT\right] \mathrm{~cm}^2 \mathrm{~s}^{-1}, \qquad (7.52)$$

based on two data points. Significant enhancement of diffusivity with increasing Se_2 pressure was observed with the pressure dependence $p_{Se_2}^{\gamma}$, with $0.4 \lesssim \gamma \lesssim 0.5$ on the Se side of the phase field, but was relatively insensitive to component pressure on the Zn side. The authors proposed diffusion via a neutral interstitial selenium, but noted that another mechanism appears to occur at lower Se_2 pressures. A summary of the results is given in figure 7.12.

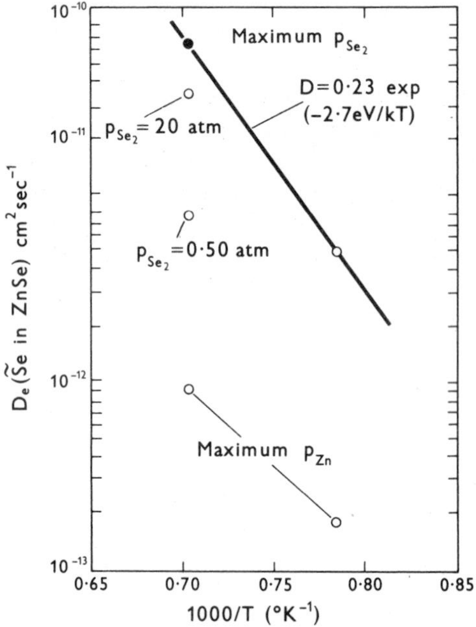

Figure 7.12. Temperature dependence of $D_e(\widetilde{Se}$ in ZnSe). The lower points (saturated Zn) were regarded as upper limits and the solid point is a value extrapolated from lower pressure data. After Woodbury and Hall (1967) [90].

Henneberg (1970) [84] measured the diffusion coefficient of Se-75 in ZnSe single crystals using the same technique described for his measurements of Zn-65 in ZnSe. The self-diffusion coefficient of Se in Se-saturated ZnSe is described by the relationship

$$D_e(\widetilde{Se}) = 0.13 \exp\left[-2.6/kT\right] \text{ cm}^2 \text{ s}^{-1} \qquad (7.53)$$

for the temperature range 860-1020° C, which is in acceptable agreement with the findings of Woodbury and Hall (1967) [90]. The dependence of $D_e(\widetilde{Se})$ on p_{Se_2} was studied at 1021° C and found to vary as $D_e(\widetilde{Se}) \propto p_{Se_2}^{0.35}$, for $p_{Se_2} < 10^{-3}$ atm, and to be rather insensitive to changes in component pressure for smaller Se_2 pressures. This is in general agreement with the results of Woodbury and Hall (1967) [90]. The authors suggested that a neutral interstitial mechanism for Se diffusion occurs on the Se side of the phase field, and that Se diffusion proceeds via a doubly ionized vacancy on the Zn side. Such a mechanism is consistent with the defect equilibrium diagram shown in figure 7.11.

Interdiffusion in ZnSe

Interdiffusion in ZnSe was investigated by Sirota and Koren (1962) [91] in the temperature range 300-577°C by measuring the rate of film growth on Zn single crystals and Zn liquid when exposed to Se vapor. Parabolic growth kinetics were observed, and the associated diffusion coefficient for the formation of dense greenish films was expressed by the equation

$$\underset{\sim}{\mathbf{D}}(\mathrm{ZnSe}) = 8.3 \times 10^{-4} \exp\,[-1.13/kT]\ \mathrm{cm}^2\ \mathrm{s}^{-1}. \qquad (7.54)$$

In addition to a dense greenish film, a loose yellow film was also observed. Both films were identified as the sphalerite structure, with the observed diffusivity in the latter case represented by a larger D_o value, viz., $1.2 \times 10^{-2}\ \mathrm{cm}^2\ \mathrm{s}^{-1}$. As would be anticipated, these values for $\underset{\sim}{\mathbf{D}}(\mathrm{ZnSe})$ are many orders of magnitude larger than $D_e(\tilde{\mathrm{Zn}})$ or $D_e(\tilde{\mathrm{Se}})$, when extrapolated to the same temperature.

7.3.4 Diffusion in ZnTe

Zinc Diffusion in ZnTe

The self-diffusion of Zn-65 in ZnTe single crystals was investigated by Reynolds and Stevenson (1969) [38] within the temperature range 693-1031°C using Zn-65 and a mechanical sectioning method. The technique was essentially the same as the one used by Henneberg and Stevenson (1971) [85] described previously in section 7.3.3. Autoradiographic evidence was presented showing enhanced grain boundary diffusion in polycrystalline samples; however, all reported values were for bulk diffusion in single crystals. The diffusion coefficient was found to be identical, within experimental scatter, for Zn-saturated crystals as well as Te-saturated crystals and was expressed as

$$D_e(\tilde{\mathrm{Zn}}) = 14 \exp\,[(-2.69 \pm 0.08)/kT]\ \mathrm{cm}^2\ \mathrm{s}^{-1} \qquad (7.55)$$

for the temperature range studied, 693-1031°C. One data point was determined at 695°C for the congruent subliming composition and was found to be in agreement with data points for Zn- and Te-saturated crystals. This demonstrated that $D_e(\tilde{\mathrm{Zn}})$ does not go through a minimum across the phase field, as would be expected for un-ionized Frenkel disorder. The addition of Al impurity in a concentration of $\sim 5 \times 10^{19}\ \mathrm{cm}^{-3}$ was shown to enhance $D_e(\tilde{\mathrm{Zn}})$ with $D_e(\tilde{\mathrm{Zn}}$ in Al doped ZnTe) $\approx 10^{-4} \exp\,[-1.7/kT]\ \mathrm{cm}^2\ \mathrm{s}^{-1}$.

The independence of $D_e(\tilde{\mathrm{Zn}})$ to p_{Zn} was explained by postulating that thermal disorder on the zinc sublattice dominates the charge neutrality condition for the crystals, with ionized Frenkel disorder proposed. The enhancement of the diffusion for Al-doped samples was explained assuming the charge neutrality condition $[\mathrm{Al}^{\cdot}_{\mathrm{Zn}}] = [V'_{\mathrm{Zn}}]$ to be dominant in the extrinsic range. Thus, the activation energy for diffusion in this range may be identified with the energy of motion of the zinc vacancy. Although the model proposed by the authors was tentative, there is a discrepancy between the observation that the $D_e(\widetilde{\mathrm{Zn}})$ is independent of component pressure and the zinc vacancy model proposed for the acceptor in ZnTe, e.g. Thomas and Sadowski (1964) [92].

A recent study performed by Secco and Yeo (1971) [93] determined the exchange ratio between a ZnTe single crystal and Zn-65 vapor for the temperature range of 760-860°C, with $p_{\mathrm{Zn}} = 0.10$ atm, using the technique previously described in section 7.3.1 (Secco and Moore, 1957) [48]. The exchange rate was found to be diffusion controlled at all times and the corresponding diffusion coefficient was expressed as

$$D_e(\widetilde{\mathrm{Zn}}) = (2.34 \pm 0.37) \exp\,[(-2.04 \pm 0.22)/kT]\ \mathrm{cm}^2\ \mathrm{s}^{-1}. \quad (7.56)$$

The results are considerably different from those of Reynolds and Stevenson (1969) [38], with no obvious reason for the discrepancy.

Secco assumed that the Zn diffusion proceeded via a doubly ionized Zn vacancy, V''_{Zn}, in accord with the model of Thomas and Sadowski (1964) [92] in which the charge neutrality condition is assumed to be $2[V''_{\mathrm{Zn}}] = p$. Expressing $D_e(\widetilde{\mathrm{Zn}})$ as $\propto \gamma^2 \Gamma [V''_{\mathrm{Zn}}]$, he computed D_o using the following values: the expression of Thomas and Sadowski giving $[V''_{\mathrm{Zn}}]$ as function of temperature and pressure; a jump frequency, Γ, taken from the Debye frequency; and an assumed jump distance, λ. The agreement was excellent: $D_o(\mathrm{calc}) = 1.67\ \mathrm{cm}^2\ \mathrm{s}^{-1}$ versus $D_o(\mathrm{obs}) = 2.34 \pm 0.37\ \mathrm{cm}^2\ \mathrm{s}^{-1}$. If Zn diffuses via a V''_{Zn}, then a very large change in $D_e(\widetilde{\mathrm{Zn}})$ should be observed when comparing Zn-saturated with Te-saturated crystals, contrary to the observations of Reynolds and Stevenson (1969) [38].

Diffusion of Te in ZnTe

Reynolds and Stevenson (1969) [38] also measured the self-diffusion of Te-123 in ZnTe single crystals using the same experimental technique as in their determination of $D_e(\widetilde{\mathrm{Zn}})$. For Te-saturated crystals,

$$D_e(\widetilde{\mathrm{Te}}) = 2 \times 10^4 \exp\,[(-3.8 \pm 0.4)/kT]\ \mathrm{cm}^2\ \mathrm{s}^{-1}, \quad (7.57)$$

for the temperature range 750-970°C. A relatively large error band was a result of the small values for $D_e(\widetilde{\mathrm{Te}})$ and the uncertainty in defining the position of the initial interface after slight surface irregularities developed during the diffusion anneal. For Zn-saturated ZnTe crystals, much lower $D_e(\widetilde{\mathrm{Te}})$ values were observed. The corresponding concentration profiles were rather erratic and appeared to be influenced by diffusion along subgrain boundaries. One could assign maximum possible values, however, and it was well established that $D_e(\widetilde{Te})$ increased with increasing p_{Te_2}. The behavior was interpreted as evidence that an interstitial Te atom of unknown electrical activity is the dominant mobile defect on the Te sublattice of ZnTe.

Interdiffusion in ZnTe

The rate of film growth (reaction diffusion) on single crystals of Zn and Zn liquid in a Te atmosphere was investigated by Sirota and Koren (1963) [94] for the temperature range 320-520° C. The rate of film growth was followed by microstructural examination of the reaction product and a parabolic rate law was observed. The use of the inert marker method established that Zn diffusion dominates the film growth process. The effective interdiffusion coefficient was represented by

$$\underset{\sim}{D}(ZnTe) = 1.1 \times 10^{-4} \exp\left[-1.08/kT\right] \ cm^2 \ s^{-1}. \qquad (7.58)$$

The higher mobility of Zn is consistent with the observed differences in the self-diffusion coefficients, i.e., $D_e(\widetilde{Zn}) \gg D_e(\widetilde{Te})$ (Reynolds and Stevenson, 1969) [38]. In addition, the relative magnitudes of $\underset{\sim}{D}(ZnTe)$ and $D_e(\widetilde{Zn})$ are as expected; at 400° C, $D_e(\widetilde{Zn}) \cong 10^{-19}$ cm^2 s^{-1} and $\underset{\sim}{D}(ZnTe) = 10^{-12}$ cm^2 s^{-1}. Using equation 7.1 and estimates of the thermodynamic factor, as discussed in section 7.2.3, it can be shown that a difference of this general magnitude is reasonable (Reynolds, 1965) [76].

7.4 DIFFUSION IN THE CADMIUM CHALCOGENIDES

7.4.1 Diffusion in CdO

Diffusion of O in CdO

The oxygen-18 exchange between gaseous oxygen and cadmium oxide crystals has been investigated by Haul and Just (1962) [95] as a function of temperature in the temperature range 630-855°C and as a function of pressure at 756° C for 70 torr $\leqslant p_{O_2} \leqslant$ 700 torr. Two types of samples were used: small crystals, 40-70 μm in size, which were prepared by oxidizing Cd metal; and tablets prepared by pressing (10,000 atm) and sintering CdO powder (300-400 h at 950° C). The latter were prepared pure and with additions of Li_2O, Al_2O_3 and In_2O_3. The sample surface areas were determined by gas adsorption methods (BET). The O-18 exchange kinetics were followed with mass spectrometric techniques (Haul, Just and Dümbgen, 1960; Haul and Just, 1958) [96, 97] and the exchange was reported to be mixed control; the phase boundary reaction was comparable to the diffusion of O in the solid. Calculated values for

$D_e(\tilde{O})$ and the phase boundary reaction rate constant, K, were obtained by suitable analysis (Haul, Just and Dümbgen, 1962) [98] and represented by

$$D_e(\tilde{O}) = 3.8 \times 10^6 \exp [(-4.00 \pm 0.17)/kT] \text{ cm}^2 \text{ s}^{-1} \qquad (7.59)$$

and

$$K = 1.3 \times 10^2 \exp [(-2.13 \pm 0.48)/kT] \text{ cm}^2 \text{ s}^{-1} \qquad (7.60)$$

for the temperature range 630-855° C.

The authors commented that the absolute values for the pre-exponential, D_o, are uncertain because the true diffusion cross-section could not be determined unambiguously.

The value of $D_e(\tilde{O})$ at 750° C was reported to decrease with increasing p_{O_2}, with the dependence expressed as

$$D_e(\tilde{O}) \approx K p_{O_2}^{-1/\gamma}, \qquad \gamma = 5 \pm 1 \qquad (7.61)$$

for 70 torr $\leqslant p_{O_2} \leqslant$ 700 torr. The time required to establish defect equilibrium at different p_{O_2} values was estimated to be of the order of minutes and the observed values of $D_e(\tilde{O})$ were the same whether or not the samples were preannealed at exactly the same pressure. The observed decrease in $D_e(\tilde{O})$ with increasing p_{O_2} was interpreted as evidence for a vacancy diffusion mechanism, with the observed pressure dependence explained by the proposed defect reaction:

$$3\text{Cd}_{\text{Cd}} + 3\text{O}_{\text{O}} \rightleftharpoons 2V_{\text{O}}^{\cdot\cdot} + \text{Cd}_{\text{Cd}}'' + 2\text{Cd}_{\text{Cd}}' + \text{O}_{\text{O}} + \text{O}_2(G), \qquad (7.62)$$

with $[V_{\text{O}}^{\cdot\cdot}] = [\text{Cd}_{\text{Cd}}'] = 2[\text{Cd}_{\text{Cd}}'']$, leading to a $p_{O_2}^{-1/5}$ dependence for $[V_{\text{O}}^{\cdot\cdot}]$. In light of the modest uncertainty in the p_{O_2} exponent, it would seem as reasonable to select the simpler defect reaction, $\text{O}_{\text{O}} \rightleftharpoons V_{\text{O}}^{\cdot\cdot} + 2e' + \tfrac{1}{2}\text{O}_2(G)$, with $2[V_{\text{O}}^{\cdot\cdot}] = n$ leading to a $p_{O_2}^{-1/6}$ dependence of $[V_{\text{O}}^{\cdot\cdot}]$.

The vacancy model was further substantiated by the reported increase in $D_e(\tilde{O})$ with Li_2O doping and a decrease in $D_e(\tilde{O})$ with In_2O_3 addition. The observed enhancement of $D_e(\tilde{O})$ with Li_2O concentration was analyzed to show the combined influence of native disorder and impurities. The atom fraction of oxide vacancies was estimated to be 4.4×10^{-4} in pure CdO at 790° C with $p_{O_2} = 0.16$ atm, in good agreement with earlier determinations reported in the literature. Estimates of thermochemical quantities were made to express the dependence of the vacancy fraction upon p_{O_2} and temperature, and the activation energy for diffusion of 4.0 eV was separated into 2.7 eV for motion and 1.3 eV for formation.

7.4.2 Diffusion in CdS

Diffusion of Cd in CdS

The diffusion of Cd in CdS was first investigated by Woodbury (1964) [99] using the etching technique described in section 7.3.3. Either Cd-109 or Cd-115m isotopes in the form of CdO or Cd metal were used as the diffusion source. The observed diffusion profiles showed an abrupt sharp decrease near the sample interface, in the first few microns, followed by a normal error function complement profile. This was interpreted as non-equilibrium at the surface arising from a slow interface step. The profiles were analyzed using an expression derived by Crank (1956) [100] for a surface reaction rate-limiting condition; however, Woodbury noted that there was only about 10% difference between values calculated from the error function complement or the surface reaction rate limited analytical solution. For undoped Cd-saturated crystals, $D_e(\tilde{\text{Cd}})$ was expressed as

$$D_e(\tilde{\text{Cd}}) = 3.4 \exp[-2.0/kT] \text{ cm}^2 \text{ s}^{-1} \tag{7.63}$$

in the temperature range 750-1000°C. When CdO was used as the tracer source, $D_e(\tilde{\text{Cd}})$ was 1/4 to 1/40 of the values given above, and this decrease was attributed to the lower p_{Cd} using this source. A further check on the dependence of $D_e(\tilde{\text{Cd}})$ on p_{Cd} was made by exposing CdS crystals to the tracer source at 800° C under saturated Cd vapor for 15 min followed by a 2-h anneal under saturated sulfur pressure at 900° C. The profile was characteristic of the short diffusion anneal at 800° C indicating that diffusion under sulfur-saturated conditions, even at higher temperatures, was slower than under Cd-saturated conditions. The results were also interpreted as 'consistent with the fact that the radioactive Cd is in normal lattice sites, since Cd in other sites (in particular as interstitials) might be expected to be removed in the sulfur firing'. This interpretation does not consider that the tracer atoms may diffuse interstitially and exchange with Cd on normal sites and, hence, are not removed by annealing in sulfur. There is evidence that interstitial-substitutional exchange is extremely rapid in some systems (Rogalla, 1968) [101]. If this is the case, there would be no significant preference for tracer atoms to out-diffuse during the sulfur anneal. In addition, the change of composition during sulfur annealing could be accommodated by the in-diffusion of Cd vacancies, which would trap the Cd interstitials. Other experiments were carried out at intermediate pressures—slight sulfur excess—with resulting $D_e(\tilde{\text{Cd}})$ values 10^{-2}-10^{-3} times those for Cd-saturation, suggesting that $D_e(\tilde{\text{Cd}})$ is a minimum somewhere intermediate in the phase field. The influence of donor solutes—In or Cl—on $D_e(\tilde{\text{Cd}})$ was studied; $D_e(\tilde{\text{Cd}})$ was not influenced by the added solutes in Cd-saturated crystals, but increased linearly in S-saturated crystals. Woodbury concluded that $D_e(\tilde{\text{Cd}})$ occurs via Cd vacancies in the presence of excess S and that the concentrations of these acceptor defects are controlled by the donor solute concentration. In Cd-saturated CdS crystals, the predominant defect was presumed to be a sulfur vacancy. The more rapid self-diffusion of Cd in the presence of excess Cd was explained by postulating that either a ring mechanism is operative and involves the four Cd atoms around a S vacancy, or the four Cd atoms surrounding the S vacancy interchange by using the free volume available due to the absence of the S atom. This latter process was termed the covacancy mechanism. Both models would result in $D_e(\tilde{\text{Cd}})$ being dependent upon the concentration of S vacancies.

The experimental results can also be accounted for if it is postulated that Frenkel disorder exists on the Cd sublattice. In this case the model for the self-diffusion of Cd in the presence of excess S would be the same as that proposed by Woodbury. The more rapid diffusion in the presence of excess Cd, however, would arise from the creation of interstitial Cd which could diffuse via the interstitial or the interstitialcy mechanism. Since the interstitial defect could be more mobile than other defects, it is not necessary to propose that the interstitial is present in a greater concentration than other native defects.

There were also a number of observations on the rate of precipitation in doped crystals. In-doped samples change color and conductivity reversibly with alternate short anneals in Cd or S atmospheres and one crystal changed color at 800°C after annealing for only a few minutes, with an estimated diffusion coefficient of $\sim 10^{-5}$ cm^2 s^{-1}. The diffusion process for this change occurs in a concentration gradient and corresponds to the chemical diffusion coefficient, $\underset{\sim}{D}$(CdS), which is expected to be orders of magnitude larger than the self-diffusion coefficients. This is one of many such observations that indicate very large values for $\underset{\sim}{D}$ in the II-VI compounds.

In summary, the study of Woodbury has contributed the following information: (1) $D_e(\tilde{\text{Cd}})$ increases with p_{Cd}, but the explicit dependence is not well established; (2) $D_e(\tilde{\text{Cd}})$ increases with donor concentration for S-saturated crystals and is independent of donor concentration for Cd-saturated crystals; (3) $\underset{\sim}{D}(\text{CdS})$ is $\gg D_e(\tilde{\text{Cd}})$. Lacking explicit information for the exact dependence of $D_e(\tilde{\text{Cd}})$ on p_{Cd}, the mechanisms proposed by Woodbury should be considered tentative.

The dependence of $D_e(\tilde{\text{Cd}})$ in CdS on component pressure was studied by Shaw and Whelan (1969) [102] at 850°C. Measurements of the electrical conductivity as a function of cadmium pressure were made on the same samples for the temperature range 660-950° C in order to provide additional information concerning the point defect structure. Diffusion was studied parallel to the c axis using Cd-115m, in the form of $Cd(NO_3)_2$, as the radioisotope. After diffusion annealing, the tracer profile was determined by mechanically sectioning the sample. Diffusion profiles invariably indicated two distinct regions corresponding to a fast and slow diffusion process which the authors interpreted as dislocation and bulk diffusion, respectively. There was no corroboration of this interpretation by autoradiography. The profiles were quite different from those of Woodbury (1964) [99]. Only the slow component was considered by Shaw and Whelan in the treatment of their data and a significant increase in the $D_e(\tilde{Cd})$ with p_{Cd} was observed—roughly a factor of 250 over the entire phase field—and three distinct regions were apparent: S-saturation and Cd-saturation with $D_e(\tilde{Cd}) \propto p_{Cd}$; and an intermediate region with $D_e(\tilde{Cd})$ independent of p_{Cd} $-D_e(\tilde{Cd}) \propto p_{Cd}^0$; cf. figure 7.13.

Electrical conductivity as a function of p_{Cd}, when combined with estimates of the mobility, gave a series of electron carrier density isotherms as shown in figure 7.14. The pressure dependence of n could be expressed as $n \propto p_{Cd}^{\gamma}$, ranging from $\gamma \cong 0.38$ at the highest temperature to $\gamma \cong 0.05$ at the lowest temperatures. The data at the highest temperature are consistent with a doubly ionized Cd_i or V_S defect, but the decrease in γ with temperature indicated the influence of an impurity defect. At the

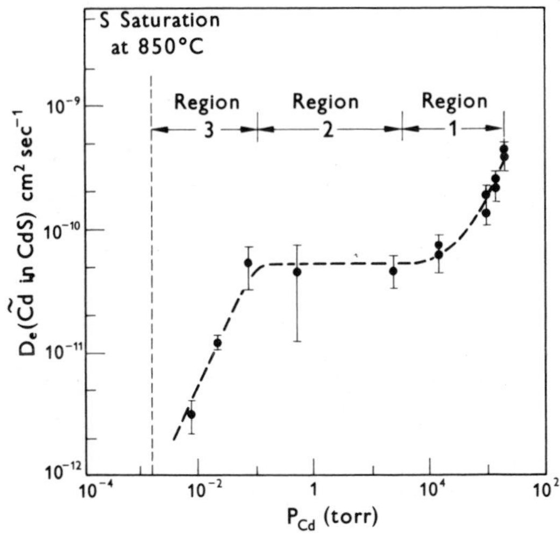

Figure 7.13. A diffusion isotherm for $D_e(\tilde{Cd}$ in CdS) at 850° C. After Shaw and Whelan (1969) [102].

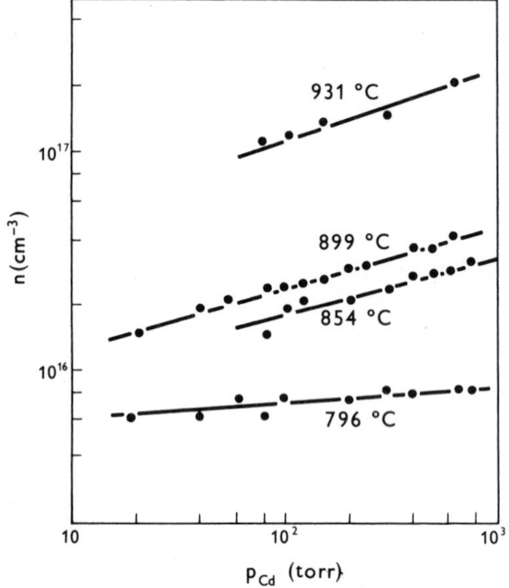

Figure 7.14. The variation of electron concentration, n, with p_{Cd} in CdS at different temperatures. After Shaw and Whelan (1969) [102].

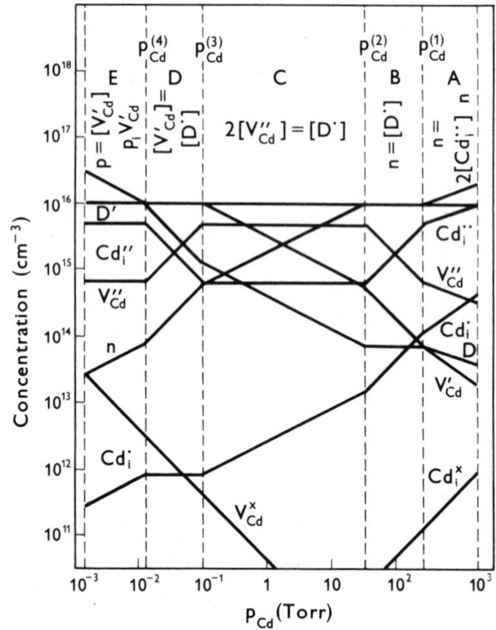

Figure 7.15. A proposed defect concentration isotherm for CdS. After Shaw and Whelan (1969) [102].

diffusion temperature of 850°C, Shaw and Whelan assumed a model involving singly and doubly ionized cadmium interstitials and vacancies and a shallow fully-ionized donor, $D^{\cdot}$, and proposed a corresponding defect equilibrium diagram, shown in figure 7.15. The following interpretation of the diagram was given to explain the observed diffusion behavior: the diffusion in the S-rich and in the center portion of the phase field proceeds via the V''_{Cd} defect ($\propto p_{Cd}$ and independent of p_{Cd} in the two regions), whereas the diffusion in the Cd-rich region proceeds via $Cd_i^{\times}$. The authors comment at length on the relatively high value of the mobility of $Cd_i^{\times}$ compared with $Cd_i^{\cdot\cdot}$ and propose an interstitialcy mechanism to explain this difference.

The work of Boyn, Goede and Kuschnerus (1965) [103], (BGK), is relevant to the previous study. BGK measured the conductivity of CdS crystals as a function of p_{Cd}, expressing their results as $\sigma \propto p_{Cd}^{\gamma}$. The values of γ were found to decrease from 0.48 at 395°C to 0.31 at 600°C. This was interpreted as evidence for ionized Frenkel disorder, with the following charge neutrality conditions at low and high temperatures respectively: $[Cd_i^{\cdot}] = n$ and $2[Cd_i^{\cdot\cdot}] = n$.

The observed behavior for γ was in sharp contrast to the reported values of Shaw and Whelan, which are compared in figure 7.16. A recent study of the carrier concentration in CdS as a function of component pressure has been reported by Hershman and Kröger (1971) [104] and the trend in γ was generally similar to that reported by Shaw and Whelan. The latter authors point out that a carrier density roughly

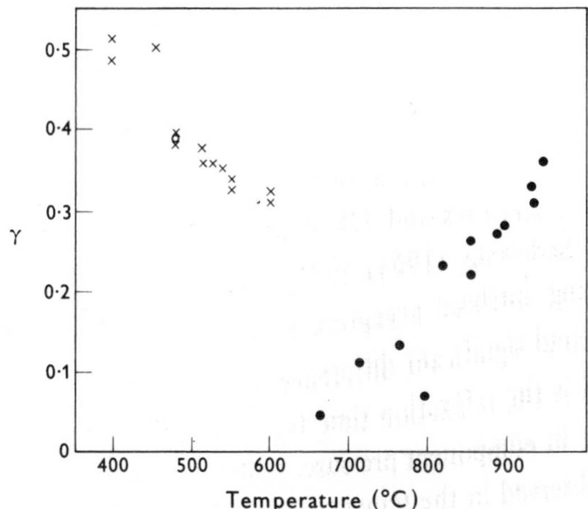

Figure 7.16. The variation of γ with temperature in CdS, where γ is the exponent in $n \propto p_{Cd}^{\gamma}$. Comparison of the results of Shaw and Whelan, ● (1969) [102], with Boyn, Goede and Kuschnerus, x (1965) [103]. After Shaw and Whelan (1969) [102].

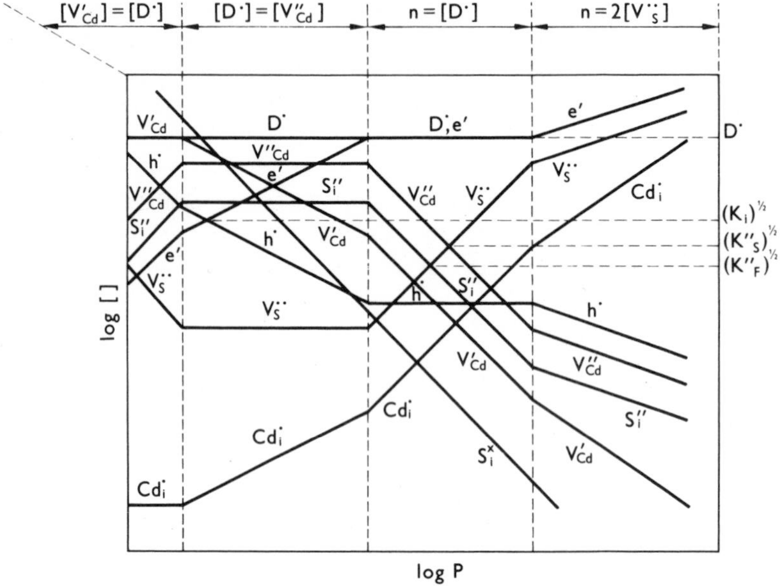

Figure 7.17. Proposed defect concentration isotherm for CdS. After Kumar and Kröger (1971) [108].

10^{14} cm^{-3} at 395°C, as reported by BGK, implies an improbably low residual impurity concentration, i.e., $\leqslant 10^{14}$. Shaw and Whelan suggest that associates involving residual donor impurities may account for the low carrier densities, particularly since evidence for such associates in CdS has been given by Boyn (1968) [105]. The low carrier density and the behavior of γ observed by BGK could also be explained by a high residual acceptor concentration. The acceptors could compensate the native donor defects causing a low carrier density, and the behavior of γ explained by including ionized acceptor impurities along with the electrons and native donors in the charge neutrality condition. Similar behavior for γ in ZnTe, with 1/3 at elevated temperatures and 1/2 at lower temperatures as reported by Thomas and Sadowski (1964) [92], was explained by Larsen (1970) [106] assuming intrinsic acceptors partially compensated by impurity donors. A second significant difference in Shaw and Whelan's results and those of BGK is the relaxation time for a change in conductivity after an abrupt change in component pressure. In the former case, relaxation times $\leqslant 60$ s were observed in the temperature range 660-950°C whereas in the latter study and in a similar study by Boer, Boyn and Goede (1963) [107], relaxation times of the order of several minutes to an hour were observed. This is only one of several such discrepancies reported for similar studies in other II-VI compounds. In general, one reports small relaxation times for systems employing the van Doorn (1961) [24] technique of dynamic refluxing, used by Shaw and Whelan, whereas much longer relaxation times are reported for investigations employing closed systems, such as used by BGK. BGK report a diffusion coefficient $D = 2.1 \times 10^{-5} \exp[-0.6/kT]$ cm^2 s^{-1} from their relaxation measurements in CdS. It should be mentioned that these relaxation measurements yield interdiffusion coefficients which are expected to be several orders of magnitude larger than self-diffusion coefficients for the same system.

Kumar and Kröger (1971) [108] have made a more recent investigation of the self-diffusion behavior of Cd in single crystal wafers of CdS as a function of p_{Cd} and doping levels in the temperature range 700-900°C. In discussing the authors' results and interpretation, it is convenient to refer to their proposed defect equilibria isotherm diagram in figure 7.17. The samples were preannealed under conditions identical to those for the diffusion anneal in cases where homogenization times would be long (for $T < 600°$C) or when out-diffusion of the excess component might alter the component pressure, near P_{min}. Diffusion isotherms were measured in the temperature range 700-1000°C on the Cd side of the stability range—$p_{Cd} \sim 0.01$ atm to $p_{Cd)SAT}$—with the diffusion equation represented by

$$D_e(\tilde{Cd}) = 7.29 \times 10^{-5} \exp[-1.2/kT] p_{Cd}^{2/3} \text{ cm}^2 \text{ s}^{-1} \text{ atm}^{-2/3} \quad (7.64)$$

Two isotherms and a comparison with Shaw and Whelan (1969) [102] are given in figure 7.18. Referring to previous work that established $n \propto p_{Cd}^{1/3}$ for this pressure range in CdS (Hershman and Kröger, 1971 [104]; Shaw and Whelan, 1969 [102]), the authors concluded that the dominant mobile defect on the Cd sublattice is not the majority defect involved in the charge neutrality condition. The charge neutrality condition, $2[V_S^{\cdot\cdot}] = n$, was selected, based on diffusion isotherms for $D_e(\tilde{S})$; $D_e(\tilde{S}) \propto P_{Cd}^{1/3}$ as discussed below. With this electroneutrality condition, $[Cd_i^{\cdot}] \propto p_{Cd}^{2/3}$, and it was proposed that $Cd_i^{\cdot}$ is the dominant mobile defect in this range, cf. figure 7.17. At lower p_{Cd}, data for $D_e(\tilde{Cd})$ showed considerable scatter, obviating a determination of the exact pressure dependence. The observed values were, however, significantly higher than values extrapolated from equation 7.64, indicating a different mechanism or a different electrical neutrality condition. The influence of In doping on $D_e(\tilde{Cd})$ isotherms provided further information. These isotherms indicated three regions, as shown in figure 7.19: (I) lowest p_{Cd}, $D_e(\tilde{Cd})$ independent of p_{Cd}; (II) intermediate p_{Cd}, $D_e(\tilde{Cd}) \propto p_{Cd}^{-1}$; and (III) highest p_{Cd}, $D_e(\tilde{Cd}) \propto p_{Cd}$. The authors proposed V_{Cd}'' as the dominant mobile defect in regions (I) and (II) and $Cd_i^{\cdot}$ in region (III) (figure 7.17).

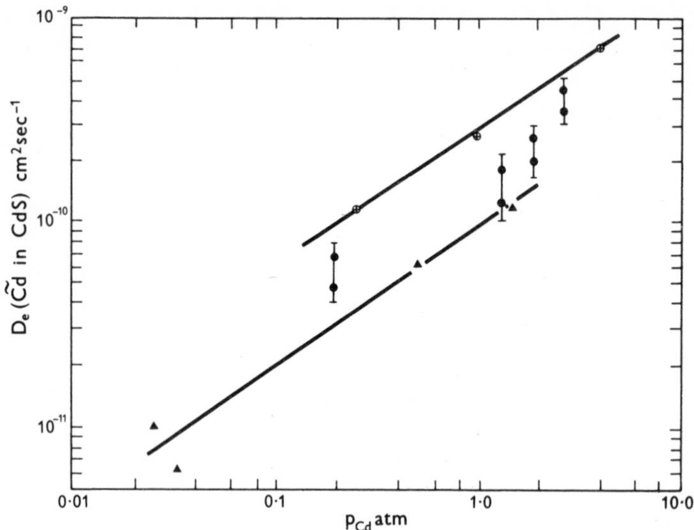

Figure 7.18. Dependence of $D_e(\widetilde{Cd}$ in CdS) upon p_{Cd} ; ▲ and ⊗, results of Kumar and Kröger (1971) [108] at 800° and 900°C, respectively and ● results of Shaw and Whelan (1969) [102]. After Kumar and Kröger (1971) [108].

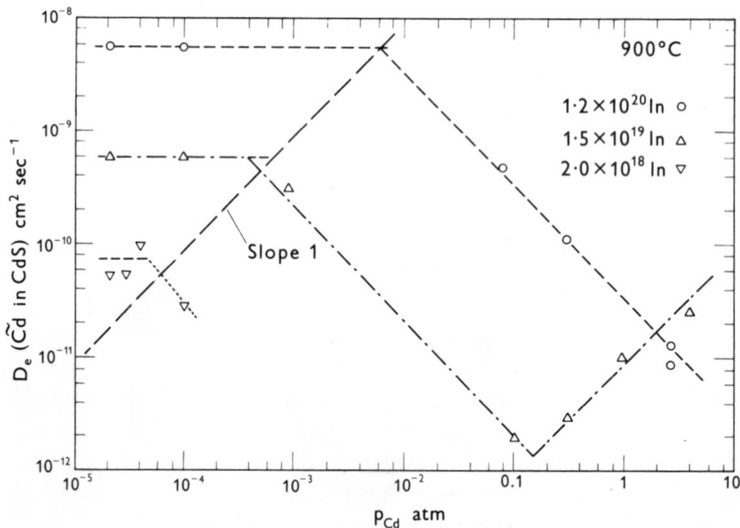

Figure 7.19. Isotherms for the dependence of $D_e(\widetilde{Cd}$ in CdS) upon p_{Cd} for crystals doped with various concentrations of In. After Kumar and Kröger (1971) [108].

Diffusion of S in CdS

Kumar and Kröger (1971) [108] also studied the self-diffusion of S in CdS. At high p_{Cd}, diffusion isotherms showed the behavior, $D_e(\tilde{S}) \propto p_{Cd}^{1/3}$, which led to the following assignments for the incorporation reaction:

$$Cd(G) \rightleftharpoons Cd_{Cd} + V_S^{\cdot\cdot} + 2e'; \qquad K_{SV} = n^2 [V_S^{\cdot\cdot}]/p_{Cd}, \tag{7.65}$$

with $2[V_S^{\cdot\cdot}] = n$, whence

$$[V_S^{\cdot\cdot}] = n/2 = \left(\frac{K_{SV}''}{4}\right)^{1/3} p_{Cd}^{1/3}. \tag{7.66}$$

From the Hall-effect measurements on the same crystals by Hershman and Kröger (1971) [104], which established n as a function of T and p_{Cd}, the variation of $[V_S^{\cdot\cdot}]$ was given by:

$$[\underset{\sim}{V}_S^{\cdot\cdot}] = 3.06 \times 10^{-3} \exp[-0.58/kT] p_{Cd}^{1/3}. \tag{7.67}$$

Combining this with an expression for $D_e(\tilde{S})$ at $p_{Cd} = 4$ atm (based on only two temperatures), $D_e(\tilde{S}) = 2.58 \times 10^{-6} \exp[-1.9/kT]$ cm^2 s^{-1}, and the relation $D_e(\tilde{S}) = D_e(V_S^{\cdot\cdot})[\underset{\sim}{V}_S^{\cdot\cdot}]$, they obtained the following expression for the defect diffusivity:

$$D_e(V_S^{\cdot\cdot}) = 5.32 \times 10^{-4} \exp[-1.32/kT] \text{ cm}^2 \text{ s}^{-1}. \tag{7.68}$$

At low p_{Cd}, a change in the pressure dependence to $D_e(\tilde{S}) \propto p_{Cd}^{-1}$ was observed, with assignment of a neutral S interstitial, $S_i^{\times}$, as the mobile defect. The neutral character of the defect was confirmed by establishing that $D_e(\tilde{S})$ was independent of In doping in this range of p_{Cd}. The temperature dependence of $D_e(\tilde{S})$ for $p_{S_2} \cong 2$ atm was represented by

$$D_e(\tilde{S}) = 1.11 \times 10^{-2} p_{S_2}^{1/2} \exp[-2.08/kT] \text{ cm}^2 \text{ s}^{-1}. \tag{7.69}$$

Interdiffusion in CdS

Kumar and Kröger (1971) [109] also measured interdiffusion in CdS single crystals in the temperature range 660-760°C by observing the relaxation times for the resistivity change attending a stepwise change in component pressure. A closed tube system was employed, with a two zone technique used to control p_{Cd}. The CdS crystal was placed in the higher temperature zone which had a flat temperature profile, whereas pure Cd was placed at the other end of the quartz tube in a cooler temperature zone that had a decreasing temperature profile. After the sample had equilibrated with the capsule in one position, the capsule was rapidly shifted so that the Cd reservoir was at a new temperature with the CdS crystal at the same temperature. The relaxation time of the resulting change in resistivity of the CdS was measured, and the interdiffusion coefficient determined from appropriate mathematical analysis (Dünwald and Wagner, 1934) [110]. Their results, which may be expressed as

$$\underset{\sim}{D}(\text{CdS}) = 8 \times 10^{-3} \exp\left[-0.68/kT\right] \text{ cm}^2 \text{ s}^{-1}, \qquad (7.70)$$

showed good agreement with a single point obtained by Whelan and Shaw at 800°C when a correction for the pressure difference in the two studies was made (figure 7.20). This is an instance where reasonable agreement is obtained for the relaxation time between results from a closed and an open system. Comparison with the results of Boyn, Goede and Kushnerus (1965) [103], however, showed the latter diffusion coefficients to be about 100 times smaller, with no obvious explanation for the discrepancy.

An analysis of the interdiffusion process was made, based on the defect model that had been presented in self-diffusion studies in CdS; namely, $V_S^{\cdot\cdot}$ is the dominant defect for controlling non-stoichiometry in CdS under these conditions of T and p_{Cd}. However, $Cd_i^{\cdot}$ is the most rapidly diffusing specie. The assumption was made that the diffusing $Cd_i^{\cdot}$ adjusted the $V_S^{\cdot\cdot}$ concentration by the reaction $Cd_i^{\cdot} \rightleftharpoons Cd_{Cd}^{\times} + V_S^{\cdot\cdot} + e'$. With this model, the following expression was derived for $\underset{\sim}{D}(\text{CdS})$:

$$\underset{\sim}{D}(\text{CdS}) = \frac{3}{[\underset{\sim}{V}_S^{\cdot\cdot}]} D_e(\tilde{\text{Cd}}). \qquad (7.71)$$

Referring to previous expressions for the dependence of $D_e(\tilde{\text{Cd}})$ and $[V_S^{\cdot\cdot}]$

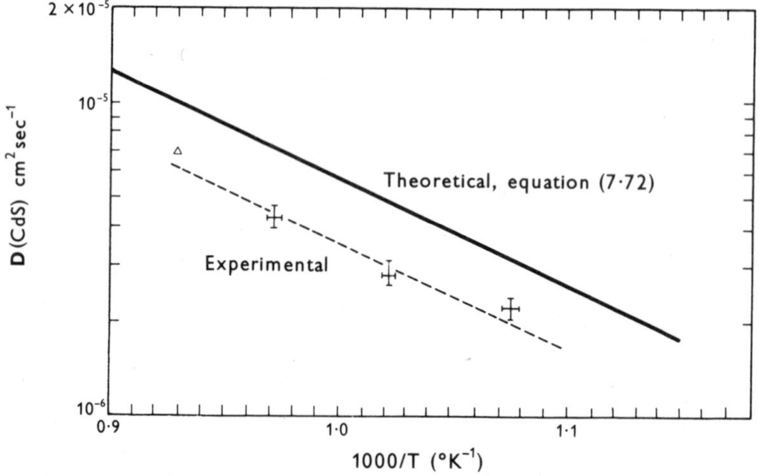

Figure 7.20. The dependence of D(CdS) upon temperature. ⊹ Kumar and Kröger (1971) [109]. △ Shaw and Whelan (1969) [102]. After Kumar and Kröger (1971) [109].

upon T and p_{Cd}, the following expression for $\underset{\sim}{D}(CdS)$ was calculated:

$$\underset{\sim}{D}(CdS) = 7.26 \times 10^{-2} p_{Cd}^{1/3} \exp[-0.68/kT] \text{ cm}^2 \text{ s}^{-1} \qquad (7.72)$$

which was in reasonable agreement with the observed values (figure 7.20).

7.4.3 Diffusion in CdSe

Diffusion of Cd in CdSe

The self-diffusion of Cd-109 in single crystals of CdSe was investigated by Borsenberger, Stevenson and Burmeister (1967) [111] using essentially the same experimental technique used by Henneberg and Stevenson (1970) [85] and described in section 7.3.3. Values of $D_e(\tilde{Cd})$ were determined for Cd-saturated crystals between 550°C and 878°C. $D_e(\tilde{Cd}) = 9.2 \times 10^{-2} \exp[-1.68/kT] \text{ cm}^2 \text{ s}^{-1}$, and for Se-saturated crystals between 760°C and 939°C, $D_e(\tilde{Cd}) = 0.33 \exp[-2.37/kT] \text{ cm}^2 \text{ s}^{-1}$. Employing the values of p_{Cd} corresponding to the limits of stability of the compound, the results were summarized in the form:

$$D_e(\tilde{Cd}) = 6.02 \times 10^{-5} \exp[-0.99/kT] p_{Cd}^{0.46} \text{ cm}^2 \text{ s}^{-1} \text{ atm}^{-0.46}.$$

The pressure dependence was confirmed by measurement of a diffusion isotherm at 798°C for five values of p_{Cd}. The data were represented as $D_e(\tilde{Cd}) = 8.05 \times 10^{-5} p_{Cd}^{0.48} \text{cm}^2 \text{ s}^{-1} \text{atm}^{-0.48}$. An exponent of 0.5 for p_{Cd} is expected if a singly ionized cadmium interstitial, $Cd_i^{\cdot}$, is the dominant mobile defect and if the charge neutrality condition for the crystal is given by $[Cd_i^{\cdot}] = n$. The agreement between the predicted and observed exponents is well within experimental uncertainty and was proposed as confirming evidence for $Cd_i^{\cdot}$ as the dominant mobile defect in CdSe on the Cd sublattice.

In the diffusion study of Borsenberger, Stevenson and Burmeister (1967) [111], the Hall effect and electrical conductivity were measured on samples taken from the same boules as the diffusion samples. Samples were annealed at elevated temperatures under varying p_{Cd}, quenched to room temperature and the electron concentration measured. The defect model proposed from the diffusion results, $[Cd_i^{\cdot}] = n$, predicts that $n = Kp_{Cd}^{1/2}$, whereas the observed electron concentration was virtually independent of p_{Cd} on the Cd-rich side of the phase field, and decreased abruptly as Se-saturation was approached, as shown in figure 7.21. The authors attributed the observed discrepancy to the failure in preserving the high temperature defect structure during the quenching operation; phenomena, such as precipitation, association or out-diffusion of defects could lower the concentration of native donors. In order to observe the electrical properties at elevated temperatures, Smith (1969) [112] and

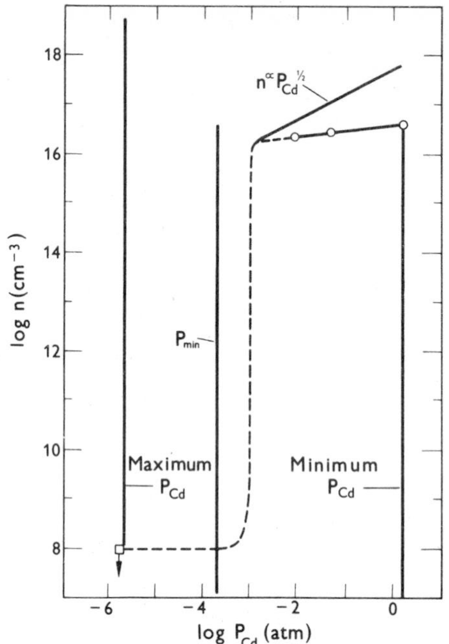

Figure 7.21. The dependence of the room temperature electron density in CdSe upon annealing conditions. Samples were annealed at 798° C under varying p_{Cd} and rapidly cooled to room temperature. After Borsenberger, Stevenson and Burmeister (1967) [111].

Callister, Varotto and Stevenson (CVS) (1970) [113] measured the high temperature conductivity and Hall coefficient of CdSe as a function of p_{Cd}. The carrier density may be expressed as $n \propto p_{Cd}^{\gamma}$. In both studies γ was found to be close to 1/3 at elevated temperatures, ~750° C and higher. This implies a doubly ionized Cd interstitial or Se vacancy, in contrast to the singly ionized Cd interstitial proposed from the measurements of $D_e(\widetilde{Cd})$. It was also observed by CVS that γ decreased with temperature for temperatures below 750° C, to values as low as 0.08, and this behavior was attributed to residual donor impurities. This assumption was further corroborated with measurements on intentionally doped samples.

A tentative defect equilibrium diagram was prepared by Callister *et al.* to explain the high temperature electrical properties of CdSe, which is given in figure 7.22. These properties are well explained by the diagram and also the behavior of $D_e(\widetilde{Se})$ discussed below; however, there is no Cd defect which varies as $p_{Cd}^{1/2}$ to explain the dependence of $D_e(\widetilde{Cd})$ observed by Borsenberger *et al.* In addition, it would be unreasonable to expect a

Cd defect to have a higher concentration-mobility product than $Cd_i^{\cdot\cdot}$, considering its proposed high concentration and probable high mobility by virtue of size. This apparent discrepancy remains unresolved.

Diffusion of Se in CdSe

Woodbury and Hall (1966 and 1967) [114, 90] investigated the self-diffusion of Se-75 in CdSe as a function of component pressure for the temperature range 700-1000° C. The value of $D_e(\widetilde{Se})$ showed a linear decrease with p_{Cd} over the major portion of the phase field, cf. figure 7.23, and was represented by: $D_e(\widetilde{Se}) = 1.3 \times 10^5 \exp[-4.43/kT] p_{Cd}^{-1}$ $cm^2\ s^{-1}$ atm. The diffusion coefficients for Se-saturation and the congruent subliming composition were, respectively, $D_e(\widetilde{Se})_{Se\ SAT} = 2.6 \times 10^{-3} \exp[-1.55/kT]$ and $D_e(\widetilde{Se})_{congr.\ sub.\ comp.} = 2.2 \times 10^{-2} \exp[-2.2/kT]$. The linear pressure behavior extended from the Se-saturation boundary beyond the minimum total pressure line (congruent sublimation composition); as p_{Cd} approached Cd-saturation, $D_e(\widetilde{Se})$ was observed to increase with p_{Cd}. The dominant mobile defect proposed was a neutral Se interstitial for the Se side of the

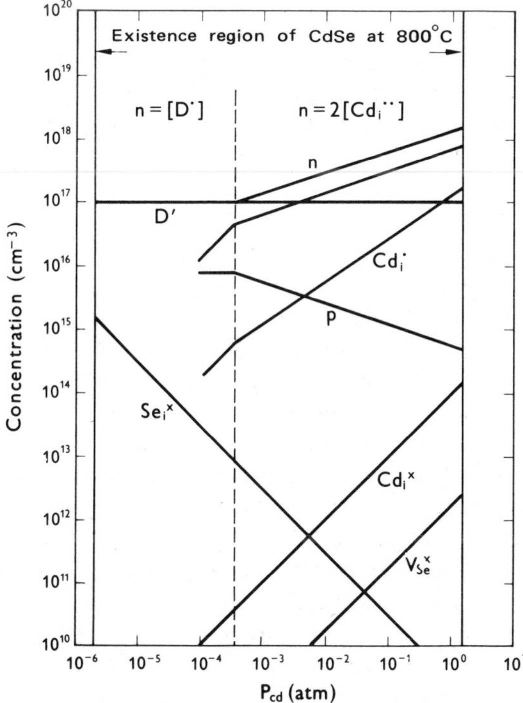

Figure 7.22. Tentative defect concentration isotherm for CdSe at 800° C. After Callister, Varotto and Stevenson (1970) [113].

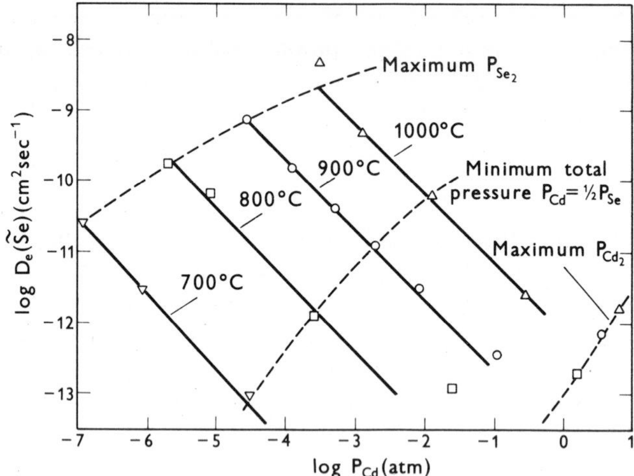

Figure 7.23. Diffusion isotherms for $D_e(\widetilde{\mathrm{Se}}$ in CdSe). After Woodbury and Hall, (1966) [90, 114].

phase diagram, with a vacancy mechanism possible as maximum p_{Cd} was approached. Further evidence for a neutral species was the observation that $D_e(\widetilde{\mathrm{Se}})$ did not vary with donor concentration; doping with up to 10^{19} cm^{-3} Cl or 7 x 10^{19} cm^{-3}. In donors produced no detectable change in $D_e(\widetilde{\mathrm{Se}})$.

Interdiffusion in CdSe

Interdiffusion in CdSe single crystals was studied by Shiozawa and Jost (1969) [115] in the temperature range 900-1100°C by measuring the electrical conductivity profile of diffusion annealed samples. Single crystal bars, 2 x 2 x 13 cm^3, were diffusion annealed in quartz ampoules with an Ar + Cd atmosphere. After quenching, the average conductivity of the bar was measured using a four-probe technique and the conductivity remeasured after successively removing the surface by hand lapping until a constant conductivity was obtained, denoted the residual conductivity, σ_r. The measured conductivity minus the residual conductivity, $\bar{\sigma}(s) - \sigma_r$, was attributed to the uptake of Cd by diffusion. Assuming that the excess Cd is fully ionized, values for $\underset{\sim}{D}$(CdSe) were obtained from the following expression, derived for the existing boundary conditions:

$$(\bar{\sigma}(s) - \sigma_r) = \frac{4\sigma_0\sqrt{\underset{\sim}{D}t}}{w_0^2}\int_x^\infty \operatorname{erfc}(x)dx, \qquad (7.73)$$

where s is the surface thickness removed, $x = s/\sqrt{\underset{\sim}{D}t}$, w_0^2 is the initial cross-sectional area of the rectangular bar and σ_0 is the conductivity at

$x = 0$. A typical profile, figure 7.24, shows a change in the homogeneous conductivity in addition to the profile near the surface. The diffusion coefficients obtained from the latter profile were represented by the expression:

$$\underset{\sim}{D}(\text{CdSe}) = 3 \times 10^6 \exp[-4.0/kT] \text{ cm}^2 \text{ s}^{-1}, \qquad (7.74)$$

for the diffusion of Cd into CdSe. The values of $\underset{\sim}{D}$(CdSe) are remarkably small for interdiffusion coefficients and are incompatible with the reported values for the self-diffusion coefficients for this compound; values of $\underset{\sim}{D}$(CdSe) are actually smaller than $D_e(\tilde{\text{Cd}})$.

Another observation of considerable interest was of the rapid bulk change when CdSe, initially in a Cd-rich state, was annealed in Se; uniform high resistivity crystals are obtained from low resistivity crystals within 90 min at 1045° C, with the exception of a shallow region near the surface. It would appear that $\underset{\sim}{D}$(CdSe) (Se into Cd-saturated CdSe) $\gg$ $\underset{\sim}{D}$(CdSe) (Cd into Se-saturated CdSe). The authors proposed a tentative mechanism for these observations assuming parallel diffusion of Cd_i and V_{Se} species, with the former diffusing much more slowly than the latter and with the precipitation of V_{Se} upon cooling to room temperatures. For diffusion in

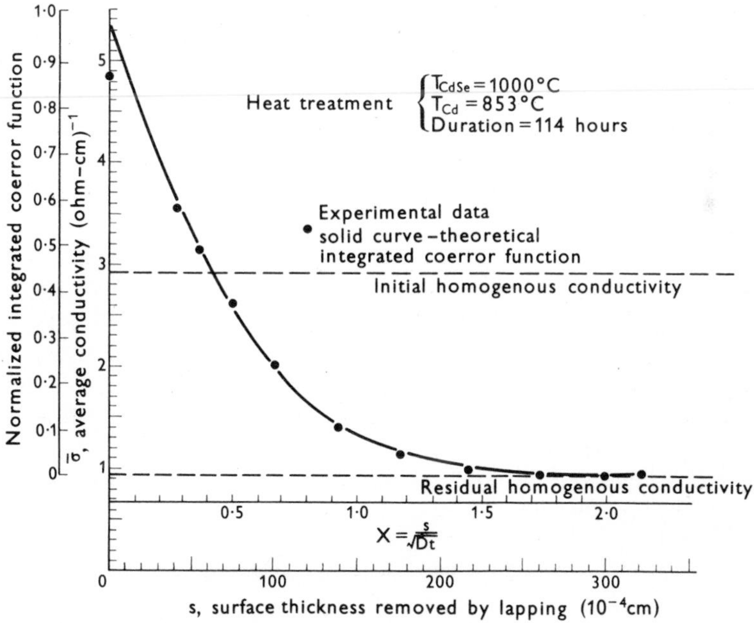

Figure 7.24. Typical conductivity profile in CdSe, in the evaluation of $\tilde{D}$(CdSe). After Shiozawa and Jost (1965) [115].

Cd-rich atmospheres, a uniform V_{Se} concentration is rapidly established in the crystal with a Cd_i profile near the surface. Upon quenching, the V_{Se} concentration is rapidly reduced by precipitation, whereas the resulting donor profile represents the Cd_i profile. For diffusion in Se-rich atmospheres, the V_{Se} concentration is quickly reduced during the diffusion anneal; however, as p_{Se_2} is reduced during the quenching process, Se vacancies diffuse into the crystal a short distance giving rise to the observed surface layer of high donor concentration. This explanation is consistent with the observations; however, for parallel and independent diffusion of Cd_i and V_{Se}, it is necessary that equilibrium between these species, e.g., $Cd_i \rightleftharpoons Cd_{Cd} + V_{Se}$, is attained in times that are large compared to the diffusion anneal. Another possibility might be the parallel diffusion of a native defect and an unknown impurity defect, with the former diffusing much more rapidly than the latter.

7.4.4 Diffusion in CdTe

Diffusion of Cd in CdTe

De Nobel (1959) [22] performed some 'preliminary experiments' on the self-diffusion of Cd in CdTe and concluded that Frenkel disorder, involving interstitial Cd and Cd vacancies, was dominant on the Cd sublattice. No details of the diffusion studies were given, however.

Whelan and Shaw (1967) [116] studied the self-diffusion of Cd-115m in single crystals of CdTe over the temperature range 700-1000°C and at various Cd pressures. The Cd-115 profile was determined by mechanically sectioning the diffusion sample and the diffusion profiles consisted of two distinct regions. The region closest to the crystal surface had a profile that could be represented by a complementary error function and yielded diffusion coefficients that were independent of time; these values were regarded as volume self-diffusion coefficients for Cd, $D_e(\tilde{Cd})$. At greater penetration depths, a second region indicated a faster diffusion process. This was interpreted as diffusion along dislocations and was represented by the expression, $c = c_o \exp(-\alpha x)$, where α decreased with increasing time and p_{Cd}. Positive identification of high diffusivity paths, using techniques such as autoradiography was not made.

The volume self-diffusion coefficient, $D_e(\tilde{Cd})$, was found to be independent of preannealing treatment and also independent of p_{Cd}, as established by isotherms at 800°C and 900°C over a range of p_{Cd} from 1 to 10^3 torr. The temperature dependence was represented by

$$D_e(\tilde{Cd}) = (1.26^{+1.07}_{-0.72}) \exp [(-2.07 \pm 0.08/kT] \text{ cm}^2 \text{ s}^{-1} \qquad (7.75)$$

for the temperature range from 700-1000°C. In discussing their results, the authors suggested the following possibilities to explain the independence of $D_e(\tilde{Cd})$ upon p_{Cd}: a ring mechanism, an exchange mechanism, or a neutral associate of a Cd vacancy and a Cd interstitial. They ruled out a neutral associate involving the Te sublattice based on the observations that Te diffusion is significantly slower and depends on p_{Cd} (Woodbury and Hall, 1967) [90]. They also ruled out any Cd vacancy or interstitial as well as any charged defect, reasoning that the concentrations of these defects would have to vary with p_{Cd}. This need not be true, however, if the appropriate charge neutrality conditions are considered. If ionized Frenkel disorder dominates the charge neutrality condition, for example $[Cd_i^{\cdot}] = [V'_{Cd}]$ and $K_{F'} = [Cd_i^{\cdot}]\,[V'_{Cd}]$, then the concentrations of these defects will depend only upon temperature, i.e., $[Cd_i^{\cdot}] = [V'_{Cd}] = K_{F'}^{1/2}$. Observations were also made on the time required to attain 'electronic equilibrium', i.e., the time necessary to establish a new uniform conductivity after a change in component pressure. This was found to be $\sim 10^3$ shorter than times necessary to homogenize the radiotracer composition. The authors explained the rapid rate of equilibration of the electronic defects by the generation of point defects internally, without mass transport from the external vapor. The crystal interior was presumed to sense the external pressure charges via dislocations and thereby cause point defect generation (annihilation) by equilibration of the jog density with p_{Cd}. It is not clear, however, how the jogs inside the crystal can equilibrate with p_{Cd} outside the crystal without mass transport along dislocations, since equilibrium implies the same value for the chemical potential of both components in the interior as in the vapor phase. This necessarily requires mass transport. The authors also showed a quantitative correlation between the rate of attainment of electronic equilibrium, an estimated dislocation diffusion coefficient and the dislocation density. By way of contrast, however, Zanio (1970) [30] showed that there is no correlation between the interdiffusion coefficient and the dislocation density in CdTe.

A large difference in the rate of attainment of electronic equilibrium and the homogeneous distribution of a tracer atom is not at all anomalous and is easily explained by the expected differences in diffusion behavior with and without a concentration gradient. The rate of attainment of electronic equilibrium for a step jump in component pressure depends upon the chemical diffusion coefficient, since a gradient in concentration is initially established by changing the external component pressure. The tracer self-diffusion coefficient, however, refers to diffusion in the absence of a concentration gradient and may be much smaller than the interdiffusion coefficient. In a binary system, the interdiffusion coefficient is related to the two self-diffusion coefficients and the thermodynamic factor, as given in equation 7.1. Since values for the thermodynamic factor are large for the II-VI compounds (see, for example, Zanio, 1970 [30] for the thermodynamic factor in CdTe) the rapid attainment of electronic equilibrium for a step jump in component pressure is to be expected.

Borsenberger and Stevenson (1968) [117] measured the self-diffusion coefficient in CdTe single crystal slices for the temperature range 510-920°C at Cd and Te saturation and for several intermediate pressures at 775°C. In addition, the influence of donor (Al) and acceptor (Au) solutes on $D_e(\tilde{\mathrm{C}}\mathrm{d})$ was investigated. The profiles showed normal behavior; they were well represented by either the error function solution or the exponential solution to Fick's law, depending on the relevant boundary conditions. The values of $D_e(\tilde{\mathrm{C}}\mathrm{d})$ were independent of p_{Cd}, within experimental error, as determined by the values at $p_{\mathrm{Cd})\max}$ and $p_{\mathrm{Cd})\min}$:

$$D_e(\tilde{\mathrm{C}}\mathrm{d})p_{\mathrm{Cd})\max} = 326 \exp\left[-2.67/kT\right] \mathrm{cm}^2 \mathrm{s}^{-1}; \tag{7.76}$$

$$D_e(\tilde{\mathrm{C}}\mathrm{d})p_{\mathrm{Cd})\min} = 15.8 \exp\left[-2.44/kT\right] \mathrm{cm}^2 \mathrm{s}^{-1}. \tag{7.77}$$

In addition, an isotherm at 775°C showed essentially no change in $D_e(\tilde{\mathrm{C}}\mathrm{d})$ for Cd pressures from 10^{-4} to 1 atm. The lack of dependence of $D_e(\tilde{\mathrm{C}}\mathrm{d})$ upon p_{Cd} agreed with the observation of Whelan and Shaw (1967) [116]; however, there was a difference in the reported values of the activation energy. The possibility of an impurity controlled defect concentration that would be independent of p_{Cd} was investigated by determining the influence of donor (Al) and acceptor (Au) solutes on $D_e(\tilde{\mathrm{C}}\mathrm{d})$. Both types of solutes enhanced $D_e(\tilde{\mathrm{C}}\mathrm{d})$ at lower temperatures which provided evidence that the behavior of the undoped crystals was characteristic of pure crystals. The lack of dependence of $D_e(\tilde{\mathrm{C}}\mathrm{d})$ on component pressure was explained by an ionized Frenkel defect on the Cd sublattice dominating the charge neutrality condition. For example, for a single ionized Frenkel defect

$$\begin{aligned}[V'_{\mathrm{Cd}}] = [\mathrm{Cd}_i^{\cdot}] &= K_a p_{\mathrm{Cd}}^{-1} \exp\left[E_F - E_{1A}/kT\right] \\ &= K_d p_{\mathrm{Cd}} \exp\left[E_{1D} - E_F/kT\right]\end{aligned} \tag{7.78}$$

where E_F, E_{1A}, and E_{1D} are the Fermi energy, the acceptor level of V_{Cd} and donor level of Cd_i, respectively, and K_a and K_d are the mass action constants for the incorporation of the neutral vacancy and interstitial defects, respectively. The above relation clearly shows that E_F may indeed increase with p_{Cd} and still leave $[V'_{\mathrm{Cd}}]$ and $[\mathrm{Cd}_i^{\cdot}]$ constant. A doubly ionized Frenkel defect was also mentioned as a possibility. The enhancement of $D_e(\tilde{\mathrm{C}}\mathrm{d})$ with electrically active impurities was interpreted as evidence for impurity controlled defect concentration imposed by self compensation. For example, with Al doping, charge neutrality may be expressed as

$$[V'_{Cd}] = [Al^{\cdot}] = [Al]_{TOTAL}.$$

Diffusion of Te in CdTe

The dependence of the self-diffusion coefficient of Te in CdTe upon p_{Cd} at 800° C was reported by Woodbury and Hall (1967) [90] using Te-123

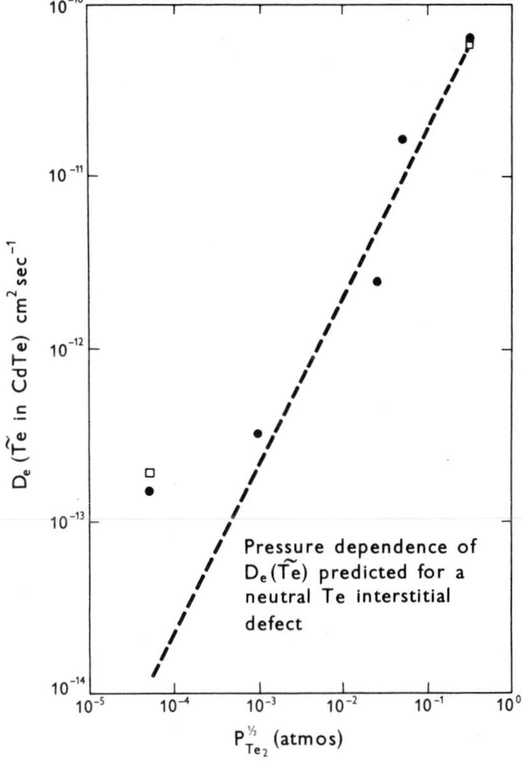

Figure 7.25. Diffusion isotherm for $D_e(\widetilde{Te}$ in CdTe) at 800° C. ● Woodbury and Hall (1967) [90]. □ Borsenberger and Stevenson (1968) [117]. After Borsenberger and Stevenson (1968) [117].

as isotope and an etching technique to determine the diffusion profile. The results, shown in figure 7.25, indicate that $D_e(\widetilde{Te})$ varies as

$$D_e(\widetilde{Te}) = 1.6 \times 10^{-14} p_{Cd}^{-1} \text{ cm}^2 \text{ atm s}^{-1}, \tag{7.79}$$

over most of the composition range, except in the region of $p_{Cd)max}$. This behavior is consistent with a neutral Te interstitial for the major portion of the phase field while another mechanism becomes operative as Cd-saturation is approached.

The behavior of $D_e(\widetilde{\text{Te}})$ was also measured by Borsenberger and Stevenson (1968) [117] who used Te-123 as tracer and determined $D_e(\widetilde{\text{Te}})$ between 510° C and 930° C by a mechanical sectioning technique. The diffusion coefficient was found to increase with p_{Te_2}, as observed by Woodbury and Hall. The agreement between the two investigations was good, as seen by comparing their results in figure 7.25. The dependence of

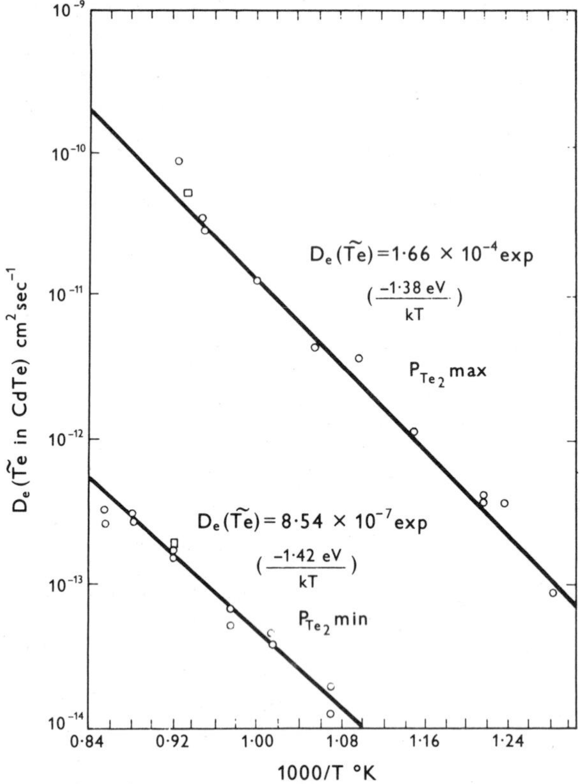

Figure 7.26. Dependence of $D_e(\widetilde{\text{Te}}$ in CdTe) upon temperature and component pressure. ○ Borsenberger and Stevenson (1968) [117]. □ Woodbury and Hall (1967) [90]. After Borsenberger and Stevenson (1968) [117].

$D_e(\widetilde{\text{Te}})$ upon temperature was also investigated for both Cd- and Te-saturated crystals, and results are given in figure 7.26. The diffusion coefficients for the respective conditions were described as

$$D_e(\widetilde{\text{Te}})_{\text{Te SAT}} = 1.66 \times 10^{-4} \exp\left[-1.38/kT\right] \text{ cm}^2 \text{ s}^{-1} \quad (7.80)$$

and

$$D_e(\widetilde{\text{Te}})_{\text{Cd SAT}} = 8.54 \times 10^{-7} \exp\left[-1.42/kT\right] \text{ cm}^2 \text{ s}^{-1}. \quad (7.81)$$

The observations showed that $D_e(\widetilde{\mathrm{Te}})$ increases with p_{Te_2} at all temperatures, and the increase was represented by the relation

$$D_e(\widetilde{\mathrm{Te}}) = D_e(\mathrm{Te}_i^{\times})K_i(T)p_{\mathrm{Te}}^{1/2} \tag{7.82}$$

over most of the region of solid stability of CdTe, where K_i is the mass action constant for the formation of $\mathrm{Te}_i^{\times}$. Near Cd saturation, however, a different diffusion mechanism appears to become operative, possibly via a Te vacancy. This behavior is similar to Se in CdSe reported by Woodbury and Hall (1966) [114].

Interdiffusion in CdTe

It is well known that undoped CdTe may be prepared n- or p-type by annealing in Cd or Te atmospheres, respectively. De Nobel (1959) [22] reported that the room temperature n-type electrical conductivity in annealed and quenched CdTe crystals varied as $p_{\mathrm{Cd}}^{1/2}$, where p_{Cd} is the pressure of Cd at the annealing temperature. From this evidence and his preliminary measurements of $D_e(\widetilde{\mathrm{Cd}})$, he concluded that a singly ionized cadmium interstitial dominates the charge neutrality condition, i.e., $[\mathrm{Cd}_i^{\cdot}] = n = Kp_{\mathrm{Cd}}^{1/2}$. In relating room temperature properties to high temperature equilibria, it is necessary to assume that the high temperature defect structure is preserved during the quenching operation. More recently there have been a number of measurements of the conductivity of CdTe as a function of p_{Cd} at elevated temperatures: Whelan and Shaw, 1968 [118]; Zanio, 1969 [119, 120]; Smith, 1970 [121]; Callister, 1971 [122]; Rud' and Sanin, 1971 [123]. In all cases the conductivity was found to vary as $p_{\mathrm{Cd}}^{1/3}$, and this was interpreted as a doubly ionized cadmium interstitial dominating the charge neutrality condition, $2\,[\mathrm{Cd}_i^{\cdot\cdot}] = n$. Smith [121] and Callister [122] confirmed the assumption that n varies directly with the conductivity from Hall mobility measurements: the Hall mobility was independent of p_{Cd} for Cd rich crystals. At high p_{Te_2}, Zanio [119] reported that the conductivity of undoped CdTe was independent of p_{Cd}, characterizing extrinsic conduction. Rud' and Sanin [123] used a range of crystals having various residual impurity concentrations. Only in the most pure crystals was the conductivity found to vary as $p_{\mathrm{Cd}}^{1/3}$. With decreasing purity the conductivity approached a dependence of $p_{\mathrm{Cd}}^{1/2}$ which was interpreted as a change in the charge neutrality condition from $n = 2\,[\mathrm{Cd}_i^{\cdot\cdot}]$ to $[\mathrm{A}'] = 2\,[\mathrm{Cd}_i^{\cdot\cdot}]$ where A' is a residual foreign acceptor.

These recent results appear to be in conflict with the reported self-diffusion behavior. The lack of dependence of $D_e(\widetilde{\mathrm{Cd}})$ upon p_{Cd} is consistent with ionized thermal disorder dominating the charge neutrality condition. In addition, the fact that $D_e(\widetilde{\mathrm{Cd}}) \gg D_e(\widetilde{\mathrm{Te}})$ and that $D_e(\widetilde{\mathrm{Te}})$ does depend on p_{Cd} suggests that the dominant disorder may involve the Cd sublattice only. There is not a simple defect model which explains both these results and the high temperature conductivity. For example, the choice $[Cd_i^{\cdot}] = [V_{Cd}']$ or $[Cd_i^{\cdot\cdot}] = [V_{Cd}'']$ leads to the electron concentration increasing as p_{Cd} or $p_{Cd}^{1/2}$, respectively, in contrast to the observed $p_{Cd}^{1/3}$.

Additional information in this system is provided by observations made on relaxation times for changes in conductivity when a step jump in pressure is made (Zanio, 1970 [30]; Whelan and Shaw, 1968 [116]). Zanio (1970) [30] obtained data for temperatures of 600°, 700° and 800° C over most of the existence range of the compound and analyzed these data to obtain the interdiffusion coefficient,

$$\underset{\sim}{D}(\text{CdTe}) = 4 \exp\,[(-1.15 \pm 0.1)/kT]\ \text{cm}^2\ \text{s}^{-1}. \qquad (7.83)$$

A calculated value of $\underset{\sim}{D}$(CdTe) was made at 700° C using equation 7.1 and the self-diffusion measurements of Borsenberger and Stevenson (1968) [117]. Values for the thermodynamic factor were taken from Zanio's measurements of $\sigma(p_{Cd})$, by extrapolating mobility values for CdTe to obtain $n = f(p_{Cd})$, assuming $2[Cd_i^{\cdot\cdot}] = n$ and assuming that all the excess Cd was in the form of $Cd_i^{\cdot\cdot}$ at 700° C. Values of the thermodynamic factor varied from 4×10^5 at the Cd-rich boundary to 3×10^6 at $p_{Cd} = 10^{-3}$ atm. At the latter point, the calculated $\underset{\sim}{D}$(CdTe) of 1×10^{-5} cm² s⁻¹ compared favorably with the measured value of 5×10^{-6} cm² s⁻¹. Since the thermodynamic factor calculated by Zanio decreased with increasing p_{Cd}, he surmised that the independence of $\underset{\sim}{D}$(CdTe) upon p_{Cd} could only be explained if $D_e(\widetilde{Cd})$ increases with p_{Cd}, contrary to the observed behavior reported by Borsenberger and Stevenson (1968) [117] and Whelan and Shaw (1968) [116]. On this basis, Zanio proposed that $D_e(\widetilde{Cd})$ must increase with increasing p_{Cd}, characteristic of an interstitial mechanism.

Relaxation measurements were also made by Whelan and Shaw (1968) [118], who obtained much shorter relaxation times and consequently much larger values for $\underset{\sim}{D}$(CdTe); at 700° C, $\underset{\sim}{D}$(Cd) was ~40 times larger than the values of Zanio. This disparity is another example of the difference in the relaxation times that have been observed if different techniques are used to control the component pressure. The refluxing technique has lead to more rapid rates of equilibration than a two-temperature zone system. One is tempted to attribute this difference to a longer apparatus time constant for the latter technique; however, Zanio reported that his $\underset{\sim}{D}$(CdTe) values were independent of sample dimensions which implies that the time constant for the apparatus does not explain the discrepancy.

7.5 DIFFUSION IN THE LEAD CHALCOGENIDES

7.5.1 Diffusion in PbO

Although lead forms four common oxides, the monoxide is the stable oxide at elevated temperatures in air and consequently has been the most studied of the lead oxides. The system is complicated by a transition from 'red lead' (low temperature modification) to 'yellow lead' at $\sim 488°C$. There are a few studies of diffusion in this system which point out rather interesting behavior; the atomic mobility on the oxide sublattice is remarkably high, leading to appreciable electrolytic conductivity via oxide ions.

Diffusion of Pb in PbO

The self-diffusion of Pb in PbO was studied by Lindner (1952) [124] in the temperature range 400-700°C using Pb-212 (Th-B) as a radioactive indicator. The 'contact method' was found suitable for the range of diffusivity values encountered; compacted pellets of radioactive and non-radioactive PbO were pressed together and the measured isotopic exchange related to the diffusion coefficient. The diffusion annealing was carried out in a nitrogen atmosphere or in a vacuum to avoid oxidation to Pb_3O_4. No mention was made of preannealing treatments or the possible influence of oxygen partial pressure on the diffusion coefficient. The results for $D_e(\widetilde{Pb})$ in yellow PbO were expressed as

$$D_e(\widetilde{Pb}) = 10^5 \exp[-2.8/kT] \text{ cm}^2 \text{ s}^{-1} \tag{7.84}$$

for the tempeature range 488-600°C, with an estimated uncertainty in the activation energy of ±10% and a much greater uncertainty in the preexponential. Values below the yellow to red PbO transition temperature of 488°C did not lead to unambiguous evidence regarding a change in the activation energy at the transition temperature.

Lindner and Terem (1954) [125] extended the measurements of $D_e(\widetilde{Pb})$ to lower temperatures using the 'recoil method' (Lindner, 1949) [126] on samples prepared by the oxidation of Pb wafers. Satisfactory agreement with the previous study (Lindner, 1952) [124] was obtained for yellow lead, and the activation energy for $D_e(\widetilde{Pb})$ in red lead was reported to be 0.58 eV, significantly lower than the 2.86 eV value for $D_e(\widetilde{Pb})$ in yellow lead. The kinetics of oxidation were also studied in the same paper and a comparison of $D_e(\widetilde{Pb})$ and the oxidation constant were cited as evidence that the oxidation rate is not controlled by the diffusion of lead ions in PbO. This suggests a high value for the oxide ion mobility.

Measurements of $D_e(\widetilde{Pb})$ were also reported by Das Gupta, Sitharamarao and Palkar (1965) [127] between 600°C and 670°C. A surface decrease method was used in which Pb-210 was employed as the tracer atom, but because of the low energy of the emitted β for Pb-210, the radiation of the Bi-210 daughter was counted. Only three individual measurements were reported and no discussion was given of the sample preparation. The limited results showed good agreement with the values given by Lindner, with a reported activation energy of 2.78 eV.

Diffusion of O in PbO

The self-diffusion of O in PbO was investigated by Thompson and Strong (1963) [128] at temperatures between 500-650°C by measuring the O-18 exchange with mass spectrometric techniques. Small spheres of PbO (90-160 μm) were prepared by dropping PbO powder through a glass blowing torch and subsequently prediffusion annealing at the diffusion temperature for 24 h in air. X-ray analysis confirmed the presence of the yellow modification. The diffusion coefficient was measured by observing the rate of isotopic exchange between normal crystals and an O-18 enriched gaseous ambient. Seven diffusion runs between 500-650°C produced values of $D_e(\tilde{O})$, which are represented by the expression

$$D_e(\tilde{O}) = 5.39 \times 10^{-5} \exp\left[-0.97/kT\right] \text{ cm}^2 \text{ s}^{-1}, \tag{7.85}$$

with probable errors of $\pm 0.45 \times 10^{-5}$ cm^2 s^{-1} and ± 0.04 eV for D_o and Q respectively. The most striking aspect of the result was the large value of $D_e(\tilde{O})$, compared to $D_e(\widetilde{Pb})$ in PbO; at least an order of magnitude larger than the values for $D_e(\widetilde{Pb})$ found by Lindner (1952) [124]. Assuming that the oxidation of lead is controlled by anion diffusion, parabolic rate constants were derived from the measured values of $D_e(\tilde{O})$. A comparison of these calculated values with experimentally observed rate constants showed good agreement, thus confirming the belief that the oxidation of Pb is controlled by the diffusion of O in PbO. The authors speculate, based on the observed activation energy, that O diffuses by an interstitial mechanism, but point to a need to study the dependence of $D_e(\tilde{O})$ on p_{O_2} to confirm this.

Heyne, Beekmans and de Beer (1972) [15] have also measured the self-diffusion of oxygen in PbO at a temperature of 514°C by following O-18 exchange with a mass spectrometer. The influence of sample type, oxygen partial pressure and doping concentration was studied. Sintered pellets, 3 mm thick x 8 mm in diameter, were densified by hydrostatic pressing at 30 kilobar while single crystals were prepared either: by pulling from a PbO melt in a platinum crucible; by slow cooling of the melt; or by melting the end of a vibrating rod. Bismuth and potassium doping was carried out with the addition of either Bi_2O_3 or K_2CO_3 (600 ppm), and it was reported that only the Bi-doped crystals retained the yellow modification to room temperature. The authors reported that the surface exchange rate was too slow to allow an accurate determination of $D_e(\tilde{O})$, in contrast to the experience of Thompson and Strong (1963) [128]. Rapid exchange between CO_2 and PbO, however, led to the utilization of enriched CO_2 and natural O_2 as isotopic source, and $D_e(\tilde{O})$ as well as the exchange reaction constant were evaluated from an appropriate analysis of mass spectrometric measurements. The values for the K-doped and pure sintered pellets showed a fair amount of scatter which was attributed to differences in the effective surface area of the samples. The authors noted the following trends in $D_e(\tilde{O})$: (1) an enhancement with Bi doping, possibly by as much as an order of magnitude; (2) a somewhat lower value for single crystals than for pellets; and (3) no significant influence of p_{O_2}, as judged by comparing results in p_{O_2} = 10 torr with those in pure CO_2. The value of $D_e(\tilde{O})$ in single crystals, $D_e(\tilde{O}) = 4 \times 10^{-11}$ cm^2 s^{-1} at 514°C, compared well with the value obtained by Thompson and Strong at the same temperature, $\sim 3 \times 10^{-11}$ cm^2 s^{-1}

Heyne, Beekmans and De Beer (1972) [15] also made a study of the electronic and electrolytic conductivity of PbO as a function of p_{O_2}, doping level and temperature using solid state electrochemical techniques and observed an appreciable electrolytic contribution to the total conductivity; the ionic transport number varied from ~0.01 to ~0.50. Although the total conductivity was observed to depend strongly upon doping content and p_{O_2}, the ionic conductivity was independent of p_{O_2} and increased with Bi doping. A comparison of the ionic conductivity, θ_{ion}, and $D_e(\tilde{O})$ was made using the Einstein relation, $D_e(\tilde{O}) = 5 \times 10^{-9}\ \sigma_{ion}\ T$. This relation would be valid if σ_{ion} is due entirely to the transport of oxygen species only. It was found that the observed value of $D_e(\tilde{O})$ was ~100 times greater than that calculated from σ_{ion}. This would suggest that a neutral form of oxygen may participate in the self-diffusion process, which is contrary to the observed independence of $D_e(\tilde{O})$ upon p_{O_2}. The authors comment rather extensively on the difficulties in explaining many observations in PbO using present defect models, particularly the relative variation of electronic and electrolytic conductivity with doping concentration and p_{O_2}.

7.5.2 Diffusion in PbS

Diffusion of Pb in PbS

Anderson and Richards (1946) [129] studied the self-diffusion of Pb-210 in pressed compacts of PbS for the temperature range 460-770°C using a mechanical sectioning technique. The authors pointed out the limitations of this study, which was one of the earliest self-diffusion studies in this type of system. They observed a dependence of $D_e(\tilde{Pb})$ on depth which could be correlated with the composition of the PbS; increasing the S concentration increased $D_e(\tilde{Pb})$, whereas increasing the Pb concentration decreased $D_e(\tilde{Pb})$. They verified this trend by diffusion annealing samples at 580°C in a vacuum and in sulfur vapor (p_{S_2} ~ 60-80 torr) and observing values for $D_e(\widetilde{Pb})$ of 7.9×10^{-12} cm^2 s^{-1} and 2.3×10^{-11} cm^2 s^{-1}, respectively. From the observed enhancement of $D_e(\widetilde{Pb})$ with increasing S activity, they inferred that Schottky disorder predominates in PbS. As pointed out by Simkovich and Wagner (1963) [130], this increase might also occur for a Frenkel defect on the S-side of the phase field, with the possibility of an increase in $D_e(\widetilde{Pb})$ with decreasing S-activity in the Pb-excess region. The study of Anderson and Richards clearly established a dependence of $D_e(\widetilde{Pb})$ upon component activity in the S-rich region of the phase field. However, as pointed out by the authors, the observation that $D_e(\widetilde{Pb})$ was a function of depth of the profile implied that isoconcentration conditions were not established, as required for true self-diffusion studies. In addition, since the samples were pressed powders, it is possible that contributions from grain boundary diffusion might be significant.

Simkovich and Wagner (1963) [130] studied the self-diffusion of Pb-210 in PbS single crystals in the temperature range 500-800° C. The defect concentrations were controlled by fixing the sulfur activity (using different H_2/H_2S gas mixtures, a two zone technique, or an inert atmosphere), and by additions of acceptor (Ag) or donor (Bi) solutes. The diffusion profiles were determined by the residual activity technique. Considerable difficulty was encountered in obtaining diffusion samples that were free of high diffusivity paths, such as internal cracks and dislocations; such defects were monitored with autoradiography. In crystals free of such defects, they observed normal profiles as predicted from Fick's law. For crystals that were annealed in an inert atmosphere (near the stoichiometric composition), an enhancement in $D_e(\widetilde{Pb})$ for both donor and acceptor doped crystals was observed. Referring to relevant equilibrium constraints for Frenkel disorder for singly ionized defects,

$$h^{\cdot} + \mathrm{Pb_{Pb}} + \mathrm{S_S} \rightleftharpoons \mathrm{Pb}_i^{\cdot} + \tfrac{1}{2}\mathrm{S_2}(G); \qquad K = [\mathrm{Pb}_i^{\cdot}]p_{\mathrm{S_2}}^{1/2}/p \tag{7.86}$$

$$e' + \tfrac{1}{2}\mathrm{S_2}(G) \rightleftharpoons \mathrm{S_S} + V'_{\mathrm{Pb}}; \qquad K = [V'_{\mathrm{Pb}}]/np_{\mathrm{S_2}}^{1/2}, \tag{7.87}$$

the enhancement in the two cases was attributed to an increase in $[V'_{Pb}]$ by the increase in n due to the donor Bi or the increase in $[Pb_i^{\cdot}]$ from the increase in p contributed by the acceptor Ag. Additional evidence was provided by the observation that the magnitude of $D_e(\widetilde{Pb})$ in undoped crystals with either excess Pb or S was higher than in crystals near the stoichiometric composition. The values of $D_e(\widetilde{Pb})$ were two orders of magnitude less than those of Anderson and Richards (1946) [129]; this difference was attributed to diffusion short circuits in the polycrystalline samples used in the earlier study.

The study of Simkovich and Wagner was consistent with the qualitative interpretation in terms of Frenkel disorder, but the composition range studied was insufficient to provide a basis for determining the degree of ionization of defects. There is evidence reported by Bloem (1955) [131] that the defects are singly ionized. To provide diffusion data over a wider range of compositions, Seltzer and Wagner (1963) [132] determined the self-diffusion of Pb-210 in PbS single crystals, undoped and doped with Bi, over the entire range of existence at 700° C using the same technique described for the previous work. The results, represented as log $D_e(\widetilde{\text{Pb}})$ versus log p_{S_2}, are shown in figures 7.27 and 7.28 for undoped PbS and

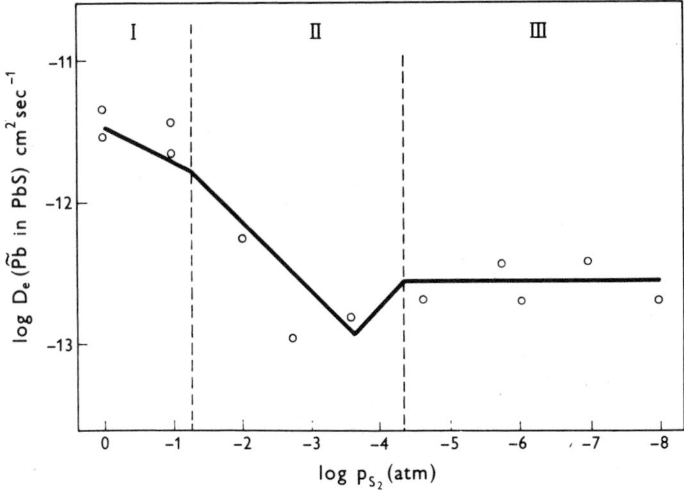

Figure 7.27. Diffusion isotherm for $D_e(\widetilde{\text{Pb}}$ in PbS) at 700° C. After Seltzer and Wagner (1963) [132].

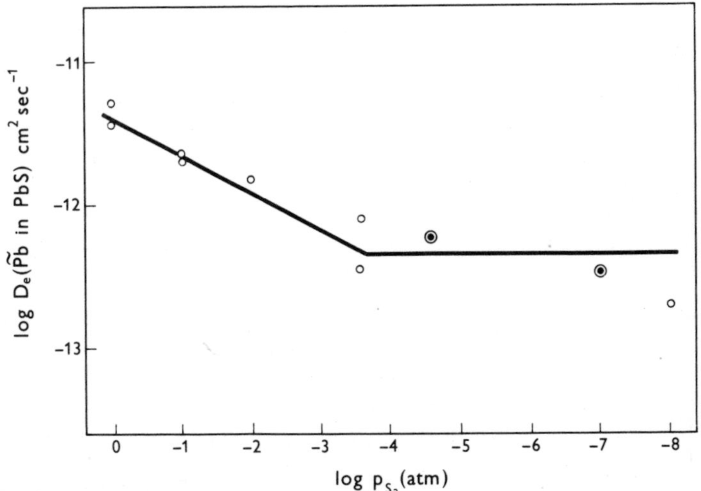

Figure 7.28. Diffusion isotherm for $D_e(\widetilde{\text{Pb}}$ in PbS -2×10^{19}Bi) at 700° C. After Seltzer and Wagner (1963) [132].

PbS doped with 1/20 mole % Bi_2S_3. For the undoped PbS, there are four distinct regions in the log $D_e(\widetilde{Pb})$ versus log p_{S_2} isotherm, with slopes of −¼, −½, +½ and 0, figure 7.27. In analysing these results, the authors constructed a defect equilibrium diagram, figure 7.29, based on the work of Bloem (1956) [131] on controlled conductivity in PbS crystals and assuming Frenkel disorder based on the observations of Simkovich and

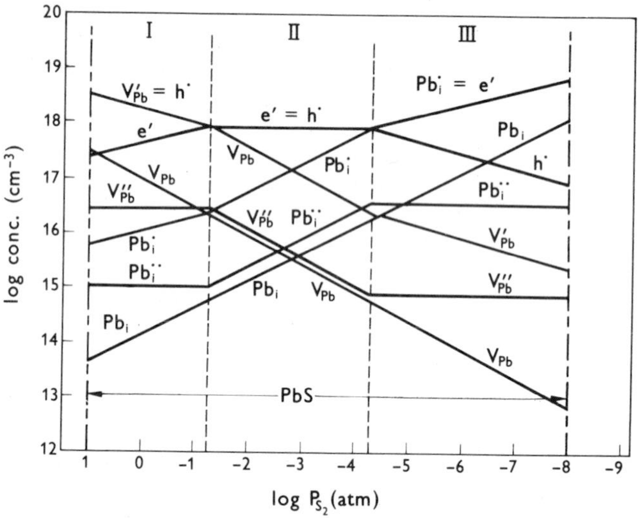

Figure 7.29. Defect concentration isotherm for undoped PbS at 700°C. e′ and h˙ denote free electrons and free holes respectively. After Seltzer and Wagner (1963) [132].

Wagner (1963) [130]. The proposed dominant mobile defects in the four respective regions were V'_{Pb} in the first two regions and $Pb_i^{\cdot\cdot}$ in the last two. It was pointed out that $[Pb_i^{\cdot\cdot}]$ is smaller than either $[Pb_i^{\times}]$ or $[Pb_i^{\cdot}]$ and would have to have a much higher mobility than the other two defects. In addition, if the second ionization level of Pb_i were smaller, the curve for log $[Pb_i^{\cdot\cdot}]$ could be much closer to log $[Pb_i^{\cdot}]$. Another explanation proposed sulfur vacancies as the dominant donors, with Pb interstitials existing at a lower, but substantial concentration.

The dependence of $D_e(\widetilde{Pb})$ upon p_{S_2} for Bi-doped samples, figure 7.28, was unexpected. It was presumed that Bi would reside on Pb sites, and ionize to produce electrons and the ionized species, $Bi_{Pb}^{\cdot}$. The charge neutrality conditions would then become $[Bi_{Pb}^{\cdot}] = [V'_{Pb}]$ for high p_{S_2} and $[Bi_{Pb}^{\cdot}] = n$ for lower p_{S_2}. For a relevant defect equilibrium isotherm, figure 7.30, one would predict that the dominant Pb defect, $[V'_{Pb}]$, would be independent of p_{S_2} at high p_{S_2}, and decrease with the p_{S_2} at lower

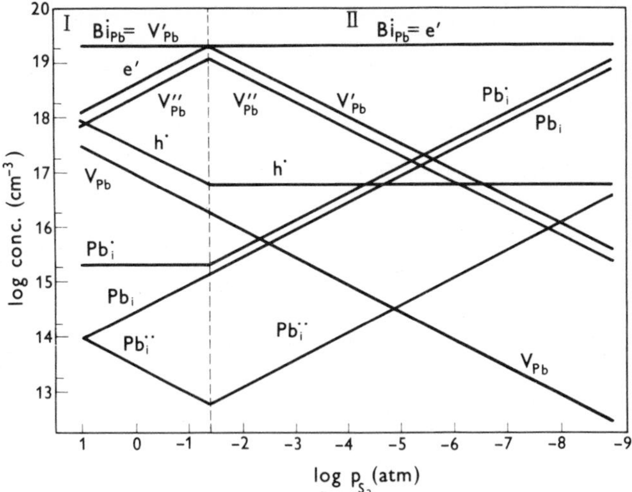

Figure 7.30. Defect concentration isotherm for PbS with 2 x 10^{19}Bi atoms per cm^3 at 700°C. e′ and h˙ denote free electrons and free holes respectively. After Seltzer and Wagner (1963) [132].

p_{S_2}. In addition, the magnitude of $[V'_{Pb}]$ would be enhanced at all values of p_{S_2}. To explain these discrepancies, the authors suggested the possibility of significant concentrations of the associate $(Bi^{\cdot}_{Pb} V'_{Pb})$, which would lower the concentration of V'_{Pb}. Furthermore, it was suggested that a significant portion of Bi_{Pb} is unionized which would allow $[V'_{Pb}]$ to decrease with p_{S_2} at high p_{S_2}. In addition, they proposed $V^{\cdot}_{S}$ as the dominant donor, instead of $Pb^{\cdot}_{i}$, and the formation of a $(V'_{Pb} V^{\cdot}_{S})$ associate at lower values of p_{S_2}, whose concentration would be independent of pressure.

Further information on the defect structure of PbS was contributed by Zanio and Wagner (1968) [133]. The simultaneous diffusion of Pb-210 and Bi-210 into single crystals of PbS was measured at 700°C to determine the tendency for the formation of $(Bi^{\cdot}_{Pb} V'_{Pb})$ associates. If an appreciable concentration of these associates exist, then the self-diffusion coefficient of Bi would be expected to be much greater than that for Pb. At a temperature of 700°C, it was reported that $D_e(\widetilde{Pb}) \approx D_e(\widetilde{Bi})$ for both undoped PbS and for PbS doped with 4 x 10^{18} cm^{-3} Bi and for various sulfur pressures.†

This suggests that there is a negligible concentration of $(Bi^{\cdot}_{Pb} V'_{Pb})$

† An example of a strong interaction of simultaneous diffusing species is reported by George and Wagner (1969) [134] for the simultaneous diffusion of Ni and Cl into PbTe.

associates, and that diffusion proceeds via the V'_{Pb} defect for $\log p_{S_2} > -3.9$. Zanio and Wagner also established that a Bi doping level of 4×10^{18} cm^{-3} did not alter $D_e(\widetilde{Pb})$, whereas 4×10^{19} cm^{-3} increased $D_e(\widetilde{Pb})$ (in the existence range $-6.7 < \log p_{S_2} < 0$). This implied a much higher native defect density than had been inferred from the room temperature electrical property studies of Bloem (1956) [131] and was taken as evidence that the high temperature defect equilibria is not preserved upon cooling to room temperature. From the doping experiments, they calculated an incorporation constant $K_R = [V'_{Pb}]p/p_{S_2}^{1/2} = 2\text{-}4 \times 10^{39}$ atm$^{-1/2}$ at 700° C.

Zanio and Wagner also commented that the independence of $D_e(\widetilde{Pb})$ upon p_{S_2} for $\log p_{S_2} < -3.9$ may be caused by micro-precipitates that are formed during growth, because of retrograde solubility. They suggest that the precipitates may not redissolve at the annealing temperature, and, therefore the bulk of the crystal does not equilibrate with the low p_{S_2} atmosphere. As a consequence, any given boule may have a fixed but non-equilibrium concentration of defects and $D_e(\widetilde{Pb})$ would be independent of p_{S_2} and could vary from sample to sample. It would be surprising for different samples to show diverse behavior since the precipitates, presumably of lead, should fix the defect concentration of the interior of the crystal corresponding to the lead saturation value for that temperature. It is also surprising that the precipitates would not be removed at a temperature of 700° C, considering the large value of the interdiffusion coefficient in this system, 2×10^{-6} cm^2 s^{-1} at 500°C, as reported by Brebrick and Scanlon (1954) [135]. There are a number of observations in the II-VI compounds showing the removal of precipitates by relatively low temperature anneals in the atmosphere of one of the components. It appears that precipitates play an important and possibly an enigmatic role in the defect chemistry of compound semiconductors.

Diffusion of S in PbS

Seltzer and Wagner (1965) [136] measured the self-diffusion of S-35 in undoped, Bi- and Ag-doped single crystals of PbS for the temperature range 500-750° C and for p_{S_2} ranging from 1 to 10^{-8} atm.

The temperature dependence of $D_e(\widetilde{S})$ was determined for undoped PbS at three compositions—10^{18} cm^{-3} excess S, stoichiometric PbS and 10^{18} cm^{-3} excess Pb—with results shown in figure 7.31. The magnitude of $D_e(\widetilde{S})$ is remarkably large, being comparable to $D_e(\widetilde{Pb})$. An increase in $D_e(\widetilde{S})$ with p_{S_2} was observed at all temperatures with $D_e(\widetilde{S})$ largest for excess S and expressed as

$$D_e(\widetilde{S})_{\text{S EXCESS}} = 4.56 \times 10^{-5} \exp[(-1.22 \pm 0.09)/kT] \text{ cm}^2 \text{ s}^{-1}. \quad (7.88)$$

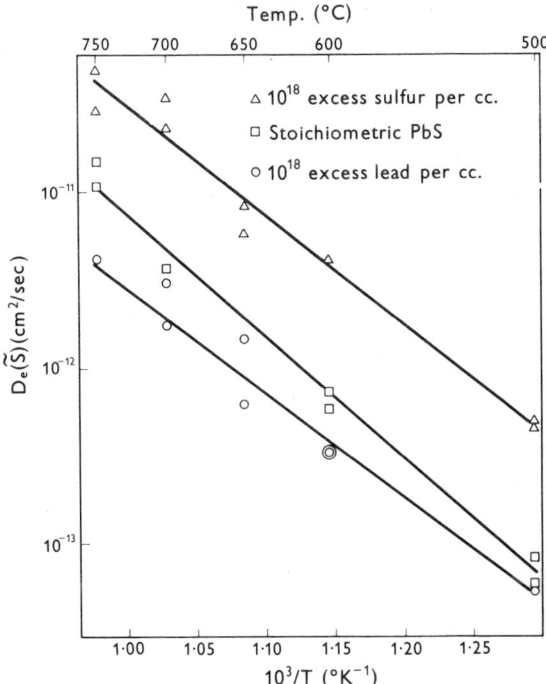

Figure 7.31. Dependence of $D_e(\tilde{\text{S}}$ in PbS) upon temperature and stoichiometry. After Seltzer and Wagner (1965) [136].

An isotherm at 700°C established two regions, $D_e(\tilde{\text{S}}) \propto p_{\text{S}_2}^{0.44}$ for $p_{\text{S}_2} > 3 \times 10^{-4}$ atm and $D_e(\tilde{\text{S}})$ independent of p_{S_2} for $p_{\text{S}_2} < 3 \times 10^{-4}$ atm. Doping with 1/20 and ½ mole % Bi_2S_3 produced a small enhancement in $D_e(\tilde{\text{S}})$ for temperatures above 600°C, whereas doping with the same amount of Ag produced essentially no change in $D_e(\tilde{\text{S}})$. It was concluded that $D_e(\tilde{\text{S}})$ does not proceed via a sulfur vacancy for two reasons: (1) $D_e(\tilde{\text{S}})$ would be expected to decrease with increasing p_{S_2}, whereas the opposite trend is observed; and (2) $D_e(\tilde{\text{S}})$ would be expected to increase with acceptor doping and decrease with donor doping, contrary to the observed behavior. It was proposed by Seltzer and Wagner (1963) [132] that $V_\text{S}^{\cdot}$ is the dominant donor in PbS. If this is indeed the case, the mobility of S vacancies must be much lower than the dominant mobile defect. It was proposed that S diffuses by a neutral interstitial or via vacancies on the lead sublattice in the S-excess region. The latter possibility, however, seems unlikely since an energetically unfavorable antistructure defect would be involved. As a consequence, diffusion of S via Pb vacancies would be expected to be energetically less favorable than diffusing Pb via Pb vacancies and one would expect $D_e(\widetilde{Pb}) \gg D_e(\widetilde{S})$, contrary to the observed behavior. In addition, diffusion of S via V_{Pb} would involve two steps, $S_S + V_{Pb} \rightarrow V_S + S_{Pb}$ and $S_{Pb} + V_S \rightarrow S_S + V_{Pb}$. The rate of the first step will be $\propto [V_{Pb}]$, but the rate of the second step will be $\propto [V_S]$, so that the p_{S_2} exponent would be expected to be less than ½. For the Pb-excess region, diffusion via a defect pair, probably $(V'_{Pb} V^{\cdot}_S)$, was proposed. Doping with Ag or Bi not only had a minor influence on $D_e(\widetilde{S})$, but also did not appear to change the mechanism of diffusion.

Interdiffusion in PbS

The interdiffusion coefficient in PbS was measured by Brebrick and Scanlon (1954) [135] at 550°C using the *p-n* junction method. Two natural PbS crystals, originally *n*-type but with different compositions, were annealed at 550° C in a sulfur atmosphere, with sulfur vapor provided by a reservoir at 123° C. After a one hour anneal, the two crystals were quenched and the location of the *p-n* junction was determined by a rectification test with a fine tungsten probe. A value for $\underset{\sim}{D}$(PbS) of 2×10^{-6} cm^2 s^{-1} was obtained for the temperature of 550°C. It was shown that an oxidized surface layer was an effective diffusion barrier; an oxidation anneal of 15 min at 500° C in air established a surface layer ~0.1 mm thick which prevented a change in carrier type below the layer upon annealing in a sulfur atmosphere.

7.5.3 Diffusion in PbSe

Diffusion of Pb in PbSe

The self-diffusion of Pb-210 in single crystals of PbSe in the temperature range 400-800°C was described by Seltzer and Wagner (1962) [137]. The diffusion samples were as-grown crystals of undoped PbSe or PbSe doped with Bi_2Se_3 or Ag_2Se (0.5 mole %). Diffusion anneals were carried out with an instantaneous tracer source of Pb-210 and a flow of gaseous nitrogen through the system which was used to develop the congruently subliming composition in all the crystals. Some of the profiles showed deviation from normal Fick's law behavior, for example, steeper profiles at the surface were caused by 'lock up' at the initial surface and more gradual profiles in the interior were due to short-circuit paths, such as dislocations. Reliable values of $D_e(\widetilde{Pb})$ however, were obtained from the dominant straight line portions of the profile.

The behavior of $D_e(\widetilde{Pb})$ for the three compositions, PbSe, PbSe + 0.5% Bi_2Se_3, and PbSe + 0.5% Ag_2Se, are shown in figure 7.32 and

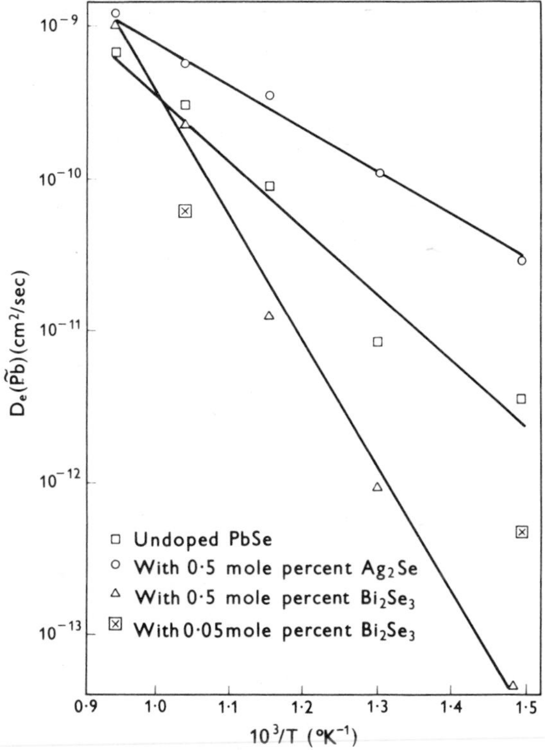

Figure 7.32. Dependence of $D_e(\tilde{Pb}$ in PbSe) upon temperature and impurity content. After Seltzer and Wagner (1962) [137].

may be represented by:

$$D_e(\tilde{Pb})_{\text{undoped}} = 4.98 \times 10^{-6} \exp[-0.83/kT] \text{ cm}^2 \text{ s}^{-1}, \quad (7.89)$$

$$D_e(\tilde{Pb})_{\text{Bi-doped}} = 4.28 \times 10^{-2} \exp[-1.61/kT] \text{ cm}^2 \text{ s}^{-1}, \quad (7.90)$$

$$D_e(\tilde{Pb})_{\text{Ag-doped}} = 4.41 \times 10^{-7} \exp[-0.55/kT] \text{ cm}^2 \text{ s}^{-1}. \quad (7.91)$$

It was noted that D_0 and Q vary in the same manner for the three crystal compositions studied. The acceptor and donor roles of Ag and Bi in PbSe were verified by room temperature measurement of the carrier type and density; Ag-doped crystals were p-type with $p = 1.75 \times 10^{19}$ cm^{-3} while Bi-doped crystals were n-type with $n = 8.56 \times 10^{19}$ cm^{-3}. For a constant value of the component pressure, as established by the congruent subliming conditions of the experiment, one would predict $[V''_{Pb}] \propto 1/p^2 \propto n^2$ and $[Pb_i^{\cdot\cdot}] \propto p^2 \propto 1/n^2$, in accordance with the

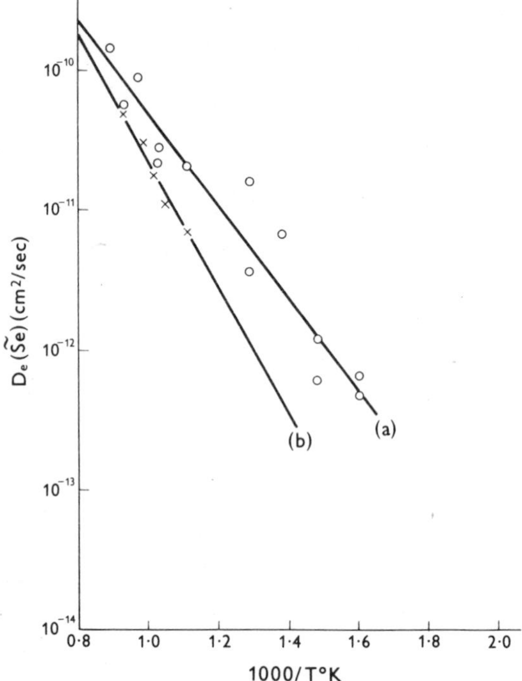

Figure 7.33. Temperature dependence of $D_e(\tilde{Se}$ in PbSe). ○ Ban and Wagner, (1970) [140]. x Boltaks and Mokhov (1958) [138, 139]. After Ban and Wagner (1970) [140].

equilibria:

$$\mathrm{Pb_{Pb}} \rightleftharpoons V''_{\mathrm{Pb}} + 2h^{\cdot} + \mathrm{Pb}(G); \qquad K = p_{\mathrm{Pb}}[V''_{\mathrm{Pb}}]p^2. \qquad (7.92)$$

$$\mathrm{Pb}(G) \rightleftharpoons \mathrm{Pb}_i^{\cdot\cdot} + 2e'; \qquad K = [\mathrm{Pb}_i^{\cdot\cdot}]n^2/p_{\mathrm{Pb}}. \qquad (7.93)$$

The authors concluded from the higher $D_e(\tilde{Pb})$ values for the p-type Ag-doped crystals, as contrasted with the n-type Bi-doped crystals, that Pb_i species are more mobile than V_{Pb} species in the diffusion of lead.

Diffusion of Se in PbSe

Boltaks and Mokhov (1958) [138, 139] and Boltaks (1963) [42] reported measurements of the tracer diffusion of Se-75 in PbSe single crystals with compositions close to the congruent subliming composition. The values of $D_e(\tilde{Se})$ for the temperature range 650-850°C, given in figure 7.33, are represented by

$$D_e(\tilde{Se}) = 2.1 \times 10^{-5} \exp[-1.2/kT] \ \mathrm{cm^2\ s^{-1}}. \qquad (7.94)$$

Ban and Wagner (1970) [140] studied the diffusion of Se-75 in single crystals of PbSe using the residual activity method. A variety of experimental conditions was used; undoped crystals and Bi- or Ag-doped crystals were measured at the congruent subliming composition over the temperature range 350-850°C, and undoped crystals were measured over the entire range of component pressures for a constant temperature of 700°C.†

The diffusion of Se-75 in undoped PbSe was given by

$$D_e(\widetilde{Se})]P_{min} = 8.74 \times 10^{-7} \exp[(-0.63 \pm 0.10)/kT] \text{ cm}^2 \text{ s}^{-1}, \quad (7.95)$$

which is in fair agreement with the values of Boltaks and Mokhov (1958) [138, 139] as shown in figure 7.33. Doping with Bi produced only a modest increase in $D_e(\widetilde{Se})$, and an unexpected result that the increase was virtually the same for 1/20 mole % and ½ mole % Bi_2Se_3 additions. The possibility of defect clusters involving Bi was discussed.

The activation energy increased upon Bi doping, similar to $D_e(\widetilde{S}$ in PbS). Crystals of PbSe doped with Ag_2Se showed essentially the same value of $D_e(\widetilde{Se})$ as undoped crystals. The value of $D_e(\widetilde{Se})$ increased by ~2 orders of magnitude over the phase field at 700°C, with three regions: $D_e(\widetilde{Se}) \propto p_{Se_2}^0$ for 10^{-7} torr $< p_{Se_2} < 10^{-4}$ torr; $D_e(\widetilde{Se}) \propto p_{Se_2}^{1/4}$ for 10^{-4} torr $< p_{Se_2} < 80$ torr; and $D_e(\widetilde{Se}) \propto p_{Se_2}$ for 80 torr $< p_{Se_2} < \sim 760$ torr. In these three regions, selenium was inferred to diffuse via the following three defects: (1) a defect pair, such as $(V'_{Pb} V^{\cdot}_{Se})$, whose concentration is a function of temperature only; (2) singly ionized Se-interstitials with $[Se^{\prime}_i] \propto p_{Se_2}^{1/4}$; and (3) neutral Se interstitials with $[Se_i^{\times}] \propto p_{Se_2}^{1/2}$.

Interdiffusion in PbSe

Butler (1964) [141] investigated interdiffusion in the PbSe system at 650°C using the *n-p* junction method. A homogeneous single crystal of *n*- or *p*-type PbSe was diffusion annealed using a Se- or Pb-rich source taken from a two phase PbSe ingot. The location of the resulting *np* junction was measured as a function of time using a thermoelectric microprobe. Referring to the initial carrier density of the crystal and the saturation value at the diffusion temperature, the following interdiffusion coefficients were calculated: $\underset{\sim}{D}(PbSe) = 5 \times 10^{-8}$ cm^2 s^{-1} for excess Se diffusing into *n*-type PbSe at 600°C, and $\underset{\sim}{D}(PbSe) = 9 \times 10^{-9}$ cm^2 s^{-1}, for excess Pb diffusing into *p*-type PbSe at 650°C. Two important features of

† In this paper, the diffusion behavior of Se-75 was also reported in crystals of PbS and PbTe at their congruent subliming compositions and the conclusion was made that non-stoichiometric and impurity defect concentrations are major factors in the diffusion behavior, in contrast to lattice parameter and atom size.

the results were emphasized. First, the values for $\underset{\sim}{D}(PbSe)$ were 2-3 orders of magnitude larger than $D_e(\widetilde{Pb})$ or $D_e(\widetilde{Se})$, as expected. Second, the value of $\underset{\sim}{D}(PbSe)$ for the diffusion of a Pb-rich source into a p-type crystal, $\underset{\sim}{D}(PbSe)_{n\to p}$, was significantly smaller than the reverse, $\underset{\sim}{D}(PbSe)_{p\to n}$. Two alternative explanations were presented for this behavior: Frenkel disorder on the Pb sublattice is dominant in PbSe, as suggested by Seltzer and Wagner (1962) [137], with the mobility of Pb_i (the dominant defect in n-type Pb-rich crystals) greater than the mobility of V_{Pb} (the dominant defect in p-type Se-rich crystals); or precipitates of Se in p-type PbSe exist, acting as sinks for diffusing Pb. The first explanation appears likely and has been analyzed in more detail by Brodersen, Walpole and Calawa (1970) [142], but in their analysis, the relative mobilities of Pb_i and V_{Pb} are reversed. With reference to the second explanation it is clear that Se precipitates are indeed sinks for Pb. This would encourage a steeper concentration gradient and, hence, a larger diffusion flux; however, the Se in the precipitates would compensate the excess Pb and, thus, require a larger relative flux of Pb to change a given region from p to n.

A similar study of $\underset{\sim}{D}(PbSe)$ was made by Brodersen, Walpole and Calawa (1970) [142] at 400° C, with the objective of experimentally

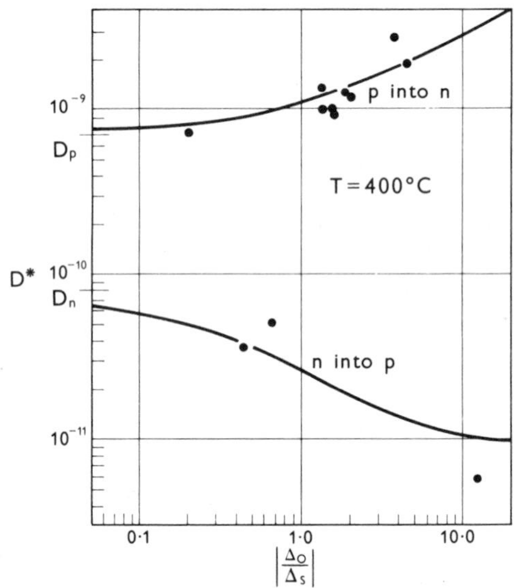

Figure 7.34. The variation of the experimental n-p junction diffusion coefficient D^* at 400° C with deviation from stoichiometry. Δ_o is the original deviation from stoichiometry and Δ_s is the deviation from stoichiometry at the surface. The points are the experimental data and the solid curve is the theoretical fit to the experimental points. After Brodersen, Walpole and Calawa (1970) [142].

determining and then theoretically analyzing the diffusion coefficients obtained by the *n-p* junction techniques with reference to their dependence upon deviation from stoichiometry, Δ. The authors make a clear distinction between the junction diffusion coefficient, D^*, which is the average value of the interdiffusion coefficient, $\underset{\sim}{D}(\Delta) = \underset{\sim}{D}(\text{PbSe})$, which may be a function of Δ. The experimental results are shown in figure 7.34. Brebrick (1959) [143] expressed $\underset{\sim}{D}(\Delta)$ in terms of deviation from stoichiometry and defect diffusivities, D_p and D_n. Taking relative values of $D_p = 9\, D_n$, where, for a Frenkel model, D_p and D_n are the diffusion coefficients of V_{Pb} and Pb_i respectively, a numerical solution for $\underset{\sim}{D}(\Delta)$ showed the following behavior:

$$\underset{\sim}{D}(\Delta) \sim D_p \quad \text{for} \quad \Delta < 0 \quad \text{and} \quad \underset{\sim}{D}(\Delta) \sim D_n \quad \text{for} \quad \Delta > 0.$$

Making the simplification that $\underset{\sim}{D}(\Delta) = D_p$ for $\Delta < 0$ and $\underset{\sim}{D}(\Delta) = D_n$ for $\Delta < 0$ and $\underset{\sim}{D}_p = 9\, D_n$, an analytical solution for D^* was obtained which provides a good fit to the observed experimental behavior (figure 7.34).

7.5.4 Diffusion in PbTe

Diffusion of Pb in PbTe

The self-diffusion of Pb-210 in PbTe was investigated by George and Wagner (1971) [144] for a temperature of 700°C as a function of component pressure and impurity content. Single crystal wafers were preannealed for 100 h at 700°C in an enclosed ampoule containing excess Te or Pb, or in a flowing argon atmosphere in order to establish compositions corresponding to Te-saturation, Pb-saturation or congruent subliming compositions, respectively. In addition, crystals containing either Ag (an acceptor) or Bi (a donor) were prepared and annealed in a flowing argon atmosphere. The tracer, Pb-210, was electrodeposited and the samples diffusion annealed at 700°C under conditions identical to the prediffusion anneal and the resulting profiles were determined by the residual activity method. The profiles obeyed the instantaneous source boundary condition with no anomalies. The resulting dependence of $D_e(\widetilde{\text{Pb}})$ upon p_{Te_2} and impurity content is illustrated in figure 7.35, which shows the following behavior: for undoped PbTe there is a minimum value of $D_e(\widetilde{\text{Pb}})$ at the congruent subliming composition; at the congruent subliming composition, both donor doped (Bi) and acceptor doped (Ag) crystals show an enhancement of $D_e(\widetilde{\text{Pb}})$. This behavior was ascribed to the predominance of Frenkel disorder on the Pb sublattice, the diffusion proceeding via Pb interstitials on the Pb-rich side of the phase field and via Pb vacancies on the Te-rich side of the phase field. The enhancement by both donors and acceptors was attributed to impurity enhanced defect

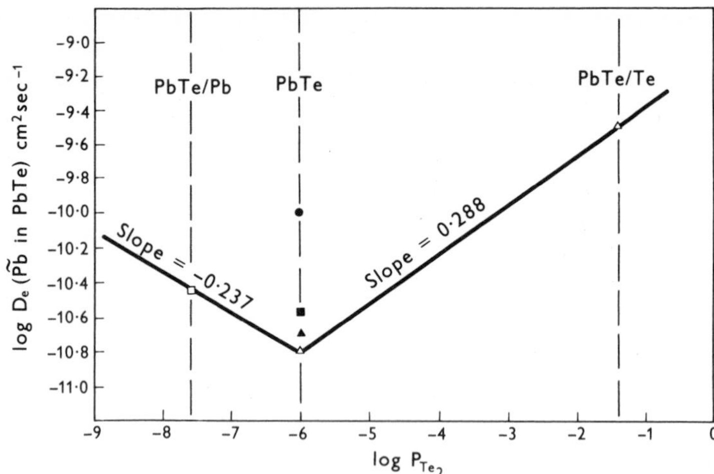

Figure 7.35. Diffusion isotherm for D_e($\widetilde{Pb}$ in PbTe) at 700° C,

□ PbTe annealed at the phase boundary PbTe-Pb.
△ PbTe annealed at the phase boundary PbTe-Te.
▲ PbTe annealed under argon corresponding to the minimum total pressure, P_{min}. This point is coincidentally very nearly the stoichiometric composition at 700° C.
■ PbTe doped with Ag[5 x 10^{18} electron holes/cm^3] annealed under argon.
● PbTe doped with Bi[9.7 x 10^{17} electrons/cm^3] annealed under argon. After George and Wagner (1971) [144].

concentrations, with the concentrations of Pb vacancies or Pb interstitials increased by donors (Bi) and acceptors (Ag), respectively.

The self-diffusion of Pb was studied as a function of crystal composition and temperature by Gomez, Stevenson and Huggins (1971) [145] in the temperature range 520-729° C. Single crystals of PbTe were grown by directional solidification (Bridgeman-Stockbarger method) and the variation in composition along the single crystal was estimated from the melt composition, the phase limits for PbTe according to Brebrick and Allgaier (1960) [146] and Brebrick and Gubner (1962) [147], and the theory of unidirectional, non-equilibrium solidification. Wafers for five compositions, varying from Te-excess to Pb-excess, were sliced from the ingot and homogenized by annealing in evacuated Vycor capsules for 120 h at 400° C. During these homogenization and subsequent diffusion anneals, a small shift in composition toward the congruent subliming composition occurred due to the small free volume of the capsules. An identical procedure using Pb enriched with Pb-210 was used to produce single crystal wafers of $Pb^{210}Te$ for the tracer source. The diffusion anneal consisted of encapsulating wafers of PbTe and $Pb^{210}Te$ of the same composition in an evacuated ampoule. The value of D_e($\widetilde{Pb}$) was obtained from the resulting Pb-210 profile, as determined by mechanical sectioning. Normal profiles for the imposed boundary conditions were obtained in all cases. The results, represented in figure 7.36, show distinctly different

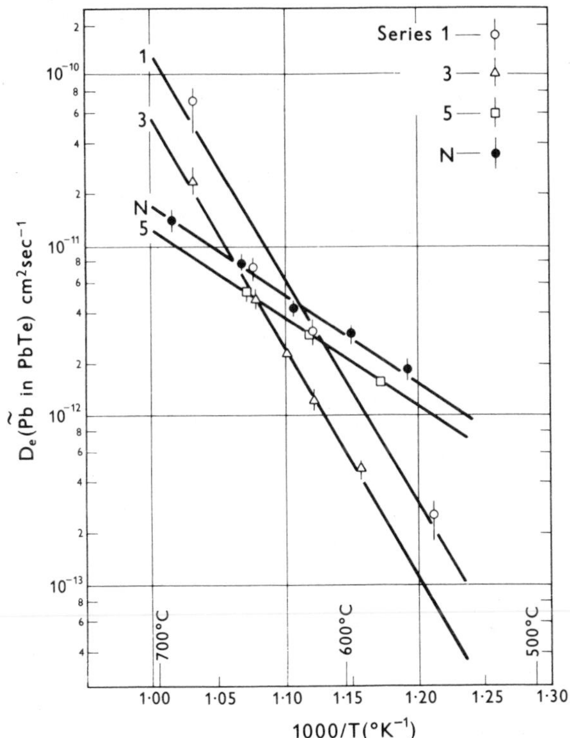

Figure 7.36. Temperature and composition dependence of $D_e(\widetilde{Pb}$ in PbTe). Concentration of Te increase from series N to 1 with N containing excess Pb and 5, 3 and 1 containing excess Te. After Gomez, Stevenson and Huggins (1971) [145].

activation energies for Pb-rich ($Q = 1.04$ eV), and Te-rich crystals ($Q = 2.65$ eV), and indicate that $D_e(\widetilde{Pb})$ goes through a minimum at intermediate compositions in the PbTe phase field. A Frenkel defect was proposed to explain these results; Pb interstitials and Pb vacancies dominate the diffusion in Pb-rich and Te-rich crystals, respectively, with the activation energy for motion of these defects given by the respective diffusion activation energies. This study appears to be in good agreement with that of George and Wagner (1971) [144], both with respect to the experimental values of $D_e(\widetilde{Pb})$, where they may be compared, and with respect to the proposed models.

Diffusion of Te in PbTe

The diffusion of Te-127 in single crystals of PbTe was investigated by Boltaks and Mokhov (1958) [138, 139] and discussed by Boltaks (1963) [42]. Single crystal samples of PbTe were diffused with Te-127 over the temperature range 500-800°C. The reported values for $D_e(\tilde{Te})$ were represented by the expression:

$$D_e(\tilde{Te}) = 2.7 \times 10^{-6} \exp[-0.75/kT] \text{ cm}^2 \text{ s}^{-1}. \qquad (7.96)$$

The diffusion of Te-127 was also studied by Gomez, Stevenson and Huggins (1971) [145] using the technique previously described. The tracer sources and sinks were pairs of single crystal wafers sliced from adjacent portions of the single crystal boule to assure identical composition and one wafer of each pair was neutron activated. Five samples, with compositions varying from Pb-rich to Te-rich, were measured; the value of $D_e(\tilde{Te})$ was found to increase as the Te content increased and to have the same

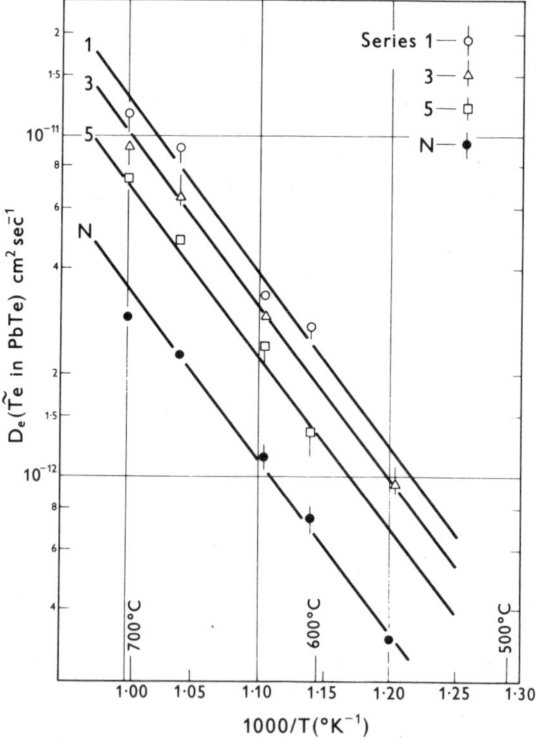

Figure 7.37. Temperature and composition dependence of $D_e(\tilde{Te}$ in PbTe). Concentration of Te increase from series N to 1 with N containing excess Pb and 5, 3 and 1 containing excess Te. After Gomez, Stevenson and Huggins (1971) [145].

activation energy of 1.04 eV for each composition (figure 7.37). The activation energy was higher than that of 0.75 eV reported by Boltaks and Mokhov (1958), and the highest value of $D_e(\tilde{Te})$—for the highest Te concentration—was lower than their diffusivity values by a factor of ~2 at 700° C and ~10 at 500° C. There is no obvious reason for this disparity. To explain the increase of $D_e(\tilde{Te})$ with Te concentration, the authors proposed an interstitial or interstitialcy mechanism.

Interdiffusion in PbTe

Brady (1954) [148] observed the time dependence of penetration depth of a *np* junction in a *p*-type PbTe specimen heated between two Pb discs at 500° C and observed a value of $5.6\text{-}9.1 \times 10^{-8}$ $cm^2\ s^{-1}$ for the interdiffusion coefficient in PbTe, $\underset{\sim}{D}(PbTe)$. The *np* junction technique was also used by Boltaks and Mokhov (1956, 1957) [149, 150] for the temperature range 250-500° C, with their results described by $\underset{\sim}{D}(PbTe) = 2.9 \times 10^{-5} \exp[-0.6/kT]$ $cm^2\ s^{-1}$ For a temperature of 500° C, Boltaks and Mokhov observed a smaller value of $\underset{\sim}{D}(PbTe)$ than Brady, $\underset{\sim}{D}(PbTe) \sim 3.5 \times 10^{-9}$ $cm^2\ s^{-1}$ in contrast to $5.6\text{-}9.1 \times 10^{-8}$ $cm^2\ s^{-1}$. Both of these results are significantly larger than the largest values of either $D_e(\tilde{Pb})$ or $D_e(\tilde{Te})$ at the same temperature, $\sim 5 \times 10^{-13}$ $cm^2\ s^{-1}$, as might be expected in a relatively sharp binary compound such as PbTe.

Butler (1964) [141] also measured the kinetics of the *np* junction penetration in PbTe using the experimental method described in the PbSe section. He obtained a value of $\underset{\sim}{D}(PbTe) = 6 \times 10^{-7}$ $cm^2\ s^{-1}$ at 650° C for diffusion of an *n* layer into *p*-type PbTe. Comparing this value with the extrapolated value of Boltaks and Mokhov (1956, 1957) [149, 150] of 2×10^{-8} $cm^2\ s^{-1}$ shows the same general disparity as the comparison of Brady with the latter authors.

7.6 SUMMARY AND CONCLUSIONS

In the preceding discussion, a review of most of the literature on the component diffusion† in the respective chalcogenides was given, with specific reference to the defect structure of those compounds. Careful study of this literature leads one to the conclusion that the present defect models for the compounds are not well established in most instances. In some cases, for example ZnO, there are numerous studies reported with considerable dispersion in the results. One of the main reasons is the variety of sample preparation and experimental conditions in the respective studies. In some systems, for example, CdO, PbO and ZnS, there are very limited data on which to base a defect model. In other systems,

† A summary is given in Table 7.1.

for example, CdTe and the remaining II-VI compounds, there is good agreement between independent investigations of the diffusion behavior on the one hand and of the electrical properties on the other; however, the defect models implied from each of these sources of information are not always mutually compatible. In this summary, a few general trends will be described and some of the more significant discrepancies will be noted. These generalizations are based on the experimental information to date and are subject to change by future studies.

In the II-VI compounds, the self-diffusion of the more electropositive component (metal, M) shows two types of behavior; in CdTe, ZnTe and ZnSe, $D_e(\tilde{M})$ is independent of component pressure over the entire phase field whereas in CdSe and CdS, $D_e(\tilde{M})$ increases with increasing p_M. Zinc oxide may be classified with the former group since $D_e(\tilde{Zn})$ appears to be

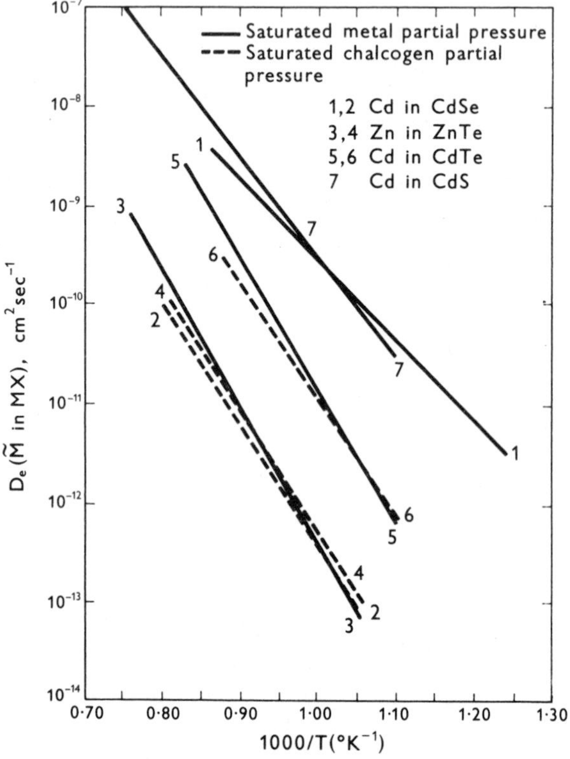

Figure 7.38. Some representative curves for $D_e(\tilde{M}$ in $MX)$ in Zn and Cd chalcogenides showing dependence upon temperature and component pressure. 1, 2; $D_e(\tilde{Cd}$ in CdSe), Borsenberger, Stevenson and Burmeister (1967) [111]. 3, 4; $D_e(\tilde{Zn}$ in ZnTe), Reynolds and Stevenson (1969) [38]. 5, 6; $D_e(\tilde{Cd}$ in CdTe), Borsenberger and Stevenson (1968) [117]. 7; $D_e(\tilde{Cd}$ in CdS), Woodbury (1964) [99].

relatively insensitive to changes in component pressure, whereas ZnS may be classified with the second group. Compounds of the first group show enhanced self-diffusion of the M component upon addition of both donor and acceptor impurity atoms. The respective behavior of $D_e(\tilde{M})$ for the two groups has been explained by postulating that ionized atomic disorder on the M sublattice controls the electrical neutrality condition for the former group with comparable mobility for the oppositely charged species, and that diffusion proceeds via an M interstitial for the latter group.† A summary of some of the reasonably well established data is given in figures 7.38 and 7.39. It will be noted that a reduced temperature plot–log $D_e(\tilde{M}$ in $MX)$ versus T_m/T as in figure 7.39–shows a rough separation into two groups. Compounds of the first group, which show dominant ionized atomic disorder, have significantly lower metal

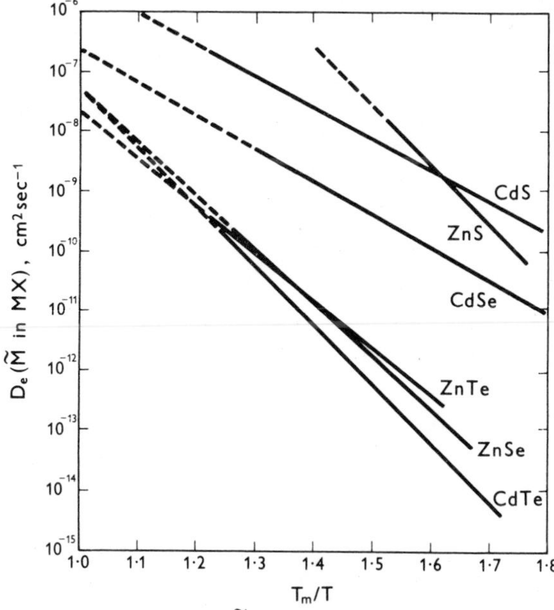

Figure 7.39. Dependence of $D_e(\tilde{M}$ in $MX)_{M\,SAT}$ upon the reciprocal reduced temperature for some representative Zn and Cd chalcogenides. CdS; Woodbury (1964) [99]. ZnS; Secco (1958) [75]. CdSe; Borsenberger, Stevenson and Burmeister (1967) [111]. ZnTe; Reynolds and Stevenson (1969) [38]. ZnSe; Henneberg and Stevenson (1971) [85]. CdTe; Borsenberger and Stevenson (1968) [117].

self-diffusion coefficients than those of the second group, which show diffusion via an interstitial mechanism, when compared at the same reduced temperature.

† $D_e(\tilde{Cd}$ in CdS) appears to proceed via a V_{Cd} specie, in S-rich crystals.

The self-diffusion of the chalcogen component is summarized in figures 7.40 and 7.41. There is a common type of behavior for all the II-VI compounds; for chalcogen rich compositions (at higher p_{X_2}, approaching X saturated crystals), $D_e(\widetilde{X}$ in $MX)$ increases with p_{X_2}. The exact pressure dependence is well established for all compounds, except ZnS and ZnTe, and may be represented as

$$D_e(\widetilde{X} \text{ in } MX) = K_T p_{X_2}^{\frac{1}{2}} = K_T' p_M^{-1}, \tag{7.98}$$

where K_T is a function of temperature only—a product of the incorporation constant of the dominant mobile defect and a defect diffusion coefficient. The above pressure dependence is consistent with a neutral interstitial chalcogen defect as the dominant mobile defect. As metal-rich compositions are approached (maximum p_M), however, an

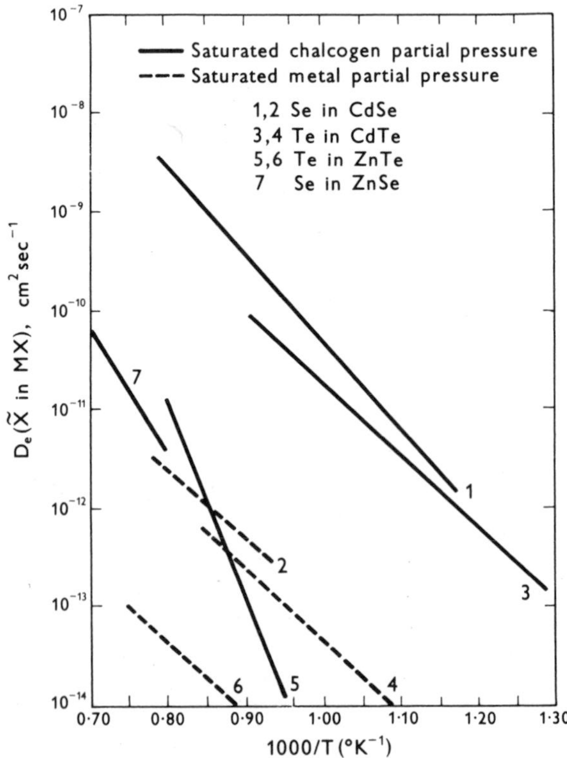

Figure 7.40. Some representative curves for $D_e(\widetilde{X}$ in $MX)$ in the Zn and Cd chalcogenides showing dependence upon temperature and component pressure. 1, 2; $D_e(\widetilde{\text{Se}}$ in CdSe), Woodbury and Hall (1967) [90]. 3, 4; $D_e(\widetilde{\text{Te}}$ in CdTe), Borsenberger and Stevenson (1968) [117]. 5, 6; $D_e(\widetilde{\text{Te}}$ in ZnTe), Reynolds and Stevenson (1969) [38]. 7; $D_e(\widetilde{\text{Se}}$ in ZnSe), Woodbury and Hall (1967) [90].

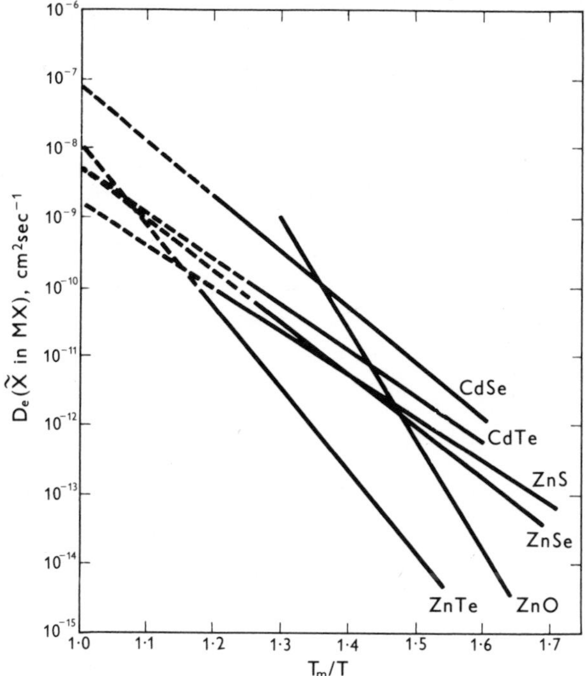

Figure 7.41. Dependence of $D_e(\tilde{X}$ in $MX)_{X\,SAT}$ upon the reciprocal reduced temperature for some representative Zn and Cd chalcogenides. CdSe; Woodbury and Hall (1967) [90]. CdTe; Borsenberger and Stevenson (1968) [117]. ZnS; Gobrecht, Nelkowski, Baars and Weigt (1967) [79]. ZnSe; Woodbury and Hall (1967) [90]. ZnO; Moore and Williams (1959) [49]. ZnTe; Reynolds and Stevenson (1969) [38].

abrupt change in the dependence upon component pressure is observed; $D_e(\tilde{X}$ in $MX)$ increases with increasing p_M in all cases, except for $D_e(\tilde{\text{Se}}$ in ZnSe) which is essentially independent of p_M for M-rich compositions. This change in behavior is explained by a chalcogen vacancy preempting the diffusion mechanism in the M-rich portion of the phase field. The corresponding pressure range for this behavior is too limited for the CdTe and CdSe systems to obtain an explicit pressure dependence, consequently the degree of ionization of the corresponding vacancy cannot be implied. In CdS, $D_e(\tilde{\text{S}}$ in $\text{CdS})_{\text{Cd-rich}} \propto p_{\text{Cd}}^{1/3}$, with the behavior ascribed to a $V_{\text{S}}^{\cdot\cdot}$ defect and a charge neutrality condition $2[V_{\text{S}}^{\cdot\cdot}] = n$, whereas the relative insensitivity of $D_e(\tilde{\text{Se}}$ in $\text{ZnSe})_{\text{Zn-rich}}$ upon p_{Zn} was ascribed to a $V_{\text{Se}}^{\cdot\cdot}$ defect with a charge neutrality condition $[V_{\text{Zn}}''] = [\text{Zn}_i^{\cdot\cdot}]$. The mechanism for $D_e(\tilde{\text{O}}$ in ZnO) is not clear at present because of conflicting results. In one study, $D_e(\tilde{\text{O}}$ in ZnO) was found to vary with $p_{\text{O}_2}^{1/2}$ with a neutral oxygen interstitial mechanism proposed, whereas a later study reported a

TABLE

Comparison of diffusion and electrical transport

Compound	Diffusion				
	Diffusing component	Defect model	Temperature range (°C)	Pressure range (torr)	Reference
ZnSe	Zn	$[Zn_i^{\cdot\cdot}] = [V_{Zn}'']$	878	8×10^{-5}-600 (Zn)	Henneberg and Stevenson [85]
	Se	$Se_i^{\times}$	1000-1150	2×10^{-3}-6×10^{3} (Zn)	Woodbury and Hall (1967) [90]
	Se	$Se_i^{\times}$, Se_i''	1021	4×10^{-8}-1.5×10^{4} (Se_2)	Henneberg and Stevenson [85]
ZnTe	Zn	$[Zn_i] = [V_{Zn}]^a$	690-1020	10^{-4}-2×10^{-3} (Zn)	Reynolds and Stevenson (1969) [38]
	Te	$Te_i^{\times}$	750-980	7×10^{-4}-10^{3} (Zn)	Reynolds and Stevenson (1969) [38]
CdS	Cd	$[V_S]$, $[V_{Cd}]^a$	700-1150		Woodbury (1964) [99]
	Cd	$[V_{Cd}'] = [D^{\cdot}]$, $2[V_{Cd}''] = [D^{\cdot}]$	850	10^{-2}-10^{3} (Cd)	Shaw and Whelan (1969) [102]
	Cd	$Cd_i^{\cdot}$, $[D^{\cdot}] = 2[V_{Cd}'']$	700-1000	Cd-rich, S_2-rich	Kumar and Kröger [108]
	S	$S_i^{\times}$, $n = 2[V_S^{\cdot\cdot}]$	700-1000	S_2-rich, Cd-rich	Kumar and Kröger [108]

^{a} = Unspecified charge state.
^{b} = Included at lower temperatures.

7.2

defect models for II-VI semiconducting compounds

Defect model	Electrical transport			
	Temperature range (°C)	Pressure range (torr)	Temperature or quench	Reference
$n = [Zn_i^{\cdot}]$ or $n = [V_{Se}^{\cdot}]$	711-1009	6-800 (Zn)	T	Smith (1969) [112]
$p = 2[V_{Zn}'']$	700-950	10-400 (Zn)	T	Thomas and Sadowski (1964) [92]
$p = 2[V_{Zn}'']$			T	Smith (1969) [151]
$n = 2[Cd_i^{\cdot\cdot}]$, $[Cd_i^{\cdot\cdot}] = [V_{Cd}'']$	395-600	2×10^{-5}-100 (Cd)	T	Boyn, Goede and Kuschnerus (1965) [103]
$n = [D^{\cdot}] + 2[Cd_i^{\cdot\cdot}]$	600-950	20-800 (Cd)	T	Shaw and Whelan (1969) [103]
$n = 2[V_S^{\cdot\cdot}] (+[D^{\cdot}] - [A'])^b$	400-1000	1.5×10^{-1}-600 (Cd)	T	Hershman and Kröger [104]

Table 7.2–(*continued*)

Compound	Diffusion: Diffusing component	Defect model	Temperature range (°C)	Pressure range (torr)	Reference
CdTe	Cd	$[Cd_i] = [V_{Cd}]^a$	775	10^{-1}-800 (Cd)	Borsenberger and Stevenson (1968) [117]
	Cd	$(Cd_iV_{Cd})^{\times}$	800-900	1-1000 (Cd)	Whelan and Shaw (1967) [116]
	Te	$Te_i^{\times}$	800	4×10^{-2}-300 (Te_2)	Borsenberger and Stevenson (1968) [117]
CdSe	Cd	$n = [Cd_i^{\cdot}]$	798	1.5×10^{-3}-10^3 (Cd)	Borsenberger, Stevenson and Burmeister (1967) [111]
	Se	$Se_i^{\times}$	700-1000	8×10^{-5}-6×10^3 (Cd)	Woodbury and Hall (1967) [90]

^{a} = Unspecified charge state.

pressure dependence $p_{O_2}^{-0.5}$ and $p_{O_2}^{-1.5}$ at different temperatures and proposed an oxygen vacancy mechanism. For $D_e(\tilde{O}$ in CdO), a pressure dependence of $p_{O_2}^{-1.5}$ was taken as evidence for a $V_O^{\cdot\cdot}$ diffusing specie.

The diffusion of the chalcogen component is significantly slower than for the corresponding metal component. All that may be implied from this observation is that the sum of the concentration-mobility products of the dominant mobile defects on the M sublattice is larger than on the X sublattice. It has generally been assumed that the actual defect concentrations are larger on the M sublattice and that corresponding

Electrical transport				
Defect model	Temperature range (°C)	Pressure range (torr)	Temperature or quench	Reference
$n = [Cd_i^{\cdot}]$, $p = [V'_{Cd}]$	700-1000	1.5-3 x 10^3 (Cd)	Q	deNobel (1959) [22]
$n = 2[Cd_i^{\cdot\cdot}]$	576-895	10-600 (Cd)	T	Whelan and Shaw (1968) [118]
$n = 2[Cd_i^{\cdot\cdot}]$ or $n = 2[V_{Te}^{\cdot\cdot}]$	600-879	10-400 (Cd)	T	Smith (1970) [121]
$n = [A'] = p$	300-1000	Te Rich	:	
$n + [A'] = 2[Cd_i^{\cdot\cdot}]$	700-950	14-2000 (Cd)	T	Rud' and Sanin (1971) [123]
$n = 2[Cd^{\cdot\cdot}]$	600-800	Cd Rich	T	Zanio (1969) [119]
$n = p + [D^{\cdot}]$	600-900	Te Rich		
V_{Se}[a]	350		Q	Tubota, Suzuki and Hirakawa (1960) [152]
$n = 2[Cd_i^{\cdot\cdot}] + [D^{\cdot}] - [A']$ or $n = 2[V_{Se}^{\cdot\cdot}] + [D^{\cdot}] - [A']$	750 and 920	10^{-1}-10^3 (Cd)	Q	Hung, Ohashi and Igaki (1969) [153]
$n = 2[Cd_i^{\cdot\cdot}]$ or $n = 2[V_{Se}^{\cdot\cdot}]$	721-913	100-600 (Cd)	T	Smith (1970) [154]

charged species will dominate the charge neutrality condition. This, however, is not necessarily the case if there is a large difference in mobility of the defects. In the analysis of CdS by Kumar and Kröger (1971) [108], the dominant electrically active atomic defect is assumed to be $V_S^{\cdot\cdot}$. The self-diffusion of S and Cd, respectively, are presumed to proceed via a $V_S^{\cdot\cdot}$ and $Cd_i^{\cdot}$ defect, respectively. The model of $V_S^{\cdot\cdot}$ dominating the charge neutrality condition and the observation that $D_e(\tilde{Cd}$ in CdS$) > D_e(\tilde{S}$ in CdS) implies that the mobility of Cd_i is much larger than that of $V_S^{\cdot\cdot}$.

A point of considerable interest is the incompatibility of the models proposed for some of the II-VI compounds based, respectively, on the diffusion behavior and the high temperature electrical properties. A tabular summary comparing these two sources of information is given in Table 7.2 for five systems for which reasonably complete and consistent data are available. Mutually compatible models have been developed for CdS and ZnSe; however, this is not the case for CdSe, CdTe and ZnTe, as discussed in the respective sections.

The large difference in interdiffusion and self-diffusion coefficients predicted from the Darken equation (equation 7.1), and the expected large values of the thermodynamic factor in these systems has been well corroborated in the literature. Interdiffusion coefficients have been obtained either by reactive diffusion (i.e., film growth kinetics), or by measurements of the relaxation time for the change in conductivity attending a step jump in component pressure. Several investigations have been made using the latter technique. Concerning the latter studies, it is important to note a significant discrepancy in the relaxation times and, hence, the values of $\underset{\sim}{\mathbf{D}}(MX)$, reported by different investigators. These differences can be correlated with the respective method of controlling component pressure; with the static two-temperature technique, relaxation times are significantly larger than those obtained using the dynamic refluxing, two-temperature zone technique. The reason for this discrepancy is not obvious.

The lead chalcogenides, PbS, PbSe and PbTe, also show some general similarities in their diffusion behavior. In all three compounds, $D_e(\widetilde{\text{Pb}}$ in PbX) is enhanced by both donor and acceptor impurities. This is interpreted as evidence for an ionized V_{Pb} and Pb_i disorder active in the diffusion process, for example V_{Pb}'' and $\text{Pb}_i^{\cdot\cdot}$. Further support for this model is given by the observation in both PbS and PbTe that the dependence of $D_e(\widetilde{\text{Pb}}$ in PbS) upon p_{X_2} shows a minimum, with $D_e(\widetilde{\text{Pb}}$ in PbX) decreasing with p_{X_2} in the Pb-rich portion of the phase field and increasing with p_{X_2} in the X-rich portion of the phase field. This is consistent with diffusion via Pb_i and V_{Pb} species in the respective regions.

The behavior of $D_e(\widetilde{X}$ in PbX) also shows similar behavior in the three compounds; there is a significant increase with p_{X_2} over the phase field, similar to the behavior in the II-VI compounds. In the case of PbS and PbSe, diffusion isotherms imply that there are distinct regions in the phase fields in which different defects, respectively, dominate the diffusion process. It appears, however, that a neutral X interstitial dominates near X saturation. For PbTe, the activation energy for $D_e(\widetilde{\text{Te}}$ in PbTe) is independent of composition which implies that only one defect mechanism dominates over the entire phase field. It is also noteworthy that the magnitude of $D_e(\widetilde{X}$ in PbX), is comparable to the corresponding value of $D_e(\widetilde{Pb}$ in PbS) in sharp contrast to the behavior of the II-VI compounds.

The values of the self-diffusion coefficients in the lead chalcogenides are much smaller than the interdiffusion coefficient, as expected from equation 7.1. Inasmuch as the Pb chalcogenides have a more extended range of stability, this difference is not expected to be as large as for the II-VI compounds.

The remarkable feature of diffusion in PbO is the high mobility of oxygen; the oxidation of Pb is controlled by the oxygen diffusion through PbO and there is appreciable electrolytic conductivity via oxide ions. The influence of component pressure and impurity concentration on the self-diffusion coefficients has not been studied on the lead sublattice, whereas the results on the oxygen sublattice lead to results which cannot be easily correlated using present defect models.

There is considerable work necessary to reliably document the diffusion behavior of this group of compounds. There are, however, several systems for which reliable data are available. Even in these systems, apparent inconsistencies appear to exist in the analysis of the defect structure using present theory. One might question the suitability of the present theory, developed for the treatment of the ionic I-VII compounds and the oxides, for treating groups of compounds having a high covalent character to their bonding. In view of the potential technological and theoretical interest in the present group of compounds and the existence of a substantial amount of reliable fundamental information on these compounds, further work on these systems will be a valuable contribution to the field of solid state chemistry.

REFERENCES

1. G. Mandel, *Phys. Rev.*, **134**, A1073 (1963)
2. M. Aven, *Appl. Phys. Letters*, **7**, 146 (1965)
3. A. R. Hutson, J. H. McFee and D. L. White, *Phys. Rev. Letters*, **7**, 237 (1961)
4. J. H. McFee, *J. Appl. Phys.*, **34**, 1548 (1963)
5. T. A. Midford, *J. Appl. Phys.*, **35**, 3423 (1964)
6. D. A. Cusano, *Solid-St. Electron.*, **6**, 217 (1963)
7. H. G. Grimmeis and R. Memming, *J. Appl. Phys.*, **33**, 2217, 3596 (1962)
8. K. Zanio, *Appl. Phys. Letters*, **14**, 56 (1969)
9. W. Akutagawa and K. Zanio, *J. Appl. Phys.*, **40**, 3838 (1969)
10. J. Kiefer and A. Yariv, *Appl. Phys. Letters*, **15**, 26 (1969)
11. A. G. Foyt and A. L. McWhorter, *Trans. I.E.E.E.*, **Ed.-13**, 79 (1966)
12. C. J. Young and H. G. Grieg, *R.C.A. Rev.*, **15**, 469 (1954)
13. G. Heiland, E. Mollwo and F. Stockman, *Solid State Physics*, **8**, 239 (1959)
14. K. Hauffe, B. Martinez, J. Range and R. Schmidt, *Photo. Korr.*, **6**, 113 (1968)
15. L. Heyne, N. M. Beckmans and A. DeBeer, *J. Electrochem. Soc.*, **119**, 77 (1972)

16. H. H. Woodbury, 'Physics and Chemistry of II-VI Compounds', p. 225 (M. Aven and J. S. Prener, eds.). (North-Holland Publishing Company, Amsterdam, 1967)
17. H. H. Woodbury, 'II-VI Semiconducting Compounds, 1967 International Conference'. (W. A. Benjamin, Inc., New York, 1967)
18. P. G. Shewmon, 'Diffusion in Solids'. (McGraw-Hill Book Co., New York, 1963)
19. W. Jost, 'Diffusion in Solids, Liquids or Gases'. (Academic Press, New York, 1960)
20. T. R. Manning, 'Diffusion Kinetics for Atoms in Crystals'. (D. Van Nostrand Co., Inc., Princeton, N.J., 1968)
21. L. S. Darken, *Trans. AIME,* **175**, 184 (1948)
22. D. deNobel, *Phillips Res. Repts.*, **14, 361 (1959)**
23. M. R. Lorenz, *J. Phys. Chem. Solids,* **23**, 939 (1962)
24. C. Z. Van Doorn, *Rev. Sci. Inst.*, **32**, 755 (1961)
25. I. Prigogine and R. DeFay, 'Chemical Thermodynamics'. (John Wiley & Sons, Inc., New York, 1962)
26. P. Goldfinger and M. Jeunehomme, *Trans. Faraday Soc.*, **59**, 2861 (1965)
27. M. R. Lorenz, 'Physics and Chemistry of II-VI Compounds', Chapter 2 (M. Aven and J. S. Prener, eds.). (North-Holland Publishing Company, Amsterdam, 1967)
28. C. D. Thurmond, *J. Phys. Chem. Solids,* **26**, 785 (1965)
29. C. Wagner, *J. Chem. Phys.*, **21**, 1819 (1953)
30. K. Zanio, *J. Appl. Phys.*, **41**, 1935 (1970)
31. C. Wagner and W. Schottky, *Zeit. f. Chemie,* **B11**, 163 (1931)
32. G. Brouwer, *Phillips Res. Rep.*, **9**, 336 (1954)
33. F. A. Kröger and H. J. Vink, *Solid State Physics,* **3**, 307 (1956)
34. H. J. Vink, 'Proc. Inter. School of Physics', Enrico Fermi, Course XXII, Semiconductors, p. 68 (R. A. Smith, ed.). (Academic Press, New York, London, 1963)
35. F. A. Kröger, 'The Chemistry of Imperfect Crystals'. (North-Holland Publishing Company, Amsterdam, 1964)
36. D. Long, 'Energy Bands in Semiconductors'. (Interscience Publishers, 1968)
37. E. Nebauer, *Phys. Stat. Sol.*, **36**, K63 **(1969)**
38. R. A. Reynolds and D. A. Stevenson, *J. Phys. Chem. Solids,* **30**, 139 (1969)
39. J. P. Roberts and C. Wheeler, *Phil. Mag.*, **2**, 708 (1957)
40. P. M. Borsenberger, Ph.D. Thesis, Stanford University, May, 1967
41. C. G. B. Garrett, *Adv. in Electronics and Electron Physics,* **14**, 1 (1961)
42. B. I. Boltaks, 'Diffusion in Semiconductors', English Translation. (Academic Press, New York, 1963)
43. P. J. Harrop, *J. Materials Science,* **3**, 206 (1968)
44. M. O'Keeffe, 'Diffusion in Oxides and Sulfides. Sintering and Related Phenomena' (C. G. Kuczinski, N. A. Hooten and G. F. Gibbon, eds.). Proc. of Int. Conference, Univ. of Notre Dame, June, 1965. (Gordon and Breach Pub. Co., New York, 1967)
45. P. H. Miller, *Phys. Rev.,* **60**, 820 (1941)
46. R. Lindner, *Acta. Chem. Scand.,* **6**, 457 (1952)
47. F. Münnich, *Naturwissenschenschaften,* **42**, 340 (1955)
48. E. A. Secco and W. J. Moore, *J. Chem. Phys.,* **26**, 942 (1957)
49. W. J. Moore and E. Williams, *Disc. Faraday Soc.,* **28**, 86 (1959)
50. E. A. Secco and W. J. Moore, *J. Chem. Phys.,* **23**, 1170 (1955)
51. D. G. Thomas, *J. Phys. Chem. Solids,* **3**, 229 (1957)
52. C. E. Wicks and F. E. Block, Bulletin 605, Bureau of Mines, U.S. Government Printing Office (1963)
53. J. C. Fisher, *J. Appl. Phys.,* **22**, 74 (1951)
54. E. G. Spicar, Thesis (Dr. rerinat.) Technische Hochschule, Stuttgart (1956)

55. E. Spicar, 'Reactivity of Solids', p. 637. 1st International Conference, Madrid, 1956, Proccedings Elsevier, Amsterdam (1957)
56. J. P. Roberts and C. Wheeler, *Trans, Faraday Soc.*, **56**, 570 (1960)
57. H. Lamatsch, Diplomarbeit, University of Erlangen, Germany (1958)
58. G. Heiland and E. Mollwo, *Disc. Faraday Soc.*, **28**, 123 (1959)
59. J. P. Roberts, *Disc. Faraday Soc.*, **28**, 123 (1959)
60. E. A. Secco, *Disc. Faraday Soc.*, **28**, 94 (1959)
61. E. A. Secco, 'Reactivity of Solids', p. 188 (J. H. De Boer *et al.*, eds.). (Elsevier Publishing Co., Amsterdam, 1960)
62. E. A. Secco, *Can. J. Chem.*, **39**, 1544 (1961)
63. E. A. Secco, 'Reactivity in Solids', p. 523 (R. W. Roberts and P. Cannon, eds.). (John Wiley and Sons, 1969)
64. R. Arneth, Erlanger Diplomarbeit, unpublished (1955)
65. R. Pohl, *Z. Physik*, **155**, 120 (1959)
66. J. W. Hoffman and I. Lauder, *Trans. Faraday Soc.*, **66**, 2346 (1970)
67. R. C. Robin, E. L. Solem, A. H. Heuer and A. R. Cooper, Fourth Semi-annual Report, ARPA Contract DAHC 1559 G-3, June (1970)
68. C. Wagner and K. Grünewald, *Z. Phys. Chem.*, **B40**, 455 (1938)
69. W. J. Moore and J. K. Lee, *Trans. Faraday Soc.*, **47**, 501 (1951)
70. J. O. Cope, *Trans. Faraday Soc.*, **57**, 493 (1961)
71. K. Hauffe, 'Oxidation of Metals', p. 202. (Plenum Press, New York, 1965)
72. K. Hauffe, 'Reaktionen in und an festen Stoffen'. (Springer-Verlag, Berlin, (1966)
73. P. Kofstad, 'High Temperature Oxidation of Metals'. (John Wiley and Sons, Inc., New York, 1966)
74. K. Hauffe and C. Z. Gensch, *Z. Physik. Chem. (Leipzig)*, **195**, 116 (1950)
75. E. A. Secco, *J. Chem. Phys.*, **29**, 406 (1958) and **34**, 1844 (1961)
76. R. A. Reynolds, Ph.D. Thesis, Stanford University, November (1965)
77. E. A. Secco, *Can. J. Chem.*, **42**, 1396 (1964)
78. T. Bansagi, E. A. Secco, O. K. Srivastava and R. R. Martin, *Can. J. Chem.*, **46**, 2881 (1968)
79. H. Gobrecht, H. Nelkowski, J. W. Baars and M. Weigt, *Solid State Comm.*, **5**, 77 (1967)
80. R. H. Blount, G. A. Marlor and R. H. Bube, *J. Appl. Phys.*, **38**, 3795 (1967)
81. N. N. Sirota and N. N. Koren, *Dan BSSR*, **6**, No. 10 (1962)
82. N. N. Sirota and N. N. Koren, *Dan BSSR*, **7**, 160 (1963)
83. E. A. Secco and C. H. Su, *Can. J. Chem.*, **46**, 1621 (1968)
84. M. M. Henneberg, Ph.D. Thesis, Stanford University (1970)
85. M. M. Henneberg and D. A. Stevenson, *Phys. Stat. Sol.*, **48**, 225 (1971)
86. W. W. Piper and S. J. Polich, *J. Appl. Phys.*, **32**, 1278 (1961)
87. V. V. Illarionov and L. M. Lapina, *Dokl. Akad. Nauk. SSSR*, **114**, 1021 (1957)
88. F. T. J. Smith, *Solid State Comm.*, **7**, 1757 (1969)
89. H. H. Woodbury, General Electric Res. Lab. Scientific Report No. 9 on Contract No. AF-19 (628)-329, November (1964)
90. H. H. Woodbury and R. B. Hall, *Phys. Rev.*, **157**, 641 (1967)
91. N. N. Sirota and N. N. Koren, *Dan BSSR*, **6**, No. 12, 760 (1962)
92. D. G. Thomas and E. A. Sadowski, *J. Phys. Chem. Solids*, **25**, 395 (1964)
93. E. A. Secco and R. S. C. Yeo, *Can. J. Chem.*, **49**, 1953 (1971)
94. N. N. Sirota and N. N. Koren, *Akad. Nauk. BSSR Minsk. Doklady*, **7**, 373 (1963)
95. R. Haul and D. Just, *J. Appl. Phys.*, **33**, 487 (1962)
96. R. Haul, D. Just and G. Dümbgen, 'Fourth International Symposium on the Reactivity of Solids', p. 65 (J. H. DeBoer, ed.). (Elsevier Publishing Company, Inc., New York, 1960)
97. R. Haul and D. Just, *Z. Electrochem.*, **62**, 1124 (1958)
98. R. Haul, D. Just and G. Dümbgen, *Z. Physik. Chem.*, **31**, 309 (1962)

99. H. H. Woodbury, *Phys. Rev.*, **134**, A492 (1964)
100. J. Crank, 'Mathematics of Diffusion', p. 34. (Clarendon Press, Oxford, 1956)
101. W. Rogalla, Dissertation, University of Göttingen (1968)
102. D. Shaw and R. C. Whelan, *Phys. Stat. Sol.*, **36**, 705 (1969)
103. R. Boyn, O. Goede and S. Kuschnerus, *Phys. Stat. Sol.*, **12**, 57 (1965)
104. G. Hershman and F. A. Kröger, *J. Solid State Chem.*, **2**, 483 (1970)
105. R. Boyn, *Phys. Stat. Sol.*, **29**, 307 (1968)
106. T. L. Larsen, Ph.D. Thesis, Stanford University (1970)
107. K. W. Boer, R. Boyn and O. Goede, *Phys. Stat. Sol.*, **3**, 1684 (1963)
108. V. Kumar and F. A. Kröger, *J. Solid State Chem.*, **3**, 387 (1971)
109. V. Kumar and F. A. Kröger, *J. Solid State Chem.*, **3**, 406 (1971)
110. H. Dünwald and C. Wagner, *Z. Physik. Chem.*, **B24**, 53 (1934)
111. P. M. Borsenberger, D. A. Stevenson and R. A. Burmeister, 'International Conference on II-VI Compounds', p. 439 (D. G. Thomas, ed.). (W. A. Benjamin, Inc., New York, 1967)
112. F. T. J. Smith, *Solid State Comm.*, **1**, 1757 (1969)
113. W. D. Callister, C. F. Varotto and D. A. Stevenson, *Phys. Stat. Sol.*, **38**, K45 (1970)
114. H. H. Woodbury and R. B. Hall, *Phys. Rev. Letters*, **17**, 1093 (1966)
115. L. R. Shiozawa and J. M. Jost, 'Research on II-VI Compound Semiconductors', p. 78. ARL-65-98 (1965)
116. R. C. Whelan and D. Shaw, 'II-VI Semiconducting Compounds', p. 451.1967 International Conference (D. G. Thomas, ed.). (W. A. Benjamin, Inc., New York, 1967)
117. P. M. Borsenberger and D. A. Stevenson, *J. Phys. Chem. Solids*, **29**, 1277 (1968)
118. R. C. Whelan and D. Shaw, *Phys. Stat. Sol.*, **29**, 145 (1968)
119. K. R. Zanio, *Appl. Phys. Letters*, **15**, 260 (1969)
120. K. R. Zanio, *Bull. Amer. Phys. Soc.*, **14**, 833 (1969)
121. F. T. J. Smith, *Trans. Metall. Soc.*, **1**, 617 (1970)
122. W. D. Callister, Ph.D. Thesis, Stanford University, May 1971
123. Y. V. Rud' and K. V. Sanin, *Soviet Phys. Semicond. (USA)*, **5**, 244 (1969)
124. R. Lindner, *Arkiv Kemi*, **4**, 385 (1952)
125. R. Lindner and H. N. Terem, *Arkiv Kemi*, **7**, 273 (1954)
126. R. Lindner, *J. Chem. Soc.*, **5**, S395 (1949)
127. A. K. Das Gupta, D. N. Sitharamarao and G. D. Palkar, *Nature*, **207**, 628 (1965)
128. B. A. Thompson and R. L. Strong, *J. Phys. Chem.*, **67**, 594 (1963)
129. J. S. Anderson and J. R. Richards, *Journal of Chem. Soc.*, **000**, 537 (1946)
130. G. Simkovich and J. B. Wagner, Jr., *J. Chem. Phys.*, **38**, 1368 (1963)
131. J. Bloem, *Philips Res. Rep.*, **11**, 273 (1956)
132. M. S. Seltzer and J. B. Wagner, Jr., *J. Phys. Chem. Solids*, **24**, 1525 (1963)
133. K. R. Zanio and J. B. Wagner, Jr., *J. Appl. Phys.*, **39**, 5686 (1968)
134. T. D. George and J. B. Wagner, Jr., *J. Electrochem. Soc.*, **115**, 1956 (1968)
135. R. F. Brebrick and W. W. Scanlon, *Phys. Rev.*, **96**, 598 (1954)
136. M. S. Seltzer and J. B. Wagner, Jr., *J. Phys. Chem. Solids*, **26**, 233 (1965)
137. M. S. Seltzer and J. B. Wagner, Jr., *J. Chem. Phys.*, **36**, 130 (1962)
138. B. I. Boltaks and Y. V. N. Mokhov, *Zhur. Tech. Fiz.*, **28**, 1046 (1958)
139. B. I. Boltaks and Y. V. N. Mokhov, *Soviet Phys. Tech. Phys.*, **3**, 974 (1958)
140. Y. Ban and J. B. Wagner, Jr., *J. Appl. Phys.*, **41**, 2818 (1970)
141. J. F. Butler, *J. Electrochem. Soc.*, **111**, 1150 (1964)
142. R. W. Brodersen, J. N. Walpole and A. R. Calawa, *J. Appl. Phys.*, **41**, 1484 (1970)
143. R. F. Brebrick, *J. Appl. Phys.*, **30**, 811 (1959)
144. T. D. George and J. B. Wagner, Jr., *J. Appl. Phys.*, **42**, 220 (1971)

145. M. P. Gomez, D. A. Stevenson and R. A. Huggins, *J. Phys. Chem. Solids*, **32**, 335 (1971)
146. R. F. Brebrick and R. S. Allgaier, *J. Chem. Phys.*, **32**, 1826 (1969)
147. R. F. Brebrick and E. Gubner, *J. Chem. Phys.*, **36**, 1283 (1962)
148. E. L. Brady, *J. Electrochem. Soc.*, **101**, 466 (1954)
149. B. I. Boltaks and Y. V. N. Mokhov, *Zhur. Tech. Fiz.*, **26**, 2448 (1956)
150. B. I. Boltaks and Y. V. N. Mokhov, *Soviet Phys. Tech. Phys.*, **1**, 2366 (1957)
151. F. T. J. Smith, *Bull. Am. Phys. Soc.*, **14**, 351 (1969)
152. H. Tubota, H. Suzuki and K. Kirakawa, *J. Phys. Soc. Japan*, **15**, 1701 (1960)
153. M. P. Hung, N. Ohashi and K. Igaki, *J. Appl. Phys. Japan*, **8**, 652 (1969)
154. F. T. J. Smith, *Solid State Comm.*, **8**, 263 (1970)

DIFFUSION IN OXIDE SEMICONDUCTORS 8

J. BRUCE WAGNER, Jr.

CONTENTS

8.1 INTRODUCTION

Oxides play an important role in certain semiconductors, as corrosion product layers on metals, as catalysts, and as materials used in the manufacture of ceramics. In this chapter an attempt will be made to point out trends in transport through oxides.† The diversity of different oxides, their structures and compositions make classification difficult. Some generalizations can be made which aid in classifying the types of diffusion in oxides. Bulk diffusion depends on temperature, oxide structure, the number of point defects in the oxide and hence the position of the metal in the periodic table and finally on the partial pressure of oxygen coexisting over a particular oxide. Special attention will be directed to the role of the partial pressure in equilibrium with the oxide.

In the case of the alkali halides, for example, both the ionic conductivity and the diffusion coefficient exhibit Arrhenius behavior and the resulting plots show a break at a certain temperature above which

† The oxides ZnO and PbO will not be treated in this chapter as they are covered in chapter 7.

intrinsic behavior predominates and below which extrinsic behavior predominates. In contrast, most oxides [1] at all temperatures exhibit deviations from the stoichiometric ratio. They only rarely exist as a stoichiometric compound. Consequently, a similar Arrhenius type plot for most oxides heated, e.g. in a vacuum, may exhibit two straight line portions in apparent analogy to the behavior of the alkali halides. However, the higher temperature portion would represent that composition, usually non-stoichiometric, which coexists with the metal† (see figure 8.1). It is therefore useful to classify oxides in terms of their dissociation pressures and because ΔG° is proportional to the logarithm of the partial pressure of oxygen one may classify these in terms of the free energy per mole of the oxide or per gram atom of oxygen.

8.2 ELLINGHAM DIAGRAMS

Ellingham [2] was the first to present information concerning the dissociation pressures of oxides in the form of a chart on which was plotted the free energy of formation of the particular oxide versus temperature. This free energy of formation, ΔG°, in turn is related to the dissociation pressure of the oxide in question. As an example, consider the metal oxide, MO, which dissociates according to

$$MO(S) \rightleftharpoons M(S) + \tfrac{1}{2}O_2(G) \tag{8.1}$$

whence

$$K_D = p_{O_2}^{1/2}\{M\}/\{MO\} \tag{8.2}$$

The activities of the pure components may be taken as unity. Therefore,

$$K_D = \exp\,(+\Delta G^{\circ}/kT) \tag{8.3}$$

and

$$\Delta G^{\circ} = +kT \ln p_{O_2}^{1/2} \tag{8.4}$$

where ΔG° is the free energy of formation of MO. The oxygen pressure,

† The term, intrinsic, has sometimes been applied to two different physical situations. It is sometimes used to denote the condition of an undoped stoichiometric crystal which exhibits only thermally generated point defects. Alternatively, it is also applied to an undoped but non-stoichiometric crystal. The second condition is also sometimes termed extrinsic in the sense that native point defects ('self-impurities') are brought about by changes in stoichiometry. For these situations, it seems better to describe crystals as stoichiometric or non-stoichiometric as the case may be and reserve the terms intrinsic and extrinsic as terms referring to defects native to the crystal, stoichiometric or non-stoichiometric, and those brought about by impurities, respectively. This nomenclature will be adopted here. The reader is cautioned of possible differences in definition in the literature.

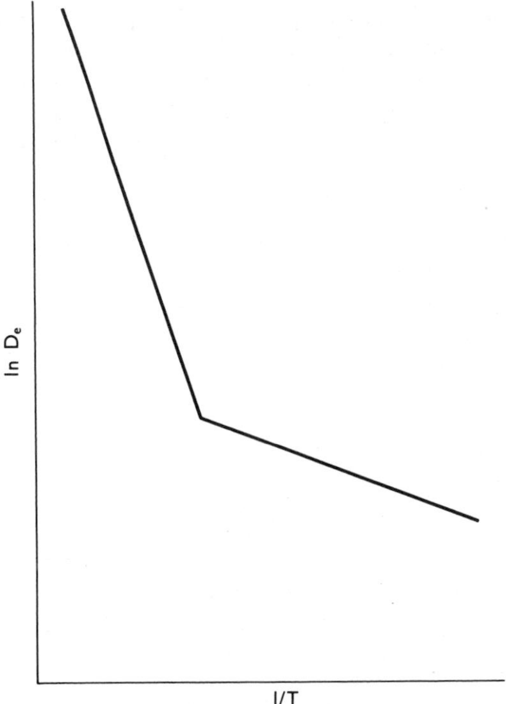

Figure 8.1. Schematic diagram of an Arrhenius plot for tracer diffusion as a function of temperature for an oxide crystal heated in a vacuum. For most oxides, both segments represent non-stoichiometric behavior because oxides only rarely exist at the stoichiometric composition when heated in a vacuum. For example, the higher temperature segment would usually represent diffusion in the composition coexisting with the metal. The lower temperature segment represents diffusion in an oxide of constant composition, i.e. constant oxygen to metal ratio, where there may also be a fixed impurity level.

p_{O_2}, is the dissociation pressure of MO. The dissociation oxygen pressure, p_{O_2}, may also be expressed in terms of gas mixtures such as CO-CO_2 or H_2-H_2O because,

$$CO_2(G) \rightleftharpoons CO(G) + \tfrac{1}{2}O_2(G) \tag{8.5}$$

and

$$H_2O(G) \rightleftharpoons H_2(G) + \tfrac{1}{2}O_2(G). \tag{8.6}$$

From the equilibrium constants for these two equations one can relate the dissociation oxygen pressure to a particular gas mixture. Richardson and Jeffes [3] used this property to prepare a nomograph as is described in the caption to figure 8.2. These data are extremely useful in relating the stability range for the oxides under consideration. It should be emphasized

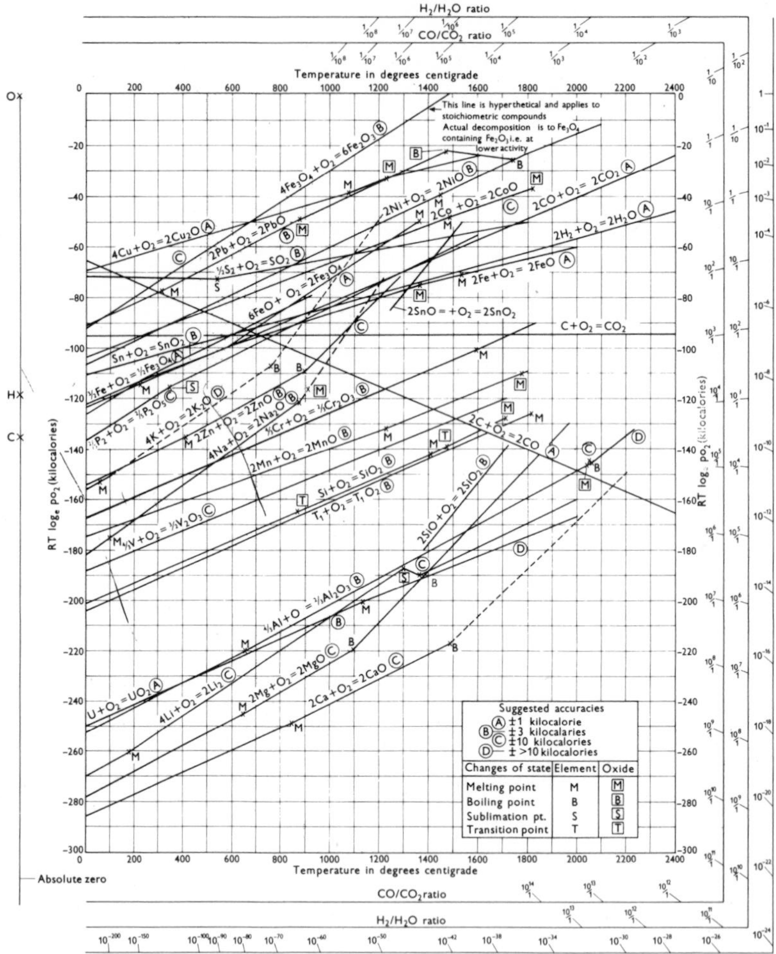

Figure 8.2. Standard free energy of formation of oxides as a function of temperature. The scales on the nomograph consist of the CO-CO_2 ratio, the H_2/H_2O ratio and the equivalent partial pressure of oxygen. In order to use the chart, a straight edge is placed on the chart to intersect the abscissa on the left at the point marked O, H, or C for the gases, oxygen, H_2/H_2O or CO/CO_2, respectively. The straight edge is adjusted to intersect the line on the chart representing the reaction under consideration at the temperature in question and the intersection on the appropriate oxygen or gas mixture scale yields the partial pressure or volume ratio directly. On this chart, the free energies are reported in cal/mole. In the text, energy values are cited in units of electron volts per particle. One calorie/mole is equivalent to 23,062 eV particle. Diagram prepared by F. D. Richardson 1961. Reproduced by permission.

that the dissociation pressure refers to that pressure for the co-existence of the metal and the metal oxide or in some cases the equilibrium pressure over a two phase mixture of two oxides. Because oxides are generally non-stoichiometric the stability range of an oxide must also be defined.

8.3 PHASE DIAGRAMS FOR OXIDES

The conventional temperature versus composition diagram is useful for oxides exhibiting very large deviations from stoichiometry. More usual are cases wherein the deviations from stoichiometry are very small and consequently the homogeneity range or limits of the existence of the single phase region are more conveniently expressed in terms of the partial pressure of oxygen co-existing with a phase. Figure 8.3 shows an example of the former case and figure 8.4 shows an example of the later case. The use of these types of pressure, temperature, composition diagrams does not specify the defect equilibria existing within the single phase field. It is a completely rigorous thermodynamic expression of the homogeneity range. In order to relate the partial pressure of oxygen to the defect equilibria, one must have a model for the type of disorder within a given oxide.

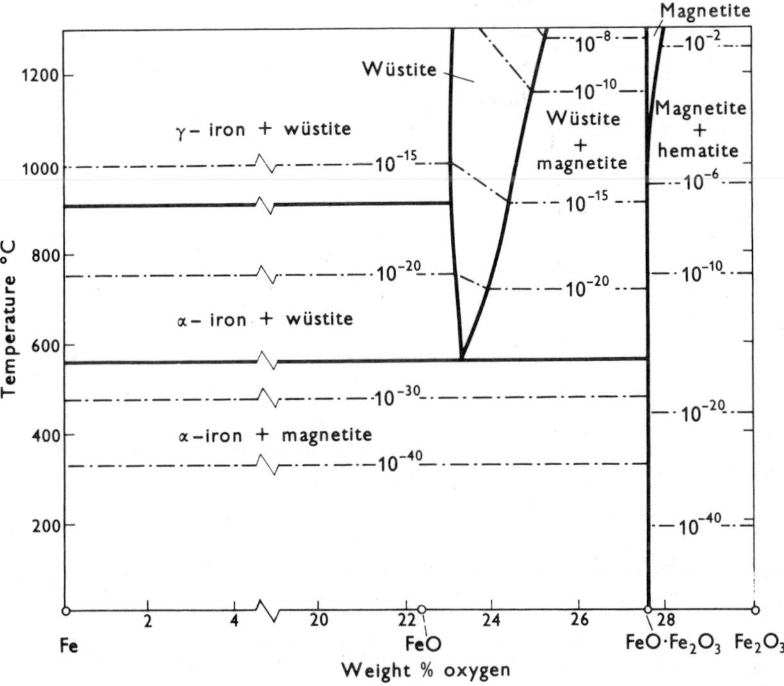

Figure 8.3. Example of a phase diagram plotted conventionally as temperature versus composition, the iron-oxygen system. The three phases, wüstite, magnetite and hematite have the nominal formulae FeO, Fe_2O_4 and Fe_2O_3, respectively. The heavy, solid lines represent phase boundaries and the light dash-dot lines denote isobars of oxygen in atm.

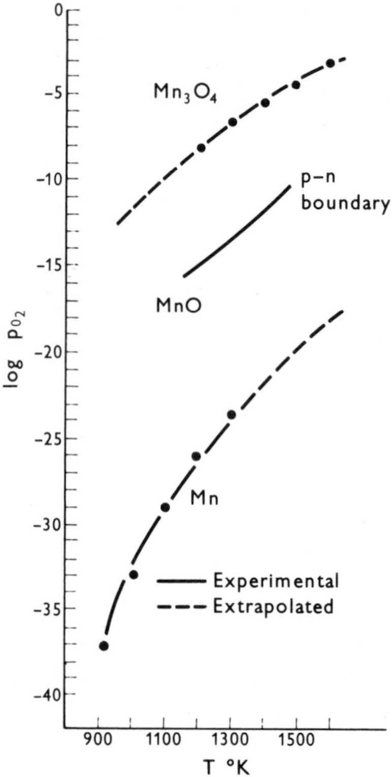

Figure 8.4. Example of a phase diagram using the logarithm of the oxygen pressure coexisting with an oxide (manganous oxide) as a function of temperature. The solid lines with the points marked ● denote the phase field limits of Mn_3O_4-MnO and MnO-Mn. The solid line marked, '*p-n* boundary' denotes the position of the apparent *p* to *n* transition, *n*-type at the lower oxygen pressures. After Price and Wagner [29].

8.4 DEFECT EQUILIBRIA, ELECTRONEUTRALITY AND MASS ACTION LAWS

When an oxide exists at the stoichiometric composition, it may have thermal disorder—either Schottky, Frenkel, anti-Schottky, anti-Frenkel or anti-structural disorder. Of course these are limiting cases and combinations are possible. One convention is to denote the mass action constant for these thermal defects as K_V, $K_F(\mathrm{M})$, K_{AV}, $K_F(\mathrm{O})$ and $K_{A\,Str.}$ respectively. When the defects are ionized, a superior prime is used to denote the state of ionization. For example,

$$K_V = [V_{\mathrm{M}}^{\times}]\,[V_{\mathrm{O}}^{\times}]\,, \tag{8.6}$$

$$K_V' = [V_M'][V_O^{\cdot}], \tag{8.7}$$

$$K_V'' = [V_M''][V_O^{\cdot\cdot}], \text{ etc.} \tag{8.8}$$

When a given oxide exists over a range of oxygen pressures, the metal or the oxygen which is added to or subtracted from the lattice (resulting in a departure from stoichiometry) must be accommodated in some way. This dissolution or exsolution, as the case may be, will form or destroy lattice defects. As examples consider the introduction of excess oxygen into a simple oxide MO. Excluding antistructural disorder, excess oxygen may enter the lattice according to

$$\tfrac{1}{2}O_2(G) \rightleftharpoons O_i^{m'} + mh^{\cdot} \tag{8.9}$$

or

$$\tfrac{1}{2}O_2(G) \rightleftharpoons O_O + V_M^{m'} + mh^{\cdot} \tag{8.10}$$

where the superior m denotes the degree of ionization. Likewise excess metal may enter the lattice of MO according to

$$M(G) \rightleftharpoons M_i^{m\cdot} + me' \tag{8.11}$$

or

$$M(G) \rightleftharpoons M_M + V_O^{m\cdot} + me'. \tag{8.12}$$

If the concentrations of defects are sufficiently low, the thermodynamic activities may be replaced by the respective concentrations in the mass action laws. Thus equations 8.9-8.12 become

$$K_{Ox} = [O_i^{m'}]\,p^m/p_{O_2}^{1/2} \tag{8.13}$$

$$K_{Ox} = [V_M^{m'}]\,p^m/p_{O_2}^{1/2} \tag{8.14}$$

$$K_R = [M_i^{m\prime}]\,n^m/p_M \tag{8.15}$$

$$K_R = [V_O^{m\cdot}]\,n^m/p_M \tag{8.16}$$

These equations may be expressed in terms of either the partial pressure of M or of O_2 because the free energy of formation of the oxide MO is related to the partial pressures of the constituents by

$$M(G) + \tfrac{1}{2}O_2(G) \rightleftharpoons MO(S) \tag{8.17}$$

whence

$$K = \frac{1}{p_M\, p_{O_2}^{1/2}}. \tag{8.18}$$

Note the metal gas phase is written in equations 8.17 and 8.18 in contrast to equations 8.1 and 8.2. The difference is of course the free energy for vaporization of the metal.

Because electrical neutrality is required in the crystals, the sum of all the positive charges must be equal to the sum of all the negative charges in the lattice. As an example, consider a crystal, MO, which exhibits doubly ionized cation vacancies and free holes. The simple electroneutrality condition would be, $2[V_M''] = p$.

Brouwer [4], and more extensively, Kröger and Vink [5], have outlined graphical procedures for describing the isothermal variation in concentration of defects as a function of the partial pressure of the metal or metalloid. Unfortunately, the experiments which are being carried out for diffusion as a function of the partial pressure of the metalloid in the chalcogenides are not yet available for oxides and by and large most diffusion studies have been carried out only in a narrow interval of composition over which only one of the limiting electroneutrality conditions is dominant. Consequently, the graphical procedure of Brouwer will not be described here.

8.5 DIFFUSION IN OXIDES

8.5.1 Tracer and Self-diffusion in Binary Oxide Semiconductors

The self-diffusion coefficient is denoted by $D_e(i)$. It represents the migration of an atomic species, i, of the homogeneous oxide due to thermal excitation only, i.e. no gradient in chemical potential exists within the oxide. It is related to the tracer diffusion coefficient, $D_e(\widetilde{i})$, of each component, i, by

$$D_e(\widetilde{i}) = fD_e(i) \tag{8.19}$$

for diffusion via a vacancy mechanism. In equation 8.19, f is the correlation factor. Expressions relating the self-diffusion coefficient to the tracer diffusion coefficient for other mechanisms are discussed in Chapter 1.

If it can be assumed that the various migration mechanisms operate independently and that successive jumps via these mechanisms are random, then the effective self-diffusion coefficient of the ith species, $\mathbf{D}_e(i)$, may be represented by,

$$\mathbf{D}_e(i) = \sum_j d_j^2 \Gamma_j c_j \tag{8.20}$$

where d_j is the jump distance, Γ_j is the jump frequency, and c_j is the defect concentration by which the ith species jumps.† Experimentally, one

† The concentrations of defects and activities of all species cited in sections 8.5.1-8.5.5 are those under equilibrium conditions of a fixed temperature total pressure and chemical composition.

tries to work in a region of temperature and oxygen pressure where only one mechanism predominates. Under such conditions, the composition of the crystal is usually within one of the limiting electroneutrality conditions as portrayed on a Brouwer diagram [4, 5] and the self-diffusion coefficient may be approximated by a single diffusion mechanism as,

$$D_e(i) = d_l^2 \Gamma_l c_l \tag{8.21}$$

where c_l denotes the majority defect lattice concentration which also is supposed to be the most mobile, for example, singly ionized cation vacancies in cobaltous oxide. In what follows, it will be assumed that only one mechanism predominates unless otherwise specified.

The temperature dependence of the self-diffusion coefficient generally obeys the Arrhenius expression

$$D_e(i) = D^*_{o,e}(i) \exp(-Q/RT) \tag{8.22}$$

Consider first un-doped, non-stoichiometric oxides. If measurements are made on crystals whereby each crystal has the same, fixed concentration of defects then the activation energy,† Q, is equal to the enthalpy of migration, ΔH_m. If measurements are made under a fixed oxygen pressure, then the value of the activation energy, Q, will be the sum of two terms. One term will consist of the enthalpy of motion, ΔH_m, and the second term will contain the enthalpy of formation of the defects, ΔH_f. It is emphasized that both sets of data represent the temperature dependence of self-diffusion coefficients. That is, each isothermal diffusion measurement is obtained on a crystal which is equilibrated with the gas atmosphere. However, in the former case, the gas atmosphere would be adjusted at each temperature in order to provide a fixed composition (oxygen to metal ratio) whereas in the second case, the oxygen pressure remains constant at each temperature and the composition (oxygen to metal ratio) is fixed but different for each temperature. Therefore, activation energies for different oxides and indeed for the same oxide must be compared under conditions where this distinction is clear. As an illustration, when ferrous oxide (wüstite) is heated under a constant oxygen pressure, it reduces (oxygen to metal ratio decreases) whereas the oxygen to metal ratio increases for cobaltous oxide and nickelous oxide. Thus the value of Q obtained from diffusion studies on ferrous oxide heated under constant oxygen pressure would reflect the enthalpy for destruction of the defects and conversely for cobaltous and nickelous oxides.

† Generally, usage has dictated the symbol, Q, to denote the activation energy obtained from an Arrhenius plot. Actually, it is the activation enthalpy. This distinction will be made in the sections in which Q will be shown to be the algebraic sum of a migration enthalpy and a formation enthalpy.

With the use of the transition state theory, the jump frequency can be expressed as

$$\Gamma_l = (\alpha kT/h) \exp(-\Delta G_m/kT) = (\alpha kT/h) \exp(-\Delta H_m/kT + \Delta S_m/k) \quad (8.23)$$

where α is a geometric factor which depends on the lattice on which the atom is migrating and $\Delta G_m = \Delta H_m - T\Delta S_m$ represents the free energy, enthalpy and entropy of migration. And from the thermodynamics of the dissolution or exsolution of oxygen, e.g. equations 8.9 through 8.12, the majority lattice defect concentration, c_l, may be expressed by an equation of the form,

$$c_l = \left(\frac{1}{m^m}\right)^{1/(m+1)} p_{O_2}^{\pm\frac{1}{2}(m+1)} \exp\left[-\Delta H_f/(m+1)kT + \Delta S_f/(m+1)k\right] \quad (8.24)$$

where the plus or minus sign as exponent over the partial pressure of oxygen represents the possibility for the defects to be created or destroyed, by the addition or removal of oxygen to the lattice, as the case may be. Hence, combining equations 8.21, 8.23 and 8.24 yields

$$D_e(i) = \left(\frac{\alpha kT}{h}\right) d^2 m^{1/(m+1)} p_{O_2}^{\pm\frac{1}{2}(m+1)} \exp\left\{\left[-\left(\frac{\Delta H_f}{(m+1)} + \Delta H_m\right)/kT\right] + \left[\left(\frac{\Delta S_f}{(m+1)} + \Delta S_m\right)\right]/k\right\}. \quad (8.25)$$

whence

$$D_e(i) = D_{o,e} p_{O_2}^{\pm\frac{1}{2}(m+1)} \exp\left\{-\left(\frac{\Delta H_f}{(m+1)} + \Delta H_m\right)/kT\right\}. \quad (8.26)$$

where $D_{o,e}$ differs from the pre-exponential in equation 8.22. In equation 8.22 the pre-exponential may contain a pressure term and of course the activation energy, Q, contains terms for the migration and formation enthalpies.

As an example, consider diffusion in an oxide MO which is metal deficient according to equation 8.12 with $m = 1$. Thus the simplified

† Several different units for concentration may be used, as, for example, defects per cubic centimeter, defects per ion pair and mole fraction. One must of course be consistent. In equation 8.21, defects per ion pair or mole fraction must be used.

In applying the mass action law, activities should be used which result in a dimensionless mass action or equilibrium constant. Consequently, the partial pressures should be relative to or normalized by a standard state partial pressure. When this *choice* is made as one atmosphere, equations of the type presented in equation 8.28 are dimensionally correct. Other choices of a standard state may be made, e.g. the equilibrium pressure co-existing over the metal-metal oxide phases. When the standard state partial pressure is not unity, an explicit constant or the standard state pressure must be included in equations of the type shown by 8.28.

electroneutrality condition is given by $[V'_{\mathrm{M}}] = p$. Hence the majority lattice defect concentration† is given by

$$c_l = [V'_{\mathrm{M}}] = K^{1/2} p_{\mathrm{O}_2}^{1/4} \tag{8.27}$$

whence

$$[V'_{\mathrm{M}}] = p_{\mathrm{O}_2}^{1/4} \exp\left[-\Delta H_f/2kT + \Delta S_f/2k\right]. \tag{8.28}$$

Thus equation 8.26 becomes,

$$D_e(\mathrm{M}) = D_{o,e} p_{\mathrm{O}_2}^{1/4} \exp\left[-\left(\frac{\Delta H_f}{2} + \Delta H_m\right)/kT\right] \tag{8.29}$$

for diffusion of the metal ion, M, via singly ionized cation vacancies, V'_{M}. Thus a plot of log $D_e(\mathrm{M})$ versus log p_{O_2} should yield a slope of +¼ and these data in conjunction with other types of data such as electrical conductivity, Hall effect, Seebeck coefficient, lattice parameter, density, and thermogravimetric analysis allow a diffusion mechanism to be inferred. There will be analogous expressions with an oxygen pressure dependence of $p_{\mathrm{O}_2}^{-1/(m+1)}$ for oxides containing excess metal. In each case, a model must be available against which the observed pressure dependence may be tested. In any such simple correlation, one must be assured that only one mechanism predominates over the composition interval studied and likewise over the temperature range studied—conditions which may not always be met.

The pre-exponential term, $D_{o,e}$ contains the entropy of formation (partial molar entropy of oxygen in MO), ΔS_f, plus the migration entropy of the atoms. From equations 8.24-8.26,

$$D_{o,e} = (d^2 \alpha kT/h) \exp\left[\left(\frac{\Delta S_f}{(m+1)} + \Delta S_m\right)/k\right]. \tag{8.30}$$

When the crystal is doped to provide extrinsic defects in excess of those which would exist at the same temperature and under the same oxygen pressure as the undoped crystal, the pre-exponential is reduced by approximately the fraction of impurity atoms added.

Next, consider undoped, stoichiometric oxides. Some oxides formed from metals on the left side of the periodic chart, for example calcium oxide, are reported to be stoichiometric when heated in an inert gas or in a vacuum. The behavior of such oxides is expected to be analogous to that for the alkali halides. The activation energy for diffusion contains a term for the formation of thermal defects plus a term for the migration of the atoms. Referring to figure 8.1, the high-temperature portion of the Arrhenius plot is supposed to represent intrinsic behavior while the low temperature portion represents extrinsic behavior. The intrinsic diffusion is represented by equation 8.22 modified as follows [6],

$$D_e(\tilde{M} \text{ in MO}) = D_{o,e} \exp\left[-(\Delta H_V/2 + \Delta H_m)/kT\right] \tag{8.31}$$

where ΔH_V denotes the enthalpy of formation of a Schottky pair and it is divided by two to account for a single vacancy and ΔH_m is the migration enthalpy. A similar expression would hold for other types of thermal disorder. In the extrinsic range, the concentration of defects created by impurities dominates the diffusion behavior and

$$D_e(\tilde{M} \text{ in doped MO}) = D_{o,e} \exp\left[-\Delta H_m/kT\right]. \tag{8.32}$$

By measuring the diffusion of a constituent in both the intrinsic and in the extrinsic range, one can thus determine the enthalpy of formation of the thermal disorder. For this approach to be valid, one must ascertain that the oxide does not change composition, that is, deviate from stoichiometry on heating.

The pre-exponential term, $D_{o,e}$ in equation 8.31 contains the entropy of formation of the thermal defects, ΔS_V, and the entropy of migration of the atoms, ΔS_m. Using the transition state approach,

$$D_{o,e} = (d^2 \alpha kT/h) \exp\left[\Delta S_V/2 + \Delta S_m)/k\right] \tag{8.33}$$

and in the extrinsic range where impurities dominate, the pre-exponential is approximately equal to the fraction of impurities times $D_{o,e}$ for the undoped crystal [6]. Thus the pre-exponential for doped crystals exhibits a much lower value than that for the undoped crystals.

8.5.2 Tracer and Self-diffusion in Binary Oxides of Nominal Formulas M_2O and MO

Almost no studies exist for oxides of metals in the first column in the periodic chart. A notable exception is cuprous oxide, Cu_2O. Cuprous oxide is a *p*-type semiconductor and the transference number of electron holes is almost unity at temperatures around 700° to 1000°C. A combination of thermoelectric effect measurements, electronic conductivity studies, radiotracer measurements and experiments involving oxidation of copper to cuprous oxide to be discussed later, have shown that the majority lattice defects are singly ionized cation vacancies with compensation via electron holes according to the simple expression

$$\tfrac{1}{2}O_2(G) \rightleftharpoons O_O + 2V'_{Cu} + 2h^{\cdot}, \tag{8.34}$$

for pressures between about 10^{-4} torr and the pressure for the formation of cupric oxide (CuO). When the point defects are in a dilute solution the mass action law can be applied using concentrations in place of the thermodynamic activities of the species. Hence the mass action constant may be written as

$$K_{Ox} = [V'_{Cu}]^2 p^2 / p_{O_2}^{1/2} \tag{8.35}$$

The simplified electroneutrality condition requires that $[V'_{Cu}] = p$. Hence

$$[V'_{Cu}] = K_{O_2}^{1/4} p_{O_2}^{1/8} \tag{8.36}$$

$$= p_{O_2}^{1/8} \exp[-\Delta G_f/4kT] \tag{8.37}$$

$$= p_{O_2}^{1/8} \exp[-\Delta H_f/4kT + \Delta S_f/4k] \tag{8.38}$$

$$= \text{const.}\ p_{O_2}^{1/8} \exp[-\Delta H_f/4kT] \tag{8.39}$$

and of course the electron hole concentration obeys the same dependence on oxygen pressure. The terms, $\Delta G_f = \Delta H_f - T\Delta S_f$, represent the free energy, enthalpy and entropy of equation 8.34. They therefore represent the partial molar quantities for the dissolution of one half mole of oxygen in solid cuprous oxide.

Radiotracer diffusion studies using Cu-63 have been carried out by Moore and Selikson [7, 8]. Their results for cuprous oxide heated in 0.1 torr of oxygen are,

$$D_e(\widetilde{Cu} \text{ in } Cu_2O) = 0.0436 \exp(-1.6/kT) \text{ cm}^2 \text{ s}^{-1}. \tag{8.40}$$

Their measurements as a function of temperature were made at constant oxygen pressure so the activation energy obtained from an Arrhenius plot of $\log D_e(\widetilde{Cu})$ versus $1/T$ contained both the enthalpy of migration plus the enthalpy for forming the defects. When the effect of constant pressure was taken into account, comparison with measurements on electronic conductivity of homogeneous samples and with parabolic oxidation experiments of a type to be described later, cation vacancies were inferred to be the dominant lattice defect according to equation 8.34. Cuprous oxide is one of the most studied oxides and its defect structure and diffusion properties are relatively well-understood for 10^{-4} torr $\leqslant p_{O_2} \leqslant p_{O_2}$ (over Cu_2O-CuO). Although cuprous oxide coexisting with copper has been suggested to contain appreciable concentrations of interstitial copper in analogy with cuprous sulfide, Hall effect, parabolic oxidation, and optical studies, suggest that the dominant defects are cation vacancies over the entire homogeneity range [9].

The diffusion of oxygen in cuprous oxide under constant oxygen pressure ($p_{O_2} = 135$ torr) has been measured using O^{18} and an exchange method [10]. The results are given by

$$D_e(\widetilde{O} \text{ in } Cu_2O) = 6.5 \times 10^{-3} \exp(-1.7/kT) \text{ cm}^2 \text{ s}^{-1}. \tag{8.41}$$

Because the data were obtained for constant oxygen pressure, the activation energy of 1.7 eV represents the enthalpy for formation of the defects plus the enthalpy for migration. It is surprising that the activation energies for the diffusion of copper and for oxygen are approximately the same. Moore and coworkers [10] suggested that the oxygen and the copper diffused by way of jumps into cation vacancies. They termed such a coupling of the oxygen diffusion and the copper diffusion the 'counter vacancy mechanism'. Moreover, there were preliminary data [10] which showed that the diffusion of oxygen increased as the partial pressure of oxygen and hence the cation vacancy concentration increased. This type of behavior, as Moore carefully pointed out, does not distinguish between oxygen diffusing via an interstitial mechanism and their proposed counter vacancy mechanism. This dilemma has also been reported for lead sulfide and for lead selenide although in the lead salts other physical measurements have been used to support the movement of the anion (sulfur or selenium, as the case may be) via interstitial jumps rather than via cation vacancies (the counter vacancy mechanism) [11-13]. In spite of many studies on Cu_2O, the detailed defect equilibria, especially at low oxygen pressures, and at elevated temperatures requires more experimental data. For example, the electronic conductivity decreases with increasing oxygen pressures at temperatures above about 1000° C. Diffusion studies as a function of oxygen pressure, especially at low oxygen pressures and on single crystals, are needed to extend our knowledge of the mechanism especially because the dependence of the electronic conductivity on oxygen pressure is not proportional to $p_{O_2}^{1/8}$ over the entire range of stability.

The series of transition metal oxides, NiO, CoO, and FeO and MnO have been studied by a variety of techniques in some detail. All are metal deficient, possess transference numbers for electrons of almost unity and exhibit the NaCl structure. The deviation from stoichiometry for the first three at temperatures around 1100° C are in the order NiO $<$ CoO $<$ FeO. The activation energies for cation self-diffusion decrease according to this same sequence. Consequently, the cation diffusion can be related to the position of the metal in the periodic chart, at least for this series of oxides having the same crystalline structure. The lower the ionization energy for a given cation (ease of formation of electron holes), the larger the deviation from stoichiometry for essentially the same values of the dielectric constant and the smaller is the activation energy for diffusion. For a more realistic comparison, one should cite the concentration of defects in each crystal. Nevertheless, this kind of trend is useful for predictions and tests should be made for oxides of common structure but different free energy of formation per gram atom of oxygen.

Cobaltous oxide is one of the best understood because its deviation from stoichiometry is large enough to be measured easily (e.g., by thermogravimetry) [14, 15] and yet small enough to allow the concentrations of defects to be used in place of thermodynamic activities. As is the case for cuprous oxide, cobaltous oxide exists over most of its homogeneity range as a p-type semiconductor with the electron holes compensating singly charged cation vacancies. Hence,

$$\tfrac{1}{2}O_2(G) \rightleftharpoons O_O + V'_{Co} + h^{\cdot} \tag{8.42}$$

and application of the mass action law using the simplified electroneutrality condition that $[V'_{Co}] = p$ yields,

$$[V'_{Co}] = p_{O_2}^{1/4}\sqrt{K_{Ox}} = \text{const.}\, p_{O_2}^{1/4} \exp\left[-\Delta H_f/2kT\right]. \tag{8.43}$$

Again, because cation vacancies are the majority lattice defects, the migration of cobalt occurs via jumps into these defects. Carter and Richardson [16] were the first to study the tracer diffusion of cobalt as a function of oxygen partial pressures. Because $D_e(\tilde{Co})$ is proportional to $[V'_{Co}]$, a plot of log $D_e(\tilde{Co})$ versus log p_{O_2} should yield a slope of 0.25. Carter and Richardson found experimentally that the slopes were 0.35 at 1000°, 0.30 at 1150° and 0.28 at 1350° C. The slopes were slightly greater than predicted. In similar experiments, the electronic conductivity, σ of single crystals of cobaltous oxide has been measured as a function of oxygen pressure [15]. Plots of log σ versus log p_{O_2} yield slopes of 0.26 at 1000-1200° C (see figures 8.5 and 8.6). Thus the assumption of the singly ionized cation vacancies as the *majority* defect in cobaltous oxide is verified for oxygen pressures between 1 atm and about 10^{-4} atm and the temperature range cited. The deviations from the predicted slope of 0.25 suggest that jumps into vacancies having ionization states other than one and/or the influence of impurities may be affecting the data. Between 950° and 1600° C, and in air,

$$D_e(\tilde{Co} \text{ in CoO}) = 5 \times 10^{-3} \exp(-1.67/kT) \text{ cm}^2 \text{ s}^{-1}. \tag{8.44}$$

The activation enthalpy contains the partial molar enthalpy of dissolution of oxygen plus the migration enthalpy. For 1 atm of oxygen Eror [15] calculated the migration enthalpy, ΔH_m, to be 1.2 eV and the enthalpy of formation, ΔH_f, to be 0.62 eV. The value of ΔH_f is composition dependent and hence depends on the partial pressure of the oxygen above the crystal.

Actually, there is evidence of appreciable concentrations of interstitial cobalt present as a minority defect as the composition approaches the phase boundary Co/CoO. Most of these data involve electrical properties [14, 17]. There have not been sufficient tracer experiments at low

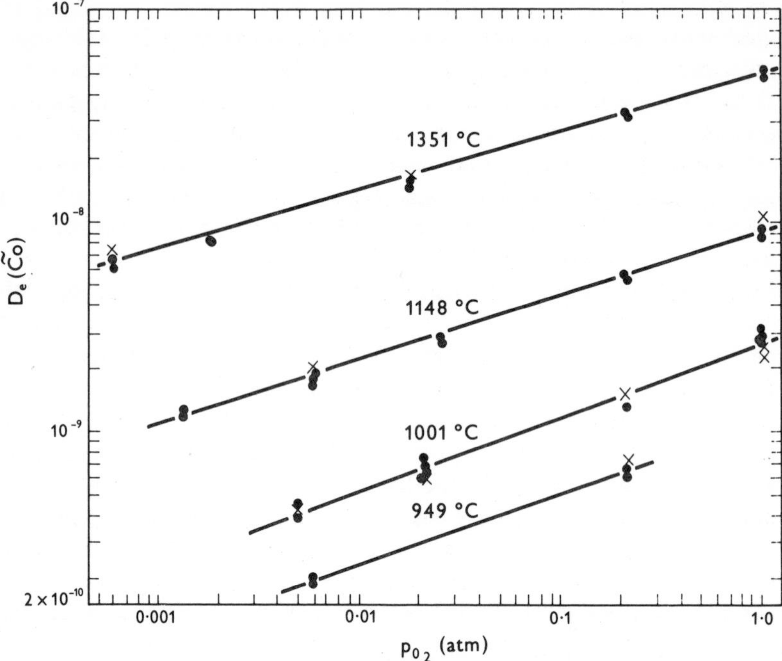

Figure 8.5. Tracer diffusion of cobalt in cobaltous oxide as a function of oxygen pressure. The symbols, *X*, denote data obtained by a sectioning technique while the filled circles denote data obtained by a surface decrease method. The slopes of the lines are approximately one fourth indicating the existence of singly ionized cation vacancies (see equation 8.43). After Carter and Richardson [16].

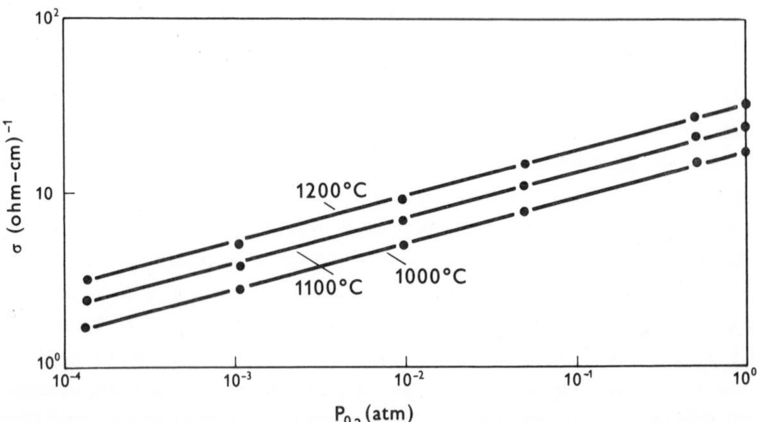

Figure 8.6. Electronic conductivity of cobaltous oxide single crystals as a function of oxygen pressure. The slopes of the lines are approximately one quarter indicating the existence of singly ionized cobalt vacancies (see equation 8.42). After Eror and Wagner [15].

chemical potentials of oxygen to test the possible contribution due to interstitials. This contribution may be important in the other oxides of this series as will be discussed below.

More recently Crow [18], and Chen [19] have studied the diffusion of cobalt in single crystals of cobaltous oxide. Their results agree with those of Carter and Richardson. Moreover, Chen and co-workers carried out a quantitative 'isotope experiment' on cobaltous oxide. These workers used Co-60 and Co-55 in a constant oxygen pressure (air, $p_{O_2} = 0.21$ atm). The strength of the isotope effect, E_j, is defined by

$$E_{60} = [1 - D_{60}/D_{55}]/[1 - m_{55}/m_{60}]^{1/2} = f\Delta K \tag{8.45}$$

where the subscripts refer to the isotope of cobalt studied, m denotes nuclide mass, f is the correlation factor = 0.78 for the NaCl structure, and ΔK denotes the fraction of the total translational kinetic energy of the jumping atom at the saddle point and in the direction of the jump. Because E_{60} was 0.582, $\Delta K = 0.746$. These data were consistent with a vacancy mechanism over this temperature range and under the highly oxidizing condition, $p_{O_2} = 0.21$ atm. These studies provide an elegant confirmation of the dominant lattice defects under high oxygen pressures. More importantly, they point to the feasibility of the method for determining diffusion mechanisms in oxides.

Thompson [20], Chen and Jackson [21], and Holt [22] have also studied the diffusion of oxygen in cobaltous oxide. Chen and Jackson used an oxygen exchange method while Holt used proton activation wherein the O-18 isotope is located using the nuclear reaction, $O^{18}(p, n)F^{18}$, and autoradiography to detect the fluorine. The diffusion data for oxygen in CoO under air (0.21 atm) are in excellent agreement between different investigators using different techniques [21, 22] and are given by

$$D_e(\tilde{O} \text{ in CoO}) = 50 \exp(-4.1/kT) \text{ cm}^2 \text{ s}^{-1}. \tag{8.46}$$

The activation energy of 4.1 eV is much greater than that for the diffusion of cobalt and the mechanism of oxygen diffusion can be clearly inferred not to occur via cation vacancies. Referring to equation 8.42 one can see the non-stoichiometric undoped oxide incorporates oxygen via singly ionized cation vacancies, and

$$K_{Ox} = [V'_{Co}]p/p_{O_2}^{1/2} \tag{8.47}$$

The introduction of aliovalent impurities changes the electroneutrality condition to

$$p = [V'_{Col}] + [Li'_{Co}] \tag{8.48}$$

for lithium as a dopant and,

$$p + [Al^{\cdot}_{Co}] = [V'_{Co}] \tag{8.49}$$

for aluminium as a dopant. Because of the Schottky constant, $K_V = [V_{Co}^{m\prime}][V_O^{m\cdot}]$, the introduction of lithium increases the concentration of electron holes, decreases the concentration of cation vacancies and increases the concentration of anion vacancies. Conversely, the introduction of aluminium decreases the concentration of oxygen vacancies. Therefore, if the oxygen diffuses via oxygen vacancies, the diffusion coefficient for oxygen should decrease in the order, lithium doped, undoped, aluminium doped corresponding to the relative order of concentration of anion vacancies. This behavior was exactly what Chen and Jackson found for CoO at 1428°C and in air. Consequently, oxygen migrates via anion vacancies. Chen and Jackson infer the dominant anion defect to be the singly ionized oxygen vacancy, $V_O^{\cdot}$. A study of the oxygen pressure dependence of the oxygen diffusion is needed to test this suggested degree of ionization. At any rate, cobalt oxide is one of the best characterized oxides from the view of diffusion and other physical measurements. The dominant disorder is of the Schottky type under highly oxidizing conditions where the oxide always exists as a metal deficient compound [23]. As with cuprous oxide, there is not complete agreement on the minority defects in cobaltous oxide. Diffusion studies of cobalt and studies of the mobilities of the electrons and electron holes at very low oxygen concentrations must be carried out to test whether there exist appreciable concentrations of interstitial cobalt.

Diffusion in nickel oxide has been less extensively studied. Because the range of homogeneity is less than that of CoO, the apparent activation energy, Q, for the cation $Q_{NiO} = 1.9$ eV [24] whereas $Q_{CoO} = 1.67$ eV. The diffusion of oxygen in NiO yields [25] an activation energy of 2.5 eV, much less than that for oxygen in CoO. Moreover the diffusion of oxygen *increases* with increasing oxygen pressure as was also reported for cuprous oxide. In these respects the model for oxygen diffusion in NiO differs markedly from that for CoO. O'Keeffe and Moore [25] infer the migration of oxygen in NiO to be via interstitial jumps.

It is most desirable to extend the measurements of both cation and oxygen diffusion to NiO doped with aliovalent impurities and to cation diffusion in CoO as a function of oxygen pressure because these oxides should exhibit similar behavior with respect to anion migration as well as cation migration. At any rate the ratio of the tracer diffusion coefficient for the cation to that of the oxygen is very large, of the order of 10^3 at temperatures of the order of 1200°C. A large ratio is also substantiated for FeO and MnO as well as for NiO and CoO by the displacement of an inert marker during the parabolic oxidation of metals and by chemical diffusion measurements to be discussed later.

The other two members of this series, manganous oxide and ferrous oxide or wüstite, both exhibit some curious anomalies which tracer diffusion data have aided in interpreting. Both these oxides exhibit apparent p-n transitions (change of sign of the Seebeck coefficient) in the center of the phase fields. The electronic conductivity of single crystalline

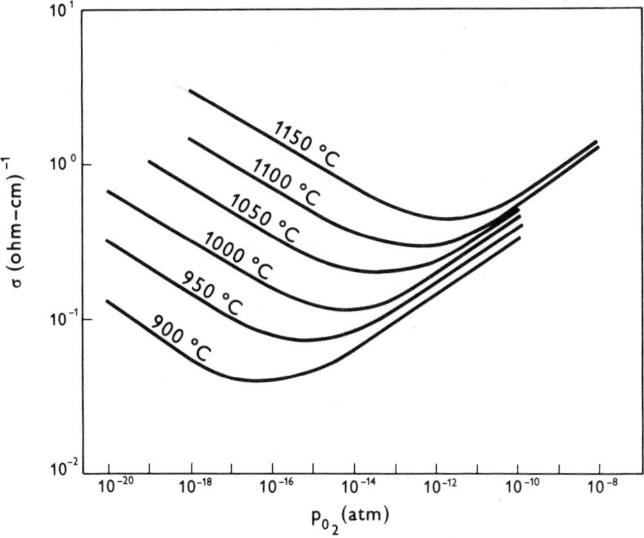

Figure 8.7. Electronic conductivity of single crystalline manganous oxide as a function of oxygen pressure. After Eror and Wagner [126].

MnO was shown to exhibit a minimum at approximately the center of the phase field—the oxygen pressures corresponding to $p_{CO_2}/p_{CO} \cong 1$ (see figures 8.4 and 8.7). At higher oxygen pressures the defects were clearly described by,

$$\tfrac{1}{2}O_2(G) \rightleftharpoons O_O + V_{Mn}'' + 2h\ . \tag{8.50}$$

On applying the law of mass action,

$$p = \text{const.}\ p_{O_2}^{1/6} \tag{8.51}$$

so that the electronic conductivity, σ, was proportional to $p_{O_2}^{1/6}$ (see figure 8.7). At lower oxygen pressures the electronic conductivity was given by

$$\sigma \propto n = \text{const.}\ p_{O_2}^{-1/6} \tag{8.52}$$

Two models were suggested to account for this behavior. One involved the reaction,

$$MnO \rightleftharpoons Mn_i^{\cdot\cdot} + 2e' + \tfrac{1}{2}O_2(G) \tag{8.53}$$

and the other [26] involved equation 8.50 but with the suggestions that the mobility of the minority carriers, the electrons, was much greater than that of the hole majority carriers. Thus in the second model, cation vacancies were predominant over the entire homogeneity range but at lower oxygen pressures, the smaller electron concentration with higher mobility dominated the electrical properties. Hall measurements [27] confirmed the fact that $\mu_p < \mu_n$. A critical experiment was to carry out diffusion across the phase field in analogy to the experiments first carried out on lead sulfide crystals [11]. If cation vacancies were the dominant defect the diffusion coefficient of manganese would monotonically decrease across the phase field with no inflection point. On the other hand, if interstitial manganese contributed to the migration, the isothermal diffusivity would at first decrease, pass through a minimum in the region near the apparent p to n transition depending on the relative mobilities of the defects and then the diffusivity would increase again. Figure 8.8 shows

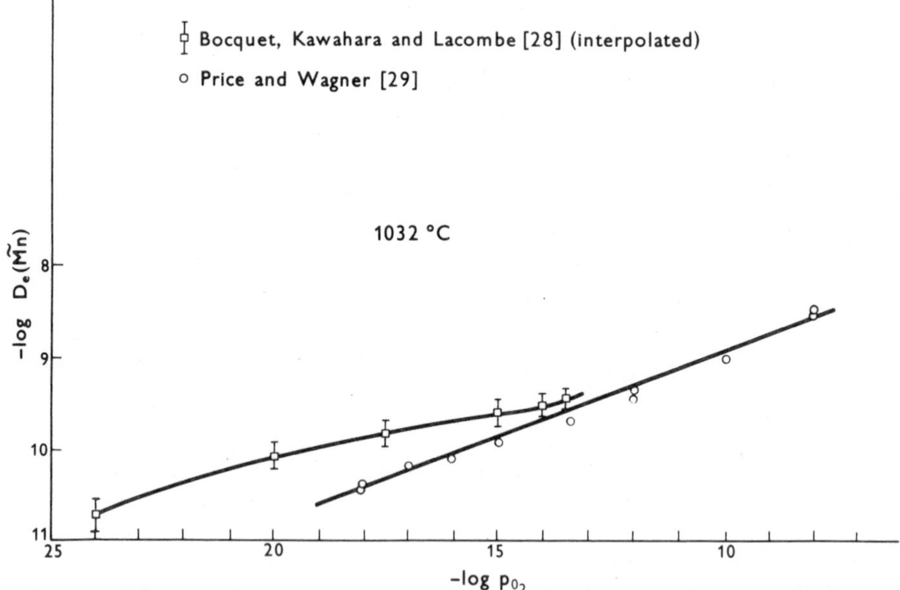

Figure 8.8. Oxygen pressure dependence of the tracer diffusion coefficient of manganese in manganous oxide. The diffusion coefficient decreases monotonically across the phase field. The slope of the line through the closed circles is 1/5.4 indicating doubly ionized manganese vacancies as the predominant defect. After Price and Wagner [29].

the data of Lacombe [28] and of Price and Wagner [29]. Because the isothermal diffusivity decreases continuously across the phase field the model involving cation vacancies as majority defects and the higher mobility of electrons compared to electron holes is preferred as was originally suggested by Hed and Tannhauser [26].

This kind of study illustrates the information necessary for a decision between different mechanisms which can be obtained from tracer diffusion studies.

Ferrous oxide or wüstite is a special case. It contains from about 5% to 11% vacancies at 1100° C. The thermodynamics of ferrous oxide or wüstite are probably better known than any other oxide owing to its wide range of stability and consequent ease of measurement and to its industrial importance. Unlike MnO, the apparent *p-n* transition as determined by the change of sign in the Seebeck coefficient is *p*-type for low oxygen pressures and *n*-type for higher oxygen pressures. At low oxygen pressures the defect equilibria have been *approximated* by

$$\tfrac{1}{2}O_2(G) \rightleftharpoons O_O + V''_{Fe} + 2h^{\cdot}. \tag{8.54}$$

However, neutron diffraction data on quenched samples and high temperature X-ray studies have indicated clustering of the vacancies and long range ordering. Galvanic cell measurements have indicated there exist three different (higher order) phases of ferrous oxide. Equation 8.54 is far too simple a model. Diffusion measurements of iron would seem to offer a good indication of the onset of ordering or of second phases. So far the data all show the tracer diffusion coefficient proportional to the total cation vacancy concentration.

Himmel, Mehl and Birchenall [30] studied the diffusion of iron in wüstite as a function of composition at 800°, 897° and 983° C. The tracer diffusion coefficient depended linearly on the total concentration of iron vacancies. The concentrations of iron vacancies were calculated assuming that all the iron vacancies were distributed randomly, i.e., no interactions and no interstitials were considered. Hembree and Wagner [31] studied the diffusion of iron across the phase field at 1100° C. They also report a linear dependence on iron vacancy concentration (see figure 8.9). These results do not rule out the possibility of 'complexes' or 'clusters' or iron vacancies and/or interstitials but the data strongly suggest that the diffusion occurs via the same mechanism across the phase field. Desmarescaux and Lacombe [32] have studied the tracer diffusion of iron in wüstite at fixed compositions as a function of temperature. The activation energies increased with increasing concentrations of cation vacancies suggestive of interactions between vacancies and ferric ions as suggested earlier by Richardson [33] and others.

Desmarescaux and Lacombe also demonstrated the existence of an isotope effect in wüstite by simultaneously diffusing two tracers, Fe-55 and Fe-59. No value for the strength of the isotope effect (E_{59} applied to equation 8.45) was reported and the data presented in the small scale figures makes an evaluation difficult. This type of quantitative data is much needed, especially in wüstite where controversies exist over the type of association between the defects.

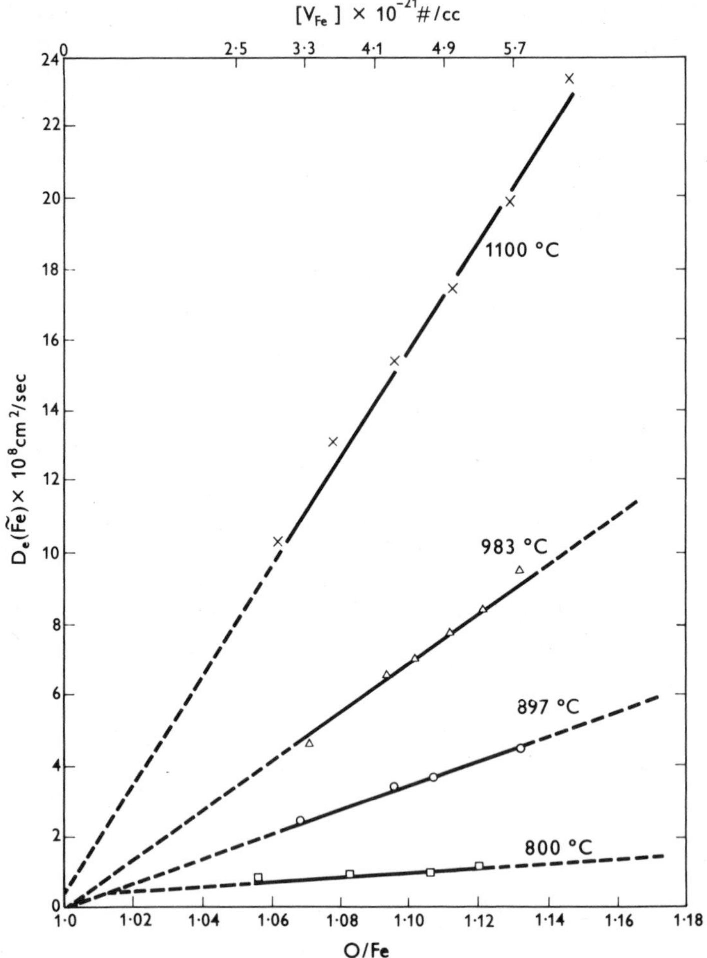

Figure 8.9. Tracer diffusion of iron in wüstite. The oxygen to metal ratio, O/Fe, is a measure of the cation vacancy concentration. The data at 1100° C are from Hembree and Wagner [31]. The data at 800°, 897° and 983° C are from Himmel, Mehl and Birchenall [30]. Straight lines (not least square lines) have been drawn through the data by the present author. The lines converge approximately at an O/Fe ratio of unity and a diffusion coefficient of approximately zero. This convergence would represent the diffusion coefficient in hypothetical samples of wüstite of ideal stoichiometric composition. The values of $D_e(\widetilde{Fe})$ at O/Fe = 1 would be finite due to the thermal disorder. However, they would probably be one or two orders of magnitude less than those in non-stoichiometric wüstite. It is puzzling that the diffusion coefficients are proportional to the total vacancy concentration as calculated for a simple random distribution of cation vacancies, a condition which does not hold in wüstite.

In summary it can be said that ferrous oxide is similar to NiO, CoO and MnO in that diffusion of the cation occurs primarily via cation vacancies. Moreover, for a given oxygen pressure a decrease in electron holes (increase in electrons or Fermi level) will result in a larger concentration of cation vacancies and hence the cation diffusivity. Thus, this series of oxides exhibits a consistent pattern. The solution of defects is so concentrated in ferrous oxide that detailed measurements using X-ray or neutron diffraction under equilibrium conditions must be made for a more complete understanding of the mechanism.

So far we have considered predominantly electronic conductors. If we turn our attention to oxides of formula MO from the Group II metals we find the deviations from stoichiometry are too small to have been experimentally detected as yet. Moreover, the activation energies for diffusion are much larger than for MnO, NiO, CoO or FeO. With the exception of BeO, the oxides of the Group IIA metals also exhibit a sodium chloride lattice as do the manganese, iron, cobalt, and nickel oxides. These Group IIA oxides are much more ionic. Because the deviations from stoichiometry are so small, trace amounts of impurities exert a very large effect on the diffusion coefficients. Moreover, in keeping with the trend developed in the preceding paragraphs, the activation energy for diffusion in this series is larger than that for the series NiO, CoO, FeO, and MnO as has been discussed by Birchenall [34]. Calcium oxide is perhaps the best characterized of the Group IIA metal oxides. Kumar and Gupta [35] have recently studied the diffusion of calcium-45 in single crystalline calcium oxide. These authors heated their crystals under argon and claimed this condition corresponded to intrinsic calcium oxide. In other words, the crystals were stoichiometric over the temperature range from 1465° to 1760° C, i.e., in figure 8.1, the upper portion of the curve would represent stoichiometric CaO. The diffusion coefficient was represented by

$$D_e(\widetilde{\mathrm{Ca}} \text{ in CaO}) = 11.25 \times 10^{-5} \exp(-2.79/kT) \text{ cm}^2 \text{ s}^{-1}. \quad (8.55)$$

The activation energy, 2.79 eV, is approximately twice that observed for the electronic conductors already discussed. Moreover, Gupta and Weirick [36] studied the tracer diffusion of calcium in calcium oxide doped with 0.001% aluminium. In these experiments the activation energy amounted to 1.5 eV and diffusion was presumably entirely extrinsic. Therefore the activation energy in the doped crystals is taken to be the migration energy for diffusion. From these data Kumar and Gupta infer that intrinsic diffusion is via cation vacancies. Using equations 8.31 and 8.32, Kumar and Gupta were able to calculate the energy for the formation of a Schottky pair in CaO to be 3.08 eV. Lindner and co-workers [37] report diffusivities which are somewhat higher than those of Kumar and Gupta. A possible explanation is the fact that Lindner used samples containing unknown impurities. So far the diffusion of oxygen in calcium oxide has not been measured directly but has been inferred from ionic conductivity measurements by Surplice [38]. The activation energy was reported to be 3.5 eV in keeping with the general trend in the sodium chloride structures where the cation is smaller than the anion, that the diffusion coefficient of the oxygen is less than the diffusion coefficient of the metal.

The diffusion of magnesium in magnesium oxide has been studied by Lindner and Parfitt [39]. The crystals were heated in air. For the temperature range, 1400° to 1600° C, their results can be expressed by

$$D_e(\tilde{\text{Mg}} \text{ in MgO}) = 0.249 \exp(-3.42/kT) \text{ cm}^2 \text{ s}^{-1}. \tag{8.56}$$

Heating MgO in a constant oxygen pressure should result in a change in composition. However, in analogy with the case of calcium oxide, the deviations from stoichiometry are extremely small [40] and it has been implicitly inferred that the MgO crystals were stoichiometric. Therefore, the activation energy of 3.4 eV represents the migration energy plus one half the energy for formation of a Schottky pair. These authors report a 'kink' in the Arrhenius plot at lower temperatures where a lower activation energy was observed. These findings suggest the onset of extrinsic behavior. Unfortunately no numerical data were reported for this low temperature range so a direct calculation of the energy terms cannot be made. Gupta [35] estimates the energy of migration to be 1.3-1.5 eV in MgO and so obtains 4.1 eV for the formation of a Schottky pair.

It is surprising that the diffusion of oxygen [41] in MgO exhibits a lower activation energy (2.7 eV) than that for the diffusion of magnesium.

Beryllium oxide, BeO, has been often studied because of its use in nuclear applications. Most of the studies have been carried out on polycrystalline samples. The activation energies [42, 43] for diffusion of beryllium vary between about 1.57 and 5.1 eV. If beryllia is somewhat similar to calcium oxide one would expect that heating the crystals in an inert atmosphere would yield stoichiometric material. Because of the large differences in the reported diffusion coefficients it is most likely that many of the data were obtained on samples containing impurities. In view of this, the higher activation energies are probably most representative of undoped, bulk diffusion. Condit and Hashimoto [44] have studied the diffusion of Be-7 in polycrystalline BeO containing known impurities. They analyze their own data and those of previous authors in terms of diffusion which is controlled by impurities. These authors conclude that an activation energy of 4 eV is representative of bulk diffusion not controlled by impurities.

The activation energy for the diffusion of barium in BaO has been reported [45] to be ~11 eV. This is an exceedingly large value even when compared to the activation energy for the cation diffusion in BeO. Barium oxide is reported to exhibit Frenkel defects and barium is reported to diffuse via a neutral interstitial and via charged cation vacancies [45]. According to Boltaks [46] the migration energy of an interstitial atom is 0.44 eV and of an atom jumping via a vacancy, 0.3 eV so that the value of enthalpy of formation of the Frenkel defects, ΔH_f, amounts to 23 ± 5 eV.

An activation energy for the diffusion of strontium in SrO has been inferred [38] from electrical measurements to be 2 eV. Thus the series, BeO, MgO, CaO, and SrO form a consistent pattern within the Group IIA oxides. The transition metal oxides already considered exhibit a trend in that the cation diffusion is smaller for oxides composed of elements which are more difficult to ionize and hence whose oxides exhibit narrower ranges of stoichiometry. Cation diffusion in the oxides of Group IIA elements is supposed to represent migration into stoichiometric crystals. The trend is, with the notable exception [45] of BaO, for a decrease in activation energy as one goes to less electronegative elements as is seen by this comparison:

Oxide	BeO	MgO	CaO	SrO	BaO
Activation energy (eV) for cation diffusion	4	3.42	2.79	(2)	11

Copeland and Swalin [47] have made conductivity, Seebeck coefficient, and thermogravimetric measurements on SrO. Their data show that SrO may deviate from stoichiometry. Excess oxygen is reported to be accommodated as singly ionized interstitials. At low oxygen pressures, these authors report an intrinsic region in which $n = p$ or $[O_i'] = [V_O^{\cdot}]$. Their data lend support to the assumption that heating the Group IIA metal oxides in an inert atmosphere yields a stoichiometric crystal. These assumptions should be verified by experiment.

Barium oxide is the anomaly in the group. Redington [45] reported a near surface diffusion coefficient for barium and one for bulk diffusion deep within the crystal. It is probable that the near surface diffusion was through a layer of hydroxide because of the strong affinity of BaO for moisture. The extremely high value of the activation energy, 11 eV, for bulk diffusion is clearly unreasonable and yields unreasonable values for the energy to form the defects. These measurements should be repeated.

Although the rate of uptake of oxygen in crystals under an activity gradient has been studied, very few self-diffusion studies of oxygen in the Group IIA metal oxides have been made. Except for beryllium, the cations have ionic radii not too different from oxygen and the oxides are of the sodium chloride structure. Consequently, one would expect the activation energy for the diffusion of oxygen to be not too different from that of the cations. O'Keeffe [48] has discussed the relative diffusivities of cations and anions in oxides in terms of the parameters used by Lidiard [49] for alkali halides. According to O'Keeffe, the general rule for oxides is that polarization (the ability of a crystal to relax around a point defect) will favor the formation and migration of the less polarizable ion. Consequently, in general, cation diffusion is more rapid than anion diffusion in oxides of nominal formulae M_2O and MO. In oxides where the cation has a higher oxidation state, e.g. MO_2, anion diffusion will generally be more rapid than cation diffusion. These arguments apply best to stoichiometric crystals although O'Keeffe suggests the same general rule may apply to non-stoichiometric oxides.

Zinc oxide was considered in a preceding chapter. The Group IIB elements, cadmium and mercury form the oxides CdO and HgO. The instability of mercuric oxide has prevented studies. However, diffusion measurements of oxygen in cadmium oxide have been made by Haul and co-workers [50, 51] to determine the predominant disorder on the anion sublattice. The diffusion of oxygen-18 has been studied in both single and polycrystalline specimens by the oxygen exchange method. The pre-exponential term $D_{o,e}$ varies from 4×10^5 to 8×10^6 $cm^2\ s^{-1}$ but the activation energies agree at 4 eV. The oxygen pressure dependence of the diffusion coefficient has been found to be proportional to p_{O_2} to the $-1/5$ or $-1/6$ power. This behavior suggests that the lattice disorder on the anion sublattice is given by,

$$O_O \rightleftharpoons V_O^{\cdot\cdot} + 2e' + \tfrac{1}{2}O_2(G) \tag{8.57}$$

In order to test this suggested disorder, Haul and co-workers made diffusion experiments on cadmium oxide doped with lithium or with indium or aluminium oxide. In analogy to the oxygen diffusion studies on CoO performed by Chen and Jackson [21] discussed earlier, the introduction of a monovalent solute, lithium, creates anion vacancies while the introduction of a trivalent solute, indium or aluminium destroys anion vacancies in CdO. As a consequence one expects oxygen diffusion to be more rapid in the lithium doped crystals than in the indium doped crystals. This behavior was found by Haul and co-workers. Thus, the disorder on the anion sublattice has been established to be the oxygen vacancy type. There are suggestions that excess cadmium may enter the lattice as interstitial atoms. Such behavior would indicate a Frenkel disorder for the stoichiometric crystals. The diffusion coefficient of cadmium in cadmium oxide is needed to test this suggestion.

In the preceding, little attention was directed to the magnitudes of the pre-exponential terms. There are alternate formulations of $D_{o,e}$ than that presented in equation 8.33 for example. Most of the formulations yield calculated values of the pre-exponential term which are close to one another in magnitude. The precision of the experimental data has not been good enough such that decisions between models can be made. Of special concern here, however, are the very large values of experimental $D_{o,e}$ and the large range of values reported for oxides ($\sim 10^{-6}$-10^{30} cm^2 s^{-1}) [43]. In the case of stoichiometric oxides the pre-exponential term using the transition state theory is given by equations 8.31 or 8.34. Using 'reasonable values' for the migration entropies ($\ll |1|$ eV/deg), enormous values for the entropy of formation of defects would have to be invoked as has been pointed out by Moore [52]. In the case of non-stoichiometric oxides, the dissolution of oxygen in the lattice results in a large change in configurational entropy as has been pointed out by Mitoff [53] in connection with electrical measurements on oxides. Haven and Wilkinson [54] are evaluating the wide range of values of $D_{o,e}$ for oxygen diffusing in oxides [$10^{-9} \leqslant D_{o,e}$ ($\tilde{O}$ in MO) $\leqslant 10^{+11}$ cm^2 s^{-1}]. They have initiated calculations to account for part of this large spread of values by considering the influence of the atmosphere with which the oxide is in equilibrium. Using the transition state theory in conjunction with statistical mechanics, they suggest the following. For non-stoichiometric intrinsic oxides, the pre-exponential factor for oxygen is $\sim 10^{-6}$ cm^2 s^{-1} for interstitial and $\sim 10^4$ cm^2 s^{-1} for vacancy diffusion. For stoichiometric, intrinsic diffusion $D_{o,e} \cong 10^{-2}$ cm^2 s^{-1}. These calculations are much needed. They should be extended to the pre-exponential terms for cation diffusion for which there exist more diffusion data and thus offer a broader range of oxides to test.

8.5.3 Tracer and Self-diffusion in Oxides of Nominal Formula $MO_{1.33}$

Most oxides of the formula $MO_{1.33}$ which have been studied belong to the spinel class and are made up of two different cations plus oxygen, e.g. $NiFe_2O_4$. One important difference from the treatment of diffusion data cited previously is to be pointed out. When the oxide consists of three components, the phase rule demands that four variables be fixed. These are usually the temperature, total pressure, oxygen activity (by means of the oxygen partial pressure co-existing with the spinel) and *the activity of one additional component.* Failure to define this additional variable has negated many otherwise useful data. In addition, the concentrations of defects are usually expressed in terms of defects per molecule of spinel (actually per unit formula because molecules may or may not exist as entities within the spinel) rather than defects per cubic centimeter.

The spinels have the oxygen ions arranged in the f.c.c. packing and there are two types of interstices, tetrahedrally coordinated and octahedrally coordinated. The occupation of these interstices depends on the size and hence on the oxidation state of the cations. The spinels are divided into so-called normal and inverse spinels and, as mentioned above, usually refer to a class of ternary oxides, AB_2O_4, where A and B denote different metal atoms. The classification is [55]

$A^{+2}B_2^{+3}O_4^{-2}$	2-3 Spinel	Normal Spinel
$A^{+4}B_2^{+2}O_4^{-2}$	4-2 Spinel	
$A^{+3}(A^{+3}B^{+2})_2O_4^{-2}$	Inverse spinel	

Because the spinels have a more complicated lattice than the simple structures considered earlier and because an additional composition variable is usually present, there are a greater possible number of types of defects. A useful notation [56] which catalogues these defects as well as atoms on normal sites is:

A_A cation A on a normal lattice site
B_B cation B on a normal lattice site
O_O anion O on a normal lattice site
$A_i^{\cdot\cdot}$ divalent cation A on an interstitial site
$B_i^{\cdot\cdot\cdot}$ trivalent cation B on an interstitial site
O_i'' divalent cation O on an interstitial site
$V_O^{\cdot\cdot}$ anion vacancy on a normal O site
A_B' divalent cation A on a normal trivalent cation B site
$B_A^{\cdot}$ trivalent cation B on normal divalent cation A site

In the above notation, a normal spinel has $\{A_A\} \cong 1$ and $\{B_B\} \cong 1$ where the curly brackets denote activities of the indicated species. The inverse spinel has $\{B_A^{\cdot}\} \cong 1$ and $\{A_B'\} \cong \{B_B\} \cong 1$. See the preceding classification.

The thermodynamics of point defects in spinels has been discussed by Schmalzried [57]. If electronic disorder can be neglected and the concentration of defects is small, Schmalzried shows that the spinel phase can be regarded as a psuedo-binary system with components as AO and B_2O_3. As examples, FeO and Fe_2O_3 react to form Fe_3O_4 and CoO and Fe_2O_3 react to form $Co_{1-x}Fe_{2+x}O_4$. The free energy of formation of the spinel from the components given by

$$AO + B_2O_3 \rightleftharpoons AB_2O_4 \tag{8.58}$$

whence

$$K = \exp(-\Delta G_s^\circ/kT) \tag{8.59}$$

where ΔG_s° denotes the free energy of formation of the spinel from the pseudo-components, AO and B_2O_3. The activities of these pseudo-components are related by

$$\{AO\}\{B_2O_3\} = \exp(+\Delta G_s^\circ/kT) \tag{8.60}$$

Schmalzried considers a spinel in which the concentration of electronic disorder is negligibly small and in which cations may not substitute in oxygen ion sites. He uses the activity of either pseudo-components, AO and B_2O_3, along with the total pressure, the temperature and the partial pressure of oxygen to fix the thermodynamic state of the spinel. For example, a spinel in equilibrium with its atmosphere is represented by

$$AO + \tfrac{2}{3}B_i^{\cdot\cdot\cdot} \rightleftharpoons \tfrac{1}{3}B_2O_3 + A_i^{\cdot\cdot} \tag{8.61}$$

whence

$$K = [A_i^{\cdot\cdot}]\{B_2O_3\}^{1/3}/[B_i^{\cdot\cdot\cdot}]^{2/3}\{AO\} \tag{8.62}$$

where the concentrations of lattice defects are expressed in terms of numbers per nominal formula of spinel and used in place of the activities as before.

The possible types of lattice defects (see notation presented above) are related to the ratio of AO to B_2O_3 which is variable for non-stoichiometric spinels. The equation,

$$(1+\alpha)AO + (1+\beta)B_2O_3 = A_{1+\alpha}B_{2(1+\beta)}O_{(4+\alpha+3\beta)} \tag{8.63}$$

expresses this deviation. A parameter, y, where

$$(1+\beta)/(1+\alpha) \equiv 1+y \tag{8.64}$$

is used to denote this deviation from stoichiometry. When $y \cong 0$, the spinel is virtually stoichiometric. When $y > 0$ the spinel contains an excess of B_2O_3 and conversely when $y < 0$ the spinel contains an excess of AO. The types of lattice disorder are shown in Table 8.1.

TABLE 8.1

Lattice disorder in spinels according to Schmalzried [57][a]

	$A_i^{\cdot\cdot}$	$B_i^{\cdot\cdot\cdot}$	V_A''	V_B'''	A_B'	$B_A^{\cdot}$	$V_O^{\cdot\cdot}$	O_i''
$A_i^{\cdot\cdot}$			$y \cong 0$	$y < 0$	$y < 0$			$y < 0$
$B_i^{\cdot\cdot\cdot}$			$y > 0$	$y \cong 0$	$y < 0$			$y > 0$
V_A''						$y > 0$	$y > 0$	
V_B'''						$y > 0$	$y < 0$	
A_B'						$y \cong 0$	$y < 0$	
$B_A^{\cdot}$								$y > 0$
$V_O^{\cdot\cdot}$								$y \cong 0$
O_i''								

[a] No distinction is made between octahedral or tetrahedral sites and electronic disorder is neglected.

In order to relate the deviation from stoichiometry to the lattice disorder, models for the predominant disorder are postulated and the influence of electrical conduction on changing the activity of one of the pseudo-components, AO or B_2O_3, at constant P, p_{O_2} and T is determined. This is quite analogous to the oxygen pressure dependence of the diffusivity discussed in the preceding sections. As an example consider an equation which expresses the formation of $A_i^{\cdot\cdot}$ and V_A'' as

$$A_A \rightleftharpoons A_i^{\cdot\cdot} + V_A'' \tag{8.65}$$

whence

$$K = [A_i^{\cdot\cdot}]\,[V_A'']/\{A_A\}. \tag{8.66}$$

Using a series of equations of this type and two series of equations for mass and site balances, Schmalzried related the predominant lattice defects to the activity of one of the pseudo-components, AO or B_2O_3. For example, combining equations 8.60 and 8.62 yields,

$$\frac{[A_i^{\cdot\cdot}]}{[B_i^{\cdot\cdot\cdot}]}\,\frac{1}{\{AO\}^{4/3}} = K \exp\,(-\Delta G_s^\circ/3RT) \tag{8.67}$$

Using a set of equations of the type expressed by equation 8.65 for all possible defects, the concentration of one type of defect is expressed as a function of the activity of one pseudo-component, usually AO, with P, p_{O_2} and T being constant. The resulting equation is written in logarithmic form and differentiated with respect to $\{AO\}$ to yield,

$$\left[\frac{\partial \log c_l}{\partial \log \{AO\}}\right]_{O,\,p_{O_2},\,T} \equiv n_i \tag{8.68}$$

where n_i is a characteristic number for each type of lattice disorder and c_l is the concentration of majority lattice defects. As an example consider a spinel with a lattice disorder consisting of predominantly $A_i^{\cdot\cdot}$ and V_B'''. The spinel contains excess B_2O_3 ($y > 0$, see Table 8.1). The reaction expressing the formation of the lattice disorder may be written

$$3\,AO + 2\,B_B \rightleftharpoons 3\,A_i^{\cdot\cdot} + 2\,V_B''' + B_2O_3 \tag{8.69}$$

and the mass action law yields,

$$K = [A_i^{\cdot\cdot}]^3\,[V_B''']^2\{B_2O_3\}/\{AO\}^3 \tag{8.70}$$

where $\{B_B\} = 1$. The electroneutrality condition is

$$[A_i^{\cdot\cdot}] = (3/2)[V_B'''] \tag{8.71}$$

combining equations 8.60, 8.70 and 8.71 yields,

$$K = \frac{[A_i^{\cdot\cdot}]^5}{\{AO\}^4}\,(2/3)^2 \exp\,(+\Delta G_s^\circ/kT) \tag{8.72}$$

whence

$$\left[\frac{\partial \log [A_i^{\cdot\cdot}]}{\partial \log \{AO\}}\right]_{P,\, p_{O_2},\, T} = 4/5 = n_A \tag{8.73}$$

Likewise, sets of equations are derived for other lattice disorder types and finally for electronic disorder types. The electroneutrality conditions are listed in Table 8.2 and the concentration dependences for interstitial cations and excess electrons are listed in Table 8.3.

TABLE 8.2

Electroneutrality conditions for spinels AB_2O_4 which exhibit disorder only on the cation sublattices. After Schmalzried [57]

$y < 0$ (excess of AO)	$2[A_i^{\cdot\cdot}] = 3[V_B''']$	$2[A_i^{\cdot\cdot}] = [A_B']$	$3[B_i^{\cdot\cdot\cdot}] = [A_B']$
$y = 0$ (stoichiometric)	$[A_i^{\cdot\cdot}] = [V_A'']$	$[B_i^{\cdot\cdot\cdot}] = [V_B''']$	$[A_B'] = [B_A^{\cdot}]$
$y > 0$ (excess of B_2O_3)	$3[B_i^{\cdot\cdot\cdot}] = 2[V_A'']$	$[B_A^{\cdot}] = [V_B''']$	$[B_A^{\cdot}] = 2[V_A'']$

TABLE 8.3

Values of the characteristic number, n_i, for interstitial cations and for electrons in AB_2O_4 as a function of activity of AO at constant p_{O_2}, T and P. After Schmalzried [57]

Disorder type	$n_A = \dfrac{\partial \ln [A_i^{\cdot\cdot}]}{\partial \ln \{AO\}}$	$n_B = \dfrac{\partial \ln [B_i^{\cdot\cdot\cdot}]}{\partial \ln \{AO\}}$	$\dfrac{\partial \ln n}{\partial \ln \{AO\}}$
$A_i^{\cdot\cdot}, V_B'''$	+4/5	−4/5	+1/10
$A_i^{\cdot\cdot}, A_B'$	+4/3	0	−1/6
$B_i^{\cdot\cdot\cdot}, A_B'$	+2	+1	−1/2
$A_i^{\cdot\cdot}, V_A''$	0	−2	+1/2
$A_B', B_A^{\cdot}$	+4	+4	−3/2
$B_i^{\cdot\cdot\cdot}, V_B'''$	+4/3	0	−1/6
$B_i^{\cdot\cdot\cdot}, V_A''$	+4/5	−4/5	+1/10
$B_A^{\cdot}, V_B'''$	+2	+1	−1/2
$B_A^{\cdot}, V_A''$	+4/5	0	−1/2

Similar expressions can be derived for the dependence of the concentration of defects on p_{O_2} when P, $\{AO\}$, and T are kept constant [57].

The lattice disorder in crystals of formula MO were inferred by studying the dependence of tracer diffusivity and electronic conductivity on p_{O_2} at constant P and T. The situation for the spinels is analogous. If only one diffusion mechanism predominates, then the ratio of the tracer diffusivity at two different values of the activity of AO (denoted by superior * and **) will be the concentration ratio of the majority lattice defects. Thus,

$$\frac{D_j^*}{D_j^{**}} = \frac{c_l^*}{c_l^{**}}\,. \tag{8.74}$$

The concentration of majority lattice defects, c_l, may be expressed in terms of the activities to the power n_i which is shown by equations of the type 8.73. Thus,

$$\frac{D_j^*}{D_j^{**}} = \left[\frac{\{AO\}^*}{\{AO\}^{**}}\right]^{n_i} \tag{8.75}$$

The activities in turn can be related to the free energy of formation of the spinel through equation 8.60 so that,

$$D_j^*/D_j^{**} = \exp\,[n_i \Delta G_s^\circ/kT] \tag{8.76}$$

The value of n_i will depend on the type of majority lattice defect and the asterisks denote the two different values of $\{AO\}$. It is experimentally convenient to work with the spinel equilibrated first with pure AO and then with pure B_2O_3. This gives the widest possible variation in activity (with P, p_{O_2} and T constant) and hence the widest variation in majority defect concentration. Provided the lattice defects considered are only those which contribute significantly to the diffusion process across the entire homogeneity range of the spinel, then for the jth species migrating via the ith defect,

$$[D_{e,j}(\{AO\}=1)]/[D_{e,j}(\{B_2O_3\}=1)] = [\exp\,(n_i \Delta G_s^\circ/kT)] \tag{8.77}$$

and

$$n_i = (kT/\Delta G_s^\circ)\,[\ln D_{e,j}(\{AO\}=1) - \ln D_{e,j}(\{B_2O_3\}=1)] \tag{8.78}$$

These types of experiments have been carried out by Schmalzried and co-workers [57]. Some results are summarized in Table 8.4.

TABLE 8.4
Experimental data for cation diffusivity in some spinels and the inferred probable disorder type according to Schmalzried [57]

Compound	Temperature (°C)	Isotope	$\{AO\}^*/\{AO\}^{**}$	D_e^*/D_e^{**}	Probable disorder
Co_2TiO_4	1200	^{60}Co	1.6	0.2	$[Co'_{Ti}] \cong [Ti^{\cdot\cdot}_{Co}]$
$Co\,Al_2O_4$	1200	^{60}Co	5	0.2	$[Al_i^{\cdot\cdot\cdot}] \cong \frac{2}{3}[V''_{Co}]$
$Co\,Cr_2O_4$	1200	^{60}Co	10	2-3	$[Co_i^{\cdot\cdot}] \cong \frac{3}{2}[Cr_i^{\cdot\cdot\cdot}]$
$Ni\,Cr_2O_4$	1200	^{51}Cr	8	0.2	$[Ni_i^{\cdot\cdot}] \cong \frac{3}{2}[Cr_i^{\cdot\cdot\cdot}]$
$Sr\,TiO_3$	1450	^{89}Sr	7	0.1	$[Ti_i^{\cdot\cdot\cdot\cdot}] \cong \frac{1}{2}[V''_{Sr}]$ $[Ti_i^{\cdot\cdot\cdot\cdot}] \cong [V''''_{Ti}]$

The spinel for which both the composition and temperature dependence has been studied is magnetite, Fe_3O_4. Because it is a two component system, fixing the oxygen pressure at constant T and P fixes the system. Nevertheless, the same arguments described above have been applied to magnetite by Schmalzried [57] who shows that, because

$$Fe_3O_4 \rightleftharpoons 3\,FeO + \tfrac{1}{2}O_2\,(G) \tag{8.79}$$

whence

$$\{FeO\} = K^{1/3}\,p_{O_2}^{-1/6} = \text{const.}\left(\frac{p_{CO_2}}{p_{CO}}\right)^{-1/3} \tag{8.80}$$

and

$$\frac{\partial \log\,[V''_{Fe}]}{\partial \log\,(p_{CO_2}/p_{CO})} = 4/3\ \text{(theo.)} \tag{8.81}$$

The oxygen pressure dependence of the tracer diffusivity of iron was studied using CO-CO_2 mixtures. The experimental value of the slope was,

$$\frac{\partial \log D_e(\tilde{Fe})}{\partial \log\,(p_{CO_2}/p_{CO})} = 2/5\ \text{(expt'l)} \tag{8.82}$$

(see figure 8.10). The experimental slope is less than the theoretically predicted value (equation 8.81) possibly due to contributions of jumps via an interstitial or interstitialcy mechanism according to Schmalzried [57]. Himmel, Mehl and Birchenall [30], have also studied the diffusion of iron in magnetite of an average composition given by $Fe_{2.993}O_4$ with the result that

$$D_e(\tilde{Fe}) = 5.2 \exp\,(-2.4/kT)\ \text{cm}^2\ \text{s}^{-1}. \tag{8.83}$$

The activation energy is higher than the migration enthalpy for iron in wüstite of composition $Fe_{0.907}O$ which amounts to 1.3 eV. This difference is expected because in general the activation enthalpies for cation diffusion [58, 59] in the spinels are larger than the migration enthalpies for cations in oxides of the formula, MO, that exhibit the NaCl lattice. The diffusion of oxygen in Fe_3O_4 has been reported by Castle and Surman [60]. An activation energy of 0.74 eV was found. This low value seems improbable for bulk diffusion in view of the presumed disorder in Fe_3O_4. Birchenall [58] has pointed out the large variation in self-diffusion coefficients for the ferrite spinels. Very probably these differences can be attributed in part to the fact that many studies have been carried out under conditions which fixed the oxygen partial pressure but not the activity of another component.

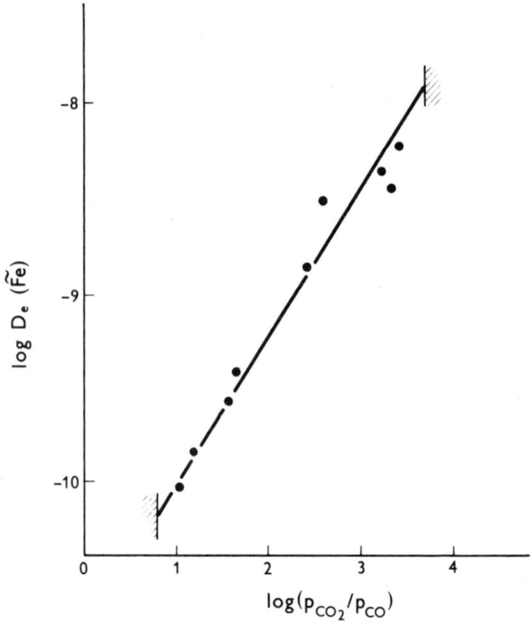

Figure 8.10. Tracer diffusion coefficient of iron in magnetite (Fe_3O_4) as a function of composition at 1115° C. The composition limits, as expressed by the CO/CO_2 ratio coexisting with magnetite, refer to the coexisting phases of wüstite (FeO) and hematite (Fe_2O_3). After Schmalzried [57].

To illustrate this point, consider the cobalt ferrite spinel, $Co_{1-x}Fe_{2+x}O_4$ which has been studied by Müller [61] and which is described in detail in the review by Schmalzried [57]. The self-diffusion coefficient of cobalt in cobalt ferrite coexisting with Fe_3O_4 passes through a minimum as a function of the oxygen pressure (see figure 8.11). Thus the mechanism for diffusion changes on either side of the minimum as also has been reported for the lead chalcogenides [11]. Therefore, if the diffusion of cobalt were carried out at constant oxygen pressure as a function of temperature and for a crystal not coexisting with another phase, e.g. the activity of Fe_3O_4 is not fixed, it is quite possible that the mechanism of diffusion would change because such a minimum (figure 8.11) might shift as the temperature changes.

Clearly more studies of the type described by Schmalzried are needed before further trends can be seen.

8.5.4 Tracer and Self-diffusion in Oxides of Nominal Formula $MO_{1.5}$

The oxides in the group include the corundum structure (e.g. α-Al_2O_3 and α-Fe_2O_3), the ilmenites (e.g. $FeTiO_3$) and the perovskites (e.g. $SrTiO_3$).

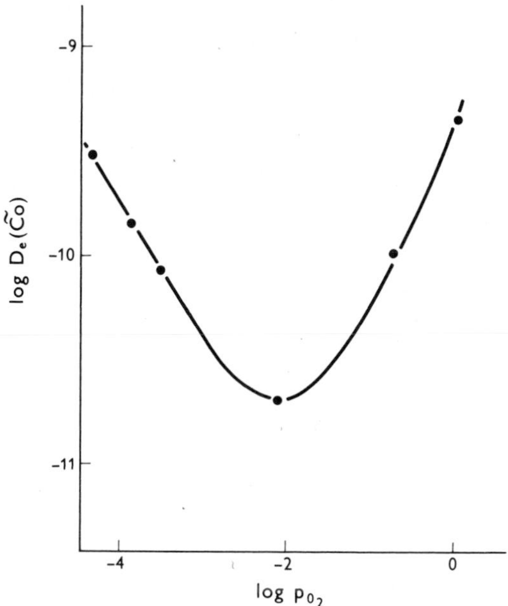

Figure 8.11. Tracer diffusion coefficient of cobalt in cobalt ferrite ($CoFe_2O_4$) as a function of oxygen pressure at 1180° C. After Schmalzried [57].

Hematite (α-Fe_2O_3) is among the most studied. Both the diffusion of iron [30, 62, 63] and of oxygen [64] have been studied.† The results for the diffusion of iron are given by

$$D_e(\tilde{Fe}) = 4 \times 10^4 \exp(-4.87/kT) \text{ cm}^2 \text{ s}^{-1} \tag{8.84}$$

for 750°-1300° C [63]. For the diffusion of oxygen, there are large discrepancies. Kingery *et al.* [64] report that,

$$D_e(\tilde{O}) = 1 \times 10^{11} \exp(-6.35/kT) \text{ cm}^2 \text{ s}^{-1} \tag{8.85}$$

for 1150°-1250° C while Hagel [65] finds,

$$D_e(\tilde{O}) = 2.04 \exp(-3.39/kT) \text{ cm}^2 \text{ s}^{-1} \tag{8.86}$$

for 900°-1250° C. Clearly another study is needed to test these conflicting data. With either set of data, the diffusivity of oxygen and of iron are the same at a temperature of about 1150° C. Below this temperature the oxygen ions appear to move more rapidly than the cations and conversely above about 1150° C.

† Oxygen has no useable radioisotope. Oxygen diffusion is usually studied using a stable isotope in conjunction with a mass spectrometer. In what follows, the stable isotope will also be labeled with a superior tilde, e.g. $\tilde{O}$.

All of these experiments were carried out in fixed oxygen potentials. Chang and Wagner have recently studied the dependence of iron diffusivity on oxygen pressure in Fe_2O_3. The diffusivity was proportional to $p_{O_2}^{-3/4}$ indicating migration via intestitial jumps [134]. It is known that α-Fe_2O_3 exhibits deviations from stoichiometry on the excess oxygen side [66-69]. The defects are probably oxygen vacancies [68]. Consequently, the activation enthalpies, Q, reported in equations 8.84, 8.85 and 8.86 contain both the enthalpy of formation of defects as well as the enthalpy of migration of the respective species. A comparison of activation energies for the diffusion of iron in the three iron oxides is instructive to this point. The parameters are as follows,

Oxide	Fe_xO	Fe_3O_4	α-Fe_2O_3
Lattice	NaCl	inverse spinel	corundum
Q(const. p_{O_2}), eV	~0	–	4.87
ΔH_m(const. O/Fe), eV	1.3($Fe_{0.907}O$)	2.39($Fe_{2.993}O_4$)	–

The trend for migration enthalpies correlates with the increased density of packing as the iron attains higher oxidation states. At first sight, it is surprising that the value for the activation energy, Q, for wüstite (FeO) should be virtually zero. This may be seen as follows. From equation 8.26, $Q = [\Delta H_f/(m+1) + \Delta H_m]$. The value of the partial molar enthalpy for the dissolution of oxygen (enthalpy for formation of the defects) has been determined by a number of investigators [70-72]. It is slightly dependent on composition but for the purposes of this calculation can be taken as −2.8 eV. The value of the degree of ionization of the defects in wüstite is a matter of controversy but in the low oxygen side of wüstite at temperatures of the order 900-1100° C, the vacancies are taken to be singly or doubly ionized (or a combination plus various degrees of association between defects). This means that $(m+1)$ is between 2 and 3. The activation energy for migration, ΔH_m, is also slightly dependent on composition [32] but may be taken as 1.3 eV. Substituting these values in the expression for Q yields virtually zero activation energy. This has been experimentally verified by Pettit [72] who oxidized iron to wüstite in a constant oxygen pressure of 10^{-13} atm using CO-CO_2 mixtures. Pettit found that $Q = 0.16 \pm 0.17$ eV. Himmel *et al.* [30] report that their activation energy for diffusion of iron in magnetite was for an average composition of $Fe_{2.993}O_4$. If the composition was constant, the value of the activation enthalpy, 2.39 eV, represents enthalpy to move the ions and not the enthalpy for creating defects. On the other hand, the hematite crystals used by Himmel *et al.* [30] were heated in oxygen. Therefore, the activation energy reported for iron diffusing in hematite (α-Fe_2O_3) represents the enthalpy for migration plus the term for the enthalpy of formation of defects. When the appropriate values of the partial molar enthalpy of oxygen, ΔH_f, are known the value of Q for Fe_3O_4 and ΔH_m for α-Fe_2O_3 may be calculated. This type of calculation emphasizes the words of caution concerning comparison of activation energies which was presented in section 8.5.1.

Al_2O_3 and Cr_2O_3 are oxides of special industrial importance because of the protection these compounds offer against oxidizing environments. However, the reported data exhibit much scatter [43]. Because of the very small deviations from stoichiometry in both, trace impurities exert a large effect. The trend for the oxides, Cr_2O_3 and Al_2O_3, is for increased activation energy (Q at constant oxygen pressure) for self-diffusion of the cation and for the anion as the melting point (bonding energies) of the corresponding compound increases. Fe_2O_3 does not follow this trend as is seen from the comparison,

Oxide	Fe_2O_3	Cr_2O_3	Al_2O_3
Approximate melting point (°C)	1565	1990	2050
Q(cation, p_{O_2} = const.; eV)	4.87	2.66	4.95
Q(oxygen, p_{O_2} = const.; eV)	3.38-6.34	4.4	6.6

Himmel *et al.* [30] state that a lower activation energy than 4.87 eV for iron diffusion in α-Fe_2O_3 is likely. An estimate from their three data points yields about 3.9 eV. Moreover, the relatively low activation energy for oxygen suggests that α-Fe_2O_3 is markedly different from alumina and chromia. This difference is apparent at elevated temperatures where relatively large deviations from stoichiometry are exhibited by α-Fe_2O_3 as compared to Al_2O_3 and to Cr_2O_3 and the fact that the anions are apparently less mobile than the cations in α-Fe_2O_3 at elevated temperatures [65].

In any case, the separation of Q into the enthalpy of formation for the defects and enthalpy of migration of the respective ions is needed. Very careful diffusion experiments using doped crystals would be very helpful in this regard because the crystals could be doped to a level to provide extrinsic behavior at the lower temperatures from which ΔH_m could be obtained. Then from the intrinsic behavior at the higher temperatures, the Arrhenius plots would yield $Q = (\Delta H_f/(m + 1) + \Delta H_m)$ from which the two terms could be separated. So far little attention has been paid to diffusion along different crystallographic directions. This is undoubtedly due to the fact that unwanted and unknown impurities in these systems still play a major role in the diffusion process.

8.5.5 Tracer and Self-diffusion in Oxides of Nominal Formulas MO_2 and $MO_{2.5}$

Oxides in this group exhibit relatively large deviations from stoichiometry and the diffusivity of the anions are usually equal to or greater than that of the cations. Because of its importance in nuclear reactor technology, more studies have been made on uranium dioxide than any other oxide in this group. The range of stoichiometry in UO_{2+x} is large. The value of x can attain a value of 0.24 and the diffusivity of oxygen varies markedly with stoichiometry. The excess oxygen has been inferred to be accommodated as interstitial oxygen. Thus, the larger the value of x in UO_{2+x} the more rapid the diffusion of oxygen. This trend is reported by Belle, Auskern and co-workers [73] and the migration energy for samples of constant composition decreased from 2.8 eV for stoichiometric UO_2 to 1.3 eV for oxygen excess $UO_{2.06}$. This suggests a complex diffusion process. Thorn and Winslow [74-76] have used a statistical mechanical model to analyze the defect structure of uranium dioxide in a series of papers in which there is formulated a model assuming a perfect cation lattice and the existence of anion vacancies and anion interstitials. Because the diffusion of uranium is so much smaller than that of oxygen, it is likely that the disorder on the cation sublattice is smaller than on the anion sublattice. Nevertheless the disorder on the two sublattices are coupled through an equilibrium constant. Thorn and Winslow consider the cation disorder as negligible compared to the anion disorder. They use an expression developed by Rice [77] and apply it to the diffusion of oxygen,

$$D_e(\tilde{O}) = \tfrac{1}{2} Z \langle a^2 \rangle \nu \theta_d (1-\theta_d) \exp(-U_e/kT) \tag{8.87}$$

where Z is the number of neighbors in the shell contained in the diffusion-complex unit, $\langle a^2 \rangle$ is the mean squared jump distance, ν is the average jump frequency and U_e is the activation energy to expand the shell so the atom can jump. According to Thorn and Winslow, the concentration of defects in terms of mole fraction, θ_d, is given by

$$\theta_d \cong \tfrac{1}{2}\left\{x + \left[x^2 + \left(\frac{8q_I}{q_V}\right) \exp\,[E_I - E_V)/kT]\right]^{1/2}\right\} \tag{8.88}$$

In equation 8.88, x is the excess oxygen in UO_{2+x}, q_i and q_v are the partition functions for the oxygen interstitial and vacancy, respectively and E_T and E_V are the energies to form an oxygen interstitial and vacancy respectively. Using this approach, Thorn and Winslow derive diffusion coefficients for oxygen which vary with temperature and composition in agreement with experimental data. If the concentrations of anion and cation defects are coupled, one would expect that increasing the concentration of oxygen (increasing x) would result in decreasing values of the tracer diffusion coefficient of uranium regardless of the type of defect. The opposite behavior has been reported by Marin and Contamin [78]. Additional studies are needed to understand the lattice disorder in UO_{2+x}.

Rutile, TiO_2, is another oxide in this group in which both the cation and anion diffusivity have been studied. The results are,

$$D_e(\tilde{Ti}) = 6.4 \times 10^{-2} \exp(-2.67/kT) \text{ cm}^2 \text{ s}^{-1} \tag{8.89}$$

for titanium [79] in the temperature range of 900-1300° C and,

$$D_e(\tilde{O}) = 2 \times 10^{-3} \exp(-2.6/kT) \text{ cm}^2 \text{ s}^{-1} \tag{8.90}$$

for oxygen [80] in the temperature range of 710-1300° C. These data presumably refer to rutile of stoichiometric composition. The very similar diffusivities suggest an almost equal concentration of cation and anion defects (assuming the mobilities of the diffusion species are about the same, of course). Rutile exhibits a deficit of oxygen, TiO_{2-x}. The defects are thought to be oxygen vacancies and/or titanium interstitials. Again, tracer diffusion measurements as a function of composition are badly needed in order to contribute to an understanding of the predominant defects as has been done with the lead chalcogenides [11-13]. Furthermore, there is evidence that in TiO_{2-x} as well as some other transition metal oxides of vanadium, niobium, tantalum, molybdenum and tungsten, there exist extended planar defects. An illustration of this type of defect is shown in figure 8.12. This defect has been termed a 'crystallographic shear' or a pseudo-phase. It is not clear whether these planar defects represent the boundaries of distinct phases but their

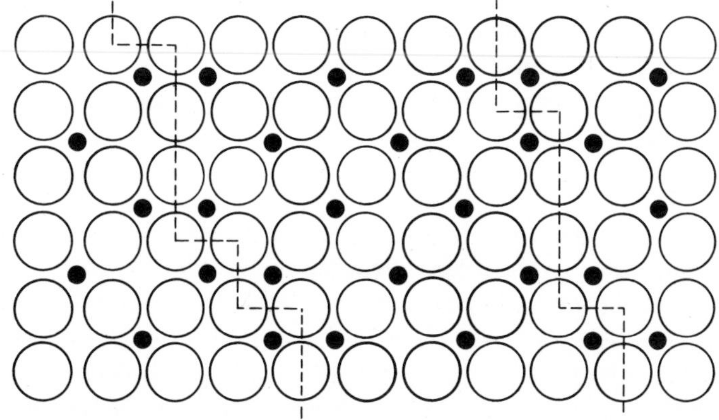

Figure 8.12. Two-dimensional model of 'crystallographic shear' in an oxide of nominal formula, MO_2. The M atoms are the small black filled circles, and the O atoms are the large open circles. There is an excess of M atoms and the arrangement of the M atoms is not random. The dotted lines divide the 'crystal' into regions in each of which the M atoms in any given region are not 'in phase' with the M atoms in a neighboring region. After Magneli [132]. The present author hypothesizes that such crystallographic shear boundaries in three-dimensional crystals may serve as rapid diffusion paths and these may be an explanation for the very rapid uptake of oxygen at low temperatures by some highly non-stoichiometric crystals.

existence has been confirmed by electron microscopy. In the case of transport properties, such defects should offer easy diffusion paths. These may be especially important in diffusion processes under a chemical gradient. It would be useful to study diffusion along these defects, for example, by autoradiographic techniques. Such studies would help in understanding divergent diffusion data for some transition metal oxides.

The oxides of the group, $MO_{2.5}$ have not been extensively studied and at the present, trends are not apparent. These oxides exist over moderate ranges of stoichiometry. One of the best known examples is α-Nb_2O_5 which has been studied by a variety of techniques. From electrical measurements [81-84], it is known that α-Nb_2O_5 is a metal excess n-type semiconductor. In analogy with studies on cobaltous oxide described earlier, a model which describes this behavior is,

$$Nb_2O_5 \rightleftharpoons Nb_2O_{5-x} + xV_O^{\cdot} + xe' + \frac{x}{2}O_2(G) \tag{8.91}$$

The mass action law gives,

$$K = [V_O^{\cdot}]^x n^x / p_{O_2}^{x/2} \tag{8.92}$$

and the electroneutrality condition is

$$[V_O^{\cdot}] = n. \tag{8.93}$$

Therefore $[V_O^{\cdot}] \propto p_{O_2}^{1/4}$ and if the diffusion of oxygen is proportional to the concentration of singly ionized oxygen vacancies, then the isothermal diffusivity will be proportional to $p_{O_2}^{1/4}$. This behavior was found by Chen and Jackson [85]. Consequently the diffusion measurements substantiated the disorder represented by equation 8.91. As pointed out previously electrical measurements alone often cannot distinguish between models. In fact, Greener [82], as well as Kofstad [83] have suggested the possibility of niobium interstitials. This defect could also account for the electrical properties. Again, measurements of the tracer diffusion of niobium as a function of oxygen pressure would help in deciding whether niobium interstitials made a significant contribution to the disorder. Because α-Nb_2O_5 is anisotropic, diffusion depends on crystallographic orientation. Sheasby and Cox [87] have reported that oxygen diffusion may be two orders of magnitude more rapid in the [010] direction than in a direction perpendicular to this. Additional data are needed on such crystals to provide limiting cases to serve as guides for understanding diffusion in anisotropic crystals and to explaining some of the apparent discrepancies between data of different investigators.

8.5.6 Diffusion in Oxides Under a Chemical Potential Gradient

In the 1930's, C. Wagner [88] developed equations for transport through oxide layers under a chemical potential gradient. Specifically, when a metal is oxidized under conditions where bulk diffusion through a thick, coherent layer is rate determining, the kinetics obey the so-called parabolic rate law, i.e., the square of the thickness is proportional to time.† C. Wagner showed that the amount of oxide in equivalents, $\tilde{n}$, formed per unit time over an area A was

$$\frac{1}{A}\frac{d\tilde{n}}{dt} = \tilde{J} = k_r/\Delta x \tag{8.94}$$

where $\tilde{J}$ is the total flux (cations plus anions) through the oxide, Δx is the film thickness at any time t and k_r is the rational rate constant expressed in equivalents/cm-s. This rational rate constant can be expressed as

$$k_r = \frac{300}{F}\frac{1}{N_0 e}\int_{\mu_M^a}^{\mu_M^i} \{(t_1 + t_2)t_3(\sigma/|z_1|)\} d\mu_M \tag{8.95}$$

where F = 96,500 coulombs/equivalent
N_0 = Avogadro's number
e = elementary electronic charge in e.s.u.
μ_M = chemical potential of the metal at the metal-metal oxide interface (superscript i) and at the oxide-gas interface (superscript a)
t = transference number
σ = total electrical conductivity
z = valence
1, 2, 3 = subscript used to designate the cation, anion and electron, respectively.

Because the electrical conductivity is related to the diffusivity, equation 8.96 may be written in terms of the diffusivity of the slow moving species (cations and/or anions) in an electronic conductor. Using the Gibbs-Duhem equation to relate the chemical potential of the metal to that of the oxygen and expressing the chemical potential in terms of the activity of oxygen yields, for an electronic conductor

$$k_r = c_{eq}\int_{\{O\}^i}^{\{O\}^a} \{(|z_M|/|z_O|)D_e(M) + D_e(O)\} d\ln\{O\} \tag{8.97}$$

† The concentration of defects is a function of position within the oxide layer. These concentrations at the metal-oxide and at the oxide-gas interface are fixed at the equilibrium value defined by the respective oxygen activities at these two interfaces.

where c_{eq} = number of equivalents of oxide per cm^3
$\{O\}$ = activity of oxygen at the metal-oxide interface (superscript i) and at the oxide-gas interface (superscript a)
$D_e(M), D_e(O)$ = self-diffusion coefficients of the metal (as the cation) and of the oxygen (as the anion). Note that the subscript e is used. The self-diffusion coefficients are composition dependent and they are behind the integral sign

In practice, it is convenient to measure either weight gain per unit area ($\Delta m/A$), as a function of time whence,

$$(\Delta m/A)^2 = k_p t \tag{8.98}$$

where k_p is the practical parabolic rate constant in units of $gm^2/cm^{-4}\,s^{-1}$ of the layer thickness per unit time (x), whence

$$x^2 = k_T t \tag{8.99}$$

where k_T is the Tamman rate constant in units of cm^2/s^{-1}. These rate constants are related to the rational rate constant by the following equations,

$$k_r = \tfrac{1}{2}\tilde{v}\left(\frac{|z_O|}{A_O}\right)^2 k_p \tag{8.100}$$

and

$$k_r = k_T/\tilde{v} \tag{8.101}$$

where $\tilde{v}$ = equivalent volume (cm^3/eq)
A_O = atomic weight of oxygen and the other terms have previously been defined.

Under conditions where the self-diffusion of one component is very much less than the other in an electronic conductor ($t_3 \cong 1$), one of the terms involving the self-diffusion coefficients may be neglected in equation 8.97. Moreover, if isothermal oxidation kinetics are obtained as a function of oxygen pressure, the self-diffusion coefficient of the more mobile species may be obtained at a definite activity of oxygen, $\{O\}'$, by differentiating equation 8.97 and using equation 8.100 to yield

$$D_e(M) = (1/f)\,D_e(\tilde{M}) = \left\{\frac{[\partial k_p/\partial \ln\{O\}]\{O\}'}{2[z_1/z_2][A_O/z_2\tilde{v}]^2}\right\} \tag{8.102}$$

This approach has been used by Pettit [89] for the diffusion of iron in wüstite, Mrowec and co-workers [90-92] for the diffusion of cobalt in CoO and by Fueki and Wagner [93, 94] for the diffusion of nickel in NiO and manganese in MnO, respectively. These calculated results are in excellent agreement with radiotracer studies (see Table 8.5). It is to be emphasized that the measurements are made of a metal being oxidized to an oxide and hence the kinetics are obtained for an oxide existing under a chemical potential gradient. The gradient is determined by the ratio of the activity of oxygen at the metal-metal oxide interface to that of the ambient gas. By this means, the self-diffusion coefficient of the more mobile species can be obtained as a function of oxygen activity (deviation from stoichiometry) across the entire homogeneity range of the oxide (see figure 8.13). Such measurements are badly needed to help elucidate the defect structure of some oxides. To this end Swaroop and Wagner [95] have initiated measurements on cuprous oxide growing on copper in order to obtain $D_e(\tilde{Cu})$ at low values of oxygen activity.

Childs and Wagner [96] have termed this kind of measurement interphase kinetics because there is a solid state phase boundary fixing the activity of oxygen during the measurement. Another type of diffusion measurement is obtained when a crystal of oxide is equilibrated with a given oxygen pressure. This gas atmosphere is quickly removed from the sample and a different but known gas atmosphere is admitted [97-99].

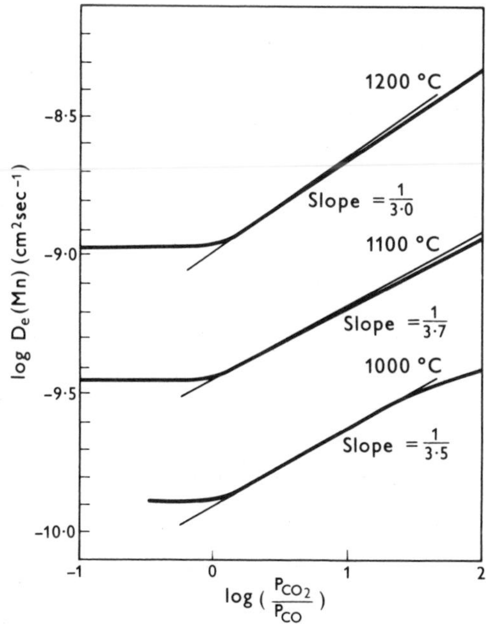

Figure 8.13. Self-diffusion coefficient of manganese in manganous oxide as a function of oxygen pressure. The diffusion coefficients were calculated from oxidation kinetics of maganese to manganous oxide in CO-CO_2 mixtures. After Fueki and Wagner [94].

TABLE 8.5

Comparison of tracer or self-diffusion coefficients[a] obtained by direct measurement, by parabolic oxidation kinetics and by chemical diffusion studies

Diffusion system	Direct measurement	Oxidation kinetics	Chemical diffusion
$D_e(\tilde{Ni}$ in NiO) (p_{O_2} = 0.21 atm, 1000°C)	2.49×10^{-11} [24]	0.198×10^{-11} [93]	1.12×10^{-11} [97]
$D_e(\tilde{Co}$ in CoO) (p_{O_2} = 1 atm, 1000°C)	2.60×10^{-9} [16]	2.84×10^{-9} [90]	4.07×10^{-9} [97]
$D_e(\tilde{Co}$ in CoO) (p_{O_2} = 0.21 atm, 1000°C)	2.51×10^{-9} [16]	1.6×10^{-9} [92]	–
$D_e(\tilde{Fe}$ in FeO) (O/Fe = 1.07 atm, 1100°C)	5.3×10^{-6} [31]	–	5.5×10^{-6} [98]

[a] D_e in cm^2/s^{-1}.

Both the initial and final gas atmospheres are such that no second phase can form. The kinetics of re-equilibration to the new equilibrium state are followed and therefrom a rate constant, the chemical diffusion coefficient, is obtained. These kinetics have been termed intraphase kinetics [95].

The chemical diffusion coefficient is defined as the proportionality constant in Fick's first law when a concentration gradient or flow of 'a deviation from stoichiometry' is propagated through the crystal. The total flux, $\tilde{J}$, in a binary oxide is given by the algebraic sum of the two fluxes as,

$$\tilde{J} = J_1 + J_2 \tag{8.103}$$

where the two subscripts, 1 and 2, refer to the cation and anion respectively, as previously. The chemical diffusion coefficient can then be defined by

$$\tilde{J} = \underset{\sim}{\mathbf{D}}(\partial \tilde{c}/\partial x) \tag{8.104}$$

where $\tilde{c}$ is the absolute value of the concentration of the excess of one of the two components. For example, for a metal deficit compound such as CoO, $\tilde{c}$ represents the excess of oxygen above that of the stoichiometric compound and is expressed in equivalents per cubic centimeter. The chemical diffusion coefficient in equation 8.104 may be written by expressing equation 8.97 in differential form [87, 97]. The chemical diffusion coefficient is

$$\underset{\sim}{\mathbf{D}} = \left\{ \frac{300}{F} \frac{(t_1 + t_2)(t_3 \sigma)}{N_0 e} \cdot \frac{1}{|z_1|} \left[\frac{\partial \mu_M}{\partial x} \right] \right\} \tag{8.105}$$

where the symbols have been previously defined. This can be related to the self-diffusion coefficients for a binary oxide by an equation due to C. Wagner [88],

$$\underset{\sim}{\mathbf{D}} = \left\{ |z_1| D_e(1) + |z_2| D_e(2) \right\} \cdot \frac{c_{eq}}{\tilde{c}} \left[\frac{1}{N_0 kT} \frac{\partial \mu_M}{\partial \ln \tilde{c}} \right], \tag{8.106}$$

or by an equation derived by Darken [102] especially for metallic alloys,

$$\underset{\sim}{\mathbf{D}} = [X_2 D_e(1) + X_1 D_e(2)] \cdot \left[1 + \left(\frac{\partial \ln \gamma_1}{\partial \ln X_1} \right) \right] \tag{8.107}$$

where the subscripts 1 and 2 denote the components of the binary alloy, γ denotes an activity coefficient and X denotes the concentration of the indicated species in terms of mole fraction.

Brebrick [103] has derived a more detailed equation applicable to a binary compound semiconductor that exhibits Schottky disorder and is in the 'exhaustion region'. His equation is,

$$\underset{\sim}{\mathbf{D}} = \Gamma_2 d^2 \left\{ 1 + (\Gamma_1/\Gamma_2 - 1)\delta \right\} \left[1 + \left(\frac{\Delta^2 + 4K_s}{\Delta^2 + 4n_i^2} \right)^{1/2} \right], \tag{8.108}$$

where again the subscripts 1 and 2 refer to the cation and anion respectively, Γ denotes a jump frequency, d denotes a jump distance, δ is the fraction of vacancies that are cation vacancies, Δ denotes a deviation from stoichiometry and is equal to the difference between the cation and anion concentrations, K_s denotes the Schottky constant and n_i the intrinsic carrier concentration. Brebrick's equation has been successfully applied to lead sulfide but not to oxides, presumably due to the lack of the necessary parameters such as the jump frequencies. More recently, Brebrick [104] has extended his equations to other systems.

In principle, then, it is possible to compute the chemical diffusion coefficients from the self-diffusion coefficients and conversely provided that one self-diffusion coefficient is much larger than the other. For example, for an electronic conductor for which $D_e(\mathrm{M}) \gg D_e(\mathrm{O})$ and the dominant defects are cation vacancies it can be shown that [97, 98],

$$\underset{\sim}{D} = (1 + z_1) D_e(\mathrm{M}) \frac{1}{X_{V_M}} \tag{8.109}$$

where X_{V_M} denotes the mole fraction of cation vacancies.

In order to obtain the chemical diffusion coefficient, $\underset{\sim}{D}$, a physical parameter is measured as a function of time. Measurements include weight change, electrical conductivity, p-n junction migration and migration of a color front [98]. The weight change or thermogravimetric method is the most straightforward because it does not involve any model for the defect structure. For example, if the oxide crystal has the geometry of a thin plate, the solution to Fick's second law is,

$$\log (1 - \Delta m/\Delta m_\infty) = \log \left(\frac{8}{\pi^2} \right) - \frac{\pi \underset{\sim}{\mathbf{D}} t}{9.2 l^2} \tag{8.110}$$

where Δm denotes the weight change at any time t, Δm_∞ denotes the weight change at the end of the experiment, i.e., after the oxide has re-equilibrated with the new gas atmosphere, t is the time, and l is the half-thickness of the plate. Similar expressions are available for brick shaped [97] and other geometries.

Experimentally one places an oxide crystal in a controlled atmosphere at a given temperature until it equilibrates. At time zero, a new gas atmosphere is admitted to the crystal (both gas atmospheres are chosen such that the oxide remains within its single phase homogeneity region). The weight change, or other physical property is followed as a function of time and equation 8.105 is used to compute the chemical diffusion coefficient. Wagner and co-workers [96-101] have used this technique to determine the chemical diffusion coefficient in MnO, NiO, CoO, and FeO. Therefrom the self- or tracer-diffusion coefficients were calculated (see

Table 8.5). The agreement between the calculated values of the tracer-diffusion coefficient and the measured values is very good (see Table 8.5). In order for this technique to be valid, chemical diffusion, which is the diffusion of the defects, must be the slowest step. In wüstite, the reaction on the surface† appears to be partially rate determining for samples about 1-2 mm thick in CO-CO_2 mixtures at 1000-1200° C. The role of volume to surface ratio has been discussed by Childs and Wagner [95]. A method to determine the effect of a surface controlled reaction is to use a fixed ratio of CO-CO_2 gases and in separate experiments successively dilute the fixed ratio with argon. Thus the chemical potential of oxygen remains fixed but the flux of reactive atoms striking the surface changes. In this way Laub [99] has confirmed that the re-equilibration kinetics of wüstite may be controlled in part by a slow surface reaction. In addition, the use of equation 8.110 implies that $\underset{\sim}{D}$ is independent of composition. J. B. Wagner and co-workers have shown that $\underset{\sim}{D}$ is a function of composition for oxides with large deviations from stoichiometry. Experimentally, one chooses increments of composition in the phase field that are sufficiently small such that $\underset{\sim}{D}$ may be taken as a constant. The procedure is then repeated across the phase field so that the composition dependence may be obtained. For example, the wüstite phase field was divided into ten equal increments of O/Fe ratio of 0.01 each [98]. This procedure should be used to test whether $\underset{\sim}{D}$ is dependent on composition for a given oxide.

There is a third type of diffusion in oxides under a chemical potential gradient. This diffusion type is the classic interdiffusion experiment where two different oxides are welded together and diffusion annealed. The diffusion profiles can be determined, for example, by an electron microprobe analysis along the diffusion axis on the quenched couple. In this way the diffusion profile may be obtained. Two limiting cases are to be distinguished. In one, the two oxides react (i.e., diffusion occurs) to form a continuous solid solution. In the second, one or more new phases (e.g. compounds) form. The former case is illustrated by interdiffusion in the MgO-FeO couple. This classic case has been studied by Rigby and Cutler [105]. According to Rigby and Cutler the interdiffusion coefficient is not constant but increases with increasing vacancy concentration. The other limiting case is illustrated by the formation of $NiAl_2O_4$ from NiO

† The rate of re-equilibration may be controlled by one or more of the following: (1) Transport of gas to the surface, (2) Reaction on the surface, (3) Diffusion into the bulk oxide and (4) Transport of gas away from the oxide. With a linear flow of gas of about 0.9 cm/s^{-1} flowing past the sample, transport to and from the surface is usually not rate controlling. However, dissociation of the gas may be slow compared to diffusion in the oxide. Then a surface reaction may control or partially control the kinetics as: $CO_2(G) \rightleftharpoons CO_2(ads)$; $CO_2(ads) + 2e' \rightleftharpoons CO(ads) + O^{2-}(ads)$.

and Al_2O_3 [106]. A review of this type of spinel formation has been presented by Armijo [56]. These very important types of interdiffusion are beyond the scope of the present chapter.

8.5.7 Diffusion in Oxides Under an Electrical Potential Gradient

Diffusion of a radio tracer into an initially homogeneous crystal to which an electrical potential is applied has been given the name of a Chemla experiment following the work of M. Chemla [107, 108] on NaCl crystals. The phenomena of altering (especially slowing or terminating) the rate of growth of an oxide layer has been known for a long time. However, only fairly recently have Chemla experiments been carried out on oxides. These types of experiments can yield correlation factors as will be discussed below.

Under the assumption of only one species moving in a binary oxide, Fick's second law becomes

$$\frac{\partial c}{\partial t} = D_e \frac{\partial^2 c}{\partial x^2} \pm \mu E \frac{\partial c}{\partial x} \tag{8.111}$$

where D_e denotes the tracer diffusion coefficient of an ion in an initially homogeneous crystal to which there is an applied potential yielding a field, E, and μ is the mobility of the ion. When the tracer diffuses into a crystal in the absence of the electrical field, the usual penetration profile is obtained. For example, an instantaneous source into two semi-infinite crystals with the source at the interface between the crystals yields

$$c(x) = \frac{W}{\sqrt{4\pi D_e t}} \exp\left(-x^2/4D_e t\right) \tag{8.112}$$

where $c(x)$ denotes the concentration of radioactive tracer at a distance x from the interface after a diffusion time, t, and W is the original quantity of tracer per unit area at the interface at $t = 0$. When a field is imposed, the penetration profile is perturbed. Experimentally one places two oxide crystals together with a tracer at the interface. The diffusion anneal is carried out while the electric field is applied between the ends of the diffusion couple. The distribution of tracer or penetration profile under the electric field is given by

$$c(x) = \frac{W}{\sqrt{4\pi D_e t}} \exp\left[-(x \pm \Delta x)/4D_e t\right] \tag{8.113}$$

where Δx is the displacement of the maximum of the radiotracer concentration due to the electric field. For example, for an oxide which has a mobile cation and essentially an immobile anion, the displacement of the maximum, Δx, is towards the negative pole of the diffusion couple.

The mobility is directly related to this displacement by,

$$\Delta x = \mu E t. \tag{8.114}$$

Therefore D_e(tracer) may be obtained from a plot of log $c(x)$ versus $(x \pm \Delta x)^2$ in the usual way while the mobility is obtained using equation 8.114. In principle the correlation factor, f, can be obtained by using the Nernst-Einstein equation to calculate the self-diffusivity as

$$D_e(\text{self}) = \mu k T \tag{8.115}$$

whence

$$f = D_e(\text{self})/D_e(\text{tracer}) \tag{8.116}$$

Cline and co-workers [109] have recently reported the results of a Chemla experiment on BeO, which is essentially an ionic conductor with the cation as the mobile ion. Lacombe and co-workers [28, 110] have applied the technique to electronic conductors such as ferrous oxide and manganous oxide. In electronic conductors, the majority of the current is of course carried by electrons. Using the total current passed, the transport number of ions may be obtained, for example, by analyzing the number of cations deposited at the cathode. Then, if the concentration of ions, c_i, which transport the charge, is known, the product of the transport number, t_i, and the total electrical conductivity, σ, may be substituted into the Nernst-Einstein relation to calculate the self-diffusion coefficient of the ion or

$$D_e(\text{self}) = (kT\sigma t_i)/(c_i z_i^2 e^2) \tag{8.117}$$

where z_i is the valence, e is the electronic charge and the other terms have been previously defined. A comparison with the experimentally measured tracer-diffusion coefficient can then be made.

8.5.8 Diffusion Under a Thermal Gradient

The unmixing of a homogeneous solid solution when placed in a thermal gradient has been known for many years and is sometimes termed the Ludwig-Soret effect. Most of the studies have been carried out on metals and alloys. Oriani [111] has published a critical resumé of these data. Relatively few data are available on solid compounds. In the 1930's, Rheinhold and co-workers [112] studied the Ludwig-Soret effect in solid solutions of salts such as CuBr-AgBr. C. Wagner presented a thermodynamic treatment of non-isothermal systems in which these types of solid salts and many other systems were treated [113, 114]. Due to the technological requirements of nuclear reactors, oxides are frequently heated under a thermal gradient and there has been a resurgence of interest in thermotransport.† Aitken and co-workers [115-117] have studied this phenomenon on UO_{2+x} and on $Pu_{0.2}U_{0.8}O_{2-x}$. They placed an initially homogeneous sample in an evacuated tube and subjected the sample to a thermal gradient. After some time, the sample presumably comes to steady state. They report for $Pu_{0.2}U_{0.8}O_{2-x}$ that the deviation from stoichiometry as expressed by x in the preceding formula is given by

$$\ln x = Q^*/kT + C \tag{8.118}$$

where Q^* is the 'heat of transport' which depends strongly on the oxygen to metal ratio, T is the temperature and C is a constant. According to Markin and Rand [118], the oxygen should concentrate at the low temperature side of the thermal gradient. Aitken and co-workers find a more complex behavior. In such oxides heated in vacuo, the oxide may lose oxygen and mass transport may occur through the gas phase. Bowen [119] has initiated a study of diffusion under a thermal gradient on ferrous oxide and he is attempting to overcome the problem of the dissociation of the condensed phase by using controlled oxygen pressures over his sample. More research is needed on this phenomenon to serve as a guide for high temperature applications of oxides.

8.5.9 Diffusion of Impurities in Oxides

While there are many studies of foreign cations in oxides, there are few systematic studies. Recently W. Crow [18] has made systematic measurements of the diffusivity of foreign cations in cobaltous oxide and in nickel oxide. Typical results are presented in figure 8.14. For NiO, the smaller the atomic number of the diffusing impurity, the more rapid is the diffusion. This trend is nearly the same in CoO except the diffusivities of Co-60 and Fe-55 are virtually the same (see figure 8.14). The reason for this apparent anomaly is not yet available.

Crow has studied the diffusion of nickel-63 and also the diffusion of iron-55 in CoO and NiO as a function of oxygen pressure and temperature. The data support diffusion of the foreign cation via cation vacancies and therefore the greater the departure from stoichiometry within a given oxide or between the two oxides, the more rapid is the diffusion coefficient of the foreign cation. Moreover, when nickel oxide is doped with chromium to provide extrinsic cation vacancies, the diffusion of cobalt-60 is greater than that in undoped crystals at the same oxygen pressure and temperature. Wuensch and Vasilos [120] have studied the diffusion of transition metals in MgO. They report that the activation

† The term, thermal diffusion, is sometimes used to denote isothermal, tracer diffusion. It seems best not to use this terminology to describe diffusion under a thermal gradient.

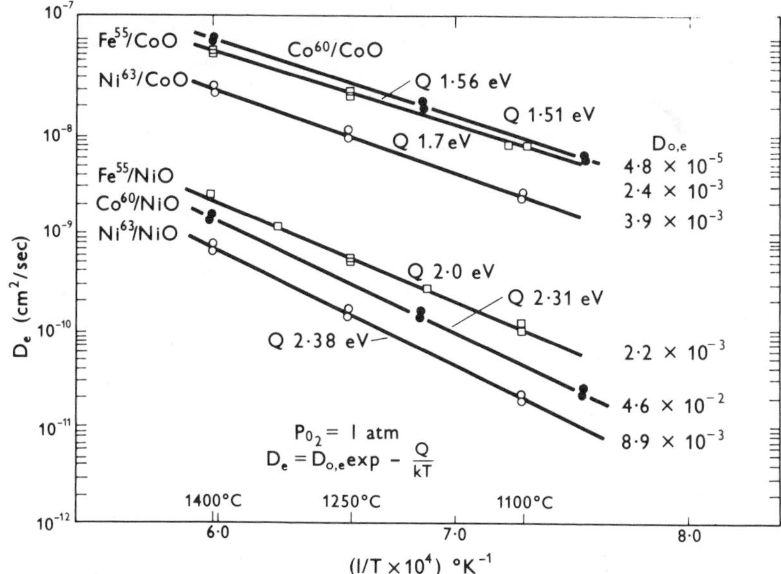

Figure 8.14. Cation self and impurity diffusion in cobaltous and in nickelous oxide. After W. Crow [18].

energy for diffusion increases exponentially with the ratio of the ionic radius of the diffusing ion divided by the polarizability for that ion and that the pre-exponential in the Arrhenius equation increases with the cation radius cubed. These studies provide very useful guides and more work of this type is needed.

Except for the diffusion of fission fragments in oxides used in nuclear technology, very few studies have been made on anion diffusion in oxides. It has been generally supposed that foreign anions are relatively immobile in simple oxides like FeO, NiO, and CoO because of the large radii of the foreign anions and the relative slow diffusivity of oxygen in these compounds. On the other hand, in the area of stress corrosion studies, it has been assumed that a chloride ion could migrate through a thin layer of oxide rapidly and thus gain access to the underlying metal. Recently, George and Wagner [121] initiated a test of this behavior by measuring the diffusivity of chlorine-36 in cobaltous oxide and in nickel oxide single crystals. Preliminary values for CoO and NiO crystals annealed in air at 1000°C yield 2×10^{-12} cm^2/s^{-1} and $2\text{-}3.4 \times 10^{-11}$ cm^2/s^{-1} respectively. Thus, the diffusivity of chlorine is about two orders of magnitude more rapid than oxygen in CoO and one order more rapid than oxygen in NiO.

Because both CoO and NiO are metal deficient, one would expect the greater the deficit of metal, the smaller the concentration of defects on the anion sublattice and hence the slower the diffusion of a foreign anion. The converse appears to be the case since cobaltous oxide has a greater concentration of cation vacancies than nickel oxide when both are annealed under the same oxygen pressure and temperature. However, these data are preliminary and should be treated as such. Further studies are under way on the diffusion of chlorine in cobaltous oxide as a function of defect concentrations by Laub. [122].

Another area of technological importance is the diffusion of sulfur in oxides. In some cases small amounts of sulfur appear to be transported through the oxide layer on turbine blades with a resulting degradation of the underlying metallic blade. Seybolt [123] recently reported the diffusion of sulfur-35 in Cr_2O_3 at 1000°C to be 1.19-2.36 x 10^{-10} cm^2/s^{-1}. His results indicate that sulfur migrates very rapidly in Cr_2O_3. The reported data in the form of the logarithm of concentration versus the square of the penetration distance exhibit two distinct segments and Seybolt's analysis is taken from data obtained deep within the crystal where the slope is smaller than that near the surface. Sandulova and Andrievskii [124] have reported a very high diffusivity of sulfur in $Cu_2O[\underset{\sim}{D}(\tilde{S}) = 8.9 \times 10^{-9}\ cm^2/s^{-1}$ at 1000°C]. It may be that sulfur migrates rapidly as an interstitial as in the case of lead salts [11-13]. This possibility should be checked on a number of different oxides. Additional data are available on NiO and CoO crystals by Chang, Stewart and Wagner [133].

8.6 SUMMARY AND CONCLUDING REMARKS

In the foregoing, an attempt was made to point out trends in diffusion in oxides. Examples were cited using simple oxides as illustrations. From the point of view of oxides with which to study model behavior of diffusion, this writer believes that cobaltous oxide, CoO, and magnesium oxide, MgO, offer the best choices for an electronic and a predominantly ionic conductor, respectively. Cobaltous oxide has a sufficiently narrow range of stoichiometry such that the concentration of defects may be treated as an ideal solution yet the range is large enough to be measured using thermogravimetric techniques. Moreover, both the cation and anion diffusion have been measured and the predominant disorder, at least under high oxygen pressures, seems established by a variety of techniques including electronic conductivity and Seebeck coefficient measurements. Magnesium oxide, on the other hand, exhibits a very narrow range of stoichiometry. Thus trace impurities exert a large influence on the diffusivity and on other properties. However, MgO can be obtained in the form of high purity crystals. Moreover, with improved materials characterization such as the use of a spark source mass spectrometric analysis for chemical analysis both before *and* after a diffusion anneal, there is the possibility of ultimately understanding the intrinsic behavior of these oxides.

From the point of view of technological importance, probably SiO_2, Al_2O_3, Cr_2O_3 and UO_2 are the most important oxides and certainly these deserve more study—again with careful attention to materials characterization.

The diffusion of impurities, especially anions deserves more attention and the simultaneous diffusion of two or more impurities where there may exist interactions such as has been reported for nickel and chlorine in lead telluride crystals [125] should be investigated. The rate and mechanism formation of so-called double oxides, more complex oxides such as spinels, from two simpler oxides needs more study. Finally, diffusion over the entire homogeneity range of some of the simpler oxides should be studied. To this end, parabolic oxidation kinetics obtained as a function of oxygen pressure offer an attractive method, especially in view of the rapid survey possible.

The reader may wish sources of diffusion data. Very useful reviews of diffusion mechanisms have been published by C. E. Birchenall [58, 127] and O'Keeffe's [48] summary is a useful guide to the diffusion behavior of oxides and sulfides. Recently, a continuous compilation, 'Diffusion Data' which is published about every four months, has become available [128]. This source provides a relatively rapid survey which gives an abstract of data and serves as a guide to the literature. The National Bureau of Standards is conpiling data in their files, 'Diffusion in Metals Data Center' for subsequent critical evaluation and inclusion in the National Standard Data Reference System [129]. When this source becomes available it will be a great help to workers in the field. The bibliographies by Harrop [43] and by Dragoo [130] are presently available and are very helpful. In the foregoing, no treatment of the studies on surface diffusion was given. This important area has been critically reviewed by Robertson [131]. Surface diffusion as a function of deviation from stoichiometry is a challenging research area for the future.

If the diffusion of a particular element in an oxide is not available in the literature, the obvious choice is to seek an analogous system with which to make a comparison. For example, it was pointed out in the foregoing that the migration enthalpy for cation diffusion in oxides exhibiting the NaCl lattice amounts to about 1.3 eV. If the partial molar enthalpy of oxygen can be obtained from independent thermodynamic data, the activation energy may be calculated. The estimation of the pre-exponential is most difficult. According to O'Keeffe [48], there is a relation between the activation energy, Q, divided by the melting temperature, T_m (°K) and the pre-exponential, $D_{o,e}$, in the Arrhenius equation such that Q/T_m versus log $D_{o,e}$ yields approximately a linear relation. The calculations of Haven and Wilkinson [54] on the pre-exponential term will be a great help. It must be re-emphasized that many studies are reported wherein the diffusion measurements are carried out under a constant oxygen pressure, especially in air. Therefore, the activation energy term, Q, is the algebraic sum of the migration energy and formation energy which is the partial molar enthalpy of oxygen in the nonstoichiometric oxide and may be positive or negative. When comparisons of energy terms are made, this distinction must be clear.

A comparison of the data on oxides with the data on the chalcogenides indicates that the former data appears more self-consistent than the latter. Part of this is due in large measure to the fact that conductivity and Hall measurements as well as diffusion measurements have been made on oxides at temperature and under a fixed oxygen pressure.

On the other hand, the chalcogenide elements attack lead wires at elevated temperatures and most auxiliary data have been obtained on samples quenched from elevated temperatures. Because of retrograde solubility in the lead salts precipitation, clustering, and association phenomena of unknown magnitude have been detrimental to a more complete understanding of these salts. On the other hand, the semiconductor industry has required and has obtained materials of high purity for many compound semiconductors which has generally not been the case for oxides. Greater purity and materials characterization are needed for oxides. Finally, oxides deserve the very careful measurements as have been illustrated by the elegant studies of Chen [21] whereby information concerning the migration mechanism may be inferred directly from the isotope effect. This type of measurement is much needed on oxides as a function of oxygen pressure.

REFERENCES

1. C. B. Alcock, *Chem. Brit.*, **5**, 216 (1969)
2. H. J. T. Ellingham, *J. Soc. Chem. Ind.*, **63**, 125 (1944)
3. F. D. Richardson and J. H. E. Jeffes, *J. Iron and Steel Inst.*, **160**, 261 (1948)
4. G. Brouwer, *Phillips Res. Rep.*, **9**, 366 (1954)
5. F. A. Kröger and H. J. Vink, Solid State Physics, Vol. 3 (F. Seitz and D. Turnbull, eds.). (Academic Press, N.Y., 1956)
6. P. G. Shewmon, Diffusion in Solids (McGraw Hill, N.Y., 1963)
7. W. J. Moore and B. Selikson, *J. Chem. Phys.*, **19**, 1539 (1951)
8. W. J. Moore and B. Selikson, *J. Chem. Phys.*, **20**, 927 (1952)
9. F. A. Kröger, Chemistry of Imperfect Crystals, pp. 577-585. (North-Holland Pub. Co., Amsterdam, 1964)

10. W. J. Moore, Y. Ebisuzaki and J. A. Sluss, *J. Phys. Chem.*, **62**, 1468 (1958)
11. G. Simkovich and J. B. Wagner, Jr., *J. Chem. Phys.*, **38**, 1368 (1963); M. S. Seltzer and J. B. Wagner, Jr., *J. Phys. Chem. Solids*, **24**, 1525 (1963); K. Zanio and J. B. Wagner, Jr., *J. Appl. Phys.*, **39**, 5686 (1968)
12. M. S. Seltzer and J. B. Wagner, Jr., *J. Chem. Phys.*, **36**, 130 (1962)
13. M. S. Seltzer and J. B. Wagner, Jr., *J. Phys. Chem. Solids*, **26**, 233 (1963); Y. Ban and J. B. Wagner, Jr., *J. Appl. Phys.*, **41**, 2818 (1970)
14. B. Fisher and D. S. Tannhauser, *J. Electrochem. Soc.*, **111**, 1194 (1964); *J. Chem. Phys.*, **44**, 1663 (1966)
15. N. G. Eror and J. B. Wagner, Jr., *J. Phys. Chem. Solids*, **29**, 1597 (1968)
16. R. E. Carter and F. D. Richardson, *J. Metals*, **6**, 1244 (1954)
17. B. Fisher and J. B. Wagner, Jr., *J. Appl. Phys.*, **38**, 3838 (1967)
18. W. Crow, Ph.D. Thesis, Ohio State University (1969)
19. W. K. Chen, N. L. Peterson and W. T. Reeves, *Phys. Rev.*, **186**, 887 (1969)
20. B. A. Thompson, Ph.D. Thesis, Rensselaer Polytechnic Institute (1962); known through reference 19
21. W. K. Chen and R. A. Jackson, *J. Phys. Chem. Solids*, **30**, 1309 (1969)
22. J. B. Holt, *Proc. Br. Ceram. Soc.*, **9**, 157 (1967)
23. F. A. Kröger, *J. Phys. Chem. Solids*, **29**, 1889 (1968)
24. J. S. Choi and W. J. Moore, *J. Phys. Chem.*, **66**, 1308 (1962)
25. M. O'Keeffe and W. J. Moore, *J. Phys, Chem.*, **65**, 1438 (1961); **65**, 2277 (1961)
26. Z. Hed and D. S. Tannhauser, *J. Chem. Phys.*, **47**, 2090 (1967)
27. M. Gvishi, I. Bransky and D. S. Tannhauser, *Solid State Comm.*, **6**, 135 (1968)
28. J. P. Boquet, M. Kawahara and P. Lacombe, *Compt. Rend.*, **265**, 1318 (1967)
29. J. B. Price and J. B. Wagner, Jr., *J. Electrochem. Soc.*, **117**, 242 (1970)
30. L. Himmel, R. F. Mehl and C. E. Birchenall, *Trans. Inst. Min. Met. Engrs.*, **197**, 827 (1953)
31. P. Hembree and J. B. Wagner, Jr., *Trans. Am. Inst. Min. Met. Engrs.*, **245**, 1547 (1969)
32. P. Desmarescaux and P. Lacombe, *Mem. Sci. Rev. Met.*, **12**, 899 (1963)
33. F. D. Richardson, *Disc. Faraday Soc.*, **4**, 256 (1948)
34. C. E. Birchenall, Physics and Chemistry of Ceramics (C. Klingsberg, ed.). (Gordon and Breach, N.Y., 1963)
35. V. Kumar and Y. P. Gupta, *J. Phys. Chem. Solids*, **30**, 677 (1969)
36. Y. P. Gupta and L. J. Weirick, *J. Phys. Chem. Solids*, **28**, 811 (1967)
37. R. Lindner, St. Austründal and A. Akerstrüm, *Acta. Chem. Scand.*, **6**, 468 (1962)
38. N. A. Surplice, *Brit. J. Appl. Phys.*, **17**, 175 (1966)
39. R. Lindner and G. D. Parfitt, *J. Chem. Phys.*, **26**, 182 (1957)
40. S. P. Mitoff, *Prog. in Ceramic Sci.*, **4**, 265 (1966). (J. E. Burke, ed.). (Pergamon Press, N.Y.)
41. Y. Oishi and W. D. Kingery, *J. Chem. Phys.*, **33**, 905 (1960)
42. H. J. DeBruin, *J. Aust. Inst. Metals*, **14**, 247 (1969)
43. P. J. Harrop, *J. Materials Sci.*, **3**, 206 (1968)
44. R. Condit and Y. Hashimoto, *J. Am. Ceram. Soc.*, **50**, 425 (1967)
45. R. W. Redington, *Phys. Rev.*, **87**, 1066 (1952)
46. B. I. Boltaks, Diffusion in Semiconductors (H. J. Goldsmid, ed., J. I. Carusso, translator). (Academic Press, New York, 1963)
47. W. D. Copeland and R. A. Swalin, *J. Phys. Chem. Solids*, **29**, 313 (1968)
48. M. O'Keeffe, 'Diffusion in Oxides and Sulfides', p. 57 in Proc. Second (1965) International Conference on Sintering and Related Phenomena (G. C. Kuczynski, N. A. Hooten and G. F. Gibbon, eds.). (Gordon and Breach Pub. Co., New York, 1967)
49. A. B. Lidiard, *Handbuch der Physik*, **20**, 246, Springer-Verlag (1957)
50. R. Haul, D. Just and G. Dümbgen, Reactivity of Solids, Proc. of the 4th International Symposium, p. 65. (J. H. DeBoer, ed.). (Elsevier Publishing Co., Amsterdam, 1961)

51. R. Haul and D. Just, *J. Appl. Phys.*, **33**, 487 (1962)
52. W. J. Moore, Diffusion in Solids, a series of published lecture notes given at Centro Bras'leiro de Pesquisas Fisican, Rio de Janeiro, Brazil (August, 1962); available as a U.S.A.E.C. Final Report No. COO, 250-41 under contract AT-(11-1)-250
53. S. P. Mitoff, *J. Chem. Phys.*, **35**, 882 (1961)
54. Y. Haven and W. Wilkinson, First Tech. Rept. for ARPA Contract DAHC-15-68-C-0179, Wake Forest University, July, 1969
55. K. Hauffe, Oxidation of Metals (Eng. Transl.). (Plenum Press, New York, 1965)
56. J. S. Armijo, *Oxidation of Metals,* **1**, 171 (1970)
57. H. Schmalzried, *Prog. in Solid State Chemistry,* **2**, 265 (1965). (Pergamon Press, H. Reiss, ed.)
58. C. E. Birchenall, *Metallurgical Reviews,* **3**, 235 (1958)
59. R. Lindner, *Z. Naturforschung,* **10A**, 1027 (1955)
60. J. E. Castle and P. L. Surman, *J. Phys. Chem.*, **71**, 4255 (1967)
61. W. Müller, Diplomarbeit, Univ. Göttingen (1963); known through reference 57
62. V. I. Izvekov, N. S. Gorbunov and A. A. Babad-Zakhryaphin, *Physics and Metallography* [Eng. Transl.], **14**, 30 (1963)
63. R. Lindner, *Arkiv Kemi,* **4**, 381 (1952)
64. W. D. Kingery, D. C. Hill and R. P. Nelson, *J. Am. Ceram. Soc.,* **43**, 473 (1960)
65. W. Hagel, *Trans. Am. Inst. Min. Met. Engrs.,* **236**, 179 (1966)
66. D. Bevan, J. Shelton and J. Anderson, *J. Chem. Soc. (London),* **1948**, 1729 (1948)
67. R. Gardner, F. Sweett and D. Tanner, *J. Phys. Chem. Solids,* **24**, 1175 (1963)
68. G. H. Geiger and J. B. Wagner, Jr., *Trans Am. Inst. Min. Met. Engrs.*, **233**, 2092 (1965)
69. D. Chang, Thesis research in progess, Northwestern Univ. (1970)
70. L. S. Darken and R. W. Gurry, *J. Am. Chem. Soc.*, **67**, 1389 (1946)
71. G. H. Geiger, R. L. Levin and J. B. Wagner, Jr., *J. Phys. Chem. Solids,* **27**, 947 (1966)
72. F. S. Pęttit, D. Eng. Thesis, Yale (1962)
73. J. Belle, A. B. Auskern, W. A. Bostrom and F. S. Susko, Reactivity of Solids (J. H. Deboer, ed.), p. 452. Proc. of the Fourth International Symposium. (Elsevier Publishing Co., Amsterdam, 1961)
74. R. J. Thorn and G. H. Winslow, *J. Chem. Phys.*, **44**, 2632 (1966)
75. R. J. Thorn and G. H. Winslow, *J. Chem. Phys.*, **44**, 2822 (1966)
76. R. J. Thorn and G. H. Winslow, Thermodynamics, Vol. II, p. 213. (International Atomic Energy Agency, Vienna, 1966)
77. S. Rice, *Phys. Rev.*, **112**, 804 (1958)
78. J. F. Marin and P. Contamin, *J. Nuclear Mat.*, **30**, 16 (1969)
79. D. A. Venkatu and L. E. Poteat, *Mater. Sci. and Eng.*, **5**, 258 (1969/70)
80. R. Haul and G. Dumbgen, *J. Phys. Chem. Solids,* **26**, 1 (1965)
81. E. H. Greener, D. H. Whitmore and M. E. Fine, *J. Chem. Phys.*, **34**, 1017 (1961)
82. E. H. Greener, G. A. Fehr and W. M. Hirthe, *J. Chem. Phys.*, **38**, 133 (1963)
83. P. Kofstad and P. B. Anderson, *J. Phys. Chem. Solids,* **21**, 280 (1961)
84. W. K. Chen and R. A. Swalin, *J. Phys. Chem. Solids,* **27**, (1966)
85. W. K. Chen and R. A. Jackson, *J. Chem. Phys.*, **47**, 1144 (1967)
86. E. N. Greener, Ph.D. Thesis, Northwestern University (1960)
87. J. S. Sheasby and B. Cox, *J. Less Common Metals,* **15**, 129 (1968)
88. C. Wagner, *Z. physik Chem.,* **B11**, 139 (1930); *Z. Physik Chem.,* **B32**, 447 (1936)
89. F. S. Pettit, *J. Electrochem. Soc.*, **113**, 1249 (1966)
90. S. Mrowec, *Bull. de L'Acad. Polonaise des Sciences, Serie des sci. chim.,* **XV**, 373 (1967)
91. S. Mrowec, T. Walec and T. Werber, *Bull. de L'Acad. Polonaise des Sciences Serie des sci. chim,* **XIV**, 179 (1966)

92. S. Mrowec, T. Walec and T. Werber, *Corrosion Science,* **6**, 287 (1966)
93. K. Fueki and J. B. Wagner, Jr., *J. Electrochem, Soc.,* **112**, 384 (1965)
94. K. Fueki and J. B. Wagner, Jr., *J. Electrochem. Soc.,* **112**, 970 (1965)
95. B. Swaroop and J. B. Wagner, Jr., unpublished research at Northwestern University
96. P. Childs and J. B. Wagner, Jr., Proc. of an International Conference in Metallurgy and Materials Science, University of Pennsylvania, September 8-10, 1969, Heterogeneous Kinetics at Elevated Temperatures, p. 269 (G. R. Belton and W. L. Worrell, eds.). (Plenum Press, 1970)
97. J. B. Price and J. B. Wagner, Jr., *Physik Chem. (N.F.),* **49**, 257 (1966)
98. J. B. Wagner, Jr., Mass Transport in Oxides, p. 65 (J. B. Wachtman and A. D. Franklin, eds.). Proc. of a Symposium held at Gaithersburg, Maryland, October 22-25, 1967. (Published as National Bureau of Standards Special Publication 296, 1968)
99. P. E. Childs, L. W. Laub and J. B. Wagner, Jr., Proceedings No. 19, *J. Brit Ceramic Soc.,* pp. 29-53 (1971)
100. R. L. Levin and J. B. Wagner, Jr., *Trans. Am. Inst. Min. Met. Engrs.,* **233**, 159 (1965)
101. J. B. Price and J. B. Wagner, Jr., Extended Abstract Number 123, Chemical Diffusion Studies on Single Crystalline Manganous Oxide. The Electrochemical Society Meeting, Detroit, Michigan, October 5-9, 1969
102. L. S. Darken, *Trans. Am. Inst. Min. Met. Engrs.,* **175**, 184 (1948)
103. R. F. Brebrick, *J. Appl. Phys.,* **30**, 811 (1959)
104. R. F. Brebrick, Chapter IX in Solid State Physics: An Introduction (P. F. Weller, ed.). (In press)
105. E. B. Rigby and I. Cutler, *J. Am. Ceram. Soc.,* **48**, 95 (1965)
106. F. S. Pettit, E. H. Randkler and E. J. Felton, *J. Am. Ceram. Soc.,* **49**, 199 (1966)
107. I. M. Chemla, *Ann. Phys. (Paris) ser. 13,* **1**, 959 (1959)
108. F. Beniere, M. Beniere and M. Chemla, *J. Phys. Chem. Solids,* **31**, 1205 (1970)
109. C. F. Cline, H. W. Newkirk, R. H. Condit and Y. Hashimoto, Mass Transport in Oxides, p. 177. (National Bureau of Standards Publication 296, 1968)
110. F. Schein, B. LeBoucher and P. Lacombe, *Comptes Rendus Acad. Sci., France,* **252**, 4157 (1961)
111. R. A. Oriani, *J. Phys. Chem. Solids,* **30**, 339 (1969)
112. H. Reinhold, Z. *phys. Chem.,* **A141**, 137 (1929); **B11**, 321 (1931); *Z. Elektrochem.,* **35**, 617 (1929); H. Reinhold and R. Schulz, *Z. physik Chem.,* **A164**, 241 (1933)
113. C. Wagner, *Annalen der Physik* (5), **3**, 629 (1929)
114. C. Wagner, *Annalen der Physik* (5), **6**, 370 (1930)
115. R. E. Frywell and E. A. Aitken, *J. Nuclear Materials,* **30**, 50 (1969)
116. S. K. Evans, E. A. Aitken and C. N. Graig, *J. Nuclear Mater.,* **30**, 57 (1969)
117. E. A. Aitken, *J. Nuclear Mater.,* **30**, 62 (1969)
118. T. L. Markin and M. H. Rand, UKAEA Report AERE-R5560 (1967); known through reference 6
119. H. K. Bowen, Research in progress at the Massachusetts Institute of Technology, private communication, August, 1970
120. B. J. Wuensch and T. Vasilos, *J. Chem. Phys.,* **36**, 2917 (1962)
121. T. George and J. B. Wagner, Jr., unpublished research, Northwestern University (1969)
122. L. Laub, research in progress, North western University.
123. A. Seybolt, *Trans. Am. Inst. Min. Met. Engrs.,* **242**, 752 (1968)
124. A. V. Sandulova and A. I. Andrievskii, *Radio Engineering and Electronic Physics (U.S.S.R.),* **12**, 1492 (1956)
125. T. George and J. B. Wagner, Jr., *J. Electrochem. Soc.,* **115**, 956 (1968); **116**, 847 (1969)

126. N. G. Eror and J. B. Wagner, Jr., submitted for publication to *J. Electrochem. Soc.*
127. C. E. Birchenall, Reactivity of Solids, p. 24. Proc. of the Fourth International Symposium (J. H. Deboer, ed.). (Elsevier Publishing Co., Amsterdam, 1961)
128. Diffusion Data, A Continuous Compilation of New Reference Data on Diffusion Processes in Inorganic Solids and Their Melts. (Published by the Diffusion Information Center, Cleveland, Ohio)
129. D. Butrymowicz, private communication, August, 1970
130. A. L. Dragoo, *J. Research of the National Bureau of Standards,* **72A**, 157 (1968)
131. W. M. Robertson, *J. Nuclear Mat.,* **30**, 36 (1969)
132. A. Magneli, Crystallographic Principle of Some Non-stoichiometric Transition Metal Oxides in Transition Metal Compounds, p. 109. Proc. of the Buhl International Conference on Materials, October 3-November 1, 1963 (E. R. Schatz, ed.). (Gordon and Breach Science Publishers, New York, 1964)
133. R. H. Chang, W. Stewart and J. B. Wagner, Jr., Proc. 7th International Symposium on Reactivity of Solids. (In press)
134. R. H. Chang and J. B. Wagner, Jr., *J. Am. Ceram. Soc.,* **55**, 211 (1972).

INDEX

Z